Evolving Toolbox for Complex Project Management

Complex and Enterprise Systems Engineering Series

Evolving Toolbox for Complex Project Management

Edited by
Alex Gorod • Leonie Hallo
Vernon Ireland • Indra Gunawan

CRC Press
Taylor & Francis Group
Boca Raton London New York

CRC Press is an imprint of the
Taylor & Francis Group, an **informa** business

AN AUERBACH BOOK

CRC Press
Taylor & Francis Group
6000 Broken Sound Parkway NW, Suite 300
Boca Raton, FL 33487-2742

© 2020 by Taylor & Francis Group, LLC
CRC Press is an imprint of Taylor & Francis Group, an Informa business

No claim to original U.S. Government works

Printed on acid-free paper

International Standard Book Number-13: 978-0-367-18591-6 (Hardback)

Visit the Taylor & Francis Web site at
http://www.taylorandfrancis.com

and the CRC Press Web site at
http://www.crcpress.com

Alex

In memory of Natalia.

Leonie

To my PhD thesis supervisor, the late Dr Frank Dalziel, Psychology Department, University of Adelaide, for believing in me.

Vernon

To my wife, Margaret, for her support over many years.

Indra

I would like to acknowledge with gratitude the support and love of my family – my parents, Suwita and Effie Gunawan; my wife, Donna; and my children, Jessica and Cynthia. I am grateful to my colleagues Alex Gorod, Leonie Hallo, and Vernon Ireland for our collaborative efforts and discussions in completing this task. The book would not have been possible without them.

Contents

Foreword

The speed of change and advancements in technology, coupled with higher expectations of consumers in the 21st century, far exceed those experienced previously. This period of the fourth Industrial Revolution has introduced cyber physical systems and is characterized by such features as ubiquitous digitization, technology blending of the biological and physical (man versus machine), continuous reinvention, autonomous systems, genetic editing, mobile super-computing, Artificial Intelligence (AI) and robotics, all coupled with the goal of supporting sustainable ecosystems. With such advancements, it is inconceivable to think we can continue only with the existing and traditional applications and approaches to project management.

If we consider the growing trend of connecting existing systems in ways they were never originally perceived, the interactions and interdependencies expand to an increased level of complexity. Likewise, the applications of new technologies across all industries have created linkages through data creation and transfer, often with non-linear effects which in turn increase complexity. This is not solely confined to technology or engineering, but also penetrates the conventional practices and tools associated with project management, contract management, supply chain, risk management, standards and other non-technical disciplines. Failure to address these "black elephants" will result in sub-optimized project deliveries. Just as systems engineering communities are grappling with these challenges, so too must project management communities to deliver value-driven quality solutions in the expected shortened timeframes today's consumer demands.

Underlying all disciplines are their principles and heuristics, which are enabled through the disciplines' practices, frameworks, tools, and applications. If the dynamics in complexity are constantly changing, then the principles and heuristics must continue to evolve or be reconfirmed. For project management, a toolbox which will grow and be able to be tailored to the conditions of the ecosystem of the solution and its delivery framework is very advantageous to its users. If not, project management, and, indirectly, systems engineering, may suffer from the misapplication or, worse, the wasted application of a process-centric approach where it is not warranted.

This book will assist project managers, project support managers, systems engineering managers and other project leaders to be equipped to navigate their path in managing and delivering complex projects. It establishes a strong foundation for the start of this journey. The book presents a multi-faceted approach leveraging expertise worldwide, contributing to a comprehensive and holistic view in complex project management. The use of case studies throughout the book, in particular the Fukushima Daiichi Nuclear Power Plant disaster recovery effort, illustrates the diversity and evolution in research, theory and practice of project management for complex solutions. The toolbox is not all-inclusive, nor is it intended to be, as recognized by the editors.

From a technical viewpoint, the toolbox presents insight into pattern recognition and its usefulness. Likewise, using systems thinking and systems theory to take a more integrated approach to the development and delivery of the solution is discussed. Emphasis on the increasing importance of visualization and the growth and fidelity of models and simulation is another element of the rich toolbox topics collated by the editors.

The interesting latitude of topics is further illustrated when considering the breadth of the Internet of Things (IoTs). IoTs increase connectedness between devices, machines, objects and people; creates communities of interest over collaboration network; and lends growth to the number of large, complex systems, composed often of other systems under a "System of Systems" arrangement. This creates challenges in security, interoperability, reliability, regulation compliance and so forth. The consumer also takes on an expendable mentality, which is further reinforced by products being developed with intentional obsolescence considerations. The toolbox explores techniques for handling this growing importance of adaptability, agility and resilience.

From the managerial viewpoint, the editors have provided guidance on the emerging practices, applications and tools best suited for various complex projects. In managing these, you may ask what are the critical success factors? Are we able to forecast reliably and accurately when the solutions we are providing are becoming more non-deterministic in nature with the adoption of increasing automated decision-making through AI and machine learning? Can we estimate costs to deliver and support such solutions? Avoidance or minimization of rigid contract milestones, documentation waste, unbalanced risk profiles, over-specifying existing product and a requirements-driven monoculture are addressed across multiple chapters. These are becoming increasingly important with the application of emerging practices such as the digital twin and virtual engineering as a means of addressing the acceleration of complexity. Such challenges are discussed, identifying the emerging tools and associated research.

Furthermore, new development life cycles will evolve adding to the existing set of life cycle frameworks. This elaboration is introduced with evolution of the toolbox required to support the next-generation life cycle framework to better integrate early and frequent releases, continuous customer visibility and the handling of changes through the incorporation of lessons learnt, evolving requirements and undocumented emergent behaviour in a faster manner. Likewise, the need to test and evaluate early and often is discussed, complementing the need to balance uncertainty levels through a trusted approach to verification and validation.

Another facet of the toolbox is the need for enterprise governance through the emerging field of Complex Systems Governance (CSG), vital for the growing number of complex cyber-physical system projects. Similarly, policy management, legal considerations and the global challenge of the protection of Intellectual Property (IP) rights have not been ignored.

With the constant increase in project complexity, the growing number of non-deterministic solutions, expanding dependencies and greater interrelationships across many projects, the discipline of project management requires ongoing review and evolution of its practices, applications and tools. Recognizing this need, the editors have created such a repository to capture these advancements in project management through the introduction of the toolbox detailed in this book. The information provided will continue to mature and evolve over time. The toolbox is a vital means for supporting the advancements in the solution offerings of the future and will be invaluable to all stakeholders involved in the delivery and sustainment of complex projects. This book will contribute to both the researcher and practitioner's library of knowledge in project management.

Kerry Lunney
Engineering Director & Chief Engineer, Thales Australia
INCOSE President (2020–2021)

Preface

On March 11, 2011, Shizuko A* opened her soba noodle shop in Futaba town on the Tōhoku coast of Japan as usual. It was a typical morning, with a steady flow of customers for breakfast and through the morning. Soon after her lunch rush, she heard a loud unusual sound, a bit like an aeroplane taking off at close quarters. Very soon, people were running in the street and shouting.

This day was like the end of the world for people living near the Fukushima Daiichi Nuclear Power Plant. An earthquake of magnitude 9.0 on the Richter scale generated a tsunami, which crashed over a wide swathe of the northeast coast of Japan. Almost 20,000 people died, and hundreds of thousands of people have been left homeless [1]. The bill for damages was immense and has been estimated at between US$250 and US$500 billion [2]. This natural disaster and the resulting nuclear crisis spelled the end of tourism in Fukushima, which was formerly known for its sake, its hot spring baths and its spring cherry blossoms [3]. The people of the affected region had their lives, plans and aspirations entirely disrupted.

While the damage from the earthquake itself was minimal because residents had learned from previous disasters and were prepared, the subsequent tsunami caused tremendous devastation to life and property. The Fukushima Daiichi Nuclear Power Plant accident following the earthquake and tsunami led to the evacuation of 160,000 people from the vicinity of the plant: many thousands are still in temporary accommodation. These individuals suffered acute physical and mental health effects from the disaster and ongoing dislocation [1]. Adding to the trauma, information sharing was neither timely nor transparent and in some cases misleading [4].

This was the world's most significant nuclear accident since the Chernobyl disaster in April 1986. The clean-up will take decades and is extremely expensive [5]. Challenges faced in the nearby regions are still enormous. Widespread radioactive contamination has been detected [6], and the recovery process is highly demanding. Tokyo Electric Power Company (TEPCO) has already collected more than 1 million tonnes of contaminated water [7]. The Japanese Minister of the Environment, Yoshiaki Harada, stated on September 10, 2019, that Japan will run out of room for storing by 2022: "The only option will be to drain it into the sea and dilute it" [7].

On Wednesday, June 22, 2016, Mr and Mrs Smith* of Lancaster, Lancashire, United Kingdom (UK), both retired, invited their neighbours, the Joneses*, in for a meal in anticipation of the result of the Brexit vote to be held the next day. The group animatedly discussed the likely outcome, and all were excited to be able to help bring Britain back out of the European Union (EU).

At the same time, UK bookmakers were finalizing their odds and had put Brexit well behind, having shortened their odds dramatically, giving the Remain decision a much higher probability of winning, with very unattractive odds for those wishing to bet on Leave [8].

On the day, the Leave vote won by 51.9%–48.1% [9]. The referendum turnout was 72.2%, with more than 30 million people voting [10]. Consequently, the Prime Minister of the UK, Theresa May, triggered Article 50 of the Treaty on European Union, starting the process of British withdrawal from the EU.

Many readers will remember hearing about the results of the Brexit vote, which were largely unexpected and sent shockwaves through the financial markets [11]. At the time of this writing, the UK was due to leave the EU on March 29, 2019. Research [12] indicates that the divide between the winners and losers of globalization was a key reason for the outcome. Since June 2017, the Prime Minister has been negotiating with the EU, but the draft withdrawal agreement was voted out in the House of Commons by 432 votes to 202. Working out the details of the future relationship between Britain and the EU is proving to be very difficult [13].

The world's cameras flashed as Saudi Crown Prince Mohammad bin Salman bin Abdulaziz Al Saud made his momentous announcement about the development of the new mega-city Neom as "The future of Saudi Arabia" and a centrepiece of the Kingdom's Vision 2030, a plan to pivot the economy away from oil [14]. The news was delivered with much fanfare on October 24, 2017 in Riyadh, as part of the Future Investment Initiative.

Neom will be located on an area of 26,500 km², which is more than twice the size of Qatar and 33 times the size of New York. The project aspires to create many unique development opportunities and serve as a global hub for trade, innovation and knowledge [15]. It will be funded initially with US$500 billion from the Kingdom's Public Investment Fund and private investors [16]. The city is intended to offer a unique and enjoyable lifestyle, aiming to outperform other major cities in terms of competitiveness and opportunities [17]. The task is referred to as "the world's most ambitious project" [18]. Similar large-scale ventures in Saudi Arabia have floundered in the past, which shows the arduousness of such an undertaking. It is also questionable whether this project will actually be completed due to a variety of emerging external factors.

What are the common elements in the three projects outlined above?

- They are all complex.
- They are difficult to manage.
- Traditional project management tools have limitations when applied to these types of situations.
- There is a lack of an evolving platform for managing tools.

All three projects are complex because of the high degree of uncertainty about the outcomes. All three consist of multiple autonomous, decentralized and heterogeneous constituents that often interact in a non-linear manner, creating emergent behaviour that is challenging to predict. Traditional top-down management practices do not work effectively in such complex cases due to a high degree of uncertainty, volatility and non-linearity. Finally, there is no readily adaptable platform for collecting, integrating, monitoring and updating appropriate tools in complex project management.

This book came about because of the interest of the editors in complex project management. In particular, over the years of teaching and researching in the field, the editors became aware of the current shortage of current project management tools for dealing with the growing number of complex projects and their increasing degree of complexity. In addition, it became evident that there was a lack of an evolving repository for the tools. The editors subsequently agreed that it would be helpful to draw from global experts their very best wisdom on the tools for use in their domains of interest. In turn, the initiative has resulted in this collection of 23 chapters, representing the latest thinking about tools for complex project managers and how to access them. Of course, this is not an exhaustive list and other chapters could have been added. The book is timely, and yet it is still a toolbox-in-progress. No doubt in this digital age, new approaches and ideas will be quickly developed and could also be part of future compilations. Nevertheless, the authors believe that the information and resources included in the book will bring significant value to project managers seeking more contemporary and effective ways to manage complex projects.

As a result of expanding complexity, today's projects often fail and project managers are faced with more complex tasks, requiring skills from across many disciplines. The pace of change is fast and accelerating, creating an urgent need for a paradigm shift. The rise in the sharing economy, now ubiquitous in our day-to-day experience, reflects the general move toward a more collaborative world. Smart cities and the Internet of Things (IoT) reflect this increase in interconnectedness, and the implications of this transformation are only now beginning to be explored.

This book presents a new approach for complex project management: the evolving toolbox. Each chapter provides an important and useful toolset for managers of complex projects. As complexity is increasing, tools must evolve and the toolbox will also be developing, offering new benefits in the future. The initial set of toolsets found in the book is just a snapshot, and additional tools will emerge to meet rising challenges. The Law of Requisite Variety [19] has been understood and accepted for a long time now: in order to address a variety of problems, a corresponding variety of solutions are necessary.

Present limitations in complex project management include:

- The range of available tools to manage complex projects is insufficient. Although new tools are being developed, complexity is growing at a faster rate.
- The existing tools are static and not self-updating or self-organizing. Given the rapid growth in complexity, there needs to be a smart mechanism for enabling self-learning and self-organization to occur.
- Traditional tools are typically used in a stand-alone fashion, and many were designed for different purposes. It is challenging to integrate them dynamically with other tools because there are no evolving repository and no common protocols.
- Many existing tools were created during a less complex time period and lack the ability to deal with systemic issues because they are reductionist in nature.

This book addresses these and other issues by bringing together a broad array of the latest tools suited to dealing with complex projects, thus providing project managers with current practical options and in-depth perspectives in one location. In addition, the book discusses the evolving toolbox concept as a virtual platform for managing tools. The platform is open to the external environment, making it possible for tools to interact and integrate as well as to evolve systemically in response to their usage and application by project managers. This system will operate as a cloud, containing multiple toolsets for various purposes. While individual tools may not be sufficient to solve a complex project, the collaborative smart cloud with continually updating toolsets would bring more resources to finding a more feasible resolution to such problems.

Emergence of novel properties and tools can naturally occur inside the toolbox because the tools are autonomous and responsive to the open environment. As an example, the interface of modelling and simulation tools with Artificial Intelligence (AI) may influence innovation. Another example could be integrating toolsets in scheduling, AI and risk management, which could potentially create a ground-breaking way of coping with risk while simultaneously revamping the process of scheduling. Furthermore, evolutionary fine-tuning is expected to take place since tools are all interrelated.

The collection of toolsets and the evolving toolbox approach discussed in this book represent a step forward in the ability of project managers to manage complex projects more successfully. In addition, the material can have a significant impact on education, helping prepare next-generation engineers and project managers to be more effective in addressing non-traditional challenges. The overall emphasis is on the importance of cultivating a mindset of greater adaptability, agility and openness to exploring the integration and evolution of tools that will be part of the future of complex project management.

ENDNOTE

* Fictional Characters

REFERENCES

1. E. Feldman, Compensating the victims of Japan's 3-11 Fukushima disaster, *Asian-Pacific Law and Policy Journal*, vol. 16, p. 127, 2015.
2. S. Starr, Costs and consequences of the Fukushima Daiichi disaster, *Retrieved January*, vol. 15, p. 2015, 2012.
3. J. S. Kim and S. -H. Park, A study of the negotiation factors for Korean tourists visiting Japan since the Fukushima nuclear accident using Q-methodology, *Journal of Travel & Tourism Marketing*, vol. 33, no. 5, pp. 770–782, 2016.

4. K. Hasegawa, Facing nuclear risks: Lessons from the Fukushima nuclear disaster, *International Journal of Japanese Sociology*, vol. 21, no. 1, pp. 84–91, 2012.

5. S. Muramatsu and K. Hanawa, Seven years on, no end in sight for Fukushima's long recovery; Japan faces myriad challenges to decommissioning and decontamination, in *Nikkei Asian Review*, ed. Japan, 2018.

6. H. Kawamura et al., Preliminary numerical experiments on oceanic dispersion of 131I and 137Cs discharged into the ocean because of the Fukushima Daiichi nuclear power plant disaster, *Journal of nuclear science and technology*, vol. 48, no. 11, pp. 1349–1356, 2011.

7. Japan may have to dump radioactive water into the sea, minister says, in *Reuters*, Ed.: Japan, 2019. https://in.reuters.com/article/japan-fukushima-water/japan-may-have-to-dump-radioactive-water-into-the-sea-minister-says-idINKCN1VV0BX.

8. C. Milas, T. Worrall, and R. Zymek, Watch out for winners and losers: Odd-implied brexit sentiment and FTSE returns, *Financial Times*, 2016.

9. H. D. Clarke, M. Goodwin, and P. Whiteley, Why Britain voted for Brexit: An individual-level analysis of the 2016 referendum vote, *Parliamentary Affairs*, vol. 70, no. 3, pp. 439–464, 2017.

10. S. B. Hobolt, The Brexit vote: A divided nation, a divided continent, *Journal of European Public Policy*, vol. 23, no. 9, pp. 1259–1277, 2016.

11. M. Jawad and M. Naz, Pre and post effects of Brexit polling on United Kingdom economy: An econometrics analysis of transactional change, *Quality & Quantity*, vol. 53, no. 1, pp. 247–267, 2019.

12. K. Neumann, 'Know Why' thinking is a new approach to systems thinking, *Emergence: Complexity & Organization*, vol. 15, no. 3, pp. 81–93, 2013.

13. M. Emerson, Which model for Brexit? in *After Brexit: Consequences for the European Union*, N. da Costa Cabral, J. Renato Gonçalves, and N. Cunha Rodrigues, Eds. Cham: Springer International Publishing, 2017, pp. 167–188.

14. I. Kouskouvelis and K. Zarras, Cairo and Riyadh, Vying for Leadership, *Middle East Quarterly*, 2019.

15. A. Farag, The story of NEOM City: Opportunities and challenges, in *New Cities and Community Extensions in Egypt and the Middle Eas*, Z. S. a. A. I. Sahar Attia, Ed.: Springer, 2018, pp. 35–49.

16. W. Qi and Z. Shen, A smart-city scope of operations management, *Production and Operations Management*, 2018.

17. L. K. B. Melhim, M. Jemmali, and M. Alharbi, Intelligent real-time intervention system applied in smart city, in *2018 21st Saudi Computer Society National Computer Conference (NCC)*, 2018, pp. 1–5: IEEE.

18. Neom. 2019, 17 March. *Welcome to Neom*. Available: www.neom.com.

19. W. R. Ashby, *Variety, constraint, and the law of requisite variety*. Chicago, IL: Aldine, 1968.

Acknowledgments

The editors would like to thank Rosemary Hallo, Annsley Kerr and Veronica Shipilov for their enormous input and support in the editing work.

The editors also express their sincere appreciation to the contributing authors for their expertise and collaborative efforts. This book would not have been possible without them.

In addition, the editors are most grateful to Professor Noel Lindsay at the University of Adelaide Business School for his vision and encouragement.

The editors would like to especially thank Kerry Lunney, INCOSE President-Elect, for her inspirational foreword, and Professor Michael Jackson, Associate Professor Benyamin Lichtenstein, and Professor Kunhui Ye for their supportive commentary.

Finally, the editors would like to acknowledge Richard O'Hanley for his continuous guidance and for the opportunity to publish this work. Special thanks are extended to Kritheka at Codemantra for major editing work.

Editors

Alex Gorod is a Professor at the University of Adelaide and an Adjunct Associate Professor at Zicklin School of Business, City University of New York. He is teaching courses in the areas of complex project management, leadership, systems engineering, engineering management, systems thinking, and operations management. He is a recipient of the Fabrycky-Blanchard Award for Excellence in Systems Engineering Research and Robert Crooks Stanley Fellowship in Engineering Management. His primary research focus is on management of complex adaptive systems. He is author or co-author of multiple publications, including scientific articles, book chapters, and patents. In addition, he has co-edited a book titled *Case Studies in System of Systems, Enterprise Systems, and Complex Systems Engineering*. His research appeared in such journals as *International Journal of Project Management; IEEE Aerospace and Electronic Systems Magazine; IEEE Systems Journal; IEEE Transactions on Systems, Man, and Cybernetics; Transportation Research Record; Supply Chain Management: An International Journal; and Systems Research and Behavioral Science*, among others. He holds a PhD in Engineering Management from Stevens Institute of Technology and actively participates in several professional organizations including INCOSE, IEEE, ASEE, SDS and ERS. He currently serves as an Associate Editor of *IEEE Systems Journal (ISJ)*. He is also the founder and managing member at SystemicNet, LLC and a partner at Mountava, LLC.

Leonie Hallo is an Adjunct Associate Professor at the University of Adelaide and a Co-director of the Complex Systems Innovation and Entrepreneurship Lab (CSIEL). Leonie holds a PhD in Psychology from the University of Adelaide (Gold Star Award). She has been a psychologist in private practice for many years specializing in coping with cancer. She has also been a consultant to public and private organizations across fields including leadership training and development, assessments of organizational climate and reporting/recommendations, support in redundancy, career guidance, 360-degree feedback and executive development. She has supervised almost 30 doctoral students in Australia and throughout Asia in areas relating to organizational behaviour, supply chain management and entrepreneurship. She is also a Key Note Speaker and Reviewer for several international conferences and journals in the complexity leadership and management field. Her current primary research interests are complex project management, cancer treatment and the relationship between leadership and emotion.

Vernon Ireland is currently a Professor and Foreign Expert in the Faculty of Construction Management and Real Estate, Chongqing University (2018–2019). Vernon was formerly an Emeritus Professor in Complex Systems, The University of Adelaide (2016); the Director of Project Management and the Director of the Master of Applied Project Management, The University of Adelaide (2003–2015); and the CEO of The Australian Graduate School of Engineering Education (1996–2002). He was the Corporate Development Director, Fletcher Challenge Construction (1991–1996), and the Dean of the Faculty of Design, Architecture and Building, University of Technology, Sydney (1987–1991). He practised as an engineer for 10 years (1962–1973) and was engaged by the Royal Commission into Productivity in New South Wales, to provide a comparison of 25 projects in Australia with 25 projects in the USA, Canada, New Zealand, the UK, Sweden and Germany (1992). He was the Chairman, Building Services Corporation of NSW (1987–1990): this is the licencing authority for domestic builders and electrical and plumbing contractors. Vernon has received three medals: Silver Magnolia Medal from the Shanghai Government (2000), Rotary International Gold Medal for his contributions to vocational education (2006) and Engineers Australia Medal (2006). He has been the President of the Sydney Division of Engineers Australia (2005) and has published many papers and book chapters.

Indra Gunawan is an Associate Professor in Complex Project Management, and the Director of Postgraduate Project Management Programme in the Entrepreneurship, Commercialisation and Innovation Centre (ECIC), Faculty of the Professions, the University of Adelaide, Australia. He received his PhD in Industrial Engineering and MSc in Construction Management from Northeastern University, Boston, Massachusetts, USA. Prior to joining the University of Adelaide, he was the Postgraduate Programme Coordinator for Maintenance and Reliability Engineering at Monash University, Melbourne, Australia. Previously, he has also taught in the Department of Mechanical and Manufacturing Engineering at Auckland University of Technology, New Zealand and worked as the Head of Systems Engineering and Management programme at Malaysia University of Science and Technology (in collaboration with the MIT, USA). His current research interests include system reliability modelling, maintenance optimization, project management, applications of operations research and operations management. He is actively involved in the Asset Management Council, a technical society of Engineers Australia. Indra's work has appeared in many peer-reviewed journals such as *International Journal of Project Management*; *International Journal of Production Research*; *Reliability Engineering and System Safety*; *IEEE Transactions on Industrial Informatics*; *International Journal of Reliability, Quality and Safety Engineering*; *Quality and Reliability Engineering International*; *International Journal of Project Organisation and Management*; and *International Journal of Quality and Reliability Management*. He also serves as an Editorial Board member for *International Journal of Project Organisation and Management*, *International Journal of Information Technology Project Management* and *International Journal of Performability Engineering*.

Contributors

Fran Ackermann is currently a Professor at Curtin University having previously worked at the University of Bath, University of Strathclyde (UK), University of Grenoble (France), University of Western Australia and University of Georgia (Athens, USA). She has an Honorary Professorship at The University of Manchester, UK, and an adjunct professorship at Strathclyde University. In the past, she has held adjunct professor positions at Tilburg University, Bordeaux Business School. In addition, she has held managerial positions such as the Head of the Department of Management Science, Strathclyde University, UK (2011–2012), and the Dean of Research, Curtin Business School (2013–2017). Fran's research interests focus on three main considerations: (1) complex project management including modelling disruption and delay for litigation, and systemic risk modelling; (2) strategic management including models for managing stakeholders, developing scenarios, leveraging competitive advantage, and identifying and agreeing goals; and (3) multi-organizational negotiation and collaboration. She has published widely producing 5 books, 29 book chapters, and over 70 journal articles. She has been awarded the INFORMS GDN Section Award 2016, four Best Paper prizes at the Academy of Management, and Best Paper prizes at Australian and New Zealand Academy of Management and at British Academy of Management. She is a Fellow of the British Academy of Management and a Graduate of the Australian Institute of Company Directors. Fran has received ~AUS$5.4+ million as research funds from both research councils and industry straddling the UK and Australia (from different fields such as Health, Defence, Energy and Engineering). She has been a consultant for a variety of fields in Australia, the UK, the USA, France and Italy. Fran's research is strongly influenced by the work of Lewin whose premise that "there is nothing as practical as good theory" resonates strongly with her. She has given talks to a range of industry bodies including the Risk Engineering Society, Australian Institute of Management, Australian Institute of Builders and National Association of Women in Construction, Scottish Executive (Edinburgh), Department of Environment (Paris) and Department of Health (Perth). Fran currently delivers, in association with The University of Manchester, a bespoke Executive Education programme for BAE Systems on Complex Project Management, Previously, she has also delivered Executive Education to Babcock International, Clydesdale Bank, Alliance Trust, MITRE (MIT Research Environment, USA), Motherwell Bridge and Shell International.

Alicia Aitken is the Chair of the International Centre for Complex Project Management and has 20 years' experience working in the field of project, programme and portfolio management. As the CEO of Human Systems International, she led a global team that assessed, benchmarked and advised on the improvement of organizational delivery capability for organizations and government departments. Her cross-industry experience includes banking, finance, telecommunications, defence, aerospace, pharmaceutical, government, engineering and construction. Alicia held the first Chief Project Officer role for Telstra, Australia's largest telecommunications company, and is now the Head of Investment Management and Delivery for ANZ Banking Corporation. Alicia is actively involved in several industry groups and peak bodies. She holds a PhD in Project Management and Psychology with a particular focus on how project managers cope with stress. She is a regular keynote speaker at conferences around the world and contributes to academic programmes at a number of universities.

Neal M. Ashkanasy, OAM, PhD, is a Professor of Management in the UQ Business School at The University of Queensland, Australia. He studies emotion in organizations, leadership, culture and ethical behaviour. He publishes in journals such as *Academy of Management Journal* and *Review, Journal of Management* and *Journal of Applied Psychology.* He served as the Editor-in-Chief of *Journal of Organizational Behavior* and an Associate Editor for *Academy of Management*

Review and *Academy of Management Learning & Education*. He is an Associate Editor for *Emotion Review* and Series Editor for *Research on Emotion in Organizations*. He is a Fellow of learned societies (ANZAM APS, ASSA, BAM, SIOP, SMA) and the 2019 Academy of Management Managerial & Organizational Cognition Division Distinguished Scholar.

Dolores A. Atallo is a trusted advisor to board members and senior management in matters of governance and risk oversight structure and roles, risk appetite, linkage to strategy and risk culture integration. She has over 25 years of Big Four, financial services industry and consulting experience helping clients build and enhance Enterprise Risk Management (ERM), Corporate Governance and Compliance programs that align strategy, objectives and performance. In these roles, she also facilitates regulatory readiness and communication with external stakeholders, including rating agencies, analysts and investors. She is currently a Managing Director and North American Leader for Enterprise Risk Management (ERM) and Governance and a member of the Global ERM Leadership team at Protiviti. She is a graduate from Rutgers University (B.A. and M.B.A). She is an adjunct professor of governance and risk management at Martin Tuchman School of Business at New Jersey Institute of Technology (NJIT). She can be reached at www.linkedin.com/in/doloresatallo.

Alexey Bakman, Esq., is an Intellectual Property Attorney who currently practices law in New York City. He obtained his Juris Doctor degree with a concentration in Intellectual Property and Mass Media Law from the Benjamin N. Cardozo School of Law. He is admitted to the New York State Bar. He is also admitted to the Patent Bar as a Patent Attorney, registered to practice before the United States Patent and Trademark Office. Over the years, he has successfully prosecuted numerous patent applications. He has also handled a number of IP-related adversarial proceedings, including trademark opposition proceedings. Alexey has been consulted for providing legal services to multiple global enterprises on issues relating to IP contracts, trademarks, copyrights, trade secrets and other IP matters in the context of complex project management.

Alon Ben-Meir is a Professor and Senior Fellow at New York University's Center for Global Affairs, and Senior Fellow and Middle Eastern Studies Project Director at the World Policy Institute. He has authored seven books related to the Middle East and is currently working on a new book about countering violent extremism.

Kon Shing Kenneth Chung is currently the Director of the Undergraduate Project Management Programme and a Senior Lecturer at the Project Management Programme, the University of Sydney. His research interest lies in behavioural and organizational dynamics with emphasis on social networks and complexity science. A social network analyst by training, with competencies in mixed methods research, his expertise lies in developing conceptual models through the triangulation of theories and methods in social networks, networks and complexity science, management science and information systems. His research in human social networks and how network-, actor- and relational-level properties of such networks influence social or organizational outcomes – such as coordination, learning and performance – has significant implications for projects and stakeholder management. He publishes in top-tier conferences and journals such as *Project Management Journal, IRNOP, SIG-Management of Information Systems, Computers in Human Behavior, Computer-Supported Cooperative Work* and *European Conference on Information Systems*. He holds numerous grants, including early career research grant (university level), Australian Research Council Discovery Project grant (national level) and Project Management Institute Research Grant (international level).

Robert Cloutier is a Professor at the University of South Alabama (USA). His research interests include model-based systems engineering, system architecting, 2D/3D concept of operations, complex patterns for systems engineering and visualizing socio-technical systems. He has published

22 peer-reviewed journal articles. His book, *Systems Engineering Simplified* with CRC Press/ Taylor & Francis Group, is targeted toward non-systems engineering undergraduates and corporate engineers. His research and mentoring has resulted in the awarding of nine doctoral degrees in Systems Engineering. Before joining the USA, he was an Associate Professor at Stevens Institute of Technology in Hoboken, New Jersey. Prior to Stevens, he spent over 20 years at Lockheed Martin, The Boeing Company (where he was an Associate Technical Fellow), and a commercial e-commerce consulting firm. His roles have included System Architect, Enterprise Architect, and Principal Systems Engineer. He served 8 years in the U.S. Navy and received his BS from the U.S. Naval Academy, his MBA from Eastern University and his PhD from Stevens Institute of Technology.

David Cole has 30 years' experience of managing projects, programmes and portfolios to deliver products, services and business change in private- and public-sector organizations. As a management consultant, he also advised these organizations on the management of their programmes and portfolios. He is a Chartered Engineer and a member of the Institution for Engineering and Technology. He is also a member of the Association for Project Management, where he was a founder member of Systems Thinking Specific Interest Group (SIG). He currently serves as a committee member of the Systems Thinking SIG. He holds a BSc in Electrical and Electronic Engineering from the University of Reading and an MSc in Computer Science from Imperial College.

George L. De Feis (DPS, Pace University) has been a Professor of Management at Baruch College (City University of New York), Pace University, and Iona College. In addition to policy management, he has recently written on service-learning initiatives, sustainable development, international trade theory and strategic management. His works have been published in the *Journal of General Management*, *International Journal of Business & Applied Sciences*, *Academy of Business Research Journal*, *Journal of Business and Economic Research*, and *Journal of Marketing Perspectives*, and has authored a textbook, *Management Science: A Practical Approach for Today's Business Management Student* (2004, 2007). In 2016, he received an Iona College Academic Innovation Grant for creating a Service-Learning Course on Civic Engagement and Public Values (2016). He was elected to the Board of Directors of the Business and Applied Sciences Academy of North America (BAASANA) in 2016, and currently sits on the Board of AirSoilWater, a sustainability not-for-profit organization in Pennsylvania. He has over 25 years of worldly professional and academic experience covering the fields of strategic management, global affairs and operations research, and he has been a consultant for the for-profit and not-for-profit worlds. He can be reached at www. linkedin.com/in/george-de-feis-0911046/.

Mahmoud Efatmaneshnik is a Researcher at the School of Information Technology and Electrical Engineering, University of New South Wales (UNSW) at Australian Defence Force Academy, in Canberra. He has a PhD in Complexity Management of Design Process (2009) and a ME in Manufacturing Engineering (2005) both from UNSW. He has a BE in Aerospace Engineering (1999) from Tehran Polytechnic University. He is the author of more than 80 book chapters, journal papers and refereed conference papers on topics such as complexity management, modularity value analysis, system robustness and resilience, and integrated navigation systems.

Michael Emes has an ME in Engineering, Economics and Management from The University of Oxford and a PhD from University College London's (UCL) Mullard Space Science Laboratory in developing ultra-low-temperature cooling systems for spacecraft. He has worked as a Strategy Consultant for Mercer Management Consulting (now Oliver Wyman) on projects in rail, retail and energy, including advising the UK Government on the restructuring of the rail industry when Railtrack was taken into administration. He is now an Associate Professor of Technology Management at UCL and the Director of UCL Centre for Systems Engineering. He conducts teaching and research in project management, risk management and decision-making in domains including transport, health and aerospace. He is the Programme Manager and a Lead Trainer for the European Space Agency's Project Manager Training Course, and the Programme Director for UCL MSc programmes in Technology Management and Management of Complex Projects. He is a Chartered Engineer, a Senior Fellow of the Higher Education Academy and Co-chair of the APM Systems Thinking Specific Interest Group.

Kristin Falk has led technology teams in startups, SME and large corporations primarily in the energy industry. She has been in the industry for more than 20 years, where she has introduced complex systems for integrated operations, subsea processing and automated drilling. She is teaching Systems Engineering at the University of South-Eastern Norway. Her research focus is "how to create systems fit for purpose in a volatile, uncertain, complex, and ambiguous world".

Joana Geraldi is an Associate Professor at Copenhagen Business School, Department of Organizations, and leads the Center for Advanced Studies in Project Organizing (CASPRO). She studies human and organizational behaviour in projects and develops behaviour-centric organizational contexts for projects. Her empirical contexts are in projects and project organizing across different industrial contexts, in particular large engineering projects, IT, construction and wind industries. Her research earned international awards and led to over 50 publications, most of which is the key project journals and conferences. In recent years, Joana came to recognize the relevance of decisions and behaviours in decisions to projects, and has since studied how decisions happen in projects and how objects, in particular visualizations, can inform (and misinform) decisions.

Thomas Grisham has over 44 years of global experience across a variety of business sectors including medical, infrastructure and utilities, finance and manufacturing. His last full-time job was running a Japanese trading company in Tokyo. Thomas has gained this experience in 81 countries and has lived in Turkey, Saudi Arabia, Thailand, Japan, Korea, China, Hong Kong, Germany, Brazil and part time in the UK, Singapore, India and Spain. His experience spans over 400 global organizations such as the UN, Nestlé, NSA and ZTE. He currently conducts global corporate training, mentors doctoral students for a university in Switzerland, is a visiting professor at universities in Germany, Mexico and Brazil, and provides public and cruise ship speaking. He has authored four books, chapters in three others, and a new book with Dr Arroyo about working in extreme locations like the Amazon.

George Farage Jergeas, PhD, P.Eng., is a Professor and the Director of the Centre for Project Management Excellence at the University of Calgary. He joined the university as a full-time academic coming from industry in 1994. Dr Jergeas' career in industry was in the construction of international infrastructure projects, management of claims and disputes, and investigating projects experiencing cost overruns and delays. His experience in industry and subsequent research has demonstrated a strong interest in improving the performance and predictability of construction projects.

Keith Joiner, CSC, was an Air Force Aeronautical Engineer, Project Manager and Teacher for 30 years before joining the University of New South Wales to teach and research test and evaluation. As Defence's Director-General of Test and Evaluation for 4 years, he was awarded a Conspicuous Service Cross, and for doing drawdown plans for the Multi-National Force in Iraq, he was awarded a U.S. Meritorious Service Medal. He is a Certified Practising Engineer and a Certified Practising Project Director, and has over 60 published articles and contributions in work.

Dmitry Katalevsky is the Director of Industrial Programme Department, Skolkovo Institute of Science and Technology and an Associate Professor of the Institute for Business Studies of the Russian Presidential Academy of National Economy and Public Administration (RANEPA).

Dmitry received an MS degree in Public Management from the Lomonosov Moscow State University (2004) and N. Rockefeller College of Public Affairs and Policy, State University of New York at Albany, USA (2005) graduating with honours. In 2008, he received PhD in Management from the Lomonosov Moscow State University. He has over 15 years' experience in consulting and project management working at internationally recognized companies and leading non-profit organizations. Over the last 6 years, he has been working at the Skolkovo Institute of Science and Technology, a leading innovative technical university of Russia responsible for technology transfer and programme with industry. He is the author of several publications and books, including a popular textbook for simulation management (in Russian) and a simulation game for entrepreneurs "Startup: Limits to Growth".

Polinpapilinho F. Katina is an Assistant Professor of Advanced Manufacturing Management at the University of South Carolina Upstate, Spartanburg, South Carolina, USA. He holds a BSc in Engineering Technology with a minor in Engineering Management, a MEng in Systems Engineering, and a PhD in Engineering Management and Systems Engineering, all from Old Dominion University (Norfolk, Virginia, USA). He received additional training from Environmental Systems Research Institute (Redlands, California), University of Edinburgh (Edinburgh, UK), and Politecnico di Milano (Milan, Italy), among others. His areas of research include critical infrastructure protection, energy systems (Smart Grids), engineering management, decision-making under uncertainty, complex system governance, infranomics, systems engineering, system-of-systems and systems theory. He has co-authored more than 50 peer-reviewed papers to international journals including *International Journal of Critical Infrastructure Protection*, *International Journal of Critical Infrastructures*, *Requirements Engineering*, and INCOSE's *Insight*, and conferences including ASEM, IEEE, INCOSE, IISE and WEFTEC. He served as a Guest Editor for *International Journal of Critical Infrastructures* (2014) and *International Journal of System of Systems Engineering* (2015). His profile includes three books: *Infranomics*: *Sustainability, Engineering Design and Governance* (Springer, 2014); *Critical Infrastructures*: *Risk and Vulnerability Assessment in Transportation of Dangerous Goods – Transportation by Road and Rail* (Springer, 2016); and *Critical Infrastructures, Key Resources, and Key Assets*: *Risk, Vulnerability, Resilience, Fragility, and Perception Governance* (Springer, 2018).

Charles B. Keating is a Professor of Engineering Management and Systems Engineering and the Director of the National Centers for System of Systems Engineering at Old Dominion University, Norfolk, Virginia, USA. His research and teaching focusses on complex system governance, systems thinking/theory, systems engineering, system-of-systems engineering and management cybernetics. A Fellow and Past President of the American Society for Engineering Management, he received the society's most prestigious award (Sarchet Award) in 2015 for pioneer efforts in the field. He has authored 110+ peer-reviewed publications, generated $20M+ in research funding, and graduated 25 PhDs. His research has spanned defence, security, aerospace, healthcare, R&D and automotive sectors. Prior to faculty appointment in 1994, he served in leadership and technical engineering management positions for over 12 years in the U.S. Army and private industry. He holds a BS Engineering (United States Military Academy, West Point), MA Management (Central Michigan University) and PhD Engineering Management (Old Dominion University). Memberships include American Society for Engineering Management, International Society for System Sciences and International Council on Systems Engineering.

Nil H. Kilicay-Ergin is an Associate Professor of Systems Engineering at Pennsylvania State University, Great Valley School of Graduate Professional Studies in Malvern, Pennsylvania. She is also a Professor in Charge of Systems Engineering and Engineering Management Programme at Penn State Great Valley. Prior to joining Penn State in 2009, she worked within the Research Institute for Manufacturing and Engineering Systems at the University of Texas at El Paso. She was also a

Postdoctoral Fellow at the University of Missouri-Rolla. She earned her PhD in Systems Engineering and MS in Engineering Management from the University of Missouri-Rolla (currently known as Missouri University of Science & Technology). She also holds a BS in Environmental Engineering from Istanbul Technical University, Turkey. Dr Ergin's teaching involves systems engineering, systems verification, validation and testing, requirements engineering, and systems and software architecture. Her research interests include system-of-systems engineering, complex adaptive systems, model-based systems engineering and multi-agent systems. She is also affiliated with the Systems Engineering Research Center (SERC), a DoD-funded university-affiliated research center where she was the investigator for Penn State University in a multi-phased collaborative research effort among multiple universities that worked on modelling different aspects of system-of-systems acquisition. She has been a reviewer for *Systems Engineering Journal*, *IEEE Systems Journal* and *Journal of Systems Science and Systems Engineering*. She has been on the organizing committee for Complex Adaptive Systems Conference series since 2011 and has served as a technical committee member on various conferences including IEEE International Conference on System of Systems, International Conference on Complex Systems Design and Management, and INCOSE international symposium. She is a member of IEEE and INCOSE.

Marianne Kjørstad is a PhD student at the University of South-Eastern Norway, where she focusses on how to better include human aspects within early-phase systems engineering to provide innovations. She holds an MSc in Product Design and Manufacturing from the Norwegian University of Science and Technology. She has worked in the oil and gas industry for over 10 years, before starting in academia focussing on research on systems engineering. She started her PhD study in the autumn of 2017.

Fredmund Malik, Professor, is an internationally acclaimed management expert, as well as chairman and member of governance and advisory boards in business enterprises and other organizations. Malik is acknowledged as a pioneer of Holistic System-Cybernetic Management. He and his organization have shaped generations of executives and organizations of all types. His management systems and methods are being applied where conventional management reaches its limits. In 1984, he founded the Malik Institute in St. Gallen, Switzerland. With its international subsidiaries and global partnerships, it is one of the leading knowledge organizations in management, governance and leadership especially for mastering complexity. Malik is the author of more than 15 books, among them several award-winning bestsellers, and of more than 300 further publications. His classic *Managing Performing Living* was selected among the best 100 business books of all time.

Thomas A. McDermott is currently the Deputy Director of the SERC at Stevens Institute of Technology. His primary work focusses on the development of innovative methods and tools to facilitate multi-disciplinary analysis and learning in complex systems. He teaches system architecture concepts, systems thinking and decision-making, programme management and engineering leadership. He has over 33 years of experience in technical and management disciplines, including 15 years at the Georgia Institute of Technology and 18 years with Lockheed Martin.

Gerrit Muller, originally from the Netherlands, received his Master's degree in Physics from the University of Amsterdam in 1979. He worked from 1980 until 1997 at Philips Medical Systems as system architect, followed by 2 years at ASML as manager of systems engineering, returning to Philips (Research) in 1999. Since 2003, he has worked as a Senior Research Fellow at the Embedded Systems Institute in Eindhoven, focussing on developing system architecture methods and the education of new system architects, receiving his doctorate in 2004. In January 2008, he became a Full Professor of Systems Engineering at University of South-Eastern Norway in Kongsberg, Norway. He continues to work as a Senior Research Fellow at the Embedded Systems Innovations by TNO in

Eindhoven in part-time. All information (system architecture articles, course material, curriculum vitae) can be found at Gaudí Systems Architecting www.gaudisite.nl/.

Nam Nguyen, PhD, is the Director (Australia & Southeast Asia) of Malik International AG in Switzerland (one of the world's leading organisations for holistic general management, leadership, governance and transformation solutions). Nam is also a recipient of the prestigious Davos Australian Leadership Award for being at the forefront of his chosen field – systems thinking and complexity management – and a Vice President (2014-16; 2018-20) of the International Federation for Systems Research (IFSR). Nam has presented as a keynote speaker at various national and international conferences and events, and has so far contributed to the knowledge base of his research fields by authoring/co-authoring more than 80 refereed publications.

Josef Oehmen is an Associate Professor at DTU Management, Engineering Systems Group, at the Technical University of Denmark (DTU). He is also the Founder and Coordinator of DTU's RiskLab. His research focusses on better managing large-scale engineering programmes, particularly the application of advanced risk management techniques to the project and strategy level. He has a background in product development and design, systems engineering and lean management.

Barbara Rapaport (PhD, The University of Adelaide) is a Lecturer and Researcher with expertise in complexity theory in the context of political negotiations within intractable conflicts. She has provided high-level advice to the Department of Trade and Economic Development on the economic implication for South Australia of various Free Trade Agreements. She is also the author of several papers in the field of complex negotiations and conflicts, and takes interest in complex project management.

Kaye Remington is a Practitioner, Researcher and Teacher, specializing in leadership of projects in contexts characterized by uncertainty and dynamicity. She has facilitated project conflicts in Australia, Scotland, Europe and the Middle East, using soft systems methods and tools. She is the author of three books and several chapters and papers in the field of complex projects.

Azadeh Rezvani is a Lecturer and Research Fellow in the School of Management at The University of Queensland, Australia. Her research interests include project management, innovation adoption, information system and organizational psychology. Her work appears in various journals and conferences including *International Journal of Project Management*, *International Journal of Project Organisation and Management*, *International Journal of Information Management*, *Team Performance Management*, *European Management Journal* and *Computers in Human Behaviors*.

Pedro Parraguez Ruiz received an MSc degree in Innovation and Technology Management from the University of Bath, in 2010, and a PhD degree in Engineering Systems from the Technical University of Denmark (DTU), where he continued as a Postdoctoral Researcher until 2018. He is currently the Co-founder and CEO of Dataverz, a technology-based startup that develops decision-support systems to tackle societal challenges. His research and applied work are focussed on complex socio-technical systems, with emphasis on network science and data-driven analyses. This includes the study and development of decision-making support for industrial clusters, complex organizations and large engineering projects.

Sergey Suslov is the Director of International Sales and Marketing for The AnyLogic Company. He received an MS degree in Computer Science from Saint Petersburg State University and has been using simulation tools and technologies until the present day. After getting a business degree and working as a software developer, he landed at The AnyLogic Company, which has become a leading vendor in the global simulation market. In his 15 years at The AnyLogic Company, he has seen hundreds of simulation projects in many different industries, helping simulation be applied from both the technical and business sides. Currently, he is leading international sales, global marketing and technical support.

Christian Thuesen is an Associate Professor at the Technical University of Denmark (DTU), where he currently is heading the ProjectLAB. He has been working with projects for more than 15 years as Teacher, Researcher, and Consultant in various engineering settings, including construction and IT. This has sparked his fundamental research interest in the role of projects in engineering work but also more broadly in today's society. His current research interest concerns linking management of projects, programmes and portfolios with societal challenges such as climate change and sustainability.

Malcolm Tutty has served in the Air Force, Public Service and Industry in a multitude of test, operations, engineering, staff, project management and command roles. This includes being a Flight Test Armament Engineer at a research and development unit, an aircraft stores compatibility engineer while on exchange with the U.S. Air Force during Gulf War I, the AP-3C Orion Chief Engineer at Tenix, Director of both ASCENG and the Woomera Test Range, and being launch authority for two hypersonic firings into space. Recently, he deployed into Afghanistan to conduct trials and field several new high-end EW systems. He is currently serving as a Research Fellow at the Air Power Development Centre, and he has been a Fellow of both the Royal Aeronautical Society and the Institution of Engineers for over a decade and half. He was included in the Who's Who of the World for Engineering and Science in 2003.

Ricardo Valerdi is a Professor at the University of Arizona (UA) in the Department of Systems and Industrial Engineering. Prior to UA, he was a Research Associate in the Engineering Systems Division at the Massachusetts Institute of Technology (MIT). His research focusses on improving our understanding of complex systems – both technical and social – by building predictive models. His research has been funded by Army, Navy, Air Force, BAE Systems, Lockheed Martin, Raytheon and the NCAA. He received a PhD in Industrial and Systems Engineering from the University of Southern California. He is a foreign member of the Mexican Academy of Engineering and was a Visiting Fellow of the UK Royal Academy of Engineering.

Brian E. White received PhD and MS degrees in Computer Sciences from the University of Wisconsin, and S.M. and S.B. degrees in Electrical Engineering from MIT. He served in the U.S. Air Force and for 8 years was at MIT Lincoln Laboratory. For 5 years, he was a Principal Engineering Manager at Signatron, Inc. In his 28 years at The MITRE Corporation, he held a variety of senior professional staff and project/resource management positions. He was the Director of MITRE's Systems Engineering Process Office, 2003–2009. He retired from MITRE in July 2010, and currently offers a consulting service, CAU←SES ("Complexity Are Us" ←Systems Engineering Strategies). Website: www.cau-ses.net.

Seyed Ashkan Zarghami is a Lecturer at Torrens University Australia. He has recently submitted his doctoral thesis at the University of Adelaide. He holds a Bachelor's degree in Civil Engineering, a Master's degree in Project Management, and a Master's degree in Civil & Structural Engineering. His current research interests include project management, system engineering, reliability analysis of infrastructure networks and operations management. Ashkan has a great passion for proposing novel ideas, and his research has appeared in top-ranked international journals such as *Reliability Engineering and System Safety, International Journal of Production Research, Engineering, Construction and Architectural Engineering*, and *System Research and Behavioral Science*. Prior to his doctoral study, he worked for several years in large-scale construction and infrastructure network projects. Ashkan has developed his expertise and interest in project management through undertaking a wide range of research and engineering duties to support a variety of complex projects.

1 Introduction to the Evolving Toolbox

Alex Gorod and Leonie Hallo
The University of Adelaide

CONTENTS

Let's consider the task of building a toolbox for complex project management. What should be inside that toolbox? What are the essential toolsets that will enable project managers to deal more effectively with complex projects? To answer these and other important questions, the editors approached subject-matter experts in a variety of topic areas and asked them to contribute the most useful tools from their own toolbox and experience. These state-of-the-art tools have been brought together for this book. Additional topics may have been included at this time; artificial intelligence and machine learning, blockchain technology, systemic quality management, feasibility, and managing stakeholders in a complex environment, among others. New concepts and technologies are constantly emerging, and the toolbox is expected to evolve over time. As a starting point, the editors gathered 22 initial toolsets, with a chapter devoted to each of these valuable toolbox components for managing complex projects.

In this chapter, as an introduction to the toolsets included and to highlight their contribution, the toolsets are described through the lens of a complex project, the Fukushima Daiichi Nuclear Power Plant disaster recovery effort. The Fukushima recovery project is a major undertaking given its magnitude, scope, longevity, and the costs involved. There are numerous stakeholders, a great number of unknowns, and no clear correlation between cleanup efforts and the results that will be achieved. Enormous funds have been allocated to the cleanup and recovery, and it is expected to be the costliest natural disaster in history [1]. Given the continuous emergence of unanticipated factors, the problem space is perpetually changing, which makes progress much more difficult. Further, the project is complex because of its inherent interconnectedness. Whatever approach is applied, there are cumulative effects due to the interrelatedness of the constituents. The sophisticated management tools needed are not readily available and there is no adaptable platform for managing them.

In addition to providing an overview of the latest toolsets for complex project management in the context of the Fukushima recovery project, the evolving toolbox approach, which is a systemic platform for managing tools, is discussed in more detail later in the chapter. Project managers, teams, and organizations can use such a system for collecting, integrating, monitoring, and updating tools to ensure that their tools are optimized in the light of contemporary developments.

The first toolset presented in this book concerns the use of case studies. Case studies have many advantages [2,3]. They are based upon real-life situations that can come from different domains and industries, revealing a range of best practices and lessons learned. They are also a source of patterns [4] and can be used as blueprints for modeling and simulation (M&S) [5]. Case studies can provide feasibility assessment of selected approaches [6]. In the Fukushima recovery project, it is expected that the authorities have established contact with international bodies, such as the International Atomic Energy Association (IAEA) in Geneva, Switzerland, to access information about previous nuclear recovery operations. From a review of existing case studies, including Chernobyl and the Three Mile Island Nuclear Accident, it would be possible to evaluate the success of previously utilized approaches, recognize best practices, identify patterns, and determine the likely most effective strategies. In addition, case studies complement modeling in gaining a more accurate understanding of the boundary of the disaster recovery project. The Fukushima recovery project itself is a case study for future disaster response planning [7]. **Chapter 2: Case Studies Toolset** presents a case study of the Boston Big Dig, a major infrastructure project, and describes case study approaches that can be used to assist managers of complex projects in extracting best practices, patterns, and other critical information from similar cases around the world.

Scheduling plays a key role in any project. However, this is particularly problematic within a complex environment because forecasting is not always reliable and accurate prediction may not be feasible. It was originally thought that the Fukushima cleanup would take 30–40 years to complete [8], but now it has been suggested that this project may take up to 200 years [9]. Yet, scheduling is fundamental to project success. Therefore, it is vital to plan and adjust forecasts in response to new impediments and developments. **Chapter 3: Scheduling Toolset** presents a range of project scheduling tools and describes how these have advanced from the mediaeval period to the present time. The chapter then looks to the future and discusses how project scheduling tools can continue to evolve in order to cope with increasingly complex projects. The toolset can offer support to project managers in forecasting and in assessing the accuracy of their forecasts.

> Scheduling is extremely difficult within a complex environment because forecasting is not always reliable and accurate prediction may not be feasible.

No matter how complex a project is, it is still necessary to plan and allocate a budget and to establish the costs associated with the project and the various activities involved. Cost estimation becomes a more demanding process when considering complex projects. **Chapter 4: Cost Estimation Toolset** presents novel modern approaches on how cost estimation is conducted under uncertainty and provides an overview of models that are relevant to complex projects. Further, the chapter offers a case study that illustrates how one of the selected models is applied in the complex project management environment.

Project managers require tools to effectively manage the human factor or the people involved. Managers may be faced with great stakeholder diversity and need to strive to understand the nature of interactions between individuals and groups, recognize ways to influence and motivate them, and foster a culture of creativity and flexibility. **Chapter 5: Human Factors Toolset** discusses the central role of people in complex project management. Both the unique characteristics of any individual and the way people relate to each other and to technology will have an enormous impact on the results. Whenever there is a human factor present, outcomes are unpredictable and uncertainty will ensue. Managers of complex projects need to have skills in the effective management of themselves, their staff, and their teams. Both internal and external stakeholders need to be identified and managed, which may not be as easy in a complex dynamic environment as it is in traditional projects. In the case of Fukushima, it is crucial to identify the thousands of different entities involved. It is the job of the project manager to integrate and streamline these stakeholders towards the overarching shared purpose as well as to instill a commitment to the project. This chapter provides an overview of the human factor in complex project management and includes several relevant tools to assist project managers.

> Whenever there is a human factor present, outcomes are unpredictable, and uncertainty will ensue.

What model of leadership should be applied in leading a complex project? There are many leadership styles, such as transactional leadership, transformational leadership, visionary leadership, participatory leadership, servant leadership, and responsible leadership, to name just a few. Each style is a tool that can be used under certain circumstances. The question of the appropriate leadership model for the Fukushima recovery effort has been debated in the literature [10], with many observers indicating that a top-down style, while consistent with Japanese culture, was not the most optimal approach under these circumstances [11]. **Chapter 6: Leadership Toolset** discusses leadership as a critical skill for managers of complex projects. Due to the growing role of globalization, technology, and the human factor, projects are becoming exponentially more complex. Project managers need tools to unite and lead disparate groups of people of different time zones, disciplines, and goals towards a common vision. The future obviously promises to present further challenges for managers in developing a shared holistic perspective and fostering collaboration, given

> Project managers need tools to effectively unite and lead disparate groups of people of different time zones, disciplines, and goals towards a common vision.

the increasing diversity of constituents. According to a study conducted by the Project Management Institute (PMI) of 697 experienced project management practitioners, leadership skills are considered to be extremely important for successfully managing highly complex projects [12].

The same report from PMI asserts that negotiation and persuasion skills are also required in working productively with complexity. When numerous stakeholders are involved, all with different objectives, needs, and views, negotiation plays a major role in reaching an agreement on a collective mission. Thus, project management necessitates constant bargaining and trade-offs that at times give rise to conflict. In the case of Fukushima, there are ongoing discussions among the many stakeholders within the project and outside, such as between Tokyo Electric Power Company (TEPCO), IAEA, and other international bodies. **Chapter 7: Systemic Negotiation and Conflict Resolution Toolset** presents tools for complex negotiation and conflict management in projects that

contain high levels of uncertainty and unpredictability. Notions of leverage, soft systems methodology, and emotional intelligence are suggested as frameworks for aligning stakeholders' interests towards resolution of an impasse in negotiation. The chapter is focused on various techniques that can be used to successfully negotiate as well as to mitigate potential conflicts.

> Project managers also need to stand back and observe the complex project in its entirety.

Project managers also need to stand back and observe the complex project in its entirety. Systems theory enables the creation of a view of the project's evolution and life in its environment as it interacts with other projects. In the case of the Fukushima recovery project, organization and systems theory can be applied in recognizing how the multiple parties are involved, such as the public, regulatory bodies, and other participants, perceive the recovery project. Gaining such a perspective can be informative in identifying more effective approaches as part of rebuilding and revitalization efforts. **Chapter 8: Organization and Systems Theory Toolset** asserts that complexity can be actively managed. According to the chapter authors, there are a variety of different characteristics of complexity. The chapter introduces several potential strategies for coping with project complexity, such as network analysis, systems dynamics, modularization, antifragility, and mindfulness.

Taking a systems theory approach raises the issue of control. It is widely understood that in complex environments, exercising control is not always feasible. In the case of the Fukushima recovery effort, attempting to control the continuously altering situation through the existing hierarchy is not necessarily the most constructive way of operating. Since it takes time for information to go through the chain of command, decisions made at the grassroots level are often more effective because they are timely. In addition, the workers at the bottom of the organization may have more information about the situation than managers at the top. Therefore, rather than using the traditional command and control structure, governance mechanisms are needed in order to influence the environment. Governance is about establishing rules and parameters within which constituents can self-organize. This makes it possible to affect the environment without controlling it, obtain feedback, and monitor the state of the project. In terms of the disaster recovery project, appropriate standards, reporting procedures, safety requirements, and other "rules of the game" will need to be set and properly communicated. It is the responsibility of the project manager to influence the project environment to foster emergent bottom-up (grassroots) behavior through providing sufficient autonomy. As a consequence, this will help to address quickly changing dynamics while simultaneously ensuring that the overall project mission is accomplished in a safe, productive, and efficient manner. **Chapter 9: Enterprise Governance Toolset** presents a variety of governance tools concerned with oversight and accountability, focusing on complex projects. A complex project can be considered

> Complex system governance centers around ways to design, carry out, and develop the functions that maintain performance.

as a system that takes and transforms inputs from the environment. It subsequently produces outputs and receives feedback about those outputs. Complex system governance centers around ways to design, carry out, and develop the functions that maintain performance. This chapter offers information and guidance for managers of complex projects to better understand and access tools for effective project governance.

Chapter 10: Critical Success Factors and Climate Toolset looks at the factors that are most likely to lead to complex project success, including open communication, project planning and monitoring, and stakeholder satisfaction. This chapter presents practical tools to assist managers of complex projects in building upon these success factors in project management. The project manager needs to understand the benefits of fostering a climate where people feel that they are receiving relevant communication, that there is a workable plan, and that they are satisfied with their job. In the case of Fukushima, reports indicate that the sharing of information with the general public was substantially inadequate, leading to widespread feelings of disempowerment and anxiety, as well as social isolation and social stigma [13]. The former head of the US Nuclear Regulatory Commission, Dale Klein, stated the following about the need for clear communication to the public in the aftermath of the Fukushima disaster: "If TEPCO does not improve their communication, it will be very difficult for them to regain the public trust" [14].

Pattern recognition and management, another important capability for managers of complex projects, is a way of analyzing data and identifying elements with a high probability of repeating. If patterns can be recognized, they can be managed. Pattern management allows identification of repeatable best practices and lessons learned and, therefore, represents an advantage for project management. Furthermore, it is instrumental in the interpretation of big data. **Chapter 11: Pattern Identification and Management Toolset** provides an overview of pattern recognition and management tools that can achieve those tasks. It further offers an example of how patterns can be documented. The chapter compares different types of patterns and their usefulness in managing complex projects based on reports from experts. In the case of Fukushima, it would be valuable to examine the patterns that can be identified from other case studies, including Chernobyl and other major projects of such scale and complexity. What are the patterns in terms of successfully managing scope, budget, and resources?

Policies are sets of clearly defined rules and standards that have an impact on individual employees, teams, contractors, and the overall project. In the case of disaster recovery efforts, policy is crucial for communicating acceptable behaviors for the personnel involved, including partners such as subcontractors. **Chapter 12: Policy Management Toolset** describes the importance of policy management across many business domains and its components of creating, approving, communicating, implementing, and changing policies and procedures in order to maintain policy sustainability. Policy management systems provide project managers with suitable methods for managing organizational policies and keeping them up-to-date. Failure to manage policies appropriately can lead to negative and expensive consequences for organizations. Several tools are used to show how effective policy management can be achieved. Additionally, the chapter discusses reputation management and data protection, where policies provide a demonstrated standard of conduct and assist in protecting organizations from litigation. Considering the Fukushima project, there are multiple critical policies around safety, affected site access, and communication, among other key components. These should be effectively managed and regularly updated. Safeguarding information and ensuring confidentiality related to the parties impacted by the disaster is another highly important policy in this situation.

Legal issues represent a major topic for complex project management. In the case of Fukushima, human lives are disrupted and many potential liabilities are involved, which may lead to costly litigation seeking compensation and sanctions, or for example the development of regulatory frameworks and government-underwritten compensation as part of the project, given the scale of the disaster. In any project, especially in a complex one, managers must ensure that all aspects of a project are handled according to the local/state/federal and international laws and regulations. **Chapter 13: Legal Aspects Toolset** offers a description of some of the legal aspects of project management with an emphasis on contractual and negligence obligations. The chapter focuses upon legal responsibilities arising from contract and tort laws, and also presents practical applications and examples of tools from current industry projects.

It is important to protect and respect intellectual property, brands, and trademarks, and the same principle certainly applies during disaster recovery efforts, when it is particularly likely that new approaches will be trialed. Therefore, managers must ensure that intellectual property is properly acknowledged and licensed. Similarly, the intellectual property generated during the Fukushima recovery project will need to be protected. **Chapter 14: Intellectual Property Toolset** discusses the importance of the protection of intellectual property rights within project management. As projects become global, tools are needed to assess the impact of intellectual property laws across the relevant jurisdictions. The chapter describes basic types of intellectual property and suggests several tools that are available to assist project managers.

As with the organization and systems theory, systems thinking presents a valuable framework to assist managers of complex projects in taking a more holistic approach. Project managers may fail to fully understand the quickly evolving and ambiguous aspects of the complexity they are faced with when planning and implementing projects. Further, they may not grasp how changes in one part of a project can have an impact on its other parts. In the case of Fukushima, systems thinking is critical because of the highly dynamic and complex nature of the post-disaster environment, the enormous scope and magnitude of the recovery efforts, and the sheer range of stakeholders involved in the project, including national and international constituents. **Chapter 15: Systems Thinking Toolset** discusses systems thinking as a framework to address problems that lack simple solutions. The significant interconnectedness and complexity of problems tackled by managers of complex projects today requires them to view the problem space beyond the individual system components and instead, to consider an expanded system structure, along with interrelationships between the project's constituents and its environment. The chapter presents tools for assisting project managers to develop this type of more inclusive and multidimensional approach.

> Systems thinking presents a valuable framework to assist managers of complex projects to take a more holistic approach.

Chapter 16: Test and Evaluation Toolset describes advances in testing and evaluation (T&E) of systems for their effectiveness and suitability. The chapter discusses the challenge of conducting T&E early enough to influence the design of the project and reduce risk. The increasing complexity of systems and the interconnectedness of systems of systems highlight the importance of undertaking T&E on time. The chapter outlines ways in which T&E processes, techniques, and management have evolved in order to meet new complexities faced by project managers. In the case of the Fukushima recovery effort, all approaches applied by TEPCO would be tested and evaluated for soundness, quality, safety, and other important considerations to increase project effectiveness and reduce overall project risk.

It has been said: "If you can't draw it, you don't understand it" [15]. In order to be able to test concepts, it is first necessary to visualize them. Due to technological advances, visualization is becoming a more and more viable tool for creating proof of concept. Visualization can reveal interconnections and allows different levels of abstraction to be seen. It is possible to zoom in on the individual or the team and, then, zoom back out to view the whole project and its interaction with other projects. Interactive maps and data visualization are useful tools in recognizing the world view. In the case of the recovery effort, visualization tools would be extremely helpful in establishing proof of concept. Furthermore, visualization can be instrumental in fostering a dialogue among stakeholders to identify possible inconsistencies and gaps in understanding. **Chapter 17: Advanced Visualization Toolset** explains the role of knowledge transfer in social networks and how to visualize knowledge in order to assist in understanding and increase the likelihood of project success. Success relies on human management and decision-making, and tools for visualizing problems are needed to help in managing the dynamics of a complex project. The chapter presents a framework that relates knowledge transfer to project complexity and improves decision-making. Several types of visualization mechanisms are presented and discussed.

Social network analysis describes the dynamics of relationships and provides a variety of tools for gaining an understanding into how different elements are interrelated and integrated. In addition, it elaborates on the process of measuring various aspects of networks. In the case of Fukushima, social network analysis would be valuable in observing how people are interconnected and identifying the structure and the type of networks. In turn, this knowledge can help increase effectiveness in terms of allocation of resources and information flow. **Chapter 18: Social Network Analysis Toolset** discusses the usefulness of network theory and methodology for complex project management in recognizing how patterns and dynamics of complexity form and evolve. Network theory is now being extended into extracting patterns from massive digital data sets.

George Box, a famous British statistician, said: "All models are wrong, but some are useful" [16]. Having visualized concepts, M&S is needed in order to observe the changing dynamics of the situation. It is the responsibility of the complex project manager to come up with models that are practical and as representative of the problem space as possible. These can then be simulated against the environment. Clearly, there are many benefits to M&S. Modeling is a common human activity. Psychologists have written about simple models in the form of mind maps for many decades. A mind map is a way of organizing information visually to show the relationships among the various segments of the whole. This process has been used in problem solving and training across a wide range of domains [17]. At a broader level, the many advantages of modeling were described by Epstein [18], who argues that people often build models that are implicit rather than explicit. It is only when models are made explicit that they can be tested. In the case of Fukushima, given the magnitude of the recovery process, individual approaches in the form of explicit models could be tested to determine their applicability before applying them in real life, thereby avoiding liability issues and minimizing financial expenditure. M&S makes it possible to identify and test how external factors might affect the project, and this can subsequently enable policy dialogues and lead to better public understanding of the scope of the project. **Chapter 19: Modeling and Simulation Toolset** presents contemporary M&S techniques as applied to complex project management. In addition, an overview of the three most prevalent modeling paradigms, system dynamics, agent-based, and discrete event, is provided using case studies in complex project management.

Adaptability, agility, and resilience can help managers to become more flexible in the face of ongoing transformation. In terms of Fukushima, this approach would help in ensuring that any activities undertaken are resilient to change, taking into consideration current and potential variability due to the numerous factors involved in the recovery effort. Adaptability, agility, and resilience can thus help project managers to behave in a proactive manner and to be better prepared to adjust to modifications in plans. Tools in this area can support the ability to bounce back if a particular approach proves unsuccessful, to navigate one's overall strategy to be more effective, and to change quickly without major consequences. This is discussed in **Chapter 20: Adaptability, Agility, and Resilience Toolset**. Properties of adaptability, agility, and resilience have emerged in complex systems, and these have direct relevance for complex project management as presented in this chapter. The chapter also describes how to measure and verify these properties and the relationships between them.

> Adaptability, agility, and resilience can help managers to become more flexible in the face of constant transformation.

Chapter 21: Cyber-Systemic Toolset discusses the current transition from the "Old World" to a "New World" and the turbulence being experienced during this development. The chapter presents an advanced method as a tool for handling complex projects and supporting stakeholder management. The introduced process of Syntegration enables a sophisticated understanding of the differing opinions of key stakeholders during an intense guided workshop experience, when the most important variables are identified and the relationships between them are mapped. In turn, this approach leads to a more informed and holistic perspective about the optimum way to transform an organization.

It is evident that risks need to be managed systemically and not just in isolation. Risk plays an enormous role in project management. Only recently has it been understood that in addition to individual risks, there is always an inherent systemic risk that could be greater than the sum of the constituent risks. Mitigating a risk can create other risks because any risk is simultaneously a risk on its own and also a component of a larger risk. The reductionist approach may not be useful in a complex environment. Risks need to be managed as part of the whole since they are integrated and interrelated. In considering the case of Fukushima, systemic risks arose when the evacuation of massive numbers of people led to many social consequences, including isolation, stigma, and great upheaval in people's lives. These social consequences resulted in additional risks, such as increases in disease, both physical and psychological [19]. Systemic risk management leads to systemic resilience and adaptability as well as to innovation, enabling project managers to be concurrently reactive and proactive. **Chapter 22: Systemic Risk Toolset: Another Dimension** discusses taking a system's perspective in risk management. Project managers are encouraged to think beyond the project design and build phases through to operations and maintenance. Furthermore, they need to consider project portfolios as opposed to focusing on a single project. Finally, project managers must include other stakeholders' priorities and not only the organization's perspective as part of their risk management strategies and decision-making. The chapter provides a range of techniques to assist project managers in analyzing and managing systemic risk. Moreover, it discusses the value of causal mapping. This is particularly relevant to the Fukushima case since it makes it possible to identify and address new risks that are otherwise unknown, including a possible correlation between high evacuation-related stress levels and social outcomes [20].

> There is always an inherent systemic risk which could be greater than the sum of the constituent risks.

Risks and opportunities are closely related. Risk may provide an opportunity for new projects, development, and innovation. Complexity requires project managers to be enthusiastic about exploring possibilities, which is challenging since it introduces uncertainty and ambiguity. In today's complex settings, traditional methods of managing projects are no longer sufficient, and project managers need to create an innovation-fostering environment. The final chapter in this book, **Chapter 23: Systemic Innovation Toolset**, offers several tools for systemic innovation that will encourage project managers to be open to new ideas and promote an innovative culture in complex project management.

The toolsets for complex project management outlined above can be accessed in the evolving toolbox cloud, which is depicted in Figure 1.1. The toolbox also includes toolsets of artificial intelligence and machine learning in addition to other tools that will appear in the future.

> The evolving toolbox conceptualization for complex project management is a paradigm shift.

The evolving toolbox conceptualization for complex project management represents a paradigm shift. Not only are there various tools available in one location, but these tools are also evolving through emergence and continuous feedback loops that take place inside the toolbox. The toolbox operates as a cloud which is open to the external environment, thus allowing adaptation to outside changes. As tools are being transformed, they will contribute to further development of the toolbox. All of the tools and the toolbox itself are self-learning and self-updating, benefitting from the emergent behavior through interconnectedness within the cloud. Toolsets for modern and future

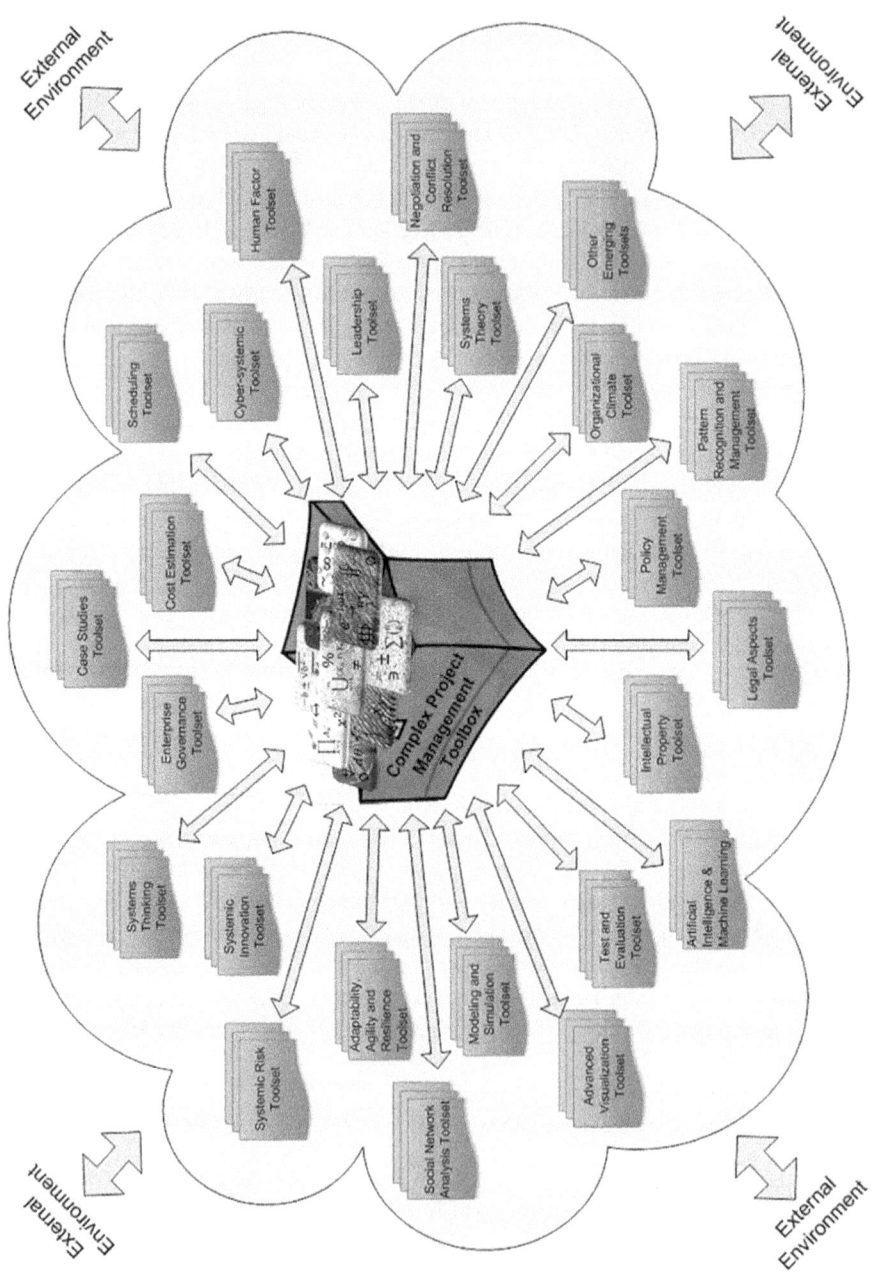

FIGURE 1.1 Evolving toolbox cloud for complex project management.

applications need to be self-evolving. In response to the tools updating themselves through their interactions inside the cloud and with external factors, the ability to manage complex projects will be enhanced and will grow more nuanced. Additional advanced tools will be added naturally over time. Gradually, an expanded and inclusive toolbox will provide broader support to project managers faced with increasingly complex projects.

The evolving toolbox approach offers several key advantages:

- It includes a variety of suitable tools for coping with many of the existing and potential challenges in complex project management.
- It enables emergence to occur within the toolbox, allowing the tools to add value through interaction.
- It can better engage and prepare students for next-generation complex project management through hands-on application of tools.
- It can foster and facilitate a policy dialogue among project stakeholders.
- Collectively, the tools provide multidimensionality and, as a result, increase the scope and capability of the toolbox.
- The tools will be smart and, therefore, will be self-updating and adaptive to change. The toolbox is also a self-learning and malleable open system, allowing its components (tools) to respond to internal and external factors. The tools can function as flexible agents, evolving more quickly as new technology and requirements arise. Through this interconnectedness, the tools can be up-to-date and optimized by searching for new possibilities.
- New algorithms can be created to increase the combinations and configurations of tools/toolsets as part of future developments.

> The tools are self-updating and adaptive to change. The evolving toolbox is also self-learning and malleable because it is open, allowing tools to respond to internal and external factors and also therefore emergent.

Even though this book is located within the project management realm, the authors believe that at the complexity level, whether dealing with a complex project, complex system, complex enterprise, or complex system of systems, fundamentally, the same type of behavior needs to be managed. Furthermore, synergy is now developing between the disciplines of software engineering, systems engineering, engineering management, and project management, leading these domains to become more interactively aligned, which will result in the availability of more appropriate and potent tools in relevant complex areas of management.

The evolving toolbox cloud offers other potential significant benefits. Presently, there are only a few standards and certifications in the complex project management world. The introduced approach can serve as a stepping stone towards the development of standardization, such as in setting requirements for proper documentation of best practices and/or obtaining a certification in systemic quality management, enterprise governance, collaborative leadership, and other practices. With time, these new developments can significantly advance the overall field of project management.

The role of a project manager will continue to be modified as projects evolve. Future attributes and requirements will reflect rapid changes in technology and the need for a more forward-thinking outlook to cultivate innovation, recognize opportunities, and be adaptive. Thus, whereas traditional project management is typically based upon a reactive mindset, project managers will need to become more future-oriented and proactive in seeking out new ideas. The sought-after skills and capabilities of project managers are already trending more towards the ability to be more inventive and to take an analytical, systemic view. Undoubtedly, creativity, holistic thinking, and integrative systemic approaches are central to tomorrow's success in complex project management.

> The role of project managers will be modified as projects evolve, and in fact, it has already been significantly altered.

Given the rapidly increasing pace of change, the usability of most traditional tools may fade as they may prove to be less effective and efficient in real-world situations. This parallels such recent developments as the closing of video stores in the film industry, traditional taxis being overtaken by Uber in the transportation domain, the dominance of Skype in telecommunications, the popularity of airbnb in the hospitality industry, and the rise of Amazon in retail. It is reasonable to view current global disruption due to swift technological progress as an opportunity to develop new tools: tools with the capability to produce relevant and timely results for the end user in tune with emerging trends. The evolving toolbox can be considered a way to build upon today's digital disruption to help project managers better navigate ever-increasing complexity and to prepare for and respond more comprehensively to such a rarely occurring but devastating event as the Fukushima Daiichi Nuclear Power Plant disaster.

> Looking to the future, given the rapidly increasing pace of change, the usability of most traditional tools may fade as they may prove to be less effective and less efficient in real-world situations.

The evolving toolbox in itself is an intelligent tool and technology consisting of an interactive and live platform of tools and toolsets for complex project management. The tools will act as agents with the autonomy to make independent choices under certain conditions in the environment. The entire system is evolving both in its totality and in the activities of the individual tools and toolsets that will have self-updating and syncing capabilities through their participation in the cloud. In turn, with new tools also being added, the toolbox will continue to expand its possibilities for overcoming unfamiliar challenges in complex project management. This will allow project managers to tap into the available tool options and integrated combinations that are at the forefront of innovation.

> The evolving toolbox in itself is an intelligent tool and technology, consisting of an interactive and live platform of various tools and toolsets for complex project management.

British Prime Minister Winston Churchill said in a worldwide broadcast on February 9, 1941: "Give us the tools, and we will finish the job" [21]. To bring this concept to the present day, the quotation could be rephrased as follows: "Give us an evolving toolbox, and we will finish the complex job." The evolving toolbox cloud described here has the potential to become one of the most powerful tools in the complex project manager's arsenal.

REFERENCES

1. B. Zhang, Top 5 most expensive natural disasters in history, *AccuWeather.com*, vol. 30, 2011.
2. K. M. Eisenhardt, Building theories from case study research, *Academy of Management Review*, vol. 14, no. 4, 532–550, 1989.
3. T. N. Srinivasan and T. S. Gopi Rethinaraj, Fukushima and thereafter: Reassessment of risks of nuclear power, *Energy Policy*, vol. 52, 726–736, 2013.
4. R. Larsson, Case survey methodology: Quantitative analysis of patterns across case studies, *Academy of Management Journal*, vol. 36, no. 6, 1515–1546, 1993.
5. N. D. Crossman, B. Burkhard, S. Nedkov, L. Willemen, K. Petz, I. Palomo, E. G. Drakou, B. Martín-Lopez, T. McPhearson, K. Boyanova, and R. Alkemade, A blueprint for mapping and modeling ecosystem services, *Ecosystem Services*, vol. 4, 4–14, 2013.
6. J.-A. S. Christiansen and T. Anderson, Feasibility of course development based on learning objects: Research analysis of three case studies, 2004.
7. S. de Saille and P. Matanle, Fukushima-The triple disaster and its triple lessons: What can be learned about regulation, planning, and communication in an unfolding emergency? 2016.
8. A. Saini and T. Koyama, Cleanup technologies following Fukushima, *MRS Bulletin*, vol. 41, no. 12, 952–954, 2016.
9. J. Green, Fukushima fallout: Updates from Japan, *Chain Reaction*, no. 126, 28, 2016.
10. Y. Funabashi and K. Kitazawa, Fukushima in review: A complex disaster, a disastrous response, *Bulletin of the Atomic Scientists*, vol. 68, no. 2, 9–21, 2012.

11. A. Nakamura and M. Kikuchi, What we know, and what we have not yet learned: Triple disasters and the Fukushima nuclear fiasco in Japan, *Public Administration Review*, vol. 71, no. 6, 893–899, 2011.
12. P. M. Institute, PMI's pulse of the profession in-depth report: Navigating complexity, ed: PMI Publishing Division, 2013.
13. C. Leppold, T. Tanimoto, and M. Tsubokura, Public health after a nuclear disaster: Beyond radiation risks, *Bulletin of the World Health Organization*, vol. 94, no. 11, 859, 2016.
14. S. Denyer, Eight years after Fukushima meltdown, the land is recovering, but public trust is not, ed. US: The Washington post, 2019.
15. B. Vandenbosch, *Designing Solutions for Your Business Problems: A Structured Process for Managers and Consultants*. John Wiley & Sons, Hoboken, NJ, 2003.
16. G. Box, Science and statistics, *Journal of the American Statistical Association*, vol. 71, 791–799, 1976.
17. J. Budd, Mind maps as classroom exercises, *The Journal of Economic Education*, vol. 35, no. 1, 35–46, 2004.
18. J. M. Epstein, Why model? *Journal of Artificial Societies and Social Simulation*, vol. 11, no. 4, 2008.
19. K. Hasegawa, Facing nuclear risks: Lessons from the Fukushima nuclear disaster, *International Journal of Japanese Sociology*, vol. 21, no. 1, 84–91, 2012.
20. E. J. Bromet, Emotional consequences of nuclear power plant disasters, *Health Physics*, vol. 106, no. 2, 206, 2014.
21. W. Churchill, Broadcast. London, 1941.

2 Case Studies Toolset

Brian E. White
"Complexity Are Us" ← Systems Engineering Strategies (CAU ← SES)

CONTENTS

BACKGROUND

In keeping with the theme of this entire book, as project situations become more complex, improved as well as additional project leadership and management tools are needed. Along with increased complexity, it is also likely that classical, conventional, or traditional project management (PM) methodologies may become even less effective. In other words, the theories of complex project management (CPM) need to be bolstered. As in the case of complex systems (CSs) engineering (CSE), in order to gain a greater understanding of what is required for successful endeavors, serious efforts in case studies are appropriate. Namely, by mounting, conducting, and analyzing case studies, much can be learned about what works and what doesn't. This nurturing of information, derived from specific cases, can inform the development of updated theories. Consequently, this chapter proposes eight easy-to-use tools that can be usefully applied in PM, particularly in connection with case studies of complex projects (CPs). These tools are defined, explained, and applied in an example case study of the ultimately successful massive and controversial Big Dig project of relocating the elevated Central Artery roadways of Boston below ground. Although some of the tools are derived from the systems engineering (SE) world, they are easily tailored for PM.

> Also included is an example case study of a hugely ambitious project where these tools, as well as CP behaviors and CPM management principles, are applied.
>
> By mounting, conducting, and analyzing case studies, much can be learned about what works and what doesn't.

HISTORICAL PERSPECTIVE AND RELEVANCE

This Case Studies Toolset consists of eight tools (listed just below), mostly developed between 2002 and 2015, all of which are considered useful in addressing the complexities associated with PM as well as SE:

1. Case Study Outline (CSO)
2. Project Management Life Cycle (PMLC)
3. Zachman Framework
4. Enterprise Project Management (EPM) Profiler
5. Project Management Activities (PMA) Profiler
6. The Enneagram
7. Soft Projects Methodology (SPM)
8. Complex Adaptive Project Management (CAPM) Methodology.

> The case studies chapter proposes eight easy-to-use tools that can be usefully applied in program management, particularly in connection with case studies of CPs.

The most important tool is CSO, a comprehensive outline for any case study to be developed, explored, and publicized. This detailed outline, conceived in 2015, is part of a comprehensive book on case studies [1] that is recommended as the overarching reference for case studies, in general, as well as most of their associated tools.

The PMLC, in the manner of [2], more properly drawn in a circle, is a cogent picture of the traditional PM process. The principal innovation is the central idea of project overview which contemplates and arbitrates various stages of the process to emphasize the iterative and feedback nature of often complex situations encountered on many projects.

Most projects are part of a larger program that, in turn, may be part of a larger enterprise that may be, indeed, a subset of an overarching CS. When it was introduced in 1987, the Zachman Framework [3] was considered an architecture for information systems; later, it was characterized as applying to enterprise architectures. This framework is an excellent way of helping one consider all aspects of a given project situation.

The original Enterprise Systems Engineering (ESE) Profiler [4], now adapted as the EPM Profiler, is a good way of characterizing the essence of any project from eight different perspectives associated with four distinct contexts. The Systems Engineering Activities (SEA) Profiler [5], now adapted as the PMA Profiler, is a way of assessing how the project team is actually approaching the project considering each stage of the PMLC and to what extent that activity is addressing the associated complexity.

Although the EPM and PMA Profilers are fundamentally qualitative, they can be quantified for comparison purposes by selecting and assigning integer (say) weights for each aspect and then computing a total numerical score. These profilers can and should also be used for intra-team and cross-team discussions to highlight common concerns and/or areas requiring more attention.

It's important to understand that qualitative tools can be as (or even more) useful than some quantitative tools where the technical details of the quantification become "rabbit holes" obfuscating the overall objectives in providing good tools. This raises the question: What are the positive attributes and potential negative pitfalls of PM tools?

First, a good tool must be useful in comparing and contrasting a particular element of the contemplated system of the relevant domain of interest. The tool must exhibit sufficient richness, diversity, or variety to at least roughly match the complexity of the system element in question. The tool should be relatively easy to apply and within a reasonably short time frame. The tool structure should facilitate discussions among protagonists that help illuminate commonalities or distinctions among different situations within the same domain. One should continually rate the usefulness of every tool in terms of achieving acceptable project solutions within the overall desired outcome space. It should be feasible to suggest possible improvements to be further tested based upon objective observations of tool usage. Thus, tool "evolvability" is a key attribute.

The Enneagram [6,7] is an important tool for ensuring more complete discussions of PM issues. Checkland's famous Soft Systems Methodology (SSM) [8] is easily adapted as the next tool, SPM.

The latest Complex Adaptive Systems Engineering (CASE) Methodology [2,9] that builds upon and emphasizes SSM, is now adapted as the CAPM Methodology. This is intended as very comprehensive guidance for applying effective and common-sense techniques to CP leadership and management.

Here are a couple of thoughts this author asserts about leadership versus management that are often ignored in SE practice. Exemplar leadership, including the positive ramifications of many basic leadership tenets [10], needs to be the fundamental core of any CP. Leadership in addressing complex change to establish an effective strategy is more important than the subsequent management of that change strategy.

REVIEW AND ANALYSIS OF TOOLSET

If anyone would like soft copies of the following tools (and their supporting artifacts) that one might use, tailor, modify, or improve upon, please contact this author at bewhite71@gmail.com.

CASE STUDY OUTLINE

The first tool in Toolset is the recommended outline for conducting and explaining CP case studies. The original full-blown outline, then oriented more toward CSs and CSE as opposed to CPs and CPM, is provided in great detail and substantial justification in Chapter 3 of Ref. [1]. Table 2.1 only provides the rudimentary outline specifically adapted for CP case studies. It doesn't matter much in

TABLE 2.1

Principal Objectives of Case Study Sections

Main Section of Case Study	This Section's Principal Case Study Objective
Case Study Elements	Provide enough concise information to enable a project leader, or team members, or general readers to decide whether this case study is of particular interest to them.
Background	Further define the case study, especially for those working on other projects that are not yet sure if this project is relevant to their interests.
Purpose	Capture the impetus behind and specific reasons for the case study, and show some of the passion that drove or is driving this project's evolution.
Constituents	Characterize people and institutions interacting in the case study, and illuminate their motivations, e.g., what incentives drove or drive them?
(Complex) Project/Program/ System/System of Systems/ Enterprise	Provide a clear and complete but focused description of this, presumably complex, project (or more generally, how it fits into a larger program, system, system of systems, or enterprise).
Challenges	Highlight the principal aspirations and difficulties of the subject project.
Development	Show just how transformational change can occur on the project.
Results	Answer the "What Happened?" and "So what?" questions.
Analysis	Provide suggestions to others in how to interpret results of this project's transformation, hopefully improvements, and what it all meant or means.
Summary	Complement mainly the Case Study Elements, Purpose, and Results sections, especially for those that only want to skim the case study and not delve into the details of the project.
Conclusions	Whet a reader's appetite to revisit the case study body for more detail.
Suggestions for Future Work	Motivate additional effort to further understand complex projects and advance exemplar PM techniques.
References	Lend credibility to the project's case study and highlight relevant literature from related bodies of knowledge.

which order these outline sections are obtained; what is important is the collective impact of their objectives.

- How did the tool arise?

    Early on in the preparation of Ref. [1], the first co-editor, Alex Gorod, who motivated the effort, recognized the need for and offered an outline for case studies. The second co-editor, the present author, expanded upon the outline to make it more comprehensive, while suggesting some aspects that might be optional.

- Why is the tool needed for PM?

    A logical and orderly outline that helps guide the development and documentation of any case study should be welcome by CP leaders (PLs), managers, and teams. Use of such an outline also facilitates the preparation and comparison of parallel case studies within and without the project domain.

- What example PM practices should be improved and how?

    More case studies should be instituted for the benefit of not only the subject project but also other projects.

- What is the most appropriate portion of a project case study where the tool should/could be applied?

    All portions (cf., Table 2.1)

- To what extent has the tool been applied?

    The CSO was advocated and utilized to some extent in Ref. [1].

- How might the tool be improved?

    With additional use, the outline can certainly be improved, e.g., toward highlighting which areas of the outline should be mandatory versus optional.

PROJECT MANAGEMENT LIFE CYCLE

A telling depiction of the traditional PMLC is provided in Figure 2.1. In addition to the distinctive and important stages of the process, typical milestones are indicated at appropriate points in the cycle as defined in the associated table of PMLC stage reviews. Most of these reviews are derived from traditional military electronic systems acquisition programs, although such reviews are prevalent in many civil PM contexts.

With respect to these reviews, it is important to understand that each review should have well-defined entry and exit criteria. This helps prevent the natural human tendency to approve a review for spurious reasons such as a contractor's need for more funding, as opposed to insisting upon a greater effort to meet the criteria.

Although this life cycle was originally labeled as SE, it is now being appropriately viewed as a PM depiction. The various stages are quite relevant to both points of view. Note that there is a Disposition stage which is often omitted from consideration in many projects because they focus more on developing or improving a given situation and not on the end state. However, it is advisable to consider the end state as one conceives and exercises the process because that can influence development or future improvements.

The project overview "centerpiece" was labeled "Solution Engineering" in the SE depiction of Ref. [2]. This aspect, as well as the double arrows connecting this centerpiece with the stages,

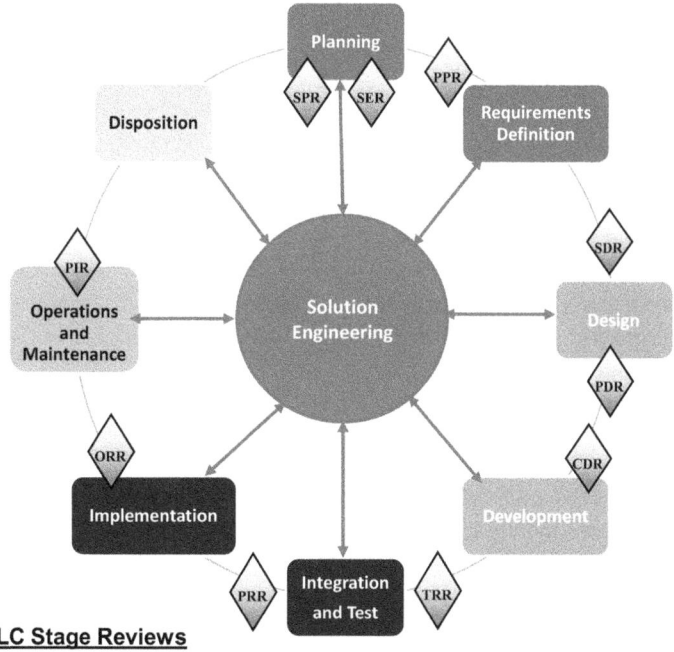

PMLC Stage Reviews

SPR: Study Plan Review
SER: Solution Engineering Review
PPR: Project Planning Review
SDR: System Definition Review
PDR: Preliminary Design Review
CDR: Critical Design Review

TRR: Test Readiness Review
PRR: Production Readiness Review
ORR: Operational Readiness Review
PIR: Post Implementation Review

FIGURE 2.1 Project Management Life Cycle (PMLC) [2].

emphasizes the nonlinear nature of most PM situations where feedback is at least as important as forward progress.

- How did the tool arise?

    The much older "waterfall" model, although originally including feedback, was misapplied for many years as a strictly serial linear process. More recently, the so-called "V" model, though an improvement over the waterfall because it shows parallelism between comparable earlier and later stages, is still largely viewed as a linear sequential process, unfortunately. This is quite misleading. Such diagrams are essentially traps for postponing prudent "look-aheads" in anticipating what may be coming, as well as temptations to proceed before the current stage has been adequately addressed, as with mistakes in prematurely approving an associated review milestone. The further to the left one has to backtrack in such situations, the faster the resulting cost grows and the longer the eventual schedule extends. Hence, it is better to enact the PMLC as the "circle" of Figure 2.1.
- Why is the tool needed for PM?

    All eight stages around the circle and the ten interspersed reviews are still recommended. The project overview portion in the center and the interactions among that activity and the stages indicated by the bidirectional arrows is what is most important. Successful PM more often results from enlightened viewpoints, continually keeping the future in mind, and working backwards as well as forwards. It is critical to recognize and prepare for the unexpected, knowing that most initial views of what is needed may be fundamentally flawed. It is largely through humble efforts that proceed with the best intentions, and that are open to change as new information becomes available, that the project can reach a satisfactory conclusion.
- What example PM practices should be improved and how?

    PM is complementary to SE. A PL is rightly concerned mainly with performance, cost, and schedule, whereas a systems engineer concentrates more on technologies and the other more technical aspects of the endeavor. The greatest successes result from a balanced approach following both PM and SE principles and recommendations. In the more ideal situations, the PL and chief engineer (CE) work well together, hopefully under a common boss that can help adjudicate unresolved differences.
- What is the most appropriate portion of a project case study where the tool should/could be applied?

    The PMLC is most relevant in the middle or heart of the case study, cf., mainly the description objective of the (complex) project in the fifth row of the CSO in Table 2.1, once the foregoing foundational aspects have been established, and before the subsequent points have been made.
- To what extent has the tool been applied?

    As with many of these tools, it is difficult to reliably assess the degree of application of this (circular) life cycle depiction. A likely reason for this may stem from the fact that practitioners will not likely inform the proponents of a given model or tool that they are using it, successfully or not.
- How might the tool be improved?

    The author, and perhaps many others, may agree that the Requirements Definition stage conveys the most egregious misconception. The true requirements of most systems, let alone projects, are extremely difficult to establish firmly at the beginning. Most stakeholders, especially users, generally are unable to precisely express the ultimate requirements. Instead, they typically try for increased capabilities that would alleviate their present problems. It is really counterproductive to insist upon detailed requirements, let alone premature specifications, until the problem at hand is better understood. It is wiser to establish and embrace a desirable outcome space, and then to help guide the project evolution in the proper direction.

ZACHMAN FRAMEWORK

A current version of the Zachman Framework [3] artifact is shown in Figure 2.2. This is likely worth copying and printing in a larger size for better readability and easier reference.

- How did the tool arise?

  John Zachman developed his ideas in the 1980s (https://en.wikipedia.org/wiki/Zachman_Framework), so little insight is available here into just what his motivations were other than he was probably looking for a good way to help guide people in addressing the entire situation surrounding a given system or project problem.

- Why is the tool needed for PM?

  One should be able to see this within a few minutes of studying Figure 2.2. The column headings express the fundamental and comprehensive set of questions one might ask about any subject, whereas the column footers characterize just how these questions often apply. The rows' left headings provide different perspectives of key stakeholders, whereas the right headings give model names appropriate to each perspective. The rows from top to bottom address identifications, definitions, representations, specifications, configurations, and instantiations, respectively. Thus, in embarking on any project, the Zachman Framework is an excellent tool for considering all aspects of the situation at hand, and the framework should be kept in mind and revisited as the work proceeds.

- What example PM practices should be improved and how?

  Too often project teams charge ahead without taking enough up-front time to thoroughly understand the problem. More will be said about this when discussing the SPM tool. Furthermore, as the project evolves, considerable effort is devoted to "getting back on track," in being disproportionally wedded to "requirements" and/or previous plans instead of being

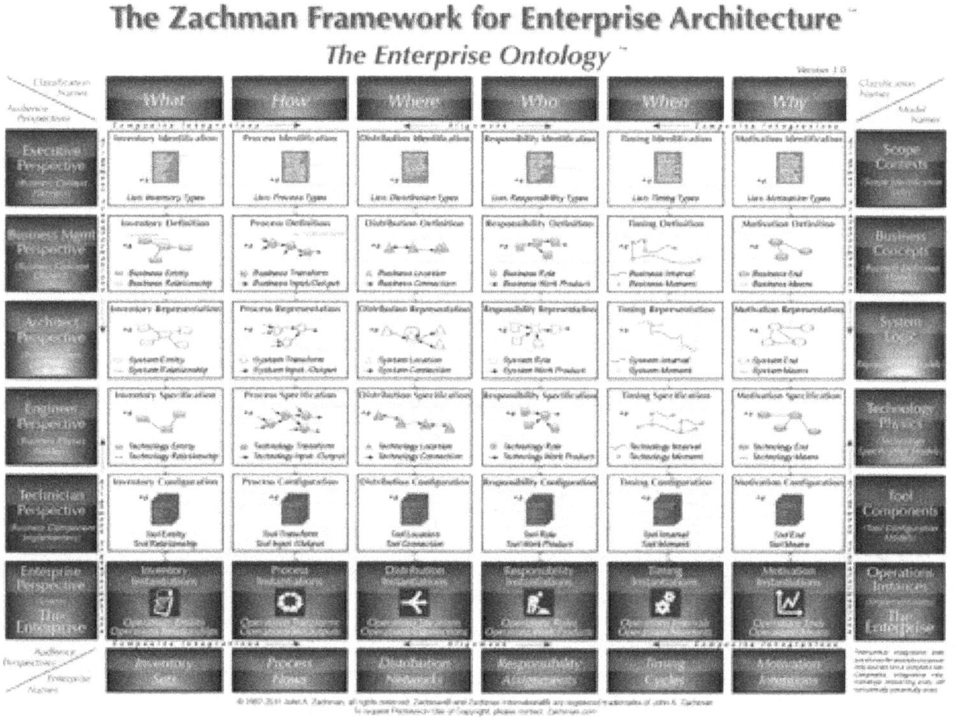

FIGURE 2.2 Current version of the Zachman Framework (www.zachman.com/images/ZI_PIcs/ZF3.0.jpg).

open to emerging *opportunities* and re-planning. This will be echoed when describing the PMA and CAPM tools. The Zachman Framework is a good remedy for both these shortfalls.

Regarding opportunities, this author asserts that in many SE projects, there is too much emphasis on identifying and mitigating risk and not enough attention to treating opportunities in the same fashion. To become successful with CPs, it is critical to balance opportunities and risks. The same methodology can be used to handle both these important dual aspects. Too often project managers focus just on risks and risk management, continually trying to get back on track with whatever project plan had been fashioned when that may not be the best course of action.

- What is the most appropriate portion of a project case study where the tool should/could be applied?

    At the beginning or very early on, as well as periodically throughout the project.

- To what extent has the tool been applied?

    Most extensively, it may be assumed, based upon the long existence and broad knowledge of this framework, although there has been no attempt here to obtain quantitative data to support this opinion.

- How might the tool be improved?

    One could study the present Version 3.0 as well as previous versions, and suggest updates in the terminology and/or cell representations, perhaps justified by the most current or projected recommended PM practices.

EPM PROFILER

This tool can be applied to characterizing the *nature* of most any CP "enterprise," depending on the size, domain, and stakeholders involved. Referring to the diagram in Figure 2.3, there are four largely disjoint contexts: strategic, implementation, stakeholder, and system, where each instance of "system" can easily be replaced by "project." There are two octants in each context quartile which subdivide important notions. Strategically, one is concerned with the project's mission, environment, and scope of the effort. Acquisition environment and scale of effort are under implementation. The PM team needs to identify and understand the key stakeholders, and their likely involvement and interrelationships. Finally, desired project outcomes must be established, and solution-oriented project behavior should be anticipated and tracked.

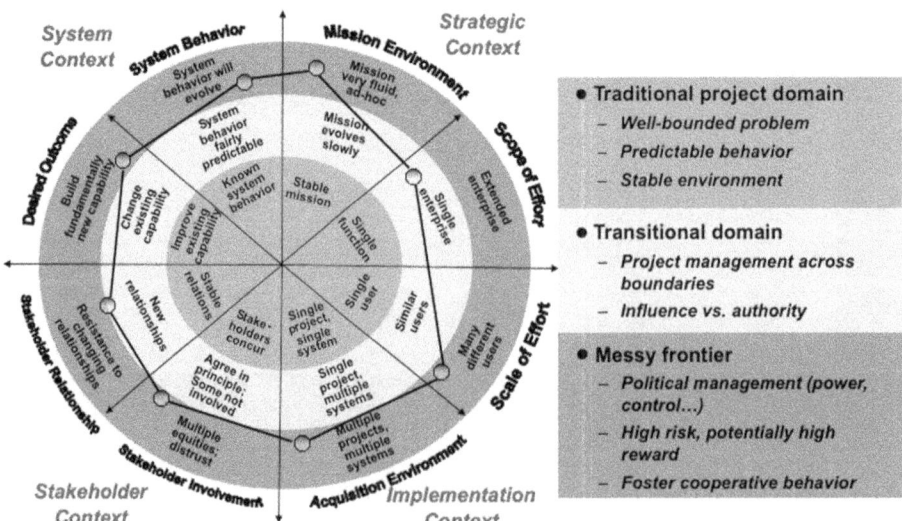

FIGURE 2.3 Enterprise Project Management (EPM) Profiler [4,11].

The diagram's inner circle purports to contain the easier characteristics of the problem posed by the given project. The most difficult environmental characteristics are represented by the outer ring, the so-called messy frontier. The octants of the middle ring lie somewhere in between, and represent characteristics of medium difficulty. It will be useful to read the descriptors and absorb the qualitative differences from less to more complex within each quadrant as one proceeds outward from the center.

Any connected line segment overlay in Figure 2.3 consisting of nodes and links is called a "spider" diagram. The basic idea is to assess in which of the three regions of each octant a project is best characterized. Once these nodes are marked on the figure, they are connected by straight lines as shown. The more this overlay impinges on or intrudes into the messy frontier, the more difficult the overall project. No doubt most people prefer projects contained in the inner region! Unfortunately, many projects involve wider concerns. For instance, in many seemingly relatively benign case studies, when viewed as an enterprise focused more on potential *external* stakeholders, the project would best be characterized by the large spider diagram shown in Figure 2.3. Alternatively, when viewed more from the point of view of *internal* stakeholders, a smaller spider diagram might better apply. The result depends not only on one's own perspective but also on the shared perspectives of the project team, for example.

- How did the tool arise?

    This profiler was invented circa 2008 by Renée Stevens, a close colleague of this author while both worked at MITRE. She astutely understood the relatively new concept of enterprise and what SE was all about in that type environment. Renée recognized the need for characterizing all aspects of the challenges presented to a project team.

- Why is the tool needed for PM?

    The three rings, four quadrants, and eight octants contain hints for addressing many of the most important issues when embarking on a given project. None of these aspects should be omitted from consideration by the project team. The PL is responsible for ensuring discussions take place as consensus is reached as to what spider diagram is appropriate, at least at the beginning of the project. As the work progresses, of course, things can change, and a new spider diagram more representative of the then current situation can be created.

- What example PM practices should be improved and how?

    The mission environment needs to be recognized as stable, evolving, or fluid. Oftentimes, this is not discussed, and project members make assumptions that do not agree. This can lead to arriving at the wrong scope of effort which has little hope in solving the real problem. A proper evaluation of this strategic context provides a good basis for discussing implementation. The appropriate scale of effort regarding the number of potential users should be determined. If this is done well, the acquisition environment, particularly the number of programs and projects involved, can be established. Sometimes, the implementation context is too narrow which can lead to not satisfying needs, or too broad which may cause eventual project failure because the effort is too ambitious. Often missing from direct confrontation is stakeholder analysis. Most times, PLs seem to accept the stakeholder situation without modification and try to deal with it as best they can. However, this is a time-consuming and often distracting task when their energies could be better spent in trying to move the project forward more efficiently and effectively. It is quite worthwhile to clearly identify and characterize the key stakeholders as early in the project as possible. Included in that analysis is a characterization of the relationships stakeholders have, not only with the project but also among themselves. This can pay dividends in learning which stakeholders are supportive, which are malleable, and which require special attention to mitigate, if not eliminate, their contrariness. The fourth quadrant relates to project context; that is, the level of desired outcomes and the anticipated behavior the project is attempting to establish. Here, it is important to not be very specific in defining success, but rather keep a healthy stash of potential solutions were possibly attractive opportunities could be pursued with informed risk.

- What is the most appropriate portion of a project case study where the tool should/could be applied?

  The EPM Profiler is probably more useful near the front end of the case study especially covering the Background, Purpose, and Constituents (cf., Table 2.1) of the project. But as with many of these tools, they can be usefully applied while addressing other stages of the case study.

- To what extent has the tool been applied?

  This author has often cited this tool in his many presentations and publications. He also recalls that Renée successfully recommended this tool in one of her US Army contracts, where the commanding officer attested to its usefulness and thanked her for this contribution. Considering that this tool was published in an important systems conference in Montréal [4] and was also mentioned in her book [11], it would not be surprising if her profiler has been applied in many other venues.

- How might the tool be improved?

  Within a given project, it might be worthwhile to consider rewording some of the aspects in the profiler to be more closely connected with the actual environment encountered. If this is done well, it could provide greater motivation for utilizing the profiler to more accurately and precisely characterize what is facing the team.

PMA PROFILER

This tool, which is meant to focus on what activities are *actually happening* on the project, is shown in Figure 2.4. Each mention of SE in the SEA Profiler [5] was replaced by PM, and every "system" was replaced by "project."

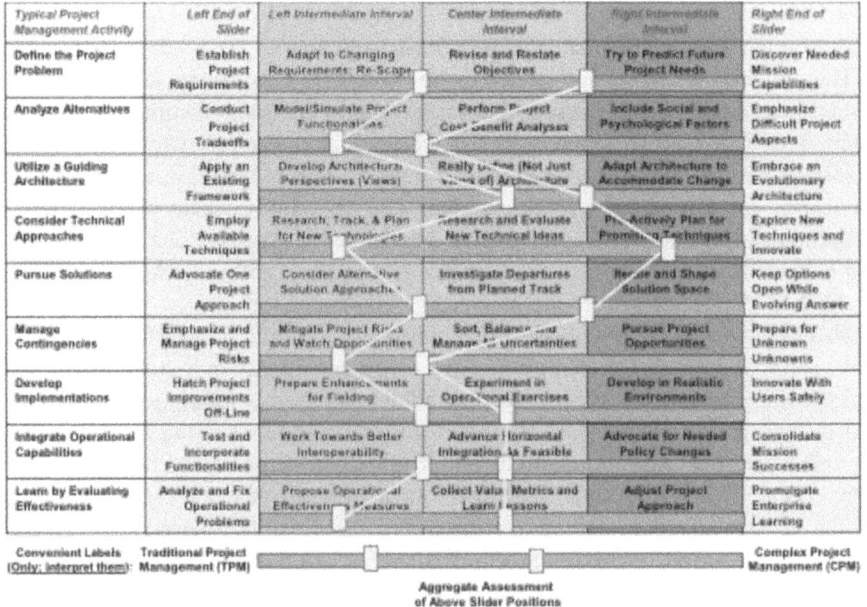

FIGURE 2.4 Project Management Activities (PMA) Profiler [5].

As seen, the first column lists PM activities that are quite typical of any project. These are listed in a more or less logical order although they are not intended to be sequential, and any particular activity can be addressed before any other one. In each row of the chart, the position of the relatively-small rectangle-shaped "slider" box is chosen to indicate where in the continual spectrum of activity the intensity of that current effort resides. The left-hand end of this slider marks the traditional minimum effort, whereas the right-hand end attempts to depict the most reasonable extreme effort for very CPs. The three intermediate intervals express interim levels of activity. The farther to the right, the more aggressive the activity in terms of solving CP problems.

The reader should consider all the entries to gain an appreciation for the intent and flexibility of this tool. The placement of the slider positions in each connected pattern represents the judgment of the project leader and/or team members regarding where activities stand. As project problems increase or intensify, at least some of the slider positions would likely be moved to the right, provided there are available resources for more vigorous activities in an attempt to make improvements. The two patterns shown represent characterizations of the activities in an unpublished case study of the development of the case study book [1]. The slider pattern on the left is made from the perspective of internal stakeholders, primarily the co-editors of the book. The pattern on the right is made more from the perspective of external stakeholders, e.g., the readers of the book.

- How did the tool arise?

  Renée Stevens was starting to use her enterprise profiler of Figure 2.3 in a before-and-after mode where on a given project the environment would be characterized at the beginning and then again as the project evolved. This would show a better understanding of the environment as reflected in an updated characterization. This author created another way to show what the project was *actually doing* using another profiler, i.e., the SEA (now PMA) Profiler.

- Why is the tool needed for PM?

  It's important for the project team to know and contribute to what is going on in the project. The PMA Profiler is oriented toward showing the relative complexity of possible approaches to each stage of the PMLC of Figure 2.1. Once the PMA Profiler is established with appropriate slider positions as a consensus effort by the project team led by the PL, it can be publicized and referred to periodically. This is a useful artifact for presentation to the key stakeholders, as well. Anyone can quickly assess the degree to which the activities match the characterized environment of the EPM Profiler, for example. The PMA can also be used to track changes in the effort.

- What example PM practices should be improved and how?

  A guiding architecture should be established early on once the project's problem is reasonably well understood. A good architecture should not change significantly unless for some reason there is a major change in problem understanding or an unexpected "bombshell" impacts the project development. Because of these possibilities, although they may be considered remote, there should be a conscious effort to plan for contingencies and develop a process in place to be exercised when such events occur. Finally, lessons learned from various decisions and accomplishments or failings on the project should be well documented and shared not only with stakeholders but with other project teams. Too often in PM lessons learned are accumulated and perhaps documented but they are soon forgotten and not applied in the next phases of the project or on other projects.

- What is the most appropriate portion of a project case study where the tool should/could be applied?

  The PMA Profiler tool is best utilized during the Challenges, Development, and Results phases (cf., Table 2.1) of a project case study. This would provide a fundamental view that may be helpful to those learning about previous projects and aspiring to incorporate effective PM approaches on their own projects.

- To what extent has the tool been applied?

    Again, it is difficult to cite specific instances where this tool has been actually utilized in practice. However, its potential utility has been explained and espoused by this author in many papers, book chapters, tutorials, and teachings. In his 2017 online Masters-level SE course at the Worcester Polytechnic Institute (WPI), for instance, four students indicated they were recommending its use to their workplace leaders and managers.

- How might the tool be improved?

    Soon after the SEA was developed, there was some criticism from the staff assistant working for the author's boss to the effect that they thought the tool was too complicated. In response, it was pointed out one must recognize that tools intended to solve complex problems must have at least the same degree of variety/diversity as the problem itself. This is a fundamental principle of CSE, i.e., keep Ashby's notion of "requisite variety" in mind [12].

THE ENNEAGRAM

The Enneagram tool, depicted in Figure 2.5, is an effective guide for having productive discussions that address all aspects of virtually any problem. The blue triangle (images are available in color in the eBook versions) theoretically represents the recommended self-organized foundation of every project team effort, consisting of the team's Identity, the Relationships among team members, and the Information needed on the project. The directed pathways represent the natural order of topics typically followed by Living Systems, i.e., people, who participate in and help govern discussions of the overall subject at hand. One starts with the general project Intention followed by consideration of the Principles & Standards to which the team endeavor to abide. Next, the Issues are addressed within the Structure/Context of the project, including considering the overseeing leadership and management players and stakeholders. The team then dives into discussing The Work in detail while learning what may or may not apply. Optionally, the team may return to the beginning of this discussion cycle, update their intentions, make refinements, and restart the process.

A potential positive result of applying the Enneagram is the ability to rapidly build strong trust among the team members and their leadership/management. Usually, trust is developed, albeit relatively slowly, through a mutual iterative process of sharing information that is credible and friendly. As the trust builds more and more is shared. Isn't this counter to many prevailing business cultures where one is *punished* instead of rewarded for sharing information?

The 2-8-5 triangle indicates the typical pattern usually imposed by command-and-control-oriented management. High performing teams should avoid the trap of considering just these three aspects which usually results in ineffective solutions and the necessity to start over in addressing the original problem.

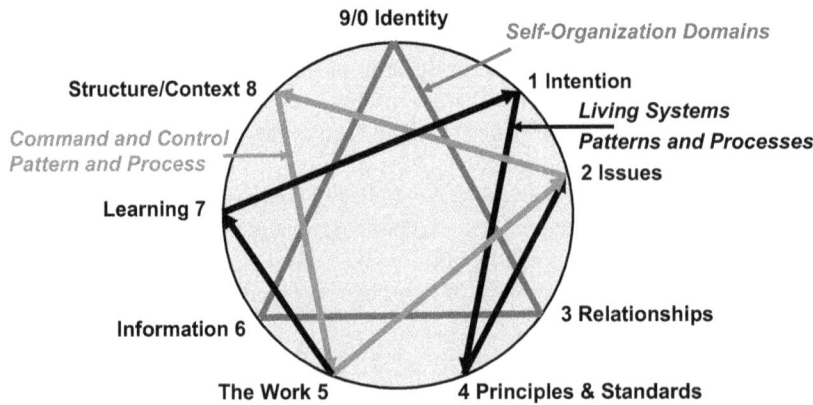

FIGURE 2.5   The Enneagram [6,7, pp. 30, 32, 33, and 39].

This author feels that the connotations invoked by the term command-and-control, particularly associated in military systems, promote a wrong mindset. CP participants must have the proper mind*sight* of *not* being in control because that state of affairs is dictated by the behaviors of CPs. CPs need to be self-organized, the conditions for which is largely up to the CP's leadership to establish, precisely because no one is truly in control. It's much better to substitute "guidance" or "direction" for "control" in this context.

Another common and misguided mistake, in this author's opinion, is for quality assurance to be treated as a separate Work Breakdown Structure (WBS) project task or quality program to be managed overtly. An effective CP leader will insist that true quality must be either built-in or integrated into everything. Everyone involved should be responsible for assuring the quality of their work and contributions.

Also, in CSs one can*not* often successfully apply reductionism and/or constructionism except perhaps within the simplest subsystems. One must take a holistic, systems thinking, and fully integrated approach to CPM. While one is working on separate pieces trying to somehow optimize their individual performance, the CS or CP has moved on through interactions of the collection of pieces among themselves and with the environment. Consequently, these optimization attempts are brought together, they will likely result in a suboptimum whole.

- How did the tool arise?

  This tool was invented by Richard Knowles based on needs for more comprehensive discussions that he uncovered through his extensive experience in PM.
- Why is the tool needed for PM?

  Many experienced practitioners would likely agree that discussions among team members and stakeholders are usually quite limited in scope, and primarily driven by individual preferences and perceptions of what might be considered important. This Enneagram tool is a good way to remedy that situation.
- What example PM practices should be improved and how?

  This tool provides a structured reminder to help facilitate a more complete interpersonal interchange about what needs to be verbalized on a project. It is intended to elicit the frank expression of all points of view so that nothing is left on the table when people leave the meeting room or teleconference. Its proper use can go a long way toward addressing the "elephants in the room" while clarifying unexpressed implicit assumptions that may be invalid.
- What is the most appropriate portion of a project case study where the tool should/could be applied?

  The Enneagram is best applied during the Purpose and Constituents phases of a project case study (cf., Table 2.1).
- To what extent has the tool been applied?

  Ref. [7] documents many instances of this tool being successfully applied. This author attests to its value during his testing of the Enneagram in first-hand applications in a few professional face-to-face meetings. It really works!
- How might the tool be improved?

  Rather than considering possible improvements, it would be better to continue trying the Enneagram in realistic environments and then properly tailor the tool for the specific domains of interest.

SOFT PROJECTS METHODOLOGY

Figure 2.6 shows the SPM which is identical to Checkland's SSM [8] that can be interpreted for PM as well as SE by replacing each instance of system(s) by project. The four upper left nodes relate to the real world, and the four lower right nodes represent the project team's "project thinking"

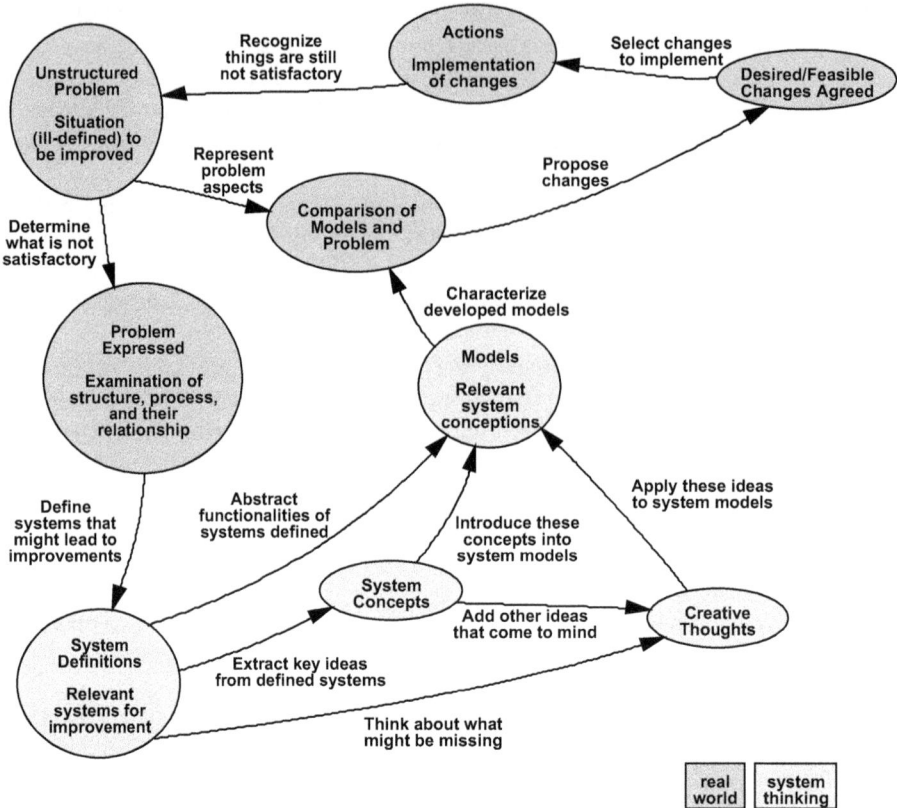

FIGURE 2.6   Soft Projects Methodology (SPM) (think of replacing each instance of "system(s)" by "project") [8].

in parallel. Note that the diagram resembles a "systemigram" [13], where the nodes are labeled with noun objects, and the links indicate actions describing the flow from one node to the next. Study this chart to better appreciate the logic and benefit of utilizing this tool. Checkland's principal concept is the importance of understanding the (project) problem. To do so, one might traverse the cycle more than once.

- How did the tool arise?

    Peter Checkland is considered an SE guru, especially within the International Council on Systems Engineering (INCOSE), and he invented the SSM (now SPM) to assist practitioners, particularly system engineers.
- Why is the tool needed for PM?

    The SPM is especially important because it emphasizes people, mainly the project team, and the importance of their discussions, as is also emphasized by the Enneagram. Great attention is focused on not only understanding the project problem but also creating a model in parallel with reality that is conducive toward creating solutions.
- What example PM practices should be improved and how?

    The SPM shows this schema should be iterated a finite number of times in ways that guides the project toward achieving more effective capabilities.
- What is the most appropriate portion of a project case study where the tool should/could be applied?

    The Challenges and Development phases of a project case study (cf., Table 2.1) will benefit most.

- To what extent has the tool been applied?

  The SSM has been widely applied in many SE environments, as much of the literature of the field, including INCOSE publications, attests.

- How might the tool be improved?

  Far be it for this author to suggest improvements. In the opinion of many practitioners, Checkland has already accomplished the close-to-ideal tool in this instance.

CAPM METHODOLOGY

Figure 2.7 summarizes 25 activities that are deemed worthwhile in attacking CP problems. Each activity is given a coded number-lowercase letter for reference purposes. Each node in the diagram is labeled by a noun and categorized by type according to the bar chart at the bottom of the figure. The links between nodes are labeled by active phrases that indicate how one node affects the next.

This chart also resembles a systemigram [13]. The reader should peruse this chart to get a feel for the methodology. Considerable further detail is available in Ref. [9].

For instance, consider 10a in Figure 2.7's upper right portion and rewarding contributors for useful results. This author advocates paying, e.g., contractors, for results *only after* being assured that progress has indeed been made toward desirable goals. Most likely, and this is usually the case with both private and government procurements, contractors/vendors are paid significant amounts of money up-front for what sponsors/customers perceive as promises. Unfortunately, this often seems to be the accepted practice where everyone pretends that the work will cost a lot less than a realistic estimate would otherwise suggest. This flies in the face of actual data showing that most contracted projects overrun by significant amounts, typically by one-third to one-half the original cost estimate.

Then when, as is often the case, proposed milestones are not reached because of inabilities to satisfy the exit criteria, additional sorely needed funds to continue are provided with the hope that the work will somehow proceed to a successful conclusion. Not only that, it is also common to grant award fees even though there is little justification for such rewards. This penchant for throwing good money after bad can then lead to cancellation of the contract with huge losses in sunk cost.

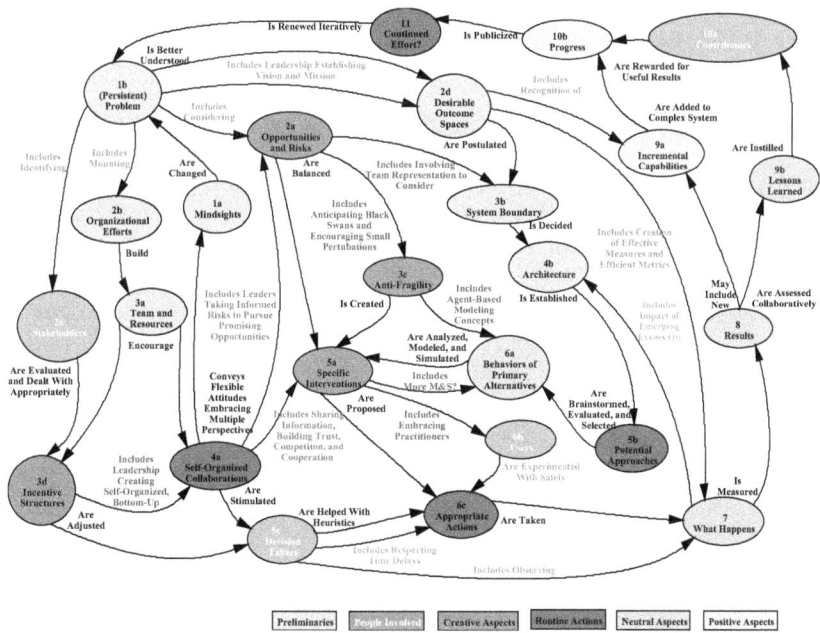

FIGURE 2.7 Complex Adaptive Project Management (CAPM) methodology [2,9].

It would be much better if procurement practice were gradually modified over time (this might take decades) to withhold most of the funding until the desired results are produced. Contractors who are successful will thus develop "deep pockets" to be able to start the next job without receiving money up-front; those contractors who cannot perform adequately will eventually go out of business. One can argue whether rewarding for results is even remotely feasible on mega-systems or mega-projects like the Big Dig. But before rejecting this rather fresh notion out of hand, shouldn't it at least be seriously tried on a significantly large project?

- How did the tool arise?

    The CAPM Methodology has evolved ever since the author became interested in CSs and CSE in 2004. This methodology was built on many of the ideas prevalent in complexity theory, CSs, and the SE of difficult electronic systems acquisition problems in the author's work experience. The behaviors of CSs have since been observed, recognized, and documented, and the principles of CSE have been codified, developed, and improved.
- Why is the tool needed for PM?

    The CAPM Methodology addresses many aspects of trying to run an efficient and effective CP.
- What example PM practices should be improved and how?

    The entire PM practice can be improved significantly by seriously applying the CAPM Methodology.
- What is the most appropriate portion of a project case study where the tool should/could be applied?

    The CAPM Methodology is most relevant to the Development phase (cf., Table 2.1).
- To what extent has the tool been applied?

    This is uncertain. The author, for one, promulgates it in his presentations and publications, and tries to follow the methodology in his consulting work, as well as his daily life.
- How might the tool be improved?

    Surely improvements can always be made, and the author has periodically done so over the years. Anyone who has studied the methodology and/or has tried to apply it is welcome to offer suggestions for improvement.

APPLICATION EXAMPLE

This section attempts to address a huge-scale project, Boston's Big Dig, and how the Toolset may (or should) have applied in that case.

Title: The Big Dig—Reviewing Boston's Hugely Complex Project: Burying Its Unsightly Overhead Central Artery

This massive project has been recognized for decades as one of the world's most ambitious, difficult, highly controversial, expensive, and yet successful undertakings. Much of this story is readily available from published accounts, and readers are encouraged to explore some of these materials, as cited in the References, to obtain additional details.

All Big Dig machinations detailed are characteristics of a CP, so, in proceeding, the focus is on the core ideas of good PM and how this case study informs the improvement of PM practices.

CASE STUDY ELEMENTS

- Fundamental Essence: improving transportation delays without unduly disrupting metropolitan operations
- Topical Relevance: courage in tackling massively CPs

- Domain: civil discourse, engineering realizations, government funding, and metropolitan transportation
- Country of Focus: the United States
- Interested Constituents: public and government stakeholders, contractors, and PM/SE practitioners
- Primary Insights: foster leadership to create target outcome spaces, stimulate self-organization, and instill quality work, know no-one is in control, avoid hierarchical and partitioned structures, take holistic non-reductionist approach, focus on improved capabilities not requirements, value open/honest communication to build trust, ensure realistic cost estimation, pursue opportunities with informed risk, and reward desired results

BACKGROUND

Boston's so-called Central (ground transportation) Artery was constructed in the early 1950s, becoming a precursor of and an adjunct to President Eisenhower's interstate highway system that was funded in 1955 [14] and more or less completed by 1960. However, this Central Artery, a set of added-on roadways, was a largely elevated eyesore through the heart of downtown that disrupted several local neighborhoods and negatively impacted residents and businesses. Moreover, rather than significantly improving traffic flow, travel times, especially for commuters, greatly increased, largely due to connecting bottlenecks that materialized on existing streets. The Artery was intended to handle 70–75k vehicles each weekday, but it soon had to support 170–200k [14,15]. Furthermore, getting to and from Boston's Logan International Airport was very problematic. One would veritably cruise into Boston on the Massachusetts Turnpike or other major thoroughfares, such as Cambridge's Storrow Drive, only to exit into a complicated but small set of optional paths that would hopefully converge on the Callahan Tunnel leading to Logan [16]. The Sumner Tunnel from Logan emptied into a similar morass.

Even with this limited and summarized information characterizing the initial state of the problem, the EPM Profiler, Tool 4 of Figure 2.3, would have been useful in helping to characterize the situation more completely. It is unknown whether those involved contemplated such a 360° overview of the problem early on but they could have.

The principal components or sub-projects of the Big Dig included the following: (1) rerouting of the Central Artery of Interstate (I)-93 into the 1.5-mile-long Thomas P. O'Neill, Jr., Tunnel; (2) Ted Williams Tunnel connecting I-90 and Logan; (3) Leonard P. Zakim Bunker Hill Memorial Bridge over the Charles River; and (4) Rose Kennedy Greenway, a 1.5-mile-long array of surface parks and public spaces just under what used to be the elevated Central Artery roadway [16]. The principal Big Dig directly connected project stakeholders probably developed a least a preliminary version of the PLMC Tool 2 of Figure 2.1, and most likely several versions, perhaps one for each of the four sub-projects.

"Bill" Reynolds [16] had the idea of replacing the Central Artery with a buried version, and he expressed this thought to a friend and colleague at MIT, Frederick "Fred" Salvucci. Fred really took this idea to heart and became the principal motivating force. The Governor of Massachusetts, Michael Dukakis, so liked Fred's notions that he made Salvucci Massachusetts's Secretary of Transportation. A few of the notable Big Dig stakeholders and their roles are listed in Table 2.2.

A few caveats are in order. First, it is rather presumptuous to suggest that the truth of what happened on the Big Dig and why is fully understood. *Au contraire*, system of systems, enterprise, and CSE, as well as CP and CPM, are still exploratory fields that will likely take further decades to mature. Second, the present author lived through the Big Dig development first-hand and therefore has some perceptions of the underlying reality of the topic, likely somewhat misguided or erroneous, but these will be shared anyway for what they might be worth. This section really offers just a guide to a Big Dig case study effort itself, particularly highlighting Virginia Greiman's enthusiastically recommended book, *Megaproject Management* [17], with the goal of exposing real difficulties and how they might be better dealt with in similar endeavors.

TABLE 2.2

Some Big Dig Stakeholder Involvements

Stakeholder	Stakeholder Type	Stakeholder Role(s)
Michael S. Dukakis	Governor of Massachusetts	Selected Fred Salvucci to be his Secretary of Transportation and oversaw the project during his tenures as Governor. Dukakis helped get Federal funding because the state had paid for the Central Artery prior to Eisenhower's funding of the interstate highway system [14]
Frederick P. Salvucci	MIT Class of 1961 civil engineering graduate (S.B. and S.M. degrees); Massachusetts's Secretary of Transportation	The principal force behind the selling and design of the Big Dig. He convinced Gov. Dukakis that it would enhance Boston's environment and that homes would not be lost [14]
Thomas P. "Tip" O'Neill, Jr.	Speaker of US House of Representatives	Helped persuade Congress to pay for 80% of the project [14,15]
William "Bill" Reynolds	Friend and colleague of Fred Salvucci at MIT	Provided Fred with the "simple" idea of tunneling the Central Artery [16]
Ronald W. Reagan	President of the US, 1981–1989	Vetoed a 1987 highway bill containing the Big Dig's first significant Federal funding [14]
Edward M. "Ted" Kennedy	Massachusetts US Senator	Worked with Tip O'Neil to garner enough political support in the Congress to override Reagan's 1987 veto [14]
Christian Menn	Swiss engineer	Suggested the ten-lane cradle, cable-stayed Leonard P. Zakim Bunker Hill Memorial Bridge [16]

CONTEXT

Much of Boston is built on silt and hauled-in landfill from US Revolutionary days in the late 1700s. This, and the high-water table, being so close to Boston Harbor, complicated the idea of digging even at relatively shallow depths, although it was necessary to go rather deep, up to 90–120 ft in places, to avoid interfering with subway lines, for example. Also, disruptions to homes and businesses needed to be avoided if possible.

Above all, during Big Dig construction, the Central Artery traffic somehow had to keep moving, and connecting streets and roadways had to be impacted as little as possible. As US Representative Barney Frank lamely joked, "It might be easier to raise the city than lower the highway!" [15].

RELEVANT DEFINITIONS

Complexity has a specific meaning in the "Context" section. A widely agreed, simple definition of complexity is elusive and may be impossible to establish. However, one can approach an understanding of varying degrees of complexity by contemplating 14 CP behaviors commonly observed, and appreciating, through their application, to what extent 14 complex PM principles seem to help [2,9,18–22]. These behaviors and principles were originally applied to CSs and SE, but they generally carry over almost exactly to CPs and CPM.

PERTAINING THEORIES

One cannot rely completely on theory nor often predict what will happen as the project evolves. Actual case study experiences, vital in showing what works and what doesn't, urgently need more attention and exposition.

EXISTING PRACTICES

Traditional (classical or conventional) PM is wedded to notions that one is in control, can know requirements, can pre-specify solutions and predict outcomes, assuming, of course, that desired performance can be achieved if there are sufficient resources, a reasonable schedule, and adequate funding. But for the more difficult problems to be addressed, of which there are many, one is rarely in complete control, especially because people, largely autonomous beings that interact with each other, are integral to impacting the target project outcome(s). This often leads to behaviors of self-organization and interpersonal relationships. Thus, requirements are unstable, and one cannot fully pre-specify or accurately predict results and must honor the principle of bringing a healthy dose of humility to bear.

Unexpected and sometimes surprising emergent behaviors occur. Typically, as is generally understood and sometimes joked about, at most two of the three characteristics, performance, cost, and schedule can be attained.

PLs frequently claim a life cycle as a necessary mechanism for achieving success. But behaviorally, there are often so many factors at play that the project can progress only so far, e.g., two steps forward, before needing to regroup, and perhaps take more than just one step back. Such backtracking abuse of Big Dig's life cycle was certainly prevalent, causing much of the cost growths and schedule slippages. Contractors were often allowed to start work on sub-projects before their designs and other key aspects were completed. The philosophy seemed to include the idea of "getting things done now and asking questions later" [14]. Instead, a life cycle view similar to the PMLC Tool 1 is advocated where one embraces the principle of taking a holistic approach.

General awareness of, respect for, and greater attention to Tool 3, the Zachman Framework of Figure 2.2, and adhering to the principle of balancing all aspects of the project undoubtedly would have helped the Big Dig manage better. Also, Tool 5, the PMA Profiler of Figure 2.4, would have shown what was actually being done to address the situation described by the EPM Profiler, Tool 4 of Figure 2.3, if that was applied, as well.

"AS IS" PROJECT DESCRIPTION

Here is a summary of the Central Artery prior to the Big Dig, the Big Dig's *initial* tasks, *estimated* cost, and schedule.

> The elevated six-lane Central Artery, which opened in 1959, linked the Southeast Expressway with points north, with plentiful offramps to access the city and its proliferation of above-ground parking garages. … By the early 1980s, the elevated roadway was hopelessly congested, and engineers pronounced its structural failure imminent [15].
>
> The Central Artery [traversed] the Financial District to downtown Boston, one of the most densely developed areas in New England. … the Central Artery had 38 lanes of traffic that fed into the 6 lanes of the main corridor carrying not only north-south traffic but much east-west traffic as well. … [17, p. 40].

The first formal estimates were a $2.5B total cost and project completion in 1998 [17, p. xvii].

This is where Tools 6 and 7, the Enneagram and SPM, respectively, of Figures 2.5 and 2.6, would have been useful. Again, the Enneagram helps ensure discussions are complete and comprehensive, while the SSM guides a process for fully understanding the problem before embarking too far into the Problem-Solving and Execution phases.

Project proponents were intent on pursuing opportunities but not in the appropriate spirit of that important principle, including being proactive in seeking better solutions but taking only informed risks. This would have suggested slipping the 1998 target date to reflect a more realistic schedule.

"To Be" Project Description

Cambridge objected to the original design of the Charles River crossing, and a Swiss engineer, Christian Menn, suggested and designed what later became one of Boston's beautiful downtown landmarks, the Zakim Bridge [16].

Other sub-projects included the Tobin Bridge approach/connector (called the City Square Tunnel) and the Storrow Drive Connector (a companion bridge to Zakim) [16].

The scope evolved and crept because of (1) technological uncertainties or advances; (2) cost increases or schedule delays in reconsidering project goals, influenced by increased public interest and variations in available funding; and (3) changes in the legal, political, or economic landscape [17, p. 166].

Ultimately, the Big Dig cost $14.8B and was essentially completed in 2006 [17, p. xvii]. A closer adherence to, say, the CAPM Methodology, Tool 8 of Figure 2.7, may have had a more positive influence on mitigating cost increases and schedule delays.

Purpose

The fundamental purpose was to increase traffic flow with a modicum of disruption to the city. The difficulties of improving Boston's Central Artery, via the Big Dig, in general, imply a spider diagram extending well into the outer ring of the EPM Profiler Tool 4 of Figure 2.3.

History

Big Dig planning began in 1982, environmental impact studies started in 1983, partial Federal funding was approved by Congress in 1987, and construction began in 1991. In response to pressure from business leaders for better access to Logan, the original project of transforming the Central Artery was augmented by Governor Dukakis and Secretary Salvucci, during their second terms, 1983–1991 [23], to include the building of a new (third) harbor tunnel [16].

There were three public transportation construction goals: (1) the Silver (bus) Line tunnel under the Fort Point Channel between South Station and Logan; (2) extension of the Green (streetcar) Line beyond Lechmere to connect the Red and Blue (subway) Lines, and the restoration of the Green Line to the Arborway in Jamaica Plain; and (3) North–South Rail Link connecting North and South (train) Stations [16].

Then Current Situation

> Early on … it was clear that depressing I-93 wouldn't by itself be enough to ease congestion. Traffic would still snarl trying to access the Callahan Tunnel to Logan Airport, East Boston, and points north. To close the deal, plans were drawn up for the extension of Interstate 90 under Fort Point Channel, the South Boston Waterfront, and Boston Harbor. The Ted Williams Tunnel became the necessary appendage … appealing to the feds because it would complete an interstate highway system. But with expanded scope, costs and complications started ratcheting up. Planning for the new world order got underway in 1982 [15].

During most of the Big Dig effort, this author lived about 20 miles west of Boston. He often traveled into the city for various entertainment and social events, and drove to and from Logan Airport for business or pleasure trips. He well remembers the difficulties and controversies of the early Big Dig days. Generally, the Boston public was highly critical of the effort, complaining about the significant disruption to the local norms and their daily habits. The author admits to being negatively influenced by this which fed into his own somewhat annoying experiences at times. It might have helped to have studied the Zachman Framework, Tool 3 of Figure 2.2, and contemplated CP behaviors and CPM principles.

KNOWN PROBLEMS

> As the most expensive highway in US history, the Central Artery and Tunnel project had a special, contradictory destiny: an engineering marvel deemed to be ill-conceived from the start [despite Salvucci's yeoman efforts]. … There was too much opportunity [a relatively rare negative use of this word] for incompetence and corruption. … [15].
>
> As the budget for the Big Dig kept going up and up … the public assumption was that there must be massive overcharging by contractors, if not outright corruption at work. To be sure, there were fraud charges, most notably surrounding the provider of flawed concrete [15].
>
> One of the most enduring critiques was that the entire public-private joint venture arrangement was inherently inadequate to control costs—that the state wasn't being hard enough on contractors, and thus failed to safeguard taxpayers' money. That was legitimate criticism [15].

In June 2006, after many of these difficulties had been dealt with, and in conjunction with the 45th reunion of the MIT Class of 1961, the author attended a lecture on the Big Dig by his classmate Fred Salvucci at MIT's Stata Center followed by a bus tour of Big Dig sites. Salvucci gave a detailed first-hand account of the difficulties and remarkable efforts to overcome obstacles and unanticipated problems that had arisen. The enormity of the tasks became more real and formidable. This author was much better able to imagine what the organizers, managers, and particularly the workers faced. Lamenting the absence of pictures from the bus tour, he recalls two unforgettable scenes: the Fort Point Channel including the tips of the sunken sections of the emplaced tunnels, and a bird's-eye view into a massive still open section of the Big Dig's sunken roadway. From then on, the Big Dig evoked more positive thoughts.

MISSION AND DESIRED OR EXPECTED CAPABILITIES

The Big Dig's mission was to vastly improve traffic flow to, fro, and through downtown Boston while beautifying and revitalizing the immediate surrounding area. Initially, at least, the envisioned capabilities engendered by the challenges of the Big Dig were rather limited, and even they were accompanied by rather skeptical opinions regarding their potential for success.

The time from the Big Dig's initial conception to when the work started in earnest was almost a decade, so it is difficult to concisely characterize a detailed, fixed description of the mission or capabilities representing that period. Moreover, because the work proceeded for another 16 years, the perception of the project, especially by the public, varied considerably, mostly in an adversarial vein, interspersed by occasional visions hoping for the best, the latter likely inspired mostly by members of the government and Big Dig contractors and workers directly involved.

TRANSFORMATION NEEDED AND WHY

The Central Artery was not only ugly but also incapable of handling existing traffic loads, let alone future demands. The need to minimize adverse impacts on the downtown Boston area was imperative. Furthermore, project costs not only needed to be contained but had to include Federal funding assistance. On top of all that, the public trust was not to be violated.

Boston is a remarkable place, uniquely representative of New England virtues, steeped in US history, and home to many respected institutions of higher education and cutting-edge or high-technology companies. Many residents and semi-permanent interlopers felt that the region deserved the Big Dig modernization and would not only survive the effort but also be proud to claim the considerable plaudits of its inevitable success.

CHALLENGES

Here are a few challenges, in addition to those already mentioned, implied, or inferred.

ANTICIPATED

There were dozens of sub-projects, albeit supposedly with well-defined interfaces among the seven heavy-builder contractors [16]. Nevertheless, the coordination problem was immense, and this surely led to anticipation of unforeseen difficulties. Tools 6 and 8, the Enneagram, and the CAPM Methodology, cf., Figures 2.5 and 2.7, respectively, would have been especially applicable in enabling and facilitating the necessary levels of coordination.

ACTUAL

As already stated, the scope of the Big Dig project expanded continually. For example, Big Dig's tasks had to be modified, as follows, and their *estimated* costs and schedules increased and extended.

- 1991: "Added $554 million for Dewey Square Tunnel, East Boston tunnel covers, landscaping, railroad relocation, material disposal program, West Virginia Fire Tunnel Test, project utilities, change in steel and underpinning designs, and miscellaneous other items"
- 1992: "Added $324 million in scope elements related to the New Charles River Crossing and miscellaneous other items"
- 1997: "Added $400 million due to bid results, noise mitigation, dust mitigation, traffic mitigation, and deletion of future air rights credit"
- 2000: "Added $1.1 billion for design development, design during construction, additional construction costs, and force account work" [17, pp. 166–7].

It's no surprise that the Big Dig costs escalated continually from $2.8B (1982 dollars), to $6.0B (2006 dollars), to $14.6B (2006 dollars), and to $22B (2018 dollars), including interest that won't be paid off until 2038. Similarly, the schedules slipped from 1998 to December 2007 [16]. This is a good example of the need to continually apply the SPM Tool 7 of Figure 2.6. Doing so could have led to a much better appreciation of the more serious problems.

DEVELOPMENT

Something akin to the PMA Profiler, Tool 5 of Figure 2.4, could have been utilized to help acknowledge development problems. However, the more useful artifact here would be the CAPM Methodology Tool 8 summarized in Figure 2.7.

> There seemed to be mistakes at every turn, … from design blueprints that didn't line up properly, to the faulty mixing of concrete, to, most tragically, a ceiling collapse that killed a car passenger in one of the new tunnels [15].
> … for every triumph, there were gaffes. Some were in design, like a miscalculation of a tunnel alignment, the guardrails that turned out to be lethal in car accidents, or the lighting fixtures that cost more than $50 million to replace. Some failures were due to problems in the construction process, such as the concrete that was not properly mixed, leading to leaks. And some were a combination of design and execution; the ceiling collapse that killed the car passenger was traced to problems in epoxy. But it turned out the heavy panels requiring the glue weren't necessary in the first place. At the drawing board, engineers thought an elaborate system was necessary to make sure the ventilation and exhaust fans operated properly, but in fact something more bare-bones worked fine. The architects of infrastructure are expected to get everything 100 percent right, but tragically, some realities only became apparent by way of failure [15].

As the author recalls, there were two main Big Dig contractors that were set to working separately, starting at opposite ends of the Central Artery, one team working essentially North to South, and the other South to North, hopefully to meet somewhere in the middle at the completion of the effort. This must have had something to do with the initial blueprint and tunnel misalignments above.

How could this separation possibly work without some reliable mechanism for ensuring these major contractors worked together in a largely integrated fashion?

> The Big Dig was led by Bechtel/Parsons Brinckerhoff, one of the largest and most experienced teams in infrastructure design and construction. Extensive environmental feasibility studies, risk assessments, and other documentation were completed prior to the project's start. Nonetheless, costs increased across all contracts throughout the project's life cycle despite enormous efforts to transfer, mitigate, or avoid risk and contain costs [24].
>
> In other words, things can go dramatically wrong despite the best efforts. Few infrastructure projects have used as many innovative tools and programs to control project risk and cost as the Big Dig. These included an owner-controlled insurance program that saved $500 million by providing group coverage for contractors, subcontractors, and designers and an unprecedented safety program; a cost-containment program that saved $1.2 billion; an integrated audit program that identified and mitigated writing and potential overruns and delays; a labor agreement that established a no-strike, no-slowdown guarantee for the life of the project; a quality-assurance program that was recognized by the Federal Highway Administration as one of five noteworthy accomplishments; and a dispute-resolution process that avoided extensive litigation costs [24].

A huge temporary dam was built/devised to dry out a Fort Point Channel basin in which to construct multi-lane tunnel segments, followed by flooding, floating these segments, weighing about 50k tons each [15], and then re-sinking them into their new underwater locations. Engineers built underground bridges to hold up operating subways, threading highway tunnels above and below various mass transit facilities. Underneath active railroad tracks, they also had to freeze the ground which was too soft to sustain construction efforts [25]. Near South (railroad) Station progress in path building was just 3 ft/day, where excavated earth had to be shoved backwards as one moved forward [26].

Here are factual descriptions of the Big Dig's principal milestones.

> On January 18, 2003, the opening ceremony was held for the I-90 Connector Tunnel, extending the Massachusetts Turnpike (Interstate 90) east into the Ted Williams Tunnel, and onwards to Boston Logan International Airport. The Ted Williams tunnel had been completed and was in limited use for commercial traffic and high-occupancy vehicles since late 1995. … [16].
>
> The next phase, moving the elevated Interstate 93 underground, was completed in two stages: northbound lanes opened in March 2003 and southbound lanes (in a temporary configuration) on December 20, 2003. A tunnel underneath Leverett Circle connecting eastbound Storrow Drive to I-93 North and the Tobin Bridge opened December 19, 2004 … . All southbound lanes of I-93 opened to traffic on March 5, 2005, including the left lane of the Zakim Bridge, and all of the refurbished Dewey Square Tunnel [16].
>
> By the end of December 2004, 95% of the Big Dig was completed. Major construction remained on the surface, including construction of final ramp configurations in the *North End* and in the *South Bay* interchange, and reconstruction of the surface streets [16].
>
> The final ramp downtown—exit 20B from I-93 south to Albany Street—opened January 13, 2006 [16].
>
> In 2006, the two Interstate 93 tunnels were dedicated as the Thomas P. O'Neill Jr. Tunnel … [16].

RESULTS

With additional time and determined development efforts, some Big Dig's problems were overcome.

> Thanks to federal and state investigations, criminal prosecutions, and other follow-up, most of the costs of the most blatant mistakes were recouped. The Globe identified approximately $1 billion in design flaws, and Bechtel ended up essentially reimbursing about half of that. … But there was no systematic corruption, at least not the kind seen in infrastructure projects elsewhere in the world. … Because of the overruns, however, Massachusetts taxpayers are still paying the bill [15].

Ultimately, the mission of vastly improving Boston's arterial traffic handling without unduly disrupting the surrounding environment was accomplished, and the capabilities exceeded most expectations.

> … 10 years after the official completion in 2006. How are we feeling now? Maybe there's room for some grudging appreciation. Rush hour brings what radio reporters refer to as heavy volume. But the traffic moves, and for 1.5 miles through downtown Boston, it moves out of sight, underground. Above those famously expensive tunnel boxes is some of the most beautiful and valuable urban real estate anywhere in the nation, if not the world. Getting to and from Logan airport has never been easier, whether picking up grandma or getting an executive to a startup in Fort Point or along Route 128 [15].

As of 2015, the Big Dig comfortably handled 536k vehicles each weekday [15].

Despite the most expensive of the three public transportation goals, cf., the "History" section, the bus tunnel was ultimately successful. As of 2015, the Lechmere extension had begun, but no funding was earmarked for the Red and Blue Line connection. The Rail Link was dropped because negotiations with the Federal Government for expanding the harbor tunnel left no room for rail lines [16].

Transformation Accomplished

> The dismantling of the elevated Central Artery was nothing short of transformational, opening up views at Faneuil Hall Marketplace across to the harbor, knitting the urban fabric back together at Hanover Street in the North End, and ushering in new cityscapes at Bulfinch Triangle — where the elevated Green Line at North Station was also depressed. Entire generations have no memory and no idea that the hulking elevated viaducts were ever even there. … So it is that the greatest success of the Big Dig is this: It established a new landscape for the city to flourish all around it. Buildings once overlooking a clogged highway now have a beautiful park at their front door. … Boston would probably be booming even without the Big Dig. But the project removed all doubt. Numerous reports have chronicled big jumps in property values. The Shawmut Peninsula is some of the most sought-after and desirable urban real estate in the country. … John Kerry, then a US senator, took a lot of flak for saying that the Big Dig would ultimately be viewed as a bargain, but he wasn't that far off the mark in this context [15].

Personally, the author relishes the much faster and easier trip to and from Logan Airport via the Massachusetts Turnpike extension and the Ted Williams Tunnel, the most direct beneficial improvements from his perspective. On the other hand, traveling into/out of Boston from/to the northeast via Route 1 or Route 93 entails some annoying delays at a fringe of the Big Dig construction, e.g., traveling in-bound, one must await in long lines of vehicles for access to Storrow Drive. No infrastructure improvement, or any CP result, is perfect!

Final System Description

Here are some impressive facts and figures from the State of Massachusetts website [27].

> Traffic: Elevated 6 lanes became underground 8-10 lanes; 161 lane-miles, about half in tunnels, and 4 major highway interchanges in a 7.5-mile corridor; 27 on-off ramps became 14 with improved associated streets.
>
> Dirt: 16M cubic yards excavated; 2/3, 541k truckloads, went to landfills, etc.; 1/3, 4,400 barge loads, went to Spectacle Island in Boston Harbor, capping an old dump to create a new park; and 3M cubic yards of clay went to New England cities and towns to cap landfills.
>
> Concrete and steel: 3.8M cubic yards of concrete; 26,000 feet of steel-reinforced concrete slurry walls [27], forming walls of underground highway and supports for elevated highway during construction; reinforcing steel would make a 1-inch bar that would wrap the Earth at the equator; and steel from the elevated demolition would make 5 Tobin (connecting Boston and the north) Bridges.
>
> First, most, biggest: Widest and deepest circular cofferdam [28] (connecting Ted Williams Tunnel and South Boston and including a ventilation building) in North America; Ted Williams Tunnel

interface in East Boston is 90 feet below the surface of Boston Harbor, the deepest such connection in North America; and the Zakim Bridge is the widest cable-stayed bridge in the world.

Parks and open space: 300 acres of new parks and open space; and 3/4 of 27 downtown acres remain open; more than 4,800 trees and 33,000 shrubs were planted.

Underground utility relocation: 29 miles of utility lines were moved; and 5,000 miles of fiber optical and 200,000 miles of copper telephone cable were installed.

Odds and ends: Boston's carbon monoxide level dropped 12% citywide; at peak, about 5,000 construction workers were on the job; deepest point of underground highway is 120 ft; highest point passes over the Blue Line subway tunnel, and the roof of the highway is the street above; and 118 separate construction contracts with 26 geotechnical drilling contracts.

Few readers would disparage these numerical results. However, the Big Dig's tangible qualitative improvements in the lives of Boston area residents is what matters most.

ANALYSIS

This section focuses on what else might be gleaned from the Big Dig experience.

ANALYTICAL FINDINGS

First, some selected quotations from Virginia Greiman's paper [24] on her learnings from the megaproject of the Big Dig are shown. This should whet the reader's appetite toward acquiring and reading her wonderful book [17] where she goes into great detail while heroically documenting the project. (Ms. Greiman was the former deputy counsel and risk manager for the Big Dig from 1996 to 2005 and an executive manager responsible for daily operations.)

> Our research on the Big Dig has shown us that no single catastrophic event or small number of contracts caused costs to escalate. Multiple decisions by project management across all contracts contributed to the increase. The critical cause was a lack of experience and knowledge about dealing with the complexity and uncertainty that giant projects bring with them.
>
> The most difficult problems on the Big Dig involved the means and methods used to address issues raised in the project's design and drawings, and the failure to properly account for subsurface conditions during the construction process. Project documents show that the challenges of subsurface conditions were substantially underestimated. … many unanticipated conditions and a large volume of claims and changes.
>
> The surprises included uncharted utilities, archeological discoveries, ground-water conditions, environmental problems, weak soil, and hazardous materials. The project faced safety and health issues, frequent design changes, and changes in schedules and milestones. The unexpected discovery of 150-year-old revolutionary-era sites and Native American artifacts was one surprise complication and source of delays, requiring approvals from yet another diverse set of stakeholders, including historical and preservation organizations and Native American groups.
>
> Because contracts were negotiated separately with designers and contractors, there was little room for collaboration among the project's most important stakeholders.
>
> If there is a single cause for the massive cost escalation on the Big Dig, it probably involves the management of the project's complex integration. Integration problems were exacerbated by the project's organizational structure, which separated design from construction through its traditional design-bid-build model and required managing thousands of stakeholders. … In the early phases of the project, there was little communication between and among many of the internal and external stakeholders, other than an impressive outreach to the local community, particularly residents living close to several of the project's major worksites. Community and social costs were vastly underestimated on the Big Dig. No one ever envisioned the full cost of dealing with the media, community interests, numerous regulatory agencies, auditors, and neighborhood stakeholders.

Avail yourself of the opportunity to peruse Greiman's great book [17] which is highly worthwhile, especially as the basic of an academic study of large CPs. The very important Chapter 3, Stakeholders [17, pp. 79–109], is particularly instructive. And there are many good discussion questions (located near the end of each chapter). There are also a lot of good quotations, some rather humorous, that encapsulate useful guidance.

Lessons Learned

What changes in policies or procedures were being or would be implemented so that lessons are *really* learned?

> Mega projects will always struggle with unforeseen events, massive regulatory requirements, technical complexities, community concerns, and a challenging political environment. What has been learned from the Big Dig can help future large projects. Of the many lessons from this huge undertaking, here are the major ones:
>
> - Project integration is critical to success.
> - Goals and incentives must be mutual and built into contracts throughout the project life cycle to ensure quality, safety, financial soundness, and a commitment to meeting budget and schedule.
> - Continuous improvement and rigorous oversight are both essential.
> - Doing things as they have always been done does not work for complex projects that require constant innovation and a culture of collaboration [24].

How Were the Biggest Challenges Met?

Unfortunately, there were few, if any, foregoing projects of the same ilk and magnitude as the Big Dig that could be used as role models to guide this massive effort. Consequently, many of the obvious challenges were met with meticulous and time-consuming planning that attempted to explore avenues for success and accompanying persistence in executing intended activities that were often somewhat *ad hoc* in nature and conducted within an environment of frustration. Other unforeseen challenges that continued to occur were dealt with in a similar fashion. It is likely that some aspects of Toolset were applied though not in a systemic, comprehensive way.

What Worked and Why?

As summarized just below, much of Boston's travel problems were alleviated.

> The wrong lesson to take from the Big Dig is that other states shouldn't bother with ambitious infrastructure. While the Big Dig's real worth will be measured in decades, its impact so far, three years after workers dismantled the Central Artery, shows its value. Travel time through downtown at afternoon rush hour is down from nearly 20 minutes to less than three, consistent with pre-construction estimates. Elsewhere on the underground highways, travel times are between one-quarter and two-thirds shorter; average speeds in some sections have shot from ten miles per hour to 43 (speed, rather than drivers' veering toward too many exits in slow traffic, is the tunnels' biggest safety problem). Airport trips are between one-half and three-quarters shorter. A 62 percent drop in hours spent on the new roads saves nearly $200 million annually in time and fuel [14].

Reasons for such success primarily come down to ambitious optimism and persistence in pursuing the overall goals and their associated opportunities while weighing and protecting against the various and looming risks involved.

What Did Not Work and Why?

The most obvious shortcomings were what might be construed as the excessive length of the project and its high cost. Up-front planning took a decade and the construction another decade and a half. Estimated costs grew by a factor of six! Again, the reasons why have much to do with the unprecedented nature of such a CP, and the need for garnering sufficient political, financial, and public support for the undertaking. Here's an example of a specific after-the-fact rationale that lends some flavor to the issues.

> ... the Big Dig's biggest pitfall was that Massachusetts never understood a basic fact: that it couldn't pay someone else to assume its own responsibility for an immensely complex, risky project. As the National

Transportation Safety Board (NTSB) would later say in its report on the 2006 ceiling collapse, Bechtel and Parsons, the state's long-term consultants, were 'performing the role that would normally be carried out by a government agency, specifically, the state highway department' [14].

What Should Have Been Done Differently?

In answer to this question, suffice it to say, so many things! In this chapter, it's too easy and self-serving to merely point to a more rigorous application of Toolset as the panacea. For instance, a more inclusive involvement of transdisciplinary experts, e.g., in philosophy, psychology, sociology, organizational change theory and management, economics, environment, politics, and morality, should have been included in order to do justice to the principle of looking beyond the most immediate set of stakeholders for additional guidance. More broadly, there also is much fodder in the other chapters of this book that could have been applied to great effect, if the Big Dig organizers and enactors had been more aware of the wisdom of these fellow authors' contributions.

CONCLUSIONS

CPs require a different set of tools than recommended by classical, traditional, or conventional PM techniques as often espoused by the Project Management Institute or INCOSE, for example. This author believes Toolset goes quite far in remedying this situation and encourages readers to further explore the use of these tools in their own CPM efforts. Consider initiating or completing and documenting other case studies utilizing the CSO tool and sharing the results with colleagues. This author would certainly welcome any of your inputs along these lines.

THE BIG DIG

By now, the Big Dig milieu is rather well known. This chapter's author hopes and trusts that the description and analysis provided herein is useful to readers concerned with CPs, even those less ambitious than Boston's recent experience. When contemplating embarking on such complex efforts, readers should be sure to avail themselves of appropriately matched complex tools as expressed in this book.

> When embarking on complex efforts, readers should be sure to avail themselves of these appropriately matched complex tools.

QUESTIONS

This section is common with other chapters of this book in focusing on academic questions that will hopefully suitably challenge and provide some mental exercise for students and practitioners of PM.

1. What (other?) tools are most recommended for application in case studies of CPs?
2. What are the relative pros and cons of the tools offered in this chapter, especially in terms of case studies?
3. What might one think of the premise that case studies are instrumental in driving complex PM theory?
4. On CPs to what extent is outstanding project leadership [29] needed in addition to exemplar PM?

Consider these questions and perform some work on possible answers before reading the Summary. Also, obtain Virginia Greiman's outstanding book [17] and peruse the many excellent discussion questions therein.

SUMMARY

Here again is the list of tools introduced in this chapter that may be needed for coping with PM challenges.

1. Case Study Outline
2. Project Management Life Cycle
3. Zachman Framework
4. EPM Profiler
5. PMA Profiler
6. The Enneagram
7. Soft Projects Methodology
8. CAPM Methodology.

In the author's opinion, Tool 1 is the most important for case studies, Tool 4 is most useful in characterizing the project environment, while Tool 5 best describes project activities; Tool 6 is invaluable and properly guiding project discussions, and Tool 8 is most important for good PM on CPs, although Tool 7 is also helpful; Tool 2 is more limited but can be useful in simpler contexts. Tool 3 is great for contemplating an essentially complete array of potential project issues and their interrelationships.

Reasons for the above assessment may be ascertained by reviewing the preceding tool descriptions and analyses. The Big Dig case study example shows how these tools can be applied in a specific project scenario. The author hopes this chapter contributes positively to the evolving body of knowledge of PM particularly with respect to case studies in CPs which are becoming more prevalent.

ACKNOWLEDGMENT

The author gives special thanks to Alex Gorod for inviting this chapter and encouraging its successful completion. The technical and personal stimulation provided over the years by Alex and other colleagues such as Vernon Ireland, Jimmy Gandhi, Brian Sauser, Beverly Gay McCarter, Sarah Sheard, Gerrit Muller, Mikhail Belov, Chuck Keating, Rob Cloutier, and Mihaela Ulieru is also gratefully acknowledged.

REFERENCES

1. A. Gorod, B. E. White, V. Ireland, S. J. Gandhi, and B. J. Sauser, *Case Studies in System of Systems, Enterprise Systems, and Complex Systems Engineering.* Boca Raton, FL: CRC Press, Taylor & Francis Group. 2015.
2. B. E. White, System of Systems (SoS) course. Developed for the Worcester Polytechnic Institute (WPI). WPI's graduate program in Systems Engineering. February 2015.
3. J. A. Zachman, A framework for information systems architectures. *IBM Systems Journal.* Vol. 26. No. 3. 1987. 276–92. www.zachman.com/images/ZI_PIcs/ZF3.0.jpg; https://en.wikipedia.org/wiki/Zachman_Framework.
4. R. Stevens, Profiling complex systems. *IEEE (Institute of Electrical and Electronics Engineers) Systems Conference.* Montreal, Quebec, Canada. 7–10 April 2008.
5. B. E. White, Systems Engineering Activities (SEA) profiler. *8th Conference on Systems Engineering Research (CSER).* 17–19 March 2010, Hoboken, NJ. 18 March 2010.
6. B. E. White, Complex adaptive systems engineering. *8th Understanding Complex Systems Symposium.* University of Illinois at Urbana-Champaign, IL. 12–15 May 2008. http://www.howhy.com/ucs2008/schedule.html.
7. R. N. Knowles, *The Leadership Dance—Pathways to Extraordinary Organizational Effectiveness.* 3rd Edition. Niagara Falls, NY: The Center for Self-Organizing Leadership. 2002.
8. P. Checkland, *Systems Thinking, Systems Practice—Soft Systems Methodology: A 30 Year Perspective.* New York: John Wiley & Sons. 1999.

9. B. E. White, A Complex Adaptive Systems Engineering (CASE) methodology—The ten-year update. *IEEE (Institute of Electrical and Electronics Engineers) Systems Conference.* Orlando, FL. 18–21 April 2016.

10. B. E. White, On leadership in the complex adaptive systems engineering of enterprise transformation. *Journal of Enterprise Transformation.* Vol. 5. No. 3. 192–217. 2015. Supplementary Material (Appendices): www.tandfonline.com/doi/suppl/10.1080/19488289.2015.1056450; ISSN: 1948-8289 (Print) 1948-8297 (Online) Journal homepage: www.tandfonline.com/loi/ujet20; doi: 10.1080/19488289.2015.1056450.

11. R. Stevens, *Engineering Mega-Systems—The Challenge of Systems Engineering in the Information Age.* Boca Raton, FL: CRC Press. 2011.

12. W. R. Ashby, Requisite variety and implications for control of complex systems. *Cybernetica.* Vol. 1. No. 2. 83–99. 1958.

13. J. Boardman and B. Sauser, *Systems Thinking—Coping With 21st Century Problems.* Boca Raton, FL: CRC Press. 2008.

14. N. Gelinas, Lessons of Boston's Big Dig—America's most ambitious infrastructure project inspired engineering marvels—and colossal mismanagement. From the Magazine. *City Journal.* Autumn 2007.

15. A. Flint, 10 years late, did the Big Dig deliver?—The $15 billion project is a road paved with failures, successes, and what-ifs. *The Boston Globe Magazine.* 29 December 2015.

16. Big Dig—The Central Artery/Tunnel Project. Wikipedia, the free encyclopedia. https://en.wikipedia.org/wiki/Big_Dig. Accessed 14 September 2018.

17. V. A. Greiman, *Megaproject Management—Lessons on Risk and Project Management from the Big Dig.* Project Management Institute. Hoboken, NJ: John Wiley & Sons. 2013.

18. B. E. White, A personal history in system of systems. Special session on System of Systems (SoS). *International Congress on Ultra Modern Telecommunications and Control Systems (ICUMT-2010).* Moscow, Russia. 18–20 October 2010.

19. B. E. White, Managing uncertainty in dating and other complex systems. *Conference on Systems Engineering Research (CSER) 2011.* Redondo Beach, CA. 15-16 April 2011.

20. B. E. White, Let's do better in limiting material growth to conserve our earth's resources. *Conference on Systems Engineering Research (CSER) 2012.* St. Louis, MO. 19–22 March 2012.

21. B. E. White, Applying complex systems engineering in balancing our earth's population and natural resources. *The 7th International Conference for Systems Engineering of the Israeli Society for Systems Engineering (INCOSE_IL).* Herzlia, Isreal. 4–5 March 2013.

22. B. E. White, On principles of complex systems engineering – Complex systems made simple tutorial. *23rd INCOSE International Symposium.* Philadelphia, PA. 24–27 June 2013.

23. Michael Dukakis, Wikipedia, the free encyclopedia. https://en.wikipedia.org/wiki/Michael_Dukakis. Accessed 12 August 2018.

24. V. Greiman, The Big Dig: Learning from a Mega Project. Home/39/ASK_Magazine/, 15 July 2010, https://appel.nasa.gov/2010/07/15/the-big-dig-learning-from-mega-project/.

25. Slurry wall, Wikipedia, the free encyclopedia. https://en.wikipedia.org/wiki/Slurry_wall. Accessed 11 August 2018.

26. F. Salvucci, Big Dig talk during MIT Class of 1961's 45th reunion. 2006.

27. The Big Dig: Facts and figures—Statistics from the Central Artery Project. State of Massachusetts. www.mass.gov/info-details/the-big-dig-facts-and-figures.

28. Cofferdam, Wikipedia, the free encyclopedia. https://en.wikipedia.org/wiki/Cofferdam. Accessed 11 August 2018.

29. A Guide to the Project Management Body of Knowledge (PMBOK® Guide). 5th Edition. Project Management Institute (PMI), 2013. www.orange.ngo/wp-content/uploads/2016/09/PMBOK-Guide-5th-Edition-PMI.pdf.

3 Scheduling Toolset

Seyed Ashkan Zarghami and Alex Gorod
The University of Adelaide

CONTENTS

INTRODUCTION

The concept of scheduling is not new. It has a history that is almost as long as the dawn of civilization. Sun Tzu, the reputed author of *The Art of War*, addressed strategy and scheduling from a military perspective nearly 2,500 years ago (Dimovski et al., 2012). The ancient cities of Mesopotamia, the pyramid of Egypt, the Great Wall of China, and the Coliseum are examples of the ancient complex projects that were successfully completed. None of these major projects could have been accomplished without careful planning and scheduling.

Since ancient time, much effort has been devoted to the development of a vast array of methods and tools for project scheduling. However, despite having numerous project scheduling methodologies, the historical records highlight a high prevalence of project failures due to delays (Herroelen, 2005). It appears to be an accepted fact that scheduling of complex projects is by nature complex. The complexity that may arise from multiple interacting activities, as well as unpredictable factors, is often considered as one of the main sources of the project's delay. This, in turn, has necessitated the development of effective scheduling methods by which to manage such challenges. In fact, project scheduling has evolved over time in a manner to master the increased projects complexity as well as the complexity inherent within project scheduling.

Given the growing number of project scheduling tools, it is important to gain an understanding of the historical developments of these tools.

This chapter provides an overview of how project scheduling tools have evolved over time and discusses how researchers have applied these tools. It is based on the division of history into seven periods, each of

> Project scheduling has evolved over time in a manner to master the increased projects complexity as well as the complexity inherent within project scheduling.

which characterizes the development of particular tools. A discussion of each tool is accompanied by the additional information as to how to obtain the tool, including links to sites, seminal books, and published articles. It also sets forth the underlying principles of various scheduling methodologies and contextualizes (1) what these tools are, (2) what motivated the development of different tools, and (3) how these tools overcome constraints in project scheduling. Finally, the chapter forecasts the future evolution of project scheduling tools, where the authors believe the forthcoming works are going.

THE SEVEN PERIODS OF PROJECT SCHEDULING

This section tracks the development of scheduling tools and attempts to provide a broadly comprehensive catalogue of various scheduling approaches. In this respect, the section divides the history of project scheduling into seven periods: pre-20th century, early 20th century, 1950s, 1960s, 1970s, 1980s, and 1990s–present.

PRE-20TH CENTURY: THE FORMATION OF VISUALIZATION TOOLS

Visualization tools for scheduling can trace their origins to the 14th century (Der and Everitt, 2014). Nichole Oresme, a philosopher of the later Middle Ages, is considered to be a prodigious contributor to the development of time series graphical diagrams in preference to words and tables. In *The Latitude of Forms*, a 14th-century publication, as shown in Figure 3.1, Oresme plotted the first bar chart in history, illustrating the velocity of constantly accelerating objects against time (Oresme, 1482).

> Visualization tools for scheduling can trace their origins to the 14th century.

Skipping ahead a few hundred years, in 1765, the British polymath Joseph Priestly published his seminal book, *A Chart of Biography*, in which he plotted lifespans of 2,000 famous people from 1200 BC to AD 1765 (Schofield, 2004). Figure 3.2 illustrates Priestley's chart of biography.

Priestley's chart of biography inspired William Playfair, a Scottish engineer and political economist, to develop a range of time series charts that first appeared in *The Commercial and Political Atlas*, published in 1786. Playfair's charts have set the basis for the development of most of the graphical forms used today (Friendly, 2008). Figure 3.3 depicts the time series chart created by Playfair, illustrating the Scotland trading partners.

EARLY 20TH CENTURY: THE EMERGENCE OF GANTT CHARTS

These previous graphical innovations provided the necessary ingredients for the development of the project-based scheduling graph. In 1896, the idea of plotting the intended sequences and timing of project activities appeared in a work by Karol Adamiecki, a polish engineer and management

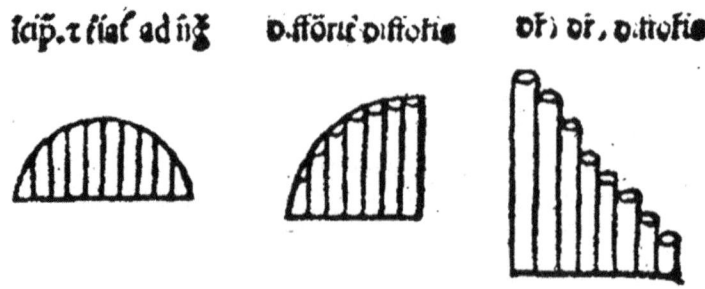

FIGURE 3.1   Oresme's bar charts (14th century). (Source: Oresme, 1482.)

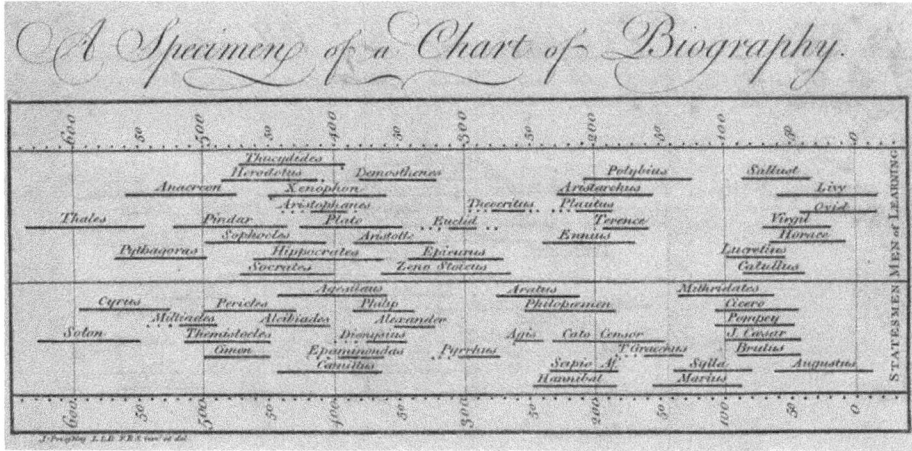

FIGURE 3.2   Priestley's chart of biography. (Source: Priestley, 1765.)

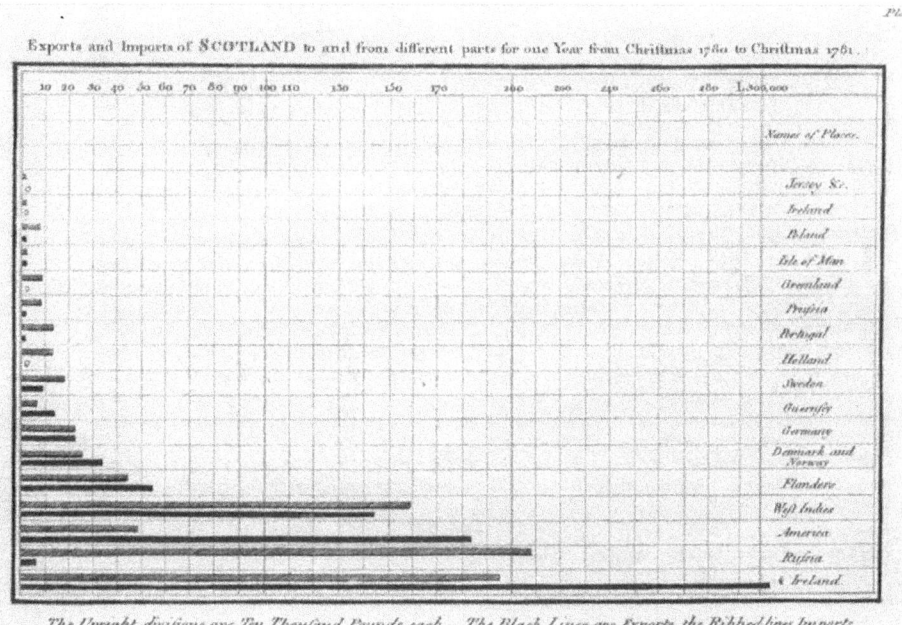

FIGURE 3.3   Playfair's chart of Scotland trading partners. (Source: Playfair, 1786.)

researcher. Adamiecki introduced a graphical analysis method termed Harmonogram, in which the activities and their durations are depicted by the position and length of paper strips. In fact, the Harmonogram is a workflow network diagram that provides solutions to production scheduling problems (Marsh, 1976). Figure 3.4 illustrates the Harmonogram designed by Adamiecki.

A few years later, an American mechanical engineer, Henry Laurence Gantt, introduced one of the most widely used charting techniques in projects today. Gantt (1903) first outlined the essence of his charts in his seminal article "*A Graphical Daily Balance in Manufacture*" (Wilson, 2003). Although Gantt charts were originally created for planning and managing batch production, shortly after, the span of application grew from the manufacturing sector to a wide range of areas, including planning and scheduling projects. As shown in Figure 3.5, the first version of the Gantt chart was designed to monitor the current productivity level of employees and their performance

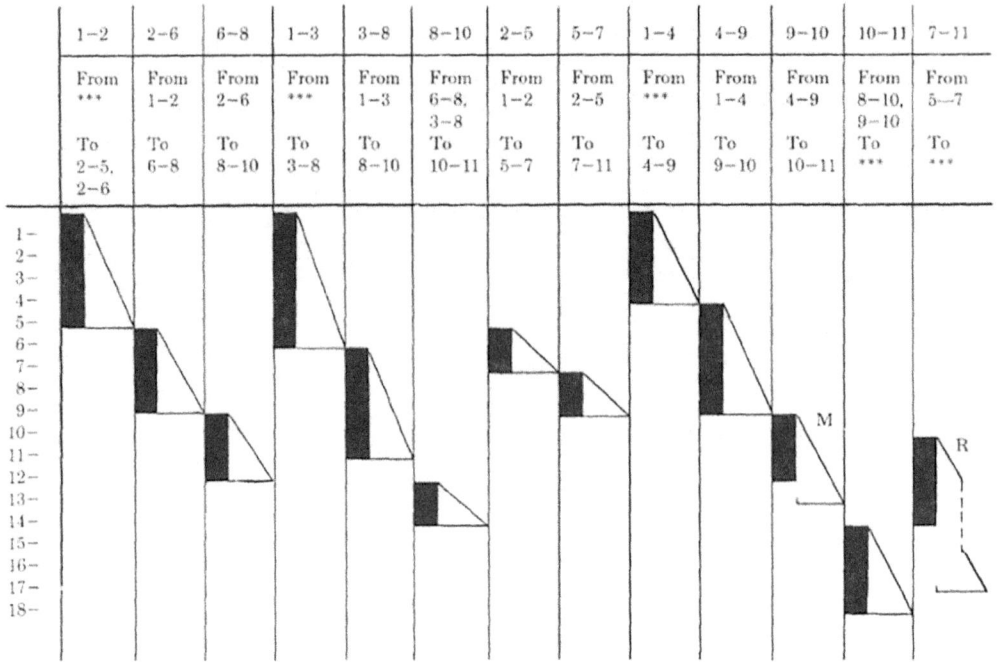

FIGURE 3.4 Adamiecki Harmonogram. (Source: Adamiecki, 1970.)

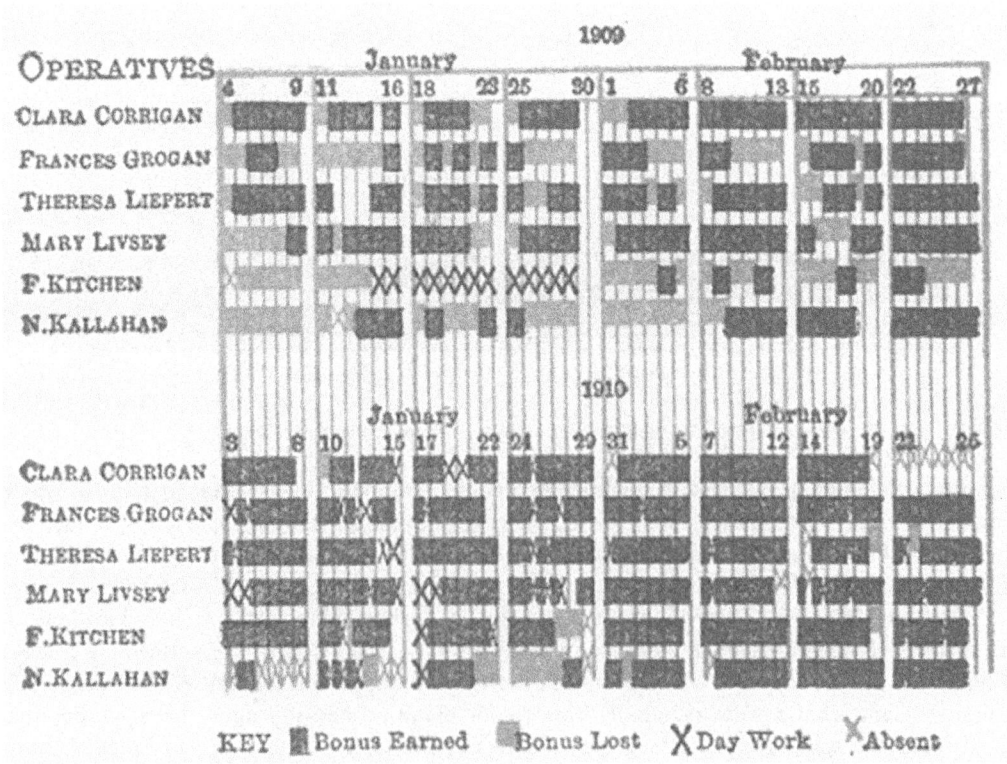

FIGURE 3.5 Original Gantt chart. (Source: Gantt, 1913.)

in manufacturing facilities, that is, focusing on the past rather than planning for the future. This visualization tool was further extended to show not only the present but also the difference between plan and actual variance (Geraldi and Lechter, 2012). In this refined version, an activity was first measured by the amount of time required for its completion, and then a space on the chart was provided to depict how long the activity was expected to take (Herrmann, 2010).

> A century after it was created, the Gantt chart is still alive and is being employed as one of the most widely used project scheduling tools.

A century after it was created, the Gantt chart is still alive and is being employed as one of the most widely used project scheduling tools. With the development of modern project management, its application has continued to grow over the past three decades. The Gantt chart has been the forerunner of many sophisticated scheduling tools. Several desktop and online Gantt chart software programs have provided superior capability for planning and tracking projects of all sizes. The following are some popular Gantt chart software packages that are currently assisting in streamlining project scheduling from planning to controlling: GanttPRO (www.ganttpro.com), GanttProject (www.ganttproject.biz), Microsoft Project (www.office.microsoft.com), ProjetLibre (www.projectlibre.com), Redbooth (www.redbooth.com), and Smartsheet (www.smartsheet.com).

1950s: The Era of Network Diagramming

Almost simultaneously in the late 1950s, different groups of engineers developed three project scheduling tools that came to be known as Critical Path Method (CPM), Project Evaluation Review Technique (PERT), and Precedence Diagram Method (PDM) (Hyatt and Weaver, 2006). The development of these three tools followed a similar pattern but with different orientations to solve project scheduling problems. In the CPM network, the primary focus is on project activities, whereas the PERT network places emphasis on project events or milestones. PDM differs from CPM and PERT along three dimensions (Wiest, 1981). First, in the PDM, activities are represented with nodes instead of arrows and the relationships between activities are shown as arrows. Second, contrary to the CPM/PERT networks that consider only Finish to Start (FS) precedence dependency between project activities, PDM takes into account three additional precedence relationships: Start to Start (SS), Finish to Finish (FF), and Start to Finish (SF). Third, PDM incorporates the concepts of lead and lag time into network diagramming. The former indicates the overlap between the first and second activities, and the latter indicates the minimum amount by which the start or finish of one activity lags the start or finish of another activity.

CPM, PERT, and PDM have attracted the attention of practitioners and academic researchers intrigued by their capability in planning, organizing, and managing projects. In academic settings, these diagramming tools have been widely discussed in books (for example, Hutchings, 2004) and journal papers (for example, Trietsch and Baker, 2012). Several project management-related journals (for example, *International Journal of Project Management* and *Project Management Journal*) have published a great number of articles addressing the latest advances in these tools. In practice, a great many commercial project management software packages containing a network chart feature, such as Edraw Max (www.edrawsoft.com), Lucidchart (www.lucidchart.com), OpenProj (www.openproj.org), PERT Chart EXPERT (www.criticaltools.com), and WBS Schedule Pro (www.criticaltools.com), are currently used for planning and managing complex real-world projects.

> CPM, PERT, and PDM have attracted the attention of practitioners and academic researchers intrigued by their capability in planning, organizing, and managing projects.

1960s: Formation of Project Scheduling under Resource Constraints

CPM, PERT, and PDM implicitly assume that unlimited resources are assigned to project activities. However, in actual reality, where resources (for example, labors, materials, equipment, and capital)

are constrained, the concept of infinite resources loses its
meaning. As early as the 1960s, researchers began to take
a new angle on project scheduling problems, which was
to become the Resource-Constrained Project Scheduling
Problem (RCPSP). RCPSP deals with the process of creating a project schedule by joint consideration of two

> As early as the 1960s, researchers
> began to take a new angle on project
> scheduling problem, what was to
> become the RCPSP.

kinds of constraints: precedence constraints and resource constraints. Precedence constraints imply
dependency between activities, and resource constraints arise when more resources are required than
available. RCPSP resolves resource conflicts (over allocations of resources) under different scheduling
objectives by examination of the possible unbalanced use of resources over time. RCPSP has received
considerable attention within the project scheduling literature, and many innovative methods and tools
have been developed since the 1960s. These mainly use combinatorial techniques to identify an optimal solution from a finite set of scheduling objectives. More specifically, the RCPSP tools aim at finding a feasible solution such that a specific scheduling objective is minimized or maximized (Artigues,
2008). As listed by Demeulemeester and Herroelen (2002), a wide variety of scheduling objectives
have been considered in the development of different RCPSP methods.

Time-based objectives: A dominant scheduling objective is the minimization of the project completion time. The basic principle is scheduling a set of project activities within a minimal project
duration and subject to precedence relations in order to reduce the risk of violating a deadline
(Vanhoucke, 2012).

Resource-based objectives: In many RCPSP methods, resource-based objectives are used as a
measure of performance. The inclusion of resource availability costs and resource-leveling problems into the method are some examples (Hartmann, 1999).

Financial objectives: In this case, the scheduling objective is assumed to be the maximization of
the net present value of the project. The underlying concept of the tools in this category is to schedule the project activities in such a way that cash inflows are obtained as early as possible and cash
outflows occur as late as possible (Kimms, 2001).

Quality-oriented objectives: Maximizing the quality of a project and its outcomes is the primary
objective for many projects. In the existing literature, different indicators have been used to measure the perception of quality in a project. For example, Icmeli-Tukel and Rom (1997) considered
quality as a function of rework in projects, whereas Kellenbrink and Helber (2016) studied the
relationships between quality attributes and project flexibility.

The last four decades of the 20th century witnessed the development of a number of RCPSP tools
and techniques. Wiest (1964) and Dike (1964) are among the earliest RCPSP works that attempted
to minimize project duration such that the precedence relations and resource availabilities were
both taken into account.

At the end of the 20th century, there was what might be perceived as a silver bullet for minimizing the total duration of resource-constrained projects. In 1997, Eliyahu M. Goldratt published
his best-selling business novel, *Critical Chain*. Goldratt set out a project scheduling tool named
"Critical Chain Project Management (CCPM)," as a new paradigm for resource feasible project
scheduling. Goldratt argued that the traditional network diagramming tools failed to tackle the
misuse of safety time created within the estimated times for each activity (Rand, 2000). Therefore,
he suggested that rather than building the safety time into each estimate, the original duration
estimates should be reduced, and subsequently, safety times should be pooled into time buffers
at control points. At its most basic, CCPM outlines the five focusing steps of the theory of constraints (Watson et al., 2007). First, CCPM identifies the constraints that limit system performance,
such as the weakest link in a chain of activities or the bottleneck resources. In the second step,
CCPM focuses on activities in the critical chain (a set of activities that results in the longest path to
project completion) to ensure the timely completion of the project. The third step of the five focusing steps subordinates the non-bottleneck resources to improve the performance of the bottleneck
resources. Should system performance be unsatisfactory, the fourth step elevates the constraint

and adds resources to reduce the constraint. Finally, the fifth step renews the improvement cycle and prevents inertia from becoming a constraint. Since the introduction of CCPM, the concept and principles underlying this method have been investigated in several books (for example, Newbold, 1998 and Vanhoucke, 2016). A number of researchers have discussed the strengths and weaknesses of CCPM (for example, Steyn, 2002 and Lechler et al., 2005), and other researchers have addressed the technical aspects of this method (for example, Tukel et al., 2006 and Zhang et al., 2016). CCPM has also attracted the attention of project management practitioners who express concerns over using non-resource-based scheduling tools. Hence, a number of software packages based on the CCPM scheduling concept have been used in practice, such as Aurora-CCPM (www.aurora-ccpm. com), BeingManagement (www.beingmanagement.com), Exepron (www.exepron.com), and Lynx TameFlow (www.a-dato.com).

1970s: A System Approach to Project Management

Despite significant efforts on the development of project scheduling tools, many researchers found continued schedule performance problems in projects. The main argument centered on the fact that most methodologies were inadequate for addressing the dynamic complexity of projects (Lyneis et al., 2001). In fact, these methods were confined to analyzing the project scheduling problems either statically or by taking a separate view of parameters involved in the scheduling analysis. However, there were several responses to this challenge. One response was to adopt System Dynamics (SD) modeling, where the dynamic interactions within a system are captured from a holistic perspective. In the early 1960s, Jay W. Forrester, a pioneering American computer engineer and system scientist, laid the foundation of SD as a potent tool to capture the dynamic interactions within a system from a holistic perspective (Zarghami et al., 2018). The SD modeling approach first highlights the underlying feedback loops that characterize the behavior of a system and then proceeds with simulating the complex interactions between constituent elements of the system, where causes and effects are indiscernible.

SD has been the focus of increasing international research over the last six decades. Since then, several influential books (for example, Wolstenholme, 1990 and Sterman, 2000) have been published. To date, most literature on system engineering advocates the use of SD to better understand and predict the dynamic behavior of complex systems. There are many SD software tools ranging from open source software to commercial ones. Most published work in this field has used these tools. Examples of software and web-based tools supporting SD are as follows: AnyLogic (www.anylogic.com), Insight Maker (www.insightmaker.com), Powersim Studio (www.powersim. com), Vensim (www.vensim.com), Wolfram SystemModeler (www.wolfram.com), and iThink and STELLA (www.iseesystems.com). For the updated list of available tools on SD, the reader is referred to System Dynamics Society web page (www.systemdynamics.org).

Shortly after the SD modeling invention, the project management community caught up with this approach. Since then, SD models have been widely used in the project management areas as diverse as software development projects (Abdel-Hamid and Madnick, 1983), the water sector (Ahmad and Simonovic, 2004), manufacturing development projects (Williams et al., 1995), construction projects (Khan et al., 2016), and infrastructure networks (Rashedi and Hegazy, 2016). Nevertheless, the literature on the application of SD modeling to project scheduling is rather sparse. Cooper (1980) employed an SD approach to quantify the impact of customer changes on the project schedule overrun in naval ship production. Jalili and Ford (2016) used the SD modeling approach to evaluate the impact of three reinforcing feedback loops generated by rework, schedule pressure, and ripple effect loops on project schedule performance. In 2017, Wang and Yuan (2017) utilized the SD approach to investigate the risks' interactions on schedule delay on infrastructure projects. Finally, Kumar and Thakker (2017) analyzed schedule and cost overrun in R&D projects.

Clearly, further continuous research into the intersection of SD application and project scheduling across various industries is needed and may prove very useful for successful project management.

1980s: Project Scheduling under Uncertainty

Uncertainty is an inherent aspect of project scheduling. The high record of scheduling disruptions due to considerable uncertainty in project activity durations has resulted in questioning the validity of deterministic and network diagramming scheduling tools. Although uncertainty in task duration was addressed by proposing a number of probabilistic methods in the 1960s and 1970s (for example, Charnes et al., 1964 and Burt and German, 1971), the lack of historical data was the most important barrier to the successful application of these methods to the project scheduling domain. Accordingly, the critics of probabilistic methods argue that a different approach that formalizes the knowledge of experts in estimating activity durations seems a more effective method. As a result, the Fuzzy Inference System (FIS) came to the attention of schedulers as early as the 1980s.

> Uncertainty in project activity durations led to the formation of several methods to tackle it.

FIS, proposed by Zadeh (1965), is a mathematical tool for modeling nonlinear systems in which membership is a matter of degree. It works in such a way that is very similar to human thinking, reasoning, and cognition. The implementation of FIS requires three consecutive steps: fuzzification, fuzzy inference, and defuzzification. The first step is fuzzification, through which the membership functions for input and output variables are constructed and the linguistic variables, as nonnumerical variables, are defined. The second step, fuzzy inference, combines the membership functions, using "IF-THEN" rules in order to derive the fuzzy output. The defuzzification step transforms the linguistic variables, generated by the previous step, into numerical values so as to make them available for real applications (Bai and Wang, 2006). During the last five decades since the inception of FIS, the number of researchers, journals, and consequently of publications addressing fuzzy set theory has increased rapidly. Successful application of fuzzy theory in many disciplines is triggering further interest and research, with currently more than 5,000 FIS-related papers being published every year (Merigó et al., 2015). Several journals such as *Fuzzy Sets and Systems*, *IEEE Transactions on Fuzzy Systems*, *Journal of Intelligent & Fuzzy Systems*, and *Fuzzy Optimization and Decision Making* are specifically devoted to the advancement and applications of fuzzy theory, to name a few. In addition, hundreds of journals publish papers on the topic of fuzzy logic (for example, *International Journal of Computational Intelligence Systems*, *International Journal of Production Research,* and *Journal of Mathematical Analysis and Applications*). A good number of FIS software design tools exist today that assist in the implementation of fuzzy-based systems. The most popular FIS software developed by the scientific community are as follows: FISPro (www.fizpro.org), FuzzME (www.fuzzme.wz.cz), FuzzyLite (www.fuzzylite.com), FuzzyTECH (www.fauzzytech.com), and MATLAB-FLT (www.mathworks.com).

Thanks to the ability of FIS in modeling nonlinear systems, as well as its compatibility with decision-making situations, it has gained recognition in the project scheduling literature since the early 1980s. Prade (1979), Chanas and Kamburowski (1981), and Ayyub and Haldar (1984) were among the first papers that tackled the project scheduling problem by adopting FIS. FIS has been widely used to deal with uncertainty specifically in RCPSP, and it is still garnering much attention in the research literature. For example, Hapke and Slowinski (1996), Kim et al. (2003), Masmoudi and Hait (2013), Chen and Zhang (2016), and Birjandi and Mousavi (2019).

1990s–Present: The Era of Optimization

Since the last decade of the 20th century, there has been a growing realization that classical mathematical approaches cannot solve the complexity inherent in the computation process of RCPSP. This reflection has led to the evolution of several optimization tools. Subsequently, many books (for example, Dréo et al., 2006 and Bozorg-Haddad et al., 2017), a great many articles (for example,

Artigues et al., 2013 and Elsayed et al., 2017), and book reviews (for example, Brucker, 2002) have been devoted to this subject. Further, many societies such as *Mathematical Optimization Society* (www.mathopt.org) and *Informs Optimization Society* (www.informs.org) have been formed with the goal of promotion and development of the optimization methods.

Over the last three decades, the increasing complexity of optimization problems has been accompanied by the advent of the modern computer that has resulted in a rise in the widespread adoption of optimization software. Recent advances in optimization problems including RCPSP would have been impossible without the use of computer-aided optimization modeling. Among the most common software packages used in optimization problems are AMPL (www.ampl.com), Artelys Knitro (www.artelys.com), and FICOXpress (www.fico.com), just to name a few. In the following section, the most widely applied optimization tools in the project scheduling domain will be discussed.

Agent-based project scheduling (ABPS): An agent is an autonomous entity that interacts with the environment where it is situated. Agent-based models are composed of multiple identifiable and discrete agents that are capable of autonomous actions and independent decisions. More specifically, an agent-based model is a simulation model of the real environment in which multiple agents operate. These agents represent different individuals from the system of interest. The power of agent-based modeling lies in its ability in demonstrating emergent phenomena. Additionally, it provides an intuitive representation of the system under investigation and offers high level of flexibility and scalability (Siegfried, 2014). ABPS is now considered to be a common simulation method for project scheduling. Five attributes have made ABPS appealing to researchers in the project scheduling domain. First, ABPS decomposes a global problem to a number of smaller local problems, thereby reducing the computational times. Furthermore, breaking down problems into sub problems facilitates the process of handling large-sized problems. Second, thanks to the agents' ability to continuously monitor the status of a project, ABPS presents higher reactiveness degrees in a project and its dynamic environment (Barbati et al., 2012). Third, an agent is adaptable and can modify its behavior in response to changing environmental conditions (Jennings and Wooldridge, 1995). Fourth, ABPS is a distributed and an autonomous modeling approach, and hence, the resource allocation and scheduling functions are performed in a decentralized manner (Adhau et al., 2012). Fifth, ABPS follows a negotiation-based decision-making process rather than a pre-planned process (Rabelo et al., 1999). ABPS has been widely employed for multi-project scheduling in the decentralized decision-making environment (for example, Li and Xu, 2018 and Song et al., 2018). Knotts and Dror (2003) and Ratajczak-Ropel (2018) investigated the application of ABPS to solve non-preemptive multimode RCPSP. Han et al. (2017) applied ABPS to solve resource conflicts and task assignment in offshore projects.

Genetic algorithm (GA): This is a heuristic exploratory procedure based on the Darwinian principle of survival of the fittest that "the strong will succeed and the weak shall perish." Several authors have implemented the GA approach to locate near-optimal solutions to RCPSP. The common underlying idea of all these works is to maintain a set of trial solutions and force them to evolve towards an acceptable solution through a four-step process (Frenzel, 1993). GA begins with the evaluation of a randomly generated population of possible solutions (individuals) to a given RCPSP optimization problem in order to investigate how well they solve the problem (Bozorg-Haddad et al., 2017). The second step takes the advantage of a survival of the fittest strategy with the aim of creating a new population of possible solutions whose members best solve the problem according to their fitness level. In the third step, a new population is generated by applying a recombination that combines two or more selected individuals to generate new offspring (Boussaïd et al., 2013). The last step of the GA is mutation through which new traits are transferred to offspring. In fact, this step promotes diversity by injecting new information into the population that results in getting new solutions. Since its discovery, GA has proved its effectiveness in the project scheduling research literature. Alcaraz and Maroto (2001) applied GA as a resource allocation tool in project scheduling. Gonçalves et al. (2008) adopted GA as an optimization tool for the resource-constrained multi-project scheduling problem. Zamani (2013) developed a modified GA by using a precedence-based permutation crossover, called magnet-based crossover, for solving RCPSP. Kadri and Boctor (2018) employed GA in order to minimize the project duration by choosing an optimum start time for each activity.

Simulated annealing (SA) algorithm: SA technique is a metaheuristic optimization algorithm proposed by Kirkpatrick et al. (1983). It emulates the annealing process in metallurgy, involving heating a metal to a specified temperature and then slowly lowering the temperature to decrease the defect in the metallic structure. In the SA algorithm, a control parameter in optimization acts as the temperature of the physical system. The algorithm begins with generating an initial random solution that represents the current state of the metal in the real annealing process. A new state (solution) is then generated and compared with the current state by using a probability distribution. In a minimization problem, the algorithm accepts all new states that lower the objective. In order to avoid being trapped in local minima, some states that increase the objective function with a certain probability are also accepted. As the system is cooled, the process is repeated by generating a certain number of new states at each temperature and is stopped when termination criteria are satisfied (Bozorg-Haddad, 2017). Since its discovery, SA has enjoyed a successful application in a wide variety of optimization problems including RCPSP. Boctor (1996) and Cho and Kim (1997) are pioneering project scheduling works that applied SA to optimize different objective functions in RCPSP. Józefowska et al. (2001) employed SA to minimize the project makespan in RCPSP with multiple execution modes. He et al. (2009) utilized SA for maximizing the net present value (NPV) of multimode project payment scheduling where the activities can be performed with one of several discrete modes.

Particle swarm optimization (PSO): In 1995, Kennedy and Eberhart (1995) introduced PSO as an optimization method inspired by flocking and schooling patterns of birds and fish. In PSO, individuals (the particles) are grouped into a swarm. The particles are then placed in the search space of an objective function. Each particle is associated with a candidate solution to the optimization problem. At each step, the algorithm evaluates the objective function of each particle at its current location. A particle then positions itself by combining the history of its own locations as well as the best positions encountered by one or more members of the swarm (Brits et al., 2002; Poli et al., 2007). Over a number of iterations, the particles are likely to converge together around an optimum of the objective function. Over the last decade, PSO has been used to solve constrained optimization problems for RCPSP. Jarboui et al. (2008) attempted to minimize the makespan for multimode RCPSP with the aid of PSO. Zhang and Li (2010) suggested a multi-objective PSO aimed at finding an optimal trade-off between time and cost for resource-constrained scheduling of construction projects. Koulinas et al. (2014) proposed an extended hyper-heuristic PSO algorithm for handling RCPSP, working as a multilevel algorithm that generates low-level heuristics. Shou et al. (2015) employed PSO for solving preemptive RCPSP where the maximum of one interruption per activity is allowed.

Ant colony optimization (ACO): This approach is a nature-inspired optimization tool that borrows features from the collective behavior of ants searching for food. Some species of ants are capable of finding the shortest path between the food source and the nest in an indirect way by means of releasing a chemical substance called a pheromone. As ants move, they deposit pheromone on the ground, marking favorable paths for colony members to find food. The more ants follow a given trail, the more attractive this trail becomes to be followed by other colony members. ACO is an algorithmic methodology that simulates the pheromone-laying and the pheromone-following behavior of ants to solve combinatorial optimization problems. In ACO algorithms, the emergent results of the cooperation and adaption among an ant colony build solutions by moving on a graph where the tour of the ants from the nest to a source of food denotes a possible solution to the optimization problems (Ünal et al., 2013 and Bozorg-Haddad et al., 2017). Several varieties of ACO have been developed for combinatorial optimization problems in general and RCPSP in particular. Most ACO works in the area of RCPSP have attempted maximizing the final NPV of the project (Chen et al., 2010), resource-leveling optimization (Geng et al., 2011), and minimizing project duration in multimode resource-constrained project scheduling in which each activity has multiple execution modes (Chiang et al., 2008 and Zhang, 2012).

Tabu search (TS) algorithm: The algorithm, developed by Fred W. Glover, is a metaheuristic algorithm to solve combinatorial optimization problems. The main concern in TS is to avoid local optimality traps by satisfying certain constraints. In fact, the TS algorithm classifies selected attributes of local modification as tabu. These restrictions placed on the algorithm are designed to solve the problem of convergence to local optima. The use of short-term memory in the algorithm prevents reversal and repetition of certain moves (Glover, 1990). A decade after its inception, TS began being successfully applied to the project scheduling field. The works of Punnen and Aneja (1995) and Tsai and Gemmill (1998) are among the first works that employed TS to schedule activities in RCPSP. Huang and Yu (2013) adopted the TS algorithm to minimize the cost of the unexpected event in RCPSP by using slack time. Skowronski et al. (2013) applied TS to multi-skills RCPSP, in which project activities require a specific amount of resources for each skill. Mika et al. (2008) and Dai et al. (2018) developed a TS model for multimode RCPSP where the processing time of project activities are affected by their setup time.

Hybrid optimization methods: There has been a revolutionary change in the optimization of the objective functions in RSPSP in recent times. Many researchers are now advocating the idea that the strengths of optimization can be significantly increased through combining different optimization methods, which in turn results in more powerful and flexible project scheduling tools. The focus of research is now shifting away from the sole utilization of an optimization method towards the hybridization of different optimization methods. A pioneering effort is the work of Shen (2002) that attempted to combine agent-based modeling and GA to address resource-constrained manufacturing scheduling problems. Shan et al. (2007) suggested a hybrid model of ACO and PSP for multimode RCPSP with a minimum time lag. The hybridization of GA and SA algorithms for RCPSP was proposed by Chen and Shahandashti (2009) and Bettemir and Sonmez (2015). In addition, Yannibelli and Amandi (2013) proposed a hybrid SA and evolutionary algorithm to solve RCPSP with the aim of minimizing the project makespan along with assigning the most effective set of human resources to project activities. Furthermore, Tran et al. (2015) coupled the artificial bee colony algorithm with the differential evolution technique to handle RCPSP. Afshar-Nadjafi et al. (2017) then developed a model by combining SA and TS algorithms to solve preemptive project scheduling problems. Finally, Sebt et al. (2017) hybridized GA and PSO for solving multimode RCPSP.

THE FUTURE OF PROJECT SCHEDULING

Based on the emerging trends in project scheduling tools, several directions for the future evolution of these tools are forecast.

First, there has been major progress in the development of project scheduling methods during the last six decades. Among these methods, optimization tools are currently pervasive. It seems clear that an integrated optimization method that hybridizes multiple algorithms will be necessary to capture complexity and diversity inherent in project scheduling problems. As suggested by Banda et al. (2014), a huge demand for building complex hybrid solutions and simplification of the use of optimization methods is likely.

> Optimization tools are currently pervasive in the project scheduling domain.

Second, optimization methods have mainly revolved around how best to achieve a set of scheduling objectives. More specifically, the available optimization tools are rooted in reductionism premised on the planning of a project in terms of required resources and individual activities. These tools attempt to minimize or maximize a specific scheduling objective by simplifying the complexity inherent in projects rather than facing up to that complexity. This trend in optimization methods is likely to continue and intensify. However, substantial time overruns in complex projects raise concerns about the effectiveness of applying optimization tools to project scheduling. It has been long understood that the dynamic interaction among multiple interacting elements of a project (for example, resources and external factors) governs the behavior of the project. The current optimization methods neglect emergent behaviors in complex projects caused by these dynamic interactions. To sidestep this weakness, the project management community has advocated a transition from reductionism to holism. On this premise, the complexity inherent in projects has been treated by adopting a systems thinking approach. As noted earlier, project scheduling has seen a rise in the use of SD modeling approach. However, the stand-alone use of the available system engineering tools does not seem yet quite able to capture complexity in project scheduling. The available system engineering tools have shortcomings in capturing the more detailed and operational issues of project management (Howick, 2003). Furthermore, uncertainty in quantifying the qualitative variables makes the output of the system analysis methods unreliable (Coyle, 2000). Coupling a system engineering technique and an optimization tool is anticipated to result in a more comprehensive and systematic project scheduling. This new approach will broaden the scope of optimization from an exclusive focus on strategic planning to operational decision-making.

Third, the optimization of a project's elements does not guarantee the project's success. The existing optimization algorithms attempt to address optimization problems with the goal of achieving the best possible outcome. However, as pointed out by Liu and Chen (2012), the optimization requirements in complex projects are hardly possible to be satisfied due to the existence of multiple interacting and dynamically evolving elements. To circumvent this difficulty, the idea of finding a satisfactory or an adequate result rather than the optimal solution, which came to be known as "satisficing theory," appeared in a work by Simon (1956). A scheduler is confronted by a plethora of choices for optimizing different objective functions. This, in turn, hinders the achievement of the best results. The satisficing theory, then, is a viable alternative for the creation of innovative project management tools, including in scheduling, by setting a good enough benchmark, rather than aiming for an illusionary best outcome in a complex and rapidly transforming modern environment (Gorod et al., 2009 and Hallo et al., 2018).

Fourth, project schedules are concerned with the prediction or projection of the amount of time required for completing each task. There are many different types of uncertainty associated with duration estimates, such as deficiencies in information, lack of familiarity with the task, and unreliability in resource acquisition. Managing uncertainty in duration estimates is a complex problem in itself and presents a major impediment to a realistic schedule. This challenge expands with an increase in the complexity of the project because of the existence of multiple interacting activities and their associated uncertainties. Despite the level of attention that uncertainty in task estimates has received in existing scheduling tools, consideration of uncertainty is often oversimplified. This overly simplistic perspective may be avoided in the development of future scheduling tools by employing forecast error methods such as mean absolute deviation (MAD) and mean absolute error (MAE) (Khair et al., 2017).

Fifth, while project autonomy has been an active area of past research in the project management literature, the influence of autonomy on project scheduling has received no attention. As pointed out by Gemünden et al. (2005), various dimensions of autonomy such as goal autonomy, resource autonomy, structural autonomy, and social autonomy may affect project performance. Future tools might seek to articulate the change in project completion time with an increasing level of autonomy.

Sixth, as stated earlier, diverse natural phenomena have inspired researchers to develop several novel methods to solve scheduling problems. This trend is anticipated to continue in the future. Indeed, it is an accepted fact that biomimicry – solving human problems via emulating nature – offers technically novel and feasible solutions to various engineering and management problems, and the scheduling domain is no exception. As an example, Duggal (2018) has recently suggested the idea of decoding the elements of management and strategy execution in a similar way to that in which DNA encodes an organism's genetic blueprint. This exemplary idea may offer new perspectives on the development of the next generation of practical project scheduling tools through finding a code containing the elements of scheduling, just as DNA contains each person's unique code that is used for the functioning of living organisms.

QUESTIONS FOR DISCUSSION

1. What are the limitations of WBS-based tools such as the Gantt chart in complex project management?
2. Identify and describe four key differences between CPM and PERT.
3. What is the purpose of CCPM?
4. What external factors might affect scheduling accuracy in complex projects?
5. What is meant by the precedence constraints in project scheduling? Provide examples.
6. How might changing requirements affect scheduling?
7. What is the role of technology and innovation in scheduling?
8. Describe why identifying an optimal solution from a set of scheduling objectives is demanding.

SUMMARY

To better follow the development of various project scheduling tools, this chapter has reviewed the evolution of these tools throughout history. It divides the history of project scheduling into seven periods, each of which chronicles advancements in the development of different methods.

As early as the 14th century, the use of visualizing tools began to be recognized. These diagramming tools included the necessary ingredients for the development of network-based scheduling techniques in the 1950s. In the 1960s, the real-world problem of limited resources motivated researchers to take a new approach to solving scheduling problems under resource constraints. In the 1970s, as a consequence of systems thinking, scheduling tools began to shift from reductionism towards holism. Uncertainty in project activity durations led to the formation of several methods. Starting from the last decade of the 20th century, project scheduling blossomed into the optimization research area to solve the complexity in the computation process of scheduling, resulting in the application of a great many optimization algorithms to this field.

> Under the influence of the system thinking approach, in the 1970s, scheduling tools began to shift from reductionism towards holism.

From this history, it can be observed that the practice of project scheduling has gradually developed over time, resulting in numerous novel methods to overcome scheduling constraints. It is hoped that schedulers will appreciate the history in more depth and be more successful in the future. The authors strongly believe that the increasing complexity of projects reinforces the need to manage this complexity through development of the next generation of practical scheduling tools.

> Increasing project complexity requires continuous development of innovative practical scheduling tools.

REFERENCES

Abdel-Hamid, T.K., and Madnick, S.E. 1983. The dynamics of software project scheduling. *Communications of the ACM.* **26**(5): 340–346.

Adamiecki, K. 1970. *O Nauce Organizacji.* Panstwowe Wydawnictwo Ekonomicze, Warsaw.

Adhau, S., Mittal, M.L., and Mittal, A. 2012. A multi-agent system for distributed multi-project scheduling: An auction-based negotiation approach. *Engineering Applications of Artificial Intelligence.* **25**(8): 1738–1751.

Afshar-Nadjafi, B., Yazdani, M., and Majlesi, M. 2017. A hybrid of tabu search and simulated annealing algorithms for preemptive project scheduling problem. *Lecture Notes in Computer Science.* **10350**(2017): 102–111.

Ahmad, S., and Simonovic, S.P. 2004. Spatial system dynamics: New approach for simulation of water resources systems. *Journal of Computing in Civil Engineering.* **18**(4): 331–340.

Alcaraz, J., and Maroto, C. 2001. A robust genetic algorithm for resource allocation in project scheduling. *Annals of Operations Research.* **102**(1–4): 83–109.

Artigues, C. 2008. The resource-constrained project scheduling problem. In Artigues, C., Demassey, S., and Neron, E. (Eds). *Resource-Constrained Project Scheduling: Models, Algorithms, Extensions and Applications*. John Wiley & Sons, London: pp. 20–35.

Artigues, C., Leus, R., and Nobibon, F.T. 2013. Robust optimization for resource-constrained project scheduling with uncertain activity duration. *Flexible Services and Manufacturing Journal.* **25**(1): 175–205.

Ayyub, B.M., and Haldar, A. 1984. Project scheduling using fuzzy set concepts. *Journal of Construction Engineering and Management.* **110**(2): 189–204.

Bai, Y., and Wang, D. 2006. Fundamental of fuzzy logic control-fuzzy sets, fuzzy rules and defuzzifications. In Wang, B., and Zhuang, H. (Eds). *Advanced Fuzzy Logic Technologies in Industrial Applications.* Springer-Verlag, London: pp. 17–36.

Banda, M., Stuckey, P., Hentenryck, P., and Wallaca, M. 2014. The future of optimization technology. *Constraints.* **19**(2): 126–138.

Barbati, M., Bruno, G., and Genovese, A. 2012. Applications of agent-based models for optimization problems: A literature review. *Expert Systems with Applications.* **39**(5): 6020–6025.

Bettemir, Ö.H., and Sonmez, R. 2015. Hybrid genetic algorithm with simulated annealing for resource-constrained project scheduling. *Journal of Management in Engineering.* **31**(5): 04014082: 1–8.

Birjandi, A., and Mousavi, M. 2019. Fuzzy resource-constrained project scheduling with multiple routes: A heuristic solution. *Automation in Construction.* **100**(2019): 84–102.

Boctor, F.F. 1996. Resource-constrained project scheduling by simulated annealing. *International Journal of Production Research.* **34**(8): 2335–2351.

Boussaïd, I., Lepagnot, J., and Siarry, P. 2013. A survey on optimization of metaheuristics. *Information Sciences.* **237**(2013): 82–117.

Bozorg-Haddad, O., Solgi, M., and Loáiciga, H.A. 2017. *Meta-Heuristic and Evolutionary Algorithms for Engineering Optimization.* John Wiley & Sons, Hoboken, NJ.

Brits, R., Engelbrecht, A.P., and Van den Bergh, F. 2002. A niching particle swarm optimizer. In *Proceedings of the 4th Asia-Pacific Conference on Simulated Evolution and Learning*, Orchid Country Club, Singapore: 692–696.

Brucker, P. 2002. *Project Scheduling under Limited Resources: Models, Methods and Applications.* Sönke Hartmann, Springer, Berlin. ISBN: 3-540-66392-4.

Burt, J.M., and German, M.B. 1971. Mont Carlo technique for stochastic PERT network analysis. *INFOR: Informational Systems and Operational Research.* **9**(3): 248–262.

Chanas, S., and Kamburowski, J. 1981. The use of fuzzy variables in PERT. *Fuzzy Sets and Systems.* **5**(1): 11–19.

Charnes, A.C., Cooper, W.W., and Thompson, G.L. 1964. Critical path analysis via chance constrained and stochastic programming. *Operations Research.* **12**(3): 460–470.

Chen, L., and Zhang, Z. 2016. Preemption resource-constrained project scheduling problems with fuzzy random duration and resource availabilities. *Journal of Industrial and Production Engineering.* **33**(6): 373–382.

Chen, P.H., and Shahandashti, S.M. 2009. Hybrid of genetic algorithm and simulated annealing for multiple project scheduling with multiple resource constraints. *Automation in Construction.* **18**(4): 434–443.

Chen, W.N, Zhang, J., Chung, H.S.S., Huang, R.Z., and Liu, O. 2010. Optimizing discounted cash flows in project scheduling- An ant colony optimization approach. *IEEE Transactions on Systems, Man, and Cybernetics, Part C (Applications and Reviews).* **40**(1): 64–77.

Chiang, C.W., Huang, Y.Q., and Wang, W.Y. 2008. Ant colony optimization with parameter adaptation for multi-mode resource-constrained project scheduling. *Journal of Intelligent and Fuzzy Systems.* **19**(4–5): 345–358.

Cho, J.H., and Kim, Y.D. 1997. A simulated annealing algorithm resource constrained project scheduling. *Journal of Operational Research Society.* **48**(7): 736–744.

Cooper, K.G., 1980. Naval ship production: A claim settled and a framework built. *Operations Interfaces.* **10**(6): 20–36.

Coyle, G. 2000. Qualitative and quantitative modelling in system dynamics: Some research questions. *System Dynamics Review.* **16**(3): 225–244.

Dai, H., Cheng, W., and Guo, P. 2018. An improved tabu search for multi-skill resource-constrained scheduling problems under step deterioration. *Arabian Journal for Science and Engineering.* **43**(6): 3279–3290.

Demeulemeester, E., and Herroelen, W.S. 2002. *Project Scheduling: A Research Handbook.* Kluwer Academic Publishers, Dordrecht, the Netherlands.

Der, G., and Everitt, B. 2014. *A Handbook of Statistical Graphics Using SAS ODS.* CRC Press, Boca Raton, FL.

Dike, S.H. 1964. Project scheduling with resource constraints. *IEEE Transactions on Engineering Management.* **11**(4): 155–157.

Dimovski, V., Maric, M., Uhan, M., Durica, N., and Ferjan, M. 2012. Sun Tzu's "The Art of War" and implications for leadership: Theoretical discussion. *Organizacija.* **45**(4): 151–158.

Dréo, J., Pétrowski, A., Siarry, P., and Taillard, E. 2006. *Metaheuristics for Hard Optimization: Methods and Case Studies.* Springer-Verlag, Berlin, Heidelberg.

Duggal, J. 2018. *The DNA of Strategy Execution: Next-Generation Project Management and PMO.* John Wiley & Sons, Hoboken, NJ.

Elsayed, S., Sarker, R., Ray, T., and Coello, C.C. 2017. Consolidated optimization algorithm for resource-constrained project scheduling problem. *Information Sciences.* **418**(2017): 346–362.

Frenzel, J.F. 1993. Genetic algorithms. *IEEE Potentials.* **12**(3): 21–24.

Friendly, M. 2008. A brief history of data visualization. In Chen, C.H., Härdle, W.K., and Unwin, A. (Eds). *Handbook of Data Visualization.* Springer-Verlag, Berlin, Heidelberg: pp. 15–56.

Gantt, H.L. 1903. A graphical daily balance in manufacture. *ASME Transactions.* **24**: 1322–1336.

Gantt, H.L. 1913. *Work, Wages, and Profits*, New York. Engineering Magazine Co.

Gemünden, H.G., Salomo, S., and Kriegar, A. 2005. The influence of project autonomy on project success. *International Journal of Project Management.* **23**(5): 366–373.

Geng, J.Q., Weng, L.P., and Liu, S.H. 2011. An improved ant colony optimization algorithm for nonlinear resource-leveling problems. *Computer & Mathematics with Applications.* **61**(8): 2300–2305.

Geraldi, J., and Lechter, T. 2012. Gantt charts revisited: A critical analysis of its roots and implications to the management of projects today. *International Journal of Managing Projects in Business.* **5**(4): 578–594.

Glover, F. 1990. Tabu search: A tutorial. *Interfaces.* **20**(4): 74–94.

Goldratt, E.M. 1997. *Critical Chain.* The North River Press, Great Barrington, MA.

Gonçalves, J.F., Mendes, J.J.M., and Resende, M.G.C. 2008. A genetic algorithm for resource constrained multi-project scheduling problem. *European Journal of Operational Research.* **189**(3): 1171–1190.

Gorod, A., DiMario, M., Sauser, B., and Bordman, J. 2009. Satisficing's system of systems using dynamic and static doctrine. *International Journal of System of Systems.* **1**(3): 347–366.

Hallo, L., Gorod, A., and Morriss, A. 2018. Towards "Satisficing" creativity effort within project management. In *Proceedings of the Fourth International Conference on Human and Social Analytic*, Venice, Italy: 31–36.

Han, D., Yang, B., Li, J., Wang, J., Sun, M., and Zhou, Q. 2017. A multi-agent-based system for two stage scheduling problem of offshore project. *Advances in Mechanical Engineering.* **9**(10): 1–17.

Hapke, M., and Slowinski, R. 1996. Fuzzy priority heuristics for project scheduling. *Fuzzy Sets and Systems.* **83**(3): 291–299.

Hartmann, S. 1999. *Project Scheduling Under Limited Resources: Models, Methods, and Applications.* Springer-Verlag, Berlin Heidelberg.

He, Z., Wang, N., Jia, T., and Xu, Y. 2009. Simulated annealing and tabu search for multi-mode project payment scheduling. *European Journal of Operational Research.* **198**(3): 688–696.

Herrmann, J.W. 2010. The perspectives of Taylor, Gantt, and Johnson: How to improve production scheduling. *Journal of Operations and Quantitative Management.* **16**(3): 243–254.

Herroelen, W. 2005. Project scheduling: Theory and practice. *Production and Operations Management.* **14**(4): 413–432.

Howick, S. 2003. Using system dynamics to analyse disruption and delay in complex projects for litigation: Can the modelling purposes be met? *The Journal of Operational Research Society.* **54**(3): 222–229.

Huang, R.H., and Yu, T.H. 2013. A novel tabu search algorithm for solving robust multiple resource-constrained project scheduling problem. *International Journal of Computer and communication Engineering.* **2**(2): 213–215.

Hutchings, J.F. 2004. *Project Scheduling Handbook.* Marcel Dekker, Inc., New York.

Hyatt, C., and Weaver, P. 2006. *A Brief History of Scheduling.* Mosaic Project Services Pty Ltd, Melbourne.

Icmeli-Tukel, O., and Rom, W.O. 1997. Ensuring quality in resource constrained project scheduling. *European Journal of Operational Research.* **103**(3): 483–496.

Jalili, Y., and Ford, D.N. 2016. Quantifying the impacts of rework, schedule pressure, and ripple effect loops project schedule performance. *System Dynamics Review.* **32**(1): 82–96.

Jarboui, B., Damak, N, Siarry, P., and Rebai, A. 2008. A combinatorial particle swarm optimization for solving multi-mode resource-constrained project scheduling problems. *Applied Mathematics and Computation.* **195**(1): 299–308.

Jennings, N.R., and Wooldridge, M. 1995. Applying agent technology. *Applied Artificial Intelligence an International Journal.* **9**(4): 357–369.

Józefowska, J., Mika, M., Różycki, R., Waligóra, G., and Węglarz, J. 2001. Simulated annealing for multi-mode resource-constrained project scheduling. *Annals of Operations Research.* **102**(1–4): 137–154.

Kadri, R.L., and Boctor, F.F. 2018. An efficient genetic algorithm to solve the resource-constrained project scheduling problem with transfer time. *European Journal of Operational Research.* **265**(2): 454–462.

Kellenbrink, C., and Helber, S. 2016. Quality- and profit- oriented scheduling of resource-constrained projects with flexible project structure via a genetic algorithm. *European Journal of Industrial Engineering.* **10**(5): 574–595.

Kennedy, J., and Eberhart, R.C. 1995. Particle swarm optimization. In *Proceedings of the IEEE International Conference on Neural Networks IV.* Perth, Australia: 1942–1948.

Khair, U., Fahmi, H., Hakim, S.A., and Rahim, R. 2017. Forecasting error calculation with mean absolute deviation and mean absolute percentage error. *Journal of Physics: Conference Series.* **930**(1): 012002: 1–6.

Khan, K.I.A., Flanagan, R., and Lu, S.L. 2016. Managing information complexity using system dynamics on construction projects. *Construction Management and Economics.* **34**(3): 192–204.

Kim, K.W., Gen, M., and Yamazaki, G. 2003. Hybrid genetic algorithm with fuzzy logic for resource-constrained project scheduling. *Applied Soft Computing.* **2**(3): 174–188.

Kimms, A. 2001. Maximizing the net present value of a project under resource constraints using a Lagrangian relaxation based heuristic with tight upper bounds. *Annals of Operations Research.* **102**(1–4): 221–236.

Kirkpatrick, S., Gellat, C.D., and Vecchi, M.P. 1983. Optimization by simulated annealing. *Science.* **220**(4598): 671–680.

Knotts, G., and Dror, M. 2003. Agent-based project scheduling: Computational study of large problem. *IIE Transaction.* **35**(2): 143–159.

Koulinas, G., Kotsikas, L., and Anagnostopoulos, K. 2014. A particle swarm optimization based hyper-heuristic algorithm for the classic resource constrained project scheduling problem. *Information Sciences.* **277**(2014): 680–693.

Kumar, S., and Thakker, J.J. 2017. Schedule and cost overrun analysis for R&D projects using ANP and system dynamics. *International Journal of Quality & Reliability Management.* **34**(9): 1551–1567.

Lechler, T.G., Ronen, B., and Stohr, E.A. 2005. Critical chain: A new project management paradigm or old wine in new bottles. *Engineering Management Journal.* **17**(4): 45–58.

Li, F., and Xu, Z. 2018. A multi-agent system for distributed multi-project scheduling with two-stage decomposition. *PLoS One.* **13**(10): e0205445: 1–24.

Liu, J., and Chen, Y.W. 2012. Toward understanding the optimization of complex systems. *Artificial Intelligence Review.* **38**(4): 313–324.

Lyneis, J.M., Cooper, K.G., and Els, S.A. 2001. Strategic management of complex projects: A case study using system dynamics. *System Dynamics Review.* **17**(3): 237–260.

Marsh, E.R. 1976. The harmonogram: An overlooked method of scheduling work. *Project Management Quarterly.* **7**(1): 21–25.

Masmoudi, M., and Hait, A. 2013. Project scheduling under uncertainty using fuzzy modelling and solving techniques. *Engineering Applications of Artificial Intelligence.* **26**(1): 135–149.

Merigó, J.M., Gil-Lafuente, A.M., and Yager, R.D. 2015. An overview of fuzzy research with bibliometric indicators. *Applied Soft Computing.* **27**(2015): 420–433.

Mika, M, Waligóra, G., and Węglarz, J. 2008. Tabu search for multi-mode resource-constrained project scheduling with schedule-dependent setup time. *European Journal of Operational Research.* **187**(3): 1238–1250.

Newbold, R.C. 1998. *Project Management in the Fast Lane: Applying the Theory of Constraints.* The St. Lucie Press, New York.

Oresme, N. 1482. *Tractatus de Latitudinibus Formarum, Pavoda.* British Library, London: IA 3Q024.

Playfair, W. 1786. *Commercial and Political Atlas.* Corry, London. Re-published in Wainer, H. and Spence, I. (Eds), *The Commercial and Political Atlas and Statistical Breviary,* 2005, Cambridge University Press, ISBN 0-521-85554-3.

Poli, R., Kennedy, J., and Blackwell, T. 2007. Particle swarm optimization: An overview. *Swarm Intelligence.* **1**(1): 33–57.

Prade, H. 1979. Using fuzzy set theory in a scheduling problem: A case study. *Fuzzy Sets and Systems.* **2**(2): 153–165.

Priestley, J. 1765. *A Chart of Biography.* British Library, London: 611.I.19.

Punnen, A.P., and Aneja, Y.P., 1995. A tabu search algorithm for the resource constrained assignment problem. *Journal of the Operational Research Society.* **46**(2): 214–220.

Rabelo, R.J., Camarinha-Matos, L.M., and Afsarmanesh, H. 1999. Multi-agent based agile scheduling. *Robotics and Autonomous Systems.* **27**(1–2): 15–28.

Rand, G.K. 2000. Critical chain: The theory of constraints applied to project management. *International Journal of Project Management.* **18**(3): 173–177.

Rashedi, R., and Hegazy, T. 2016. Holistic analysis of infrastructure deterioration and rehabilitation using system dynamics. *Journal of Infrastructure Systems.* **22**(1): 04015016: 1–10.

Ratajczak-Ropel, E. 2018. Experimental evaluation of agent-based approaches to solving multi-mode resource-constrained project scheduling problem. *Cybernetics and Systems.* **49**(5–6): 296–316.

Schofield, R.E. 2004. *The Enlightened Joseph Priestly: A Study of His Life and Work from 1773–1804.* The Pennsylvania State University Press, University Park, PA.

Sebt, M.H., Afshar, M.R., and Alipouri, Y. 2017. Hybridization of genetic algorithm and fully informed particle swarm for solving the multi-mode resource-constrained project scheduling problem. *Engineering Optimization.* **49**(3): 513–530.

Shan, M., Wu, J., and Peng, D. 2007. Particle swarm and ant colony algorithms hybridized for multi-mode resource-constrained project scheduling problem with minimum time lag. In *Proceedings of 2007 International Conference on Wireless Communications, Networking and Mobile Computing,* Shanghai, China: 58989–5902.

Shen, W. 2002. Genetic algorithms in agent-based scheduling systems. *Integrated Computer-Aided Engineering.* **9**(3): 207–217.

Shou, Y., Li, Y, and Lai, C. 2015. Hybrid particle swarm optimization for preemptive resource-constrained project scheduling. *Neurocomputing.* **148**(2015): 122–128.

Siegfried, R. 2014. *Modeling and Simulation of Complex Systems: A Framework for Efficient Agent-Based Modeling and Simulation.* Springer Vieweg, Wiesbaden.

Simon, H.A. 1956. Rational choice and the structure of the environment. *Psychological Review.* **63**(2): 129–138.

Skowronski, M.E., Myszkowski, P.B, Adamski, M., and Kwiatek, P. 2013. Tabu search approach for multi-skills resource-constrained project scheduling problem. In *Proceedings of the 2013 Federated Conference on Computer Science and Information Systems,* Krakow, Poland: 153–158.

Song, W., Xi, H., Kang, D., and Zhang, J. 2018. An agent-based simulation system for multi-project scheduling under uncertainty. *Simulation Modelling Practice and Theory.* **86**(2018): 187–203.

Sterman, J.D. 2000. *Business Dynamics: System Thinking and Modeling for a Complex World.* McGraw-Hill/Irwin, New York.

Steyn, H. 2002. Project management applications of the theory of constraints beyond critical chain scheduling. *International Journal of Project Management.* **20**(1): 75–80.

Tran, D.H., Cheng, M.Y., and Cao, M.T. 2015. Solving resource-constrained scheduling problems using hybrid artificial bee colony with differential evolution. *Journal of Computing in Civil Engineering.* **30**(4): 04015065: 1–10.

Trietsch, D., and Baker, K.R. 2012. PERT 21: Fitting PERT/CPM for use in the 21st century. *International Journal of Project Management.* **30**(4): 490–504.

Tsai, Y.W., Gemmill, D.D. 1998. Using tabu search to schedule activities of stochastic resource-constrained projects. *European Journal of Operational Research.* **111**(1): 129–141.

Tukel, O.L., Rom, W.O., and Eksioglu, S.D. 2006. An investigation of buffer sizing techniques in critical chain scheduling. *European Journal of Operational Research.* **172**(2): 401–416.

Ünal, M., Ak, A., Topuz, V., and Erdal, H. 2013. *Optimization of PID Controllers Using Ant Colony and Genetic Algorithms.* Springer-Verlag, Berlin Heidelberg.

Vanhoucke, M. 2012. *Project Management with Dynamic Scheduling: Baseline Scheduling, Risk Analysis and Project Control.* Springer-Verlag, Berlin Heidelberg.

Vanhoucke, M. 2016. *Integrated Project Management Sourcebook: A Technical Guide to Project Scheduling, Risk and Control.* Springer International Publishing, Switzerland.

Wang, J., and Yuan, H. 2017. System dynamics approach for investigating the risk effects on schedule delay in infrastructure projects. *Journal of Management in Engineering.* **33**(1): 04016029: 1–13.

Watson, K.J., Blckstone, J.H., and Gardiner, S.C. 2007. The evolution of management philosophy: The theory of constraints. *Journal of Operations Management.* **25**(2): 387–402.

Wiest, J.D. 1964. Some properties of schedules for large projects with limited resources. *Operations Research.* **12**(3): 395–418.

Wiest, J.D. 1981. Precedence diagramming method: Some unusual characteristics and their implications for project managers. *Journal of Operations Management.* **1**(3): 121–130.

Williams, T., Eden, C., Ackermann, F., and Tait, A. 1995. Vicious cycles of parallelism. *International Journal of Project Management*. **13**(3): 151–155.

Wilson, J.M. 2003. Gantt charts: A centenary appreciation. *European Journal of Operational Research*. **149**(2): 430–437.

Wolstenholme, E. 1990. *System Enquiry: A System Dynamics Approach*. John Wiley & Sons, New York.

Yannibelli, V., and Amandi, A. 2013. Hybridizing a multi-objective simulated annealing with a multi-objective evolutionary algorithm to solve a multi-objective project scheduling problem. *Expert Systems with Applications*. **40**(7): 2421–2434.

Zadeh, L.A. 1965. Fuzzy sets. *International Journal of Information and Control*. **8**(1965): 338–353.

Zamani, R. 2013. A competitive magnet-based genetic algorithm for solving the resource-constrained project scheduling problem. *European Journal of Operational Research*. **229**(2): 552–559.

Zhang, H. 2012. Gantt charts: Ant colony optimization for multimode resource-constrained project scheduling. *Journal of Management in Engineering*. **28**(2): 150–159.

Zhang, H., and Li, H. 2010. Multi-objective particle swarm optimization for construction time-cost tradeoff problems. *Construction Management and Economics*. **28**(1): 75–88.

Zhang, J., Song, X., and Dias, E. 2016. Project buffer sizing of a critical chain based on comprehensive resource tightness. *European Journal of Operational Research*. **248**(2016): 174–182.

Zarghami, S.A., Gunawan, I., and Schultmann, F. 2018. System dynamics modelling process in water sector: A review of research literature. *System Research and Behavioral Science*. **35**(6): 776–790.

4 Cost Estimation Toolset

Ricardo Valerdi
University of Arizona

CONTENTS

COMPLEX PROJECT MANAGEMENT AND COST ESTIMATION

It is clear that we have been living in the Systems Age for some time as evidenced by the role of technologically enabled systems in our everyday lives. Most of our everyday functions are dependent on, or enabled by, large-scale socio-technical[1] systems that provide useful technological capabilities. The advent of these systems has created the need for Complex Project Management.

The function of Complex Project Management – coupled with engineering disciplines – enables the creation and implementation of systems of unprecedented size and complexity. However, these disciplines differ in the way they create value. Traditional engineering disciplines are value-neutral; the laws of physics control the outcome of electronics, mechanics, and structures. Tangible products serve as evidence of the contribution that is easily quantifiable. In contrast, Complex Project Management has a different paradigm in that its intellectual output is often intangible and more difficult to quantify. Common work artifacts such as project plans, schedules, cost estimates, requirements, architectures, designs, verification, and validation plans are not readily noticed. Yet without these artifacts, it is difficult to imagine the success of any complex venture. For this reason, Complex Project Management is better suited for a value-based approach where the contributions are evident in better decisions and better project outcomes.

Traditional views about technology focus on performance, but in practice, performance is one of many attributes of a project. Equally important are the project's cost and schedule. The discussion of quality is not deliberately left out; in fact, for purposes of this chapter, it is considered to be embedded in performance.

Rather than optimizing on performance, successful projects jointly optimize on cost simply because of the fact that a solution that is too expensive is not a solution. This is best illustrated in one of the pillars of Project Management: the iron triangle, also known as the triple constraint. This is where performance, cost, and schedule are considered in parallel and represented in the form of a triangle to further demonstrate that a change in one dimension will cause a change in the other two.

> Optimizing on cost is a necessary criterion. The mantra should be emphasized: a solution that is too expensive is not a solution.

For example, decreasing the schedule of a project is likely to result in an increase in cost. Sometimes, however, a project's schedule is fixed because of its commitments. Events like the Olympic Games or the World Cup are excellent examples of fixed schedule projects that happen every 4 years and cannot be moved. Since their schedule is fixed, only the other two dimensions – cost and time – can be adjusted to absorb changes around it.

COST ESTIMATION APPROACHES

Many useful cost estimation techniques are available to Complex Project Management practitioners. They vary in both maturity and sophistication. Subsequently, some are more easily adaptable to the changing environment, while others take more time to develop. The logic behind these approaches is fundamentally different, leaving only their outcomes as measures of merit. It is believed that a hybrid approach that borrows from each method is the best way to capture cost phenomena that a single approach may miss. Six estimation techniques are presented here in order of sophistication.

> It is believed that a hybrid approach [to cost estimation] that borrows from each method is the best way to capture cost phenomena that a single approach may miss.

Heuristics and rules of thumb. Heuristic reasoning has been commonly used to arrive at quick answers to Complex Project Management questions. Project Managers, through education, experience, and examples, accumulate a considerable body of contextual intuition. These experiences evolve into instinct or common sense that are seldom recorded. These can be considered insights, lessons learned, and rules of thumb, among other names, that are brought to bear on certain situations. Ultimately, this knowledge is based on experience and may provide valuable results. Cost estimation heuristics and rules of thumb have been developed by researchers and practitioners (Boehm et al. 2000; Honour 2002; Rechtin 1991). One such rule of thumb, provided by Barry Horowitz, retired CEO of MITRE Corporation, adopts the following logic for estimating systems engineering effort (Horowitz 2004):

> If it is a custom-developed system (mostly) or an Off-the-Shelf (OTS) integration (mostly),
> > Then the former gets 6%–15% of the total budget for systems engineering, and the latter gets 15%–25% of budget (where selection of OTS products as well as standards is considered systems engineering).
> The following additional rules apply:
> > If the system is unprecedented,
> > > Then raise the budget from minimum level to 50% more.
> > If the system faces an extreme requirement (safety, performance, etc.),
> > > Then raise the budget by 25% of minimum.
> > If the system involves a large number of distinct technologies, and therefore a diversity of engineering disciplines and specialties,
> > > Then raise the budget by 25% of minimum.
> > If the priority for the system is very high compared to other systems also competing for resources,
> > > Then add 50% to the base.
> Note that the % of systems engineering is larger for OTS, but since the budgets for these projects are much lower, so are the numbers for systems engineering.

Expert opinion. This is the most informal of the approaches because it simply involves querying the experts in a specific domain and taking their subjective opinion as an input. This approach is useful in the absence of empirical data and is very simple. The obvious drawback is that an estimate is only as good as the expert's opinion, which can vary greatly from person to person. However, many years of experience is not a guarantee of future expertise due to new requirements, business processes, and added complexity. Moreover, this technique relies on experts, and even the most highly competent experts can be wrong. A common technique for capturing expert opinion is the Delphi method (Dalkey et al. 1969) which was improved and renamed Wideband Delphi (Boehm 1981).

Case studies and analogy. Recognizing that organizations do not constantly reinvent the wheel every time a new project comes along, there is an approach that capitalizes on the institutional memory of an organization to develop its estimates. Case studies represent an inductive process, whereby estimators and planners try to learn useful general lessons by extrapolation from specific examples. They examine in detail elaborate studies describing the environmental conditions and constraints that were present during the development of previous projects, the technical and managerial decisions that were made, and the final successes or failures that resulted. They then determine the underlying links between cause and effect that can be applied in other contexts. Ideally, they look for cases describing projects similar to the project for which they will be attempting to develop estimates and apply the rule of analogy that assumes previous performance is an indicator of future performance. The sources of case studies may be either internal or external to the estimator's own organization. Homegrown cases are likely to be more useful for the purposes of estimation because they reflect the specific engineering and business practices likely to be applied to an organization's projects in the future. Well-documented case studies from other organizations doing similar kinds of work can also prove very useful so long as their differences are accounted for.

Top down/design to cost. This technique aims for an aggregate estimate for the cost of the project based upon the overall features of the project. Once a total cost is estimated, each subcomponent is assigned a percentage of that cost. The main advantage of this approach is when a certain cost target must be reached regardless of the technical features. However, the top-down approach can often miss the low-level nuances that can emerge in large complex projects.

Bottom up/activity based. Opposite the top-down approach, bottom-up begins with the lowest level cost component and rolls it up to the highest level for its estimate. The main advantage is that the lower-level estimates are typically provided by the people who will be responsible for doing the work. This work may be represented in the form of a Work Breakdown Structure (WBS), which makes this estimate easily justifiable because of its close relationship to the activities required by the project elements. This can translate to a fairly accurate estimate at the lower level. The disadvantages are that this method is labor-intensive and is typically not uniform across entities. In addition, every level folds in another layer of conservative management reserve which may result in an over estimate at the end.

Parametric cost estimation models. This method is the most sophisticated and most difficult to develop. Parametric models generate cost estimates based on mathematical relationships between independent variables (e.g., lines of code, requirements, and square feet of a building) and dependent variables (e.g., effort). The inputs characterize the nature of the work to be done, plus the environmental conditions under which the work will be performed and delivered. The definition of the mathematical relationships between the independent and dependent variables is the heart of parametric modeling. These relationships are commonly referred to as cost estimating relationships (CERs) and are based upon statistical analyses of large amounts of data. Regression models are used to validate the CERs and operationalize them in linear or nonlinear equations. The main advantage of using parametric models is that, once validated, they are fast and easy to use. They do not require a lot of information and can provide relatively accurate estimates. Parametric models can also be tailored to a specific organization's historical projects. The major disadvantage of parametric models is that they are difficult and time-consuming to develop, and require a lot of clean, complete, and uncorrelated data to be properly validated.

STATE OF THE PRACTICE

The origins of parametric cost estimating date back to World War II (NASA 2015). The war caused a demand for military aircraft in numbers and models that far exceeded anything the aircraft industry had manufactured before. While there had been some rudimentary work to develop parametric techniques for predicting cost, there was no widespread use of any cost estimating technique beyond a bottom-up build-up of labor-hours and materials. A type of statistical estimating was suggested in 1936 by T. P. Wright in the *Journal of Aeronautical Science* (Wright 1936). Wright provided equations which could be used to predict the cost of airplanes over long production runs, a theory which came to be called the learning curve. By the time the demand for airplanes had exploded in the early years of World War II, industrial engineers were using Wright's learning curve to predict the unit cost of airplanes. Today, parametric cost models are used for estimating software development (Boehm et al. 2000), unmanned satellites (USCM 2002), hardware development (PRICE-H 2002), and many other areas.

A parametric cost model is defined as a group of CERs used together to estimate entire cost proposals or significant portions thereof. These models are often computerized and may include many interrelated CERs, both cost-to-cost and cost-to-non-cost. The use of parametric models in Complex Project Management serves as a valuable tool for project managers to estimate the effort and cost needed to successfully complete a project. Developing these estimates requires a strong understanding of the factors that affect the effort and costs of projects.

> A parametric cost model is defined as a group of CERs used together to estimate entire cost proposals or significant portions thereof.

There are dozens of cost models available, some for free while others require a paid license (as noted in Table 4.1). Cost models have been an essential part of Complex Project Management for decades. Hardware models were the first to be developed followed by software models in the 1980s (Ferens et al. 1999). The corresponding owner/developer and domain of applicability for a selected subset of models are provided in Table 4.1.

The aforementioned models were compared in five key areas relevant to systems engineering:

1. Model inputs for software or hardware size
2. Definition of systems engineering
3. Model inputs for systems engineering
4. Life cycle stages used in the model
5. Domain of applicability.

TABLE 4.1
Cost Models with Systems Engineering Components

Model Name	Owner/Developer	Domain
COCOMO II[a]	University of Southern California	Software
TruePlanning[b] (PRICE-S and PRICE-H)	PRICE Systems, LLC	Hardware and Software
Software Evaluation and Estimation of Resources (SEER) for Hardware[b]	Galorath, Inc.	Hardware
SEER for Software[b]	Galorath, Inc.	Software
Small Satellite Cost Model (SSCM)[a]	The Aerospace Corporation	Hardware
Unmanned Space Vehicle Cost Model (USCM)[a]	Los Angeles Air Force Base	Hardware
COSYSMO[a]	University of Arizona	Systems Engineering

[a] Denotes free.
[b] Denotes paid license required.

These areas provided valuable information on the applicability of each model to Complex Project Management. The increasing frequency and number of programs that have run significantly over budget and behind schedule (GAO-03-1073 2003) because systems engineering problems were not adequately understood should, by itself, be reason enough to press for improvement in forecasting systems engineering resource needs. However, even if the history of systems engineering problems is ignored, the future paints an even more demanding picture. The undeniable trend is toward increasingly complex systems dependent on the coordination of interdisciplinary developments where effective system engineering is no longer just another technology, but the key to getting the pieces to fit together. It is known that increasing front-end analysis reduces the probability of problems later on, but excessive front-end analysis may not pay the anticipated dividends. The key is to accurately estimate early in a program the appropriate level of systems engineering in order to ensure system success within cost and schedule budgets.

Most of the estimation tools listed in Table 4.1 treat systems engineering as a subset of a software or a hardware effort. One exception, the Constructive Systems Engineering Cost Model (COSYSMO), treats systems engineering as separate from hardware and software. Because many functions can be implemented using either hardware or software, systems engineering is becoming the discipline for selecting, specifying, and coordinating the various hardware and software designs. Given that role, the ideal path is to forecast systems engineering resource needs based on the tasks that systems engineering must perform and not as an arbitrary percentage of another effort. Hence, systems engineering estimation tools must provide for aligning the definition of tasks that systems engineers are expected to do on a given project with the program management's vision of economic and schedule cost, performance, and risk.

Tools that forecast systems engineering resources largely ignore factors that reflect the scope of the systems engineering effort, as insufficient historical data exists from which statistically significant algorithms can be derived. To derive CERs from historical data using regression analysis, one must have considerably more data points than variables, such as a ratio of 5 to 1. It is difficult to obtain actual data on systems engineering costs and on factors that impact those costs. For example, a typical factor may be an aggressive schedule, which will increase the demand for systems engineering resources. The result is a tool set that inadequately characterizes the proposed program and therefore inaccurately forecasts resource needs. Moreover, the tools listed in Table 4.1 use different life cycle stages, complicating things even further. The names of the different life cycle stages and a mapping to each other are provided in Figure 4.1. The three software models have different life cycle stages than the five hardware models. As a result, only models with similar life cycle phases are mapped to each other.

As the parallels between hardware and software estimation models are drawn and the relationships between these and systems engineering are defined, it is easy to identify the pressing need for a model that can estimate systems engineering as an independent function. The fundamental approach for developing a model that meets this demand relates back to the area of software cost estimation from which the theoretical underpinnings of COSYSMO are derived. This area of research is described in the next section.

COST MODELING SUITE

In the late 1970s and early 1980s, as software engineering was starting to take shape, software managers found they needed a way to estimate the cost of software development and to explore options with respect to software project organization, characteristics, and cost/schedule. Along with a number of commercial and proprietary cost/schedule estimation models, one of the answers to this need was the open-internals Constructive Cost Model (COCOMO). This and the other models allowed users to "reason" about the cost and schedule implications of their development decisions, investment decisions, established project budget and schedules, client negotiations and requested changes, cost/schedule/ performance/functionality trade-offs, risk management decisions, and process improvement decisions.

By the mid-1990s, software engineering practices had changed sufficiently to motivate a new version called COCOMO II, plus a number of complementary models addressing special needs of the software estimation community. Figure 4.2 shows the variety of cost models that have been

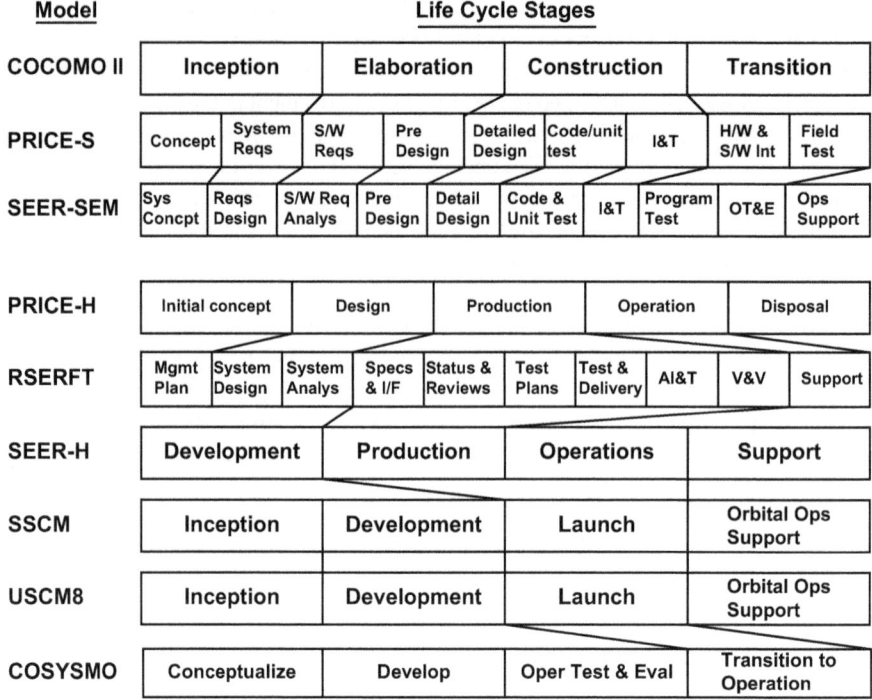

FIGURE 4.1   Model life cycle phases compared.

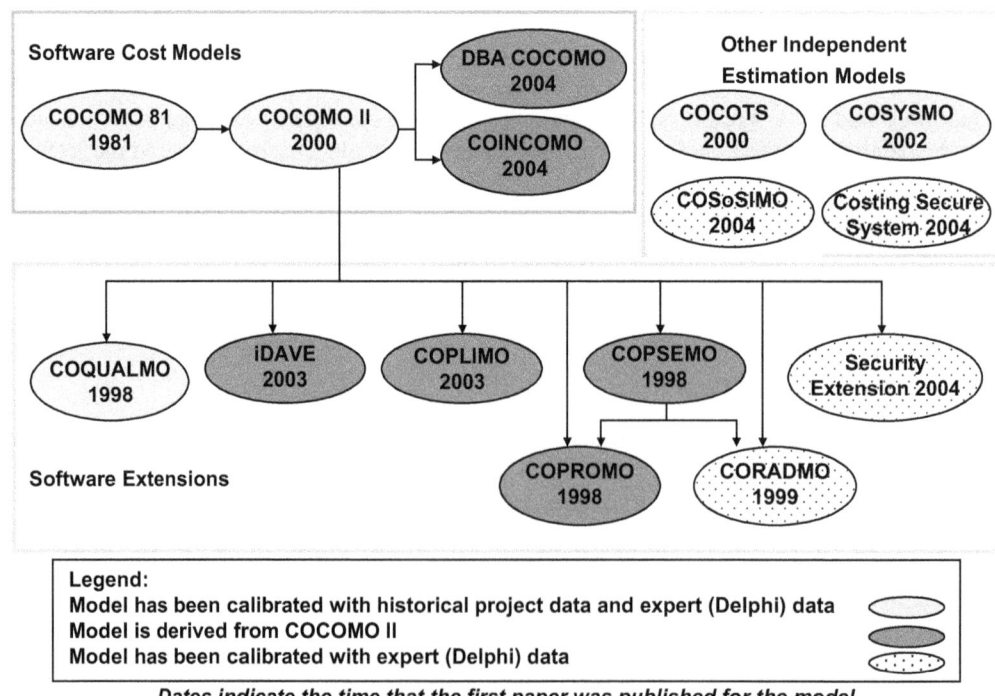

FIGURE 4.2   Historical overview of COCOMO suite of models.

developed at the University of Southern California to support the planning and estimation of software-intensive systems as the technologies and approaches have evolved since the development of the original COCOMO in 1981 (Boehm 1981).

Figure 4.2 shows the evolution of the COCOMO suite categorized by software models, software extensions, and independent models. The more mature models have been calibrated with historical project data as well as expert data via Delphi surveys. The newer models have only been calibrated by expert data. For more detail regarding these models, see Boehm et al. (2005).

Equally important as COSYSMO's lineage is its development method. It provides valuable context of the state of the practice surrounding it while informing users of the genesis of the model.

SYSTEMS ENGINEERING COST MODEL DEVELOPMENT

A constituency of practitioners familiar with the benefits provided by the COCOMO in the realm of software engineering proposed the development of a similar model to focus on systems engineering (Boehm et al. 1998). No formal approach to estimating systems engineering existed at the time, partially because of the immaturity of systems engineering as a formal discipline and the lack of mature metrics. The beginnings of systems engineering can be traced back to the Bell Telephone Laboratories in the 1940s (Auyang 2004). However, it was not until almost 30 years later that the first U.S. military standard was published (MIL-STD-499A 1969). The first professional systems engineering society, the International Council on Systems Engineering, was not organized until the early 1990s and the first commercial U.S. systems engineering standards, ANSI/EIA-632 and IEEE 1220, followed shortly thereafter. Even with the different approaches of defining systems engineering, the capability to estimate it is desperately needed by organizations. Several heuristics are available, but they do not provide the necessary level of detail that is required to understand the most influential factors and their sensitivity to cost.

Based on the previous support for COCOMO II, COSYSMO was positioned to leverage off the existing body of knowledge developed by the software community. The synergy between software engineering and systems engineering is intuitive because of the strong linkages in their products and processes. Researchers identified strong relationships between the two disciplines (Boehm 1994), opportunities for harmonization (Faisandier and Lake 2004), and lessons learned (Honour 2004). There have also been strong movements toward convergence between software and systems as reflected in two influential standards: ISO 15504 *Information technology—Process assessment* (2004) and the Capability Maturity Model Integration (2010). Organizations went as far as changing their names to reflect their commitment and interest in this convergence. Some examples include the Software Productivity Consortium becoming the Systems & Software Consortium, the Software Technology Conference becoming the Software & Systems Technology Conference, and the Center for Software Engineering at the University of Southern California becoming the Center for Systems & Software Engineering. Despite the strong coupling between software and systems, they remain very different activities in terms of maturity, intellectual advancement, and influences regarding cost.

COST MODELING OBJECTIVES

The process involved with cost modeling can help people reason about the cost implications of the decisions they make. User objectives include the ability to make the following:

- Investment decisions. A return-on-investment analysis involving effort needs an estimate of the cost or a life cycle effort expenditure profile.
- Budget planning. Managers need tools to help them allocate project resources.
- Trade-offs. Decisions often need to be made between cost, schedule, and performance.
- Risk management. Unavoidable uncertainties exist for many of the factors that influence project costs.

- Strategy planning. Setting mixed investment strategies to improve an organization's capabilities via reuse, tools, process maturity, or
other initiatives.
- Process improvement measurement. Investment
in training and initiatives often need to be measured. Quantitative management of these programs can help monitor progress.

> The process involved with cost mod-
> eling can help people reason about
> the cost implications of the decisions
> they make.

To enable these user objectives, cost modeling can provide decision support capabilities. Among these include, a model that is

- Accurate. Where estimates are close to the actual costs expended on the project.
- Tailorable. To enable ways for individual organizations to adjust the model so that it reflects their business practices.
- Simple. Understandable counting rules for the drivers and rating scales.
- Well-defined. Scope of included and excluded activities is clear.
- Constructive. To a point that users can tell why the model gives the result, it does and helps them understand the systems engineering job to be done.
- Parsimonious. To avoid the use of highly redundant factors or factors which make no appreciable contribution to the results.
- Pragmatic. Where inputs to the model correspond to the information available early on in the project life cycle.

To aid in these objectives, industry standards serve as collective experiences that help shape the field as well as the scope of some cost models.

SYSTEMS ENGINEERING AND INDUSTRY STANDARDS

The synergy between software engineering and systems engineering is evident by the integration of the methods and processes developed by one discipline into the culture of the other. Researchers from software engineering (Boehm 1994) and systems engineering (Rechtin 1998) have extensively promoted the integration of both disciplines but have faced roadblocks that result from the fundamental difference between the two disciplines (Pandikow and Törne 2001).

The development of systems engineering standards has helped the crystallization of the discipline as well as the development of COSYSMO. Table 4.2 includes a list of the standards most influential to this effort.

The first U.S. military standard focused on systems engineering provided the first definition of the scope of engineering management (MIL-STD-499A 1969). It was followed by another standard that provided guidance on the process of writing system specifications for military systems

TABLE 4.2
Notable Systems Engineering Standards

Standard (Year)	Title
MIL-STD-499A (1969)	Engineering Management
MIL-STD-490-A (1985)	Specification Practices
ANSI/EIA-632 (1999)	Processes for Engineering a System
CMMI (2010)	Capability Maturity Model Integration
ANSI/EIA-731.1 (2002)	Systems Engineering Capability Model
ISO/IEC 15288 (2002)	Systems Engineering – System Life Cycle Processes

(MIL-STD-490A 1985). These standards were influential in defining the scope of systems engineering in their time. Years later, the standard ANSI/EIA-632 *Processes for Engineering a System* (ANSI/EIA 1999) provided a typical systems engineering WBS. This list of activities was selected as the baseline for defining systems engineering in COSYSMO. The standard contains five fundamental processes and thirteen high-level process categories that are representative of systems engineering organizations. The process categories are further divided into 33 activities which help answer the *what* of systems engineering and helped characterize the first significant deviation from the software domain covered by COCOMO II. The five fundamental processes are (1) Acquisition and Supply, (2) Technical Management, (3) System Design, (4) Product Realization, and (5) Technical Evaluation. These processes are the basis of the systems engineering effort profile developed for COSYSMO.

This standard provides a generic industry list which may not be applicable to every context. Other types of systems engineering WBS lists exist such as the one developed by Raytheon Space & Airborne Systems (Ernstoff and Vincenzini 1999). Lists such as this one provide, in much finer detail, the common activities that are likely to be performed by systems engineers in those organizations, but are generally not applicable outside of the companies or application domains in which they are created.

Under the integrated software engineering and systems engineering paradigm, or Capability Maturity Model Integration® (CMMI 2010), software and systems are intertwined. A project's requirements, architecture, and process are collaboratively developed by integrated teams based on shared vision and negotiated stakeholder concurrence. A close examination of CMMI process areas – particularly the staged representation – strongly suggests the need for the systems engineering function to estimate systems engineering effort and cost as early as CMMI Maturity Level 2. Estimates can be based upon a consistently provided organizational approach from past project performance measures related to size, effort, and complexity. While it might be possible to achieve high CMMI levels without a parametric model, an organization should consider the effectiveness and cost of achieving them using other methods that may not provide the same level of stakeholder confidence and predictability. The more mature an organization, the more benefits in productivity they experience (ANSI/EIA 2002).

After defining the possible systems engineering activities used in COSYSMO, a definition of the system life cycle phases is needed to help establish the model boundaries. Because the focus of COSYSMO is systems engineering, it employs some of the life cycle phases from ISO/IEC 15288 *Systems Engineering – System Life Cycle Processes* (ISO/IEC 2002). These phases were slightly modified to reflect the influence of the aforementioned model, ANSI/EIA-632, and are shown in Figure 4.3.

Life cycle models vary according to the nature, purpose, use, and prevailing circumstances of the system. Despite an infinite variety in system life cycle models, there is an essential set of characteristic life cycle phases that exists for use in the systems engineering domain. For example, the *Conceptualize* stage focuses on identifying stakeholder needs, exploring different solution concepts, and proposing candidate solutions. The *Development* stage involves refining the system requirements, creating a solution description, and building a system. The *Operational Test & Evaluation* stage involves verifying/validating the system and performing the appropriate inspections before it is delivered to the user. The *Transition to Operation* stage involves the transition to utilization of the system to satisfy the users' needs. These four life cycle phases are within the scope of COSYSMO. The final two were included in the data collection effort

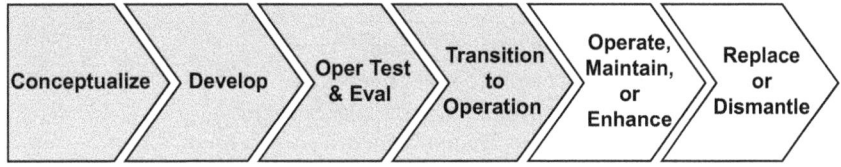

FIGURE 4.3   COSYSMO system life cycle phases.

but did not yield enough data to perform a calibration. These phases are *Operate, Maintain, or Enhance*, which involves the actual operation and maintenance of the system required to sustain system capability, and *Replace or Dismantle*, which involves the retirement, storage, or disposal of the system.

Life cycle models vary according to the nature, purpose, use, and prevailing circumstances of the system.

Each stage has a distinct purpose and contribution to the whole life cycle and represents the major milestones associated with a system. The stages also describe the major progress and design reviews of the system through its life cycle. These life cycle stages help answer the *when* of systems engineering and COSYSMO. Understanding when systems engineering is performed relative to the system life cycle helps define anchor points for the model.

System-of-interest. The ISO/IEC 15288 standard also provides a structure that helps define the system hierarchy. Systems can be characterized by their architectural structure or levels of responsibility. Each project has the responsibility for using levels of system composition beneath it and creating an aggregate system that meets the customer's requirements. Each particular subproject views its system as a system-of-interest within the grand scheme. The subproject's only task may be to deliver their system-of-interest to a higher level in the hierarchy. The top level of the hierarchy is then responsible for integrating the subcomponents that are delivered and providing a functional system. Essential services or functionalities are required from the systems that make up the system hierarchy. These systems, known as enabling systems, can be made by the organization itself or purchased from other organizations (Blanchard and Fabrycky 2010).

The system-of-interest framework helps answer the *where* of systems engineering for use in COSYSMO. In the case where systems engineering takes place at different levels of the hierarchy, organizations should focus on the portion of the system which they are responsible for developing and testing. Identifying system test responsibility helps crystallize the scope of the systems engineering estimate at a specific level of the system hierarchy.

The diversity of systems engineering standards can be quite complex (Sheard 2001); therefore, only the applicable standards have been mentioned here.

ANATOMY OF A COST MODEL

COSYSMO is an "open-source" model that contains eighteen parameters: four size drivers and fourteen effort multipliers. It is built on a framework similar to its well-known predecessor, COCOMO II, and integrates accepted systems engineering standards to define its scope.

Supported by the prime contractors in the U.S. aerospace and defense industry, the model focuses on large-scale systems for military applications that employ a disciplined approach to systems engineering. Data were collected from six aerospace companies in the form of expert opinion and historical project data to define and calibrate the model. In reduced form, the model yields a PRED(30) of 50% for programs within a defined productivity range.

Each parameter in the COSYSMO algorithm is part of the CERs in the form:

$$\text{PM}_{\text{NS}} = A \cdot \left(\sum_k \left(w_{e,k} \Phi_{e,k} + w_{n,k} \Phi_{n,k} + w_{d,k} \Phi_{d,k} \right) \right)^E \cdot \prod_{j=1}^{14} \text{EM}_j$$

where

PM_{NS} = effort in person-months (Nominal Schedule)
A = calibration constant derived from historical project data
k = {Requirements, Interfaces, Algorithms, Operational Scenarios}
w_x = weight for "easy", "nominal", or "difficult" size driver
Φ_x = quantity of "k" size driver
E = diseconomies of scale

EM = effort multiplier for the j_{th} cost driver. The geometric product results in an overall effort adjustment factor to the nominal effort.

The size of the system is the weighted sum of the Requirements, Interfaces, Algorithms, and Operational Scenarios parameters and represents the additive part of the model, while the EM factor is the product of the 14 effort multipliers and represents the multiplicative part of the model which are listed in Table 4.3.

After determining the size of the system by assigning values to the size drivers, the user must give an assessment of the system-of-interest with respect to the understanding, risk, implementation difficulty, development environment, and people capability with respect to systems engineering. This is done through the cost drivers, also referred to as effort multipliers, because of their multiplicative effect on the systems engineering effort calculation. Assigning ratings for these drivers is not as straightforward as the size drivers mentioned previously. The difference is that most of the cost drivers are qualitative in nature and require subjective assessment in order to be rated.

In addition to a definition, each driver has a corresponding rating scale that describes different attributes that could be used to rate the degree of impact on systems engineering effort. Rating levels include Very Low, Low, Nominal, High, Very High, and in some cases Extra High. The Nominal level is assigned a multiplier of 1.0 and therefore represents no impact on the systems engineering effort estimate. Levels above and below nominal are assigned multipliers above or below 1.0 according to their individual impact on systems engineering effort. The incremental impact of each step along a multiplier's rating scale depends on the polarity of each driver. For example, the requirements understanding multiplier are defined in such a way that *Very Low* understanding will have a productivity penalty on systems engineering. As a result, it will have a multiplier of greater than 1.0, such as 1.85, to reflect an 85% productivity penalty. The rating scale values for the cost drivers are provided in Table 4.3.

For example, the *Requirements Understanding* driver is worded positively since there is an effort savings associated with High or Very High understanding of the requirements. This is indicated by multipliers of 0.77 and 0.60, respectively, representing 23% and 40% savings in effort compared to the nominal case. Alternatively, the *Technology Risk* driver has a cost penalty of 32% for High

TABLE 4.3
Rating Scale Values for Cost Drivers

	Very Low	Low	Nominal	High	Very High	Extra High	EMR
Requirements understanding	1.85	1.36	1.00	0.77	0.60		*3.08*
Architecture understanding	1.62	1.27	1.00	0.81	0.65		*2.49*
Level of service requirements	0.62	0.79	1.00	1.32	1.74		*2.81*
Migration complexity			1.00	1.24	1.54	1.92	*1.92*
Technology risk	0.70	0.84	1.00	1.32	1.74		*2.49*
Documentation	0.82	0.91	1.00	1.13	1.28		*1.56*
# and diversity of installations/platforms			1.00	1.23	1.51	1.86	*1.86*
# of recursive levels in the design	0.80	0.89	1.00	1.21	1.46		*1.83*
Stakeholder team cohesion	1.50	1.22	1.00	0.81	0.66		*2.27*
Personnel/team capability	1.48	1.22	1.00	0.81	0.66		*2.28*
Personnel experience/continuity	1.46	1.21	1.00	0.82	0.67		*2.18*
Process capability	1.46	1.21	1.00	0.88	0.77	0.68	*2.15*
Multisite coordination	1.33	1.15	1.00	0.90	0.80	0.72	*1.85*
Tool support	1.34	1.16	1.00	0.85	0.73		*1.84*

and 74% for Very High. Not all rating levels apply to all of the drivers. Again, it is a matter of how the drivers are defined. The *Migration Complexity* driver, for example, only contains ratings at Nominal and higher. The rationale behind this is that the more complex the legacy system migration becomes, the more systems engineering work will be required. Not having a legacy system as a concern, however, does not translate to a savings in effort. The absence of a legacy system is the Nominal case which corresponds to a multiplier of 1.0.

The cost drivers are compared to each other in terms of their range of variability or Effort Multiplier Ratio (EMR). The EMR column in Table 4.3 is representative of an individual driver's possible influence on systems engineering effort. The four most influential cost drivers are *Requirements Understanding, Level of Service Requirements, Technology Risk,* and *Architecture Understanding.* The least influential *Documentation, # of Installations, Tool Support,* and *# of Recursive Levels in the Design* were kept because users wanted to have the capability to estimate their impacts on systems engineering effort. The relatively small influence of these four drivers does not mean that they are insignificant. Their existence gives users the ability to quantify their impact on systems engineering.

CASE STUDY USING COSYSMO

The scope of COSYSMO can be best illustrated through a case study. The assumptions in this example mirror a typical case in which a customer provides a system specification and requests an estimate of systems engineering effort from a contractor. It is also assumed that the information needed to populate the COSYSMO inputs is readily available; although this is the exception rather than the rule.

Assume that a customer has provided a system specification that contains 200 requirements, 5 interfaces, 4 algorithms, and 5 operational scenarios. This information must be used to obtain an estimate of systems engineering effort. The first step is to translate the requirements given into the appropriate level for systems engineering. It is determined that, after some requirements decomposition, 225 requirements can be derived from the system specification provided by the customer. The next step is to allocate the derived requirements into complexity levels. Through additional dialog with the customer, it is determined that the 225 derived requirements can be allocated as follows: 100 *easy*, 50 *nominal*, and 75 *difficult.*

The five interfaces provided by the customer must also be allocated into complexity levels. After review of system specification and discussion with experts, it is determined that two of the interfaces are *easy* and three are *difficult.* Similarly, the four easy algorithms are reviewed and assigned a complexity level of *easy.* Finally, the five operational scenarios are determined to be *nominal.* These quantities are entered into the COSYSMO model as size drivers.

At this stage, the model provides an initial systems engineering person-month estimate based exclusively on the COSYSMO size parameters. But additional information about the program is available that can be used to adjust the estimate. In particular, three things are known to be unique about this project. The first is that the contractor has done similar projects and therefore has a *high* degree of requirements understanding. Second, it is determined that a critical technology used in the project is relatively immature and requires a significant degree of research and development to make it usable. This translates to a *high* technology risk. Third, the contractor responsible for this project has relatively mature systems engineering processes and recently obtained a CMMI rating of 3. This translates to a *high* process capability. Together, these three characteristics of the project being estimated can be captured in the COSYSMO cost drivers by selecting the appropriate settings on the following three effort multipliers: requirements understanding, technology risk, and process capability. This additional project information also provides deeper insight into the project's potential performance and possible risk factors that may introduce schedule or cost variation.

In summary, the information obtained from the system specification – and other project documentation – combined with the project characteristics yields an estimate of 192 person-months of systems engineering as shown in Figure 4.4.

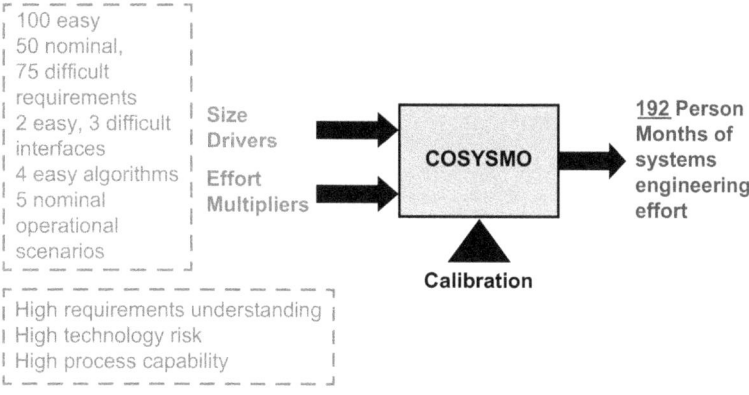

FIGURE 4.4 COSYSMO example.

This single-point estimate can be further decomposed into specific life cycle phases and systems engineering activities. For purposes of this example, it is assumed that the project being estimated includes the standard systems engineering life cycle phases described earlier (Conceptualize, Develop, Operational Test and Evaluation, and Transition to Operation) and a standard systems engineering WBS.

COMPARISON BETWEEN COCOMO II AND COSYSMO

On the surface, COCOMO II and COSYSMO appear to be similar. However, there are fundamental differences between them that should be highlighted. These are obvious when the main assumptions of the model are considered:

- Sizing. COCOMO II uses software size metrics such as lines of code, while COSYSMO uses metrics at a level of the system that incorporates both hardware and software such as requirements and interfaces.
- Life cycle. COCOMO II, based on a software tradition, focuses exclusively on software development life cycle phases defined by model-based systems engineering (Boehm and Port 1999), while COSYSMO follows the system life cycle provided by ISO/IEC 15288.
- Cost drivers. Each model includes drivers that model different phenomena. The overlap between the two models is minimal since very few of the COCOMO II parameters are applicable to systems engineering. One appreciable overlap is the software-related systems engineering effort estimated by both models which may be minimal to non-existent in many cases.

Additional differences between the two models such as the number of parameters and the model's functional form are summarized in Table 4.4.

COCOMO II serves as a natural starting point because of its useful and mature framework. The scope of this chapter is to present the relevant parameters in systems engineering while building from the lessons learned in software cost estimation. As much synergy as exists, software engineering and systems engineering must be treated as independent activities. This involves measuring them independently and identifying metrics that best capture the size and cost factors for each.

TABLE 4.4

Differences between COCOMO II and COSYSMO

Estimates	COCOMO II Software Development	COSYSMO Systems Engineering
Estimates size via	Thousands of Software Lines of Code (KSLOC), function points, or application points	Requirements, interfaces, algorithms, and operational scenarios
Life cycle phases	MBASE/RUP Phases: (1) Inception, (2) elaboration, (3) construction, and (4) transition	ISO/IEC 15288 Phases: (1) Conceptualize, (2) develop, (3) operation, test, and evaluation, (4) transition to operation, (5) operate, maintain, or enhance, and (6) replace or dismantle
Form of the model	One size factor, five scale factors, and eighteen effort multipliers	Four size factors, one scale factor, fourteen effort multiplier
Represents diseconomy of scale through	Five scale factors	One exponential system factor

DISCUSSION QUESTIONS

Why is the sequence of performance, cost, and schedule important?

Which cost estimation methodology is more reliable and why?

Why is it common for multiple estimation methodologies to be used?

Of the many objectives of cost modeling, which ones might be the most important for your organization?

What are examples of size drivers?

Think of a project you are familiar with and consider what drives its cost.

SUMMARY

One of the toolsets of Complex Project Management is the ability to estimate costs early in the project life cycle. A variety of cost estimation methods exist, some more appropriate than others based on the phase of the life cycle and the data available to decision makers. In any case, Complex Project Management cannot ignore cost, schedule, and performance planning. The example cost models provided, with special emphasis on COSYSMO, shed light on a methodology that can be extended to other contexts.

ENDNOTE

1 Socio-technical systems are an approach to complex organizational work design that recognizes the interaction between people and technology in workplaces (Cherns 1976).

REFERENCES

ANSI/EIA. 1999. *ANSI/EIA-632–1988 Processes for Engineering a System*. New York: American National Standards Institute.

ANSI/EIA. 2002. *EIA-731.1 Systems Engineering Capability Model*. New York: American National Standards Institute.

Auyang, S. Y. 2004. *Engineering – An Endless Frontier*. Cambridge, MA: Harvard University Press.

Blanchard, B. and Fabrycky, W. 2010. *Systems Engineering and Analysis*, 5th Edition. Upper Saddle River, NJ: Prentice Hall.

Boehm, B. W. 1981. *Software Engineering Economics.* Upper Saddle River, NJ: Prentice-Hall.

Boehm, B. W. 1994. Integrating software engineering and systems engineering, *The Journal of NCOSE, 1(1),* 147–151.

Boehm, B. W., Abts, C., Brown, A. W., Chulani, S., Clark, B., Horowitz, E., Madachy, R., Reifer, D. J., and Steece, B. 2000. *Software Cost Estimation with COCOMO II.* Upper Saddle River, NJ: Prentice Hall.

Boehm, B. W., Egyed, A., and Abts, C. 1998. Software engineering and system engineering, *Proceedings Focused Workshop #10-Center for Software Engineering,* Los Angeles, CA.

Boehm, B. W. and Port, D. 1999. Escaping the software tar pit: Model clashes and how to avoid them, *ACM Software Engineering Notes, 24(1),* 36–48.

Boehm, B. W., Valerdi, R., Lane, J., and Brown, A. W. 2005. COCOMO suite methodology and evolution, *CrossTalk - The Journal of Defense Software Engineering, 18(4),* 20–25.

Cherns, A. 1976. The principles of sociotechnical design, *Human Relations, 29(8),* 783–792.

CMMI. 2010. *Capability Maturity Model Integration (Version 1.3).* Pittsburg, PA: Carnegie Mellon - Software Engineering Institute.

Dalkey, N., Brown, B., and Cochran, S. 1969. *The Delphi Method, III: Use of Self Ratings to Improve Group Estimates,* RM-6115-PR. Santa Monica, CA: RAND Corporation.

Ernstoff, M. and Vincenzini, I. 1999. Guide to products of system engineering, *INCOSE International Symposium,* Las Vegas, NV.

Faisandier, A. and Lake, J. 2004. Harmonization of systems and software engineering, *Incose Insight, 7(3),* 21–23.

Ferens, D. V., Brummert, K. L., and Mischler, P. R. Jr. 1999. A comparative study of model content and parameter sensitivity of software support cost models, *Proceedings of the 1999 Joint ISPA-SCEA Conference,* San Antonio, TX.

GAO-03-1073. 2003. *Defense Acquisitions Improvements Needed in Space Systems Acquisition Management Policy.* Washington, DC: Government Accountability Office.

Honour, E. C. 2002. Toward an understanding of the value of systems engineering, *1st Annual Conference on Systems Integration,* Hoboken, NJ.

Honour, E. C. 2004. Understanding the value of systems engineering, *INCOSE 14th Annual International Symposium: Systems Engineering Managing Complexity and Change,* Toulouse, France.

Horowitz, B. 2004. Systems Engineering Estimation Rules of Thumb. R. Valerdi. Charlottesville, VA: Personal communication.

IEEE. 2005. Standard for Application and Management of the Systems Engineering Process, Institute of Electrical and Electronics Engineers, September 9, 2005.

ISO/IEC. 2002. ISO/IEC 15288:2002(E) Systems Engineering - System Life Cycle Processes, Geneva, Switzerland.

ISO/IEC 15504. 2004. Information technology — Process assessment, Geneva, Switzerland.

MIL-STD 490-A. 1985. *Specification Practices.* Washington, DC: U.S. Department of Defense.

MIL-STD-499A. 1969. *Engineering Management.* Washington, DC: U.S. Department of Defense.

NASA. 2015. *Cost Estimating Handbook (Version 4),* NP-2015-03-1473-HQ. Washington, DC: NASA.

Pandikow, A. and Törne, A. 2001. Integrating modern software engineering and systems engineering specification techniques, *14th International Conference on Software & Systems Engineering and their Applications,* Vol. 2. Paris, France.

PRICE-H. 2002. *Your Guide to PRICE-H: Estimating Cost and Schedule of Hardware Development and Production.* Mt. Laurel, NJ: PRICE Systems, LLC.

Rechtin, E. 1991. *Systems Architecting: Creating & Building Complex Systems.* Upper Saddle River, NJ: Prentice Hall.

Rechtin, E. 1998. System and software architecture, Proceedings USC Center for Software Engineering Focused Workshop #10: Software Engineering and System Engineering, Los Angeles, CA.

Sheard, S. A. 2001. Evolution of the frameworks quagmire, *Computer, 34(7),* 96–98.

USCM. 2002. *USCM8 Knowledge Management System.* Goleta, CA: Tecolote Research, Inc.

Wright, T. 1936. Factors affecting the cost of airplanes, *Journal of Aeronautical Science, 3(4),* 122–128.

5 Human Factors Toolset

Alicia Aitken
The International Centre for Complex Project Management

CONTENTS

BACKGROUND

As the previous chapters have described, complexity can exist in many forms and be driven by a number of different factors. However, the human factors in complexity are common across every complex project. The toolset required to successfully navigate the human factors is broad. In many ways, the toolkit for managing the human factors requires you to have learnt all the tools from the previous chapters and bring them to bear on the people in and around the project you're a part of.

There are two parts to the human side of complexity: yourself and others. The toolset required to master the human factors is drawn from a diverse set of sources ranging from the theory of psychology through to the pragmatism of using a time-zone planner to manage geographical spread.

The toolset required to not only survive but to thrive is an evolutionary one. It grows with our collective understanding of human psychology and our growth as individuals.

Before you can truly lead through complexity, you must know yourself and develop the behavioural capabilities to lead others. This chapter will delve into both sides of the human factors equation and offer practical tools to help navigate this aspect of complexity.

> Before you can truly lead through complexity, you must know yourself and develop the behavioural capabilities to lead others.

HUMAN FACTORS TOOLSET

Historically projects and project management in particular have focused on the processes required to manage time cost and quality (TCQ). The concept of human factors has been gaining a greater focus over the last decade; as project success ratings have remained static on TCQ, we have looked to find alternative levers for improving project success. The Project Management Institute (PMI) developed the PMBoK® Guide to lay out the knowledge required to successfully manage projects and was first published in 1996 (PMI Standards Committee, 1996). Although stakeholders were always mentioned, it took until version 5 (PMI, 2013) for the human factors associated with managing successful projects (beyond the basics of human resource management processes) to be called out on its own in a chapter called "Stakeholder Engagement". This chapter is oriented to the human factors associated with the people in and around the project but less so on the project manager themselves.

The International Project Management Association (IPMA) produced the International Competence Baseline (ICB) publishing the first version in 1998 (Caupin, Knopfel, Morris, Motzel, & Pannenbacker, 1999) that focused on project management as a practice. In ICB 2, the "sunflower" version began to see the inclusion of human factors with two elements for leadership and teamwork (Caupin & IPM Association, 2000). However, it wasn't until ICB 3 was published that behaviours were called out as one of the three competency domains (Caupin et al., 2006). In the ICB, the focus is on the human factors associated with the behaviours required of the project manager, and for the first time, we see the concept of self-reflection begin to emerge. In 2002, the PMI released the first version of the Project Manager Competency Development Framework which addressed the competencies for individuals. Chapter 3 focuses on the personal competencies of communicating, leading, managing, cognitive ability, effectiveness and professionalism (PMI, 2002), but still the concept of knowing yourself and building the emotional intelligence required to successfully navigate the world of projects was not present.

The Global Alliance for the Project ProfessionS (GAPPS) provides a mapping of a number of global standards showing the overlap and unique features of each. Figure 5.1 shows a relatively high level of coverage of the human factors associated with managing project stakeholders across the various standards. However, there is only a very brief reference throughout the analysis to the management of oneself, element 3.3.1 (Figure 5.2) seeking feedback on personal performance, and for this, there is relatively little to be found in any of the standards except for the ICB, the APMBoK and a partial coverage in the ISO 21500 standard.

The GAPPS has also produced a mapping of the major program management standards and the two units related to human factors are shown in Figures 5.3 and 5.4. As the complexity increases from projects to programs, the risks related to human factors increase; however, interestingly the willingness of the available standards to address this difficult topic becomes erratic with some standards such as the ICB and the Project Management Association of Japan's Program & Project Management for Enterprise Innovation (P2M) having far more coverage of human factors relating to leadership and stakeholder engagement than the others.

Morris and Pinto provide a detailed analysis of a number of international standards that clearly demonstrate a difference in focus between those that are oriented towards project, organisations and people. They found that there is a clear bias towards standards for project and organisations compared to those that cover people. Even then, within the standards that are mentioned in the people's domain, they are focused on the competencies people need to apply, predominately, process. (Morris & Pinto, 2010). For those interested in a more detailed understanding, Golpira has also

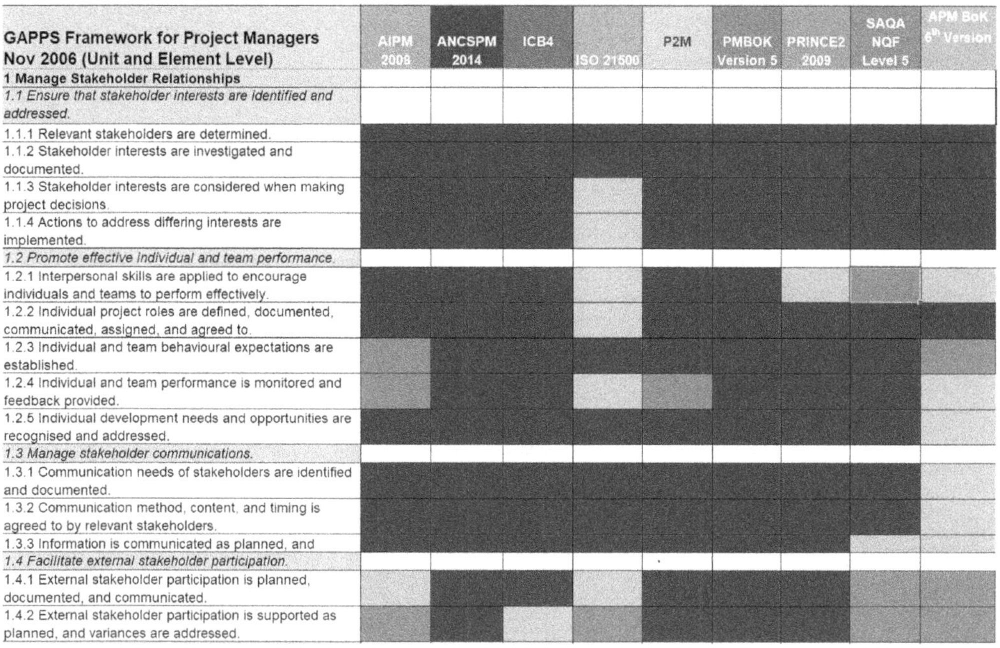

GAPPS Framework for Project Managers Nov 2006 (Unit and Element Level)	AIPM 2008	ANCSPM 2014	ICB4	ISO 21500	P2M	PMBOK Version 5	PRINCE2 2009	SAQA NQF Level 5	APM BoK 6th Version
1 Manage Stakeholder Relationships									
1.1 Ensure that stakeholder interests are identified and addressed.									
1.1.1 Relevant stakeholders are determined.									
1.1.2 Stakeholder interests are investigated and documented.									
1.1.3 Stakeholder interests are considered when making project decisions.									
1.1.4 Actions to address differing interests are implemented.									
1.2 Promote effective individual and team performance.									
1.2.1 Interpersonal skills are applied to encourage individuals and teams to perform effectively.									
1.2.2 Individual project roles are defined, documented, communicated, assigned, and agreed to.									
1.2.3 Individual and team behavioural expectations are established.									
1.2.4 Individual and team performance is monitored and feedback provided.									
1.2.5 Individual development needs and opportunities are recognised and addressed.									
1.3 Manage stakeholder communications.									
1.3.1 Communication needs of stakeholders are identified and documented.									
1.3.2 Communication method, content, and timing is agreed to by relevant stakeholders.									
1.3.3 Information is communicated as planned, and									
1.4 Facilitate external stakeholder participation.									
1.4.1 External stakeholder participation is planned, documented, and communicated.									
1.4.2 External stakeholder participation is supported as planned, and variances are addressed.									

FIGURE 5.1 GAPPS project manager standards mapping (Unit 1 – Manage stakeholder relationships).

GAPPS Framework for Project Managers Nov 2006 (Unit and Element Level)	AIPM 2008	ANCSPM 2014	ICB4	ISO 21500	P2M	PMBOK Version 5	PRINCE2 2009	SAQA NQF Level 5	APM BoK 6th Version
3.3 Reflect on practice.									
3.3.1 Feedback on personal performance is sought from relevant stakeholders and addressed.									
3.3.2 Lessons learned are identified and documented.									

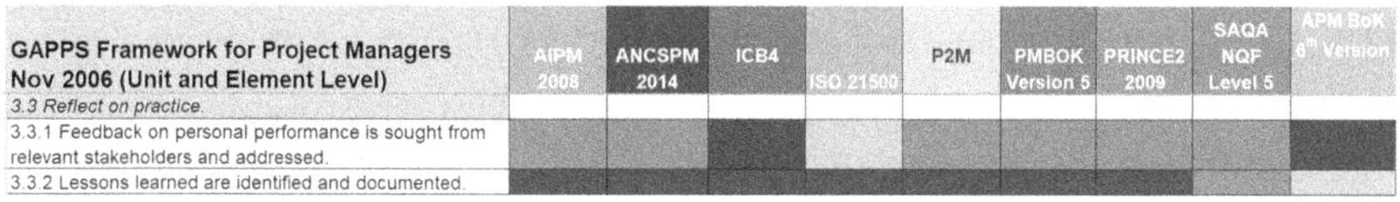

FIGURE 5.2 GAPPS project manager standards mapping (Element 3.3 – Reflect on practice).

created a summary of the major project management standards in their work to apply a method for evaluating project management standards that can be useful in understanding how the myriad of standards fit together (Golpîra, 2014).

The International Centre for Complex Projects (ICCPM) is the custodian of the first complex project manager standards which came into being in 2005 and was the first to include a unit called "Being Human" which addressed the competency required for managing both the self and others in complex projects. As a relative newcomer to the scene of project management standards, the shifting focus from process to include human factors was a welcome change. In the 2012 revision, there are two views as they refer to them, View 7: Project Leadership and View 8: Culture and Being Human. View 7 simply put, states "Good Leadership is the most important competence of a Project Manager" followed closely by View 8 which "specifies the competencies required to understand culture, cognition, personality and human life cycle and to use them in the design and operation of the project organisation and its systems. Being human refers to the psychological realities of being human and its impact on how we think, make decisions and hold memory and values. It also includes issues such as our personality and aging" (ICCPM, 2012). This distinction between the management of others and the management of oneself is the clearest distinction in the currently available standards.

These project management standards form part of the toolset for managing the human factors of complex projects, but on their own, they are not enough. To fully understand what tools leaders need, we must break down the problem into its component parts of managing yourself and managing others and look at them one at a time.

GAPPS (2011) A Framework for Performance Based Competency Standards for Program Manager	ICB3	MSP 2011	The Standard for Program Management, 2nd Edition	P2M
Provide Leadership for the Program				
1.1 Promote the program vision.				
1.1.1 Alignment of the program vision with the vision, mission, principles, and values of the sponsoring organization is maintained.				
1.1.2 Engagement with the program vision is stimulated through ongoing review, monitoring, communications, and negotiations with pertinent stakeholders.				
1.1.3 Commitment to the program vision is demonstrated by the program manager.				
1.2 Build an environment of confidence and trust within the program.				
1.2.1 Stakeholders are treated fairly and equitably.				
1.2.2 Open discussion is encouraged and facilitated.				
1.2.3 Differences are managed constructively.				
1.2.4 Issues and concerns are attended to in a timely manner.				
1.2.5 Interpersonal and leadership styles are chosen and applied based on the circumstances.				
1.2.6 Personal commitments are realistic and honoured.				
1.3 Embed socially responsible practice into the program.				
1.3.1 Expectations for socially responsible practice are made explicit and communicated to constituent projects and other pertinent stakeholders.				
1.3.2 Policies and procedures are designed to allow individuals to safely report breaches of socially responsible practice without fear of retaliation.				
1.3.3 Threats to socially responsible practice within the program are identified and addressed.				
1.4 Develop the potential of program staff.				
1.4.1 Individual behavioural expectations for constituent project managers are established.				
1.4.2 Individual program roles are defined, documented, communicated, assigned, and agreed to.				
1.4.3 Desirable behaviours are encouraged, and undesirable behaviours are discouraged.				
1.5 Support a learning environment.				
1.5.1 Program planning and program plan implementation are viewed as a learning process.				
1.5.2 Errors, mistakes, and expressed concerns are treated as learning opportunities.				
1.5.3 Plans for identifying, capturing, disseminating, and exchanging knowledge are developed and maintained.				
1.5.4 Program knowledge is identified, captured, disseminated, and exchanged as planned.				
1.5.5 Reflection on and review of practice is encouraged as a basis for learning.				

FIGURE 5.3 GAPPS program manager standards mapping (Unit 1 – Provide leadership for the program).

GAPPS (2011) A Framework for Performance Based Competency Standards for Program Manager	ICB3	MSP 2011	The Standard for Program Management, 2nd Edition	P2M
Facilitate Stakeholder Engagement				
2.1 Communicate effectively with stakeholders.				
2.1.1 Program stakeholders and their communication needs are identified and documented.				
2.1.2 Communication approaches are agreed to by pertinent stakeholders.				
2.1.3 Information is shared as planned, and variances are identified and addressed.				
2.1.4 Communication interfaces among constituent projects are monitored.				
2.2 Cultivate stakeholder commitment.				
2.2.1 Interests and expectations of pertinent stakeholders are investigated, documented, and considered when making program decisions.				
2.2.2 Approaches to influence ongoing stakeholder commitment are developed and implemented.				
2.2.3 Actions are taken to accommodate differing stakeholder interests and expectations.				
2.2.4 Evolving stakeholder interests and expectations are shared across the program.				

FIGURE 5.4 GAPPS program manager standards mapping (Unit 2 – Facilitate stakeholder engagement).

MANAGING YOURSELF

First let's look at the toolkit for managing yourself. As a leader, the first step is understanding yourself, your preferred working styles, how you react under stress, your leadership style, learning style, values and purpose. By truly understanding yourself, you will be better able to understand your impact on others, interpret their reactions to your approach and, where useful, adjust your approach to improve your chances of effecting the outcome you desire.

SELF-AWARENESS: DATA INPUTS

Reflection is the key to successfully mastering self-management, but there must be inputs to feed this practice. Data is required. There are a myriad of assessment tools available to help. Each assessment tool uses a slightly different approach and provides data from a slightly different perspective so it's important to investigate the right one for the right situation. There is no commonly agreed "best" assessment, they each have

> Reflection is the key to successfully mastering self-management.

something to offer. It is recommended that as and when you have the opportunity you use them all. It is common these days when applying for roles companies will ask you to participate in their preferred assessment process, if you do, always ask for a copy of your results as each will provide you with more data to fuel your journey of self-discovery.

Some of the most commonly used tools include the Myers Briggs Type Indicator (MBTI), DISC, Gallup Builder Profile 10 (BP10), Sixteen Personality Factor Questionnaire (16PF), the Herrmann Brain Dominance Instrument (HBDI) and the Life Style Inventory (LSI).

Personality Profile Assessments

MBTI: This assessment is based on Jungian psychology and his notion of psychological types. The assessment uses four dimensions to create 16 personality types.

- Favourite world: Do you prefer to focus on the outer world or on your own inner world? Extraversion (E) or Introversion (I).
- Information: Do you prefer to focus on the basic information you take in or do you prefer to interpret and add meaning? Sensing (S) or Intuition (N).
- Decisions: When making decisions, do you prefer to first look at logic and consistency or first look at the people and special circumstances? Thinking (T) or Feeling (F).
- Structure: In dealing with the outside world, do you prefer to get things decided or do you prefer to stay open to new information and options? Judging (J) or Perceiving (P).

The MBTI is a good general psychological assessment based on your preferences rather than your abilities or traits. The data is helpful for understanding how you prefer to work which can assist you in understanding how to manage your time, how you might approach decision-making and managing stressors throughout the endeavour, project or program. Having some form of general assessment such as this one is a good starting point for getting to know yourself and starting on the managing yourself pathway.

16PF: The 16PF assessment uses a similar construct to the "Big 5" personality trait model (Cattell, 1973). The original model breaks personality down into 16-pair traits that group across the five factors of extroversion, agreeable, conscientiousness, neuroticism and openness to experience. The 16-pair traits include items such as reserved vs outgoing, concrete thinking vs abstract thinking and undisciplined vs self-controlled.

Similarly, the assessment measures 16 different primary personality characteristics grouped into five global factors: extraversion, tough-mindedness, self-control, anxiety and independence. The language has been altered to be less clinical and more appealing to a wider audience, but the concepts are the same. The 16PF focused on understanding the way people work and interact with others.

This assessment is another generalist psychological assessment but with a stronger focus on the application to the work environment as compared to the MBTI which is designed to apply to whole of life.

Behavioural Style Assessments

DISC: The DISC profile is based on the work of Dr. William Marston, originally published in 1928 in a book titled *Emotions of Normal People* (Marsten, 1928). It focuses on the assessment of preferred behavioural styles. It can be used effectively as an individual assessment on its own or for understanding how different team members' behavioural styles are likely to interact with one another.

The model uses four dimensions: dominance, influence, steadiness and conscientiousness. These are defined as:

Dominance: Person places emphasis on accomplishing results, the bottom line, confidence
Influence: Person places emphasis on influencing or persuading others, openness, relationships
Steadiness: Person places emphasis on cooperation, sincerity, dependability
Conscientiousness: Person places emphasis on quality and accuracy, expertise, competency

The aspect of human factors that makes working in complexity so difficult is the lack of visibility and tangibility of the materials you're working with i.e. humans. The actions you can see on the surface are the product of a myriad of underlying emotions, behaviours and psychological predispositions. They are also lagging indicators. The purpose of using assessments such as the DISC profile is to find leading indicators for behavioural conflict that could derail your endeavour, project or program, whether those behaviours be your own and their effect on the team or between team members.

By increasing your self-knowledge with regard to how you respond to conflict, what motivates you, what causes you stress and how you solve problems you'll be able to improve your working relationships, manage team dynamics more proactively and better communicate with your stakeholders in a way that will maximise the chances of them hearing and understanding your message.

LSI: The Life Styles Inventory™ (LSI) is a behavioural assessment that has both a self-assessment and colleague feedback component so you can see not only what your perceptions are but those of the people working with you. The analysis identifies strengths, unrecognised strengths, blind spots and stumbling blocks.

The results identify 12 styles grouped into three general clusters: constructive, passive and aggressive/defensive. They are as follows:

- Constructive styles
  - Achievement
  - Self-actualising
  - Humanistic-encouraging
  - Affiliative

Constructive styles draw on the positive aspects of building self-satisfaction and strong relationships with people to help attain goals, develop teams and promote adaptability and effectiveness.

- Passive styles
  - Approval
  - Conventional
  - Dependent
  - Avoidance

Passive styles are characterised by self-protecting thinking and behaviour centre around protecting the individual. These styles can have negative consequences such as stifling creativity and innovation which can lead to the organisations not driving forward.

- Aggressive/defensive styles
  - Oppositional
  - Power
  - Competitive
  - Perfectionistic.

Aggressive or defensive styles lead people to focus on their own needs rather than those of the organisation. For the individual they can lead to stress, turnover and inconsistent performance. For the organisation or group, they can put the achievement of goals at risk.

Thinking Style Assessments

HBDI®: This assessment focuses on your thinking style(s). It identifies preferences for emotional, analytical, structural or strategic thinking as defined by Ned Herrmann's Whole Brain® Thinking Model. The model uses a four quadrant framework divided by colour: blue (analytical), red (relational), yellow (experimental) and green (practical) and serves as an organising principle for how the brain works. Each of the four quadrants is made up of a number of subthemes as shown in Figure 5.5.

The model is based on the premise that everyone has access to all four quadrants but to varying degrees. Most people (58%) will have a preference for two quadrants, while 34% have a preference for three. Very few people have a preference for just one or all four.

When working in complexity, there are a myriad of decisions to make and problems to solve. Having a strong understanding of how you prefer to think will help you do both of these things better at scale and at pace which is needed when working at this level.

Talent Assessments

BP10: The BP10 by Gallup is designed specifically around the ten innate talents they say "builders" possess: confidence, delegator, determination, disruptor, independence, knowledge, profitability, relationship, risk and selling. Gallup uses the term builder to describe a specific kind of talent, similar to entrepreneurship that folds in creating start-ups, building teams and innovation. As with much of the Gallup toolset, the focus is on strengths and how to maximise the value you can derive from what you're inherently good at rather than looking for deficiencies that need to be bridged. This positive approach to self-development is uplifting and motivating, and can be applied to the management of both yourself and others.

When working in complexity, the need for innovation, building something out of nothing and creating economically viable outcomes is magnified. A specific self-assessment such as the BP10 can give you a focused lens on your talents in this area and help you build and grow what you need to successfully lead your project. There are many other skill-specific assessments that you might choose to use at different times, the BP10 is one example.

Analytical	Experimental
- Logical	- Holistic
- Analytical	- Intuitive
- Fact-Based	- Integrating
- Quantitative	- Synthesising
Practical	Relational
- Organised	- Interpersonal
- Sequential	- Feeling-Based
- Planned	- Kinesthetic
- Detailed	- Emotional

FIGURE 5.5 HBDI quadrants.

HEALTH: STRESS AND COPING

The risk of stress and burnout is inherent in complex endeavours, projects and programs. Managing your own stress through coping mechanisms as well as the team is essential. The European Agency for Safety and Health at Work's literature review on the cost of stress provides a comprehensive summary of the statistics on societal, organisational and sector impacts of stress. The cost globally is in the hundreds of billions of dollars and the impacts on individual health in the form of burnout, cardiovascular disease and other life-limiting conditions is well documented (Hassard et al., 2014). As project managers, we tend to approach coping with stress in a uniform way across our personal health, home and work stressors, using acceptance, planning and active coping strategies. As the application of project management competencies increases so does the use of these coping strategies (Aitken, 2011). This may have its roots in classical project management training, based on the standards described above, the notion of solving challenging problems by applying the process of making a plan and taking action to solve the problem is inherent in how we go about our business. However, when managing the human factors associated with complexity a one-size-fits-all approach to managing stressful situations can do more damage than good. A focus on a strong mind and good mental health is vital to working in complexity. Although there a myriad of tools and techniques for this, the three essentials for the human factors toolkit are exercise, support networks and mindfulness.

Exercise needs to be regular and designed around what works for you. Endorphins released from exercise work to release tension, help you sleep better, stabilise moods, maintain energy levels and a whole host of additional benefits. There are often claims from poor leaders that they do not have the time to exercise; however, given we are all gifted exactly the same amount of time each day, it is less a matter of availability and more of priority. A clever solution – walking meetings. For your 1-2-1 meetings with team members or other stakeholders – go for a walk and talk through the issues of the day and you'll be able to combine work and exercise in an efficient solution.

Support networks are essential, life is a team sport and you need a strong personal team to see you through the good and challenging times. A team you can talk to, seek advice and counsel from, not just with surface level issues but the underlying hopes, fears and dreams you have that drive your decisions and actions. Your team may be big or

> Life is a team sport, and you need a strong personal team to see you through the good and challenging times.

it may be small, size doesn't matter, what matters is the depth of vulnerability you can take yourself to with your team. Complexity is by definition hard and you will fail at something if not a lot of things along the way. Without your team, you will struggle to build and maintain the resilience you need to see your project through. Thinking through who is on your team now and who should be on your team, consider family, friends, mentors and sponsors as a starting point for team selection. Once you have your team in place, it is important to maintain the health of that team, dedicating sufficient time and effort to maintain the relationships is key to ensuring your team is there when you need them.

And finally mindfulness, this is a simple practice of taking time to calm and still the mind. You can do this on your own without the aid of any specific tools, but if you are like me, you may need a little guidance to get your to focus. There are apps available to help such as Head Space which take you through guided meditations and moments of stillness that can start as short as 1 min (Headspace, n.d.).

QUESTIONS TO CONSIDER

When applying the toolset for managing yourself, there are a number of questions to consider. The first set relates to the selection of a self-assessment tool:

- **What is it about yourself that you are seeking to understand?** If you are seeking a general starting point for understanding yourself and how that may be different to others, a personality profile is a good choice. If you are experiencing difficulties or have noticed times when your interactions with others have not been as positive as you'd have liked, a behavioural or thinking styles assessment where you get deeper in to not only how you think but also how you act and how this may interact well or not so well with others of a different type is a better choice.

- **Should I use a self-assessment or 360° assessment?** The personality assessments tend to be self-assessments validated through scientific research and can be relied upon as a starting point for getting to know yourself. If you are seeking to understand in more depth how you can improve the interactions you have with others in a team, the behavioural assessments such as DISC and HBDI can be used by everyone in the team. A facilitator will debrief each individual and then run a team workshop to learn together about how the different styles complement or hinder each other. If you're seeking an in-depth analysis of your interactions with others, a full 360° assessment such as LSI provides both your self-view and the view of others. Where these two materially differ can provide next-level insights for use in adjusting your behaviours.

Questions related to health and stress that you might consider asking include the following:

- **How important is mental and physical health to me?** This question is about the prioritisation of the toolset for stress and coping. The options here are simple but require the prioritisation of time and the discipline to apply the toolset to achieve the long-term benefits of mental and physical health.
- **What is the risk to the endeavour, project or program if I am not well?** Another angle to the prioritisation is that of risk. It can be easy to think if it's just for myself, I won't or can't dedicate the time to regular exercise, mindfulness etc., but what if by choosing to not prioritise your health, you are putting the endeavour, project or program at risk?

MANAGING OTHERS

The second aspect to the human factors in complexity is the toolkit for managing others. In complex projects, the human factors that make up the project team and the wider stakeholder group are, like all things, exponentially more difficult to manage. With complexity, inevitably comes some or all of the problems from the sheer volume of individual people involved, number of teams, dependencies, diversity (of all kinds), capability levels relative to what's needed and geographical dispersion. To be able to manage the human factor at a scale in complexity project, leaders need to have an over-flowing toolkit at their disposal from which they can draw the right combination of tools to suit the changing needs of their project.

Team Assessments: Understanding the Team

As with managing oneself, the starting point for managing teams is to get to know them and then get them to know one another. Team building workshops are a common starting point, but in complexity, the volume of individuals requires a more analytical approach. Self-assessments such as the ones described above can be used to get a sense of individual styles, but beyond that the individual leaders in complexity should be looking to the use of team assessments such as the Belbin Team Roles to understand how the interaction of those individuals in a team is likely to work or not.

Belbin Team Roles: The Belbin assessment is based on the work of Dr. Meredith Belbin. The assessment uses nine clusters of behaviour called "Team Roles". These include:

- Resource investigator
- Team worker

- Co-ordinator
- Plant
- Monitor evaluator
- Specialist
- Shaper
- Implementer
- Completer–finisher.

The theory states that every team needs access to each of the nine Team Role behaviours to become a high performing team. Most people have two to three team roles they draw on, and these may change over time so you don't need to have nine people each of a different type in each team you run!

When working in complexity, it is important to keep track of what kind of team and team dynamics you need at any given time as this will shift over the course of the endeavour, project or program. Using a team assessment such as the Belbin team profile provides you with the data you need to form a solid basis for developing strategies for managing the human factors.

TEAM ASSESSMENTS: MANAGING THE HEALTH OF THE TEAM

Throughout the project, teams need regular health checks, particularly on complex projects that often have long durations and can have prolonged periods of high stress. An easy-to-use tool for framing regular check-ins is the Atlassian Team Playbook, www.atlassian.com/team-playbook (Atlassian, 2018b). No technology or special training is needed, just follow the steps starting with a Health Monitor workshop. The Health Monitor is "part cure, part preventative medicine", by regularly using a 360° assessment, you'll be able to honestly assess what's working and what's not and adjust as you go. When working in complexity, the environment is constantly evolving, and as a result, teams and their dynamics are shifting landscapes. Leaders in complexity need to apply their team assessment tools regularly.

Another emerging tool available to project and program directors is an artificial intelligence (AI)-powered stress monitor from Pioneera, www.pioneera.com (Owen Witford, 2018). The algorithm works to monitor business communications and behaviours such as emails or chatter on Slack for stress indicators. The system can be configured to provide both in-the-moment interventions for individual team members to help them manage their stress, and/or it provides de-identified data and insights to leaders on stress patterns in the team. As the system gets to know you and your team, it can predict stressors, so you know if a specific meeting or workshop is going to cause stress and design it in a way that prevents stress and maximises productivity. Knowing when and for how long stress events are occurring in the project provides invaluable data for leaders looking to maximise the effectiveness of their interventions to manage stress across the team. A tool such as this can also help you understand how those receiving your project are feeling, which is very useful. Pioneera is one of the 2018 cohort of tech start-ups that came through the BlueChilli SheStarts incubator sponsored by Google for Startups, LinkedIn, MYOB, ANZ, Microsoft and Herbert Smith Freehills.

DIVERSITY AND INCLUSION

Diversity is the lifeblood complexity needs for innovative solutions to difficult problems. Diversity of all kinds whether it's cognitive diversity, cultural, gender or any other form adds value. However, the greater the diversity, the harder it is to foster inclusion, and in complexity, the leadership challenge on inclusion at scale requires additional effort.

> Diversity is the lifeblood complexity needs for innovative solutions to difficult problems.

One of the greatest assets in complexity is cognitive diversity. Different people think differently about problems and possible solutions. Earlier in the chapter, we discussed the use of the HBDI® assessment for understanding thinking styles. When thinking about diversity in your team, this tool can provide insight into a type of diversity that has no externally visible signs.

> One of the greatest assets in complexity is cognitive diversity.

Much of the toolset required for actively managing diversity and inclusion is an extension of the assessments and health monitoring described above. As you get to know your team through individual and team assessments and monitor the health of the team, the inclusion challenges will surface allowing you to develop and apply management strategies for ensuring all members of the team are able to feel included.

Atlassian are constantly playing around with ways to build an inclusive and effective organisation as they scale from start-up to world domination. Dom Price, Atlassian's work futurist, scours the world in search of ideas that Atlassian can weave into their practices. He has five simple steps to inclusion from diversity that form the basis of a great inclusion toolkit.

1. Hire for values, not cultural fit. "Culture fit gives you more what you already have, which is not diversity. Values give you culture add, people who'll help evolve your culture" (Price, 2017).
2. Recognise (not reward) the good. Break the link between reward, which is usually financially based, and good behaviour. Instead, develop ways in which to recognise and celebrate the positive behaviours you want to see more of.
3. Educate your people on how bias, conscious and unconscious affect their actions and the flow on effect to inclusion.
4. Shout out. Be heard on the topic by writing blogs, post on LinkedIn or your company message system. Speak at company events so your team knows you take inclusion seriously. If they know you care, they're more likely to feel included. Dom stepped on the digital stage with his first blog post, scared to talk about diversity? I'll go first (Price, 2017), it's a good read for anyone getting started.
5. Make it part of how you do business. The practical part of inclusion then comes down to applying some simple techniques in meetings to build inclusion. The playbook on inclusive meetings (Atlassian, 2018a) is a great starting point to build the habit of inclusive leadership. www.atlassian.com/team-playbook/plays/inclusive-meetings

If you're looking for a more data-driven way to understand the dynamics of your team over time, there is a tool that has emerged in recent times that provides analytics on the quality of meetings (McGrath, 2014) being held across the project. The process is a simple survey of the participants in key project meetings such as steering committees, team meetings and retrospectives. AI is used to calculate the intensity of anger, fear, sadness and joy (see Ekman and Plutchick below) exhibited by attendees, and then social network analysis is used as the analytical framework for mapping the quality of interactions within and between meeting participants of each of the various project meetings. The real-time combination of relationships, emotions and perceptions across all meetings in a project is used to predict the project success.

Also provided is a real-time Jungian tool similar to DISC and iMA (http://ima-pm.co.uk) which maps project team meeting attendee profiles on a recurring basis as part of the same survey. The toolset consists of meeting mapping for projects, customer, safety, coaching, interviews and many more. The combination of the relationships, emotions and perceptions across all of these meeting types provides insights into the culture of an organisation.

To operate, you simply add the meeting invite email to your meeting calendar invite (Outlook etc.), and a 30-second survey is sent to all invitees once the meeting has ended. The tool provides leading rather than lagging indicators of potential project success or failure based on the data. As part of the cultural insights, the effects of lack of inclusion and diversity are able to be surfaced through the data making what can be a hidden issue transparent and the next steps can be actioned.

Emotions as Drivers

Emotional intelligence is the ability to identify and manage one's own emotions and those of others. The literature on emotional intelligence identifies motivation and empathy as two of the key elements of the construct which provide the pathway for identifying the toolsets needed for leading in complexity. There is no assessment for emotional intelligence. It is not a single construct but rather a collection of skills honed over time. Historically, the world of business was one where emotions were frowned upon and leaders were taught to "leave their emotions at the door" when walking into work as they were considered to cloud decision-making and rational thought.

As our understanding of complexity and the human factors within it has grown, we now perceive emotions as an integral part of how humans behave, make decisions and adapt to change. As such, leaders in complexity need to have a detailed understanding of the types of emotions humans have, the way they emerge and how they can be used to lead people through the ambiguity of complexity to successfully deliver outcomes. There are many different models; two of the most common include Paul Ekman, best known for his work on micro-expressions and the detection of deceit from facial expressions, identified the six basic emotions (Ekman, 1999) that underpin human behaviour, anger, disgust, fear, happiness, sadness and surprise. Robert Plutchick developed the eight-factor model that grouped emotions into four pairs of polarities, joy–sadness, anger–fear, trust–distrust and surprise–anticipation (Plutchik, 1980). There are many other variations that have been published by academics that provide variations on these that leaders can use as a framework for understanding the emotions that underpin the way our teams and stakeholders behave.

Once you have the ability to identify the emotional drivers behind the decisions and behaviours you see manifesting in your stakeholder network, you will have the power to bring your stakeholders down the paths you desire. If we look back at Aristotle's work on the roads of logic and rhetoric, we find that in modern business, the first road of logic is well defined and applied far more often than rhetoric. In the world of complex endeavours, the first road of logic is revered, science, engineering, design and mathematics play significant roles in the creation of the right solution. The emphasis of logic in the development of leadership skills that rely on building logical arguments has come at the expense of the art of rhetoric. It is rhetoric, the ability to argue a case using a wider array of tools including persuasion through emotional means that can offer leaders in complexity an edge when it comes to keeping the complex web of human factors moving in the same direction. It is for this reason that taking the time to understand the range of human emotions, how they manifest in stakeholders and developing the skills required to use this knowledge that will allow you to follow the second road is a worthwhile investment.

Grow People Selflessly

In the world of projects, we have come to rely heavily on the argument that project teams are temporary, therefore, the project manager is not responsible for the development of team members as they usually have a line manager in their parent company that is responsible for this. On the very complex endeavours, projects and programs, a special purpose vehicle (SPV) will often be created that overcomes the home manager scenario, but the temporary nature of SPVs can result in a lack of culture and process to support growing capability in the team. Either way, career development and planning is nearly always the sole responsibility of the individual, but skill development has to be within the human factors managed by the project or program manager as it is skills or a lack thereof that will introduce risk to the project. It is through the lens of risk that the notion of growing people selflessly moves from being a "nice thing to do" to an essential activity for project success. If all of the tools described above are applied to the understanding of self, others and team dynamics, you will have all the data you need as input to helping others to grow and develop the skills your project needs as it evolves. From there, it is about the application of a continuous development conversation with distinct markers to allow for the measurement of progress. At a minimum, set annual performance goals and associated development goals have a midpoint check-in at the 6-month mark to review progress on both performance and development goals and course correct if necessary.

Continuing with the theme of a data-based approach to the management of human factors, there are assessment tools that can assist in the selection of new candidates for roles and support your ongoing efforts to grow people selflessly. One of the most commonly used tools is the Caliper Profile which has been in use for over 50 years. This assessment is designed specifically to assess the likelihood of candidate-job fit.

Caliper Profile: The Caliper Profile is designed to assess an individual's job-performance potential. It looks at over 25 personality traits that relate to job performance. It is a tool that helps you find out which person is best suited for a given job based on their intrinsic motivations relative to the role's responsibilities.

Clifton Strengths Assessment: The Clifton Strengths Assessment is part of the Gallup suite of tools. It is designed to identify an individual's strengths from the 34 themes within the tool. These themes are broken down into four domains: Strategic Thinking, Executing, Influencing and Relationship Building. By understanding the strengths of your team members, you'll be able to support them in finding opportunities to exploit their strengths that will help them to grow and also add value to your project.

When you need the team to focus on what could be, those with strengths in strategic thinking themes will be key. When you need to implement a solution, you'll look to bring forward those team members with strengths in Executing themes. When it's imperative that the project team can sell their ideas inside and outside the organisation, you will turn to people with strengths in the Influencing themes, and finally, when the team needs to achieve more than is seemingly possible, you'll rely on your team members with strengths in Relationship Building to bring the team together.

GEOGRAPHICAL DISPERSION

Geographical spread of project teams can occur in all kinds of projects, in complex projects, the chances are simply higher that this will occur and that the number of geographies involved will be greater. This spread not only adds to the complexity of scheduling work but has a direct impact on inclusion and team cohesion. Although it seems simple, a time-zone meeting planning tool such as www.timeanddate.com/worldclock/meeting.html is an essential part of everyone's toolkit.

By simply being able to see all the cities on one page with a view of the best crossover times booking meetings that optimise the use of team member's reasonable working time is easily achieved. Inevitably there will be at least one group that joins at an unreasonable hour, implementing a rotation system such that each group takes a turn at staying up late or getting up early will pay dividends in terms of team health in the long run.

QUESTIONS TO CONSIDER

When applying the toolset for managing others, there are a number of questions to consider as you need to ensure the right tools are selected for the team you have. Not all of the tools are needed or appropriate for all teams.

- **Is the team in a place of psychological safety?** Before embarking on a round of team assessments, it's important to ensure the team you have is in a safe place psychologically. There are no external forces such as wider organisation restructures taking place that may make team members feel uneasy taking an assessment. Even the most well-meant developmental assessment in a time of job uncertainty will increase stress and not achieve the desired outcome.

- **Analogue or digital?** A number of the emerging tools for analysing team health such as Pioneera and MeetingQuality are exciting. They use AI and machine learning to process large volumes of data to provide insights. However, if you work in a large corporation, it is important to ensure that your IT security team is engaged early to assist. Another question to ask is how well does this technology fit within our current culture or are we introducing a new form of digital communication that team has not used before. For example, Pioneera can deploy through email scanning (pretty common) or Slack, if you haven't already chosen a team chat platform yet the introduction of Pioneera and Slack could be ideal. If, however, you've already selected a different platform such as WhatsApp moving to Slack may not be an ideal choice, better to go with Pioneera on email.
- **Is my team ready for an assessment?** When it comes to growing people selflessly, the title reflects all the best intentions; however, this is a highly personal and often challenging conversation for people to have. Using an assessment can provide a neutral basis for discussion, diffusing any personalisation that may make the more difficult aspects of the conversation easier. However, if the team is distrustful or unsure of the privacy of their data and how it may be used, you may be better off starting with 1-2-1 conversations rather than leap into a full digital assessment with benchmark comparisons.

ETHICAL DECISION-MAKING

The study of ethics has traditionally been the domain of philosophers dealing with morality and the principles needed to guide humanity through this quagmire. However, in recent times, ethics has become popular with scholars from a wider range of disciplines leading to the creation of different streams of ethical study including behavioural and applied ethics. Complexity gives rise to ethical dilemmas, and it for this reason that we need rules or guidelines established for what conduct is right and wrong for the team and wider stakeholders to use when facing into these issues.

For the most part, the decisions individuals and teams need to make on any given complex endeavour, project or program are relatively obvious in terms of right and wrong. They may be technically complex or require a complex set of human interactions, but at their heart, they are not ethically complex. The ethically complex decisions are the outliers, but when we encounter them, we need to have a method for doing so consistently and systematically.

Before getting into the design of ethical frameworks, it's good to cover some of the basics to provide context. Moral philosophy has three branches: Meta-Ethics which investigates the big questions such as "what is morality?" and "Is there truth?"; Normative Ethics which provides a framework for deciding what is right and wrong; and finally Applied Ethics which looks at specific practical issues of moral importance such capital punishment. For the purposes of developing an ethical framework for project or program teams to help them address the ethical dilemmas, they will encounter it is Normative Ethics that provides the path.

There are three common frameworks within Normative Ethics that can help to establish your approach to ethical behaviour, deontology, utilitarianism and virtue ethics. Deontology relies on the establishment of rules and regulations and the absolute application of those rules, regardless of consequence, is seen to be the ethical path. Although a valid approach to right and wrong in some instances in the complex space, the consequences of rigid application of rules can be wildly out of line with community standards. Utilitarianism is a form of consequentialism that relies on the assessment of the consequence to determine whether the action is ethical. The choice that holds the greatest good for the greatest number is deemed to be the best. This is a form of ethics that is often applied in business, particularly in the assessment of risk; however, there is a complication in that it is impossible to predict the future, therefore, impossible to know with certainty the actual outcomes, and in some situations, this could prove the undoing of the ethical reasoning applied. Virtue Ethics was borne out of the Greek philosophers such as Aristotle and looks to the ways in which a person can live a life of moral character. By practicing or applying the virtues of honesty, bravery etc., a person will develop virtuous habits and when confronted with an ethical choice will be more likely to make an ethical choice. The reality is that for each complex endeavour, project or program, you will need to find the right approach for your situation which may be a hybrid of these rather than a simple application of one over the other.

The contemplation of ethics is a deep and complex subject which could drive any curious person down an all-consuming rabbit hole. When we step back to look at how we can help our teams and stakeholders approach ethically complex decisions within our endeavours, projects or programs, we can draw from the above with a few practical steps.

Step 1: Ensure the mission or purpose to which the team is working is clear, true and compelling. This will create the "north star" for your endeavour, project or program.

Step 2: Surface the set of shared values that the collective agrees to work with.

Step 3: Develop a set of principles for decision-making. A useful form for this is a set of questions to be applied to any ethically complex decision that the team can work through to arrive at their answer.

When approaching the above, it is worth considering that "**Values** are an expression of what we think to be good... **Principles** are an expression of what we think to be 'right'" (Longstaff, 2016). Together they create a common platform to facilitate the process of ethical decision-making.

Although hopefully not too frequent, the ethically complex decisions that arise in complexity are potential stressors for the team. The decisions themselves are difficult, but the process of arriving at a decision tests each team member's personal values. To ensure that your team are supported through this, it is important that these decisions are made as transparently as possible. It is not possible in all circumstances to let the whole team into the full details but to the extent that it is possible as a leader explaining why you made the call you made helps people put ethics into practice, particularly with grey areas and challenging stakeholders. If you're leading by example in this way, your team is more likely to make a similar ethical call when they need to, but they'll also be comfortable coming to talk to you when the decisions get hard (as they will) or they make a mistake (which we all do!).

Questions to Consider

When developing your ethical framework, there are a few key questions to consider:

- What kind of ethical questions would be raised as part of this endeavour, project or program?
- How frequently would they occur?
- What is the likely consequence of getting ethical questions wrong?
- Who should be involved in ethical question decision-making?
- Do we need external assistance to help us see our world from the outside?

These questions are the basis for helping you decide the path you want to go down to develop your framework. In particular, the last question is the one that should be considered carefully. The tendency to argue that subject matter or industry experience is needed to fully appreciate the details of the most ethically complex questions is common; however, the risk in that is that from the inside, it is that you can only see the problem from the inside, and an external view can be invaluable especially when it comes to sense checking what the world outside your industry will see and hear about your decisions.

HUMAN FACTORS TOOLS INDEX

This chapter has described many different tools for managing the human factors in complexity. The following provides details on where to find more information. I have grouped them by topic as a reference back to the section in which they were discussed (Table 5.1).

TABLE 5.1

Human Factor Toolset Resources

Topic	Tool	Reference
Project management standards	ICCPM – Complex Project Manager Standard	https://iccpm.com/cpm-competency-standards/
	PMI – PMBoK®Guide	www.pmi.org/pmbok-guide-standards/foundational/pmbok
	PMI – Project Manager Competency Development Framework	www.pmi.org/pmbok-guide-standards/framework/pm-competency-development-3rd-edition
	IPMA – ICB	www.ipma.world/individuals/standard/
	P2M – Program and Project Management for Enterprise Innovation	www.pmaj.or.jp/ENG/p2m/p2m_guide/p2m_guide.html
	GAPPS – Standards Mappings	https://globalpmstandards.org/tools/comparison-of-global-standards/comparison-of-project-program-management-standards/
Managing Yourself		
Self-assessment	MBTI	www.myersbriggs.org/my-mbti-personality-type/mbti-basics/
	16PF	www.16pf.com/en_US/
	DISC	www.discprofile.com/
	LSI	www.human-synergistics.com.au/change-solutions/change-solutions-for-individuals/diagnostic-tools-for-individuals/life-styles-inventory
	HBDI	www.herrmann.com.au/the-hbdi/
	Gallup BP10	www.gallup.com/builder/225332/builder-profile-10.aspx
Mindfulness	Head Space	www.headspace.com/headspace-meditation-app
Managing Others		
Team assessments: Understanding the team	Belbin Team Roles	www.belbin.com/about/belbin-team-roles/
Team assessments: Managing the health of the team	Atlassian Team Playbook	www.atlassian.com/team-playbook
	Pioneera – Stress Analyser	www.pioneera.com
Diversity & inclusion	Atlassian Inclusive Meeting Play	www.atlassian.com/team-playbook/plays/inclusive-meetings
	Meeting Quality	www.meetingquality.com
Geography	Time and Date	www.timeanddate.com/worldclock/meeting.html
Grow people selflessly	Caliper Profile	www.calipercorp.com/caliper-profile/
	CliftonStrengths	www.gallupstrengthscenter.com/home/en-us/cliftonstrengths-themes-domains

When considering the toolkit that's right for you and your current complex endeavour, it is important to ask yourself the following questions:

1. What do I already know about myself and my team?
2. What do I need to know about myself or my team?
3. What decisions or actions am I going to take based on the data I collect?
4. Have I created a culture of psychological safety to make using the data collection tools effective in my team(s)?
5. How can I bring in diversity and then how can I foster inclusion?

SUMMARY

The landscape painted by humanity is itself complex, within the individual, between individuals and across teams. The toolset required to not only survive but to thrive is an evolutionary one. It grows with our collective understanding of human psychology and our growth as individuals.

For too long, the concepts of stakeholder engagement and human resource management processes have been the only lens we have provided project leaders to see the human element of their work. To be effective in complexity, the human factors must be faced into through multiple lens of self and others and draw heavily on our understanding of human psychology. To truly master this element of the toolset for complexity, we must start with a genuine curiosity about ourselves, others and the world around us. Empathy and compassion will smooth the journey, and the data will be your guide. As intangible and ephemeral as the constructs of the human psyche can be, we can apply tools and techniques that help us to gather and process the information available to us. The use of psychological assessments, team health checks, AI-driven stress monitors and social network analysis all help to create the data lake needed to draw insights and make human factor decisions.

There is no simple recipe for understanding and managing ourselves or others. Much of it is trial and error as we reach into our toolkit and try different techniques with each person, team or situation we encounter. What works with one person in a given situation will have no effect on another. And so, to be a master at this takes time and practice, like all other skills. Good luck and I wish you well on your journey of self-discovery!

REFERENCES

Aitken, A. (2011). Coping strategies of project managers in stressful situations. *Doctoral Dissertation*, Bond University.

Atlassian. (2018a). Inclusive meetings playbook. Retrieved July 1, 2018, from www.atlassian.com/team-playbook/plays/inclusive-meetings.

Atlassian. (2018b). Team playbook. Retrieved December 2, 2018, from www.atlassian.com/team-playbook.

Cattell, R. (1973). Personality pinned down. *Psychology Today*, (July), 40–46.

Caupin, G. & IPM Association. (2000). IPMA competence baseline: ICB; Version 2.0. Retrieved from https://books.google.com.au/books?id=69h1AAAACAAJ.

Caupin, G., Knoepfel, H., Koch, G., Pannenbäcker, K., Pérez-Polo, F., & Seabury, C. (2006). *IPMA ICB 3* (3rd ed.). Amsterdam, The Netherlands: International Project Management Association.

Caupin, G., Knopfel, H., Morris, P., Motzel, E., & Pannenbacker, O. (1999). IPMA competence baseline, Bremen: International Project Management Association.

Ekman, P. (1999). Basic emotions. In T. Dalgleish & M. Powers (Eds.), *Handbook of Cognition and Emotion* (pp. 45–60). New York: John Wiley & Sons.

Golpîra, H. (2014). A method for evaluating project management standards, based on EFQM criteria, using FTOPSIS method. *International Journal of Research in Industrial Engineering*, *3*(3), 1–12.

Hassard, J., Teoh, K., Cox, T., Dewe, P., Cosmar, M., Grundler, R., … Van den Broek, K. (2014). Calculating the cost of work-related stress and psychosocial risks. European Agency for Safety and Health at Work. doi: 10.2802/20493.

Headspace. (n.d.). Head space mindfulness app. Retrieved January 7, 2019, from www.headspace.com/headspace-meditation-app.

ICCPM. (2012). *Complex Project Manager Competency Standards (4.1)*. Canberra: ICCPM.

Longstaff, S. (2016). Values & principles are your organisation's DNA: Get to know them. Retrieved January 20, 2019, from www.ethics.org.au/on-ethics/blog/october-2016/values-principles-are-your-organisations-dna.

Marsten, W. (1928). *Emotions of Normal People*. Taylor & Francis Ltd.

McGrath, K. (2014). Meeting quality. Retrieved December 2, 2018, from www.meetingquality.com.

Morris, P. & Pinto, J. K. (Eds.). (2010). *The Wiley Guide to Project Organization*. Hoboken, NJ: John Wiley & Sons.

Owen Witford, D. (2018). Pioneera stress tool. Retrieved July 1, 2019, from www.pioneera.com/.

Plutchik, R. (1980). A general psychoevolutionary theory of emotion. In *Emotion: Theory, Research, and Experience* (pp. 3–33). Cambridge, MA: Academic press.

PMI. (2002). *Project Manager Competency Development Framework*. Upper Darby, PA: Project Management Institute.

PMI. (2013). *PMBOK® Guide* (5th ed.). Upper Darby, PA: Project Management Institute.

PMI Standards Committee. (1996). *A Guide to the Project Management Body of Knowledge*. Upper Darby, PA: Project Management Institute.

Price, D. (2017). Scared to talk about diversity? I'll go first. Retrieved December 9, 2018, from http://collectivehub.com/2017/06/scared-to-talk-about-diversity-ill-go-first/.

6 Leadership Toolset

Thomas Grisham
SMC University

CONTENTS

A QUICK LOOK BACK

The need to manage projects began long ago—Mesopotamia, Egypt, and the Indus Valley around 3200 BCE. Early man undertook to build things, and that consumed financial resources, and required people to do the work. Many of the projects like Mohenjo Daro in what is now Pakistan constructed practical works that people needed. In this case, it is the first known public sewer system. Early projects were often conducted by rulers who had access to the necessary funds and labor. Being labor-intensive projects, it helped if slave labor was abundant. In these cases, the ruler was like a design-build-construct-operate entity. One stop shopping, with few if any outsourced services. All of the risk was taken by the owner. There are limited communication requirements and few disputes. Who would argue with the king?

Fast forward a bit into Ottoman times, 1500 CE, and enter people like Mimar Sinan the architect. He worked for the Sultan as an outside consultant, who was an early sort of design-build service provider. The Sultan provided the financial resources and people and materials, and looked at Sinan to design and direct the work. Simple success criterion is you make the customer unhappy or you die. Sinan's works, like the Selimiye Mosque in Edirne and the Suleiman Mosque in Istanbul, are legendary for their acoustics and beauty. He is often described as the Michelangelo of the Eastern world. He served as an apprentice to learn his craft, through practice, and built a reputation before engaging in projects. Not much had changed, one more communication channel was added.

Then, there were projects like the Duomo di Milano, which was started around the time Sinan worked, but was not completed until 1805, when Napoleon was to be crowned King of Italy. In some ways, this set the stage for 20th-century Project Management to define projects as having a beginning and an end, it did, and to have multiple Project Managers, customers, and stakeholders. It started with the Archbishop undertaking a religious edifice, funded by the church, and first ended with Napoleon wanting to showcase the extent of his power and control. There were 78 architects and Project Managers over the course of the construction, which was ongoing until 1988.

The foundations for Project Management were constructed by Frederick Taylor in the early 20th century and his students including the well-known father of the Gantt chart. Taylor was an engineer and devoted to scientific principles of efficiency. This systems approach led to critical path method (CPM) and Program and Evaluation Review Technique (PERT) in the 1950s. In 1969, Project Management Institute (PMI) was formed to provide a platform for the professional practice of Project Management. The systems thinking approach continued through Earned Value Method (EVM), and Kerzner's initial work—A Systems Approach to Project Management (Kerzner, 2006). Other Project Management systems and organizations sprang up with PRINCE and International Project Management Association (IPMA) being two of the more prominent.

Also during this same time frame, Agile Project Management began to gain popularity. Failure rates for projects, according to the Standish Group, were over 70%, and in other industries, the rates were at eye-watering rates above 90% when using the Project Management Body of Knowledge (PMBOK) waterfall approach—especially for IT projects. This PMBOK systems approach was too slow, subject to failure, and costly for the fast-paced world of technology; enter Agile. Still a systems approach but with a shift toward more empowerment, frequent customer interaction, limited sprints or project durations, and little or no paperwork. The focus is on having products that work rather than documentation. Have professionals do the work rather than plan and report on it. Organizations are trying to adapt it in non-IT industries, with sporadic success.

As a practitioner, we saw the need for another view of Project Management. So we started with our 40+ years of experience and began to research global leadership. We had seen numerous failures that came from too much emphasis on systems and processes, and too little on the people that perform the work. And, the application of 20th-century systems to a world of rapid change, global resources, multiple disciplines, value chains, and virtual teams was not a recipe for success. When we began our research work, PMI and IPMA had been publishing journals since 1995. Out of 1,173 articles published when we started the review, only 77 dealt with leadership and culture, which led us to study cross-cultural leadership in depth and was culminated in our first book (Grisham, 2009a). The purpose was to develop a model for leadership that could be applied to any endeavor regardless of the culture, industry, or location.

With this work as a background, we decided to explore the relationship between leadership and global Project Management. The second book (Grisham, 2009b) was written to provide a more people-based view of global Project Management, founded on leadership. We will talk more about that in the next section, so for now we want to trace out the trajectory of the profession. As of 2009, there were few books, that we know of, that addressed leadership and global Project Management. We decided to use the PMBOK outline for topics in which we work so that those who became Project Management Professionals (PMPs) would not have to learn another set of processes to use this new approach. But, as we saw the need for speed, virtual teams, greater productivity, a reduction in scope disagreements, and overallocated resources, we decided to set out a way for all organizations to use Agile and Waterfall Project Management techniques (Grisham, 2016). Projects have always been complex and risky endeavors, and subject to numerous outside influences. Dr. Arroyo and I addressed many of these outside influences in our work on extreme projects (Arroyo & Grisham, 2017). Political, local societal, and market conditions frequently have a large influence on global projects. Almost all Project Managers have no training or skills in these areas, which creates big opportunities for failure.

More recently, Dr. Arroyo and I were invited to write about the future of Project Management (Arroyo & Grisham, 2018), and we called it CPE (Collaborative Project Enterprise) 2050. In this work, we took a view looking backward from 2050, imagining where technology will lead and how global Project Management might anticipate the changes coming. The CPE means Collaborative Project Enterprise and is taken from our second book, as we think it represents the future of the profession. We will discuss CPEs in the next section.

It summarizes the concepts of thinking of global projects as programs and used the concept of a CPE to view these endeavors as if a new business.

WHAT IS A CPE?

CPE (Grisham, 2009b) is a way of looking at leadership on complex projects. First, perhaps we should define what a complex project is. From our experience, a complex project involves teams that include multiple organizations, cultures, disciplines, time zones, goals, languages, and time frames in markets that can change rapidly. Projects like new mining facilities are more stable than cutting-edge technology projects, but both are subject to rapid changes, just of different types—perhaps a change in government for the former, and a change in the technology for the latter. In the 21st century, most organizations outsource knowledge, because it is expensive to own it. This means that most projects are a collage of organizations that have quite different goals. The complexity comes from this and the need for massive amounts of coordination and communication. As we do training around the globe, participants estimate the amount of time they spend communicating to be 6 h/day, and in most countries, the hours per day average about 10.

People are generally not very good at communicating efficiently, it takes a lot of practice and cultural understanding, which we will describe later in more detail as our XLQ model. We have sat through corporate presentations by finance people, supported by IT people, wanting to persuade the boss, from a manufacturing background, that this bitcoin idea is really the future. All intelligent highly trained people have trouble communicating concepts or tacit knowledge (Fernie, Green, Weller, & Newcombe, 2003; Nonaka & Teece, 2001). From an academic perspective, they have MBAs and were taught to communicate within their discipline. Now take this dilemma and export it to 12 other organizations coming together to perform a project. The communications alone are complex.

In part, complexity also comes from the need to drive down labor costs, which can have overhead multipliers of 3.5, so organizations are not inclined to keep expertise on staff that is not used regularly. For organizations at the cutthroat end of the scale, the impetus is to shed headcount and increase profits by having people work on multiple projects concurrently. There are plenty of studies that show multitasking reduces productivity (DeVaro & Gürtler, 2016), yet many organizations continue the practice. Over the last 10 years, we have found few organizations where Project Managers work on one project at a time. Most often we are told the number of projects is between five and ten. And, this is not a successful strategy, for overallocation of resources is one of the major causes of project failure.

Again the idea is to reduce headcount, and one popular way is to utilize outsourced services. The son of Uber for projects that use only what they need. The basic concept is good, to create more efficiency, yet it comes with some challenges. Having dozens of partners, subcontractors, vendors, and agents means that to be successful, a CPE Lead must spend much of her time communicating, building, and maintaining relationships and dealing with conflict. The work itself of managing often must be ceded to others. This adds complexity and in turn requires more communication.

The concept of a CPE first began to take shape for us in 1980. The project was an EPC (engineer, procure, construct) venture where we worked for the EPC contractor. This was still in the days of unions, where jurisdiction over work was a constant battle. The jurisdictional disputes slowed the progress, required a lot of effort to communicate and negotiate with the unions, and could lead to violence. We still remember union business agents bringing guns to work. Then, in 1987, we were managing a US$400 million project which had numerous organizations involved from acquisition of real estate to public education, political management, numerous lawsuits, and of course the actual work itself. What we began to see was that all of the groups turned to our organization for guidance. Sometimes, they would accept the suggestions, sometimes not. There were dozens of different disciplines, cultures, goals, etc.

It became clear to us that we could reduce the number of problems by helping other organizations solve theirs. We found that by empathizing and going out of our way to see the world through their eyes, we could more easily persuade them to take positions and actions that were not solely in their own best interest. There was no contractual vehicle that required cooperation, and as with the unions, often personality conflicts created costly diversions. But, by stepping in between warring factions, we found that we could reduce future conflicts. In short, we saw that leadership could ease many of the conflicts that occur when multiple organizations and disciplines were present.

What we began to see, in a very general way, was that when we acted as if this was one organization, and we were the CEO, all of the organizations were more successful. At this point in time, the concept was just beginning to form, and we could not distinguish leadership from management as fully as we wished. We just knew that most of our duties required for success were not the ones on our employment contract. And, not surprisingly, when we put on this CEO hat, our employer was unhappy with us (on many projects, we were a consultant), as the employer thought short term and believed that we should be devoting all of our time to Project Management processes. Don't go outside this border was the message to us. Even then we knew that there was much more required than the narrow borders of Project Management theory.

During many training sessions with organizations globally, we have seen the schism between the work of Business Analysts, Project Managers, and Warranty support. Project Managers are seldom engaged during the initial creation of projects, or if they are, it is fleeting and superficial. Political and societal issues take time to research and understand. A project we did for a new power facility was located in very close proximity to high-end residential property. We were told that there had been engagement with the community, and no concerns were voiced—the feedback for us lasted about two minutes. One year later when we became involved and the work began, there were many concerns and a political firestorm. We had to create a community group to channel the complaints and to chair regular public meetings to explain what was occurring on the project, what was planned for the next month, and what the community could expect as far as noise, dust, or other disruption. Had we been involved earlier, we could have anticipated the issues and created more of a partnership relationship with the community. As it turned out, we were forced to react, and the community saw this as our intention. It took a very long time to persuade them that this was not our intention.

So, that is a very brief view of how the concept began. Over the next couple dozen years, we were fortunate enough to experiment, explore, research, adjust, and test the concept. We were able to define what a CPE is and how it can improve success on complex projects. A CPE is collaborative because the leader pulls together the diversity, harnesses the energy and creativity, and shows all of the organizations that acting like one single organization can increase success. A CPE is a project because the organizations come together for a limited time to create something unique. And, a CPE is an enterprise because it is led as an initial public offering, whose job is to be creative, efficient, and successful—everyone makes a profit and enhances their reputation. So that is the metaphor. Figure 6.1 provides an example of a normal CPE. If you do not work outside of your home country, you can replace the country names with states or regions. One other layer of complexity is shown for short- and long-term relationships. In the case of the financial alliance, it would consist of organizations with a more intimate and long-term relationship, more like a marriage. For the operations joint venture (JV), the relationship may have been created specifically for this endeavor and will dissolve afterwards.

The complexity in Figure 6.1 is obvious. It shows differences in relationships, long term versus transactional, difference in corporate and individual culture, different goals, etc. And, consider that each group like procurement in Frankfurt will have a third dimension of subcontractors who can often be invisible. The CPE Lead would possibly have contracting relationships with the other entities, but no other power. In the case of the financial long-term alliance, it is likely that the CPE Lead would communicate with a spokesperson for the alliance, not all the members. Contracts may certainly tell one party with whom they should communicate, but do not deal with the openness or effectiveness of those communications. That is one example of where leadership can make a very big difference.

We were involved in a project in India that had a similar structure to Figure 6.1, where our focus was on the financial groups, located in Japan. There was a consortium, not alliance, of banks with a Japanese import–export bank involved as well. The banks (there were seven of them) communicated

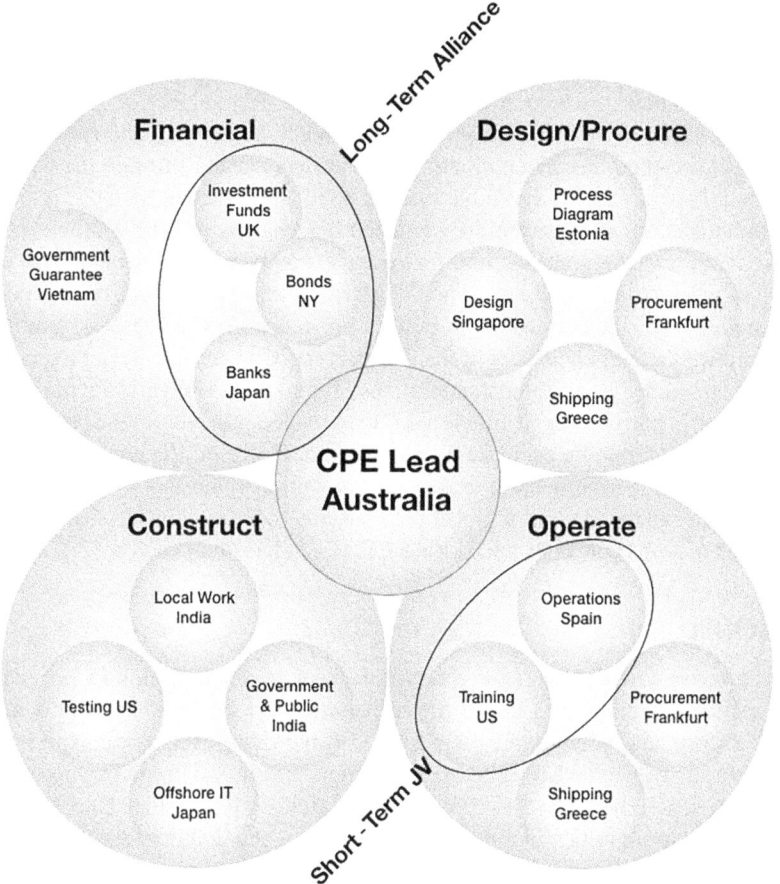

FIGURE 6.1   Collaborative program enterprise.

with the remainder of the CPE and the Japanese import–export bank, through our organization. We then communicated with the rest of the CPE including the teams in the United States, Eastern and Western Europe, the Middle East, Asia, and India. The CPE Lead's long-term goal was fast completion so that the facility could produce revenue, a goal shared by the financial institutions, and almost everyone else. Short-term goals of course varied, but the differences could be managed, if they were recognized as valid. The biggest problem for the CPE Lead was a lack of trust between the parties. If you were to draw lines between all of the organizations in Figure 6.1, you would have a map of mistrust.

What we saw was an opportunity to increase the trust for the CPE Lead by providing transparency and dependable communication channels. There was no contract that required us to do this; our interest was seeing the project succeed, so that the financial groups continued their support and were repaid promptly. So we offered a communication link between the Japanese organizations and CPE Lead in India. Because of the import–export bank participation, numerous Japanese suppliers were involved in the project. These were those subcontractors we mentioned earlier as the third dimension. The result was an increased level of trust between the organizations in the CPE, just because of the additional information and perception of transparency. We tell you this to illustrate what we mean by leadership, in this case Servant Leadership. Doing more than is required by contract to increase the probability of success for the entire CPE, by building trust.

There is one other short story about this project. There were conflicts between the organizations on this project, as we are sure you can imagine. The value of the project was billions of US dollars, so the conflicts had potentially large consequences. To finance such a project, with payments made in multiple currencies, it is necessary to develop hedges to cushion the effect of losses due to the relative imbalances between currencies at any particular time. They mean there must be a reasonably good estimate of the funding needed each month. As a result of the lack of trust, the funding agencies expected financial perfection from an industry that is subject to large amounts of risk and change. There was a significant disciplinary schism between two of the major groups and the potential shortfall of capital any given month. The conflicts rippled through the CPE, as multiple organizations could not pay their bills. What we described above helped with this issue as well. The education, communication, and transparency reduced the tensions and the conflict.

In our experience and research, we find that using a company as a metaphor for a program is more accurate of what actually happens. The CPE Lead is an entrepreneur on an endeavor that is unique and complex. The conventional definitions of projects and programs are somewhat misleading when one applies them to endeavors in the 21st century. Conventionally, a project would be the view that a subcontractor for the Spanish operations organization shown in Figure 6.1 would have and that is dangerously narrow. A program is a group of related projects, so the organizations shown in Figure 6.1 could be seen more accurately as a program. However, the conventional concept of a program is that of a single organization looking at the world. Again that is far too narrow a view. We saw a need for an expanded view of global projects, and thus the idea (Grisham, 2009b) to better describe the range of responsibilities and skills a CPE Lead has and needs.

THE XLQ MODEL

As we said above, we saw the need for a richer view of leadership. Having decades of Project Management experience globally, I began to see patterns and similarities, in how actual leaders went about their work, and why people seem to get behind them. Anecdotal experience is useful, but not adequate, so we decided to take an academic route to study leadership and try to connect it to our experience, which led to 3 years of research on leadership. One of the first things we saw was the need to a multidisciplinary approach—thus the lengthy research. Leadership was frequently equated to management and ignored philosophy, psychology, sociology, politics, music, art, literature, technology, anthropology, and much more—thus the 3 years. What we found in the Project Management literature was not broad enough to deal with the complexity in global markets. In fact, we found it was not rich enough to deal with the complexity in domestic markets as well—particularly over the last 15 years. We then tested the model that resulted, with a diverse group of global people from different backgrounds and disciplines. The results are shown in Figure 6.2 and

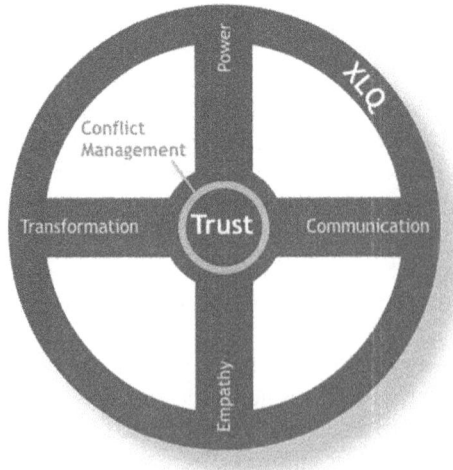

FIGURE 6.2   The XLQ model.

our second book (Grisham, 2009a). It is a blend of 40 years of experience with our research and testing. Since then, we have tested the model globally with numerous organizations and see that it is verifiable in practice, effective, and durable.

We chose to tell you briefly about trust and conflict and Servant Leadership to illustrate a few of the skills that a CPE Lead needs from our earlier work (Grisham, 2009a). Figure 6.2 provides what we call the XLQ model (cross-cultural leadership intelligence). Using the metaphor of the wheel, borrowed from Hindu mythology,

> Our definition of leadership is the ability to inspire the desire to follow and to inspire achievement beyond expectations.

trust is the axel and critical, but the wheel needs lubricant to turn and deal with the conflicts that are natural—as noted in the stories above. Our definition of leadership is *the ability to inspire the desire to follow and to inspire achievement beyond expectations.* Leaders are determined by the people who choose to follow them. And, leaders will strive to help those who follow to become more than they believe themselves capable of achieving. As Lao Tzu (Pheng, 1995) said millennia ago: "The Sage [CPE Lead] is self-effacing and scanty of words. When his task is accomplished and things have been completed, All the people (CPE Organizations) say, We ourselves have achieved it!"

The XLQ model is based on the idea that there are leadership attributes that are effective globally. The competence of a leader is to know which ones are needed and in what degree for each individual. A strong individualist in Germany and a strong individualist in the United States may have quite different personalities and backgrounds. One may need 100% empowerment, the other something less. The key is to understand each of them and provide an individualized type of leadership for each. Using a metaphor, imagine a salad of leadership, with different dressing for different individuals—personalized leadership.

The dimensions of the XLQ model, shown in Figure 6.2, start with trust. The descriptors of trust (characteristics or attributes) are integrity and ethics, truth and justice, care and concern, fearlessness, character, dependability, humaneness, integrator, and competence. Take fearlessness. In our definition of leadership (to inspire achievement beyond expectations) in a complex fluid environment, no matter how much anticipation one does, there will be the unexpected. A CPE Lead can attempt to do it all herself, but that is a poor use of expertise. Far better to empower others to take responsibility for various tasks and changes that may occur. This builds adaptability, diversity, creativity, and challenges people to be more than they think themselves capable of becoming.

Fearlessness comes in because people learn by making mistakes. So as a CPE Lead, one is in effect encouraging mistakes and must take responsibility for the mistakes of others. We do not mean mistakes springing from a lack of interest or concern or effort. Mistakes from people stretching themselves to be come more. This is not for the weak of heart and is difficult within one's own organization. Now extend the idea to other organizations, and the challenge becomes even more acute. Yet, the rewards can be far greater. Under the dimension of empathy, there is Servant Leadership (Greenleaf, 1997): to serve others without expecting reciprocation. The CPE Lead is demonstrating empathy by offering another person the opportunity to achieve something that they perhaps did not think they were capable of doing and to do so without fear failure.

Even better, there is also the gain from mirror neurons (Iacoboni et al., 2005). Other members of the CPE see the CPE Lead doing this, and the good feelings it provides. They in turn may choose to copy the behavior, and it can ripple through the CPE. People copy behavior is the point, especially the behavior of someone they respect—the CPE Lead in our example. The beauty is that these theories work in practice. We have practiced them on many occasions. Please do not believe, however, that it works 100% of the time. Some people are not wired in a way that they are interested in empathy, either by nature or by nurture.

Conflict is the lubricant that keeps the wheel turning. Conflicts in personalities, disciplines, organizations, cultures, gender, sexuality, income, goals, and idea are present on all human endeavors. We see conflict as a positive thing, not a negative thing, as it is a normal part of life and is inescapable. It is positive because it offers the opportunity for the CPE to see how the CPE Lead deals with it. One Chinese parable says that you don't know someone until you have fought with them. We can attest to that. Adapting it to projects, the CPE does not know the CPE Lead until they have seen how he deals with conflict. Imagine a conflict between an Israeli and a Palestinian on a project and that the CPE Lead needs them to work with him to solve a problem being confronted by the project. It will not be smooth, there will be arguments, and the two parties may not want to dine together at the end. The point is the CPE witnesses if the CPE Lead steps into the conflict, without judging it, and works to get the parties to cooperate. It tells the CPE, if you have a conflict, there will be support.

Back to the dimension of empathy. The descriptors are cultural intelligence, humaneness, servant leadership, compassion, and competence. For compassion, we follow the advice of the Dalai Lama: I understand you, I feel for you, I want to help you. To understand another person requires some time, even for people who are gifted with the ability to read others. To feel for another is a personal characteristic which some have and some do not. And, to want to help is a measure of emotional intelligence or EQ in our mind (Goleman, 1998). As with all the dimensions of the XLQ model, people have varying degrees of ability and can certainly learn to increase their competence. Empathy is showing compassion, and easy ways of doing so are simply saying good morning in another person's language—I cared enough to bother to find out about your culture.

There is a good deal of research showing that empathy also produces oxytocin in people. Oxytocin is a chemical like serotonin, though not as strong, which gives another person a feeling of connection or happiness (Barraza & Zak, 2009). The point being that all Homo sapiens are constructed this way, so it works everywhere. The challenge is if one can do it, genuinely. If you do not care and fake empathy, it will have a very different result. This idea of empathy is a measure of one's EQ in our view. The CPE Lead must be genuine first, always. If you don't care about others, don't pretend. To be a CPE Lead, one must know him or herself, and certainly their capacity for empathy.

Power is a more fundamental dimension with the descriptors being referent, knowledge, position, power distance, reward/punish, and competence. Referent means respect, and we see as the strongest version of power. Our definition of leadership begins with the ability to inspire the desire to follow. Think about a person like Mahatma Gandhi, who had no position, no special knowledge, and could not reward or punish, yet millions chose to follow him. Because he inspired them to do so because of their respect for his soul. His Sabarmati Ashram near Ahmedabad is a small block structure with a concrete floor simple, basic, human scale. He empathized with his fellow countrymen and showed them the dignity in living a simple and spiritual life. The other dimensions are less durable, yet some require understanding for a CPE Lead like Hofstede's Power Distance (Hofstede, 2001). Some societies such as Korea and Nigeria place significant importance on position. A CPE Lead needs to understand such things to show empathy for those who come from these backgrounds.

Transformation has the descriptors of inspiration, risk change, vision and curiosity, and charisma. We discussed the issue of encouraging others to become more, and this means a person will risk changing themselves. As with corporate takeovers and mergers, change is inevitable on projects. So the effort devoted to building trust and inspiring the desire to follow takes support from the transformation descriptor. Vision is important for a CPE, because of the variety of goals that are always present. The CPE Lead can help pull the parties together by setting forth a vision for the project that all organizations can accept, which then becomes the consistent goal that grounds everyone when things change. As in constructive conflict, vision can create a common enemy that organizations can team up against—not me against you, me and you against the dragon, the failure to reach our vision.

Then, there is the dimension of communication with the descriptors of wisdom, adaptability, competence, creativity, patience, and sensitivity. This is the most complex and difficult of the dimensions that a CPE Lead must acquire, for it requires pieces of all of the others, EQ, XLQ, and CQ or cultural intelligence (Earley & Ang, 2003). For this chapter, we would like to focus only on wisdom.

By wisdom, we mean the person appropriate application of knowledge. For example, consider Abdul: he is a 30-year-old man, born in Nigeria, educated in the UK with a degree in finance, and living in New York. He is the Project Manager for the Bonds organization in Figure 6.1. Think about communicating the difficulties of building something in India, with the poor infrastructure, and why that will cause problems with precision in the amount of draw that will be needed in rupees each month. On the knowledge side, what do you as the CPE Lead know about Bonds, what does he know about construction and India, what do you know about how Nigerians communicate, what does he know about how those from Australia communicate, does he prefer paper or electronic format, how can you show empathy, do you have anything in common, etc. It takes time—a lot.

Last but not least is the cultural XLQ that is the tire for the wheel. Here, we spent about 3 years of research looking at the thinking that preceded us regarding leadership, culture, and cross-cultural leadership. The work of Hofstede, the GLOBE Survey (House & Javidan, 2004), Trompenaars (Trompenaars, 1993), Thompson (Thompson, 1998), and many, many more directed my research and thinking. Staying with Hofstede, who is the most prominent when it comes to culture, our XLQ model includes his dimensions of culture: power distance, masculinity, indulgence, individualism, long-term orientation, and uncertainty avoidance. If you are not familiar with his work, I highly recommend that you have a look at his webpage hofstede.com. Hofstede did his initial work in the 1980s and later added in countries that he could not access during that time. The world has changed significantly, and the generational changes show up in differences that are obvious today, like in China.

Take power distance as an example. From his webpage (hofstede.com), he defines this dimension as "the extent to which the less powerful members of institutions and organizations within a country expect and accept that power is distributed unequally." So countries such as South Korea, Malaysia, Nigeria, and Saudi Arabia score high on this dimension, and countries such as Nordic Europe low. When the GLOBE Survey (House & Javidan, 2004) came along, it asked two questions on this dimension: How are things in your country today? and How should they be in the future? Interestingly, not a single country of the 60 studies said that they wanted the same level of power distance in the future, not one. Thinking about empowerment of underlings, everyone wants to make their own decisions in the future.

So, here is an opportunity for a CPE Lead to empower people. Some cultures will still be very tentative and fearful of taking on this challenge, but want to do it. As we discussed earlier, inspiring achievement beyond expectations is an opportunity. A CPE Lead can provide a safe place to land for those who want to make their own decisions, but are fearful of failure. Think of a time in your life that you were afraid to do something, but another person helped confront your fear and protected you. Remember that feeling? It is not easy, quick, or painless to do this, but the returns are large and durable far better than lengthy detailed contracts. It is what the Chinese call *Guanxi*—building and maintaining a relationship. For those of you interested in further research on the leadership dimensions have a look at (Csikszentmihalyi, 2001) for transformation, for communication and trust (Stewart, 2001), and for emotional intelligence (Goleman, Boyatzis, & McKee, 2002)

The people side of the equation requires the majority of CPE Lead's time. It is critically important, and we believe it determines whether an endeavor is successful or not. But for now, there are the managerial things that must be attended to until the technology arrives. The key is not to be seduced into emphasizing these because it is easier and requires less time. Another look at trust comes from leader profiles (Goldsmith, Greenberg, Robertson, & Hu-Chan, 2003). Some other authors that we find prescient (Yukl, 1998; Bartlett & Ghoshal, 1998; Bass & Avolio, 1994; Earley & Singh, 2000; Heil, Bennis, & Stephens, 2000; Marquardt & Berger, 2000; Bass & Stogdill, 1990; Sendjaya, Sarros, & Santora, 2008) offer other perspectives on leadership. And, globally one excellent tool and skill is storytelling (Denning, 2005). We believe that 80% of the time a CPE Lead spends each day must be devoted to people. If a CPE Lead is working, doing stuff, she is not doing her job. Her job is to motivate and inspire others to do the work. But there are some tools that can be used to facilitate the inspiration and communication of tacit knowledge.

There are many tools out there to help teams become more efficient and effective: leadership, emotional intelligence, cultural intelligence, cross-cultural leadership intelligence, Myers Briggs, and more. The idea is to understand ourselves by looking within from as many angles as possible. A review of the details is beyond the scope of this article, but you can look above at the references to get a starting point. The absolute key to becoming a leader is to know yourself. There are no shortcuts, and while you may find it difficult to accept what you see, there is no other way to become a leader. As we said earlier, the XLQ model is designed to be a tool belt that requires knowledge to use. There are currently 7.5 billion souls on this planet, and each one of them is unique. No short cuts, just hard work to get to know each key member of a CPE, individually.

For more prosaic tools, we find that the Nominal Group Technique (Potter et al., 2004) works quite well for diverse groups—culture/discipline/organization/age. It is a basic brainstorming approach that asks participants to prioritize the items that are developed from basic brainstorming. We find that using an Ishikawa diagram (Corno, Reinmoeller, & Nonaka, 1999) after the initial round is quite useful, especially for virtual teams. Graphical interfaces are particularly useful for global teams where the languages often get in the way. Other tools such as SWOT (strengths, weaknesses, opportunities, and threats) analysis or systems thinking (Maani & Maharaj, 2004) provide ways to improve the efficiency of decision-making, collaboration, and concurrence in complex environments. We also like affinity diagrams as a way to inspire creativity and help people to think outside the box. Such tools provide a leader with tangible ways to show a team how to come together, but are just that processes (Cavaleri & Reed, 2008).

THE DEVIL IS IN THE DETAILS

Regardless of whether you do all waterfall, all agile, or some mix, the details matter as you all know. The PMBOK provides a good checklist of requirements that need to be addressed on all projects—cost, risk, schedule, stakeholders, communications, and quality. In addition, we would add in politics, markets, cultures, and currencies. Organizations that do projects regularly have the data to learn from performance over time that is essential to improve productivity and increase profits. So lessons learned are as important long term as are the shorter term KPIs and profitability. For this example, assume that all of the contracts are negotiated fixed price type and that the CPE Lead is the ultimate customer.

Using Figure 6.3, we would like to sketch out a process for one of the requirements noted above, risk, using the financial consortium as an example. In Figure 6.3, the Tier 1 contracts are shown with heavy gray lines. These are the primary foundations for the CPE and create the key relationships. The other relationships are then negotiated between the Tier 1 organizations and the Tier 2 organizations, which are called subcontractors. So, for example, in the financial circle, each organization shown (Bonds NY, Banks Japan, Government Guarantee Vietnam) has a Tier 2 subcontract with the Tier 1 organization—Financial Risk Funds UK. As with the management concept of span of control, it is necessary to minimize so-called direct reports for purposes of efficiency. Otherwise, there are simply too many communication paths, which lead to confusion and misunderstandings.

The CPE Lead needs to include a requirement in the Tier 1 contracts that the terms and conditions would be passed along to each Tier 2 subcontract. In this way, the terms and conditions are symmetrical and consistent. One of these conditions should require the Tier 1 organization to obtain a risk management plan. The format and level of detail are provided in the contracts, again for consistency.

The Financial Tier 1 organization should be required to acquire and scrutinize the plans from each Tier 2 organization, and combine them into a single risk plan submitted to the CPE Lead. The CPE Lead then combines all of the Tier 1 organizations into a CPE risk plan and shares that with the Tier 1 organizations. The basic risk template is in a spreadsheet format and includes

- Risk description
- Risk profile (RP) of the organization=RP (see below)
- Probability from 0.0 to 1.0
- Impact from 0.0 to 1.0
- Probability times impact=P×I
- Risk response plan
- Cost of the plan
- Assumptions.

The RP is an assessment of the level of risk the organization is comfortable accepting from 0.0 to 1.0. At a level of 0.0, the organization would be risk-averse and not willing to take any risk. The RP is then used to determine if the identified risk is below the RP. If so, the organization takes the risk, and it becomes a passive risk. That means the organization monitors the risk to see if the probability or impact change over time. If the risk is equal or above the RP, then the organization must make a specific risk response plan.

Imagine the Bonds organizations in NY prepare their plan and they have an RP of 0.50. They identify a risk of currency devaluation and a cost of US$1 million to mitigate the risk. The banks in Japan have an RP of 0.40 and identify a potential US$0.5 million to mitigate the risk. The Vietnamese Government Guarantee has an RP of 0.20, but has no requirement to convert currencies so no risk. But, they provide the guarantee of capital and have identified a risk of bankruptcy by the local Indian organization if the currency devaluation of rupees is extreme. That risk is dependent upon the financial group having put a US$-¥-Indian Rupee hedge in place. In this case, the risk of a Tier 2 organization can be created if another organization does not take some action that is where the CPE Lead comes in.

In this case, imagine the CPE Lead has eight risks that are above its RP. Remember, this organization may choose to take some of the risks that a Tier 2 organization cannot either accept

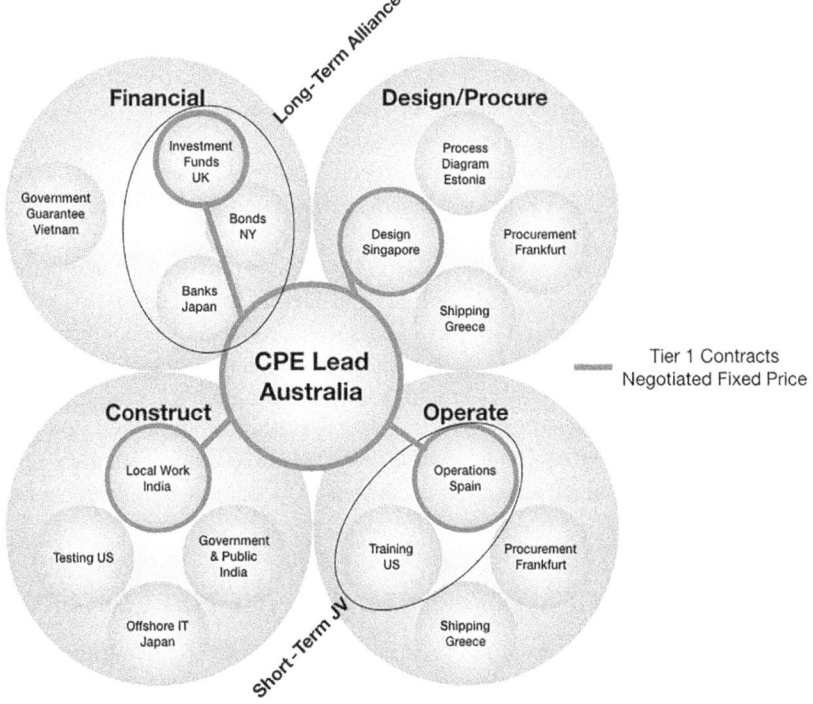

FIGURE 6.3 CPE contract structure.

or manage. Let's assume that they chose to do so for this risk. For the CPE Lead to manage this risk, they will need to put currency hedges into place and to do that they will need information from the Construct group for the expected dates and amounts of progress. That could be gleaned from the detailed schedule that the Construct group provides and their risk plan, which would likely include the risk of not being paid. Complicated indeed, but manageable if the facts are presented and shared, openly.

If everyone does a proper job of planning for risks, they will include contingency funds in their estimates to cover risk mitigation. If the CPE Lead permits this to happen in a non-structured way, the extent of the contingencies included by all parties cannot be known, and the cumulative effect can be significant. And, some organizations will assume certain risks will be taken by others and not build a plan to manage them. Thus, when the risk manifests itself, it can create conflict and place multiple organizations in a reactionary mode. We have been told that such transparency is counterproductive, because it may inform an organization that they have overlooked a risk and thus increase their price. This strategy hopes the risk does not occur, and if it does, then the parties will try to deal with it.

There is more, however. In embarking on such an undertaking, the CP Lead has an opportunity to build trust, and a culture of transparency and teamwork, by moving discussions about difficult topics out into the open. Imagine a subcontractor to one of the Tier 2 organizations that is a local organization, say in Estonia. The local organization may have never worked with international organizations, and may be less sophisticated in their technology and operations. Here again, there is an opportunity for the CPE Lead to build trust and show empathy, by taking on the education of organizations in the CPE—not by lecturing or judging, but by educating and helping. In the literature, and as we mentioned earlier, it is called Servant Leadership.

> In embarking on such an undertaking, the CPE Leads has an opportunity to build trust, and a culture of transparency and teamwork, by moving discussions about difficult topics out into the open.

So for each plan that is needed for a project, there is an opportunity to build what we call a microculture. The CPE Lead should establish the vision and goals for the entire CPE, and the values and rituals to be followed. Organizations will be transparent, cooperative, collectivistic, of their own choosing, not because it is required by contract or penalty, but because they chose to do so. If the CPE Lead builds such a microculture, she leads by example and inspires the desire to follow. This must be done by offering assistance without any expected payback. It needs to be a gift that does not need to be returned. If the CPE Lead does this, the gift will actually be repaid in a multitude of other ways.

The Internet has enabled information to be shared globally in a matter of seconds. The reputation of organizations can now be assessed easily online. Consider an organization that is predatory by culture and engages with others through lengthy unbalanced contracts, withholding of information, and self-interest. If your organization has an opposite set of values and must engage with them on a project, what do you do? Certainly, your organization should assess the damage that the project may suffer and include costs to avoid, mitigate, or remedy problems that could flow from them to you. In economic terms, organizations may increase prices based on one's reputation. Or, as we have seen, more aggressive organizations may underprice their work purposely to reap big benefits from changes and disputes.

WHY LEADERSHIP IS CRITICALLY IMPORTANT IN THE 21ST CENTURY

If you follow sites like TED.com, there are ample warnings of the accelerating trends in global business. According to a variety of speakers, four exabytes of unique data will be transmitted in 2019 which is estimated to be more than the previous 5,000 years combined. Facebook will be the third largest country, and as an educator, we are preparing students for jobs that do not yet exist.

The videos estimate that people entering the workforce in 2018 will have 14 or more jobs by the time they are 38. The speed of change is the point we would like to emphasize. Think of describing the Ganges River in the Himalayas; what you see now is not what you see at the end of a sentence. Training people with ideas published 5 years ago might as well as have been written in the last century.

Change is not easy for people and certainly not for organizations. A number of my doctoral students have researched change in organizations, and while the topic is somewhat complex, there are a few findings that bear on our topic here. It is essential to communicate the changes well in advance and to engage employees in the design and implementation. Especially if the organization has somewhat inflexible processes and procedures—banks are an easy example because of regulatory issues. Organizations that are the opposite, say an IT organization, often suffer from change fatigue. There is so much daily change that processes and procedures are impossible to maintain. In such circumstances, change is the process, and there must be a greater degree of freedom to adapt individually. Waterfall projects are more like the banks, agile projects more like the opposite.

Change that comes from the flood of information from scholarly publications, blogs, podcasts, news, global organizations, tweets, emails, Facebook is also staggering. And it has pushed us to further compartmentalize expertise. For example, we asked our primary care physician how much time it takes him to keep current in his reading. He tells us it would take him 6 h/day to read just the more well known. That is not possible for him, so he must rely on specialists to keep current in areas of medicine, while he only looks at general information. The same is the case in most all disciplines in today's global markets. Organizations have increasingly complex value chains that supply the expertise and resources to make products and services available to global markets.

As with make-buy decisions in the past, organizations must decide what fundamental capacity they must have to run their business and to minimize the number of people on payroll. Take an easy example of shipping. A shipping company needs ships, fuel, cargo, deckhands, captains, customs support export and import ends, financial services for payments, bill of lading support, a webpage, tax information, and much more. Organizations look at the required support needed, decide which are critical for operations, and then outsource the rest. What is outsourced requires more communications and results in less control. Outsourced organizations have different goals, long- and short-term needs, training programs, hiring criteria, financial needs (like cash flow), insurance, and much more. An organization that is outsourcing places itself at risk by ignoring these things.

A recent article in *The Economist* (The digitization of trade's paper trail may be at hand, 2018) said in part:

> Maersk, the world's biggest container-shipping line, found that a shipment of avocados from Mombasa to Rotterdam in 2014 entailed more than 200 communications involving 30 parties. A giant container vessel may be associated with hundreds of thousands of documents. "A Venetian merchant…would recognize some of our documentation," says John Laurens….

We have been shipping things for thousands of years and still use antiquated systems. But think about the difference between docking in Singapore or Mombasa. The facilities are wildly different, as are the technology, corruption, processes, governments, and much more. Developed economies had the benefit of the industrial revolution to assimilate technology and adapt to the changes it brings in all aspects of life. Undeveloped or developing countries may have completely skipped over 20th-century changes and jumped directly into the 21st century. Imagine trying to adjust to 100 years of change in a generation. And, imagine trying to do a project in such a location.

Another current favorite is to acquire some or all of these outsourced organizations. In 2009, we did some training in Germany for a merger. One organization was a conservative manufacturer with mature product lines and very mature individuals. They all sat on one side of the room in suits and ties, and took notes on paper. The other organization was an IT group, with fast-changing product lines. They sat together in jeans, T-shirts, and sneakers, and took notes on their iPads. The cultures of the organizations were like night and day, and we wondered how long it would take them to find a balance. Turns out it never happened. We were with the same group about 5 years later, and things had not changed. A few years after this second session, the merger was ended. The literature is rife with examples of how difficult it is to merge two companies. Running one is difficult enough. So, imagine what a CPE is like where the CPE Lead is attempting to marry up multiple organizations with multiple cultures, for a one time endeavor.

Technology changes exponentially faster than we can evolve. We are now able to use CRISPR to edit genes, change species, and edit Homo sapiens. Most scholars believe that with natural evolution, changes happen every 10,000 years, but even if they are badly wrong and it is 500 years, we can now edit our genetic makeup in one generation and avoid natural evolution. Children can now be tested for diseases and treated long before they become a problem. The thing is we have not yet evolved to the point that we can see the effects of this over generations. We are still short-term thinkers by design, and even trans-generational perspectives are hard for us. Some global organizations, while well along the path of adopting technology, are still struggling to get all of their systems integrated. Common with banking acquisitions, where legacy systems may be too difficult to integrate and too expensive to replace.

In today's developed economies, there are still big differences. In Estonia, the entire government is digital—everything. We were told by our students that the entire country is "backed up" in Switzerland in case of a hack. Compare that to the United States where the government is maybe 50%–60% digital and the rest paper. We still have paper voting ballots, tax forms, and a lot more. Yet, India is ahead of the United States in electronic identification for citizens, so just because the country is developing does not mean it is behind. In the United States, many people still cannot vote because they have no acceptable ID. And in my state of Florida, news of voting problems sound like a new broadcast in Uganda.

In the 21st century, there are still many governments that do not provide for their citizens, have dictators at their head, do not follow the rule of law, or still persecute different groups of people—many. If an organization needs to do work in the Democratic Republic of Congo, one such place, would it be best to invest in building hard assets, or hire a local organization to do this? Most would hire locals. The risk of the government extracting bribes for customs, or wanting to be hired as a member of the board of directors, or countless other means of extracting money for services is against the law in many countries like the United States and Germany. As a Western organization, say in the United States, with laws about such things, it becomes a real challenge to do business in such locations. It may not be feasible to walk away from entire markets, so partners are needed.

There are two major accounting systems, around eight legal systems, and well over a hundred monetary systems on this planet. If you look back at Figure 6.1, you can see many of them. Institutions such as the United Nations, World Bank, International Monetary System, and World Court (International Court of Justice) were created to provide a semblance of consistency across the diverse systems in the world, but their power and acceptance are not consistent across countries. In the 21st century, there is still no recognized law that regulates the Internet, genetics, governance, environment, and immigration. We do not mean to imply that it is lawless out there, but that it is complicated and risky. The work that Dr. Arroyo (Arroyo & Grisham, 2017) and I did on extreme projects lays out the range of risks that need to be addressed and provides suggestions on how to minimize the exposure organizations face.

Consider the expertise one would need to just determine the legal requirements of a complex project, or how to pay everyone in their currency, or how to get proprietary technical expertise across borders. These are things that the PMBOK does not address. It was written for a different time, with different rules, different conditions, and different technology. In a global environment, leaders need to have the skills that we summarized in Figure 6.2. For example, when we worked in Saudi Arabia, bribing customs officials was standard practice. Working for a US firm, we operated under the Foreign Corrupt Practices Act, which forbids US organizations from bribing foreign officials. At that time, our German competitor could pay bribes and take tax deductions as a business expense. Fast forward today and the German Strafgesetzbuch forbids paying bribes. As we said a number of times, change is normal in the 21st century, and it requires people to stay current on many topics, some seemingly unrelated to the project itself.

Back when oil prices were quite high, we were doing training in Trinidad & Tobago. The country was awash in capital, and everyone was building. The training was for Project Management, and we were discussing risk. One of the students described a problem that his company faced while building a new high rise. They could not get limestone for concrete. With the large demand, limestone had to be barged in from the United States at eye-watering prices. We asked how long it took the student's organization to plan and design and acquire the permits—years was the answer. Oil prices are very well published, yet his organization neglected to look at the broader economy in assessing their risks. Not to suggest that they could have been precise in assessing the impact, but they certainly could have anticipated a problem. And that is the point, anticipating market conditions needs to be part of a CPE Lead's portfolio.

On a project in Thailand, we had an important customer, who was very unhappy. We were asked to lead the project and repair the relationship. Clearly, the first order of business was to determine the problem. As we engaged with the customer, we discovered that the arrogance and disrespect shown by our organization was at the root of the problems. Our organization had left the customer with the impression that they were a group of jungle people, and needed to be Westernized. That took some time to correct and is beyond the scope of this chapter. The more superficial problem was that the customer wanted to unpack shipping containers of our product, inventory the contents, and warehouse the parts. Trouble was that there were missing packing lists, different formats, different languages, different levels of detail, and different technical descriptions. It was a global value chain issue produced by having sourced product in dozens of countries. When we put Western experts on the task, they were unable to sort it out as well. We needed to fix the global procurement system for a very large organization. You probably guess that did not happen, and we spent large piles of money to remedy the situation on the ground. Had the organization spent the time to consider the impact of a global value chain, we could have built a work-around for a fraction of the cost and without damaging the relationship.

Reputation is a serious concern in this century. Think of the last time you bought something without a follow-up survey or wanted to know something about a potential service provider. The Internet enables people around the world to find out about you and your organization, in seconds. Imagine that you have a person in Cambodia interested in finding a global partner who is "flexible" when it comes to bribes and corruption. You can find them with some work on the web or with less work on the dark web. Or if you are a Norwegian organization that wants to find a partner in Niger who has Nordic European values. Again, you can find them.

During a recent course in Tokyo, a student asked if the idea of quality is different in different cultures. We told him it was and that unlike Germany and Japan where precision is cherished, most cultures and organizations are comfortable with accuracy. If you live in a place where potable water is hard to find, one does not really care if the protective cap is perfect. And, in cultures where people work to live, precision is an unnecessary distraction—kit will be accepted if it does not require added effort. We also explained that in the 21st century, it is very possible for organizations to have thousands of organizations in their supply chains. This means that the supply chain (the children) actions are directly attributed to the contracting organization by the customer. So if your great-great-great grandson provides a product or service that is not even accurate, the father and mother are held to account.

ISO (International Standards Organization) describes quality as "…meeting the implied and stated needs of the customer…." Our question for the student was how would the great-great-great grandson know what the implied needs of your customer are? The burden of course is that the father and mother must communicate this throughout the value chain. The responsibility for quality is with the parent, certainly from the customer's point of view. Quality is determined by the customer, and it may be very different from that of your organization. And, it manifests itself at all touch-points with the customer, from reception, to emails, to meetings, to the value chain, to accounting, to the product itself, and to the Project Management services provided. This is another place where the CPE Lead must provide leadership.

In the 20th century, a business would last on average about 30–50 years—in the 21st century, it is about 14 years. Artificial Intelligence (AI) and machine learning now enable robots to more accurately read X-ray films than physicians with 20+ years of experience. The AI algorithms for the X-ray test were written by IT people, not physicians. AI can process masses of data and learn so quickly that it is somewhat indifferent to disciplinary expertise. Currently, AI uses human neurology as a model, and human neurology is being reconsidered based upon what we learn from developing AI. Experts disagree on the timing for the arrival of AI mainstream, but most seem to be as early as 2023, or as late as 2030. So, people who graduate in 2023 will not have been trained on fully developed AI, but will be interfacing with it fully before they are 30 years old.

Many experts believe that AI will not be able to replace human capabilities like management or leadership, or emotions before 2050 and perhaps not by then. So conservatively let us assume by the end of the century. That means that starting with the people who will graduate university in 2023, three or more generations will work in a world where people are still needed to do some things, like leadership. And, those people will provide the skills needed to train AI to do leadership. With that heads-up, let us look more closely at the future of projects.

WHAT ABOUT THE FUTURE?

Think about the time you spend on a project estimating, scheduling, planning, communicating data, assessing quality, and then updating it all. This is the sort of work that AI can perform far more quickly, accurately, and efficiently than can humans. Consider estimating the amount of effort that is necessary to do a design for a roadway, or write and test code, or develop a database for medical records, or send a man to the moon. Most of us have worked with person-hours and perhaps Earned Value (EV) for the human effort, because hours are more easily measured. But some tasks such as shipping, or communications, or testing, or setting steel beams in place for a high-rise building do not lend themselves to comparisons with person-hours. What if there was a database of calories consumed for various tasks (being a common measure of energy consumed)? It is very easy to crunch the numbers about progress and efficiency, and relate them back to currency.

Using calories, all tasks can be compared, at a level of detail that is subject to little uncertainty. Imagine progress updates where one knows, exactly, what effort has been expended and the level of productivity. The numbers are indifferent to geographic location, discipline, culture, or training. Quick comparisons to large databases would easily show where the productivity is low, compared to benchmarks. We will know the efficiency of different cultures and different disciplines, in different countries for multiple years. Soon, databases with more information than has been created since mankind began recording things can be statistically assessed in microseconds. That data can then be analyzed in conjunction with cultural, market, technical, and political data to adjust it to current conditions in a few more microseconds.

With such detailed data, an organization can use financial algorithms to determine what labor in what location can provide the optimum financial benefit at any moment of a project. It enables such things as currency swaps for payments to be optimized through other algorithms running in the financial and book-keeping systems. A CPE Lead (Project Manager will become a remnant of the past) can determine the progress within a few seconds for a multimillion-dollar global enterprise, from her sailboat. AI will keep track of everything that is occurring real time. No need for paper, or reports, or schedule updates to be done by a Project Manager. All data will be available 24/7 at any time desired, through the cloud. So, the Executive for the customer can get a summarized dashboard while having a martini on a flight from Singapore to New York.

For a US$1.6 billion (2018 dollars) project in Thailand, it took six experienced people and a Project Leader a month to build a preliminary schedule and work breakdown structure (WBS). As preliminary work was ongoing concurrently, by the time this planning was completed, it was out of date and needed to be refreshed. On a similar size project in India, the constructor had a team of three people whose only job was to prepare monthly reports. We have no clinical data to support this, but from our experience, scheduling, estimating, risk assessment, quality control, and planning represent at least 50% of what a CPE Lead does in 2018. Add to this the communications of information, updates, and answering questions about the work, and most all of a CPE Lead's time is used up. Then, there are stakeholders to consider politics, market changes, technology changes, and competition changes to address.

As a result of the mass of details on complex projects, CPE Leads must spend a majority of their time dealing with details that are continually changing. This is a perfect place for AI, Augmented Reality (AR), and robots to help out. Soon, there will be an app for projects that will fit comfortably onto your iPhone, which will connect to the cloud and the Project Central Page. All of the organizations involved in the project will be able to view

> As a result of the mass of details on complex projects, CPE Leads must spend a majority of their time dealing with details that are continually changing. This is a perfect place for AI, AR, and robots to help out.

all of the project information for all tasks, at all times. Each organization posts its data each minute as it is obtained by the AI systems that collect and analyze it. For example, the CPE Lead wants to know the progress of wire pulling on the project. The lineal footage of the cable being removed from a spool would be continuously monitored, and the video images of the work analyzed for unusual or unnecessary activity, and the graphic displayed immediately. Then, the progress would be compared to plan, what is needed to recover lost time, and the communications through the CPE back to the person pulling the wire in a few microseconds.

Each organization would have a secure area on the Project Central Page so that it could conduct confidential business, and deal with proprietary information. They could then upload public information required to the public area from a predetermined checklist that is downloaded and supervised by the AI system. "Mr. Khan, it is time to upload your daily report," says Alexa. All the data would be automatic. Real-time video would be available of work locations or computer screens for such things as architecture or code or testing. Each organization will be able to create their own dashboards, reports, and analysis features from a central list of options. Each option would have available a tutorial on how to use it, and would be provided with 24 h support. Each Project Central Page will be able to be viewed in the languages of the project. So if there are Japanese, Chinese, Vietnamese, Indian, German, Egyptian, Brazilian, and US organizations, each can retrieve the information or post the information in their mother tongue, and it will automatically be translated into the other languages of the project.

The value chain will use new versions of the QR code on all products. From a dump truck filled with rock to a chip, to a few hundred lines of code, and to meeting minutes, the QR code would be attached to each WBS activity. Remember, WBS activities could easily be 30 min of work as level of detail is really not important as it is all produced, measured, and analyzed through AI systems. The QR code can be read through the Project Portable Portal app and would then provide all of the details associated with that WBS activity. For things like high-pressure piping for steam, it would provide the details of where it was manufactured, by whom, when it was shipped, who shipped it, where and when it was unloaded, which organization is responsible for installation and testing, what the current status is, etc.

Meetings would be video-recorded and the videos available to those involved or effected. The meetings could be virtual or face-to-face, or some combination. The Project Central Page would adjust the timing for the meetings to minimize the inconvenience across time zones and to modify future meeting times to spread the inconvenience around. No more meetings that question about progress, test results, cost, risk, contingency, unfulfilled requests, or finger-pointing about who has the ball. The system will produce the entire record of discussions by pulling from emails, searches, and more. No more time wasted arguing over occurrences and facts. It is all tracked.

You may be wondering at this point, what will our people do? Well, what we do well and AI will not be able to do for 30–40 years, see a project holistically. AI will not be able to empathize, or be creative, or be ethical, or understand emotions. As human beings, we will not have evolved out of the need for human contact. What we need is the time for these things and that is currently being used up on the details. Einstein once said that creativity is the residue of time wasted. The problem in 2018 is that people are way overstretched with the details to do these important things. AI, AR, and robots can do a far better job with the details.

In complex program environments, there are simply too many details for humans. There are multiple organizations, disciplines, governments, taxes, laws, and risks—the list is long. All must be dealt with, so currently, the issue is who should deal with each one and to what level, and then who should communicate the necessary information to whom. Currently, CPEs must decide where to put their effort and where to get support. The more people, the more complexity, and the more complexity, the more people—it is a self-perpetuating system. And, Homo sapiens have some difficulty with communications, even the most adroit of us.

To close, what we see around the globe, in all countries and organizations, is the overallocation of resources. People are playing whack-a-mole in a reactionary mode, crisis by crisis. Studies show people have 2 min of uninterrupted time during a work day, which spans up to 10 h/day and 6 days/week in many cultures. Productivity is abysmal, creativity is absent, tempers are short, stress is high, and people are not happy. There is a lot of attention and emphasis on reducing costs, but most often, it seems focused on short-term profits. Our hope is that technology comes to the rescue soon and to give people a break.

CONCLUSION

In the early 21st century, think back to the early 20th century and the industrial revolution. Then, think how far developed economies came in 100 years. Now, imagine we are currently in the same place as far as technological development, and we will cover the same improvements in 30 years. That is the world that CPE Leads occupy. There are still about 2 billion people living on less than US$1 per day, and in places like India, 50% of the population does not have access to clean water, education, power, and all the rest. Doing a project in conditions where there are no benchmarks, guide posts, or stability and providing leadership to others, how does one prepare to do that?

In our view, it requires the skills and traits that we have sketched out here. The Internet is both a blessing and a curse, for it provides access to unlimited information, but not access to knowledge. It enables quick communications and explicit information, but less tacit knowledge. It permits instantaneous global visibility and opinions posing as facts, foments strife, and intolerance, and helps people educate themselves and doctors treat patients in remote locations. Our idea is like an unbroken horse, wild and untamed. We are certain that the future is filled with rapid continual change, and incredible challenges to Homo sapiens. We may not have evolved fast enough to keep pace with the technological change.

Yet, we believe that leadership is our hope to navigate these changing times. And, since Project Management has always been a necessity for mankind in undertaking complex endeavors, it is the place that needs leadership now. Processes and technology are necessary to provide time to lead, they are only that. We hope that we have convinced you of the importance to lead in our brief work here.

QUESTIONS TO PONDER

How much should a CPE Lead feel responsible for when implementing a project in a developing country likes social well-being, filling the absence of infrastructure, education of the local population? Is it your responsibility to act like a government?

What do you do when in a society where class due to economic segregation or caste discrimination is prevalent? Is it your job to correct it, modify it, ignore it, pretend it does not exist? What might you do?

Leadership is the ability to inspire the desire to follow. How does one do this? Were Mao and Hitler leaders?

Nordic domiciled. Try this "You are the CPE Lead for a Nordic Domiciled organization doing a project in China. You are surprised at how lax the laws regarding water pollution are. You can greatly increase profits by following Chinese law and are encouraged to do so by the Chinese organizations. But, the laws in the EU are much more stringent, and if you follow them, the Nordic partners will be more comfortable, your organization will make less, and the Chinese partners will be very unhappy. Take the lead, what do you do?"

REFERENCES

Arroyo, A., & Grisham, T. (2017). *Leading Extreme Projects - Strategy, Risk and Resilience in Practice.* London: Gower.

Arroyo, A., & Grisham, T. (2018). *CPE 2050 The Gower Handbook of Project Performance.* Pending: Gower.

Barraza, J. A., & Zak, P. J. (2009). Empathy toward strangers triggers oxytocin release and subsequent generosity. *Annals of the New York Academy of Sciences, 1167,* 182–189. doi: 10.1111/j.1749-6632.2009.04504.x.

Bartlett, C. A., & Ghoshal, S. (1998). *Managing Across Borders - The Transnational Solution.* Boston, MA: Harvard Business School Press.

Bass, B. M., & Avolio, B. J. (1994). *Improving Organisational Effectiveness through Transformational Leadership.* London: Sage Publications.

Bass, B. M., & Stogdill, R. M. (1990). *Bass & Stogdill's Handbook of Leadership: Theory, Research, and Managerial Applications* (3rd ed.). New York; London: Free Press; Collier Macmillan.

Cavaleri, S., & Reed, F. (2008). Leading dynamically complex projects. *International Journal of Managing Projects in Business, 1*(1), 71–87.

Corno, F., Reinmoeller, P., & Nonaka, I. (1999). Knowledge creation within industrial systems. *Journal of Management & Governance, 3*(4), 379–394.

Csikszentmihalyi, M. (2001). The context of creativity. In W. Bennis, G. M. Spreitzer, & T. G. Cummings (Eds.), *The Future of Leadership - Today's Top Leadership Thinkers Speak to Tomorrow's Leaders* (pp. 14–25). San Francisco, CA: Jossey-Bass.

Denning, S. (2005). *The Leader's Guide to Storytelling - Mastering the art and Discipline of Business Narrative.* San Francisco, CA: Jossey-Bass.

DeVaro, J., & Gürtler, O. (2016). Strategic shirking: A theoretical analysis of multitasking and specialization. *International Economic Review, 57*(2), 507–532. doi: 10.1111/iere.12166.

Earley, C. P., & Ang, S. (2003). *Cultural Intelligence, Individual Interactions Across Cultures.* Stanford, CA: Stanford University Press.

Earley, P. C., & Singh, H. (2000). *Innovations in International and Cross-Cultural Management.* Thousand Oaks: Sage Publications.

Fernie, S., Green, S. D., Weller, S. J., & Newcombe, R. (2003). Knowledge sharing: Context, confusion and controversy. *International Journal of Project Management, 21*(3), 177–187.

Goldsmith, M., Greenberg, C. L., Robertson, A., & Hu-Chan, M. (2003). *Global Leadership - The Next Generation.* Upper Saddle River: Prentice Hall.

Goleman, D. (1998). The emotional intelligence of leaders. *Leader to Leader, 1998*(10), 20–26.

Goleman, D., Boyatzis, R., & McKee, A. (2002). *Primal Leadership: Realizing the Power of Emotional Intelligence.* Boston, MA: Harvard University Press.

Greenleaf, R. (1997). *Servant Leadership.* New York: Paulist Press.

Grisham, T. (2009a). *Cross-Cultural Leadership (XLQ).* Saarbrücken: VDM Verlag.

Grisham, T. (2009b). *International Project Management - Leadership in Complex Environments.* Hoboken, NJ: John Wiley & Sons.

Grisham, T. (2016). Leading Agile Projects. Kindle eBook Amazon. ASIN: B01N5FSWHT.

Heil, G., Bennis, W., & Stephens, D. C. (2000). *Douglas McGregor, Revisited.* New York: John Wiley & Sons.

Hofstede, G. (2001). *Culture's Consequence*: Thousand Oaks, CA: Sage Publications.

House, R. J., & Javidan, M. (2004). Overview of GLOBE. In R. J. House, P. J. Hanges, M. Javidan, P. W. Dorfman, & V. Gupta (Eds.), *Culture, Leadership, and Organizations - The GLOBE Study of 62 Societies*. Thousand Oaks, CA: Sage Publications.

Iacoboni, M., Molnar-Szakacs, I., Gallese, V., Buccino, G., Mazziotta, J. C., & Rizzolatti, G. (2005). Grasping the intentions of others with one's own mirror neuron system. *PLoS Biology, 3*(3), e79.

Kerzner, H. (2006). *Project Management A Systems Approach to Planning, Scheduling, and Controlling* (9th ed.). Hoboken, NJ: John Wiley & Sons.

Maani, K. E., & Maharaj, V. (2004). Links between systems thinking and complex decision making. *System Dynamics Review, 20*(1), 21–48.

Marquardt, M. J., & Berger, N. O. (2000). *Global Leaders for the Twenty-First Century*. Albany: State University of New York Press.

Nonaka, I., & Teece, D. (2001). *Managing Industrial Knowledge - Creation, Transfer and Utilization*. London: Sage Publications.

Pheng, L. S. (1995). Lao Tzu's Tao Te Ching and its relevance to project leadership in construction. *International Journal of Project Management, 13*(5), 295–302.

Potter, M., Gordon, S., & Hamer, P. (2004). The nominal group technique: A useful consensus methodology in physiotherapy research. *New Zealand Journal of Physiotherapy, 32*(3), 126–130.

Sendjaya, S., Sarros, J. C., & Santora, J. C. (2008). Defining and measuring servant leadership behaviour in organizations. *Journal of Management Studies, 45*(2), 402–424.

Stewart, T. A. (2001). Trust me on this - Organizational support for trust in a world without hierarchies. In W. Bennis, G. M. Spreitzer, & T. G. Cummings (Eds.), *The Future of Leadership - Today's Top Leadership Thinkers Speak to Tomorrow's Leaders* (pp. 14–25). San Francisco, CA: Jossey-Bass.

Thompson, L. (1998). *The Mind and Heart of the Negotiator*. London: Prentice Hall.

Trompenaars, F. (1993). *Riding the Waves of Culture: Understanding Cultural Diversity in Business*. London: Economics Books.

Yukl, G. (1998). *Leadership in Organisations* (4th ed.). Sydney: Prentice-Hall.

7 Systemic Negotiation and Conflict Resolution Toolset

Barbara Rapaport
The University of Sydney

Kaye Remington
The University of Sydney

Alon Ben-Meir
New York University

CONTENTS

INTRODUCTION

The project management domain has traditionally explored technical and administrative aspects of project management (Crawford, Morris, Thomas, & Winter, 2006). The constantly evolving field of project management, however, prompts project managers to respond to the changes in the environment by expanding its knowledge to accommodate for systemic negotiations. From this perspective, project managers need not only focus on acquiring technical skills but also consider gaining knowledge and building their core competencies in the areas of motivation, stakeholder relationships, culture, systems of belief and values, learning processes and the transfer of knowledge (Cicmil & Hodgson, 2006; Cicmil, 2006).

This chapter begins with a brief overview of conventional approaches to negotiations. It then proceeds to explore concepts of emotional intelligence, leverage and Soft Systems Methodology (SSM). These concepts are then applied to a case study – the hydrocarbon exploration projects in the Eastern Mediterranean region.

CONVENTIONAL (DETERMINISTIC) APPROACHES TO NEGOTIATIONS

Depending on the complexity of the project, project managers may employ different negotiating approaches throughout the life cycle of the project. It is, however, important to emphasize that there is a distinct difference between conventional negotiations settings and complex negotiations.

Predominantly, the rationale underpinning conventional approaches to negotiations is that the outcomes can be clearly predefined, and processes are sequential, provided that the mindset of the counterparties is fully established and revisited throughout the negotiations where project management processes, procedures and tasks follow logically, given the right set of instructions. These contexts benefit from approaches, such as the principled negotiations approach by Fisher, Ury, and Patton (1991), which predominantly relies on facts, logic and figures and prompts the conflicting stakeholders to focus on resolving a defined problem. The underlying assumption of this approach is that the stakeholders will ultimately behave rationally, when confronted with facts and evidence.

Recognition of each side that the other side has a better alternative to the current negotiated one is referred to as BATNA (best alternative to a negotiated agreement) (Fisher, Ury, & Patton, 1991). This concept has an origin in game theory and has been extensively analysed by the scientific community. The main idea is to identify alternative courses of action (Sebenius, 2017); however, each party is prompted not to take a course of action or agree on a deal that does not serve his or her best interests. The negotiators are encouraged to understand the 'minimal acceptable set of terms in negotiations'; that is in a buy–sell deal, the BATNA refers to the minimum (the least attractive option) and maximum (the most attractive option) price the buyer or seller is ready to accept (Sebenius, 2017, p.2). Understanding the options at hand and knowing that you can walk away from the deal at any given time as you understand the alternatives places you in a strong position versus the opponent. It also important to get other stakeholders on board to increase the perception of your case in the eyes of the negotiating side. The main question when applying the BATNA approach would be - what is your best alternative to a negotiated agreement? (Sebenius, 2017). Cooperation between the conflicting sides is highly desirable pre and post settlement, otherwise, feelings of resentment and mistrust may take over and interfere in mutual cooperation (Fisher, Ury, & Patton, 1991), which of necessity requires a mutually gainful outcome.

Whilst there is an increased emphasis on stakeholder management in the most recent edition of the PMBoK Guide (PMI, 2017), there appears to be little change to recommended approaches for conflict negotiation, from those discussed in papers presented to Project Management Institute (PMI) conferences by Guan (2007) and Englund (2010). These authors consistently focus on the following approaches (Table 7.1).

TABLE 7.1
Summary of Negotiation Approaches

Resolution Type	Common Result
Forcing	Win–lose
Smoothing	Yield–lose
Withdrawing	Lose–leave
Compromising	Moderate lose–moderate lose
Problem solving or confrontation	Integrative

Source: Guan (2007).

According to Guan (2007), the most effective approach is the *problem-solving* or *confrontation* approach. This approach strongly coincides with the accepted, conventional mode of negotiations, which is based on maximizing gains, minimizing losses or focussing on yielding absolute equality of the outcomes (interest-based negotiations). Like the other approaches, described above, this approach is fundamentally deterministic, emphasizing win–win solutions (Leahy, 2011).

As this chapter will illustrate, the logic behind these deterministic approaches does not address, in any meaningful way, egocentric biases and emotional dissonance that characterize complex negotiations (Aquilar & Galluccio, 2008; Babcock & Loewenstein, 1997; Babcock, Loewenstein, Issacharoff, & Camerer, 1995; Bazerman, Tenbrunsel, & Wade-Benzoni, 1998; Forgas, 1998).

Conventional approaches to negotiations need to be redefined for complex environments as projects are fuelled with high levels of uncertainty, unpredictability and dynamicity (Atkinson, Crawford, & Ward, 2006). Recent research in this field clearly indicates that the forces in complex environments contribute to disintegration of stakeholders' vision and lead to an intractable state of affairs, hostility and conflict between the parties (Rapaport, 2016). Therefore, complex projects, being enacted in uncertain and dynamic environments, require different, often non-conventional approaches to negotiations.

The prevailing models on negotiations, which employ rational and deterministic approaches, are designed to work in straightforward projects. In complex, contested environments, conventional conflict and negotiation models fail to provide project managers with effective tools for negotiation. The purpose of this chapter is to explore the nonconventional approaches to negotiations in which facilitators and key stakeholders orchestrate a multilevel agreement. The approaches expand over the use of leverage, the application of SSM and the employment of emotional intelligence. The following chapter will provide a framework of tools for practitioners to utilize in negotiating intractable disputes, where no clear solution can be agreed upon or envisaged by the key parties involved. That said, compelling interest-based negotiations can overcome complex issues (especially political and emotional) as long as the stakeholders can envisage clear benefits that outweigh such difficulties. The negotiating process, under such circumstances, requires more time while undertaking confidence-building measures in which the parties develop a continuing vested interest.

> In complex, contested environments, conventional conflict and negotiation models fail to provide project managers with effective tools for negotiation.

COMPLEX SYSTEMS AND THEIR CHARACTERISTICS

The field of complex systems has gained popularity in social sciences (Urry, 2003), where it has been used to explore how various elements within the system, through their emergence and interaction, give rise to the patterns and collective behaviour of the system as a whole (Rapaport, 2016;

Bar-Yam, 2002). This field of study also explores how systems adapt, survive and connect in constantly changing environments (Checkland, 2011).

The concept of a 'system' was coined by biologist Ludwig von Bertalanffy in the late 1940s. It was initially associated with the biological organism, which was predominantly characterized by metabolism and self-production (Checkland, 2011). Later, the meaning of a biological system, which in essence is a 'hard' system, has been extended and applied to many other fields (von Bertalanffy, 1968; Gray & Rizzo, 1973) including social problems (Checkland, 1983, 2001).

The core nature of a complex system is its unpredictability, uncertainty, dynamicity and lack of a centralized and hierarchical structure (Urry, 2003). Due to these key characteristics, complex systems are more susceptible to strong interactions of elements within the system, resulting in emergence of new patterns and subsystems, where the component elements cannot be disintegrated due to the established entanglement and interconnectivity with each other (Rapaport, 2016; Richardson, 1984; Urry, 2003).

Complexity can be found in biological, social, economic and political systems (Auyang, 1999). For example, planet Earth is a complex biological system, consisting of subsystems such as the biosphere, atmosphere and lithosphere with continuous interactions of matter and energy among them (Marwan, Donges, Zou, Donner, & Kurths, 2009; Richardson, 1984). The socioeconomic systems consist of citizens, households, governments, organizations, businesses and marketplace platforms with ongoing commodity trading, investments and transactions (Ostrom, 2010). An example of a complex socioeconomic and political system is the exploration of hydrocarbon reservoirs in the Eastern Mediterranean region (Rapaport, 2016), as it interconnected with regional political economies, social systems of the neighbouring countries, ethnic territorial disintegration, religious extremist agendas, historical and international dimensions (Cilliers, 1998; Rapaport, 2016).

CHARACTERISTICS OF A COMPLEX PROBLEM

Individuals have different depths of understanding, levels of perception as to the world around them. Each of these levels of perception are further tempered by emotional state, moral circumstances and deeply held belief systems. Each and every perception contributes in some way to the larger whole. Given the vast variety of modes of individual understanding and expression (thought, speech or action) fuelled by indeterminate positions, it can be challenging to explicitly articulate the nature of the problem. Also stakeholders and stakeholder groups may hold multiple positions, which may be conflicting.

Complex problems are interdependent and multidimensional, with multiple causes, relationships, conflicting objectives and goals. Due to intertwinedness of the issues, any attempt to provide a solution will likely to alter and change the dynamics of the system in unpredictable ways, often leading to unintended consequences. It is, therefore, highly unlikely that a perfectly clear or definitive solution to a complex problem will be found. Any proposed solution can only be classified as better or worse than the current state (Cilliers, 1998; Rapaport, 2016).

> Complex problems are interdependent and multidimensional, with multiple causes, relationships, conflicting objectives and goals.

Case Study: Oil and Gas Exploration in the Eastern Mediterranean Region

A BRIEF HISTORY OF THE CYPRIOT CONFLICT

The Cypriot conflict, which is central to this study, has a long and complicated history (Michael, 2011; Papadakis, Peristianis, & Welz, 2006). The island of Cyprus was under the rule of the Ottoman Empire between 1571 and 1878 and then under the dominion of the British until 1960 (Pericleous, 2009). In 1960, a Treaty of Guarantee (Article IV, Treaty of Guarantee, 1960) was drawn up, naming Greece, Turkey and the United Kingdom of Great Britain as guarantor powers who reserve the right to intervene by taking action in re-establishing the state of affairs in Cyprus. Article I of the

Treaty prohibits Cyprus from participating in 'any political or economic union with other state', while Article II 'requires other parties to guarantee the independence, territorial integrity and security of Cyprus'. In 1974, Turkish intervention into Cyprus led to the de facto division of the island of Cyprus (Diez & Tocci, 2009).

Following the de facto division of Cyprus, Greek Cypriots fled to the south, while Turkish Cypriots to the north of the island (Fouskas & Tackie, 2009; Joseph, 1997). The division created geographical, ethnical, economic and political partition of the island (Varnava & Faustmann, 2009; Vassiliou, 2010), which gradually crystalized into the Republic of Cyprus and the Turkish Republic of Northern Cyprus (TRNC). The Republic of Cyprus has been internationally recognized, while the TRNC has never been given such recognition (Papadakis, Peristianis, & Welz, 2006).

Given the long-standing conflict between the Greek Cypriots and the Turkish Cypriots, the elections in 2013 of Nicos Anastasiades, as President of the Republic of Cyprus, prompted the sides to resume political negotiations (Rapaport, 2016; Grigoriadis, 2014). Political negotiations resumed in 2014 with optimism that the exploration for natural gas in the Eastern Mediterranean might become a major catalyst of change, the stabilizing force and the leverage in negotiation process (Rapaport, 2016). Nevertheless, experts believe that 'the de facto division of Cyprus poses a risk of completion of any project, the legality of natural gas exploration by the Republic of Cyprus has been contested by Turkey' (Grigoriadis, 2014, p.127). The rights of Turkish Cypriots to energy resources have been contested by the Greek Cypriots.

THE EXPLORATION AND CONSTRUCTION OF A GAS PIPELINE IN EASTERN MEDITERRANEAN REGION

The discovery of large quantities of natural gas (hydrocarbons) in the seabed between Cyprus and Israel led to the proposed construction of a gas pipeline from the Eastern Mediterranean region to Europe. The project is being negotiated within the climate of a long-standing conflict between the Greek and Turkish Cypriots. It is understood that the immediate beneficiaries from the natural gas exploration are Cyprus and Israel, as they would (also) become energy exporters (Grigoriadis, 2014), while benefits for Turkey would be the transportation of the extracted hydrocarbons (natural gas) via pipeline to the European market and the world (Grigoriadis, 2014). However, an ongoing unresolved Cypriot conflict and dispute 'over the delineation of the EEZ (exclusive economic zone) and Turkey's frozen relations with Israel have deterred regional cooperation' (Grigoriadis, 2014, p. 124). Hence, discerning an interest-based approach in this particular case could change the dynamics of negotiations (Rapaport, 2016).

The initiation of the project would not be without its financial challenges. Beneficiaries need to consider 'the commercial viability of gas prices, project costs, regulatory stability and certainty, risks and the ability to ensure uninterrupted exports over the contract period' (Ellinas, 2016). Moreover, the high costs of offshore development prompt Israel to justify its investment and meet the self-consumption demands along the European market demands in the future. Cyprus, on the other hand, is looking to profit from the Aphrodite gas field given the low-price market conditions. The anticipated sales of the liquefied natural gas (LNG) could be challenging given Qatar's long-term contractual agreements and US Cheniere LNG exports to Europe (Ellinas, 2016).

For example, the agreement signed by Lebanon, with a consortium comprising France (Total), Russia (Novatek) and Italy (ENI), allows for these companies to explore gas territory claimed by Israel. Given the magnitude of the project and politico-economic ramifications for the region, this move is perceived by Israel to be provocative.

THE POTENTIAL PROJECT SCENARIOS AND THEIR IMPLICATIONS

1. The first potential project scenario is development of the natural-gas liquefaction facilities in Israel and the Republic of Cyprus. Israel's interest lies in building its own natural-gas facility

(LNG) onshore and offshore or a possible joint LNG project with the Republic of Cyprus on the Cypriot territory. The consortium aims at constructing a natural-gas pipeline by 2025 which would be between Israel, Italy, Cyprus and Greece. The pipeline would carry gas from Leviathan field off Israel and Aphrodite field off Cyprus. The project is in its pre-initiation phase as discussion are underway (Dunnahoe, 2018).

2. The second project scenario perceived to be an economically (export) viable option is the construction of a pipeline from Israel (from Leviathan) via Cyprus to Turkey. The pipeline network has been already developed by Turkey in collaboration with Russia, Azerbaijan and Iran. The transportation of the natural gas would be through a pipeline to Europe, while Turkey will consume it as well. However, without the comprehensive settlement on the Cyprus issue, in accordance with the provisions of UN Security Council resolutions, the project might not be successful (to note: the UN Security Council urges Greek and Turkish Cypriot sides to reach a comprehensive settlement based on bicommunal, bizonal federation with political equality).

3. The third possible scenario is the construction of an underwater pipeline from Israel and Cyprus to Crete and then to Greece. The cost of the project would likely be high; however, at the same time, it would consolidate strategic, economic and cultural relations between Greece and Cyprus and most likely to prompt an emerging partnership with Israel. The success of the project might be impacted by the market rationale, as Turkey's demand for the natural gas is much bigger than that of Greece.

EXPORT OPTIONS

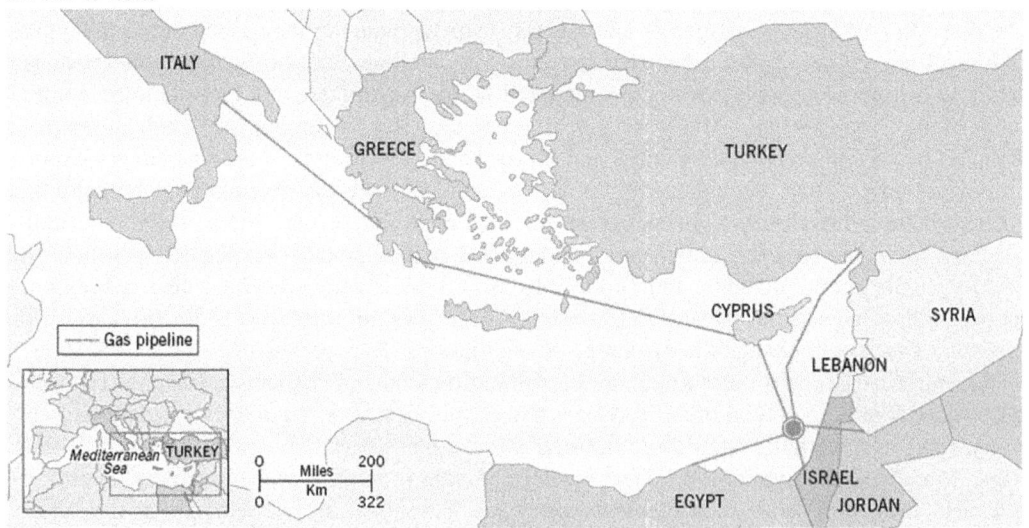

Source: Dunnahoe (2018).

Additional complicating factors relate to the internal politics of the Lebanon which serve to stagnate the development of gas. The development of the 'Gaza Marine field off the Gaza strip remains blocked by the de facto Israeli government action' (Ellinas, 2016, p.8). Terrorism in the region contributes to economic vulnerability, and shutting down the East Mediterranean Gas Company (EMG) pipe from Egypt to Israel contributed to Israel losing confidence in gas imports from Egypt, resulting in compensation to Israel Electric Corporation (arbitration decision by the International Chamber of Commerce).

NEGOTIATIONS IN COMPLEX PROJECT CONTEXTS

SYSTEMIC NEGOTIATION TOOLS

These conflicts are complex partially because there is an asymmetrical distribution of power between the key stakeholder groups. The asymmetrical power distribution between the negotiating parties may play a role in how negotiations are conducted (Zartman & Berman, 1982; Luterbacher & Norrlöf, 1999; Raiffa, 1982).

Leverage Points

There is increasing recognition of the importance of achieving transformational shift in a stagnated and conflict-ridden system, and therefore, the use of *leverage points* in negotiation is worth exploring (Meadows, 2008; Rapaport 2016). Leverage points are viewed as hypothetical places within a system (e.g. an organization, company, economy, conflict) which can alter the behavioural patterns of that system. The most prominent leverage points are considered to be *Paradigms: the mindset out of which the system arise*s (its goals, structure, rules, delays, parameters) and *Transcending Paradigms* (Meadows, 2008, pp. 147–165). Transcending paradigms provide the highest form of systems' intervention. The act of transcending paradigms involves the awakening of all the (human) senses: wisdom, insight and nonattachment to any of the existing worldviews (held by the stakeholders). Understanding leverage points in a system can have a powerful effect, assisting stakeholders in understanding the anomalies and failures of existing patterns in a given system (Meadows, 2008; Kuhn, 2009). The other top leverage points introduced by Meadows are rules, self-organization and goals:

Rules – Incentives, Punishments, Constraints

Establishing rules, laws, boundaries and the degrees of freedom of any given economy is paramount when identifying places for intervention. The presence of an ongoing Cypriot conflict enhances the complexity of the emerging disputes in the Eastern Mediterranean region and affects the economic viability of multiple economic systems. The regional conundrum introduces fragmentation of multiple economic systems by national boundaries. This further deepens and complicates the resolution of the Cypriot conflict and the future of the exploration of gas and construction of the pipeline. Many of the nation-state stakeholders involved in the international oil and gas conundrum focus singularly on their economic autonomy and policies that are bound to national boundaries. This promotes competitive advantage of one economy and potentially may have regional economic and political implications (Barnes & Ledebur, 1997). The dispute over the oil and gas enhances uncertainty of the region, contributes to regional disintegration and pulls the nation-states into potential international arbitration disputes (Morando & Joesten, 2005).

EXAMPLES OF RULES

Identifying direct and indirect benefits to those who are not directly associated with the project but who the main stakeholders need to satisfy (incentive)
Settling the border dispute, not easy but absolutely doable (boundary constraints)
Determining cost and benefit of reaching an agreement or the lack thereof (incentive)
Identifying different project scenarios that have positive regional implications (incentive).

Identifying direct and indirect benefits to those who are not directly associated with the project but who the main stakeholders need to satisfy.

Self-Organization – The Power to Add, Change or Evolve System Structure

In economies, this refers to technical advances or social revolution or self-organization. The ability to self-organize is a strong form of resilience, where political and economic structures and systems evolve and reinvent themselves. The system that is resilient is capable of evolving and self-evolving. It changes itself and develops new responses to unpredictable circumstances and situations. Self-organization is governed by the system itself and therefore is self-adaptable: it can add or subtract conditions. Self-organization is the power to change, and to put it into the context of the Cypriot conflict, this is something that the Greek and Turkish Cypriot sides need to understand. Any system that cannot evolve is doomed for failure.

EXAMPLE OF SELF-ORGANIZATION

Parties to self-organize to reach agreement on different conflicting issues unrelated to the big projects at hand to create goodwill (Ben-Meir, 2019).

Parties to self-organize to reach agreement on different conflicting issues unrelated to the big projects at hand to create goodwill.

Goals – The Purpose or Function of the System

The goal of the system is not less important than *self-organization* and *rules*. Turkey has set the goal of having independence in energy. However, Turkey's goal is not entirely shared by regional stakeholders who are also interested in the benefits of the hydrocarbon wealth (Kahraman, 2018). Given the notion of leverage, one may pose a question, what is the goal of nation-state stakeholders? What is the purpose of the system as a whole and what does it aim to achieve? Is the goal of the system to quantify the volume of hydrocarbons for the singular economy and reap the financial rewards? Or is the goal of the conflicting sides beyond the extrapolation of the hydrocarbon reserves? Can the nation-states redefine the goals to benefit all, without compromising the sovereignty of the states?

The authors emphasize other leverage points; however, the points outlined above are the most significant and relevant for the case study analysis.

EXAMPLES OF GOALS

Improve the bilateral political relations between Israel and Turkey
Develop and agree on side projects between the parties, such as military cooperation and sustainable development projects
Establish mechanism for rapid correction in case of unpredictable adverse development throughout the implementation of the project (Ben-Meir, 2019).

Case Study: How Understanding Leverage Points May Assist Negotiations

The current conflict between Cyprus and Turkey and the overall Eastern Mediterranean political dynamic has been and will continue to be an obstacle in any future negotiations. Leverage may assist in harmonizing disintegrated visions of the diverse regional forces mediated through deep appreciation of history. Leverage points in the process of systemic negotiations might include the rules and laws by which the nation-state stakeholders need to abide, politico-economic constraints, and understanding that the hydrocarbon reserves are an immense incentive for the Greek and Turkish Cypriot sides to solve the long-standing conflict (Morando & Joesten, 2005). Leverage is also about the ability for the negotiating sides to adapt and change throughout the course of negotiations as well as exhibit resilience and self-organization through change (Rapaport, 2016).

Establishing goals and purpose of the negotiated system and what the sides aim to achieve beyond the benefit of quantification of the financial rewards for the singular economy would be advantageous for the international community as a whole.

The most powerful leverage points are *paradigms* and *transcending paradigms*. This is the commitment of the stakeholders to achieving the ultimate harmonizer of the diverse socioeconomic and political forces and perseverance through the convulsions of history (Rapaport, 2016). However, even if all sides are committed to cooperate, they must not only be prepared to make the necessary concession to reach an agreement but must also engage in a process of reconciliation to mitigate distrust, and adjust positions in light of the concessions required, while collaborating on other issues or projects not necessarily related to the discovery of the sizable hydrocarbon reserves from which all sides can benefit (Ben-Meir, 2019).

Gradually, shared understanding of the mutual economic benefit, through appropriation and monetization of natural resources, derived by the stakeholders may overcome existing conflicting issues. In this case, however, the leverage that either side might have may not be sufficient alone to overcome such a deeply held animosity and distrust (Ben-Meir, 2019). Therefore, leverage needs to be considered in the combination of other systemic tools discussed in this chapter.

SOFT SYSTEMS METHODOLOGY (SSM)

The notion of different and changing world views combined with purposeful action underpins a comprehensive set of tools and approaches, known as Soft Systems Methodology or SSM (Checkland, 2011; Wang, Liu, & Mingers, 2015; Freeman & Reed, 1983). Developed over 20 years by Peter Checkland and others, SSM recognizes that before any attempt to solve a complex situation can be made, the problem situation must first be understood by all the key stakeholders. SSM assists in increasing appreciation of the problem situation between key groups of stakeholders. As argued earlier in this chapter, the complexity of many problem situations prevents clear definition of the real problems underpinning a conflict. SSM provides a framework for tackling such situations (Checkland & Scholes, 1990; Naughton, 1984; Eden & Radford, 1990).

SSM researchers and practitioners argue that problems are directly correlated to individuals' different interpretations of the world. Typically, problems do not exist in a vacuum, and solutions are merely intellectual constructs that may be held and believed by only some stakeholders. As a rule, problems are interrelated, interconnected and 'messy'. There are usually multiple problems in a complex situation. It is therefore important to consider the whole situation or as many of the interrelating systems as possible in order to comprehend underlying dimensions and root causes of problems. SSM practitioners also assert that improvements in systems are more likely to occur through perception sharing, persuasion and debate. Hence, facilitators should provoke interaction rather than take an expert position. Facilitators must also recognize that they are not divorced from the problem. Through interaction they become a part of it (Checkland & Scholes, 1990; Naughton, 1984; Eden & Radford, 1990).

Initial work on the problem situation might involve interviews and meetings to gain some understanding of the breadth of the problem situation and use of collaborative enquiry techniques, such as 'rich pictures'. Concepts of hierarchy, communication, control and emergent properties are used to identify systems and boundaries which might provide insight into the problem situation.

This section will describe in detail, one of the tools that forms part of the SSM suite of tools, 'rich pictures'. Developing collaborative 'rich pictures' captures the human activity of a system and defines the underlying reasons of a problem. This deceptively simple tool can be used over and over again to reveal subtle and unvoiced aspects of an issue. Since it is a pictorial tool, its effectiveness is not dependent upon language. Therefore, it is an ideal tool to be used in a negotiation setting.

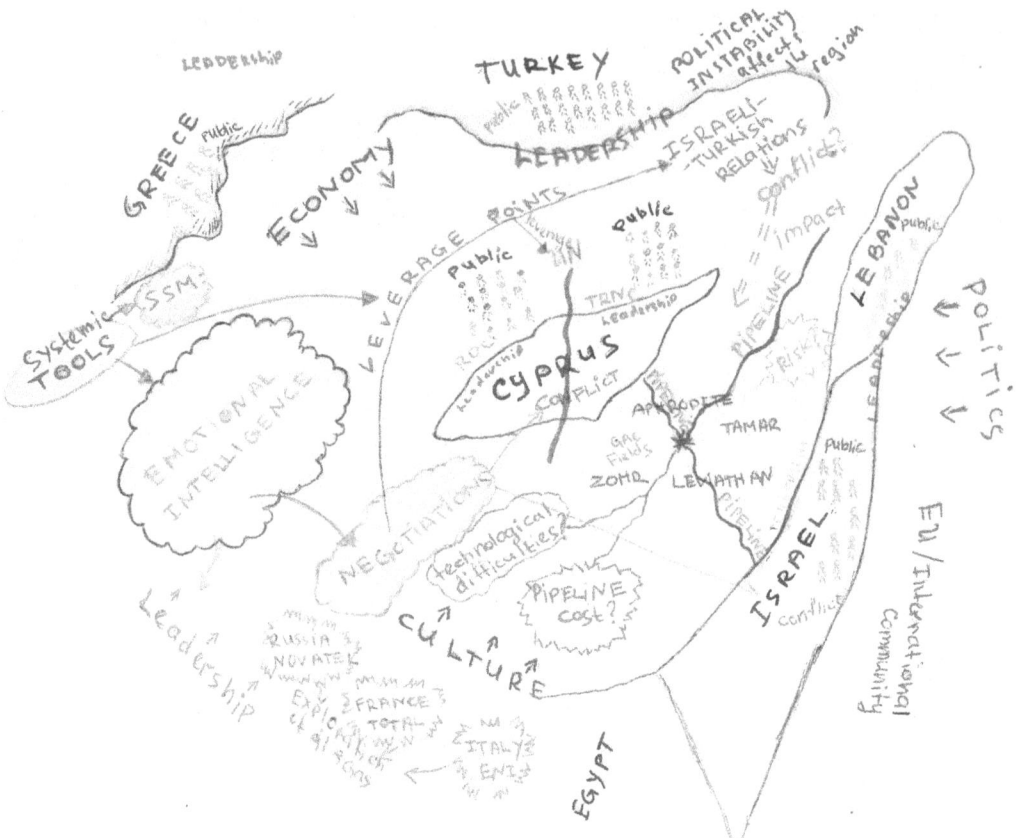

A 'Rich Picture' developed around conflict and systemic negotiations in Eastern Mediterranean region. (Image courtesy by Rapaport, 2019).

Monk and Howard (1998) describe rich pictures as a way of representing what we know about a messy situation – the issues, actors, problems, processes, relationships, conflicts and motivations. The use of this tool involves a group of people drawing pictures that best represent the stakeholders, their interaction as well as identifying any issues that may be of concern in the delivery of the project. A rich picture is most effectively used in the formulative stage of the project when the desire to remain in the creative mind set is most required. One of the very useful effects of this tool is its ability to provoke laughter among stakeholders who might otherwise be hostile. Laughter, which helps to break the tension, is invariably stimulated by the odd-looking drawings that people produce. Humour is now turned to creativity (Branagan, 2007; Eliav, Miron-Spektor, & Bear, 2017). This tool, when properly facilitated, leads to increased levels of ideation, which helps stakeholders to move from past focussed negativity to increased positivity and future thinking.

Some practical steps are needed to get the most out of this tool. A very large table should be set out, enough space for each participant to comfortably occupy a position, away from adversaries. Enough large sheets of paper should be taped together to allow each participant easy access to the paper. Taping the paper to a wall is not recommended as some people will decline to contribute if they must bend down or stretch. The question should be clearly written in the centre of the paper. Each person should be given a thick marker and encouraged to draw any thoughts they have about the question. Some will be reluctant to start, saying that they 'can't draw'. They must be gently encouraged. Participants should be allowed to draw in silence until they have all finished. Finally, participants are encouraged to explain their drawings, without interruption.

Because this tool promotes sharing of separate world views and increased levels of ideation, it is an ideal tool to use in conflicted situations, especially at the beginning of the project or at the beginning of a project phase. However, its simplicity means that it can be used whenever the project manager detects a level of conflict or lack of a shared vision.

Once a shared understanding of the problem situation has been gained, using a tool such as 'rich pictures', the relevant, active systems can be identified by constructing 'root definitions' which are then used to generate 'conceptual models' of the selected systems (Smyth & Checkland, 1976). Conceptual models representing contrasting viewpoints can become the basis of a debate, which through an appreciative enquiry process can lead to possible and agreed change and action. The mnemonic 'CATWOE' is often used to identify the perspectives that must be considered. CATWOE describes the separate world views of the C (customer), A (actor), T (transformation), W (Weltanschauung – worldview), (Owner), and E (environment), according to Armstrong (2018).

SSM AND THE CASE STUDY

Cyprus's attempts at exploring hydrocarbons in the Eastern Mediterranean region must be perceived as part of the bigger picture, solving the Cypriot conflict. Systems thinking requires the parties to understand that there is a layered multidimensional structure (Checkland, 2011). In the case of the gas pipeline project, the parties would need to find a way to adapt and respond to environments by evolving through more effective communication and control processes (Checkland, 2011). Obtaining a robust analysis of the complex entities as a whole is challenging. If a project manager working on any of the hydrocarbon projects in the region is to avoid failure, they must become fully aware that the project is a small part of a very large complex system. The application of SSM may assist practitioners to express and construct the problem and apply into the context of the real-case scenario (Armstrong, 2018).

RECOGNITION OF THE PART PLAYED BY EMOTIONS IN CONFLICT NEGOTIATION

Scholars have long focussed on the part played by *emotion* and the fact that human emotions play a vital role in human interactions and communications (Rapaport, 1942; Leeper, 1948; Schachtel, 1959). It is equally worthwhile recognizing the importance played by emotions in systemic negotiations.

Individuals experience emotional reactions in various ways (Storbeck & Clore, 2005). The same emotion can have a different effect on different people (Izard, 2013) and influence how muscles/facial muscles, circulatory and respiratory systems respond during various states of emotions (Rusalova, Izard, & Simonov, 1975). The 'neurophysiological systems and subsystems' are continuously engaged with emotions (Izard, 2013, p.46). Interestingly emotions and perceptions are interlinked, by way that emotions influence how human beings perceive the world around them (Izard, 2013). There are strong linkages between cognition and emotions, action and emotion and personality development. It therefore can be said that humans are complex entities and are far from easy to understand. At the same time, understanding, the extent possible, of the emotional intelligence of the counterparty encourages a more open dialogue with an enlarged sense of comfort, which is conductive to engagement with a greater potential for success.

It is also worthwhile noting that the history of a human kind and its prime emotion is entrenched in aggression. Aggression, as a factor, is viciously present and can be often detected in all human beings and not to be tailored to any one specific individual and/or complex situation or system. This very factor disturbs human relations in a civilized society and 'perpetually menaced with disintegration through this primary hostility of men towards one another' (Freud, 1963, p.86).

In project negotiations, emotional responses arise as social interactions through which certain patterns emerge (Izard, 2013; Clarke & Howell, 2009). As a rule, positive emotions support innovative strategies, agreements and positive outcomes (Carnevale & Isen, 1986; Kumar & Oliver, 1997) as well as flexibility in negotiations (Druckman & Broome, 1991), while negative emotions contribute to low levels of productivity and cooperation, increased aggressive behaviour (Bell & Baron, 1990) and negative outcomes (Allerd, 1999). Although there is a substantial literature on conflicts in projects (Li & Chen, 2006; Porter & Lilly, 1996; Tarr, 2007), there is limited research on how emotions play out and impact the project or project outcomes (Clarke & Howell, 2009).

EMOTIONAL INTELLIGENCE

Emotional intelligence is a term that has been associated with cognitive abilities emerging in early years of a human being to 'perceive emotion', 'to integrate emotion to facilitate thought' and 'to understand and manage emotions' (Salovey & Mayer, 1990; Mayer & Salovey, 1997; Mayer, Roberts, & Barsade, 2008). Goleman's thesis, for example, argues that emotional intelligence describes 'the capacity for recognizing our own feelings and those of others, for motivating ourselves and for managing emotions effectively in ourselves and others' (Goleman, 1998, p.317). High levels of emotional intelligence have been associated with team learning and knowledge transfer in projects. Emotional intelligence is also considered to be an important leadership attribute (Sosik & Megerin, 1999). The research suggests that leaders who are able to define and distinguish between various emotional states of their team members are more likely to succeed in an organization (Humphrey, 2002).

Leaders with vision who can evaluate, understand and manage their emotions, are the ones who are also able to emotionally connect with stakeholders and team members. Effective stakeholder management involves establishing deep connections, engendering trust, formulating sound decision-making processes and strategies, and gaining stakeholders' commitment to achieving positive results (Jones & George, 1998; Sosik & Megerian, 1999; Leban & Zulauf, 2004).

> Leaders with vision who can evaluate, understand and manage their emotions, are the ones who are also able to emotionally connect with stakeholders and team members.

The concept of emotional intelligence is strongly associated with projects (Butler & Chinowsky, 2006; Leban & Zulauf, 2004) and our understanding, as practitioners, of the part played by emotional intelligence in negotiations is limited (Turner & Muller, 2005).

EMOTIONAL INTELLIGENCE AND SYSTEMIC NEGOTIATIONS

In the case of hydrocarbon exploration, Eastern Mediterranean seabed, various stakeholders and interested sides have their own politico-economic vested interest/s in the natural resources. Others may strive to fulfil their economic interests and strengthen their politico-economic presence in the region and form economic allies or achieve an economic prosperity for the region. Understanding the various views and positions in an emerging conflict may assist the sides in moving forward.

From the standpoint of systemic negotiations, emotional intelligence serves to be a tool which may assist stakeholders and negotiators in understanding emotionally evocative interactions, interrelationships, (emotional) patterns of behaviour and perceptions of the opponent side. Systemic negotiations can offer a framework and be helpful in orchestrating and aligning stakeholders' conflicting interests for the purpose of achieving consensus and overcoming an impasse.

The hydrocarbon exploration is perceived by the Greek and Turkish Cypriots to be an impetus and catalyst in the long-standing conundrum. However, both sides must recognize that emotional baggage, born of long-standing cultural differences, contributes to preventing the sides reaching a comprehensive settlement (Rapaport, 2016). Transference and reinforcement of entrenched emotional positions to the hydrocarbon projects further generate stagnation.

Continuing lack of trust contributes to deterioration in strategic long-term partnership between Israel and Turkey and negative impact on hydrocarbon exploration in Eastern Mediterranean region. The question of trust which is lacking and may well endure as long as the current leadership is in power. This alone can torpedo any prospective negotiations, regardless of cost and benefit. Even if all sides are committed to cooperate, they must not only be prepared to make the necessary concession to reach an agreement but must also engage in a process of reconciliation to mitigate distrust and adjust positions in light of the concessions required, while collaborating on other issues or projects not necessarily related to the discovery of the sizable hydrocarbon reserves from which all sides can benefit (Rapaport, 2016).

It is important to distinguish between emotional intelligence in the context of negotiations between the stakeholders as states, versus emotional intelligence in negotiations between individuals who have personal stakes in the outcome. Negotiators who represent their country are governed differently in their negotiations in both style and substance than negotiators negotiating on behalf of themselves (Ben-Meir, 2019).

Although emotional intelligence, particularly on the part of the chief negotiators, plays a critical role in negotiations, the prevailing psychological dimension must be factored in. The psychological dimension is formed by complex historical narrative and experiences, and informed by current events. The problem in analysing this aspect is certainly subjective, something that cannot be negotiated unless it is preceded by confidence-building measures (Ben-Meir, 2019).

SUMMARY

The geopolitical environment of the Eastern Mediterranean region plays a paramount role in the global strategic politics. The discovery of the sizable hydrocarbon reserves generates vested interest from world leaders, as it significantly adds to the complexity of already intertwined international politics and existing disputes.

The exploration is also heavily embedded within the context of the long-standing conflict between the Greek and Turkish Cypriot sides (since 1974) and invigorates the debate about the delineation of the EEZ. Therefore, the feasibility of the pipeline's construction, connecting the Eastern Mediterranean's newly discovered gas fields to Europe is being questioned given the enormous cost of the pipeline and technological difficulties.

Despite the common interests in hydrocarbon agreements, the political volatility in Turkish–Israeli relations has deterred regional cooperation (Ben-Meir, 2018; Grigoriadis, 2014). Moreover, blame, mistrust and inept leadership contributes to hostility, accusations and imbalance in diplomatic relations between Turkey, Israel, Greece and Cyprus and deepens regional disintegration (Ben-Meir, 2018; Rapaport, 2016).

Nonetheless, the exploration of reserves has been perceived by experts in the energy security to be a stabilizing force which could stimulate the prospects for fully fledged negotiations between the Greek and Turkish Cypriots of achieving a sustainable peace (Rapaport, 2016). The main question to ask would be whether exploration of the hydrocarbons would stimulate economic cooperation and political synergies among the regional players or further contribute to the intractability of the long-standing conflict/s?

This chapter has identified three tools or approaches that may assist project leaders and negotiators to find shared ways that lead to action in a complex conflict situation. SSM provides a comprehensive set of tools that focus on moving from shared understanding of the problem situation to action. One tool from SSM has been described in detail – 'rich pictures'. This tool is used to develop shared understandings of world views and is generally applied at the beginning of the project or project phase. 'Leveraging' relies upon finding critical aspects in the negotiation that offer control for the parties involved. Leveraging is a dynamic process that relies on understanding the interrelated systems associated with the problem situation. Finally, application of any of these tools requires recognition of the part played by emotions. High levels of emotional intelligence are needed by key players, especially facilitators, to be able to use these tools to advantage.

> Leveraging is a dynamic process that relies on understanding the interrelated systems associated with the problem situation.

QUESTIONS FOR DISCUSSION

1. What would be the main advantages in using an SSM tool such as 'rich pictures' in a complex environment, such as the one described in the case study?
2. What do negotiators need to understand in relation to the impact of emotions in a complex negotiation environment? Use the case study as an example.
3. What kinds of leverage may assist stakeholders to enable the exploitation of the natural gas reserves?
4. How would you go about mixing different tools and approaches to develop a more customized approach to a particular negotiation?

ACTIVITY

METHODOLOGY

The outlined tools are aimed to stimulate innovative thinking and analysis which may result in alternative interpretations and (potential) approaches assisting practitioners in the field of negotiations. The purpose of the tools is to stimulate higher levels of systemic inquiry which would enable a deeper understanding of complex problems. The following section summarizes the tools, perspectives, and elements and proposes the following methodology:

Understand Boundaries of the Analysed System

a. Identify the environment, factors, patterns, limitations and scope.
b. Identify objectives, goals and overall purpose.
c. Explore the complexity of the system.
d. Identify stakeholders and conduct stakeholder analyses by identifying politico-economic interests of the various stakeholders while considering the boundaries of the system.

Data Collection Unique to the Case Study

a. Collect qualitative data. For example, through interviews with key political and commercial stakeholders, bearing in mind the viability of the data due to high confidentiality levels.
b. Collect quantitative data, bearing in mind that quantitative data will also be limited by the highly confidential nature of operations.
c. Develop a set of agreed measurable criteria.

TEAM DESIGN

a. Design a unique negotiations team which would focus on the complex problem, considering the skills, capabilities, knowledge and emotional intelligence levels.
b. Establish agreed roles and responsibilities for the team based on individuals' key strengths.
c. Develop team building, training and review mechanisms.

PREPARE FOR NEGOTIATIONS

a. Analyse the case study and design the negotiation strategy.
b. Define implementation goals and objectives.
c. Select appropriate negotiation tools and approaches to design negotiation strategy.

IMPLEMENT NEGOTIATION STRATEGY

a. Team to apply strategy with regular review (the process is dynamic, and teams should be prepared to change the strategy at short notice in accordance with feedback and current political or environmental situation).
b. Regularly evaluate the positions of the conflicting parties to assess progress (the team should be alert to additional side conflicts that might be constructed for political gain or hidden agendas that might influence the negotiations).
c. Reassess and make changes to strategy based on feedback. This is an iterative, nonlinear process.

REGULAR REPORTING AND AUDITING

a. Collate, discuss and report to key stakeholders at regular intervals.
b. Provide recommendations to stakeholders for comment.
c. Revise to accommodate disparate stakeholder views. There might be several revisions.

REPORT

a. Report the findings.
b. Conduct internal evaluation of systemic negotiations.
c. Understand the impact of negotiations.

REFERENCES

Allerd K.G., 1999. Anger and retaliation: Toward an understanding of impassioned conflict in organizations. In: R. J Bies, R. J. Lewicki, & B. H. Sheppard (Eds.), *Research on negotiation in organizations.* JAI, Greenwich, vol 7, pp 27–58.

Aquilar, F. and Galluccio, M., 2008. *Psychological Processes in International Negotiations: Theoretical and Practical Perspectives.* New York: Springer.

Armstrong, R., 2018. Systemic Practice and Action Research. doi: 10.1007/s11213-018-9466-7.

Atkinson, R., Crawford, L., and Ward, S., 2006. Fundamental uncertainties in projects and the scope of project management. *International Journal of Project Management, 24*(8), 687–698.

Auyang, S.Y., 1999. *Foundations of Complex-System Theories: In Economics, Evolutionary Biology, and Statistical Physics.* New York: Cambridge University Press.

Babcock, L. and Loewenstein, G., 1997. Explaining bargaining impasse: The role of self-serving biases. *Journal of Economic Perspectives, 11*(1), 109–126.

Babcock, L., Loewenstein, G., Issacharoff, S., and Camerer, C., 1995. Biased judgments of fairness in bargaining. *The American Economic Review, 85*(5), 1337–1343.

Barnes, W.R. and Ledebur, L.C., 1997. *The New Regional Economies: The US Common Market and The Global Economy* (Vol. 2). Thousand Oaks, CA: Sage Publications.

Bar-Yam, Y., 2002. *General Features of Complex Systems. Encyclopedia of Life Support Systems (EOLSS)*. Oxford: UNESCO, EOLSS Publishers, p. 1.

Bazerman, M.H., Tenbrunsel, A.E., and Wade-Benzoni, K., 1998. Negotiating with yourself and losing: Making decisions with competing internal preferences. *Academy of Management Review*, 23(2), 225–241.

Ben-Meir, A., 2018. Erdogan's reelection and its dire consequences, Alon Ben-Meir online website, 2.

Ben-Meir, A., 2019, Personal communication, February 16.

Branagan, M., 2007. The last laugh: Humour in community activism. *Community Development Journal*, 42(4), 470–481.

Bell, P.A. and Baron, R.A., 1990. Affect and aggression. In B. S. Moore & A. M. Isen (Eds.), *Studies in Emotion and Social Interaction. Affect and Social Behavior*. New York: Cambridge University Press, pp. 64–88.

Butler, C.J. and Chinowsky, P.S., 2006. Emotional intelligence and leadership behavior in construction executives. *Journal of Management in Engineering*, 22(3), 119–125.

Carnevale, P.J. and Isen, A.M., 1986. The influence of positive affect and visual access on the discovery of integrative solutions in bilateral negotiation. *Organizational Behavior and Human Decision Processes*, 37(1), 1–13.

Checkland, P., 1983. OR and the systems movement: Mappings and conflicts. *Journal of the Operational Research Society*, 34(8), 661–675.

Checkland, P., 2001. Soft systems methodology. In J. Rosenhead and J. Mingers (Eds.), *Rational Analysis for a Problematic World Revisited*. Chichester: John Wiley & Sons.

Checkland, P., 2011. Autobiographical retrospectives: Learning your way to 'action to improve' – The development of soft systems thinking and soft systems methodology. *International Journal of General Systems*, 40(05), 487–512.

Checkland, P. and Scholes, J., 1990. *Soft Systems Methodology in Action*. Chichester: Wiley.

Cicmil, S. 2006. Understanding project management practice through interpretative and critical research perspectives. *Project Management Journal*, 37, 27–37.

Cicmil, S. and Hodgson, D., 2006. New possibilities for project management theory: A critical engagement. *Project Management Journal*, 37(3), 111–122.

Cilliers, P. 1998. *Complexity and Postmodernism: Understanding Complex Systems*. Hoboken, NJ: Taylor and Francis.

Clarke, N. and Howell, R., 2009. *Emotional Intelligence and Projects*. Newtown Square, PA: Project Management Institute.

Crawford, L., Morris, P., Thomas, J., and Winter, M., 2006. Practitioner development: From trained technicians to reflective practitioners. *International Journal of Project Management*, 24(8), 722–733.

Diez, T. and Tocci, N. (Eds.), 2009. *Cyprus: A Conflict at the Crossroads*. Manchester: Manchester University Press.

Druckman, D. and Broome, B.J., 1991. Value differences and conflict resolution: Familiarity or liking? *Journal of Conflict Resolution*, 35(4), 571–593.

Dunnahoe, T., 2018. Israeli outreach aims to further exploration, development. *Oil & Gas Journal*, 116(5), 32–34.

Eden, C. and Radford, J., 1990. *Tackling Strategic Problems: The Role of Group Decision Support*. Thousand Oaks, CA: Sage Publications.

Eliav, E., Miron-Spektor, E., and Bear, J.B., 2017. *Humor and Creativity. The Psychology of Humor at Work: A Psychological Perspective*. New York: Routledge.

Ellinas, C., 2016. East Med Gas Export Risks, Cyprus Weekly, http://www.iene.eu/east-med-gas-export-risks-p2362.html

Ellinas, C., 2016. *Hydrocarbon Developments in the Eastern Mediterranean*. Washington, DC: Atlantic Council.

Englund, R. L., 2010. Negotiating for success: are you prepared? *Paper presented at PMI® Global Congress 2010—EMEA, Milan, Italy*. Newtown Square, PA: Project Management Institute.

Fisher, R., Ury, W., and Patton, B., 1991. *Getting to Yes: Negotiating Agreement without Giving In (No. E71202)*. Penguin.

Freud, S., Riviere, J., and Strachey, J., 1963. *Civilization and Its Discontents* (Rev. ed. / revised and newly edited by James Strachey). London: Hogarth Press.

Forgas, J.P., 1998. On feeling good and getting your way: Mood effects on negotiator cognition and bargaining strategies. *Journal of Personality and Social Psychology*, 74(3), 565.

Fouskas, V. and Tackie, A. O., 2009. *Cyprus: The Post-Imperial Constitution*. London: Pluto Press.

Freeman, R.E. and Reed, D.L., 1983. Stockholders and stakeholders: A new perspective on corporate governance. *California Management Review*, 25(3), 88–106.

Goleman, D., 1998. *Working with Emotional Intelligence*. New York: Bantam Books.

Gray, W.I.L.L.I.A.M. and Rizzo, N.D., 1973. Ludwig von Bertalanffy and the development of modern psychiatric thought. *Unity Through Diversity*, 169–184.

Grigoriadis, I., 2014. Energy discoveries in the eastern mediterranean: Conflict or cooperation? *Middle East Policy*, *21*(3), 124–133.

Guan, D., 2007. Conflicts in the project environment. *Paper presented at PMI® Global Congress 2007*—Asia Pacific, Hong Kong, People's Republic of China. Newtown Square, PA: Project Management Institute.

Humphrey, R.H., 2002. The many faces of emotional leadership. *The Leadership Quarterly*, *13*(5), 493–504.

Izard, C.E., 1977. *Human Emotions*. New York: Plenum Press.

Jones, G.R. and George, J.M., 1998. The experience and evolution of trust: Implications for cooperation and teamwork. *Academy of Management Review*, *23*(3), 531–546.

Joseph, J. S., 1997. *Cyprus: Ethnic Conflict and International Politics: From Independence to the Threshold of the European Union*. New York: St. Martin's Press.

Kahraman, S., 2018. Turkey is determined, but alone in the Mediterranean to hunt for oil and gas. Ahval News, accessed 1 March 2019, https://ahvalnews.com/east-mediterranean/turkey-determined-alone-mediterranean-hunt-oil-and-gas>.

Kuhn, L., 2009. *Adventures in Complexity: For Organisations Near the Edge of Chaos*. Chicago, IL: Triarchy Press Ltd.

Kumar, A. and Oliver, R.L., 1997. Special session summary cognitive appraisals, consumer emotions, and consumer response. ACR North American Advances.

Leahy, R.L., 2011. Personal schemas in the negotiation process: A cognitive therapy approach. In F. Aquilar and M. Galluccio (Eds.), *Psychological and Political Strategies for Peace Negotiation*. New York: Springer.

Leban, W. and Zulauf, C., 2004. Linking emotional intelligence abilities and transformational leadership styles. *Leadership & Organization Development Journal*, *25*(7), 554–564.

Leeper, R.W., 1948. A motivational theory of emotion to replace 'emotion as disorganized response'. *Psychological Review*, *55*(1), 5.

Li, L. and Chen, J.H., 2006. Emotion recognition using physiological signals. In *International Conference on Artificial Reality and Telexistence* (pp. 437–446). Berlin, Heidelberg: Springer.

Luterbacher, U. and Norrlöf, C., 1999. The new political economy of trading and its institutional consequences. *International Political Science Review*, *20*(4), 341–358.

Marwan, N., Donges, J.F., Zou, Y., Donner, R.V., and Kurths, J., 2009. Complex network approach for recurrence analysis of time series. *Physics Letters A*, *373*(46), 4246–4254.

Mayer, J.D., Roberts, R.D., and Barsade, S.G., 2008. Human abilities: Emotional intelligence. *Annual Review Psychology*, *59*, 507–536.

Mayer, J.D. and Salovey, P., 1997. What is emotional intelligence. *Emotional Development and Emotional Intelligence: Educational Implications*, *3*, 31.

Meadows, D., 2008. Thinking in systems: A primer. In D. Wright (Ed.), *White River Junction*, VT: Chelsea Green Publishing.

Michael, M.S., 2011. *Resolving the Cyprus Conflict: Negotiating History*. Basingstoke: Palgrave Macmillan.

Monk, A. and Howard, S., 1998. Methods & tools: The rich picture: A tool for reasoning about work context. *Interactions*, *5*(2), 21–30.

Morando, J. W. and Joesten, N.E. Leverage Points in International Arbitration, IPLITIGATOR, May/June 2005, available at www.fbm.com/index.cfm/fuseaction/publications.detail/object_id/a053f2f3-5924-46f8-bcb8-14ae4b4f0880/LeveragePointsinInternationalArbitration.cfm.

Naughton, J., 1984. *Soft Systems Analysis: An Introductory Guide. Block: the Soft Systems Approach*. Open University Press.

Ostrom, E., 2010. Beyond markets and states: Polycentric governance of complex economic systems. *American Economic Review*, *100*(3), 641–72.

Papadakis, Y., Peristianis, N., and Welz, G., 2006. *Divided Cyprus: Modernity, History, and an Island in Conflict*. Bloomington, IN: Indiana University Press.

Pericleous, C., 2009. *Cyprus Referendum: A Divided Island and the Challenge of the Annan Plan* (Vol. 26). London and New York: IB Tauris.

PMI, 2017. *A Guide to the Project Management Body of Knowledge (PMBOK Guide)*. Newtown Square, PA: Project Management Institute.

Porter, T.W. and Lilly, B.S., 1996. The effects of conflict, trust, and task commitment on project team performance. *International Journal of Conflict Management*, *7*(4), 361–376.

Raiffa, H., 1982. *The Art and Science of Negotiation*. Cambridge, MA: Harvard University Press.

Rapaport, D., 1942. Emotions and memory.

Rapaport, B., 2016. Exploring the contribution of complexity theory to the system of political negotiations: a case study of the intractable conflict in Cyprus/Barbara Rapaport.

Richardson, J. (Ed.), (1984). *Models of Reality: Shaping Thought and Action.* Mt Airy, MD: Lomond Publications.

Rusalova, M.N., Izard, C.E., and Simonov, P.V., 1975. Comparative analysis of mimical and autonomic components of man's emotional state. *Aviation, Space, and Environmental Medicine, 46*(9), 1132–1134.

Salovey, P. and Mayer, J.D., 1990. Emotional intelligence. *Imagination, Cognition and Personality, 9*(3), 185–211.

Schachtel, E.G., 1959. *Metamorphosis: On the Development of Affect, Perception, Attention, and Memory.* Basic Books.

Sebenius, J.K., 2017. BATNA s in negotiation: Common errors and three kinds of "No". *Negotiation Journal, 33*(2), 89–99.

Smyth, D.S. and Checkland, P.B., 1976. Using a systems approach: The structure of root definitions. *Journal of Applied Systems Analysis, 5*(1), 75–83.

Sosik, J.J. and Megerian, L.E., 1999. Understanding leader emotional intelligence and performance: The role of self-other agreement on transformational leadership perceptions. *Group & Organization Management, 24*(3), 367–390.

Storbeck, J. and Clore, G.L., 2005. With sadness comes accuracy; with happiness, false memory: Mood and the false memory effect. *Psychological Science, 16*(10), 785–791.

Tarr, P., 2007. Epupa dam case study. *Water Resources Development, 23*(3), 473–484.

Turner, J.R. and Müller, R., 2005. The project manager's leadership style as a success factor on projects: A literature review. *Project Management Journal, 36*(2), 49–61.

Urry, J., 2003. *Global Complexity.* Cambridge: Polity.

Varnava, A. and Faustmann, H. (Eds.), 2009. *Reunifying Cyprus: The Annan Plan and Beyond.* London: IB Tauris.

Vassiliou, G., 2010. *From the President's Office: A Journey Towards Reconciliation in a Divided Cyprus.* London: IB Tauris.

Von Bertalanffy, L., 1968. General system theory. *New York, 41973*(1968), 40.

Wang, W., Liu, W., and Mingers, J., 2015. A systemic method for organisational stakeholder identification and analysis using Soft Systems Methodology (SSM). *European Journal of Operational Research, 246*(2), 562–574.

Zartman, I.W. and Berman, M.R., 1982. *The Practical Negotiator.* New Haven, CT: Yale University Press.

8 Organization and Systems Theory Toolset

Joana Geraldi
Copenhagen Business School
Technical University of Denmark

Josef Oehmen and Christian Thuesen
Technical University of Denmark

Pedro Parraguez Ruiz
Technical University of Denmark
Dataverz

CONTENTS

"For every complex problem, there is an answer that is clear, simple and wrong." – Popular (and clear, simple and wrong) adaptation of a statement by H.L. Mencken (1917)[i]

INTRODUCTION

We have learned to accept that most projects are complex, at least in some aspects. Scholars studying the complexities involved in projects have made significant progress in understanding the different aspects of complexity in projects, programs and portfolios, and this work is often used as project typologies in organizations to classify projects and tailor project management governance, process, competences and resources. Yet, there is still significant work to be done in bridging complexity concepts to organizing tools that enable managers to accept, embrace and manage the complexity of their projects.

> We create complexities not complexity. Three different kinds of complexity can co-exist in projects in various degrees: Structural complexity, socio-political complexity and uncertainty. Each complexity requires different organizing tools.

Drawing upon inspiration from recent streams of research in Organization Theory and prior studies on complexity in projects (Maylor et al. 2013; Geraldi et al. 2011), we take an organizational approach towards the complexity in projects. We firmly suggest that complexity is created and can be organized, and more importantly, the way it is organized will influence the project and its potential ability to deliver useful benefits to organizations and society. Therefore, we strongly believe that organizing complexity matters to projects, project practitioners and project management.

> This chapter proposes five organizing tools for dealing with project complexity based on recent streams of work in Organization Theory and the wide field of organizing: Network analysis, systems dynamics, modularization, antifragility and mindfulness.

After presenting our mental model for project and complexity, the chapter describes three different facets of complexity: structural complexity, uncertainty and socio-political complexity. We then introduce five strategies for dealing with project complexity based on recent streams of work in Organization Theory and the wide field of organizing: network analysis, systems dynamics, modularization, antifragility and mindfulness.

WHAT KIND OF PROJECTS

Not all projects are the same. Much of the confusion in conceptual and theoretical developments in project studies starts as scholars discuss projects in general terms, void of context. While we believe that there is room for theorizing at the project level and that a project like the London Olympics can learn from a complex IT project, neither the learning nor the theorizing would be possible through unreflected use of ideas and concepts from one context to the other. Thus, before we start the discussion of complexity and complex projects, we would like to clarify how we see projects.

Our mental model of a project here is a large, socio-technical project designing society's engineering systems. We are adopting the view articulated by De Weck et al. (2011) that our modern lives are governed by "engineering systems" that fulfil central societal functions – for example, our modern communication, transportation, healthcare or energy generation and distribution systems. These systems are not "just" technical systems – they are socio-technical systems where people and technology are intertwined, and have become dependent on one another. Project, programme and portfolio management are used as vehicles to design and change these systems. Examples of such projects (or programmes and portfolios of projects) include airports, bridges, sewage systems, subway stations, large manufacturing sites, new hospitals, and installation of smart energy grids.

COMPLEXITY IN PROJECTS: BACKGROUND AND RELEVANCE

Having defined what we mean by projects, we turn now to a non-less-disputed concept: complexity. In lay terms, "complex" and "complexity" are concepts often used to describe what is considered to be intricate or complicated. However, in order to advance our understanding of complexity, it is important to draw a clear distinction between these two ideas – complexity and complicated.

Although a complex system might also be considered complicated, it does not need to be so. Complicated systems have a large number of interdependent elements, yet they are predictable and linear. Complex systems, in turn, cannot be predicted and have a high degree of self-organization. Complex systems are a key enabler of emergent value creating processes, and when appropriately designed and managed, become the source of significant competitive advantages for their owners. Such emergence is what generates and allows us to explain "higher-order" behaviours and functions, like self-reproduction, self-organization, intelligence and communication (Alexiou 2010; Holland 1997). Consequently, the understanding of how a complex structure of interactions can generate useful (or harmful) behaviours is crucial to improve the design, production and management of human-made engineering systems (Calvano & John 2004). The focus of this chapter is on complexity. For additional insights on complexity, see, for example, Holland (1997), Johnson (2007), Cilliers (2000), De Weck et al. (2011) and Geraldi et al. (2011).

Projects can be seen as a complex system composed by a large number of interdependent tasks executed by different people. For example, the London Olympics 2012 is composed of the four different objectives that are reached by the coordinated work of innumerous tasks, executed by people across several organizations. If we choose to see projects as such, we will accept that project practitioners and stakeholders will *create and manage* this complexity (Geraldi et al. 2011; Geraldi & Terrikangas 2011).

Classic project management would suggest managing complexity by breaking down the system into manageable pieces and integrating different pieces in a coherent way. Such a management approach is in line with classic understanding of complexity management, as suggested by the following fragment of a seminal work on complexity called the Architecture of Complexity, written by Simon in 1962:

> There once were two watchmakers, named Hora and Tempus, who manufactured very fine watches. Both of them were highly regarded, and the phones in their workshops rang frequently -new customers were constantly calling them. However, Hora prospered, while Tempus became poorer and poorer and finally lost his shop. What was the reason? The watches the men made consisted of about 1,000 parts each. Tempus had so constructed his that if he had one partly assembled and had to put it down-to answer the phone say-it immediately fell to pieces and had to be reassembled from the elements. The better the customers liked his watches, the more they phoned him, the more difficult it became for him to find enough uninterrupted time to finish a watch. The watches that Hora made were no less complex than those of Tempus. But he had designed them so that he could put together subassemblies of about ten elements each. Ten of these subassemblies, again, could be put together into a larger subassembly; and a system of ten of the latter subassemblies constituted the whole watch. Hence, when Hora had to put down a partly assembled watch in order to answer the phone, he lost only a small part of his work, and he assembled his watches in only a fraction of the man-hours it took Tempus.

> **Simon (1962, p. 470)**

Although in principle, complexity will require some kind of division of work, for example, we cannot manage projects like the London Olympics single-handed, such a management approach needs to be used with immense care. As a complex system is emergent, its different parts will be developing in the course of the project. It is therefore hard to integrate the parts back together as well as ensure that the parts will still cover the purpose of the project.

Indeed, the vast majority of current project, programme and portfolio management processes consider projects to be "simple" or complicated systems. We assume that we can follow a staged and deterministic process by defining requirements, investigating alternative solutions, evaluating solutions and implementing them. The reality however is characterized by "wicked problems", complex projects where true requirements are unknown (or unknowable) before the project start and develop in parallel with the solution. Treating them as simple often turns them into chaotic projects.

Acknowledging the social-technical complexity of projects, programmes and portfolios requires a development of a managerial sensitivity towards the potential wickedness of the problems (Rittel & Webber 1973).

The term "wicked problems" was originally outlined within the area social policy planning which resonates with the intrinsic element of human behaviour, uncertainty and complexity in project, programme and portfolio management. The following elements characterize a wicked problem (Rittel & Webber 1973). Note that it is now customary to use the term "wicked problem" loosely to problems that fulfil some but not necessarily all of the following characteristics at the same time:

1. There is no definitive formulation of a wicked problem.
2. Wicked problems have no stopping rule.
3. Solutions to wicked problems are not true-or-false, but good or bad.
4. There is no immediate and no ultimate test of a solution to a wicked problem.
5. Every solution to a wicked problem is a "one-shot operation"; because there is no opportunity to learn by trial and error, every attempt counts significantly.
6. Wicked problems do not have an enumerable (or an exhaustively describable) set of potential solutions, nor is there a well-described set of permissible operations that may be incorporated into the plan.
7. Every wicked problem is essentially unique.
8. Every wicked problem can be considered to be a symptom of another problem.
9. The existence of a discrepancy representing a wicked problem can be explained in numerous ways. The choice of explanation determines the nature of the problem's resolution.
10. The planner has no right to be wrong (i.e. planners are liable for the consequences of the actions they generate).

To address a wicked problem is thus a never-ending process where every proposed solution reveals aspects of the problem, which in turn cause a revision of the solution.

This does not mean that we cannot address a wicked problem; indeed, we can and should. However, wicked problems require local trial and error and learning, and hence, we cannot provide recipes to manage wicked problems but instead aim to create a more nuanced understanding of the kind of issues we could face when we are practicing project management, and propose novel concepts and strategies for dealing with project complexity.

CHARACTERIZING COMPLEXITY IN PROJECTS

Based on current and past developments in project studies and related fields, we discern three characteristics of complexity in projects: structural complexity (often termed complicatedness), socio-political complexity and uncertainty, which correspond to core practices of project managing, namely coordinating, collaborating and adapting.

> Socio-political complexity is addressed through network analysis and mindfulness, yet much more research is required in this area.

Structural Complexity

Structural complexity is related to a large number of distinct and interdependent elements, which is close to the original concept of complexity as a set of interrelated entities (Baccarini 1996), as well as the Oxford Dictionary definition of the term. Not surprisingly, structural complexity is the most common aspect of complexity, and it refers to what we now define as complicatedness.

There has been an overall lack of appreciation of the importance and difficulty of dealing with highly complicated systems. Complicated is not easy. In such an environment, mistakes are almost inevitable, but they are significantly expensive. Activities with high structural complexity demand delegation and are supported by project planning and controlling tools, and computeraided instruments. Employees lack time to collect, to analyse and to internalize information, and have to make decisions and act without properly understanding required information. The challenge here is to keep a

> The 'complexity of a project' is a choice, it is not given nor fixed, and hence it can be negotiated and shaped across project stakeholders.

holistic view and not get lost in the immense amount of details. Indeed, Green (2004) showed in an experiment that project managers tend to be more successful if they have high cognitive integration and low cognitive differentiation. Project managers with a high cognitive integration are good to see the connections and integrate parts of the project into larger umbrellas, which improves flexibility and makes the project more manageable. Thus, what Green suggests is that project managers able to maintain an overview and not micro-manage are more able to deal with the challenges arriving from highly structured complex projects. Project managers with high cognitive differentiation, in contrast, look for differences and thereby increase the fragmentation of the projects. Such project managers would divide the projects into too small chunks, search for very specialized professionals, would struggle to coordinate their work and ensure good collaboration. Following this line of thought, project managers with higher tendency for fragmentation would also tend to fragment the project even further and hence increase the structural complexity of the project, making it even harder to manage.

These insights go beyond experimental settings. For example, the automotive industry increased engineering project efficiency (i.e. the number of car models and variants brought to market per unit of engineering manpower) dramatically over the last 15–20 years to address both a customer need for differentiation and a business need for competitiveness. Real customization is now reserved to a fraction of around 20% of car components, which drive customer perception—such as the shape of the chassis. This radical simplification of technical complexity allowed a reduction of the organizational complexity.

Socio-Political Complexity

Projects are about people, and yet most tools and concepts related to projects only partially address "people"-related challenges. This trend has started to shift from the 2000s onwards, and we have seen a more research and practice focussing on people. It has also become clearer to practitioners and academics alike that the classic project management tools, such as work breakdown structures (WBSs) or Gantt charts are there to manage behaviours and people. In line with this work, we suggest that a facet of complexity in projects derives from the socio-political nature of projects.

Socio-political complexity is easy to broadly conceptualize but yet exasperatingly difficult to precisely define. We see socio-political complexity as a combination of political aspects and emotional aspects involved in projects. This complexity is expected to be high in situations such as mergers and acquisitions, organizational change or where a project is required to unite different interests, agendas or opinions.

One relatively new aspect of this complexity though is the very way in which we conceive complexity in projects. We find this perspective promising and will therefore spend a few paragraphs conceptualizing it and discussing implications of such complexity.

Independent of the actual complexity of a project is the way to perceive and manage this complexity. Arguably, regardless of its "real" complexity, what counts is our perception and reaction to it.

While our brains are incredibly adaptable and able to learn, fundamentally, our ability to make sound decisions in the face of complexity is limited by three factors: (1) Our access to information will never be perfect and completely and accurately represent the full current state; (2) we have cognitive limitations, for example, regarding the number of factors we can consider in parallel or the amount of information we can take up or speed we can process it at; and (3) the time we have to make a decision. This leads to the phenomenon that (possibly endless) information consumes our (limited) attention, and we are forced to make decisions before we are "ready". Given the fact that with raising complexity, the amount of information tends to increase and therefore the time needed to understand and process it, we are faced with a fundamental challenge of finding the "right" way of compressing complexity without sacrificing key aspects that are relevant for decision-making.

Adding to these cognitive limitations, our subconscious or built-in decision-making models and rules (e.g. heuristics) are great for ensuring our survival in the savannah but of limited help when managing complex projects. This leads to a number of challenges when we try to "intuitively" understand and deal with complexity:

- Number and diversity of items: While complex systems often consist of tens, hundreds or even thousands of elements and their relationships (e.g. project tasks), we can "keep in mind" about five to seven items at any given time.
- Dynamic behaviour and change: We subconsciously extrapolate any change as "linear" – we lack an "emotional appreciation" of what exponential growth means (one of the reasons why population growth and climate change has not lead to a global panic).
- Cherry picking: We reduce perceived complexity by selective attention – a subconscious "cherry picking" of information, that fits our existing worldview and theories (selective attention). If we just bought a red car, all we see on the road is other red cars. The availability of information on the Internet has made the phenomenon worse – whatever my opinion is, I will find "facts" to support it. In a project environment, this leads to a lack of appreciation of complexity. Decision-makers get stuck in a set frame of mind, and in its worst case disconnect between the decision-maker and reality.
- Even pessimists overestimate their abilities (overconfidence and optimism bias): When we analyse a problem or project task, we will subconsciously focus on those aspects that we are familiar with, good at and certain about. We will also prefer to start execution with "easy" tasks. This will lead to overly optimistic assessments of cost and schedule requirements, and chances of success. We are also built-in overconfident in our abilities (even pessimists): Studies show that 93% of drivers believe their driving skills are above average (i.e. in the top 50%). We also like to select our leaders from people that exhibit strong confidence in themselves and the success of their people.
- We were right after all: Reinforcing our overconfidence is a mechanism that adjusts the memory of our opinion to actual developments (hindsight bias). We typically believe (after the match) the thought that the winning team was the stronger one. In complex project environments, that creates a false trust in the quality of our "intuition" to foresee problems and choose the right solutions.
- Our estimates are terrible (anchoring bias): We are not very good (some say terrible) at making estimates based on our "gut feeling". When estimating without some level of prior data, our brains rely on the first number they can remember (typically the last number they came across, for example, the day of the month) to define the order of magnitude for the estimate (is it in the 10s, 100s or 1,000s?). Untrained estimation completely ignores even the limited information that may be available, and can be worse than using random numbers (Based on Kahneman, 2011).

Uncertainty

The concept of uncertainty and its intrinsic relationship with risks have been core to the organization and project studies. Uncertainty is often discussed in terms of aleatory uncertainty (the probability and chance of an event) and epistemic uncertainty (lack of information, and lack of agreement over current and future situation or ambiguity). In project management, common concepts involve the uncertainty in goals and methods (Turner and Cochrane 1993), as well as the level of (un)predictability of a project: variation, foreseen uncertainty, unforeseen uncertainty and chaos (DeMeyer et al. 2002).

> System dynamics, antifragility and mindfulness are tailored towards the management of uncertainty.

Uncertainty is unescapable in the development of something unique and the solving of new problems. Geraldi and Adlbrecht (2007) termed this complexity as "complexity of faith", as project managers would act as "priests", convincing the team and stakeholders to have faith in the project but not necessarily be closed to criticism.

The challenge of uncertainty is particularly true in complex projects, which typically have long life cycles, increasing the timeframe through which planning assumptions have to be projected and forecasts be made. This increases the number of factors that are seriously affected by uncertainty: For example, while a project manager can be fairly certain of the regulatory environment for the next 2–3 years, a plane or infrastructure project with a start of construction in 15 years must consider uncertainties from this domain. Second, complex projects or systems in general will display so-called emerging behaviour: We can observe behaviours that are impossible (or at least very difficult) to predict by just understanding the single factors of the project but are created by their mutual dependence. While the various shapes of snowflakes are pretty examples of emergence, emergent behaviour is typically negative or even opposed to the original intention. For example, making bike helmets mandatory (in order to reduce severe injuries) can lead to an increase of severe injuries in bicyclists due to dynamic overcompensation effects arising from human and other factors.

NOVEL APPROACHES TO ORGANIZE COMPLEXITY

Borrowing from Organization Theory in its most generic understanding, the following section will discuss some selected concepts to deal with complexity in projects. This is not an exhaustive list but is intended to introduce some lesser-known concepts with a particular focus on complexity. They focus on the three areas discussed above: structural complexity (network analysis, system dynamics and modularity), uncertainty (Antifragility) and socio-political complexity (mindfulness).

NETWORK ANALYSIS

In the last decades, we have moved from an understanding of organizations as pre-established hierarchical structures, to a more fluid forms of organizing, which do not necessarily follow hierarchical structures, but instead, its structures are an empirical question. Network

> Network analysis and modularization are particularly useful to address the structural complexity of projects.

analysis became a powerful perspective and tool to understand the structure of organizations, in our case, projects, and to capture potential forms to configure and reconfigure it. From such a perspective, we choose to understand projects as systems. A system of interconnected elements can be modelled through its network architecture. It provides a quantitative and/or graphical representation of the "interconnectedness" between the elements, as well as describing characteristics for each of the constituent elements and quantifying their diversity.

As a minimum, a network-based approach will consider a list of elements and a binary indication about the existence or not of a relationship between each of the elements. Despite the simplicity of such an elementary way of modelling a complex system, even basic network models allow us to gain insights about key features and properties of complex systems. For example, through network-based approaches, we can measure interdependence and decomposability, and describe modularity in systems. This allows technical modularization ("sub-systems") as well as organizational modularization (e.g. departments and teams). Moreover, certain types of network configurations reliably suggest particular strengths and weaknesses of a complex system or project.

Two approaches are common: matrix-based and graph-based network analyses. While the most evident difference between them is representational: Matrix-based approaches use square or rectangular matrices, and graph-based approaches use network graphs; their representational differences are rooted in different analytical methods, needs and assumptions (Wyatt et al. 2013). One of the most widely utilized matrix-based approaches in Engineering Design and Systems Engineering is the design structure matrix (DSM). DSM is a flexible method based on square matrices (also known as influence or adjacency matrices) that allows to make explicit the connections between two elements of the same domain (Steward 1981; Eppinger & Browning 2012). Typical applications are product architecture DSM, analysing dependencies/interactions between components; organization architecture DSM, analysing communication/interactions between people; and process architecture DSM, analysing dependencies and information flows between activities. In addition to DSM, the multi-domain matrix (MDM) allows us to map connections between domains (i.e. mapping organization to process, process to product, etc.).

In contrast to matrix-based approaches, graph-based approaches are more diverse, varying widely in terms of analytical capabilities and focus. Nonetheless, all graph-based approaches share a representation based on nodes and edges (although they often use different naming conventions). On one extreme, simpler graph-based approaches do not have a quantitative intent; instead, their emphasis is only on graphically summarizing information about an architecture, and this is the case of organizational charts, workflow diagrams and basic abstract representations of a product's architecture. On the other extreme, petri nets, different variants of social network analysis, IDEF0 and IDEF3 diagrams, programme evaluation and review technique (PERT) and graphical evaluation and review technique (GERT) diagrams, etc. are intended to not only visualize but also explicitly quantify the network at one or more levels of analysis (Browning & Ramasesh 2007).

Example 8.1: Understanding True Communication Needs

In a power plant engineering project, the project manager wanted to better understand what the "real" communication pathways between his project organization and to outside suppliers were. There were official rules and protocols in place, but it was unclear to what degree they were implemented, and if they were the most useful configuration for a project involving several thousand people. Using network analysis, it was possible to understand what communication pathways were the most important, who were "key communicators" in the organization and what groups of people were communicating among themselves intensively. As a result, the communication policy could be adapted to the real project needs, for example regarding the composition, frequency and content of regular coordination meetings, newsletter-type information dissemination, establishment of direct communication between key partners and review of the activity critical communication interfaces. Figure 8.1 shows a graphical representation of communication partners (nodes) and communication between partners (edges). The sizes of the nodes indicate communication intensity, and the shades of gray of the nodes indicate clusters of communicating parties (i.e. a group of people with a high degree of communication within).

Example 8.2: Aligning Organizational and Technological Structures

This next application of network analysis does not rely on a big data set but is based on a manual modelling of the information flow between processes (project management and engineering), as well as the modelling of the dependence between technical components. The system being designed is a biomass power plant. Based on a model of the technical dependencies of the system, the engineering activities were aligned in such a way that each development team focussed on well-defined sub-systems. This minimized the need for coordination between teams (as technical sub-systems were defined through a network analysis of their dependencies to have minimal dependencies at their interfaces and maximum dependencies within). The same way, project management or procurement processes can be designed to minimize the need for cross-boundary

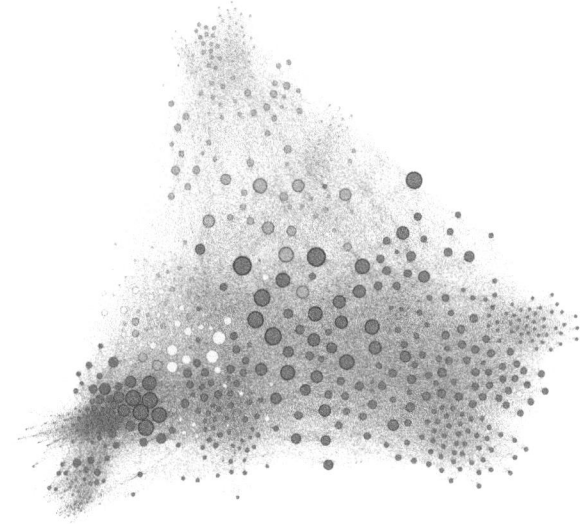

FIGURE 8.1 Communication network and clusters, based on the analysis of emails in a biomass plant engineering project. Gray shades represent clusters. Node size represents centrality.

communication. In Figure 8.2, it is possible to see the results of modelling the design process architecture as a DSM and as a graph. Through these representations, it is possible to, for example, identify modular and integrative processes, understand which subsystems are the ones with the highest influences if an engineering change were introduced, and help to redesign the process to better align with the architecture of the product or the organization.

Systems Dynamics

Following such network and system views of project organizing, we move to another adjacent approach to address structural complexity coming from systems dynamics. System dynamics is an analysis and simulation technique developed to target socio-technical systems. It is targeted at overcoming key issues in working with complex systems, such as bounded rationality, flawed and oversimplified mental models, and thinking and decision-making biases.

It consists of a number of simple building blocks, such as the variables and their positive (reinforcing) and negative (balancing) relationships (see simple example in Figure 8.3). Developing a complete and reasonably accurate system dynamics model of any concrete problems will require expert support. However, in the first step, it can be used to develop a qualitative model of the system, highlighting key factors, their relationship, and some fundamental dynamic behaviours of the system. For example, a system model that only shows reinforcing (positive) relationships between its elements will be prone to runaway effects – either exponential growth or crash. A stable system – and project – will always require a balancing influence that will stabilize its performance after a growth or decline phase.

System dynamics models can be crucial to gain clarity regarding both the structural complexity and, through simulation runs, the dynamic complexity of a project. System dynamics models for projects typically focus on

- Understanding particular project elements (e.g. decision-making processes at milestones)
- The impact and optimization of rework cycles (particularly for development projects)
- The optimization of control processes to adjust project execution to stay within budget, scope and schedule
- The root causes and countermeasures to "emerging behaviour", such as policy resistance, unintended consequences and ripple effects (Sterman 2000).

(a)

Process Architecture (matrix form)

	Air and Flue Gas	Boiler and Equipment Design	COMOS Data	Combustion System	Electrical Control and Instrumentation	External Piping	Load Plan and Layout	Overall Project Management	PFD and P&ID	Pressure Parts Design	Procurement	Design of steel structures
Air and Flue Gas		0.0	0.7	0.0	0.3	0.0	1.0	0.3	0.7	0.0	0.7	0.7
Boiler and Equipment Design	1.0		0.7	1.0	1.0	1.0	0.0	0.3	0.0	1.0	0.0	0.0
COMOS Data	0.7	0.0		0.7	1.0	0.7	0.0	0.3	0.7	0.7	0.0	0.0
Combustion System	0.0	0.0	0.7		0.3	0.0	1.0	0.3	0.0	0.0	0.7	0.0
Electrical Control and Instrumentation	0.3	0.0	0.7	0.3		0.3	0.0	0.3	0.7	0.3	0.7	0.0
External Piping	0.0	0.0	0.3		0.3		1.0	0.3	0.3	0.3	0.7	0.7
Load Plan and Layout	0.0	0.0	0.0	0.0	0.0	0.0		0.7	0.0	0.0	0.0	1.0
Overall Project Management	0.3	0.3	0.3	0.3	0.3	0.3	0.3		0.3	0.3	1.0	0.3
PFD and P&ID	0.7	0.3	1.0	0.7	1.0	0.7	0.0	0.3		0.7	0.0	0.0
Pressure Parts Design	0.0	0.0	0.7	0.0	0.3	0.0	1.0	0.3	0.7		0.7	0.0
Procurement	0.3	0.0	0.0	0.3	0.3	0.3	0.0	0.3	0.0	0.3		0.3
Design of steel structures	0.0	0.0	0.0	0.0	0.0	0.0	0.7	0.3	0.0	0.0	0.7	

(b)

Process Architecture (graph form)

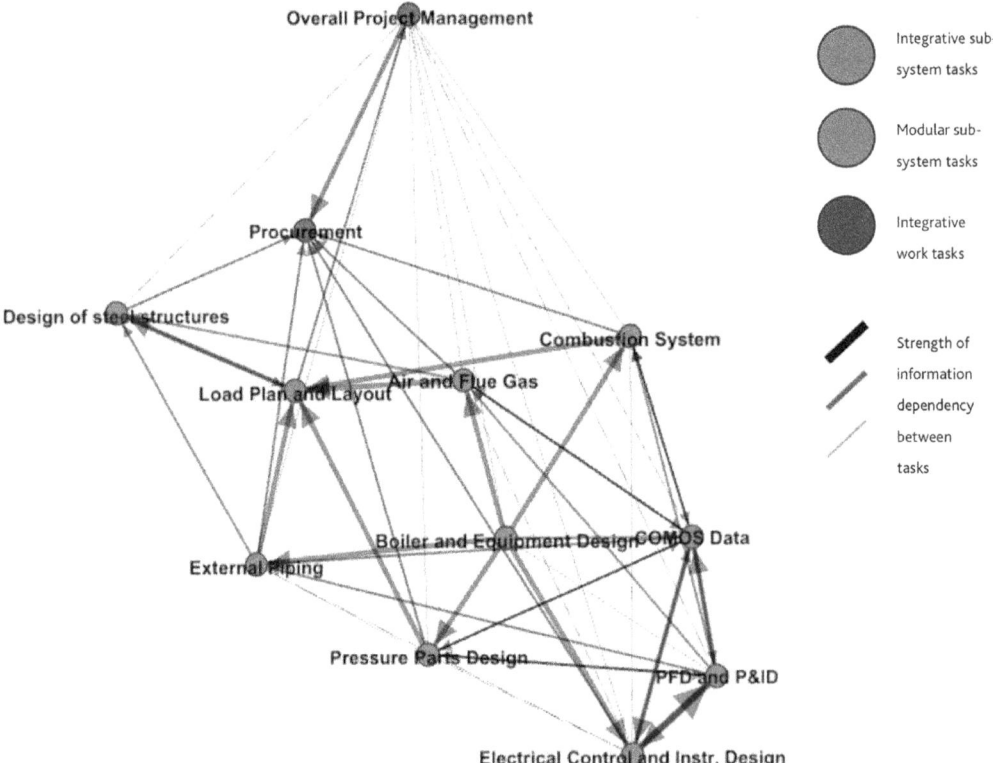

FIGURE 8.2 (a) Process architecture in matrix and (b) graph form as a simplified illustration of network complexity.

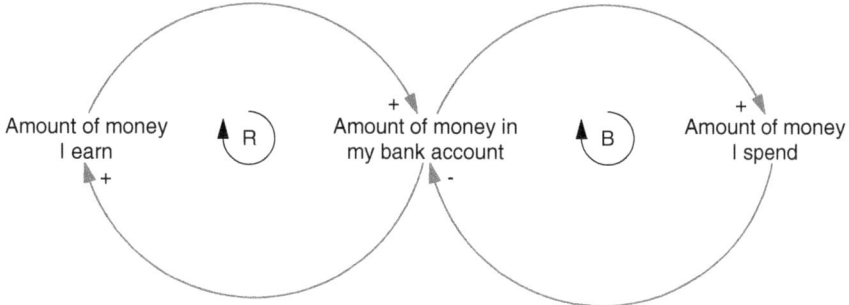

FIGURE 8.3 Example of system dynamics with financial flows.

Example 8.3: How Decreasing Project Delays Increases Them

In the late 1980s, managers and researchers became interested in better understanding the processes of developing increasingly complex software. They used a systems dynamics approach to study and model why most companies were so unsuccessful in bringing projects back on track once they were delayed. Even more puzzling, projects seemed to recover after the first intervention but then degraded quickly into failures. Figure 8.4 shows a small slice of the model that was developed to understand the problem and articulate to project managers where they went wrong: The intention was to increase labour quantity by increasing work intensity and overtime, based on the forecasted delay. While we would expect this to solve the problem, a "background mechanism" is at work that causes the opposite effect: While fatigue is low at the beginning and suggests the strategy is successful, its exponentially increasing negative effect on productivity will soon negate any positive effect of increased work intensity.

Similar models have been developed to explain a range of other counterintuitive phenomena, for example why increasing the staffing level of a project (to increase work intensity) typically also results in a decrease of work output (the most experienced and productive people are suddenly busy training the new arrivals). These models help understand counterintuitive results of our problem solving strategies.

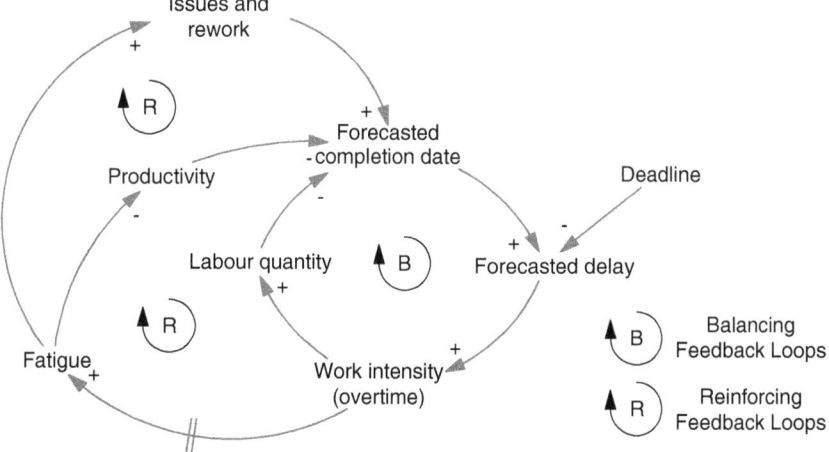

FIGURE 8.4 Why bringing a project back on track delays it more. (Adapted from Abdel-Hamid & Madnick 1991.)

Example 8.4: Video

Fluor uses system dynamics models to transform its project management and change management – www.youtube.com/watch?v=OeQcRjhhfcM.

MODULARITY

Modularity is a crucial strategy for managing complexity, enabling organizations to create products and services meeting individual customers' needs while still leveraging the benefits of similarity and standardization (Thuesen 2012).

Modularity is a general systems concept, typically defined as a continuum describing the degree to which a system's components may be separated and recombined (Schilling 2000). Usually, it refers to both the tightness of coupling between components, and the degree to which the "rules" of the system architecture enable (or prohibit) the mixing and matching of components. Given the open-ended nature of the concept, it is used in a variety of fields such as biology, nature, ecology, mathematics, cognitive science, industrial design, manufacturing, programming, and art and architecture. Within management studies, Campagnolo and Camuffo study (2010) identify three different units of analysis in (a) product design modularity, (b) production system modularity, and (c) organisational design modularity (p. 260). For simplicity, we refer to these as product, process and organisational modularity guided by the three fundamental questions (What? How? and Who?).

Product modularity (What?)

Product modularity (What?) has so far received the greatest attention. With the outset in platforms thinking, Meyer and Lehnerd (1997) described the architecture of a product as being the combination of subsystems and interfaces. They argued that every product is modular and that the goal is to make architecture common across many variants. Product modularity can also be viewed as the scheme by which the functions of the product are mapped towards the physical components, thus defining the product architecture as the arrangement of functional elements, the mapping from functional elements to physical components and the specification of interfaces between these (Ulrich 1995).

Process modularity (How?)

Process modularity (How?) "within and among organizations mirrors the degree of product modularity, with the main consequence that independent companies (e.g. suppliers) may develop, produce and deliver self-contained modules consistent with the scope and depth of their core competences" (Campagnolo & Camuffo 2010, p. 269). Thereby, modularity is not only a characteristic of a product but also the processes/task/activities for producing it. One of the consequences of focussing on modular processes is that the end product might be intangible like a service or experience (Pine & Gilmore 1999).

Organizational modularity (Who?)

Organizational modularity (Who?) can be understood as the way organizations are structured. Significant effort has been spent on developing new organizational paradigms "characterized by flatter hierarchies, decentralized decision-making, greater capacity for tolerance of ambiguity, permeable internal and external boundaries, empowerment of employees, capacity renewal, self-organizing units, and self-integrating co-ordination mechanisms" (Campagnolo & Camuffo 2010, p. 274).

A clever combination of the different modularity types creates superior performance. The use of product architecture with well-defined modules has proved to contribute to significant increases in industrial productivity, since implementation of product architecture with well-defined interfaces maintained over many years makes it possible to deliver effective (customized) products with efficient processes. One reason is that the well-defined interfaces make it considerably simpler to coordinate the individual sub-processes that are typically carried out by different organizational groups.

Particularly interesting is the relation between product and organizational modularity: "Integral products should be developed by integral organizations (tightly connected organizational units to maximize ease of communication and minimize the risk of opportunism). Modular products should be developed by autonomous, loosely coupled, easily reconfigurable organizations" (Campagnolo & Camuffo 2010, p. 274).

The abovementioned categories of modularity can all be applied in the management of project, programme and portfolios. In fact, since modularity is an attribute of a complex system, every project is modular to some extent and some of the well-known tools and practices such as WBS (What?), activity planning (How?), and organizational charts (Who?) are all developed for designing and managing the modularity of projects.

From a modularity perspective, a project might be seen as a puzzle where the pieces fit more or less together. An advantage of this perspective is that while the interface between the modules should fit well together, the core of the modules can be developed independently in the course of the project. They can potentially build on similar modules developed in prior projects, increasing economies of repetition, and hence project efficiency. If the pieces fit well together, the project will be characterized by order, efficiency and high reliability, but if the pieces do not match, resources are needed to negotiate and align the interfaces resulting in high complexity, inefficiency and uncertainty.

Example 8.5: Standardization and Modularization of Installation Shafts in Building Projects

Installation shafts are an exemplary case of complexity of construction products and practices. Following the general technological development, the shafts have got increasingly complex. Consequently, today an average installation shaft requires around 300 operations by nine to ten technical crafts, done in a space of 0.6 * 0.8 m with one-sided access and very difficult working conditions (Thuesen & Hvam 2013). Although every profession in the project has a share in the design and production of the shaft, nobody takes full responsibility for the realization of the shaft, and, as a consequence, the executing contractors usually face a significant project risk. Facing this challenge, the Scandinavian contractor Nordic Construction Company (NCC) decided to develop a flexible shaft based on the ideas of modularity as a way of managing complexity (see Figure 8.5 depicting the shaft before and after the application of modularity). The underlying idea of the developed solution is that all vertical installations of the main routes are concentrated in a shaft, which is split horizontally into factory-produced units corresponding to each floor. The units are produced off-site in an industrial process and transported to the building site in order to be installed concurrently with the erection of the base building/main structure as illustrated. Some of the effects of new modularity are

1. Significant reduction in assembly time on-site from 3 weeks to 7 minutes for each.
2. Assembly of the shaft by one craft (concrete worker) works particularly well – significantly reducing the number of crafts involved during the on-site production.
3. The in situ pouring of concrete after the assemblage results in tight slabs – an effect which usually has been impossible to achieve with the traditional construction practices.
4. Buy-in from the project workers, making the further implementation less challenging.
5. The designing project team was "forced" to develop integrated data models, clear communication interfaces and collaboration processes – a process standardization and modularization driven by a technical one. Despite theoretically a more complex design task, the quality and productivity of the design process was actually improved.

ANTIFRAGILITY

Antifragility is a management concept that seeks to establish organizations that thrive in uncertain and volatile environments. Fragile organizations (e.g. parts of our banking system) can suffer significant harm through volatility. Robust organizations (e.g. a project with ample buffers,

(a)   (b)

FIGURE 8.5 Installation shafts (a) before and (b) after the application of modularity as a tool for managing complexity.

management by deep pockets) are able to survive volatility and absorb their interference. Antifragile organizations, however, are designed to take maximum advantage of volatility and grow and improve as they are exposed to it. A classic example is evolution, where "fitter" species survive and over time fill every niche in an ecosystem. The antifragile organization derives more benefit than harm from volatility.

The antifragile mindsets take a rather radical approach and advocate much less top-down planning in favour of the ability to re-organize as a response to a disturbance or failure. At the core of antifragility is the conviction that planning is, to a large extent, futile, as unforeseeable events keep destroying our carefully laid out plans and detailed schedules. The following overview is an adaptation of advice to create antifragile organizations (Taleb 2012a, b):

- **Antifragility Rule 1: Think of projects as human bodies, not as machines.** Sophisticated machines rely on a central plan, expert operators and continuous outside maintenance. Human bodies self-heal and actually require "disorder" to survive (exposure to viruses to keep our immune system intact, stress on our bones to keep them healthy, and if they break, they grow back stronger). We run projects aiming at stability. Instead, we must allow natural fluctuations to continuously show weaknesses in our organization and have a decentralized ability to fix them as they occur. Central interventions should be saved for life-and-death situations, to allow the projects decentralized repair mechanisms to develop and strengthen.
- **Antifragility Rule 2: Create project portfolios that can collectively learn from the others' mistakes.** The airline industry has an aspect of antifragility: A plane crash triggers a thorough system-wide response. Root causes are analysed, lessons learned are shared globally, and the implementation of identified remedies is mandatory. In the project

management community, knowledge sharing and incorporating "lessons learned" are considered old news – but we are still not good at it. In order to be antifragile, this cross-project learning must be organic and not lead to an unwieldy (and fragile) centralized monstrosity.

- **Antifragility Rule 3: Small projects and project teams are efficient.** Large projects and organizations promise to deliver value by utilizing economies of scale. But they also introduce fragility: "To see how large things can be fragile, consider the difference between an elephant and a mouse: The former breaks a leg at the slightest fall, while the latter is unharmed by a drop several multiples of its height. This explains why we have so many more mice than elephants" (Taleb 2012b). Large projects are inherently more risky than small projects – the maximum possible loss is bigger. They also tend to overrun their planned cost more frequently. In project management, the exponential increase in "transaction cost" quickly offsets any economies of scale. The goal is to use a small team what a large team failed to do – not watch an even larger team fail again. This lesson has been learned at Lockheed Martin's Skunk Works in the 1950s, that designed ultra-modern military planes with a handful of people, and in many software development companies today, where large failed development teams are replaced with a small team around a gifted software architect. This also requires that we as project managers abandon the idea that a bigger team and bigger budget is bigger status.
- **Antifragility Rule 4: Fail often, fail cheaply.** Tinkering and trial-and-error approaches are central to any innovation. Almost all big leap forwards – from Edison's lightbulb to Job's iPod – rest on a pile of mostly invisible failures. The key requirement is that experiments and risk-taking come cheap, so you can keep at it until a solution emerges – as opposed to betting the house (typically once). In project environments, this idea can mean anything from exposing end-users to hundreds of cardboard prototypes (or a software interface consisting of flipcharts and post-its), to creating virtual simulation and testing environments that allow us a semi-automatic testing of more complex technical challenges.
- **Antifragility Rule 5: Project managers must have real skin in the game.** Taking big risks with someone else's resources is very different from taking big risks with one's own. We typically provide incentives for success – in fact creating a reward for excessive risk taking, as the "upside risk" is much more attractive than the "downside risk" is off-putting (we can always blame a failure on someone else). Decision-making and reward structures must align with the long-term objectives of a project or organization. This includes continued responsibility of project managers throughout the extensive life cycle of a project (and possibly the resulting system the project creates, calling for a job change).

MINDFULNESS

We now move to a behavioural view of the organization, to address some of its socio-political complexity. In particular, we borrow from the ideas of mindfulness from Weick and Sutcliffe.

The behaviour resulting from our thinking and decision-making biases are inherently human but constitute a risk to any project. Yet, we can discipline our minds to think sharper, quicker and clearer. Internalization of mindfulness principles offers one approach to do so. Mindfulness is understood as "a rich awareness of discriminatory detail. By that we mean that when people act, they are aware of context, of ways in which details differ (in other words, they discriminate among details), and of deviations from their expectations" (Weick & Sutcliffe 2001, p. 32). In other words, mindfulness is about keeping a high level of alertness and awareness of context and using this awareness in our decisions and actions. Mindful project managers know that they do not know everything and that despite their experience, new types of problems can happen. They are aware of their cognitive biases and fight against their wish to confirm their initial assumptions. They seek for deviations, engage with different perspectives of the project and attempt to create a more comprehensive understanding of the current problem and ways to solve it.

Weick and Sutcliffe (2001) developed five principles of mindfulness. The first three principles explore ways in which project managers can anticipate problems. Yet, as we mentioned above, we are fallible, and managers should be concerned not only with prevention but also with cures. In this regard, the last two suggest principles focus on reacting to unexpected events and its negative consequences. The objective of these principles is to develop more reliable organizations, that is, organizations that can reduce devastating impacts of unexpected events and recover rapidly. Therefore, in project management domain, mindfulness is particularly instrumental to manage in the intersection between human behaviour and uncertainty. The principles were developed for repetitive operations and in particular organizations such as healthcare or nuclear power plants, or aircraft. Yet, they are also applicable to project management context (see, for instance, Denyer et al. 2011). Weick and Sutcliffe (2001) provide us some pointers and a starting point in our journey to improve our cognition (Table 8.1).

QUESTIONS FOR DISCUSSION

The following questions are examples that can be used in classroom to discuss the complexities involved in projects as well as the different forms of perceiving it:

- What are the different ways of dividing the work in a specific project? What are the implications of these different ways of dividing work?
- What kind of person are you: someone how likes to see the differences or the forms of integration between different parts of a project?
- Why are projects complex?
- Did you make your current project into a complex system or was your project complex? Why?
- Can you shape the nature of complexity of your project? When? Why? How?

TABLE 8.1

Examples of Mindfulness in Project Management

Principle	Description
Reluctance to simplify	Simplify slowly, reluctantly and mindfully. Be careful with simplified explanations and categories, such as make or buy, or right or wrong. Problems faced in projects are more nuanced and often offer more options.
Preoccupation with failure	"Pay close attention to weak signals of failure that may be symptoms of larger problems within the system" (Weick and Sutcliffe 2001).
	Develop practices to preclude mistakes that have strategic negative impact on the project, such as clearly articulate and communicate these potential unacceptable mistakes to project stakeholders.
Sensitivity to operations	Be responsive to the messy reality inside of projects. This involves, on the one hand, to control project progress, and to look for potential deviations and their implications to the project as a whole, and on the other, be mindful to potential unexpected events that go beyond what you would usually control in project context.
Commitment to resilience	Accept that unexpected events will occur and that project managers need to react to them quickly. Resilience involves (1) the ability to absorb strain and preserve the functioning of the project – the show must go on, (2) ability to recover quickly and 3) ability to learn from the unexpected event and how they impacted the project.
Deference to expertize	Deference to expertize is not about assigning an expert to a problem, for instance, a risk practitioner, who will then be doing risk management, so project manager can manage other aspects of the project. Such behaviour can do more harm than good. This is about giving voice to experts actively involved in the project. They are the ones who first notice early weak signals of mistakes but are also often powerless, afraid of speak up, and even may not realize how important some deviations may be to the entire project.

SUMMARY AND CONCLUDING NOTES

Complexity is integral to organizing projects, programmes and portfolios. Project managers have a significant influence on how complexity will impact projects, for good or bad. For example, a project manager's decisions on how to organize work can create difficult and political interfaces that are hard to manage. Hence, managing projects involves the active management of complexity. The topic has received wide attention of project management practitioners and academics alike. Yet, we still have a poor understanding of complexity embedded in projects, and above all, we still insist on denying or playing down the complexity of our projects. We assume that we can follow a staged and deterministic process by defining requirements, investigating alternative solutions, evaluating solutions and implementing them. The consequence is the use of inappropriate organizational responses, leading to inefficiencies, frustration and ultimately failure.

This chapter offered some concepts and organizational approaches to understand, accept and manage the complexity of projects, programmes and portfolios appropriately. Our intention was to move outside of project management literature, in an attempt to bring new insights to the area. Our emphasis lay on the complexity involved in projects and programmes that shape engineering systems, for example energy generation and distribution, transportation infrastructure and healthcare. By looking at projects that shape engineering systems and borrowing from some streams of research within a broadly conceived array of Organization Theory, we bring difficult and abstract concepts to the realities of organizing projects and articulate implications to practice.

The chapter contributes to literature and practice in three ways. First, it builds on the existing research on complexity in general and in projects (Williams 2005; Geraldi et al. 2011; Maylor et al. 2013; Snowden & Bonne 2007; Rittel & Webber 1973), and concurring with their approaches, we define complexity as a combination of three concepts: structural complexity, uncertainty and sociopolitical complexity. What this chapter adds is the analysis of how these elements are intertwined and with what consequences to the management of projects. The second contribution is to offer innovative concepts developed outside of project management domain to embrace and manage complexity in projects. Finally, the chapter makes these abstract concepts more concrete and by doing so more visible to practitioner and academics. We do this by illustrating the concepts and potential implications and management approaches with concrete examples. We thereby contribute to closing the gap between our understanding of complexity and our ability to organize and manage it.

What we are not saying is that we have a silver bullet to organize complexity. Instead, we hope to have raised awareness of the difficulty to embrace the complexities of projects.

We are at the cutting edge of organizing practices, and therefore, approaches are still under development. Application of these concepts demands reflective practitioners who are willing to experiment, and sensitivity to context to discover what works and what does not. We are interested in moving this agenda forward, and in order to do so, the close collaboration between academics and practitioner is crucial. We are looking forward to engaging with practitioners willing to discover them together with their counterparts in academia.

Project managers have a significant influence on how complexity will impact projects, for good or bad. For example, a project manager's decisions on how to organize work can create difficult and political interfaces that are hard to manage. Hence, managing projects involves the active management of complexity.

We need more organizing tools that enable managers to accept, embrace and manage the complexity of their projects.

ACKNOWLEDGMENTS

This chapter is adapted, and many parts of the text are identical to an unpublished conference contribution: Oehmen et al. (2015). We thank the project management institute (PMI) for providing us with the copyright of the text and allow us to reuse it in this text.

ENDNOTE

1 He actually wrote: "There is always a well-known solution to every human problem – neat, plausible, and wrong."

REFERENCES

Abdel-Hamid, T. K., & Madnick, S. E. (1991). *Software Project Dynamic – An Integrated Approach.* Englewood Cliffs, NJ: Prentice Hall.

Alexiou, K. (2010). Coordination and emergence in design. *CoDesign*, 6(2), 75–97.

Baccarini, D. (1996) The concept of project complexity--A review. *International Journal of Project Management*, 14, 201–204.

Browning, T. R., & Ramasesh, R. V. (2007). A survey of activity network-based process models for managing product development projects. *Production and Operations Management*, 16(2), 217–240

Calvano, C. N., & John, P. (2004). Systems engineering in an age of complexity. *Systems Engineering*, 7(1), 25–34.

Campagnolo, D., & Camuffo, A. (2010). The concept of modularity in management studies: A literature review. *International Journal of Management Reviews*, 12(3), 259–283.

Cilliers, P. (2000). *Complexity and Postmodernism: Understanding Complex systems.* London: Routledge.

DeMeyer, A., Loch, C. H., & Pitch, M. T. (2002). Managing project uncertainty: From variation to chaos. *MIT Sloan Management Review*, 43(2), 60–67.

Denyer, D., Kutsch, E., Lee-Kelley, E. L., & Hall, M. (2011). Exploring reliability in information systems programmes. *International Journal of Project Management*, 29(4), 442–454.

De Weck, O. L., Roos, D., & Magee, C. L. (2011). *Engineering Systems: Meeting Human Needs in a Complex Technological World.* Cambridge, MA: MIT Press.

Eppinger, S. D., & Browning, T. R. (2012). *Design Structure Matrix Methods and Applications.* Boston, MA: MIT Press.

Geraldi, J. G., & Adlbrecht, G. (2007). On faith, fact, and interaction in projects. *Project Management Journal*, 38(1), 32–43.

Geraldi, J., Maylor, H., & Williams, T. (2011). Now, let's make it really complex (complicated) A systematic review of the complexities of projects. *International Journal of Operations & Production Management*, 31(9), 966–990.

Geraldi, J., & Terrikangas, S. (2011). From project management to managing by projects: Learning from the management of M&As. *International Network of Organizing by Projects (IRNOP) Biannual Conference,* Montreal, Canada, June 19–22.

Green, G. C. (2004). The impact of cognitive complexity on project leadership performance. *Information and Software Technology*, 46(3), 165–172.

Holland, J. H. (1997). *Emergence: From Chaos to Order.* Oxford: Oxford University Press.

Johnson, N. F. (2007). *Two's Company, Three is Complexity: A Simple Guide to the Science of All Sciences.* London, UK: Oneworld Publications Ltd.

Maylor, H., Turner, N., & Murray-Webster, R. (2013). How hard can it be? Actively managing complexity in technology projects. *Research-Technology Management*, 56(4), 45–51.

Mencken, H. L. (1917). The divine afflatus. *New York Evening Mail*, November 16. Republished in Prejudices: Second series. New York: Knopf, 1920; A Mencken chrestomathy. New York: Knopf, 1949.

Meyer, M. H., & Lehnerd, A. P. (1997). *The Power of Product Platforms - Building Value and Cost Leadership.* New York: The Free Press.

Oehmen, J., Thuesen, C., Parraguez, P., & Geraldi, J. (2015). *Complexity Management for Projects, Programmes, and Portfolios: An Engineering Systems Perspective.* Newtown Square, PA: Project Management Institute, Thought Leadership Series. PMI.

Pine, B. J., & Gilmore, J. H. (1999). *The Experience Economy.* Boston, MA: Harvard Business School Press.

Rittel, H. W., & Webber, M. M. (1973). Dilemmas in a general theory of planning. *Policy Sciences*, 4(2), 155–169.

Schilling, M. A. (2000). Towards a general modular systems theory and its application to inter-firm product modularity. *Academy of Management Review*, 25, 312–334.

Simon, H. (1962). The architecture of complexity. *Proceedings of the American Philosophical Society*, 106(6), 467–482. www.jstor.org/stable/985254

Snowden, D. J., & Boone, M. E. (2007). A leader's framework for decision making. *Harvard Business Review*, 85(11), 68.

Sterman, J. D. (2000). *Business Dynamics: Systems Thinking and Modeling for a Complex World*, Vol. 19. Boston, MA: Irwin/McGraw-Hill.

Steward, D. V. (1981). The design structure system: A method for managing the design of complex systems. *IEEE transactions on Engineering Management*, (3), 71–74.

Taleb, N. N. (2012a). *Antifragile: Things that Gain from Disorder*. 544 p. New York: Random House.

Taleb, N. N. (2012b). Learning to love volatility. *The Wall Street Journal*. www.wsj.com/articles/SB10001424 127887324735104578120953311383448

Thuesen, C. (2012). Understanding project based production through socio-technical modularity. In *The Academy of Management 2012 Annual Meeting, Boston, MA*.

Thuesen, C., & Hvam, L. (2013). Rethinking the business model in construction by the use of off-site system deliverance: Case of the shaft project. Journal of Architectural Engineering, 19(4), 279–287.

Turner, J. R., & Cochrane, R. A. (1993). Goals-and-methods matrix: Coping with projects with ill defined goals and/or methods of achieving them. *International Journal of Project Management*, 11(2), 93–102.

Ulrich, K. (1995). The role of product architecture in the manufacturing firm. *Research Policy*, 24, 419–440.

Weick, K. E., & Sutcliffe, K. M. (2001). *Managing the Unexpected: Assuring High Performance in an Age of Complexity*. San Francisco, CA: John & Wiley Sons.

Williams, T. (2005). Assessing and moving from the dominant project management discourse in the light of project overruns. *IEEE Transactions on Engineering Management*, 52(4), 497–508.

Wyatt, D. F., Wynn, D. C., & Clarkson, P. J. (2013). A scheme for numerical representation of graph structures in engineering design. *Journal of Mechanical Design*, 136(1). doi:10.1115/1.4025961

9 Enterprise Governance Toolset

Charles B. Keating
Old Dominion University

Polinpapilinho F. Katina
University of South Carolina Upstate

CONTENTS

BACKGROUND

Complex projects are challenged to operate under conditions of increasing complexity, scarce resources, and escalating demands to deliver higher levels of performance. This is not new. However, for project practitioners, the relentless nature of internal system flux and external environmental turbulence for projects are not likely to diminish in the near future. To more definitively examine the nature of this problem domain, we have focused on providing a more detailed accounting of challenges facing practitioners charged with managing complex projects. In this section, we specifically examine the complex system (project) problem domain and the specific resulting project governance challenges. Arguably, the degree to which practitioners master this domain will determine the effectiveness of their projects but also the continuing success of the larger enterprise served.

This exploration for application of Complex System Governance (CSG) to complex projects is focused on four primary objectives:

1. *Articulate the complex problem domain which must be confronted by modern complex project practitioners* – this examination will more precisely articulate the essence of the problem domain confronting project practitioners and which CSG has been specifically developed to confront. The historical nature and need for a 'different response' provided by CSG is placed in the context of this problem domain.

2. *Explore CSG as an emerging approach to more effectively confront the complex problem domain* – this exploration provides the conceptual foundations upon which CSG has been developed. Additionally, the details that delineate CSG as an appropriate response to the complex problem domain facing project practitioners are provided.

3. *Examine CSG application scenarios and introduce several tools to aid practitioners in discovery of systemic inadequacies in complex project design, execution, and development* – this briefly demonstrates how CSG might contribute to a more informed and effective mastery of the complex problem domain facing project practitioners.

4. *Provide development implications for practitioners performing systemic intervention to enhance CSG* – the pragmatic concerns and guidance for application and deployment of CSG are examined. Emphasis is placed on the assumptions, limitations, and considerations that practitioners should consider as they move forward in the implementation of CSG-based initiatives.

GOVERNANCE CHALLENGES FOR COMPLEX PROJECTS

PROBLEM DOMAIN FOR COMPLEX PROJECTS

The domain of the project practitioner is truly 'holistic' in nature. This spans the entire gamut of dimensions that practitioners must both consider and contend. Among these dimensions, we include social, educational, political, technology, resources, demographics, economics, culture, infrastructure, legal, regulatory, environmental, and information. A dizzying array for even the most proficient project practitioner. To exacerbate this, these dimensions are subject to shifts in understanding, criticality, and priority over the project life cycle. At first glance, the magnitude and complexity of these considerations appear insurmountable. While some may claim that projects have always been challenged to deal with these conditions, we argue that the present and future conditions are more acute for practitioners. These conditions delineate the shifting landscape of the project practitioner and can be characterized by several dominant characteristics. Following previous recitations of this landscape from recent works (Jaradat and Keating 2014; Keating, Katina, and Bradley 2015; Keating 2014; Keating and Katina 2011; Jaradat, Keating, and Bradley 2017), Table 9.1 provides a summary of characteristics and their nature for the domain faced by project practitioners who must deal with the domain.

The problems emanating from this domain continue to proliferate into all aspects of the world of project practitioners. The complex problems stemming from this domain do not have a precise cause–effect relationship that would make understanding and resolution easy or reducible to the precision demanded by rigorous mathematical formulation and solution. Instead, the landscape for project practitioners might be best characterized as a 'complex problem space'. This problem space is holistic in nature, ranging across the entire spectrum of dimensions previously mentioned that both impact, and are impacted by, complex enterprises. Of particular concern are the 'products' that emanate from the enterprise interaction with this problem space. The different aspects of this increasingly complex problem space have been previously articulated (Keating, Katina, and Bradley 2015; Jaradat and Keating 2014; Keating 2014; Keating and Katina 2011; Naphade et al. 2011) as being characterized by producing such conditions as identified in Figure 9.1. In sum, we suggest that the enterprises experience many of the recognized conditions that plague complex systems. To project practitioners, this listing is likely recognizable and represents nothing that is not, or has not been, faced on a routine basis with varying results.

Dealing with this problem space is not insurmountable. However, effectively dealing with this problem space requires a different level of thinking, consistent with Einstein's proclamation that 'We cannot solve our problems with the same thinking we used when we created them'. CSG is an emerging field designed to address this increasingly hostile landscape, which represents a 'new normal' for project practitioners. We now shift focus to an overview of governance challenges facing project practitioners.

TABLE 9.1

Domain of System (Project) Practitioners

Characteristics	Nature
Complexity	• Exponential magnitude, availability, and accessibility of information coupled with the increasingly large number of richly interconnected elements. • Incomplete, fallible, and dynamically evolving system knowledge. • High levels of uncertainty beyond current capabilities to structure, order, and reasonably couple decisions, actions, and consequences. • Emergence of behavior, performance, and consequences that cannot be known or predicted in advance of their occurrence.
Contextual dominance	• Unique circumstances, factors, patterns, and conditions within which a system is embedded – influencing the system and influenced by the system. • Enabling and constraining to decisions, actions, and interpretations made with respect to the system. • Multiple stakeholders with different worldviews, objectives, and influence patterns.
Ambiguity	• Instabilities in understanding system structure, behavior, or performance. • Potential lack of clarity in system identity, boundary conditions, delineation of system constituents, and understanding of a system and its context.
Holistic nature	• In addition to technical/technology aspects of a system, consideration for the entire influencing spectrum of human/social, organizational/managerial, policy, political, and information aspects central to a more complete (holistic) view of a system. • Behavior and performance as a function of interactions in the system – not reducible or revealed by understanding individual constituents.

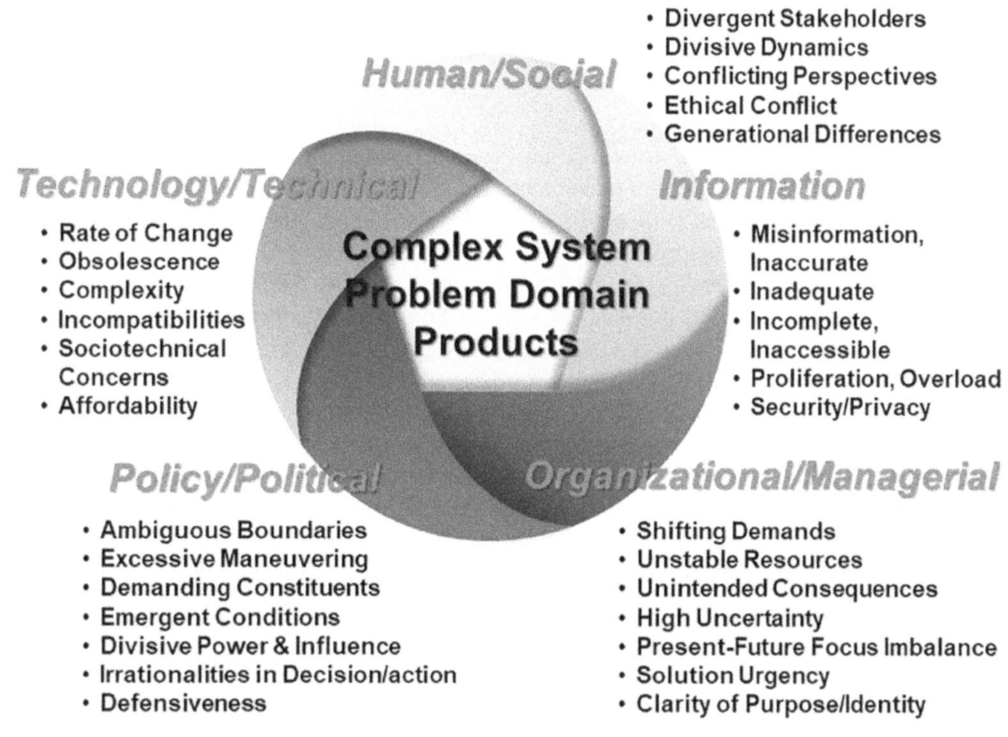

FIGURE 9.1 Products from the complex system problem space.

At the most fundamental level, 'governance', whether for a nation, system, enterprise, or project, is concerned with the provision of direction, oversight, and accountability. CSG is focused on helping enterprises (systems/projects), and their practitioners (owners, operators, designers, performers) deal more effectively with increasing complexity. It is no revelation that enterprises are under pressure to increase effectiveness in delivery of products, services, and values – with reduced resources while operating under increasingly complex conditions. These conditions are not going away. To ignore them is shortsighted, leaving project practitioners in a precarious position. At best, systems being governed under these conditions will continue to perform at a level that will assure their continued existence. At worst, these systems will continually disappoint, discourage, and decline. In the project world, this means continued difficulties in meeting cost, schedule, and performance expectations. This is not intended as a dire warning but rather an invitation to more purposefully engage in the development of systems (projects) that must operate under these conditions. CSG is focused on successfully navigating these conditions to produce higher performing systems and ease the burden of practitioners. In the following section, we delve more deeply into CSG as a response to enhance project practitioner effectiveness in dealing with increasingly complex problem domains.

COMPLEX SYSTEM (ENTERPRISE/PROJECT) GOVERNANCE

This section provides a detailed overview of CSG from a project perspective. The emphasis is directed to examination of the conceptual basis for the field, identification of the central themes upon which CSG is based, and providing a summary of the paradigm defining CSG. CSG has been previously identified as a framework to guide design, assessment, and evolution of nine essential functions that are required to sustain and evolve system performance (Keating, Katina, and Bradley 2014). We will develop these functions in detail later in this chapter. However, important to the present discussion is that the CSG functions enable systems and their practitioners to excel in the midst of constant flux, disorder, and environmental turbulence.

CONTRIBUTING FIELDS

CSG development and application draws upon a strong conceptual base found in Systems Theory, Management Cybernetics, and System Governance (Figure 9.2). In essence, *Systems Theory* offers a set of propositions (principles, laws, concepts) that have been continually developed and applied over the past eight decades. The propositions have withstood the test of time and application, serving to define the structure, behavior, and performance of all systems. Systems Theory propositions are nonnegotiable and have real consequences for systems and practitioners that, knowingly or unknowingly, 'violate' them. The strong influences of Systems Theory are found in the emphasis on integration and coordination for CSG.

Management Cybernetics provides a strong conceptual foundation for communication and control essential to CSG. In particular, Management Cybernetics offers CSG design cues for control through the model of a 'metasystem'. The 'metasystem' is a set of functions that stand above/beyond the particular systems/entities that it seeks to 'steer' – in the cybernetic sense of providing control. This 'cybernetic' concept of control found in Management Cybernetics also provides a set of communication channels associated with the 'steering' functions of the metasystem.

System Governance brings a uniquely broad and long-term view to complex systems development. Based on the spectrum of governance perspectives suggested primarily by Calida (2013) and supplemented by others, we can draw several important themes. These themes serve to inform a systems perspective of governance critical to CSG. At a most basic level, governance requires continuous achievement of:

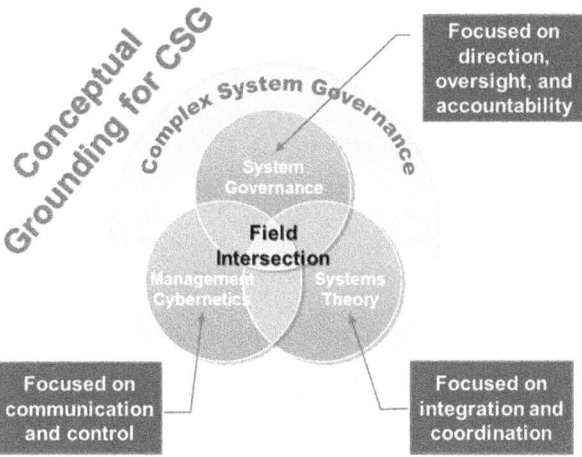

FIGURE 9.2 CSG at the conceptual intersection of three fields.

1. *Direction* – sustaining a coherent identity and vision that supports consistent decision, action, interpretation, and strategic priorities
2. *Oversight design* – providing control and integration of the system and corresponding initiatives
3. *Accountability* – ensuring efficient strategic resource utilization, performance monitoring, and resolution of aberrant conditions.

For complex systems (e.g. projects), it is important to make a distinction between governance and management, since the two are often loosely interchanged. In essence, we arguably distinguish governance from management along three dimensions (time horizon, focus, and action-response proximity). While these dimensions are certainly debatable in degree, we present them as central to our development of the different perspective intended for CSG. First, the governance time horizon is directed to the 'long view' and future. This is opposed to a management emphasis that is much more near term and operationally centered. Second, the governance focus is directed to development and pursuit of effectiveness related to vision. This suggests a more global focus, emphasizing long-term effectiveness derived from setting an appropriate vision and maintenance of trajectory toward that vision. In contrast, the management time horizon is more limited, with a dominant local emphasis on near-term objectives and performance – rooted in efficient use of resources. Third, governance is tolerant of higher levels of uncertainty in pursuit of aims. The proximity of response to governance decisions may unfold over a long-time horizon and evolve with changing circumstances. On the contrary, management, while not ignoring the long term, is more focused on detailed planning, immediate decision results impacting performance, and specific measurable objectives. While these governance–management distinctions are debatable in essence and degree, we suggest that the difference in perspective is substantial and intentionally moves CSG to a different logical level than traditional management formulations. This different logical level is critical to our further formulation of the emerging CSG paradigm.

THE CSG PARADIGM

From this conceptual grounding in Systems Theory, Management Cybernetics, and System Governance, we have the conceptual foundations to begin the definition of CSG. There are two important aspects that underpin CSG. These include (1) how it is defined and (2) what it does. For conciseness in presenting CSG themes, we will examine each of these aspects.

CSG is defined as 'Design, execution, and evolution of the [nine] metasystem functions necessary to provide control, communication, coordination, and integration of a complex system, (Keating,

Katina, and Bradley 2014, p. 274). At a high level, the following elements of the definition are elaborated as an essential foundation (noting that an enterprise is a form of system):

- **Design** – purposeful and deliberate arrangement of the governance system to achieve desirable system performance and behavior.
- **Execution** – performance of the system design within the unique system context, subject to emergent conditions stemming from interactions within the system and between the system and its external environment.
- **Evolution** – the change of the governance system in response to internal and external shifts as well as revised trajectory.
- **Metasystem** – the set of nine interrelated higher level functions that provide for governance of a complex system.
- **Control** – invoking the minimal constraints necessary to ensure desirable levels of performance and maintenance of system trajectory, in the midst of internally or externally generated perturbations of the system.
- **Communication** – the flow, transduction, and processing of information within and external to the system that provides for consistency in decisions, actions, interpretations, and knowledge creation made with respect to the system.
- **Coordination** – providing for interactions (relationships) between constituent entities within the system, and between the system and external entities, such that unnecessary instabilities are avoided.
- **Integration** – continuous maintenance of system integrity. This requires a dynamic balance between autonomy of constituent entities and the interdependence of those entities to form a coherent whole. This interdependence produces the system identity (uniqueness) that exists beyond the identities of the individual constituents.
- **Complex system** – a set of bounded interdependent entities forming a whole in pursuit of a common purpose to produce value beyond that which individual entities are capable.

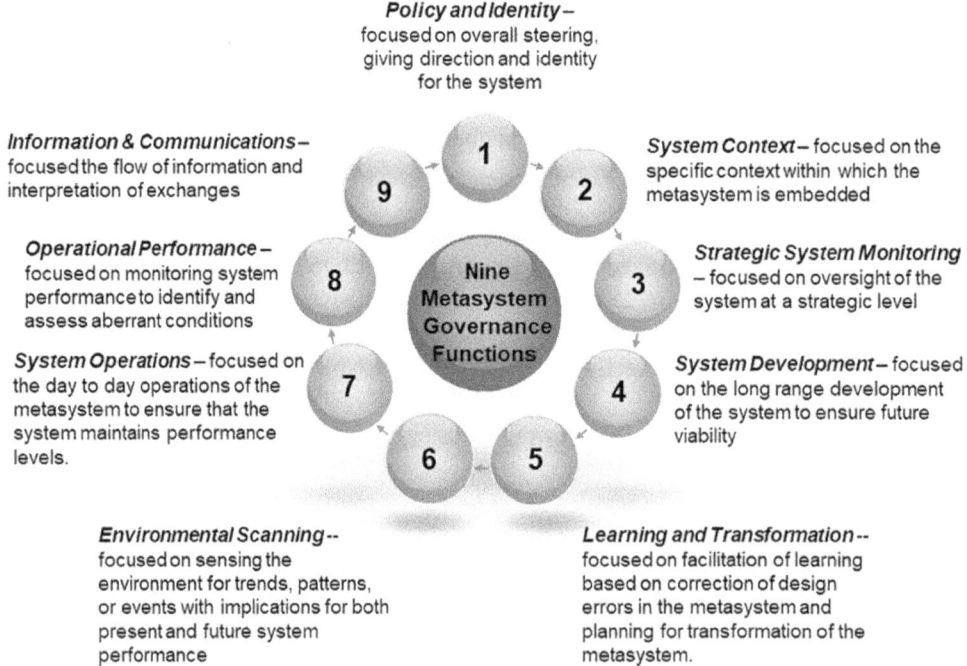

FIGURE 9.3   Nine CSG metasystem functions.

In the examination of what CSG does, our focus turns to the performance of nine essential (metasystem) functions (Figure 9.3).

Instrumental to the formulation of CSG is the unique role of the 'metasystem'. The metasystem construct brings several important considerations for the CSG, including the following: (1) the metasystem operates at a logical level beyond the elements that it must integrate; (2) the metasystem has been conceptually grounded in the foundations of Systems Theory and Management Cybernetics; (3) a metasystem is a set of interrelated functions – which only specify 'what' must be achieved for continuing system viability (existence), not 'how' those functions are to be achieved (i.e. mechanisms of enactment); (4) the metasystem functions must be performed if a system is to remain viable (continue to exist) – this does not preclude the possibility that a system may be poorly performing, yet still continue to be viable; and (5) a metasystem can be purposefully designed, executed, and maintained or left to its own (self-organizing) development. The performance of metasystem functions, required by all existing systems, supports achievement of control, communications, coordination, and integration as defined earlier in this section. Table 9.2, following earlier CSG works (Keating and Bradley 2015; Jaradat and Keating 2014; Keating 2014; Keating and Katina 2011), provides a concise overview of the metasystem functions and what they provide with respect to governance of complex systems.

TABLE 9.2
Metasystem Functions for a Complex System

Metasystem Function	Primary Role of the Function
Metasystem Five (M5): Policy and identity	To provide direction, oversight, accountability, and evolution of the system. Focus includes policy, mission, vision, strategic direction, performance, and accountability for the system such that: (1) the system maintains viability, (2) identity is preserved, and (3) the system is effectively projected both internally and externally.
Metasystem five star (M5*): System context	Identifies and monitors the system context (i.e. the circumstances, factors, conditions, or patterns that enable and constrain the system). Assesses the potential impacts (negative and positive) of contextual considerations on the system.
Metasystem five prime (M5′): Strategic system monitoring	To monitor measures for strategic system performance and identify variance requiring metasystem-level response. Ensures early identification of variabilities that might require system redesign or performance measurement reassessment. Particular emphasis is on variability that may impact future system viability in a negative or positive manner.
Metasystem four (M4): System development	To provide for the analysis and interpretation of the implications and potential impacts of trends, patterns, and precipitating events in the environment. Develops future scenarios, design alternatives, and future-focused planning to position the system for future viability.
Metasystem four star (M4*): Learning and transformation	To provide for identification and analysis of metasystem design errors (second-order learning) and suggest design modifications and transformation planning for the system.
Metasystem four prime (M4′): Environmental scanning	To provide the design and execution of scanning for the system environment. Focus is on patterns, trends, threats, events, and opportunities for the system that might enable or constrain present and future system performance.
Metasystem three (M3): System operations	To maintain operational performance control through the implementation of policy, resource allocation, and design for accountability. Facilitates modifications necessary to adjust to near-term variabilities with impact on system operations/production of goods/services.
Metasystem three star (M3*): Operational performance	To monitor measures for operational performance and identify variance in system performance requiring system-level response. Particular emphasis is on variability and performance trends that may impact system operations, enabling or constraining efficient system production of goods/services.
Metasystem two (M2): Information and communications	To enable system stability by designing and implementing architecture for information flow, coordination, transduction, and communications within and between the metasystem, the environment, and the system being governed.

Ultimately, effectiveness in purposeful design, execution, and evolution of the nine 'metasystem governance' functions determines system performance.

Consistent with the metasystem functions are the 'communication channels' that provide for the flow and interpretation of information in the system. The ten communication channels are adapted from the work of Beer (1972, 1979, 1985) and extensions of Keating and Morin (2001). To succinctly capture the communication channels, Table 9.3 provides a concise overview of the channels.

Implementing mechanisms is the final element that forms a CSG triad (Figure 9.4), complementing conceptual foundations and metasystem functions (and their corresponding communication channels). Conceptual foundations help to explain and understand 'why' systems behave and perform as they do, based on the axioms and propositions of Systems Theory and Management Cybernetics. These axioms and propositions are immutable and cannot be negotiated away – in effect providing

---

TABLE 9.3

Metasystem Communication Channels and Roles

Communications Channel and Function	CSG Metasystem Role
Command (Metasystem 5)	• Provides nonnegotiable direction to the metasystem and governed systems • Primarily from Metasystem 5 and disseminated throughout the system
Resource bargain/accountability (Metasystem 3)	• Determines and allocates the resources (manpower, material, money, information, support) to governed systems • Defines performance levels, responsibilities, and accountability for governed systems • Primarily an interface between Metasystem 3 and governed systems
Operations (Metasystem 3)	• Provides for the routine interface focused on near-term operational focus • Concentrated on direction for system production (products, services, processes, information) consumed external to the system • Primarily an interface between Metasystem 3 and governed systems
Coordination (Metasystem 2)	• Provides for metasystem and governed systems balance and stability • Ensures that information concerning decisions and actions necessary to prevent disturbances are shared within the metasystem and governed systems • Primarily a channel designed and executed by Metasystem 2
Audit (Metasystem 3*)	• Provides routine and sporadic feedback concerning operational performance • Investigation and reporting on problematic performance issues within the system • Primarily a Metasystem 3* channel for communicating between Metasystem 3 and governed systems concerning performance issues
Algedonic (Metasystem 5)	• Provides a 'bypass' of all channels when the integrity of the system is threatened • Compels instant alert to crisis or potentially catastrophic situations for the system • Directed to Metasystem 5 from anywhere in the metasystem or governed systems
Environmental scanning (Metasystem 4')	• Provides design for sensing of the external environment • Identifies environmental patterns, activities, or events with system implications • Provides for access throughout the metasystem as well as governed systems
Dialog (Metasystem 5')	• Provides for examination of system decisions, actions, and interpretations for consistency with system purpose and identity • Directed to Metasystem 5' from anywhere in the metasystem or governed systems
Learning (Metasystem 4*)	• Provides detection and correction of error within the metasystem as well as governed systems, focused on system design issues as opposed to execution • Directed to Metasystem 4* from anywhere in the metasystem or governed systems
Informing (Metasystem 2)	• Provides for flow and access to routine information in the metasystem or between the metasystem and governed systems • Access provided to entire metasystem and governed systems

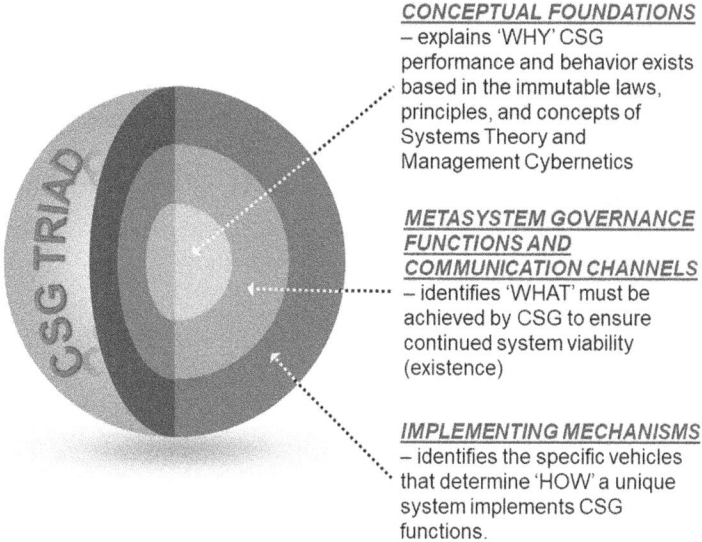

CONCEPTUAL FOUNDATIONS
– explains 'WHY' CSG
performance and behavior exists
based in the immutable laws,
principles, and concepts of
Systems Theory and
Management Cybernetics

**METASYSTEM GOVERNANCE
FUNCTIONS AND
COMMUNICATION CHANNELS**
– identifies 'WHAT' must be
achieved by CSG to ensure
continued system viability
(existence)

IMPLEMENTING MECHANISMS
– identifies the specific vehicles
that determine 'HOW' a unique
system implements CSG
functions.

FIGURE 9.4   CSG triad of foundations, functions, and mechanisms.

the 'laws' to which all systems are subject. The consequences for violation of these system 'laws' are real and will impact system viability. The metasystem functions and their communication channels (second element of the triad) identify 'what' must be achieved to ensure continued system viability.

ALL systems must perform the metasystem functions and communications channels at a minimal level to maintain viability. However, viability is not a 'guarantee' of performance excellence. On the contrary, viability simply assures that the system continues to exist. There are degrees of viability, the minimal of which is existence. The third element of the triad is implementing mechanisms. Implementing mechanisms are the specific vehicles (e.g. processes, procedures, activities, practices, plans, artifacts, values/beliefs, customs, traditions) that implement metasystem governance functions and their communication channels for a specific system of interest (enterprise/project). These mechanisms may be explicit/tacit, formal/informal, routine/nonroutine, effective/ineffective, or rational/irrational. However, all mechanisms can be articulated in relationship to the metasystem governance functions and corresponding communication channels they enact and support.

In summary thus far, the CSG paradigm (Figure 9.5) can be stated succinctly as:

From a systems theory based conceptual foundation, a set of nine interrelated metasystem functions is enacted through mechanisms. These mechanisms invoke metasystem governance to produce the communication, control, coordination, and integration essential to ensure continued system viability.

Critical to understanding the metasystem is the positioning of the metasystem for CSG in relationship to the environment, context, and system of interest (Figure 9.6).

The following descriptions are provided to focus our discussion of the CSG paradigm:

- **Environment**: The aggregate of all surroundings and conditions within which a system operates. It influences and is influenced by a system.
- **Context**: The circumstances, factors, patterns, conditions, or trends within which a system is embedded. It acts to constrain or enable the system.
- **System(s)**: The set of interrelated elements that are subject to immutable system axioms and propositions and are governed to produce that which is of value and consumed external to the system.
- **Metasystem**: The set of functions that are invoked through mechanisms to govern a system such that viability (existence) is maintained.

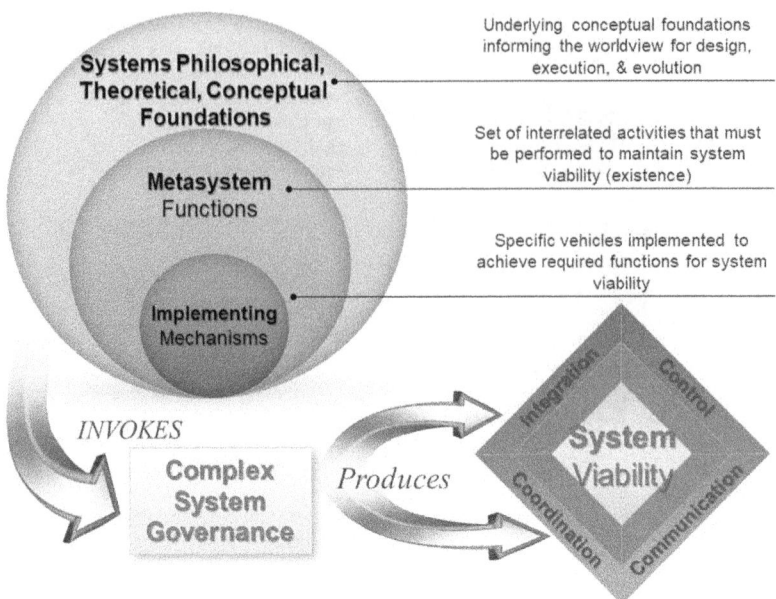

FIGURE 9.5   CSG paradigm.

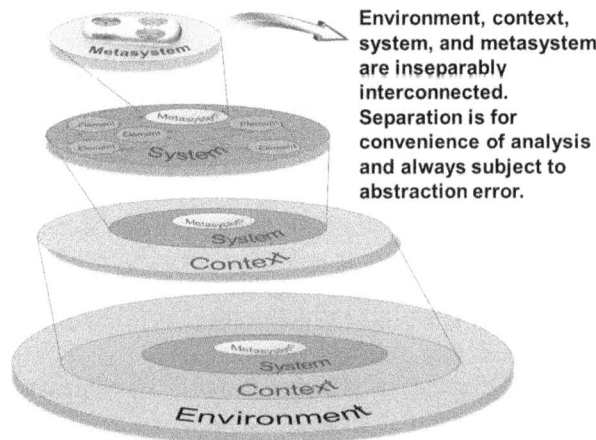

FIGURE 9.6   Interconnection and embedded nature of metasystem, system, context, and environment.

There are four important points concerning the relationship of these four elements in the CSG paradigm. First, the metasystem, system, and context are embedded in the larger environment. This implies that separation can only occur through a process of 'abstraction'. The process of abstraction is essential for analysis but also carries with it inevitable abstraction errors. There is no perfect abstraction, meaning all abstractions have some level of error. Care must be taken to account for the choices (assumptions, judgments) made to support abstraction. Second, the metasystem exists as 'meta' (beyond/above) to the system(s) that it seeks to govern. While it serves the objective of purposeful analysis and development, the metasystem simply imposes a viewpoint to examine the interconnected mechanisms that perform the functions necessary for integration, coordination, communication, and control of the system. Therefore, the metasystem is a construct that allows organization of mechanisms by essential functions to support analysis of a complex system/project. Third, the system is separated from the environment by the system boundary. The system boundary

163

is established by the criteria that define inclusion and exclusion with respect to what constitutes the system. Although the boundary is imposed for purposes of analysis, care must be taken to be conscious of both the initial establishment as well as shifts in the boundary conditions over time. Fourth, the system and metasystem are embedded within the context. In essence, the context acknowledges conditions that are more closely coupled to the system/metasystem than those in the environment (e.g. system leadership style).

The separation of the environment, context, system, and metasystem is for convenience and permits analysis. In reality, these four elements exist as an inseparable whole. The separation of these elements always requires judgments. Judgments of boundaries, relevant aspects of the environment, contextual definition, and articulation of the metasystem are always subject to 'abstraction error'. Therefore, CSG requires purposeful and accountable decisions with respect to abstraction of the context, system(s), and metasystem from the environment.

This completes the development of the CSG paradigm. We now shift our attention to the examination of where deficiencies in performance of CSG (pathologies) might occur and their relationship to failure in complex systems/projects.

DEFICIENCIES IN PERFORMANCE OF CSG

Certainly, understanding of system performance involves discovery of conditions that might act to limit that (i.e. project management system) performance. Previous research related to Systems Theory and Systems Theory-based methodologies offers insights that provide explanation for aberrant conditions affecting system performance (Keating and Katina 2012). These aberrant conditions have been labeled as pathologies, defined as, 'A circumstance, condition, factor, or pattern that acts to limit system performance or lessen system viability <existence>, such that the likelihood of a system achieving performance expectations is reduced' (Keating and Katina 2012, p. 214). Pathologies have a rich development and have been anchored in Systems Theory (the set of laws and principles that govern behavior of all complex systems) (Whitney et al. 2015; von Bertlanffy 1968; Skyttner 2005) and Management Cybernetics (the science of system structural organization) (Beer, 1979, 1985).

For grounding our present exploration, we introduce two key points related to the nature and role of pathologies in complex systems – pathologies and their relationship to Systems Theory. First, pathologies have been extensively developed for application to the design, execution, and development (governance) of complex systems (Keating and Katina 2012; Katina 2015). CSG functions provide a set of 'coordinates' to locate the existence of a pathology in relationship to performance of a system. This location is identified to one or more of the nine different functions essential to continued viability of a complex system (previously discussed, Table 9.2). Given this brief introduction to pathologies in complex systems, following the recent work of Katina (2015, 2016a, b) and earlier work of Keating and Katina (2012), a set of 53 pathologies have been developed in relationship to the metasystem functions provided earlier (Table 9.4). These pathologies are organized around the nine metasystem functions and serve to identify aberrations to normal (healthy) functioning of a complex system (e.g. project).

With respect to application of the metasystem pathologies, there are three qualifications necessary. First, the pathologies are of a generalized form. Therefore, their manifestation in 'different' complex systems may be evidenced by surface level 'symptomatic' conditions. The pathologies are rooted in the underlying dysfunctions of a system that produce 'observable' surface symptomatic conditions. For example, a missed project schedule might be a surface manifestation of the deeper underling system pathology that produced it. Thus, pathologies are not directly observable but rather are inferred from observable/demonstrable conditions (undesirable behavior, outputs, outcomes, performance) in a system. Second, pathologies have a degree of existence. Pathologies are not binary reducible and thus have a 'degree of existence', rather than a binary present/not present attribution. Third, pathologies represent 'deficiencies' in the system design (structural organization of a system to achieve desired

TABLE 9.4

Pathologies Corresponding to Metasystem Functions

Metasystem Function	Corresponding Set of Pathologies
Metasystem five (M5): Policy and identity	M5.1. Identity of system is ambiguous and does not effectively generate consistency system decision, action, and interpretation.
	M5.2. System vision, purpose, mission, or values remain unarticulated or articulated but not embedded in the execution of the system.
	M5.3. Balance between short-term operational focus and long-term strategic focus is unexplored.
	M5.4. Strategic focus lacks sufficient clarity to direct consistent system development.
	M5.5. System identity is not routinely assessed, maintained, or questioned for continuing ability to guide consistency in system decision and action.
	M5.6. External system projection is not effectively performed.
Metasystem five star (M5*): System context	M5*.1. Incompatible metasystem context constraining system performance.
	M5*.2. Lack of articulation and representation of metasystem context.
	M5*.3. Lack of consideration of context in metasystem decisions and actions.
Metasystem five prime (M5′): Strategic system monitoring	M5′.1. Lack of strategic system monitoring.
	M5′.2. Inadequate processing of strategic monitoring results.
	M5′.3. Lack of strategic system performance indicators.
Metasystem four (M4): System development	M4.1. Lack of forums to foster system development and transformation.
	M4.2. Inadequate interpretation and processing of results of environmental scanning – nonexistent, sporadic, limited.
	M4.3. Ineffective processing and dissemination of environmental scanning results.
	M4.4. Long-range strategic development is sacrificed for management of day-to-day operations – limited time devoted to strategic analysis.
	M4.5. Strategic planning/thinking focuses on operational level planning and improvement.
Metasystem four star (M4*): Learning and transformation	M4*.1. Limited learning achieved related to environmental shifts.
	M4*.2. Integrated strategic transformation not conducted, limited, or ineffective.
	M4*.3. Lack of design for system learning – informal, nonexistent, or ineffective.
	M4*.4. Absence of system representative models – present and future.
Metasystem four prime (M4′): Environmental scanning	M4′.1. Lack of effective scanning mechanisms.
	M4′.2. Inappropriate targeting/undirected environmental scanning.
	M4′.3. Scanning frequency not appropriate for rate of environmental shifts.
	M4′.4. System lacks enough control over variety generated by the environment.
	M4′.5. Lack of current model of system environment.
Metasystem three (M3): System operations	M3.1. Imbalance between autonomy of productive elements and integration of whole system.
	M3.2. Shifts in resources without corresponding shifts in accountability/shifts in accountability without corresponding shifts in resources.
	M3.3. Mismatch between resource and productivity expectations.
	M3.4. Lack of clarity for responsibility, expectations, and accountability for performance.
	M3.5. Operational planning frequently pre-empted by emergent crises.
	M3.6. Inappropriate balance between short-term operational and long-term strategic focus.
	M3.7. Lack of clarity of operational direction for productive entities (i.e. subsystems).
	M3.8. Difficulty in managing integration of system productive entities (i.e. subsystems).
	M3.9. Slow to anticipate, identify, and respond to environmental shifts.

(Continued)

TABLE 9.4 (*Continued*)
Pathologies Corresponding to Metasystem Functions

Metasystem Function	Corresponding Set of Pathologies
Metasystem three star (M3*): Operational performance	M3*.1. Limited accessibility to data necessary to monitor performance.
	M3*.2. System-level operational performance indicators are absent, limited, or ineffective.
	M3*.3. Absence of monitoring for system and subsystem-level performance.
	M3*.4. Lack of analysis for performance variability or emergent deviations from expected performance levels – the meaning of deviations.
	M3*.5. Performance auditing is nonexistent, limited in nature, or restricted mainly to troubleshooting emergent issues.
	M3*.6. Periodic examination of system performance largely unorganized and informal in nature.
	M3*.7. Limited system learning based on performance assessments.
Metasystem two (M2): Information and communications	M2.1. Unresolved coordination issues within the system.
	M2.2. Excess redundancies in system resulting in inconsistency and inefficient utilization of resources – including information.
	M2.3. System integration issues stemming from excessive entity isolation or fragmentation.
	M2.4. System conflict stemming from unilateral decisions and actions.
	M2.5. Excessive level of emergent crises – associated with information transmission, communication, and coordination within the system.
	M2.6. Weak or ineffective communications systems among system entities (i.e. subsystems).
	M2.7. Lack of standardized methods (i.e. procedures, tools, and techniques) for routine system-level activities.
	M2.8. Overutilization of standardized methods (i.e. procedures, tools, and techniques) where they should be customized.
	M2.9. Overly ad hoc system coordination versus purposeful design.
	M2.10. Difficulty in accomplishing cross-system functions requiring integration or standardization.
	M2.11. Introduction of uncoordinated system changes resulting in excessive oscillation.

behavior/performance), execution (performance of the system design), or development (evolution of the design and design/execution interface). As such, pathologies produce real consequences related to system performance which can be measured across a range of possible impacts for a system.

A second essential and fundamental grounding for development of pathologies is their linkage to Systems Theory-based laws/principles. For our present purposes, the nature of pathologies in complex systems can be captured in the following critical points and their suggested relevance to project practitioners and system development:

1. *All systems (projects) are subject to the laws of systems.* Just as there are laws governing the nature of matter and energy (e.g. physics law of gravity), so too are our systems subject to laws. These system laws are always there, always on, nonnegotiable, non-biased, and explain system performance. Project practitioners must ask, 'do we understand systems laws and their impact on our enterprise design and performance?'

2. *All systems (projects) perform essential system functions that determine system performance.* These functions are performed by all systems, regardless of sector, size, or purpose. These functions define 'what' must be achieved for maintaining viability of a system.

Every system invokes a set of unique implementing mechanisms (means of achieving system functions) that determine 'how' system functions are accomplished. Mechanisms can be formal–informal, tacit–explicit, routine–sporadic, or limited–comprehensive in nature. These functions serve to produce system performance which is a function of previously discussed communication, control, integration, and coordination. Project practitioners must ask, 'do we understand, and can we define, the mechanisms that determine how our system performs essential system functions to produce performance and maintain viability?'

3. *Governance functions can experience pathologies (deviations from 'healthy' system conditions) in performance of functions.* There is no perfect system in execution. Regardless of the noble intentions of a system design, execution includes too many variabilities to 'guarantee' realization of design intentions. The effectiveness of governance is evident in the efficacy of identification, assessment, response, and evaluation to inevitable pathologies. These pathologies represent errors in design, execution, or development of a complex system. Governance provides the degree of resilience and robustness to withstand and persevere in the midst of external environment turbulence and internal system flux as sources of error. Good systems deal with pathologies as they occur – great systems continually design out pathologies before they escalate into crises. Practitioners must ask, 'do we purposefully design and redesign our system to address current, and preclude future, pathologies?'

4. *Violations of systems laws/principles in design, execution, or development of a system (project) carry consequences.* Irrespective of noble intentions, ignorance, or willful disregard, violation of system laws carries real consequences for system performance. In the best case, violations degrade performance. In the worst case, violations can escalate to cause catastrophic consequences or even eventual system collapse. Project practitioners must ask, 'do we understand problematic system performance in terms of violations of fundamental system laws?'

5. *System (project) performance can be enhanced through development of essential system functions.* When system performance fails to meet expectations, deficiencies in governance functions can offer novel insights into the deeper sources of failure. Performance issues can be traced to governance function issues as well as violations of underlying system laws. Thus, system development can proceed in a more informed and purposeful manner. Project practitioners must ask, 'how might the roots of problematic performance be found in deeper system issues and violations of system laws, suggesting different development directions?'

Figure 9.7 provides a pictorial summary of the relationships involved in CSG functions. In essence, violations of underlying system principles (laws) generate system (project) pathologies (aberrations from normal/healthy system conditions). These pathologies are not directly observable but can be mapped to the nine CSG functions. In addition, the pathologies generate observable system failure modes (e.g. cost overrun, missed schedule, performance deficiency). These failure modes are generated from deep-seated system deficiencies and are observable as undesirable behavior/performance from the project system. Thus, while the underlying systemic deficiencies are not directly observable, the external (observable) manifestation (e.g. missed performance targets) are observable as 'symptomatic' evidence of underlying systemic deficiencies (pathologies). Also, only addressing the surface manifestation, without understanding and addressing the underlying 'systemic' source of the deficiency, can only offer temporary 'relief' of the observable (symptomatic) condition.

UTILITY AND IMPLICATIONS FOR INTERVENTION TO IMPROVE SYSTEM GOVERNANCE

Ultimately, CSG offers significant contributions to help project practitioners address some of the most vexing current, as well as future, system problems they must confront. CSG is not suggested as a panacea for all problems facing the project systems. On the contrary, CSG is advocated as an

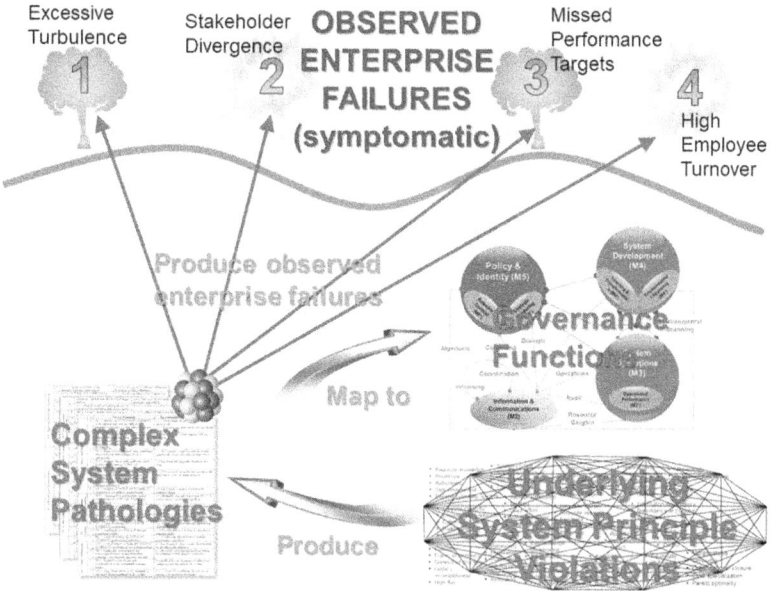

FIGURE 9.7 Symptomatic failure rooted in violation of underlying system principles (laws).

emerging field and approach with significant opportunity to provide value in the following areas (Figure 9.8). These value-adding areas range across practitioner, local system/entity, context, support infrastructure, and overarching enterprise.

- **Rigorous Guided 'Self-Study'** into CSG can provide significant insights into how the system (local entity, project, enterprise) actually functions. Although projects and their systems function routinely and successfully on a daily basis, as a matter of course, practitioners are not particularly skilled nor do they engage in deep reflection as to why, how, and what they do from a systems point of view. The gains to be made by reflective self-examination, from a systemic point of view, can reveal insights far beyond traditional

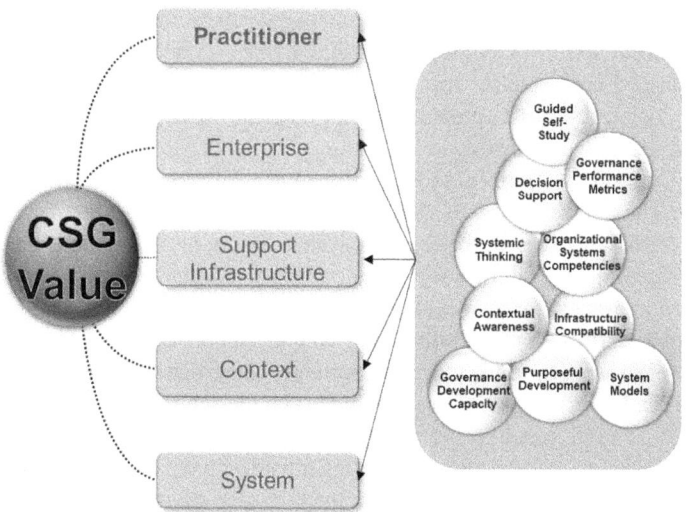

FIGURE 9.8 CSG value across a range of enterprise areas and activities.

methods of examination (e.g. strategic planning, SWOT [strengths, weaknesses, opportunities, threats] analysis, and maturity modeling). Thus, practitioners can examine a different level of analysis through 'self-study' and experience insights in a 'safe-to-fail' setting. Additionally, self-study might suggest the level of systemic education/training that might be necessary for individuals and the enterprise to increase individual capacity and project-level competence necessary to engage in higher levels of systems thinking (essential to CSG deployment).

- **Coherent Decision Support** can be achieved by the 'big picture' view of the governance landscape. This includes identification of the highest leverage strategic impact areas and their interrelationship to the larger CSG performance gaps (through pathologies assessment). Thus, decisions for scarce resource allocation can be better targeted. This allows steering away from activities that are simply 'intriguing' without demonstrating the highest substantial benefit to the larger 'systemic' governance concerns (e.g. pathologies high in existence, impact, and feasibility to address). In light of CSG development priorities, low contribution efforts can be eliminated or resources shifted appropriately.

- **Governance Performance Metrics** can be established to identify the present state of CSG functions as indicated by the pathologies. The set of 'unique' indicators developed for a specific system of interest (e.g. project) can provide a baseline that can be used to longitudinally establish the continuous progression of governance improvement. In effect, the degree of improvement stemming from initiatives undertaken to improve CSG can be established. Therefore, the state and shifts in governance can be purposefully and actively planned, deployed, monitored, and adjusted as necessary.

- **Systems Thinking Capacity** (ST-Cap) of individual practitioners to engage in the level of systems thinking necessary to more effectively deal with the entire range of complex system problems can be enhanced through CSG application. These problems are a by-product of modern enterprises and their systems (projects). Effectiveness is achieved through development and propagation of CSG language, methods, and tools to assist practitioners in their efforts to design, analyze, execute, and evolve complex systems and their associated problems (Jaradat 2015).

- **Organizational Systems Competencies** at the enterprise, or project, levels for dealing with complex systems and their derivative problems can be enhanced. This involves generation of knowledge, development of skills, and fostering abilities beyond the individual level to embrace problems holistically. For CSG application to projects, 'holism' suggests competency development that expands beyond narrow technology-centric infusions and the limiting cost-schedule-technical performance paradigm that dominates project management conversations. Instead, enhanced competencies that span the entire range of sociotechnical considerations endemic to complex systems (projects) are an outcome from CSG engagement to identify, analyze, and address pathologies.

- **Enhanced Contextual Awareness** is a direct by-product from the examination of system pathologies. The context exists as the circumstances, factors, conditions, or patterns that serve to enable or constrain performance of system functions. Thus, the wider consideration of system impediments provided by the pathology examination can open the aperture of consideration of aspects for the development of the enterprise.

- **Assess Infrastructure Compatibility** necessary to support systems-based endeavors. This compatibility is necessary to formulate contextually consistent (feasible) approaches to problems; create conditions necessary for governance system stability; and produce coherent decisions, actions, and interpretations at the individual and project levels. The most exceptional system solutions, absent compatible supporting infrastructure, are destined to outright fail in the worst-case scenario and underachieve in the best-case scenario.

- **Governance Development Capacity** can be determined to help establish the feasibility of initiatives that can be undertaken with a higher probability of successful achievement.

This does not minimize the degree of CSG inadequacies discovered through detailed examination of an enterprise. However, it does take into account the current sophistication in System Governance, the limiting/enabling context, and the individual ST-Cap that will influence what can be reasonably engaged with confidence of success. Minimally, consideration of feasibility for addressing pathologies can provide new insights into past successes/failures as well as cautions for impending future endeavors.

- **Enhanced System Models** generated through CSG exploration efforts can provide insights into the structural relationships, context, and systemic deficiencies that exist for governance of a system of interest. These insights can accrue regardless of whether or not specific actions to address issues are initiated. The models can be constructed without system modification. Therefore, alternative decisions, actions, and interpretations can be selectively engaged based on the consideration of insights and the understanding generated through CSG modeling efforts.

- **Purposeful Development** can provide focus for targeted advancement of the CSG functions. This accrues through resolution of priority pathologies derived from examination of issues in performance of system functions necessary to maintain system viability. While all viable (existing) systems perform the CSG functions and have pathologies, it is rare that they are purposefully articulated, examined, or addressed in a comprehensive fashion. Purposeful CSG development to resolve identified pathologies can produce a 'blueprint' against which development can be achieved by design, rather than serendipity. This includes establishment of the set of 'dashboard indicators' for CSG performance. These performance indicators exist beyond more 'traditional' measures of system/enterprise performance.

Ultimately, CSG seeks to increase the probability of achieving and maintaining higher levels of desirable system performance (viability, growth, etc.) in the midst of internal enterprise flux and external environmental turbulence.

THREE APPLICATION SCENARIOS AND SUPPORT TOOLS DEMONSTRATION FOR CSG AIDING DEPLOYMENT OF CSG

To illuminate the applicability of CSG and potential contributions, we examine three scenarios of CSG application. Each scenario is presented as a problematic situation with a corresponding CSG perspective and response. These scenarios are representative of completed and ongoing applications of different applications of CSG.

SCENARIO 1: WORKFORCE CAPACITY FOR SYSTEMS THINKING

Situation: A project workforce is continually behind in producing innovative thinking to effectively respond to complex demands of their environment and stakeholders – resulting in crises, surprises, or inefficiencies. The errors continue to mount with increasingly deficient performance, discontent in the workforce, and the seeming inability to effective function in relationship the demands of the complex environment within which the system and practitioners must function.

    CSG Perspective Discussion: A critical element of CSG is the dependence on the capacity of the workforce to engage at a level of thinking necessary to realize the inherent value in CSG. Without the correct frame of reference (ST-Cap), the results desired from CSG are not likely to be achieved. In essence, if the workforce does not have the necessary systemic thinking skills, then CSG is just another approach that an enterprise might grasp at for relief. Regardless of how dire the enterprise circumstances might be, there is no shortcut to having the requisite capacity in individuals to effectively engage any systems-based endeavor. There are two primary drivers for this situation. First, as previously mentioned, is the capacity of the workforce to think systemically. Second is the degree to which the environment demands with respect to ST-Cap can be met by the workforce, supporting infrastructure, and the system (project). Performance will largely be determined by the degree that there is a sufficient 'match' between the ST-Cap that exist in the workforce/project to that demanded by the environment within which the enterprise must operate.

CSG Response Discussion: ST-Cap and environment complexity demand assessment instruments have been used to identify gaps between the system workforce capacity and the demands of their environment. 'Critical' areas for enhancing ST-Cap have been identified along seven different dimensions that comprehensively define ST-Cap. Figure 9.9 depicts this gap along the seven dimensions of systemic thinking used to examine capacity. The focused closure of these gaps is essential to match the demands of the environment as well as effectively engage CSG as a systems-based endeavor.

A Simplified ST-Cap Profile Assessment Tool: Jaradat and Keating (2016) have elaborated the concept and approach for determining and using ST-Cap. However, lacking a more elaborate examination of ST-Cap, a simple tool can be used by project managers to assess the degree of systems thinking that exists in their project team. For each of the seven dimensions of systems thinking (see, for example, Figure 9.10), an assessment profile is created to delineate the state of systems thinking for a project team in contrast to the level demanded by the project being undertaken. The project manager is able to determine potential 'trouble spots' in the capacity of the project team to respond to the demands of the project. In the example, it becomes evident that for the 'Interaction' dimension, the demand of the project far exceeds the assessed capacity held within the team to meet the demand. Thus, the project manager for this project would need to be sensitive to this disparity. The response might be the inclusion of mitigation strategies and the actions to alleviate the potential pitfalls that this disparity might produce throughout the project life cycle.

SCENARIO 2: System (Project) Governance Pathologies Exploration

Situation: A focal system is experiencing continual failures (e.g. cost overruns, schedule delays, missed performance targets) that are resistant to improvement efforts. The external manifestations of failures are evident in either product/service quality, missing milestones, or required customer completion schedules, or conflicts in the adequacy, utilization, or outcomes achieved for the resources

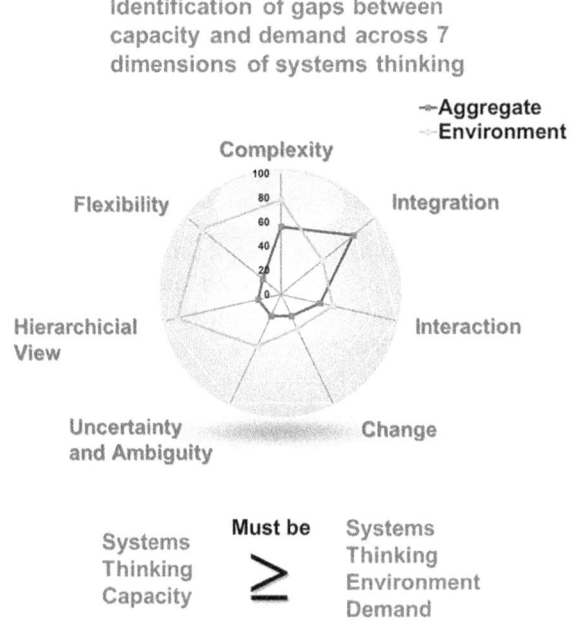

FIGURE 9.9   Gaps between ST-Cap and environment complexity demands.

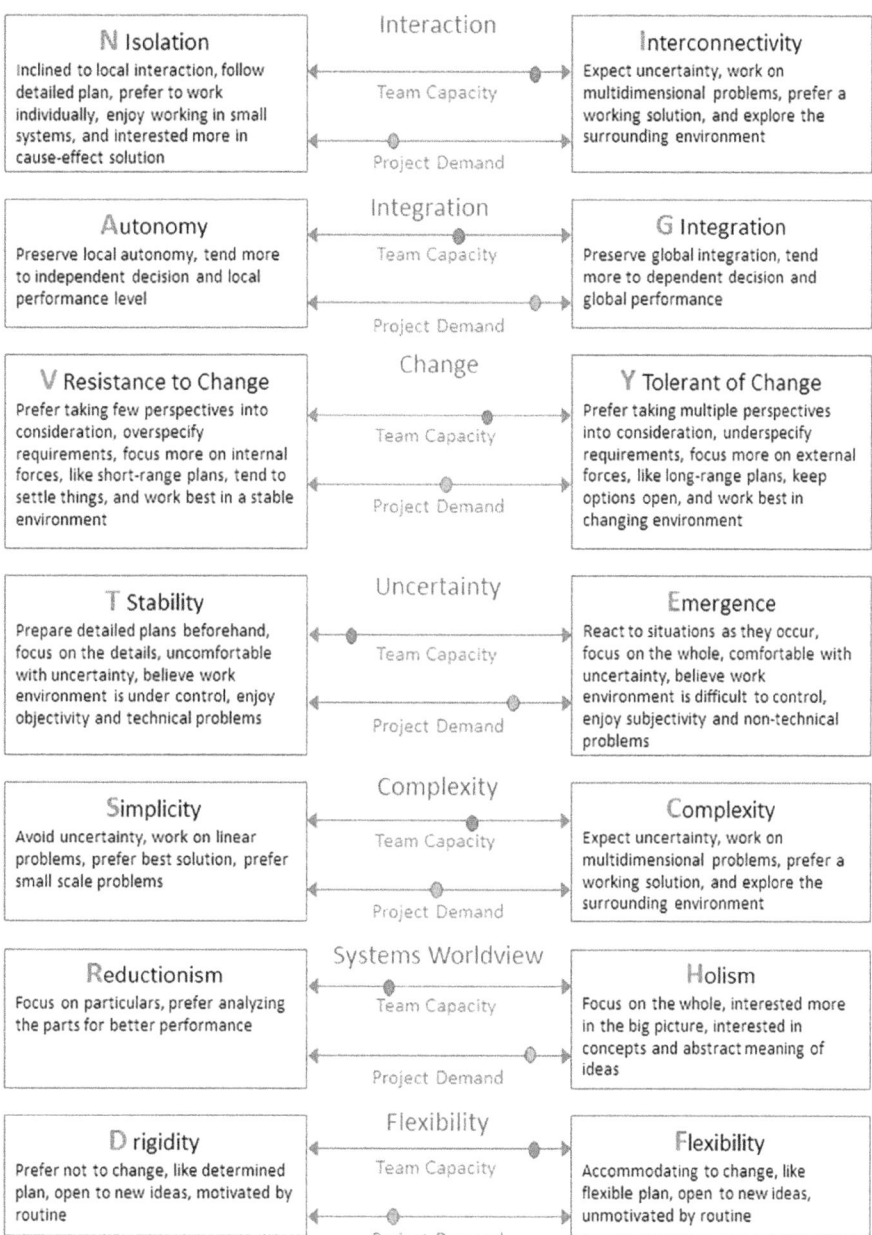

FIGURE 9.10   Profile for seven dimensions of systems thinking for CSG.

consumed. There have been several failed attempts to locate the source of the deficiencies, but there does not appear to be a singular root cause to which failures can be attributed and addressed. The result is a sagging customer confidence, resource scarcity, and a diminished workforce from the anxiety and frustration being experienced without an apparent path forward or end in sight.

CSG Perspective Discussion: It is quite easy to identify the outward observable results for the violation of underlying systems principles (evidenced as pathologies) – experienced as surface manifestations symptomatic of deeper systems issues (pathologies). Observable system deficiencies are the outward manifestation of underlying system design, execution, or development issues (attributable to underlying pathologies and existence of system law violations). Being able to properly trace systemic issues requires a 'deeper dive' into the actual system producing the performance issues. In essence, a system (project) can only produce what it produces, nothing more and nothing less. If the system (project) performance is not consistent with that we desire, we must understand the underlying system that is producing the undesirable behavior/performance. Focusing only on the outward signs of the underlying systemic issues can at best provide a temporary fix. At worst, more damage than good might accrue from superficial treatment of symptoms of underlying system deficiencies (violations of systems laws recognized as pathologies).

CSG Response Discussion: A cross section of the enterprise (project) completes a System Governance Pathologies assessment instrument. Deep system pathologies (aberrations from healthy system conditions) across the nine governance functions are identified, mapped, systemically explored, and prioritized for response feasibility (ranging from low to high feasibility). This approach provides an opportunity to discover the underlying source of deficiencies in a system. These underlying sources are not necessarily observable from the inspection of their superficial deficiencies (symptoms), recognizable as observable deficiencies in desirable behavior/performance of a system (project). Figure 9.11

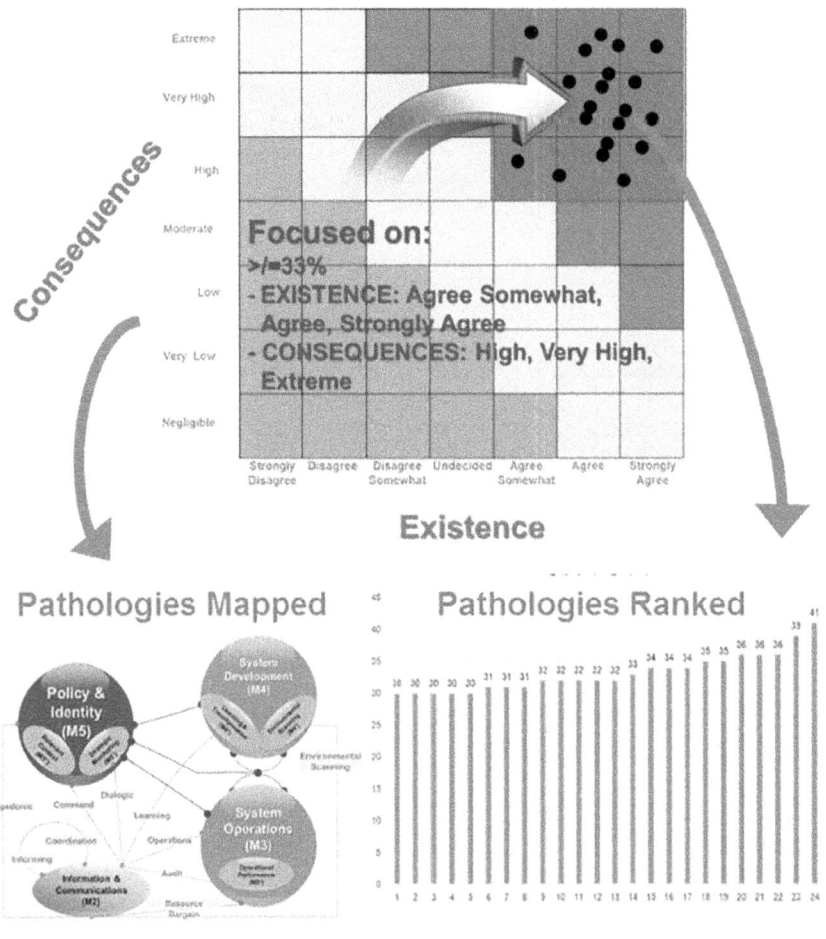

FIGURE 9.11   Mapping and ranking of CSG pathologies.

shows a mapping of one particular pathology (of 53 different potential pathologies) in an enterprise system. Based on the results, the highest leverage pathologies for system development impact were identified, traced to system law violations, and prioritized for further exploration and response strategy development.

Modified Force Field Analysis – a Tool to Identify Positive and Negative Forces Related to Project Issues (Including Pathologies). Force Field Analysis is a simple but effective tool used to identify positive (enabling) forces that are supportive of advancing movement in the situation and negative (restraining) forces that act as impediments to movement of the situation. While there have been numerous variations of the tool over the years, it was originally conceived by Kurt Lewin (1943) and has been successfully used, and adapted through years of application, to help understand forces at play in complex issues. For a situation of interest, positive forces (those contributing constructively to the situation) and negative forces (those contributing in to the situation in an adverse fashion) are identified and assessed for impact. The result allows a project manager to focus attention of the aspects of the situation (both positive and negative) to respond to the situation. In the case of pathologies, strategies/ activities can be developed to address the situation. This adaptation of Force Field Analysis begins by identifying a particular situation, event, or focus topic (e.g. a particular suspected pathology). The focus on a specific situation permits identification and capture of the nuances that can be instrumental in successfully dealing with the situation at hand. The development of the forces is free flowing and can be used to represent a multitude of different influential aspects of the situation such as environment, patterns, relationships, issues, stakeholders, geography, time, processes, people, and perspectives. These forces might be influential in analysis, design, deployment, operation, maintenance, or evolution of the situation in focus (e.g. a pathology from Table 9.4 used as a focus for Force Field Analysis in Figure 9.12). There are no limitations to the force field diagram with respect to what can be included or excluded. The primary utility of a force field diagram is the collaborative process of constructing the diagram and the 'conceptual' appreciation that is generated by the project team constructing the diagram. In effect, it provides a 'conceptual' leveling and brings out different viewpoints for consideration. Thus, participants gain a more holistic perspective and appreciation of aspects (forces) in play for the situation and hopefully a foundation to support better achievement of objectives.

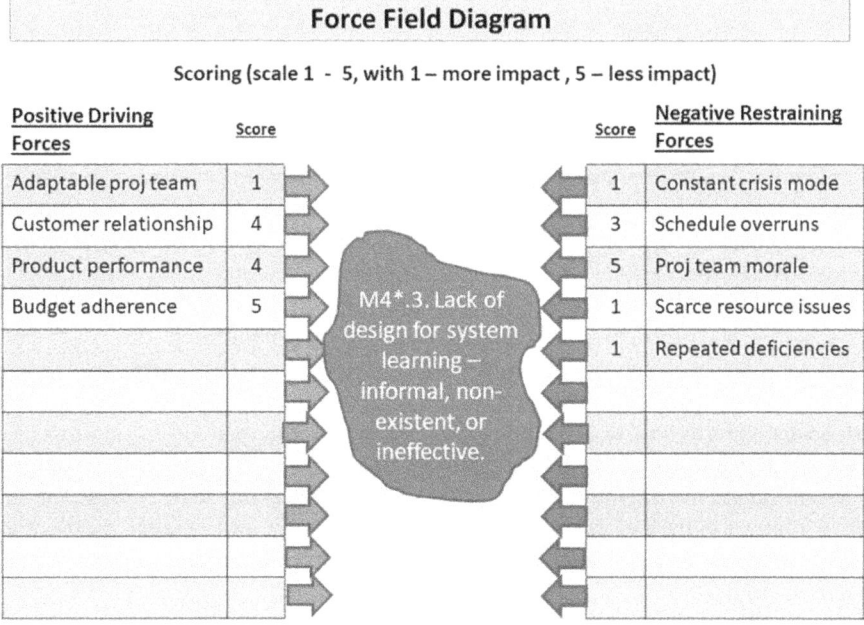

FIGURE 9.12   Force Field Analysis diagram example.

SCENARIO 3: System (Project) Governance Development

Situation: An enterprise (project) has difficulty in providing a clear, coherent, and accountable enterprise-/project-level innovation strategy to address persistent criticisms from oversight bodies/stakeholders. External forces are continually challenging the enterprise/project to provide information, performance indicators, and reasons for major decisions and strategies being pursued. Lacking a defensible design logic permits excessive regulatory force (control/constraints) to be invoked on the enterprise/project from external entities. The excessive external controls are creating a burden by clogging effective innovation thinking, slowing responsive decisions to identified issues/opportunities, slowing ability for decisive actions in response to issues, and creating confusion and ambiguity in interpretation of requirements and expectations. Needless to say, the workforce is becoming increasingly stressed, tired, and angry with the circumstances.

CSG Perspective Discussion: Although it is common to receive 'oversight' for performance of the enterprise/project mission, care must be taken to understand the degree to which the system is designed, executed, and developed such that oversight is not a burden but rather a welcomed opportunity to 'demonstrate' system proficiency. As most systems (projects/enterprises) are not purposefully and holistically designed, it is not uncommon to look at external 'oversight demands' as an annoyance at best and at worst an impediment to performance. The need to constrain a system/project may in fact stem from perceived inadequacies in the design or execution of a system by external agents with some level of authority, responsibility, and accountability for performance. This may accrue from the perception that the enterprise/project, absent detailed oversight, is incapable of mounding a response consistent with demands of the environment. Without a robust design against which to reference external control/regulatory demands, it is not likely that a system (project) will generate sufficient resilience to effectively deflect external 'meddling'. System development should emphasize generation of robustness in the design such that externally control challenges can be better understood in relationship to system design/execution. Thus, more integrated and appropriate responses to external probing can be mounted. Absence of this capability to interpret external demands against the purposeful system design/execution renders defense of the system to likely be suspect, piecemeal, and ineffective.

CSG Response Discussion: Mapping of the CSG landscape provides visualization for analysis of the most critical challenges facing CSG development (peaks in Figure 9.13). Past, ongoing, and future planned system development initiatives can be mapped against the existing governance landscape, pathologies, and system criticisms. 'Holistic' analysis provides clarity and focus for integrated system development response strategies to both internal and external challenges related to establishment, necessity, and execution of external enterprise controls. Figure 9.13 shows a mapping of a CSG landscape for a system and a depiction of the mapping of initiative against the effectiveness of CSG. As the figure indicates, there are continuing CSG effectiveness issues, even though initiatives have been targeted at CSG challenged areas.

Radar Chart – a Graphical Tool to Help with CSG Applications: The radar chart has a long history and has been recognized as an effective tool to provide a graphical representation to examine relationships in multivariate data. The various names of this tool include radar chart, web chart, spider chart as well as Kiviat diagram (Kolence and Kiviat 1973). Figure 9.13 shows a depiction of a radar chart from actual application of CSG. The chart maps each of the nine different metasystem functions ranging in effectiveness along a 100-point scale for a set of enterprise initiatives in contrast to their effectiveness for CSG. The essence of the radar chart is the ability to graphically show an array of data and support multiple different data sets on a single two-dimensional diagram.

The preceding three scenarios demonstrate the potential of CSG to enhance project performance, ranging from the individual to enterprise levels.

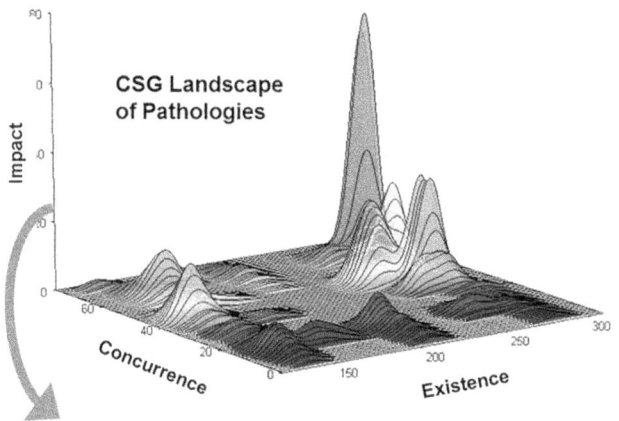

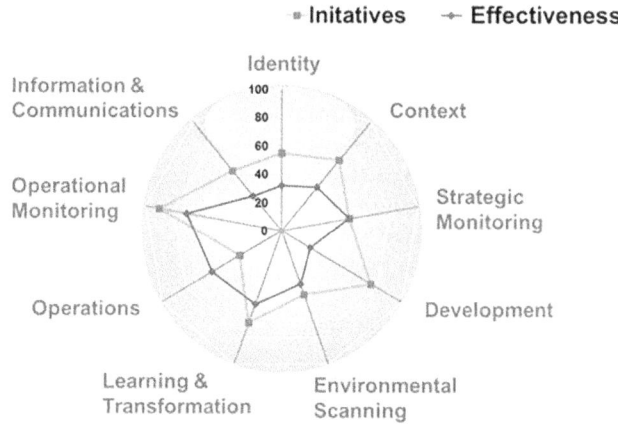

FIGURE 9.13 Mapping CSG landscape to identify the highest priority development areas.

ADDITIONAL TOOLS TO SUPPORT APPLICATION OF CSG FOR COMPLEX PROJECTS

There are several tools that might be used to support the system representation and analysis of CSG for project practitioners. While the detailed explanation of these tools is beyond the scope of this chapter, Table 9.5 provides a summary description of several readily available without cost to project practitioners.

QUESTIONS FOR DISCUSSION

The following questions are offered for discussion/exploration for project practitioners as they contemplate potential application of CSG.

1. Refer to Table 9.1 and Figure 9.1. For a familiar project which of the complex problem domain attributes/conditions might apply to that project? What examples can you provide that support the existence of the attribute/condition?

TABLE 9.5

Additional No Cost Tools to Support CSG Applications for Project Practitioners

Tool Name	Description/Utility	Access Link
NodeXL	NodeXL is an Excel Spreadsheet add-in. It permits the mapping and examination of a network (e.g. CSG network of metasystem functions). It provides capabilities for network visualization and analysis using the familiarity of a spreadsheet environment. It permits visualization to be quickly created to support analysis.	NodeXL is a free and open project and can be found at www.codeplex.com/nodexl. In addition, a network tutorial, based in social networks and NodeXL, can be accessed at http://casci.umd.edu/NodeXL_Teaching
SystemiTool	SystemiTool is based on the SystemiGram approach developed by Boardman and Sauser (2013). It offers support to help in modeling system relationships. The structure lends itself to mapping relationships in the performance of the CSG functions.	SystemiTool is downloadable via the following website http://boardmansauser.com/order/index.html
Mental Modeler	Assists in capturing knowledge in a format that can support the building and examination of scenarios. The software is based in Fuzzy-logic Cognitive Mapping to support semiquantitative models. For CSG, it can support the definition of relationships among the different system functions and examination of scenarios.	Mental Modeler is available for free download at www.mentalmodeler.org/#home
X-Mind Mindmapping Software	A powerful mindmapping software that permits definition of relationships among entities and the structuring of different ideas. Can support brainstorming as well as other creative endeavors.	Free downloadable version of X-mind is available at www.xmind.net/
yEd	A Java-based drawing tool that permits the construction of complex diagrams. Exceptional for support of CSG mapping endeavors. Provides powerful capabilities for representation of complex CSG relationships.	yEd is available for free download at www.yworks.com/products/yed
Netdraw	A network visualization program to show interrelationships in a complex network. Can support CSG efforts by visually articulation of the complex set of interrelationships in the metasystem.	Netdraw is available for free download at https://sites.google.com/site/netdrawsoftware/home

2. Identify a problematic situation for a familiar project. From Table 9.4, select at least two pathologies that you suspect exist for the project. What evidence might you offer that demonstrates the existence of the pathology?

3. For one of the pathologies identified in Question 2, construct a force field diagram that identifies both positive (enabling) forces and negative (constraining) forces that are influential in explaining the existence/state of the pathology.

4. Refer to Figure 9.10. For a familiar project, how would you distinguish between project team capacity and project demand for each of the seven dimensions? For the dimensions with the greatest gap, what would you suggest as explanations for the gap and what might be the potential impact of the gap? What would you suggest might be necessary to close the gap?

5. Consider a potential project for which CSG might be deployed. Given the considerations identified in Figure 9.14, what top three items might limit potential successful deployment of CSG? What strategies/actions might be pursued to enhance the prospects for successful deployment of CSG for the project?

CONCLUSION: CONSIDERATIONS FOR PROJECT PRACTITIONERS DEPLOYING CSG

As an emerging field, there is much that remains unknown about CSG, particularly with respect to deployment to improve CSG in projects. Much of the unknown for CSG stems from the unique demands for intervention in complex systems. CSG is somewhat unique in relationship to other systems-based approaches in three primary ways. First, CSG makes an *explicit* mapping to Systems Theory (Whitney et al. 2015) as a grounding basis for the field. This is not to suggest that other systems-based approaches are not 'born' out of an underlying Systems Theory base. However, CSG is explicit in the delineation of the Systems Theory conceptual basis. Second, engagement in CSG is metered by the degree of ST-Cap of the participating group and the state of System Governance that currently exists for the system in focus (project). Therefore, the directions and engagement will be driven by the existing ST-Cap of individuals as well as the system 'fitness' to participate across a range of CSG development activities. This range of fitness determines the nature, depth, and expectations for the level of CSG system improvement activities that might be effectively engaged. Thus, different projects will have different levels of possible engagement options with respect to CSG. Third, CSG is not equivalent to introduction of a new program or initiative (e.g. lean six sigma, Total Quality Management, balanced scorecard, Customer Relationship Management) that will be engaged 'in addition to' what is already being done by the individuals/projects. Instead, all viable (continuing to exist) systems (projects) are already performing the nine CSG metasystem functions, irrespective of acknowledgment, language, or purposeful development. Irrespective as to whether or not these functions are purposefully explored for development, they are, and will continue to be, performed if the system continues to exist. Thus, CSG is not a temporary endeavor that exists beyond the normal scope of system (project) activities/initiatives being engaged by the system (project).

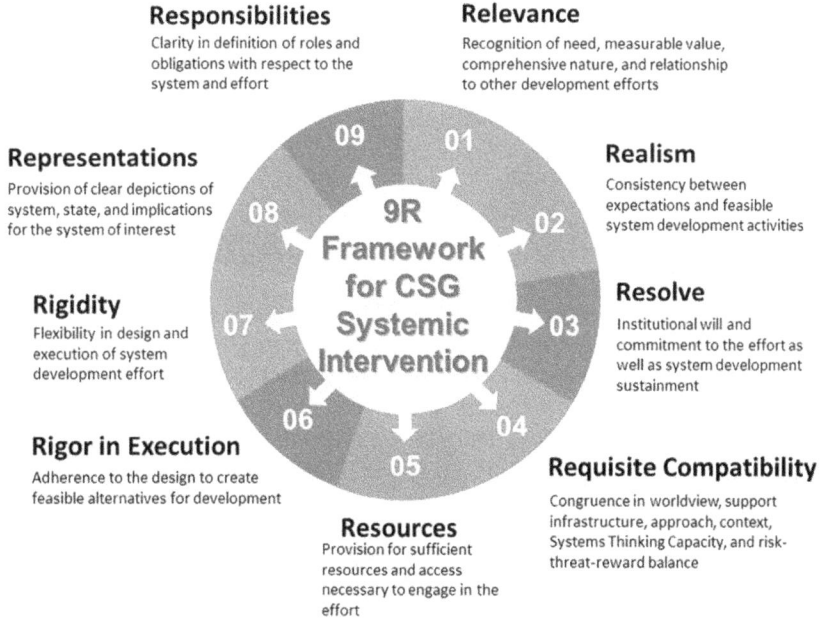

FIGURE 9.14 9R Framework of systemic intervention for CSG.

With respect to engaging CSG, following the work of Keating and Katina (2017), Pyne et al. (2018), and to a lesser degree Midgley (2000), we have identified several implications from our experiences with initial efforts to deploy CSG. These implications include the following:

1. *CSG is not the Entry Point:* As promising as CSG might be for advancing system (project) understanding and performance, it is not the highest priority for those who might be considering engagement. Instead, the priority for project practitioners is focused on 'their problems'. Thus, first understanding their problems and then drawing the linkage to potential CSG value contributions is essential. Making this connection is critical to draw attention to the possibilities that CSG might bring related to their most vexing issues.

2. *CSG Engagement is not a Binary (all or nothing) Proposition:* Following introductory CSG efforts and the implications that might be suggested from the results, there are many developmental paths that might be pursued. It is incorrect to have CSG postured as an all or nothing alternative. Instead, there is a spectrum of activities (training, development, modeling, etc.) and levels (practitioner, system, project, context, enterprise, problem) that might be pursued in the development path to enhance CSG.

3. *CSG is not an* 'In Addition To' *Endeavor:* Unlike more traditional system interventions that seek to address a new concern by introduction of a totally new initiative (e.g. Lean, Six Sigma, Total Quality Management, Customer Relationship Management, EVMS [earned value management system]), CSG functions are already being performed by a system (project) that is viable (exists). Thus, CSG is focused on understanding and potentially improving that which is already being performed by a system of interest (project). Therefore, the language, thinking, and explorations of CSG are applied to existing system execution of CSG functions which are already being 'tacitly' performed.

4. *CSG Systemic Intervention Time and Risk Should Initially Fall on the Facilitator:* It is unrealistic to expect participants to fully engage a CSG initiative in terms of investment of time and acceptance of 'risk of failure'. Instead, the CSG facilitator should bear the burden of time and risk until the value of investment (time) and utility of CSG engagement can meet an acceptable *risk-value-cost* trade-off. In effect, CSG should be conducted in a 'safe-to-fail' mode.

To elaborate implications for CSG development, also following the work of Pyne et al. (2018) and Keating and Katina (2017), a systemic intervention framework was established following initial applications of CSG. This framework, titled the *9R Framework for Systemic Intervention* (Figure 9.14), identifies nine areas of concern that project practitioners would be advised to consider as they design and execute CSG initiatives for enterprise development.

Each of the *9R Framework for Systemic Intervention* elements has been identified as having potential impact on the success of systemic interventions undertaken to improve performance of complex systems. Each element provides an area that should be considered when looking to undertake an intervention into a complex system. The following discussion elaborates each of the nine elements targeted to CSG. It is important to note that the different elements do not operate independently and are not mutually exclusive of one another. Also, the nature and priority of the elements should be expected to change over time as the intervention evolves. The nine elements include:

Relevance – Systemic intervention is undertaken in response to a recognized need or problem situation which is unresolved and persists in a system. However, CSG is not targeted to specific problems but rather to the 'underlying system' that must address problems. These problems might be more recognizable as surface manifestations stemming from deficiencies in underlying system functions. While the true value of CSG is in addressing the underlying system deficiencies, the value is most recognizable as addressing the surface 'symptomatic' conditions immediately recognizable by practitioners. Thus, systemic intervention must focus on (1) translation of surface problems to the capabilities of CSG to discover the deep seated 'systemic roots' of the problem and offer a different frame of reference for understanding potential alternative paths to resolution; (2) casting CSG in relationship to past, ongoing, and future development initiatives to better position CSG as a 'meta-initiative' that provides an integrating perspective of system development; (3) exploration of systemic intervention as *'solving the system'* such that in the future the *'system can solve the problem(s)'*; and (4) projection of CSG as system development for functions that the system is already performing, without the benefit of the CSG framing of those functions, and is therefore not 'in addition to' ongoing system work but rather amplification of existing work.

Realism – Although CSG holds great promise to identify insights into systemic deficiencies, this identification must not be confused with the underlying capability of practitioners and their system to apply those insights to fully engage systemic issues. CSG development is metered by the level of ST-Cap that exists within system participants and the current state of performance of CSG functions. Thus, expectations for system development must be appropriately metered such that the capabilities are commensurate with an appropriate level of improvement activity undertaken for system development. This defines the region of feasible engagement for system development. Knowing issues and having capabilities to address those issues must be congruent. Otherwise, the development is likely to fail and, perhaps, leave the system in a worse state than before the systemic intervention was initiated.

Resolve – Commitment of resources (manpower, material, money, methods, minutes, information – M5I) is necessary for engaging in systemic intervention. However, they are not sufficient. Sufficiency is also determined by institutional will and commitment to sustainment of system development following systemic intervention. *Institutional will and commitment* are not easily determined or measured. However, the willingness to increase engagement beyond simple resource allocations should be evident and escalate throughout the intervention. Thus, will and commitment should be congruent with increasing recognition of the value accrued from CSG initiatives.

Requisite Compatibility – Systemic intervention for CSG is not necessarily the right approach or fit to every problematic circumstance or every system. The determination of 'fitness' for CSG appropriateness should consider compatibility with existing system/enterprise: (1) *predominant worldview* recognized as the prevailing paradigm(s) which drive decision, actions, and interpretations related to system circumstances; (2) *support infrastructures* (e.g. procurement, human resources) that influence, and will be influenced by, system development stemming from CSG intervention discoveries; (3) *contextual factors* (e.g. policy, power, politics, culture, management style) that influence the prospects for conducting systemic intervention and implementing modifications; (4) approach taken to conduct the systemic intervention (e.g. level of participation) with preferred development modes in the enterprise; and (5) *risk-threat-reward balance* that indicates willingness to engage rigorous self-examination in hopes of finding deeper sources of system development. Lacking these compatibilities, CSG is not likely to achieve the desired results.

Resources – Provision for sufficient resources and access necessary to engage in the effort. This must consider the time investment of participants as well as the more mundane aspects related to sufficient levels of funding necessary to engage the desired depth of systemic intervention. Resource allocations should be consistent with expectations of value to be accrued from the effort. Additionally, shifts in resources necessary due to 'discoveries' during systemic intervention activities should be expected, scrutinized, and embraced where appropriate. Incongruence between resource allocation and expectations of value are likely to disappoint the best systemic intervention intentions.

Rigor in Execution – Systemic intervention should have sufficient detail and clarity such that in can be executed with precision. Detailed design related to data collection, analysis, and interpretation should be thorough and explicit such that *what must be done, how it will be done, who will do it, when it will be done, where it will occur, why it is necessary, and what are reasonable expectations should* be clearly delineated. This does not preclude shifts in design or execution. However, the shifts in approach, execution, and interpretations should be clearly articulated, with the underlying assumptions and supporting logic made explicit and capable of withstanding scrutiny.

Responsibilities – Each systemic intervention is unique in the specific roles that will be played and responsibilities that are allocated to those roles. Responsibilities range across the spectrum of intervention design, execution, and implementation of decisions/actions stemming from systemic intervention activities. Sufficient clarity must exist such that accountability for achievement of different aspects of the systemic intervention can be clearly assigned. This is not to support a punitive dimension for systemic intervention but rather to ensure that expectations for completion of assignments are unambiguous. Additionally, the pursuit of system changes stemming from a CSG endeavor should have clarity in responsibilities as well.

Rigidity – Systemic intervention follows a particular 'script' that lays out the design for execution. Although there might be emergent understanding that suggests alteration of the initial design, modifications should be purposeful rather that arbitrary or fickle. Execution of systemic intervention is always dynamic, emergent, and subject to shifts in direction. Reasonable and measured changes in systemic intervention should be expected and embraced, allowing for flexibility in design, execution, expectations, and trajectory of an effort. This is not an excuse to do systemic intervention by a 'seat of the pants' mode, making seemingly arbitrary shifts in direction.

Representations – The approach, results, and implications for CSG development must be provided with accuracy, brevity, and clarity. The representational forms selected must concisely and unambiguously convey the 'invitation to dialog' by those participating in the endeavor. CSG impact will be limited by the selection of representational forms that are cumbersome and difficult to easily grasp. The primary concern must be to provide a pathway for engaging in rich dialog, absent the deflections that might occur from weak representations that confuse rather than clarify.

CSG for project governance is not offered or pursued as yet another approach to system improvement. Instead, CSG provides a conceptually grounded, application-driven, and practitioner-oriented approach to enhance prospects to enable project practitioners to better engage complex system (project) development. While not presented as a panacea, CSG has shown promise to enhance system (enterprise/project) development and professional practice by (1) development of a Systems Theory-based approach to engaging complex system (enterprise) development and (2) providing a frame of reference for more rigorous examination of enterprise performance. Future development of CSG and practitioner-enabling capabilities to effectively deploy CSG are poised to provide enterprises and practitioners with a range of new and novel ways of thinking, deciding, acting, and interpreting enterprise complexities and their emergent problems.

REFERENCES

Beer, S. 1972. *Brain of the Firm: The Managerial Cybernetics of Organization*. London: Penguin Press.

Beer, S. 1979. *The Heart of Enterprise*. New York: John Wiley & Sons.

Beer, S. 1985. *Diagnosing the System for Organizations*. Hoboken, NJ: John Wiley & Sons

von Bertalanffy, L. 1968. *General System Theory: Foundations, Development, Applications* (Rev. ed.). New York: George Braziller.

Boardman, J. and B. Sauser. 2013. *Systemic Thinking: Building Maps for Worlds of Systems*. Hoboken, NJ: John Wiley & Sons.

Calida, B. Y. 2013. System governance analysis of complex systems. *Doctoral Dissertation*. Old Dominion University.

Jaradat, R. M. 2015. Complex system governance requires systems thinking. *International Journal of System of Systems Engineering*, 6(1/2), 53–70.

Jaradat, R. M. and C. B. Keating. 2014. Fragility of oil as a critical infrastructure problem. *International Journal of Critical Infrastructure Protection*, 7(2), 86–99.

Jaradat, R. M. and C. B. Keating. 2016. Systems thinking capacity: Implications and challenges for complex system governance development. *International Journal of System of Systems Engineering*, 7(1–3), 75–94.

Jaradat, R. M., C. Keating and J. Bradley. 2017. Individual capacity and organizational competency for systems thinking. *IEEE Systems Journal*. doi: 10.1109/JSYST.2017.2652218.

Katina, P. F. 2015. Emerging systems theory–based pathologies for governance of complex systems. *International Journal of System of Systems Engineering*, 6(1–2), 144–159.

Katina, P. F. 2016a. Metasystem pathologies (M-Path) method: Phases and procedures. *Journal of Management Development*, 35(10), 1287–1301.

Katina, P. F. 2016b. Systems theory as a foundation for discovery of pathologies for complex system problem formulation. In A. J. Masys (Ed.), *Applications of Systems Thinking and Soft Operations Research in Managing Complexity* (pp. 227–267). Geneva, Switzerland: Springer International Publishing.

Keating, C. B. 2014. Governance implications for meeting challenges in the system of systems engineering field. *Paper presented at the IEEE System of Systems Engineering Conference*, Adelaide, SA.

Keating, C. B. and J. M. Bradley. 2015. Complex system governance reference model. *International Journal of System of Systems Engineering*, 6(1), 33–52.

Keating, C. B. and P. F. Katina. 2011. Systems of systems engineering: Prospects and challenges for the emerging field. *International Journal of System of Systems Engineering*, 2(2–3), 234–256.

Keating, C. B. and P. F. Katina. 2012. Prevalence of pathologies in systems of systems. *International Journal of System of Systems Engineering*, 3(3/4), 243–267.

Keating, C. B. and P. F. Katina. 2017. Systemic intervention for complex system governance development. *Proceedings of the American Society for Engineering Management*, October 18–21, 2017, Huntsville, Alabama, pp. 85–95.

Keating, C. B., P. F. Katina, and J. M. Bradley. 2014. Complex system governance: concept, challenges, and emerging research. *International Journal of System of Systems Engineering*, 5(3), 263–288.

Keating, C. B., P. F. Katina, and J. M. Bradley. 2015. Challenges for developing complex system governance. *Paper presented at the Proceedings of the 2015 Industrial and Systems Engineering Research Conference*, Anahiem, CA.

Keating, C. B. and M. Morin. 2001. An approach for systems analysis of patient care operations. *Journal of Nursing Administration*, 31(7/8), 355–363.

Kolence, K. W. and P. J. Kiviat. 1973. Software unit profiles & Kiviat figures. *ACM SIGMETRICS Performance Evaluation Review*, 2(3), 2–12.

Lewin, K. 1943. Defining the 'field at a given time'. *Psychological Review*, 50(3), 292.

Midgley, G. 2000. Systemic intervention. In *Systemic Intervention* (pp. 113–133). Boston, MA: Springer.

Naphade, M., G. Banavar, C. Harrison, J. Paraszczak and R. Morris. 2011. Smarter cities and their innovation challenges. *Computer*, 44(6), 32–39.

Pyne, J. C., C. B. Keating, P. F. Katina, and J. M. Bradley. 2018. Systemic intervention methods supporting complex system governance initiatives. *International Journal of System of Systems Engineering*, 8(3), 285–309.

Skyttner, L. 2005. *General Systems Theory: Problems, Perspectives, Practice* (2nd ed.). Singapore: World Scientific Publishing Co. Pte. Ltd.

Whitney, K., J. M. Bradley, D. E. Baugh and C. W. Chesterman, Jr. 2015. Systems theory as a foundation for governance of complex systems. *International Journal of System of Systems Engineering*, 6(1), 15–32.

10 Critical Success Factors and Climate Toolset

Azadeh Rezvani and Neal M. Ashkanasy
The University of Queensland

CONTENTS

INTRODUCTION

Beginning in the late 1970s, governments in both developed and developing countries have become increasingly interested in issues related to large-scale projects (Kwak and Smith 2009). This interest stems in part from the fact that such projects hold the potential to generate high social returns; to transform the social fabric of local communities; and to contribute to private sector activity, employment, and government revenues. Consistent with this interest, researchers worldwide (e.g., see Davis 2014; Rezvani, Khosravi, and Ashkanasy 2018; Mazur, Pisarski, Chang, and Ashkanasy 2014) have sought to understand the factors that lead to successful implementation of large-scale projects.

Critical success factors (CSFs) studied include communication (Yu, Shen, Kelly, and Hunter 2006; Ogunlana 2008), stakeholder relationships (Dvir, Ben-David, Sadeh, and Shenhar 2006; Lech 2013; Wu, Liu, Zhao, and Zuo 2017), troubleshooting (Mazur et al. 2014; Rezvani, Chang, Wiewiora, Ashkanasy, Jordan, and Zolin 2016), and meeting user/owner requirements (Ferratt, Ahire, and De 2006; Fortune and White 2006; Nicolini 2002; McGillivray, Greenberg, Fraser, and Cheung 2009; Rezvani, Khosravi, and Dong 2017).

Given this accumulation of evidence, it seemed appropriate for us to conduct a systematic and critical review of the literature on large-scale project CSFs. We sought in particular to understand the best ways to deliver large-scale projects by addressing four objectives: (1) to ascertain the annual publication trends of CSFs for large-scale projects; (2) to identify countries with most published articles on the CSFs for large-scale projects; (3) to summarize, compare, and contrast the findings of CSF studies; and (4) to identify the associated tools in large-scale complex projects (LSCPs). Additionally, we hoped that our review would provide a checklist of CSFs and associated tools for large-scale projects that could be adopted by researchers and practitioners for successful delivery of large-scale projects and future empirical research in this field.

In this chapter, we begin with an overview based on previously published reviews of CSFs in project management literature. Next, we outline the methodology we employed in our review. We then present the findings reported in articles published in selected academic journals. In addition, we recommend a list of particular project management tools that should assist project organizations, project leaders, and project team members to complete their project efficiently, effectively, and successfully. We conclude with a discussion of ideas for future research.

LITERATURE REVIEW

CRITICAL SUCCESS FACTORS

We define a CSF as *a critical project activity where a favorable outcome is a prerequisite for the project manager to conclude s/he has reached her or his project goal.* As such, a project manager must identify a set of CSFs that encompass all major goals necessary to ensure the success of the project overall (Ika 2009; Rezvani and Khosravi, 2018). Given the centrality of CSFs, it is unsurprising to find that researchers have been studying project management CSFs since the 1970s (see Davis 2014). Moreover, Pinto and Slevin (1987) showed how the CSF approach could be applied to identify best ways to achieve success in large-scale projects (which is the focus of this chapter) across a wide range of industry sectors, project stages, and stakeholders. For instance, Mazur et al. (2014) and Rezvani et al. (2016) investigated CSFs for large-scale military projects in Australia. Other sectors' researchers have studied transportation, engineering, infrastructure, telecommunication, energy, and housing (Wu et al., 2017; Khosravi, Rezvani, Subasinghage, and Perera 2012; Rezvani, Barrett, and Khosravi 2018). With specific application to project stages, Jugdev and Müller (2005) examined success factors across all project stages, while Turner and Zolin (2012) focused on different stakeholders at the briefing stages of large-scale projects. In addition, Ika (2009) sought to identify success factors related to CSF forecasting. Other researchers (see, e.g., Toor and Ogunlana 2009a; Dimitriou, Ward, and Wright 2013; Ferratt et al. 2006; McGillivray et al. 2009) studied CSFs in developed and developing countries with a view to foster growth and success within large-scale projects.

Nonetheless, given the wide spectrum and coverage of studies on CSFs for large-scale projects, it is difficult for both practitioners and researchers to identify the most important CSFs for successfully implementing large-scale projects (irrespective of sector, stages, or country). It therefore seemed to us to be important to identify, to analyze, and to summarize key findings identified in previous studies in order to facilitate understanding of the most important CSFs within large-scale projects (irrespective of sector, stage, or country).

LARGE-SCALE COMPLEX PROJECTS

Although a unified definition of LSCPs remains elusive (Vidal, Marle, and Bocquet 2011), we nonetheless define them as projects that are largely ill-defined (Qureshi and Kang 2015; Thomas and Mengel 2008), chaotic (Thomas and Mengel 2008), interrelated (Browning 2014; Qureshi and Kang 2015; Yang, Shen, Ho, Drew, and Xue 2011), fast-paced, erratic, ambiguous, and nonlinear (PMI 2008; Qureshi and Kang 2015; Senescu, Aranda-Mena, and Haymaker 2012; Yang, Huang, and Wu 2011). A further problem in defining this construct derives from the fact that complexity has increased in the "hypercompetitive economy" (Assudani and Kloppenborg 2010, p. 67), resulting in organizations putting in place multiple success parameters. Just like (noncomplex, smaller) traditional projects, organizations undertaking LSCPs still require them to add value, however. More specifically, and following guidelines set out by the *Global Alliance for Project Performance Standard* (GAPPS 2006) and prior studies in complex projects (e.g., see Ahern, Leavy, and Byrne 2014; Locatelli, Mancini, and Romano 2014; Rezvani et al. 2016; Bosch-Rekveldt, Jongkind, Mooi, Bakker, and Verbraeck 2011), we consider a project as being an LSCP if it has at least one of seven key characteristics (see Table 10.1).

In spite of the interest by governments worldwide in LSCPs, their implementation nonetheless is subject to numerous impediments that need attention. These include lack of transparent communication, lengthy time periods, lack of top-management support, lack of detailed plan, and lack of user/owner satisfaction (Fortune and White 2006; Ogunlana 2008; Rezvani and Khosravi 2019a; Yu et al. 2006). In addition, large and complex projects are infamous for their high failure rates (Rezvani and Khosravi 2019b). Indeed, the failure of LSCPs has been disastrous to many organizations. According to PMI (2013), only 38% of projects meet their business goals and objectives. In view of this, identification of CSFs in LSCPs should enable practitioners to be more enlightened about the key success factors for delivering successful future LSCPs.

METHOD

Following guidelines offered by Tranfield, Denyer, and Smart (2003), we conducted a systematic review for all the available research evidence of sufficient quality over the three stages of an LSCP: planning, executing, and reporting.

TABLE 10.1
Seven Characteristics of LSCPs

LSCP Property	References
(1) A high degree of uncertainty and mixture of joined organizations and subcontracting	Ahern et al. (2014)
(2) Rapid change of technology	Davies and Mackenzie (2014)
	Bosch-Rekveldt et al. (2011)
(3) A high degree of interdependency between a number of system parts and organizations involved	Locatelli and Mancini (2012)
(4) Strong legal, social, or environmental implications from undertaking the project	Chang (2013)
(5) Clear strategic importance of the project to the organization or organizations involved	Mazur et al. (2014)
	Bosch-Rekveldt et al. (2011)
(6) Stakeholders with conflicting needs regarding the characteristics of the project's product	Locatelli et al. (2014)
(7) Newness of technology	Robinson Fayek, Revay, Rowan, and Mosseau (2006)

SEARCH STRING

We employed a three-step search strategy in our literature search. In the first step, we conducted an initial search of the *Science Direct* (Elsevier), *Wiley*, and *ABI/INFORM* databases in order to determine optimal search terms. Following discussions between the coauthors and focusing on the key articles related to large-scale project success, we next identified appropriate search terms relating to (1) project size and (2) project success. We then developed and adopted the Boolean search string: ("mega project*" OR "large-scale project*" OR "large project*" OR "major project*") AND ("success" OR "project success factor*" OR "project performance" OR "performance" OR "project success" OR "critical success factor"). We used this string to query publication keywords, titles, and abstracts in the databases; resulting in identification of 2,324 potentially relevant articles.

At the next stage, we filtered the list of articles by discarding any not published in peer-reviewed journals or that failed to provide empirical evidence regarding the success of LSCPs. We also excluded studies in languages other than English, conceptual articles, conference papers, unpublished full-text documents, and review articles. We next dropped any reports that failed to describe the research method used (cf. Savolainen, Ahonen, and Richardson 2012; Jørgensen and Moløkken-Østvold 2006). At the end of this stage, we retained 236 articles for further examination.

Finally, and following the two-stage search procedure adopted by Yang, Shen, Ho, Drew, and Xue. (2011) (see also Mok, Shen, and Yang 2015), we reviewed the abstracts and introductions of the 236 articles to identify those that specifically addressed project success factors. This process left us with a subset of 86 articles for further review. Consistent with the second stage of the Yang et al. approach, we then carefully read the remaining articles and excluded any we deemed irrelevant. In the end, we identified a subset of 63 articles that we considered to offer high quality and substantiated insights about project success in LSCPs (based on the inclusion criteria).

THEME IDENTIFICATION

Once we had identified the 63 articles, and after noting the year and country of publication, we used descriptive and thematic methods to ensure the validity and reliability of our literature analysis (Levy and Ellis 2006; Morgan and Smircich 1980; Ritchie, Lewis, Nicholls, and Ormston 2013). Using this method, we synthesized the main outcomes extracted from the literature and identified a subset of content-related categories of qualitative data (Braun and Clarke 2006; Guest, MacQueen, and Namey 2011). We then determined the major facets of our dataset by counting the number of times each topic appeared (cf. Neuendorf 2016; Levy and Ellis 2006).

At the final stage, and following Ritchie et al. (2013), we conducted a definitive thematic analysis in four steps. First, we read the literature in order to identify a set of suitable categories. This led us to identify recurring themes from the collected literature with specific reference to various large-scale project types and their success factors. After we had identified the themes, the first author arranged the main categories and their attributes using a Microsoft Excel spreadsheet to compare the identified categories. At the next stage, we condensed the results of the complete categorization set into the results shown in Table 10.2. Finally, we discussed and settled any discordances to arrive at a final set of categories.

RESULTS

OVERVIEW OF RELEVANT PEER-REVIEWED PUBLICATIONS

As can be seen in Figure 10.1, while the research on CSFs in LSCPs first appeared in a peer-reviewed journal in 1987, concerted interest in the peer-reviewed literature only began in earnest in 1996. Since then, articles have appeared in all but 2 years (2003, 2011), with more than one article published in all subsequent years except 1998, 2000, 2007, 2010, and 2014. Peak publication years

TABLE 10.2
CSFs in LSCPs

CSFs	References	Frequency
(1) Open communication	Hughes (1986); Ashley et al. (1987); Pinto and Slevin (1988); Ng and Mo(1997); Egbu (1999); Turner (1999); Chan, Ho, and Tam (2001); Chua, Kog, and Loh (1999); Mazur et al.(2014); Rezvani, Barrett, et al. (2018a); Cooke-Davies (2002); Rezvani, Khosravi, et al. (2018b); Duy Nguyen, Ogunlana, and Lan (2004); Belout and Gauvreau (2004); Phua (2004); Nicolini (2002); Fortune and White (2006); Yu et al. (2006); Ogunlana (2008); Al Nahyan et al. (2012); Chang (2013); Rezvani et al. (2016); McGillivray et al. (2009)	23
(2) Project planning and control	Belassi and Tukel (1996); Chan et al. (2001); Duy Nguyen et al. (2004); Belout and Gauvreau (2004); Fortune and White (2006, Cook-Davies (2002); Nicolini (2002); Ashley et al. (1987); Pinto and Slevin (1988); Toor and Ogunlana (2009a); Dimitriou et al. (2013); Ferratt et al. (2006); McGillivray et al. (2009); Turner and Zolin (2012)	14
(3) Clearly defined goals, mission, and priorities	Mazur et al. (2014); Rezvani et al. (2016), Hughes (1986); Pinto and Slevin (1988); Songer and Molenaar (1997); Jang and Lee (1998); Clarke (1999); Chua et al. (1999); Chan et al. (2001); Gale and Luo (2004); Yu et al. (2006); Fortune and White (2006); Nicolini (2002); Clarke (1999)	14
(4) Project manager and project team competence	Belassi and Tukel (1996); Egbu, (1999); Chua et al. (1999); Chua et al. (1999); Mazur et al.(2014); Duy Nguyen et al. (2004); Nicolini,(2002); Fortune and White (2006); Rezvani et al. (2016); Pinto and Slevin (1988); Rezvani, Barrett, et al. (2018a); Rezvani, Khosravi, et al. (2018b); Belout and Gauvreau (2004); Ogunlana (2008)	14
(5) Top-management support	Belout and Gauvreau (2004); Duy Nguyen et al. (2004); Pinto and Slevin (1988); Chua et al. (1999); Mazur et al. (2014); Rezvani et al. (2016); Yu et al. (2006); Fortune and White (2006); Nicolini (2002); McGillivray et al. (2009); Ferratt et al. (2006); Rezvani, Khosravi, and Dong (2017)	13
(6) Stakeholder satisfaction	Wu et al. (2017); Lech (2013); Ng and Mo (1997); Dvir et al. (2006); Chua et al. (1999); Phua (2004); Yu et al. (2006); Zhang and Fan (2013); Al Nahyan et al. (2012); Williams (2016); Turner and Zolin (2012); Liu and Wang (2016)	12
(7) Software use and selection	Sommerville and Craig (2004, 2006); Songer and Molenaar (1997); Ferratt et al. (2006); Duy Nguyen et al. (2004); Chan et al. (2001); Phua (2004); Fortune and White (2006); Khosravi, Rezvani, and Wiewiora (2016); Khosravi, Rezvani, and Ahmad (2013)	10
(8) Trust	Rezvani et al. (2016); Chua et al. (1999); Chan et al. (2001); Yu et al. (2006); Phua (2004); Duy Nguyen et al. (2004); Rezvani, Barrett, et al. (2018a); Rezvani, Khosravi, et al. (2018b)	9
(9) Problem solving	Eriksson et al. (2017); Belout and Gauvreau (2004); Mazur et al. (2014); Rezvani et al. (2016); Pinto and Slevin (1988); Rezvani, Barrett, et al. (2018a); Rezvani, Khosravi, et al. (2018b); Chang (2013)	8
(10) Availability of resources	Turner (1999); Belassi and Tukel (1996); Songer and Molenaar (1997); Chua et al. (1999); Phua (2004); Yu et al. (2006); Fortune and White (2006); Duy Nguyen et al. (2004)	8
(11) Staff commitment	Ogunlana (2008); Phua (2004); McGillivray et al. (2009); Songer and Molenaar (1997); Fortune and White (2006); Yu et al. (2006)	6
(12) Technical capabilities	Al Nahyan et al. (2012); Ashley, Lurie, and Jaselskis (1987); Adoko, Mazzuchi, and Sarkani (2015); Pinto and Slevin (1988); Nicolini (2002); Belout and Gauvreau (2004)	6
(13) Effective change management	Phua (2004); Arain and Low (2005); Fortune and White (2006); Cooke-Davies (2002); Yu et al. (2006); Chan et al. (2001)	6

(Continued)

TABLE 10.2 (*Continued*)
CSFs in LSCPs

CSFs	References	Frequency
(14) Meeting user's/ customer's/ owner's requirement	Toor and Ogunlana (2009a); Zhang and Fan (2013); Ogunlana (2008); Dvir et al. (2006); McGillivray et al. (2009)	5
(15) Involvement of client	Toor and Ogunlana (2009a); Pinto and Slevin (1988); Belout and Gauvreau (2004); Yu et al. (2006); Fortune and White (2006)	5
(16) Training	Chang (2013); McGillivray et al. (2009); Ferratt et al. (2006)	3
(17) Health and safety	Zhang and Fan (2013); Toor and Ogunlana (2009a); Williams (2016)	3
(18) Cleary defined contract	Phua (2004); Chua et al. (1999); Duy Nguyen et al. (2004)	3
(19) Absence of conflicts	Rezvani, Barrett, et al. (2018a); Zhang and Fan (2013); Rezvani, Khosravi, et al. (2018b)	3
(20) Technical support	McGillivray et al. (2009); Lech (2013); Khosraviet al. (2016)	3
(21) Team contributions	Ferratt et al. (2006); Duy Nguyen et al. (2004)	2
(22) Learning from past experience	Yu et al. (2006); Duy Nguyen et al. (2004)	2
(23) Project member well-being	Chang (2013)	1
(24) Achieve business/ organizational goals	Chua, Lim, Soh, and Sia (2012)	1
(25) Consulting capability	Ferratt et al. (2006)	1
(26) Defense capability	Chang (2013)	1
(27) Risks management	Zhang and Fan (2013)	1
(28) Claim management	Zhang and Fan (2013)	1
(29) Standardization of the project delivery	Locatelli and Mancini (2012)	1
(30) Project efficiency	Toor and Ogunlana (2009a)	1

were 1999, 2004, and 2012, which saw the appearance of five or six peer-reviewed publications. Overall, however, scholarly interest in exploring the best ways to deliver successful large-scale projects has tended to remain active since 1996 (Williams 2016).

In order to ascertain countries with the most research on large-scale projects in our final pool, we counted the articles in each country. Note, however, that we found several that we could not attribute to any specific country. We categorized these as "international" articles. In Figure 10.2, we list a count of research into LSCPs by country. As can be seen in the figure, the research emanated from 16 discrete countries. Most single-country research (22 articles or 35%) originated in the United States and Australia, however; although we classified a significant proportion (8 articles or 13%) as "international."

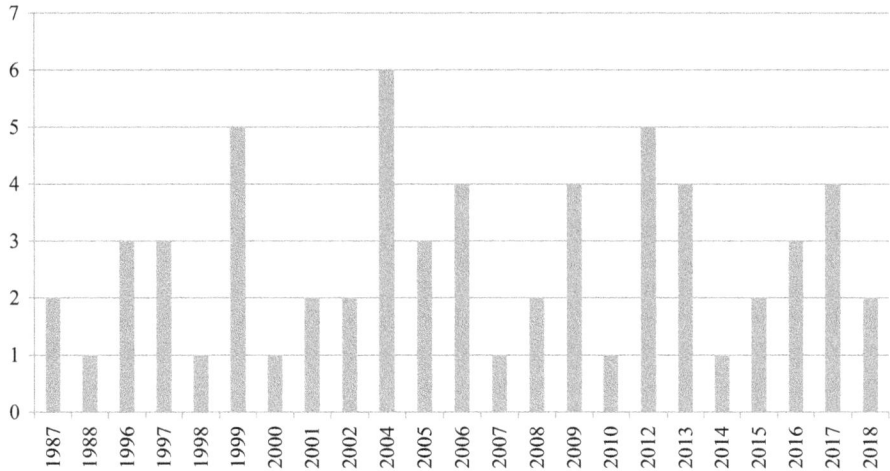

FIGURE 10.1 Publications per year.

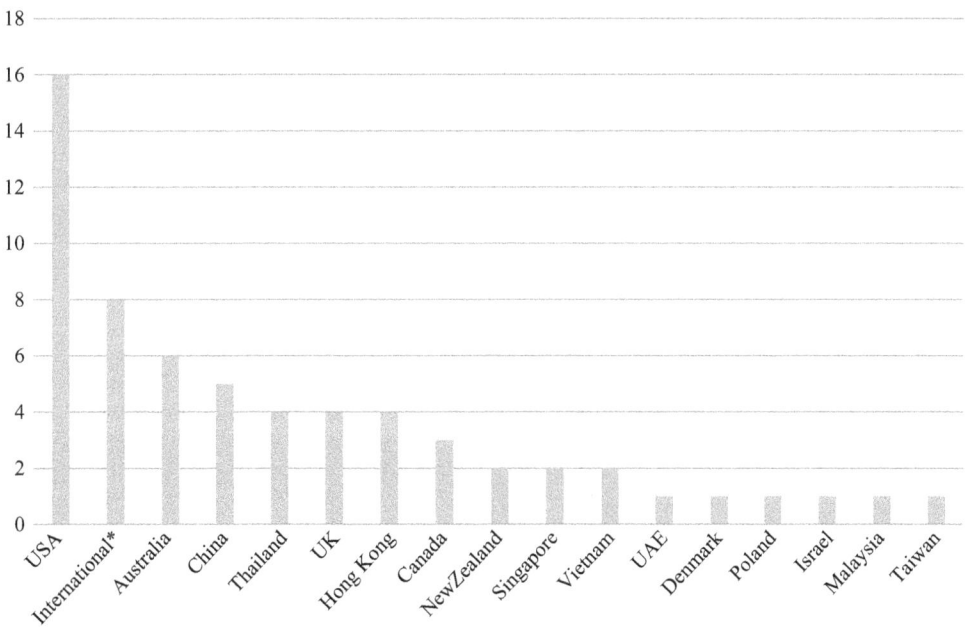

FIGURE 10.2 Projects per country. (* Contains studies that were focused on multiple countries.)

CSF Categories

Our main aim in this research was to identify which categories of CSF appeared in the peer-reviewed literature since the first one in 1987. In this regard, we list in Table 10.2 the publications in each category, ordered by frequency of appearance. In the end, we identified 30 CSFs (across the 63 publications). The top six factors we found were *open communication* (23 articles); *project planning and control* (14); *clearly defined goals, mission, and priorities* (14); *project manager and project team competence* (14); *top-management support* (13); and *stakeholder satisfaction* (12).

DISCUSSION

As we noted earlier, governments across the world are increasingly implementing LSCPs. In view of this, it is not surprising to find that project management researchers worldwide are increasingly looking into the CSFs for implementing such projects. To identify the trend in this line of research, we conducted a three-stage review of publications from 1987 to 2018 based on journal titles listed in the *Science Direct* (Elsevier), *Wiley*, and *ABI/INFORM* databases. We focused on identifying and categorizing CSFs for LSCPs. We found evidence that, over the 40 years since inception of the LSCP concept, researchers' interest in exploring the CSFs for such projects has been increasing steadily but includes three noticeable surges (in 1999, 2004, and 2012), with between one and six articles published each year in the peer-reviewed literature. In the following sections, we discuss our findings with regard to two key sets of findings: (1) the national origins of the studies we reviewed and (2) the categories of CSF we found in our review.

Country Differences

We found that researchers from the United States and Australia were the principal source of CSF publications in the period we reviewed, followed by China, the United Kingdom, Thailand, Hong Kong, and Canada. At the same time, however, it is disappointing to see so few publications on CSFs from developing countries. Clearly, there remains an untapped potential for additional research in this respect.

CSF Categories

As can be seen in Table 10.2, we identified 30 categories of CSF in the 63 peer-reviewed articles we reviewed. We identified in particular that six CSF categories were the most frequently cited. In the following, we discuss each of the six in more detail, followed by a brief summary of the remaining 24.

Open Communication

We found open communication present in 23 of the 63 articles we examined. This shows that communication is the clearly most frequently used CSF in LSCP. Interestingly, project management researchers appeared largely to have overlooked the centrality of communication as a CSF. Open communication is especially important insofar as it allows project participants to understand the requirements of project stakeholders, which in turn result in reductions of cost and time (Cooke-Davies 2002; Duy Nguyenet al. 2004; Rezvani et al. 2016). The overarching message here is that successful LSCP management requires transparency and open communication throughout the entire project. Moreover, the authors of these articles stressed that open communication must be applied in both formal and informal communication throughout the delivery of the LSCPs. Formal communication refers to the type of communication that requires much planning and includes written status reports. Informal communication refers to spontaneous communication, such as talking in the doorway, chatting, and talking in front of the screen (Chanet al. 2001; Chuaet al. 1999).

Surprisingly, no previous studies appear to have identified this as the one key CSF above all others. The importance of open communication could be found in the evidence that large projects usually have longer lifespans, so that open communication among all project participants becomes vital to share knowledge and information (Pinto and Slevin 1988; Chan et al. 2001; Cooke-Davies 2002; Duy Nguyen et al. 2004). Transparency in communication rests on the shoulders of all key internal and external stakeholders (Fortune and White 2006; Ogunlana 2008; McGillivray et al. 2009). On the other hand, lack of open communication is likely to hinder the sharing of relevant knowledge and experiences.

PROJECT PLANNING AND CONTROL

We found that project planning and control appeared as a CSF in 14 of the 63 articles. In this regard, the authors stressed that the complex nature of large-scale projects makes it very difficult for a single and specific plan from the beginning to execute the project (Chan et al. 2001; Duy Nguyen et al. 2004; Toor and Ogunlana 2009a, b). These authors stressed that project plans need to be consulted and, if necessary, updated during each stage of an LSCP. The authors also commented that the structure and compatibility of this entity influence the success of the project. In this case, we note organizations to be clear about expected outcomes and benefits (Dimitriou et al. 2013; Ferratt et al. 2006; McGillivray et al. 2009; Turner and Zolin 2012). For instance, a detailed and specific plan helps in understanding requirements, objectives, and priorities in LSCPs (Ferratt et al. 2006; Duy Nguyen et al. 2004). It looks like a weak and poorly managed plan during the project life cycle could easily result in difficulties and eventually a failure to undertake the project successfully. Consequently, project plans must be equipped with strong technical, operational, and managerial capacity to be able to execute the LSCP. Additionally, several authors (e.g., Dimitriou et al. 2013; Duy Nguyen et al. 2004; Toor and Ogunlana 2009a) found that, especially in countries where project plans tend to be underdeveloped (e.g., in developing countries), governments need to assist in strengthening local companies both financially and technically (i.e., to build their capacity to be able to compete with international project companies).

CLEARLY DEFINED GOALS, MISSION, AND PRIORITIES

Fourteen of the articles we examined identified clearly defined goals, mission, and priorities as CSFs for LSCPs. Mission clarity refers to "initial clarity of goals and general directions" (Pinto and Slevin 1987, p. 31). This is important, especially because LSCPs feature high levels of complexity and ambiguity (Dvir et al. 2006). Rezvani et al. (2016) point out further that, in large and complex defense projects, it is common for projects to have vague goals (e.g., "increase defense capability") at the beginning of a long-term project. In particular, if not enough time is allocated for clearly defined goals and mission, projects are more likely to exceed budgets and to be perceived eventually to be failures (Chua et al. 1999; Chan et al. 2001; Gale and Luo 2004; Yu et al. 2006, Fortune and White 2006).

PROJECT MANAGER AND PROJECT TEAM COMPETENCY

The authors of 14 of the articles reported that project manager and project team competency constitutes an important CSF for LSCPs. As Toor and Ogunlana (2009a) and Rezvani, Barrett and Khosravi (2018) note, project managers and project team members come across many problems and issues that need expert skills if a solution is to be found. If the project manager is not competent or does not know how to solve the problem, however, it is very difficult to achieve CSFs successfully. In this regard, the authors (see Cooke-Davies 2002; Nicolini 2002; Pinto and Slevin 1988; Toor and Ogunlana 2009a, Dimitriou et al. 2013) consistently report that skill and competency deficiencies cause excessive delays. In view of this evidence, it seems likely that incompetent project managers of failed LSCPs (as well as their project team members) simply did not know how to make decisions, how to negotiate problems, or how to resolve conflicts constructively (Toor and Ogunlana 2009a).

TOP-MANAGEMENT SUPPORT

The authors of 13 of the articles concluded that LSCP top-management support is a CSF. Top-management support refers to "the willingness of top management to provide the necessary resources and authority/ power for project success" (Pinto and Slevin 1987, p. 31). McGillivray et al. (2009) note in particular that this level of support is a CSF across all phases of project planning and execution. Similarly, Rezvani et al. (2016) found top-management support to be crucial in achieving project success in large defense projects. The support from top management needs to be clear in terms of the roles and responsibilities of project team members as well as clarity of strategic direction (Fortune and White 2006; Pinto and Slevin 1988; Rezvani, Chang and Wiewiora 2015; Yu et al. 2006).

STAKEHOLDER SATISFACTION

To round out the "top six," we found that the authors of 12 of the articles reported that LSCP stakeholder satisfaction is a key determinant of CSF. The Project Management Institute (PMI 2008) defines project stakeholders as those individuals and organizations who are actively involved in a project or whose interests depend upon successful project execution or completion. Several authors (Dvir et al. 2006; Lech 2013; Ng and Mo 1997; Wu et al. 2017) found that successful completion of LSCPs also depended upon the satisfaction of key stakeholders in large-scale projects. Studies have shown that the necessary support from all key stakeholders attracts more investors to a particular economy (Yu et al. 2006; Fortune and White 2006; Nicolini 2002; McGillivray et al. 2009). This is consistent with the findings of the OECD (2008), who concluded that a lack of investor support leads to high risk that, in turn, limits competition in the tendering process. This shows that management of all key stakeholders and meeting the concerns and expectations of project stakeholders constitute the key CSFs for LSCPs.

OTHER CATEGORIES

The remaining 24 CSF categories listed in Table 10.2 relate to three broad aspects of project planning: (1) project personnel, (2) technical issues, and (3) organizational factors. With regard to project personal, article authors identified trust, problem solving, staff commitment, effective change management, absence of conflict, team contribution, project member well-being, and learning from experience. Researchers who identified these factors include Rezvani et al. (2016), Chua et al. (1999), Chan et al. (2001), Yu et al. (2006), and Pinto and Slevin (1988). Organizational factors identified in the article we reviewed include availability of resources, software use and selection, defense capabilities, achieve business/organizational goals, consulting capabilities, and claim management (see Belassi and Tukel 1996; Chang 2013; Ferratt et al. 2006; Khosraviet al. 2013; Rezvani, Chang and Wiewiora 2016a, 2016b). Risk management, project efficiency, standardization of project delivery, and technical support were categorized as technical factors by several studies (e.g., Khosraviet al. 2016; McGillivray et al. 2009; Lech 2013; Locatelli and Mancini 2012).

To sum up, the frequency used of CSFs must be interpreted carefully. The low frequency used of the remaining CSFs does not imply that they are not important. Clearly, more research is required to pay deserved attention to the remaining CSFs.

PROJECT MANAGEMENT TOOLS

In this section, and based on our literature survey, we identify project management tools applicable to the top three categories of CSF we identified in our survey.

COMMUNICATION TOOLS

Each LSCP has a unique set of individuals and circumstances that dictate which forms of communication will work best and will help team members to make timely and appropriate decisions. As we found in our survey, it is important in LSCPs to provide open channels of communication so information can flow back and forth as needed to make correct and effective decisions. Thus, it is important to use the right tools to communicate effectively and make sense for the information that project team members share with other team members. LSCPs require communication across different levels, and choosing the right tool for each level ensures that the information exchange happens as smoothly as possible. Recognizing that different types of communication need to happen, and that each is carried out using the right tool, helps to keep the project moving forward with less friction and misunderstanding. We discuss two tools relevant for improving communication in LSCPs.

Basecamp (https://basecamp.com) is a discussion tool to enable shared, asynchronous, archived conversations about the project (Featherstone 2009). The tool provides a way to have conversations that are more useful than email because everyone on the project can participate without having to be consciously included. Moreover, newcomers to a project can go back through Basecamp log and see discussions that occurred before they joined. This is also a good place to document, to summarize, and to archive discussions that can later be retrieved to assess project performance, especially when things do not go as planned.

Slack and HipChat (https://slack.com). These (and similar) tools enable dispersed teams to have short, less formal project discussions (Perkel 2017). An especially useful feature of these tools is that conversations are archived. This enables team managers and members who were unable to be involved in real time to catch up on what they discussed earlier. It is important to make every attempt to include clients and key stakeholders in daily/weekly/monthly discussions about their project, and this is an excellent way to do just that. In LSCPs, project leaders should try to link up periodically with all project stakeholders and team members so that all project members and key stakeholders have a clear picture about what is going on with the project. These tools allow them to accomplish this.

PROJECT PLANNING AND CONTROL TOOLS

Gantt Charts, the Program Evaluation Review Technique (PERT), and the Critical Path Method (CPM) comprise the most commonly used project planning and control tools (Wilson 2003). Project managers may implement any or all of these project management tools, either manually or with commercially available project management software (e.g. www.projectmanager.com/software).

> PERT and Gantt Charts are recommended project management tools for use in LSCP management.

Gantt Charts constitute a useful tool for analyzing, planning, and controlling LSCPs.[1] Project managers can employ them to accomplish five critical tasks: (1) to identify the tasks and sub-tasks to be undertaken, (2) to lay out the tasks that need to be completed, (3) to assist in scheduling when these tasks will be carried out and in what order, (4) to assist in planning resources needed to complete the project, and (5) to assist in working out the critical path for a project where it needs to be completed by a particular date. When a complex or multitask project is under way, managers can use Gantt Charts to assist in monitoring whether or not the project is on schedule. If not, the Gantt Chart allows managers to identify what actions they need to take in order to put a delayed project back onto schedule. An essential concept behind project planning is that some activities depend upon prior completion than other activities (the "critical path"). For example, it is not a good idea to start building the walls in an office block before you have laid the foundations. These in effect comprise dependent activities – that need to be completed in a sequence, with each stage being more or less completed before the next stage can begin. As such, activities on the critical path are "sequential." Nonsequential activities are not dependent on the completion of any other tasks. Managers may schedule completion of these activities at any time before or after a particular stage in the project is reached, provided there is enough associated "slack time," which is the time left before the task affects the critical (sequential) path. These activities are called are nondependent or "parallel" tasks.

Eight known benefits of using Gantt Charts are that they

a. Provide an easy to understand visual display of the scheduled time of a task or activity
b. Make it easy to develop "what if" scenarios

c. Enable better project control by promoting clearer communication
d. Become a tool for negotiations
e. Show the actual progress against the planned schedule
f. Report results at appropriate levels
g. Allow comparison of multiple projects to determine risk or resource allocation
h. Reward the project manager with more visibility and control over the project.

The *PERT and the CPM* constitute an alternative planning and control tool that project managers can employ for defining and controlling the tasks necessary to complete a project. Note that managers often use PERT and CPM methods together; the only difference is how task times are computed. Both charts display the total project with all scheduled tasks shown in sequence. The displayed tasks show which ones are in parallel, those tasks that managers can schedule at the same time. Managers usually employ a graphic representation called a "Project Network" or "CPM Diagram" to portray graphically the interrelationships of the elements of a project and to show the order in which the activities must be performed. PERT planning involves two key steps:

1. Identify the specific activities and milestones (i.e., that mark the beginning and the end of each activity) that define the project tasks.
2. Determine the proper sequence of activities. This step may be combined with step 1 since the activity sequence is evident for some tasks. Other tasks may require some analysis to determine their exact sequencing.

Benefits to using the PERT or the CPM include

a. Improved planning and scheduling of activities
b. Improved forecasting of resource requirements
c. Identification of repetitive planning patterns which can be followed in other projects (thus simplifying the planning process)
d. Ability to identify and thus to reschedule activities (activity milestones) to reflect inter-project dependencies and resource limitations following known priority rules
e. Ability to see and control the expected project completion time, including the probability of completion before a specified date, identification of the critical path activities that impact completion time as well as the activities that have slack time and that can lend resources to critical path activities.

STAKEHOLDER SATISFACTION TOOL

As an example of a tool that can be employed to manage and meet the concerns and expectations of key stakeholders in LSCPs, we now outline the *Stakeholder Circle Tool* (https://stakeholder-management.com). As Bourne and Walker (2008) note, this tool provides a useful visual representation of all key stakeholders in

> The Stakeholder Circle Tool is recommended to manage and meet the concerns and expectations of key stakeholders in LSCPs.

LSCPs. The Stakeholder Circle is based on the premise that the success of a project business enterprise can only occur with the informed consent of its stakeholder community.

This tool can be especially useful for project managers trying to understand and to remain alert to the nature of stakeholder impact. The developers of the tool intended it to offer a mechanism for assessing the relative influence of each of the key stakeholders and for planning ways to engage with and manage their expectations/contributions. The benefit of using this tool derives in part from the analysis process itself as well as from the ease with which the influence of key stakeholders.

DISCUSSION QUESTIONS

1. Briefly describe the term complex project. What are the most frequently cited CSFs in LSCPs?
2. Provide an example of a tool that can be employed to manage and meet the concerns and expectations of key stakeholders in LSCPs?
3. Describe the importance of project planning and control in LSCPs and identify the relevant project planning and control tools.

CONCLUSIONS AND FUTURE RESEARCH

The findings revealed in the study we describe in this chapter provide a solid foundation for future research on large-scale CSFs – in terms of the scope and method, researchers may wish to adopt for subsequent research into the factors that determine LSCP success. Such research in turn has the potential to inform governments and international private developers seeking to implement and enter into the LSCP market as to the most important CSFs for engaging in complex projects irrespective of country and sectors. We found that open communication is the most cited CSF for LSCPs, followed by project planning and control; clearly defined goals, mission, and priorities; project manager and project team competence; top-management support; and stakeholder satisfaction. We provide a sample of project planning tools that can help address the top three CSFs we identified. We grouped the remaining 24 CSFs into three broad categories: project personnel, technical issues, and organizational factors.

> Open communication is the most cited CSF for LSCPs, followed by project planning and control, and clearly defined goals.

Our findings identify in particular that nearly all research in this field emanated from two developed countries (the United States and Australia). This suggests that there is potential for researchers in other, less developed, countries to contribute to this literature. We suggest in this regard that researchers in developing economies would do well to try to understand the nature of LSCP CSFs; as well as the nature of organizations, management strategies, norms, socioeconomic factors, and local cultural values in these settings. We also need to understand more about how LSCPs in developing countries differ from those in developed countries in terms of challenges, requirements, or management styles or what unique characteristics or specific factors arise from different infrastructures, local cultural values, or languages.

From a more practical perspective, via knowledge of the most cited CSFs in LSCPs, project managers can determine improvement measures to raise the probability of success and to reduce the chances of any setbacks in their own projects. There may also be practical benefits to policy development in improving the way we assess project success for LSCPs. This review may thus help organizations to divert their resources effectively to where maximum success lies while helping project leaders to accomplish their objectives. In conclusion, we recommended that project management researchers should benefit through use of the checklist of CSFs for LSCPs we developed in this study – and use it as a basis for further empirical analysis in this important field.

> Via knowledge of the most cited CSFs for LSCPs, project managers can determine improvement measures to raise the likelihood of project success.

ENDNOTE

1 For a list of commercial Gantt chart tools, see www.capterra.com/sem-compare/project-management-software?gclid=Cj0KCQiAgf3gBRDtARIsABgdL3mq4dKrTjsDoE-O-X6d2vTdvk5Gc0eXcILjt5VcgISqKLXalcsfvE8aAn-VEALw_wcB&gclsrc=aw.ds.

REFERENCES

Adoko, Moses T., Thomas A. Mazzuchi, and Shahram Sarkani. Developing a cost overrun predictive model for complex systems development projects. *Project Management Journal* 46, no. 6 (2015): 111–125.

Ahern, Terence, Brian Leavy, and P. J. Byrne. Complex project management as complex problem solving: A distributed knowledge management perspective. *International Journal of Project Management* 32, no. 8 (2014): 1371–1381.

Al Nahyan, Moza T., Amrik S. Sohal, Brian N. Fildes, and Yaser E. Hawas. Transportation infrastructure development in the UAE: Stakeholder perspectives on management practice. *Construction Innovation* 12, no. 4 (2012): 492–514.

Arain, F. M., and S. P. Low. Lesson learned from past projects for effective management of variation orders for institutional building projects. In *Proceedings of the MICRA 4th annual conference*, Kuala Lumpur, Malaysia: University of Malaya, pp. 10–1. 2005.

Ashley, David B., Clive S. Lurie, and Edward J. Jaselskis. *Determinants of Construction Project Success*. Newtown Square, PA: Project Management Institute, 1987.

Assudani, Rashmi, and Timothy J. Kloppenborg. Managing stakeholders for project management success: An emergent model of stakeholders. *Journal of General Management* 35, no. 3 (2010): 67–80.

Baccarini, David. The concept of project complexity—A review. *International Journal of Project Management* 14, no. 4 (1996): 201–204.

Belassi, Walid, and Oya Iemeli Tukel. A new framework for determining critical success/failure factors in projects. *International Journal of Project Management* 14, no. 3 (1996): 141–151.

Belout, Adnane, and Clothilde Gauvreau. Factors influencing project success: The impact of human resource management. *International Journal of Project Management* 22, no. 1 (2004): 1–11.

Bosch-Rekveldt, Marian, Yuri Jongkind, Herman Mooi, Hans Bakker, and Alexander Verbraeck. Grasping project complexity in large engineering projects: The TOE (Technical, Organizational and Environmental) framework. *International Journal of Project Management* 29, no. 6 (2011): 728–739.

Bourne, Lynda, and Derek H. T. Walker. Project relationship management and the Stakeholder Circle™. *International Journal of Managing Projects in Business* 1, no. 1 (2008): 125–130.

Braun, Virginia, and Victoria Clarke. Using thematic analysis in psychology. *Qualitative Research in Psychology* 3, no. 2 (2006): 77–101.

Browning, Tyson R. Managing complex project process models with a process architecture framework. *International Journal of Project Management* 32, no. 2 (2014): 229–241.

Chan, Albert P. C., Danny C. K. Ho, and C. M. Tam. Design and build project success factors: Multivariate analysis. *Journal of Construction Engineering and Management* 127, no. 2 (2001): 93–100.

Chang, Chen-Yu. Understanding the hold-up problem in the management of megaprojects: The case of the Channel Tunnel Rail Link project. *International Journal of Project Management* 31, no. 4 (2013): 628–637.

Chua, Cecil Eng Huang, Wee-Kiat Lim, Christina Soh, and Siew Kien Sia. Enacting clan control in complex IT projects: A social capital perspective. (2012).

Chua, David Kim Huat, Yue-Choong Kog, and Ping Kit Loh. Critical success factors for different project objectives. *Journal of Construction Engineering and Management* 125, no. 3 (1999): 142–150.

Clarke, Angela. A practical use of key success factors to improve the effectiveness of project management. *International Journal of Project Management* 17, no. 3 (1999): 139–145.

Cooke-Davies, Terry. The "real" success factors on projects. *International Journal of Project Management* 20, no. 3 (2002): 185–190.

Davies, Andrew, and Ian Mackenzie. Project complexity and systems integration: Constructing the London 2012 Olympics and Paralympics Games. *International Journal of Project Management* 32, no. 5 (2014): 773–790.

Davis, Kate. Different stakeholder groups and their perceptions of project success. *International Journal of Project Management* 32, no. 2 (2014): 189–201.

Dimitriou, Harry T., E. John Ward, and Philip G. Wright. Mega transport projects—Beyond the 'iron triangle': Findings from the OMEGA research programme. *Progress in Planning* 86 (2013): 1–43.

Duy Nguyen, Long, Stephen O. Ogunlana, and Do Thi Xuan Lan. A study on project success factors in large construction projects in Vietnam. *Engineering, Construction and Architectural Management* 11, no. 6 (2004): 404–413.

Dvir, Dov, Arie Ben-David, Arik Sadeh, and Aaron J. Shenhar. Critical managerial factors affecting defense projects success: A comparison between neural network and regression analysis. *Engineering Applications of Artificial Intelligence* 19, no. 5 (2006): 535–543.

Egbu, Charles O. Skills, knowledge and competencies for managing construction refurbishment works. *Construction Management & Economics* 17, no. 1 (1999): 29–43.

Eriksson, Per Erik, Johan Larsson, and Ossi Pesämaa. Managing complex projects in the infrastructure sector—A structural equation model for flexibility-focused project management. *International Journal of Project Management* 35, no. 8 (2017): 1512–1523.

Featherstone, Robin. Basecamp. *Journal of the Medical Library Association: JMLA* 97, no. 1 (2009): 67.

Ferratt, Thomas W., Sanjay Ahire, and Prabuddha De. Achieving success in large projects: Implications from a study of ERP implementations. *Interfaces* 36, no. 5 (2006): 458–469.

Fortune, Joyce, and Diana White. Framing of project critical success factors by a systems model. *International Journal of Project Management* 24, no. 1 (2006): 53–65.

Gale, Andrew, and Jun Luo. Factors affecting construction joint ventures in China. *International Journal of Project Management* 22, no. 1 (2004): 33–42.

GAPPS. Lynn (ed) Duncan, and Crawford (Bill (ed)). *A Framework for Performance Based Competency Standards for Global Level 1 and 2 Project Managers.* London, UK: Global Alliance for Project Performance Standards, 2006.

Guest, Greg, Kathleen M. MacQueen, and Emily E. Namey. *Applied Thematic Analysis.* London: Sage Publications, 2011.

Michael William Hughes. Why projects fail-The effects of ignoring the obvious. *Industrial Engineering* 18, no. 4 (1986): 14.

Ika, Lavagnon A. Project success as a topic in project management journals. *Project Management Journal* 40, no. 4 (2009): 6–19.

Jang, Young, and Jinjoo Lee. Factors influencing the success of management consulting projects. *International Journal of Project Management* 16, no. 2 (1998): 67–72.

Jørgensen, Magne, and Kjetil Moløkken-Østvold. How large are software cost overruns? A review of the 1994 CHAOS report. *Information and Software Technology* 48, no. 4 (2006): 297–301.

Jugdev, Kam, and Ralf Müller. A retrospective look at our evolving understanding of project success. *Project Management Journal* 36, no. 4 (2005): 19–31.

Khosravi, Pouria, Azadeh Rezvani, and Mohammad Nazir Ahmad. Does organizational identification lead to information system success. *World Applied Sciences Journal* 21, no. 3 (2013): 402–408.

Khosravi, Pouria, Azadeh Rezvani, Maduka Subasinghage, and Melville Perera. Individuals' absorptive capacity in enterprise system assimilation. In *ACIS 2012: Location, location, location: Proceedings of the 23rd Australasian conference on information systems 2012*, pp. 1–7. ACIS, 2012.

Khosravi, Pouria, Azadeh Rezvani, and Anna Wiewiora. The impact of technology on older adults' social isolation. *Computers in Human Behavior* 63 (2016): 594–603.

Kwak, Young Hoon, and Brian M. Smith. Managing risks in mega defense acquisition projects: Performance, policy, and opportunities. *International Journal of Project Management* 27, no. 8 (2009): 812–820.

Lech, Przemysław. Time, budget, and functionality?—IT project success criteria revised. *Information Systems Management* 30, no. 3 (2013): 263–275.

Levy, Yair, and Timothy J. Ellis. A systems approach to conduct an effective literature review in support of information systems research. *Informing Science: The International Journal of an Emerging Transdiscipline* 9 (2006): 181–213.

Liu, Shan, and Lin Wang. Influence of managerial control on performance in medical information system projects: the moderating role of organizational environment and team risks. *International Journal of Project Management* 34, no. 1 (2016): 102–116.

Locatelli, Giorgio, and Mauro Mancini. Looking back to see the future: building nuclear power plants in Europe. *Construction Management and Economics* 30, no. 8 (2012): 623–637.

Locatelli, Giorgio, Mauro Mancini, and Erika Romano. Systems engineering to improve the governance in complex project environments. *International Journal of Project Management* 32, no. 8 (2014): 1395–1410.

Mazur, Alicia, Anne Pisarski, Artemis Chang, and Neal M. Ashkanasy. Rating defence major project success: The role of personal attributes and stakeholder relationships. *International Journal of Project Management* 32, no. 6 (2014): 944–957.

McGillivray, Sue, Amy Greenberg, Lucina Fraser, and Ophelia Cheung. Key factors for consortial success: Realizing a shared vision for interlibrary loan in a consortium of Canadian libraries. *Interlending & Document Supply* 37, no. 1 (2009): 11–19.

Mok, Ka Yan, Geoffrey Qiping Shen, and Jing Yang. Stakeholder management studies in mega construction projects: A review and future directions. *International Journal of Project Management* 33, no. 2 (2015): 446–457.

Morgan, Gareth, and Linda Smircich. The case for qualitative research. *Academy of Management Review* 5, no. 4 (1980): 491–500.

Neuendorf, Kimberly A. *The Content Analysis Guidebook*. London: Sage Publications, 2016.

Ng, Lee Young, and Jei Kuang Weiyi Mo. Hospital procurement by design and build: A case study in Hong Kong. *CIB Report* (1997): pp. 545–554.

Nicolini, Davide. In search of 'project chemistry. *Construction Management & Economics* 20, no. 2 (2002): 167–177.

Ogunlana, Stephen O. Critical COMs of success in large-scale construction projects: Evidence from Thailand construction industry. *International Journal of Project Management* 26, no. 4 (2008): 420–430.

Perkel, Jeffrey M. How scientists use Slack. *Nature News* 541, no. 7635 (2017): 123.

Phua, Florence T. T. Modelling the determinants of multi-firm project success: A grounded exploration of differing participant perspectives. *Construction Management and Economics* 22, no. 5 (2004): 451–459.

Pinto, Jeffrey K., and Dennis P. Slevin. *Critical Success Factors across the Project Life Cycle*. Drexel Hill, PA: Project Management Institute, 1988.

Pinto, Jeffrey K., and Dennis P. Slevin. Critical factors in successful project implementation. *IEEE Transactions on Engineering Management* 1 (1987): 22–27.

Project Management Institute (PMI). *A Guide to the Project Management Body of Knowledge* (4th Edition). Newtown Square, PA: PMI, 2008.

Project Management Institute (PMI). PMI's Pulse of the Profession™. The High Cost of Low Performance (2013). Retrieved from www.pmi.org//media/pmi/documents/public/pdf/learning/thought-leadership/pulse/pulse-of-the-profession-2013.pdf.

Qureshi, Sheheryar Mohsin, and ChangWook Kang. Analysing the organizational factors of project complexity using structural equation modelling. *International Journal of Project Management* 33, no. 1 (2015): 165–176.

Rezvani, Azadeh, Artemis Chang, and Anna Wiewiora. Emotional intelligence, work attitudes and project success: An examination among project managers in complex projects. (2015).

Rezvani, Azadeh, Artemis Chang, and Anna Wiewiora. A taxonomy of project barriers in complex projects. (2016a).

Rezvani, Azadeh, Artemis Chang, and Anna Wiewiora. Project success in complex projects: A systematic literature review. (2016b).

Rezvani, Azadeh, Artemis Chang, Anna Wiewiora, Neal M. Ashkanasy, Peter J. Jordan, and Roxanne Zolin. Manager emotional intelligence and project success: The mediating role of job satisfaction and trust. *International Journal of Project Management* 34, no. 7 (2016): 1112–1122.

Rezvani, Azadeh, and Pouria Khosravi. A comprehensive assessment of project success within various large projects. *The Journal of Modern Project Management* 6, no. 1 (2018).

Rezvani, Azadeh, and Pouria Khosravi. Identification of failure factors in large scale complex projects: An integrative framework and review of emerging themes. *International Journal of Project Organisation and Management* 11, no. 1 (2019a): 1–21.

Rezvani, Azadeh, and Pouria Khosravi. Emotional intelligence: The key to mitigating stress and fostering trust among software developers working on information system projects. *International Journal of Information Management* 48 (2019b): 139–150.

Rezvani, Azadeh, Pouria Khosravi, and Linying Dong. Motivating users toward continued usage of information systems: Self-determination theory perspective. *Computers in Human Behavior* 76 (2017): 263–275.

Rezvani, Azadeh, Pouria Khosravi, and Neal M. Ashkanasy. Examining the interdependencies among emotional intelligence, trust, and performance in infrastructure projects: A multilevel study. *International Journal of Project Management* 36, no. 8 (2018): 1034–1046.

Rezvani, Azadeh, Rowena Barrett, and Pouria Khosravi. Investigating the relationships among team emotional intelligence, trust, conflict and team performance. *Team Performance Management: An International Journal* 25, no. 1/2 (2018): 120–137.

Ritchie, Jane, Jane Lewis, Carol McNaughton Nicholls, and Rachel Ormston, eds. *Qualitative Research Practice: A Guide for Social Science Students and Researchers*. London: Sage Publications, 2013.

Robinson Fayek, Aminah, Stephen O. Revay, Doug Rowan, and Donald Mousseau. Assessing performance trends on industrial construction mega projects. *Cost Engineering* 48, no. 10 (2006): 16.

Savolainen, Paula, Jarmo J. Ahonen, and Ita Richardson. Software development project success and failure from the supplier's perspective: A systematic literature review. *International Journal of Project Management* 30, no. 4 (2012): 458–469.

Senescu, Reid Robert, Guillermo Aranda-Mena, and John Riker Haymaker. Relationships between project complexity and communication. *Journal of Management in Engineering* 29, no. 2 (2012): 183–197.

Sommerville, James, and Nigel Craig. Information processing using a digital pen and paper. In *International Conference on Construction Information Technology (INCITE 2004): Managing Projects through Innovation & IT Solutions*, Langkawi, Malaysia, February, pp. 18–21. 2004.

Sommerville, James, and Nigel Craig. *Implementing IT in Construction*. London: Routledge (Taylor and Francis), 2006.

Songer, Anthony D., and Keith R. Molenaar. Project characteristics for successful public-sector design-build. *Journal of Construction Engineering and Management* 123, no. 1 (1997): 34–40.

Thomas, Janice, and Thomas Mengel. Preparing project managers to deal with complexity–Advanced project management education. *International Journal of Project Management* 26, no. 3 (2008): 304–315.

Toor, Shamas-ur-Rehman, and Stephen O. Ogunlana. Construction professionals' perception of critical success factors for large-scale construction projects. *Construction Innovation* 9, no. 2 (2009a): 149–167.

Toor, Shamas-ur-Rehman, and Stephen O. Ogunlana. Ineffective leadership: Investigating the negative attributes of leaders and organizational neutralizers. *Engineering, Construction and Architectural Management* 16, no. 3 (2009b): 254–272.

Tranfield, David, David Denyer, and Palminder Smart. Towards a methodology for developing evidence-informed management knowledge by means of systematic review. *British Journal of Management* 14, no. 3 (2003): 207–222.

Turner, J. R.. Company-wide project management: The planning and control of programmes of projects of different type. *International Journal of Project Management* 17, no. 1: 55–59.

Turner, Rodney, and Roxan Zolin. Forecasting success on large projects: Developing reliable scales to predict multiple perspectives by multiple stakeholders over multiple time frames. *Project Management Journal* 43, no. 5 (2012): 87–99.

Vidal, Ludovic-Alexandre, Franck Marle, and Jean-Claude Bocquet. Measuring project complexity using the Analytic Hierarchy Process. *International Journal of Project Management* 29, no. 6 (2011): 718–727.

Williams, Terry. Identifying success factors in construction projects: A case study. *Project Management Journal* 47, no. 1 (2016): 97–112.

Wilson, James M. Gantt charts: A centenary appreciation. *European Journal of Operational Research* 149, no. 2 (2003): 430–437.

Wu, Guangdong, Cong Liu, Xianbo Zhao, and Jian Zuo. Investigating the relationship between communication-conflict interaction and project success among construction project teams. *International Journal of Project Management* 35, no. 8 (2017): 1466–1482.

Yang, Jing, Geoffrey Qiping Shen, Manfong Ho, Derek S. Drew, and Xiaolong Xue. Stakeholder management in construction: An empirical study to address research gaps in previous studies. *International Journal of Project Management* 29, no. 7 (2011): 900–910.

Yang, Li-Ren, Chung-Fah Huang, and Kun-Shan Wu. The association among project manager's leadership style, teamwork and project success. *International Journal of Project Management* 29, no. 3 (2011): 258–267.

Yu, Ann T. W., Qiping Shen, John Kelly, and Kirsty Hunter. Investigation of critical success factors in construction project briefing by way of content analysis. *Journal of Construction Engineering and Management* 132, no. 11 (2006): 1178–1186.

Zhang, Lianying, and Weijie Fan. Improving performance of construction projects: A project manager's emotional intelligence approach. *Engineering, Construction and Architectural Management* 20, no. 2 (2013): 195–207.

11 Pattern Identification and Management Toolset

Robert Cloutier
University of South Alabama

CONTENTS

BACKGROUND

Growing up, I would watch my mother create clothing from what almost appeared to be thin air. I was brought up in what was a traditional family – a mother, father, and siblings. In our case, it was three boys and a girl. My mom made virtually all her clothes (less undergarments and socks) and all the shirts my dad and we boys wore. And, she made all my younger sister's clothes. Needless to say, she was popular at the fabric store in our town of Highland, California. Once home, she would clear the kitchen table, lay out some cloth, and then take out very flimsy paper from an envelope, and begin laying that paper out on the cloth. Once done, she would pin the paper to the cloth and begin cutting the cloth – being very careful to follow the lines – like the connect-the-dots coloring pages she would make for us.

What fascinated me most about this process is that many times, especially the dresses and pant-suits she made for herself, the final product did not look exactly like the picture on the pattern envelopes. Sure, they were a different fabric, but there were other differences. The picture on front of the pattern envelope might have shown the dress having sleeves, yet her dress did not. Or, she added a front pocket or two where the pattern had none. Same can be said about collars, hemlines, etc.

When I would ask her why she did not follow the pattern exactly, her response was that she did not like sleeves or collars (we did live in Southern California, where the outside temperatures could easily top 100°F). Her answer was that she used the pattern to get the body right – the general form, but then she liked to make changes based on what she wanted the final product to look like. Thus, began my fascination with the notion of a pattern.

As I got older, I was fascinated with airplanes. Though we could not afford a gas-powered engine for an airplane model, I still built one from scratch – well actually, I used a pattern to cut each balsa wood strut and each frame of the fuselage. I spread the tissue paper across the wings and applied the airplane dope. That airplane never flew, but it hung in my room for years.

Later in life, as a young systems engineer, I heard the software engineers talking about using software patterns. The first book about patterns I ever saw was the classic *Design Patterns: Elements of Reusable Object-Oriented Software* [1]. When I had the opportunity to study for a doctoral degree, it was very natural for me to study patterns in Systems Engineering – and down the rabbit hole[1] I proceeded.

Turns out, I was not the only one interested in using patterns to improve design to further a cause. There are many "pattern communities" today, and in the following pages, we will explore these as we develop a better understanding of patterns as they can be applied to project management.

The existence of patterns is almost universal [2]. We find them everywhere. But what constitutes a reusable pattern? One that can be followed by others? It is normal for an author to go to the dictionary first, to understand the accepted use of a word. In this case, we will look at the Merriam-Webster online dictionary [3].

1. A form or model proposed for imitation: exemplar
2. Something designed or used as a model for making things – a dressmaker's pattern.

So here we learn that a pattern can be thought of as a kind of model – an exemplar. A typical example, suitable for imitation. Looking at Merriam-Webster's learner's dictionary, one of the definitions for patterns states:

2a: The regular and repeated way in which something happens or is done
   • They are studying behavior *patterns* among high-school students.
   • Analysts are noticing different spending *patterns* by consumers.
   • The trees followed a characteristic *pattern* of growth.

It seems there is a good general understanding of what a pattern is and how it is used. However, as the problem becomes more complex, the use of patterns is less understood. In fact, many patterns are known implicitly by subject-matter experts. But those patterns have not been documented by those subject-matter experts, and that makes it difficult for others to benefit from that knowledge – why? Patterns can be very explicit or rather abstract. A cake recipe is an example of a very explicit pattern. Do the following steps, in the specified order, and your cake will turn out just fine. Deviate from the recipe at your own risk. However, as we will see later in this chapter, the more abstract the issue of task, the harder it is to provide usable guidance.

While the concept of patterns can be found throughout history, Christopher Alexander is considered the father of modern pattern thought. Alexander's seminal work *Notes on the Synthesis of Form* the design notion of design has two primary elements – form and context [4]. Form is that which can be modified in a design, and context is that part of the problem we cannot change. Alexander's early example of a teapot sitting on a stove demonstrates this in that the stove in the kitchen cannot be changed and, therefore, is the context of the problem. The design, or the form, is represented by the teapot. It is not unusual for folks to own more than one teapot, and therefore, we can decide, or change, the teapot we choose to use. Alexander goes on to explain that the context – that which we cannot change – exerts forces on our design, and those forces must be resolved or balanced in a good design. Finally, he noticed that many times these forces tend to keep reappearing in similar patterns, and that similar designs resolved these forces – and his concept of patterns emerged.

ALEXANDER'S PATTERNS

Alexander has been a prolific writer. The number of pages he has written about patterns in architecture, and nature, may approach 10,000. Many are classics [5–7] and have been cited over and over again. His patterns exist to solve common problems in architecture. One of his most commonly cited patterns might be "Light on two sides of every room". In this pattern, he is trying to address the "forces" in which people will always gravitate to those rooms which have light on two sides and leave the rooms which are lit from only one side unused and empty [5]. This is the context. The form or design solution (pattern) is to locate each room so that it has outer space outside it on at least two sides and then place windows in these outer walls so that natural light falls into every room from more than one direction.

Alexander has proposed hundreds of patterns for architecting buildings and communities. He documented patterns for the design elements of a home (such as the lights on two sides of every room pattern above), for the organization of houses to create community, for the arrangement of businesses around communities, etc. Patterns that complement one another are referred to as a pattern language [5,8]. As an example, Alexander defined a pattern language for the space between the boundaries of neighborhoods and the communities which should contain local centers. The patterns that might be used to design that space include Eccentric Nucleus, Density Rings, Activity Nodes, Promenade, Shopping Street, Night Life, and Interchange. It does not take a lot of imagination to visualize how these terms form exciting intersections between communities. This brings up another important aspect to patterns – naming the pattern matters. We will come back to this concept later in the chapter.

THE SPREAD OF THE PATTERNS GOSPEL

So how did the notion of Alexander's patterns spread outside the architecting community? Two software engineers, one from Apple Computing and the other from Tektronix, submitted a paper [9] to the 1987 OOPSLA-87 workshop on the Specification and Design for Object-Oriented Programming.

In this paper, Beck and Cunningham stated

> We propose a radical shift in the burden of design and implementation, using concepts adapted from the work of Christopher Alexander, an architect and founder of the Center for Environmental Structures. Alexander proposes homes and offices be designed and built by their eventual occupants. These people, he reasons, know best their requirements for a particular structure.

They went on to describe a pattern language "as providing workable solutions to all of the problems known to arise in the course of design…which leads a designer to ask (and answer) the right questions at the right time." Their paper is still considered the beginning of patterns in the software community.

Another software engineer, Linda Rising was expanding the idea of patterns in software at the same time as Beck and Cunningham while working at AT&T. In her book *The Pattern Almanac* [10], she created a comprehensive collection of patterns. She compiled them from conferences, papers, and association with pattern authors leading up to the publication of this body of work.

Later in her career, Rising extended her pattern's knowledge and experience into organizational dynamics. The books *Fearless Change* [11] and *More Fearless Change* [12] captured the organizational patterns which Rising observed over her career.

In 2016, Cloutier took the notion of patterns and showed they could be used to architect complex systems [2].

DESCRIBING A PATTERN

Up to this point, we have discussed the general notion of patterns. With that understanding, it is now time to discuss how to document patterns in such a way that they can be reused. After all, if someone else cannot make use of your knowledge, why take the time to document the pattern? Alexander used a specific format (thus the term form) for documenting patterns. The minimum set of information necessary to document a pattern proposed by Alexander was:

> Name: a descriptive, meaningful name of the pattern
> Context: what is the problem the pattern is meant to address
> Forces: what are the issues (forces) that are causing the problem
> Solution: the proposed solution to the problem
> Sketch: a simple sketch of the problem/solution

If we return to the pattern LIGHT ON TWO SIDES OF EVERY ROOM, Alexander documented the pattern this way [8]:

> When they have a choice, people will always gravitate to those rooms which have light on two sides and leave the rooms which are lit only from one side unused and empty.
> Therefore:
> Locate each room so that it has outdoor space outside it on at least two sides, and then place windows in these outdoor walls so that natural light falls into every room from more than one direction.

Dissecting this description, we have the name first – Light on two sides of every room. While this is a somewhat awkward name, it is very descriptive. In the patterns community, it is important to create a name that is descriptive to the person looking for an appropriate pattern. Next, there is a description of the problem being addressed, commonly called the context – people tend to gravitate to rooms with better lighting. The forces are light coming from only one side, darkness, unused space, etc. And finally, the solution is to design the room with light coming from two directions (Figure 11.1).

One of the seminal books for documenting patterns in the software community was *Design Patterns: Elements of Reusable Object-Oriented Software* [13]. Gamma et al. stated that a pattern has four essential parts:

- Pattern name: A descriptive name that will provide some context for the pattern
- Problem: What is the problem and context of the problem to be solved
- Solution: An abstract description to resolving the problem
- Consequences: The trade-offs to using this pattern – good and bad.

This is a rather elegant approach with an Occam's Razor simplicity.[2] Therefore, we will adopt this format for documenting patterns throughout the remainder of this chapter.

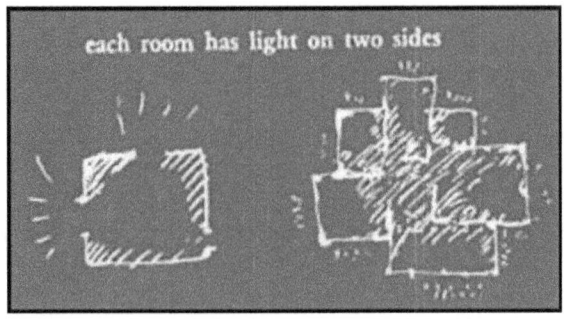

FIGURE 11.1   Sketch of light on two sides pattern.

Patterns Are Not Invented

Another concept for patterns is that they are typically not invented. Patterns, as envisioned and discussed by Alexander, are mined from previous solutions. The commonly accepted guideline is that if you can find a solution is used to solve three or more unique instances of a problem, it is probably a pattern. A pattern to be discussed later in this chapter is called "Management Reserve". I have found a common approach to addressing budget risks is to put 10% of the overall budget into a "management reserve" for that day when some unexpected event or cost arises. If and when it does arise, I have the management reserve to deal with the problem, and I do not have to go back to my management or the customer and say "oops". While some will call this a best practice, I choose to identify it as a pattern – something that works most of the time. But it does not always work. The problem/challenge may require considerably more than 10% to solve. But, as a rule, it avoids a lot of embarrassing situations.

The problem with inventing, creating, or just writing new patterns is that there is no assurance that it is a valid pattern. It may very well be, but there is no assurance. It could also be said that there is no assurance that a mined pattern will work either – but there is some history that says it may be a valid approach due to the previous experiences where it did solve the issue at hand.

The Problem with Patterns

Up to this point, we have discussed the value and benefits of patterns. However, as is true throughout most of life, there is always a downside to everything. In the case of patterns, they should not be viewed as the answer to all things. Patterns might be considered heuristics or rules of thumb. They are solutions that have worked to solve related problems but are not guaranteed to work all of the time or on your specific project. They are not a recipe in a cookbook that will provide perfect results if you simply follow the steps.

When looking for patterns to address a project challenge, one must decide whether the problem being addressed is similar to the problem the pattern is intended to solve. There is the possibility that the pattern selected may not apply to the new situation – but there is some history that it might work. When trying to solve a problem or address a challenge, it is good to have options. Patterns provide viable but not guaranteed options.

Challenges to Using Patterns

A common challenge in the patterns community is that they need to be used and implemented to be useful. There are those that have discovered the value of patterns and want to use them as often as possible. But that takes some discipline. Later we will discuss the

> A common challenge in the patterns community is that they need to be used and implemented to be useful.

"What gets measured, gets done" pattern. Basically, that pattern supports the notion that if you want to change a behavior, it requires reinforcement. Measurement is a form of reinforcement. It does not have to be formal. If management, when reviewing a project schedule, asks – "What patterns did you use?" – that is a form of measurement. It is reinforcing to the project manager that patterns are important and should be utilized where appropriate.

The next challenge with patterns is that it is not enough to simply document the patterns. Many of the patterns are such that they are there to solve a problem. Typically, on a project, that is not the time to go look for a pattern – it is a time for action. However, experienced trouble shooters know that when there is a problem that is exactly the time to take a little time to reflect and determine what in their experience bag can they call on to address the current challenge – what pattern could be applied to help the project move forward? Design patterns used by software engineers, process engineers, organizational planners are deliberate, and one can "see" the pattern. A project management pattern will many times be less obvious or visible.

Finally, if you have looked for patterns in the wild (e.g. Internet), you may have found it difficult to locate may useful patterns. What has happened over the past decade is that companies that have discovered patterns have decided they provide a competitive edge. Therefore, they lump them into that bag called Intellectual Property and do not want to share them. Unfortunately, this goes against the very heart of patterns – patterns are meant to be shared and used. Just realize that this is a possibility – it is probably a pattern "pattern".

TOOLS FOR FINDING PATTERNS

We have become a tool-driven society – tools, or apps, have made life easier and in many cases more fun. So, it is reasonable to hope for an app or tool for finding patterns. Let us first discuss how project management artifacts are constructed.

> We have become a tool-driven society – tools, or apps, have made life easier and in many cases more fun.

The Project Schedule

The first artifact we will discuss is the project schedule. It typically is represented in either a Gantt chart, consisting of a task name, a "best guess" duration, a "worst case" duration, or an "expected duration" that falls somewhere in between the first two. Each task also has a beginning date. The task will have a link back to the previous task and to the next task. That task name might be anything – pour concrete, arrange for flowers, or begin module one. However, language is a strange thing. While you may use the term "pour concrete", another planner might instead describe it as "lay concrete". Arrange for flowers may be stated as purchase flowers. And "begin module one" might be "begin first module" or "begin module 1", or it may use the actual module name.

Depending on your preference and experience, you may instead use a chart made up of nodes and links, called a network chart. However, while the graphical representation will be different from the Gantt chart, the underlying information will be the same.

We have established that patterns are mined – that is, they are harvested from previous projects. While one can find pattern recognition software, this is normally used to find numerical patterns. Research continues to discover patterns in photographs, but even that is based on digitizing (turning the photograph into a collection of numbers representing colors and densities) and then finding patterns in the numbers.

We did say that a project plan is a collection of nodes and graphs and those do have numbers, so let us assume for a minute that a complex software could be created to handle the ontology/lexicon challenge. The next challenge is that durations for like tasks might vary wildly. For instance, the time it might take me to pour the concrete needed for a small porch off the back of my house will be significantly different from that time required to pour an ice skating rink.

Another challenge is just collecting past project plans/schedules/artifacts so that they can be mined. Most times, these are found on individual project manager hard disks or locked away in corporate repositories. My personal experience is that these are very hard to come by, and most individuals and corporations treat this information as proprietary information and/or intellectual property.

Finally, let us look at the fourth challenge with creating a tool to mine patterns. This challenge may be the easiest to solve. There are dozens of software applications for managing projects: Microsoft Project and Primavera to name a few. However, there are those that use Microsoft Excel or even Microsoft PowerPoint. How would a mining tool be able to correctly read the format of each of these file formats and place the data into a common data set, so the information can be mined for common project patterns?

Hopefully the reader can now begin to understand the difficulty in creating a tool to mine project management patterns. To date, the best tool is still the human mind. It can process this dissimilar information better than any existing computing system. Maybe in the future, a quantum leap in computing power will allow that, but for now, it is a manual process.

Hopefully the reader can now begin to understand the difficulty in creating a tool to mine project management patterns. To date, the best tool is still the human mind.

We have also established that patterns are mined – that is, they are harvested from previous projects – successful or not. So, what would we be looking for? Similar nodes and lengths? Similar task names? We have already established.

PATTERNS FOR PROJECT MANAGEMENT

With that introduction, the rest of the chapter will be dedicated to patterns this author has identified over a 25-year career of managing engineering, IT, and software projects. They are broken into various categories to provide some context to their application.

SCHEDULING PATTERNS

Creating a project plan is an art form. The larger the project, the more variables arise to cause projects to overrun. The experienced project manager will work with his/her team members to develop the project plan, gaining their input and leveraging their experience. This has the added benefit of having the teams "buy-in" in relation to the scheduling of task durations and flows. However, some team members habitually underestimate the time to complete a task. This can happen for any number of reasons, but the two primary reasons are (1) to under promise and over deliver and (2) lack of confidence in the ability to complete the task. The opposite of this is the team member that will always underestimate the task at hand because they want to be a good team player and are afraid to voice their honest opinion. In any case, if the project manager is familiar with the team, the following pattern can be useful.

Pattern Name	Over/Under
Problem	A project team is made up of multiple individuals and personalities. Some individuals tend to overestimate how quickly they will accomplish a task. Others underestimate the amount of time a task will take.
Solution	Develop an over/under value for each significant team member. Use this number to increase or decrease the estimate they provide as the project plan is built. This should result in more accurate project plans.
Consequences	Even the best of estimators can underestimate a task. While this has served many a project manager well, it is not foolproof. This pattern is best used in conjunction with the "Use Buffers" pattern. In this manner, even if the project manager is off in the adjusted estimate, hopefully the buffers provide additional error protection.

Theory of Constraints Pattern

The theory of constraints was made popular by Cox and Goldratt [14]. Initially, it was used to manage the manufacturing flow of raw materials through a factory. However, a more generalized pattern exists in which the project manager schedules small buffers in the project plan to anticipate problems. This is a form of management reserve but is related to the passage of time rather than the flow of resources.

Pattern Name	Use Buffers
Problem	It is hard to determine what the impact of any individual task will have in relationship to the project end. Micromanaging each task completion individually is time consuming.
Solution	Build buffers into the schedule. A buffer is not a task but rather a timing mechanism. However, on the project schedule, it will appear as a task. Initially, buffer size was related to replenishment time of a specific machine. More generally, buffers can be built to be any length of time, and it is better to use multiple smaller buffers rather than one single large buffer. If there is remaining buffer time, the project will complete on time.
Consequences	Excessive buffers will add time to the project to the point that the scheduled time may be unacceptable.

BUDGETING PATTERNS

One of the major tasks a project manager will perform is creating a budget for the project. This effort requires estimating costs of necessary resources (salaries, paid hours, paid overtime, etc.), costs of materials used during the project, and travel expenses are among the most common budget items. But these are always done as best estimates.

Pattern Name	Management Reserve
Problem	When project budgets are constructed, they inherently contain some degree of risk. Particularly for longer projects. If the project is a few weeks in duration, there is less chance for budget-busting events to occur. However, as the project length and/or complexity grows, so does the risk of events causing a budget "miss". Costs can rise, tasks can take longer, personal issues may arise, etc.
Solution	Build a management reserve into your budget. The normal value is a 10% reserve. If your estimated budget is 100 units, then add a 10-unit reserve, making your new total budget 110 units.
Consequences	As the project increases in size and scope, 10% may make the budget too large. It may be acceptable to use a smaller percentage depending on the size of the project.

PROJECT COMMUNICATIONS PATTERNS

Effective communications are critical to project success. Often the project manager finds themselves in the role of "universal translator" between disciplines (e.g. business and engineering). While measurement is normally associated with metrics, there are actually many fewer formal means of measurement. Simply requiring weekly status reports is a form of measuring. The status report can be used to "measure" or "assess" team progress.

Pattern Name	What Gets Measured, Gets Done
Problem	Understanding and assessing the progress of various project teams is often difficult.
Solution	Design a measurement for each item you want to track. While one normally considers measurement a number, a status report is a form of measurement – it conveys interest and importance.
Consequences	If too many reports are required, the reviewing of these reports by the project manager or team leader can become onerous. If too much time is required for the team to produce the reports, the burden on the project can become measurable and may impact budgets.

RESOURCE MANAGEMENT PATTERNS

Resources can be thought of as any number of items, to include budget, machines, locations, etc. However, this pattern is geared toward personnel. My first introduction to this pattern was on a software development project. The team was falling behind and was not going to meet our contract deliverables. So, I did what any project manager would do – go to leadership and ask for more resources – in this case, programmers. After reviewing the project status and progress with the leadership, they came to a different conclusion – and they cut my team by 30%. To my surprise, we delivered the project on time and on budget. This seemed counterintuitive at the time, but I have applied this pattern numerous times since to successful outcomes.

Pattern Name	Too Many Team Members
Problem	In the face of a challenging project, with what appears to be a right-sized project team, the project continues to fall behind. Communications are good, the team is working together to address issues. In fact, they are helping each other. But they are still missing milestones.
Solution	Reduce team size. It is not uncommon that there is too much communication occurring. Communication – links between nodes – takes time and effort. That time and effort may be in the way of real work being accomplished.
Consequences	Excess communications may not be the reason for falling behind. It might simply be that the effort was underestimated.

RISK MANAGEMENT PATTERNS

Risk can be measured in many ways. Technical risk, budget risk, and schedule risk are common measures.

Pattern Name	Incremental Deliveries
Problem	The longer the project is in duration, the higher the risk of a black swan event.[a] If an event of this magnitude occurs, it might swamp the project or company.
Solution	Consider breaking the project into smaller projects and incrementally deliver the final project. This practice is common on projects considered to be agile but for a different reason. In this case, if a deliverable is broken into smaller subprojects, a black swan event only sets the project back to the last major delivery.
Consequences	This approach does add some management overhead as the project manager has multiple schedules and releases to manage – but, it is a risk management pattern. The intent is not optimization but rather risk management.

[a] A black swan event is an event that might be considered a catastrophic event. Stock market crashes and tsunamis might be considered black swan events. The theory was first proposed by Nassim Nicholas Taleb in his book *The Black Swan: the impact of the highly improbable* (2007, 2010).

Pattern Name	Schedule Management Reserve
Problem	When project budgets are constructed, they inherently contain some degree of risk, particularly for longer projects. If the project is a few weeks in duration, there is less chance for budget-busting events to occur. However, as the project length and/or complexity grows, so does the risk of events causing a budget "miss". Costs can rise, tasks can take longer, personal issues may arise, etc.
Solution	Build a management reserve into your schedule. The normal value is a 10% reserve. If your task duration is 100 units, then add a 10-unit reserve, making your new task duration 110 units.
Consequences	Doing this on each task may make the project duration too long and unacceptable. If this is the case, consider using the Buffers pattern.

PROJECT EXECUTION PATTERNS

This pattern is effective, though not popular with many white-collar teams that are used to traditional 8:00 – 5:00 workday. It is, however, very useful when a service is being provided to other white-collar workers. An example is when computing resources used by those who work a traditional workday must be updated without impacting their productivity. For instance, a very large engineering department may require upgraded workstations, but their normal work cannot be interrupted due to contract commitments. So, how does one backup the local data, upgrade the machine, restore the local files, and have it ready to go the next day? You may recognize this pattern from the legendary Federal Express company.

Pattern Name	Three-Shift Solution
Problem	Work must be performed on a mechanical/electrical system without impacting normal working hours' productivity. Normally for performing upgrades to many devices.
Solution	This pattern requires the project team to have some number of upgraded stock on hand. The upgraded stock is prepositioned for upgrade during normal working hours.
	In the second shift, existing stock is removed, and refreshed (or new) stock is put in place.
	In the third shift, the new or refreshed devices are reconfigured so that in the first shift of the next day, devices are upgraded and ready for normal operations. Also, equipment removed during the second shift are returned to centralized location to be upgraded during the first shift of the next day.
Consequences	This pattern requires talent during all three shifts of the workday. Depending on cultural norms, it may also require the project to pay that talent a shift differential – an extra amount to be willing to work the other shifts. This must be included in planning budgets in order to avoid budget overruns.

Project Management Anti-Patterns

Antigravity is a force that opposes gravity (at least hypothetically). Every science fiction reader knows that antigravity is bad and should be avoided, and anti-patterns are patterns that should be avoided. They are those activities or actions that have demonstrated bad results time and time again. It is interesting to note that while there has not been a book dedicated to project management patterns, there is a manuscript dedicated to anti-patterns in project management [15]. This book was written about anti-patterns in software development.

> Every science fiction reader knows that antigravity is bad and should be avoided, and anti-patterns are patterns that should be avoided.

Some of their patterns are titled:

- The Standards
- Micro-Management
- Corporate Craziness
- Size Isn't Everything
- Wherefore Art Thou Architecture

This also demonstrates the earlier point on naming patterns. Just by reading the anti-pattern names, one can make an educated guess regarding the intent of the anti-pattern.

Another good source of anti-patterns can be found at Sourcemaking.com [16]. As many references to project management patterns, this site is geared toward software development. However, anti-patterns such as the following are completely self-explanatory:

- Analysis Paralysis
- Viewgraph Engineering
- Death by Planning
- Fear of Success
- Irrational Management
- Throw It over the Wall

RELATED CONCEPTS

There are some related terms and concepts that are sometimes used when discussing patterns.

> *Heuristics*. A heuristic is sometimes referred to as a rule-of-thumb. Rechtin and Maier stated that heuristics are very general, spanning domains and categories of guidance [17,18].
>
> *Templates*. [2a] Document or file having a preset format, used as a starting point for a particular application so that the format does not have to be recreated each time it is used [19].
>
> *Frameworks*. Logical, organizing structure used to classify information, concepts, data, etc. They may also provide mechanisms to transform information from one form to another.

A heuristic can also be thought of as generally accepted knowledge. An example might be the saying "location, location, location" or the term "Murphy's law". Both are cited many times but are not always right. Another heuristic is that you turn a screw or bolt to the right to tighten. While this may work most of the time, it does not universally work – the left pedal of a bicycle and some plumbing fittings are examples where that heuristic does not apply.

A template is related to the pattern form. It is a device to ensure consistent capture of information. However, the template does not contain the information until it is completed.

Finally, the framework is used to organize collected information or data. Common frameworks include the Zachman Framework [20], DoDAF [21], etc.

QUESTIONS FOR DISCUSSION

1. Discuss the notion of a pattern and why patterns are useful.
2. Name four of the five information items found in a pattern as proposed by Alexander.
3. Name the four items used in this chapter to define a project management pattern.
4. What is an anti-pattern, and why are they important?
5. Discuss the difference between a pattern and a heuristic.

CHALLENGES AND CONCLUSIONS

As we wrap up this chapter, the reader should have noticed that many of the patterns and anti-patterns cited were generated by the software community. Why is this? The author can only speculate on this.

First, the software community, more than any other community, has embraced patterns. As discussed earlier in the chapter, Beck and Cunningham were on the leading edge, followed by the computer scientist, James Coplien, Linda Rising, and others. It is interesting to note that Cunningham was also the creator of the modern Wiki and wrote the software that the current Wikipedia runs on today.

Second, many of those in the original software pattern communities matured into the process arena as process management became popular with the advent of the software capability maturity model which was centered around consistent and repeatable process – the things patterns were also good for.

Pattern usage is not without challenges. Within business, the use of patterns may be thought of as a matter of intellectual property. This point warrants further discussion. Early in the process definition era, companies marked all their process work as proprietary. Second, I tried to begin a patterns repository for systems engineering – many told me they would like to contribute, but their corporate lawyers stopped them from contributing because they saw these as potential competitive advantages.

> Pattern usage is not without challenges.

This was not surprising as many companies have also deemed their general processes as competitive advantages.

OTHER SOURCES FOR PATTERNS THAT MAY BE HELPFUL

There are several sources on patterns in general or as applied to other disciplines and/or practices. Below are other references that may be of value in exploring the pattern community. Many of them are rather old but still available. Some are newer.

Portland Pattern Repository. This is the original wiki for patterns. It was primarily geared toward the software community, but there is much that can be learned from this site. It appears that it has not been updated since 2013.

http://c2.com/ppr/
http://wiki.c2.com/?PatternIndex

Project Management Patterns on the Portland Pattern Wiki. Alistair Cockburn is well known in the software community. He added his software project management patterns to the Portland Pattern Wiki.

http://wiki.c2.com/?ProjectManagementPatterns

Project Management Anti-patterns for software development. While the website considers these unique to software development, many are more broadly applicable to project management in general.

https://sourcemaking.com/antipatterns/software-project-management-antipatterns

Process Patterns from the same Portland Pattern Repository.

http://wiki.c2.com/?ProcessPatterns

Design Patterns. This site is even older. It was created by Professor Doug Lea from the Computer Science Department, State University of New York at Oswego, Oswego.

http://g.oswego.edu/dl/pd-FAQ/pd-FAQ.html

Use of Patterns for Scenario Development for Large-Scale Aerospace Projects.

www.academia.edu/1354848/PROJECT_ICARUS_STAKEHOLDER_SCENARIOS_FOR_
AN_INTERSTELLAR_EXPLORATION_PROGRAM

Lean Startup Business Model Pattern.

http://torgronsund.wordpress.com/2010/01/06/lean-startup-business-model-pattern/

Characteristics of Group Facilitation Patterns.

https://groupworksdeck.org/what-we-mean-by-pattern

Pattern Languages as Applied by the Center for Environmental Structure.

www.patternlanguage.com

Behavioral Patterns. International Project Leadership Academy.

http://calleam.com/WTPF/?page_id=555

Patterns of Effective Project Management in Virtual Projects an Exploratory Study.

www.pmi.org/learning/academic-research/patterns-of-effective-project-management-in-virtual-projects-an-exploratory-study

ENDNOTES

1 Go **down the rabbit hole**. To enter into a situation or begin a process or journey that is particularly strange, problematic, difficult, complex, or chaotic, especially one that becomes increasingly so as it develops or unfolds. (An allusion to Alice's Adventures in Wonderland by Lewis Carroll.) https://idioms.thefreedictionary.com/go+down+the+rabbit+hole.
2 Occam's Razor is the belief that the simplest, most straightforward solution to a problem is most generally the correct solution.

REFERENCES

1. Gamma, E., Helm, R., Johnson, J., Vlissides, J. (1995). *Design Patterns: Elements of Reusable Object-Oriented Software*. Boston, MA: Addison-Wesley.
2. Cloutier, R. (2016). Applicability of patterns to architecing complex systems. *Doctoral Dissertation*. Stevens Institute of Technology. Hoboken, NJ.
3. www.merriam-webster.com/dictionary/pattern. Downloaded 9/1/2018.
4. Alexander, C. (1964). *Notes on the Synthesis of Form*. Cambridge, MA: Harvard University Press.
5. Alexander, C. (1977). *A Pattern Language, with Ishikawa and Silverstein*. New York: Oxford University Press.
6. Alexander, C. (1979). *The Timeless Way of Building*. New York: Oxford University Press.
7. Alexander, C. (2002). *The Nature of Order Book 1: The Phenomenon of Life*. Berkeley, CA: Center for Environmental Structure.
8. Patternlanguage.com. www.patternlanguage.com/. Last accessed 9/5/2018.
9. Beck, K., Cunningham, W. (1987). Using Pattern Languages for Object-Oriented Programs. Technical Report No. CR-87-43. September 17, 1987. Submitted to the OOPSLA-87 workshop on the Specification and Design for Object-Oriented Programming. Last downloaded 9/5/2018 from http://c2.com/doc/oopsla87.html.
10. Rising, L. (2000). *The Pattern Almanac*. Boston, MA: Addison-Wesley.
11. Rising, L. (2008). *Fearless Change: Patterns for Introducing New Ideas*. Boston, MA: Addison-Wesley.
12. Rising, L. (2015). *More Fearless Change: Strategies for Making Your Ideas Happen*. Boston, MA: Addison-Wesley.
13. Gamma, E., Helm, R., Johnson, J., Vlissides, J. (1995). *Design Patterns: Elements of Reusable Object-Oriented Software*. Boston, MA: Addison-Wesley
14. Cox, J., Goldratt, E. M. (1986). *The Goal: A Process of Ongoing Improvement*. Croton-on-Hudson, NY: North River Press. ISBN: 0-88427-061-0.
15. Brown, W., McCormick, H., Thomas, S. (2000). *Anti-Patterns in Project Management*. New York: John Wiley & Sons.
16. https://sourcemaking.com/antipatterns/software-project-management-antipatterns.
17. Rechtin, E. (1991). *Systems Architecting: Creating and Building Complex Systems*. Upper Saddle River, NJ: Prentice Hall.

18. Rechtin, E., Maier, M. (1997). *The Art of Systems Architecting*. New York: CRC Press.
19. American Heritage Dictionary. Last accessed 10/25/2018. https://ahdictionary.com/word/search. html?q=template.
20. Zachman, J. (1987). A framework for information systems architecture. *IBM Systems Journal*. 26(3), 276–290.
21. Department of Defense Architecture Framework. Last accessed 10/25/2018. https://dodcio.defense.gov/ Library/DoD-Architecture-Framework/.

12 Policy Management Toolset

George L. De Feis
Stockton University

Dolores A. Atallo
New Jersey Institute of Technology

CONTENTS

BACKGROUND

As organizations grow in size, abilities, geography, breadth and depth, they also grow with immense policies, guidelines, rules and regulations. Thus, policy management has grown cumbersome with documents, spreadsheets, emails, websites and intranet. Maintaining some semblance of order is essential to be effective and efficient. Policy management today must eliminate waste caused by redundancy and overlap; excessive emails, documentation, and paper trails which are dated and not maintained; poor reporting with varied quality reporters; overwhelming complexity; and the lack of accountability. Policies, however, are critical for organization success, whether they are for-profit, not-for-profit, governmental, or educational enterprises, as they establish "boundaries of behavior" for individuals, groups/teams, divisions, departments, strategic alliances, relationships and transactions. Policies stem from the vision, mission, and code of ethics/conduct, which all trickle down to govern the enterprise and processes. Also, policies must be consistent and transparent, transfer knowledge, and be sustainable.

With *Governance, Risk Management and Compliance* (GRC) as an all-encompassing trium-virate for the way organizations deal with these three independent yet related activities, complex project management will help organizations achieve their goals and objectives. When properly man-aged, supervised, and enforced, GRC can establish a framework for governance, objectively identify and weigh risks, oversee compliance, and provide a framework for governance.

The content in this chapter, indeed, aligns itself well with the goals of *complex project management* as "the goal is no longer a luxury but a necessity if we are to address the high rate of project failure, which is made worse in recent years by the increasing volume and complexity of the tasks demanded of businesses" (Kilpatrick, 2006, p. 1). "As professional service firms grapple with an array of problems, delivering projects profitably has become more difficult. Globalization adds to the problem, as firms increasingly deliver in other countries via multiple subcontractors" (Deltek website). When firms are expected to deliver more for less, ultimately, this calls for the streamlining (to save costs) of the management of projects, and it calls for complex project management. "Complex projects are characterized by a degree of disarray, instability, evolv-ing decision-making, non-linear processes, iterative planning and design, uncertainty, irregularity, and randomness" (Shane et al., 2012, p. 2). Such projects require management beyond the "triple constraint" (Project Management Institute, 2017), as the amount of failure exceeds beyond twice the amount of success (Flyvberg et al., 2003; Pavlak, 2004; Standish Group, 2001). However, the PMBOK Guide (2017), which is in its sixth edition, provides a well-organized foundation for managing complexity.

> When properly managed, supervised and enforced, GRC can establish a framework for governance, objec-tively identify and weigh risks, oversee compliance, and provide a framework for governance.

CORPORATE CULTURE

Most organizations do not connect the idea of policy and policy management to the establishment of a corporate culture. Without a policy, there is no written standard for acceptable and unacceptable conduct. An organization can quickly become something it never intended, and it is too late to stop that moving train. The policy also attaches a legal duty of care to the organization, so it cannot be approached haphazardly. Mismanagement of policy can introduce risk, liability and exposure, includ-ing noncompliant policies. These gaps may be used against the organization in legal (civil or crimi-nal) and regulatory proceedings, as well as by regulators, prosecuting and plaintiff attorneys, and the public-use policy violations and noncompliance to place culpability. An organization must establish a policy it can and will enforce, but it also must clearly train and communicate the policy to make sure that individuals understand what is expected. Over time, an organization may evolve negatively even though good policies and procedures are in place. A strong positive culture, however, can rarely be achieved without an established strong risk culture and associated policies, procedures and training.

POLICIES ACROSS THE ORGANIZATION: THE GOOD AND THE BAD

Just like obeying the law, policies and procedures matter. Without these "laws" of acceptable business practice, the business world would be run like the "Wild West." Consequently, when you consider an organiza-tion operating in its business environment – whether for-profit, not-for-profit, education, government, etc. – strong business policies are not irrelevant nor a nuisance. A sampling of organizations typically will have the following in the policy area:

> Without *policies and procedures* of acceptable business practice, the business world would be run like the "Wild West."

Policies managed in documents and file shares – Such policies are informally managed such as legacy text documents, which are bound in books or annotated in procedures, emails, websites, local hard drives, and mobile devices. Some organizations will not have fully adopted centralized online management or publishing with universal access to policies and procedures. Generally, no single place existed where one could view all policies and procedures of an organization. During the 1980s, one of the authors worked for The Port Authority of New York and New Jersey (PANYNJ), and he was told all policies and procedures about working for the PANYNJ could be found in one of the three lengthy volumes of PAIs – *Port Authority Instructions*. Albeit, there were always some smaller scale policies posted, which were only on display in the particular division where one worked and only applicable within that division (e.g., tunnels, bridges, and terminals (TBT) or the rail department (PATH)).

Reactive and inefficient policy programs – Organizations of size may lack coordinated policy training or procedure management and communication programs but instead may have ineffective bureaucratic hierarchical or vertically-focused approaches to develop and coordinate training without thought for the bigger picture. Organizations can suffer from policies that are reactive and inefficient, particularly when addressing changes in laws and regulations. Think about the breadth of issues addressed within the hundreds of pages of The Sarbanes–Oxley Act of 2002 ("Sox"). As organizations persevered to meet the regulatory deadlines, policies and procedures were hurriedly put in place that are still in contention several years later. Some of these policies may have been more reactive than useful to some of their respective organizations and stakeholders of failed financial institutions still question their value.

Policies and procedures misaligned with corporate style or corporate culture – The policies of this organization do not conform to a uniform style guide; a standard template that would require that policies be presented clearly and concisely in the same style (e.g., active, passive, concise, language, level).

Policy "hackers and hijackers" (unofficial creators of policies) – Without the appropriate governance, anyone can create a policy, particularly if no rules exist on how policies are created, modified and approved. As policies are created and evolve over time, they may or may not serve their intended purpose and may become an unnecessary activity that squanders resources or even directly impact the organization in negative way. And, in some cases, the "hackers and hijackers" may use policies that result in unintended consequences. Think about "squatter's rights" or the "Homestead Act of 1862," and one can envision what could be yours if you take it.

Dated policies – When policies are not reviewed, maintained, or renewed for applicability, appropriateness, and effectiveness, and the "author" of that policy is long gone, is the policy still valid? For example, policies that don't keep pace with system or procedural changes don't maximize the tools used to support business decisions. Many laws exist in the United States and elsewhere that are archaic. In Gainesville, Georgia, an ordinance was passed to make it illegal to consume fried chicken any other way than by hand, causing a 91-year-old woman to be arrested in 2009 for consuming fried chicken with a fork. There are many examples of these types of outdated laws and their business equivalent of an outdated policy that does not match the culture or current business practices are thriving in organizations that don't invest in a policy tool kit.

Policies and procedures without lifecycle management – Many organizations maintain an ad hoc approach to creating, approving, and maintaining their policies and procedures, without much concern until problems occur. Starbucks had its policy of providing its patrons with the restroom key, which was seemingly not controversial until the day some African–American patron's request for a key was denied. The media coverage that followed triggered a nationwide mandatory training effort at Starbucks that closed business for a full day to provide compliance training and demonstrate its commitment to remediating its policies and practices.

Mapping exceptions or unusual incidents – Many times, organizations do not have a procedure in place to manage policy exceptions that are driven by new incidents or issues that have arisen. These exceptions may be a bellwether for emerging risks that may trigger the need for a policy change if the pattern is recognized. Therefore, it is important to consider why the policy wasn't sufficient to identify and/or elevate the exception.

Failure to cross-reference standards, rules, and regulations – Without the maintenance of a historical record of policy creation and maintenance for legal, regulatory, compliance use, it becomes overly burdensome and inefficient (time consuming, labor intensive, error prone) for auditors, regulators, and other primary users and consumers of this information to manage. The demand for accurate data and alignment with key standards, such as policies and regulations puts pressure on the policy tool kit to stay relevant to serve its intended purpose.

FAILURES IN POLICY MANAGEMENT

Organizations often lack a coordinated enterprise strategy for policy management, which covers policy development, approval, maintenance, communication, confirmation, training and renewal. When the intense detailed requirements for policy management are done in an ad hoc way, the organization can be exposed to significant liability. Unless policies reflect management's responsibility and accountability to oversee the enterprise, it undermines the importance of the role of policies in governing the organization. The importance of policy management is further highlighted in compliance, ethics, and other critical programs that delineate both the top-down and bottom-up accountability of roles and responsibilities in preserving the value of the organization.

To address the challenging problems of policy management, the organization must be able to show a detailed history of what policy was in effect, when, the duration, its impact, how it was communicated, and what exceptions were granted. If policies are applied haphazardly or incorrectly they do not fulfill their intended purpose or provide a dividend for the organization's investment in policy management. Subsequently, they will not provide transparent and consistent guidance for behaviors or support the desired cultural messages to advance the organization. In today's complicated business world with its complex operations, technical challenges, and ever-changing legal regulations, a well-scripted policy management program is vital.

The long and the short of policy management is clear: Time is imperative for organizations to approach policy management with a well-conceived and formulated strategy "to manage the ecosystem of policy programs throughout the organization with real-time information about policy conformance and how it impacts the organization" (Tallyfy website).

GOVERNANCE, RISK MANAGEMENT AND COMPLIANCE (GRC)

"Governance, Risk Management, and Compliance, also known as GRC, is an umbrella term for the way organizations deal with three areas that help them achieve their objectives" (Tallyfy website). The main idea behind GRC is that business practice growth in size, activities and complexity causes a potential impact to all GRC areas, mitigating repetition of tasks and ensuring that approaches used are effective and efficient (Figure 12.1).

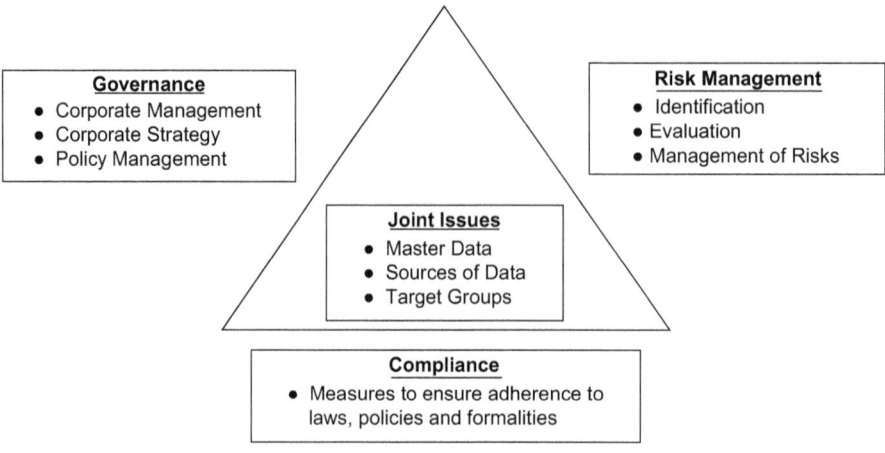

FIGURE 12.1 Interplay of GRC.

Governance

Governance has to do with the organization's TMT (top management team) or the "upper echelon" of the corporation, that was used by the military but now means the "rungs of the ladder" leading to the top of management structure. In other words, how the firm is managed from the senior levels, which includes: functional chiefs or C-Suite (chief financial officer (CFO), chief operating

> There is no way of knowing how the governance of a firm will act and interact, especially when the concept of *self* is critical and subject to individual interpretation.

officer (COO), chief information officer (CIO), etc.), the chief executive officer (CEO), the Board of Directors, the Chairman of the Board impacts the shareholders. When considering the relationship of the Board of Directors with the CEO, who is just an "agent" of the firm, the concept of "agency theory" (Jensen & Meckling, 1976) becomes an issue. There is no way of knowing how the governance of a firm will act and interact, especially when the concept of self is critical and subject to individual interpretation. While policies should not and will not be a substitute for management judgement and Board oversight, they provide guidance and boundaries for how companies are managed and judged at the highest levels, including processes and relations that allow for considerations of the rights and responsibilities of the decision makers within the operation.

Risk Management

Many organizations view policy management and the context for managing risk based on past experience and estimation of future events. Since changes to the business environment (both internal and external) can change the relevance of policies, it is critical to align and refresh policies based on strategy and risk appetite: understanding the desired outcome helps gauge what the organization is willing to put at risk. Therefore, policies should guide the risk taking as well as report-

> What changes in the *'mix of policy management and the context of managing risk'* is the business environment (both internal and external), so it is critical to align and refresh policies based on strategy and risk appetite.

ing and management of risks. This is key to policies playing a vital role in safeguarding reputation and value.

Compliance

While compliance is not the primary driver for business, businesses must comply with various standards, laws, and regulations. Over time, strategy, business climate, and other factors may change the alignment of existing policies with actual business practices exposing an organization to are penalties, financial, and otherwise from noncompliance. Therefore, considering compliance when reviewing and/or updating policies is an important step to include in the refresh efforts.

GRC Benefits and Costs

There is much to gain in the control process of your company by centralizing corporate policies, processes and procedures, though it may require more red tape and bureaucracy. However, GRC is not about adding complexity to an already overly complex process, it is about streamlining and condensing, really clarifying, an approach to enable smooth function. Indeed, consider the key benefits and costs of utilizing GRC capabilities:

Costs – Cutting some, but adding some – An integrated approach to GRC brings real financial benefit, since duplicative and unnecessary spending can be cut, coupled with more focused revenue enhancements that can boost the bottom line. Expenses to trigger revenue enhancements, which are unknown and unguaranteed, can increase costs so they must be monitored. Sometimes one has to "cut their losses."

Less redundancy – Cost cutting can be made and money can be saved when duplicate work is reduced. When businesses are set up as "divisional" or "matrix" organizations, similar processes undertaken within each strategic business unit (SBU) calls for redundancy of activities, which are enormously inefficient. GRC can streamline these operations towards more cost-efficient modes, where time consciousness and cost consciousness are paramount. Determining which operations to streamline is crucial, as redundancy is sometimes built-in for a reason.

Less losses – Having too many procedures can be cumbersome and difficult to control, causing waste and inefficiency across the entire operation. Efficient and effective approaches to GRC, communicated well throughout the organization, cut down on documentation and red tape, while boosting productivity and increasing morale. However, knowing what to cut, and when, is critical.

Greater quality on facts and figures – With centralized and coordinated GRC, the processes for gathering the needed information are accelerated with improved quality. Resulting decisions will be made quickly with greater confidence, though one must be concerned about excessive self-confidence in decision-making.

Economies of scale and of scope – When processes are repeated across different product areas, or learned and excelled at in one product area and applied to another, benefits can occur due to this "standardization" of activity for more consistent and aligned results. Sometimes, though, one may think there is repetition. But there is not a situation where applying the standard would not be appropriate.

Character security – GRC addresses the necessary components to secure your business's character, so needless to say managing your GRC in an effective and efficient process improves your reputation.

Optimal division of resources – Obtaining detailed information of your possessions (financial and otherwise) and understanding when duplicating work is or is not required can help direct your overall business operation. Scarcity of funds and resources is what business is all about, so how one maintains optimality is essential.

Breaking down "silos" – The larger a business enterprise grows in human resources and other specialized entities, the more there is staff dedicated to their specialized activity ("silos"). Getting specialized information into and out of a silo is time consuming. GRC can work to minimize, but not eliminate, the effect of the "silo structure."

THE ROLES PLAYED ON THE GRC FUNCTIONS

In the upper echelon of a business operation, there exist the GRC functions all operating towards advancing the company. These players, all of whom are critical to the GRC effort, include:

- Board Level
- CEO
- Finance Chiefs
- Risk Managers
- Compliance Officers
- Human Resource Managers
- IT Managers.

Each party holds a distinct interest, though sometimes overlapping with other key players.

CEO level and Board level – Any Board Member or CEO needs to provide strategic oversight and executive decision-making capacities, coupled along with timely and clear communication down the hierarchy to enable colleagues to fulfill their roles effectively.

Financial Chiefs – The CFO has the overall responsibility for financial operations of the business enterprise and plays a key role in the GRC implementation. The CFO is critical to accurately reporting the financial drivers for changes in the organization to management and the Board.

Risk Managers – All large organizations should have incumbents at managerial levels who are responsible for risk management, so their roles in GRC are quite complex. Knowing their organizations' strengths (and weaknesses), they need to identify the opportunities (threats) to determine what strategies to implement and those to avoid. Constant monitoring of risk managers' decisions and their impact and outcome is a crucial role for all risk personnel.

Compliance Officers – Personnel responsible for compliance need to be part of all planning decisions, which impact current operations in both the short term considering the long term alignment of compliance with evolving strategies

Human Resource Managers – To implement GRC fully throughout a business operation, HR managers must communicate vertically, horizontally and diagonally so that everyone, up and down and across, is "rowing in the same direction" relative to GRC. The HR department must be effective and efficient, and must track progress and problems or GRC is destined for failure.

IT Managers – Information technology staff have the responsibility for all technological solutions to meet the needs of the GRC strategy, so they are critical for the decision-making process. Also, they have the responsibility for methods employed in gathering information across the business enterprise and how it is used and protected.

IMPORTANCE OF GRC IMPLEMENTATION

Once key players have been identified, there is much to consider making your GRC project a success; and of course, there are indeed goals to accomplish.

Define what you aim to achieve – All decision-making leads you down a decision path, starting with *defining the problem* (i.e., why is a decision needed?) – usually the hardest step – one must know what they want to accomplish, e.g., decrease the time required to introduce a new product, decrease the overhead cost of supervision, increase our domestic market share and develop policies for international expansion into the European Union (EU). Including the primary stakeholders is key to aligning the policies with the business decision-making. Therefore, it is important to first decide what you want to achieve, set targets towards progress and measure progress (or shortcomings), so your progress can be assessed.

Take stock of your current situation – Knowing where you are strong, where you are weak and what assets or resources you have are essential tasks needed to establish potential short- and long-term actions towards achieving your goals. This key step is vital to understanding the current fields of GRC before you undertake change. A checklist to survey your regulatory activities will give you a better understanding of what you will gain from GRC and what problem areas need to be addressed.

Pick a trial entry point – Monitor progress all along the way, watching carefully over the implementation of your formulated or decided action. For smaller companies, choosing a trial entry point is essential. Indeed, larger companies have more "slack" with which to protect themselves. All companies should focus their energies on the gradual implementation of their GRC plan.

Demonstrate the progress (or benefits) – Long-term gains from progress will be shown, but only after short-term gains (early wins) are achieved, often if possible. Communicating these successes throughout the organization regularly is important to show, "in unity there is strength." So demonstrating the clear benefits of effective GRC to key staff and managers is imperative.

Define success – Defining what would represent success before the process begins is an objective way of assessing and determining success (or failure). Really one wants to determine that the project was worthwhile. The process of assessing is always a continuing process which has no end.

In short, undertaking these five tasks will be critical to accomplishing your GRC strategy. Thus, doing your research, taking an incremental approach, working cohesively as a team, communicating well, and providing proper and adequate resources are the best ideas to achieve success.

Reputation Management

The management of one's reputation is vital to control-ling or influencing how an organization is perceived now and in the future. For self-preservation in the long run, one must maintain and improve its reputation to stay ahead of the competition, who could benefit from a competitor's reputational problems. Since all is not fair in the business world, nor any world with competition, firms must strategize to keep ahead. Public relations (PR) firms are employed to help organizations create campaigns to change people's views. One practice is to create positive content pieces about a business to counter any negative content from an actual issue. Reputation managers establish a series of positive pieces in such sufficient quantities to make negative commentary appear to be less pervasive.

> For self-preservation in the long run, one must maintain and improve its reputation to stay ahead of the com-petition, who could benefit from a competitor's reputational problems.

Measuring Customer–Client Management and Authority

Customer–client management and authority are important aspects of any policy management activity, and the concept of "customer relationship management" (CRM) is part of the process. CRM focuses on creating, storing, and managing prospect and existing customer information, e.g., customer location, leads and sales opportunities (now or the future) in a central location for easy transmittal. CRM's software, apps and customizable programs change the playing field for cus-tomer development, retention and future leads in fields including retail, manufacturing, real estate, engineering, and more.

The result of state-of-the-art CRM can help increase leads; close more deals efficiently; and increase retention, loyalty, and satisfaction.

General Data Protection Regulation

Outside of the United States, other entities are charged with controlling information to the exter-nal world. Of these, the European Commission not only intends to strengthen and unify data protection for individuals within the EU but also addresses issues pertaining to releasing personal information to those outside the EU. The personnel collecting personal data must clearly disclose their collection means and must declare the lawful basis and purpose of data processing, stating its purpose, use, how long the data will be retained, and if it will be shared by third parties and those outside the European Economic Area (EEA). Persons releasing data have the right to request a copy, and the right to have their data erased under certain circumstances.

General Data Protection Regulation (GDPR) consists of 99 *articles*, grouped into 11 chapters, and an additional 171 *recitals* with explanatory remarks. The chapter headings are as follows:

- I – General provisions
- II – Principles
- III – Rights of the data subject
- IV – Controller and processor
- V – Transfers of personal data to third countries or international organizations
- VI – Independent supervisory authorities
- VII – Cooperation and consistency
- VIII – Remedies, liability, and penalties
- IX – Provisions relating to specific processing situations
- X – Delegated acts and implementing acts
- XI – Final provisions

The GDPR was adopted in April 2016 and became enforceable beginning May 2018. As the GDPR is a regulation and not a directive, it does not require national governments to pass any enabling legislation and is directly binding and applicable.

HISTORICAL PERSPECTIVE

The Industrial Revolution began in the United States around the 1860s, as before this time, most people worked for themselves in some kind of agribusiness. The means and methods of societal work were very agrarian. By the late 1800s, many in the agribusiness lifestyle had given way to industrialization and factories. The "captains of industry" at the turn of the century were John D. Rockefeller (oil), James B. Duke (tobacco), Andrew Carnegie (steel), J.P. Morgan (banking) and Cornelius Vanderbilt (steamships and railroads). These wealthy entrepreneurs started with just a few resources but had the foresight to see beyond the current day and amassed huge fortunes. Policy management was in its infancy.

As industries grew, trusts split and the Interstate Commerce Act of 1887 was passed to regulate the railroad industry and guard against its monopolistic tendencies. This act, for instance, forced railroads to publish rates and fares, and did not allow them to be changed without notification; thus, the act supervised the railroad industry. In 1890, the Sherman Antitrust Act was passed, making it illegal for companies to create monopolies. Policy management was growing.

Frederick Winslow Taylor, the "father of scientific management," developed four principles to increase productivity in the workplace, and establish policies, for instance:

1. Replace "rule-of-thumb" work methods with approaches and principles based on the scientific practices and tasks.
2. Scientifically select, train, develop and coach employees, rather than passively allow them to train themselves.
3. Provide specialized instruction and supervision of each worker in the performance of that specific worker's tasks.
4. Divide work equitably to managers and laborers, so that managers apply scientific management principles to planning the work and the way workers perform their tasks.

These policies and procedures to increase productivity put policy management into full bloom. Managers examined the relationship between the hierarchy of organizations, the working conditions of workers, and the resultant productivity and therefore created more and more policies and procedures.

The "Hawthorne Studies" of the 1920s, which had first seemed a failure, led to the conclusion that productivity could be enhanced with empathetic supervision. Thus, there was no longer Taylor's "one best way" to manage, as the era of organizational behavior, motivation, leadership and teamwork became part of the realm of management. Policy management was indeed blooming.

With motivation, leadership, teamwork and the like, workers were caused to produce more and more output. Indeed, more policies and guidelines were amassed as part of this growth. Psychologists led the way with Maslow's Hierarchy, Alderfer's ERG Theory, Herzberg's Two-Factor Theory, McGregor's Theory X–Theory Y, Ouchi's Theory Z, McClelland's Three Needs Theory, Adam's Equity Theory, Vroom's Expectancy Theory, and Locke and Latham's Goal-Setting Theory. And with them, more and more policies and procedures came into the business world.

Management's Three Eras: Executive, Expertise and Empathy

Organizations, symbolized as "machines" with "inputs, processing, and outputs" as the common way to consider them, cast a description on the way organizations and their management are thought of today. Managers face environments that range from simplicity to complexity and stability to instability (or dynamism), and this leads from a world with certainty to uncertainty. Organizations that detect, negotiate and quickly "attack" the exploitation of existing short-term advantages for gain survive well and advance. "The Capitalist's Dilemma," written by Clayton M. Christensen and Derek van Bever in the Harvard Business Review (June 2014), discusses "building and sustaining a successful enterprise" in this way. Corporations focused too narrowly on short-run shareholders' gains have found the result to be chronic decline in the long run (some company examples are Kodak and Blockbuster).

Management thought, practice and teaching have evolved to where there are three eras of management: Executive, Expertise and Empathy. Prior to the industrial revolution (late 1800s), handling tasks like planning, coordination, rewarding and controlling was the way of management and, really, the "way of the world." *Wealth of Nations* by Adam Smith in 1776 gave insight to management, with division of labor, efficiency, effectiveness and productivity, and Chester Barnard's *The Functions of the Executive* (1938) gave further insight to management.

Giving rise to the industrial revolution and the new means of production, organizations were able to "scale" operations, requiring others beyond the family, referred to as "agents" to be in charge. This period was indeed the "executive" stage of management, as it focused on execution of mass production, with managerial issues like specialization, standardization, quality control and project management. Besides Taylor, there were the time-and-motion studies of Frank and Lillian Gilbreth and Henry L. Gantt, with his emphasis on efficiency, consistency of production, low variation and statistical measurement. The goal was to optimize outputs that could be produced with limited inputs.

With the advent of business vehicles and business schools (e.g., the *Harvard Business Review* formed in 1922; the founding of the Wharton School in 1881), management was indeed changing focus from the executive to the "expertise" era. The evidence was that much progress had been made enlarging the field of management to become a growing and evolving theory in different approaches. By the mid-20th century, new management theories, beginning with Barnard but ending with the likes of Chris Argyris, Peter Drucker and Edward Deming, were being tested with empirical data.

Finally, with Douglas McGregor's *The Human Side of Enterprise* (1960), bringing to life Theory X- and Theory Y-management, we moved to the empathy era, with management now trying to incorporate emotional intelligence into the arena of management.

Management has evolved quite a bit since Frederick Winslow Taylor.

RELEVANCE TO COMPLEX PROJECT MANAGEMENT

A preponderance of data and information exists, so managers need to regularly assess, catalog, and monitor updates, and retain and discard different data to help them with complex project management. The next section will illustrate just some of the software that exists in the project management field of policy management.

A preponderance of data and information exists, so managers need to regularly assess, catalog, and monitor updates, and retain and discard different data to help them with complex project management.

CURRENT PRACTICAL TOOLS AND OBTAINING TOOLS

Many examples of tools for complex project management exist, which illustrate more than the triple constraint (cost, time and scope) of traditional project management. Below are five current examples, with brief analyses to describe their strengths and applications, along with the websites where you can find them.

  i. *Advantage* software by MetaCompliance policy management software company has features needed to automate, deliver, track and fully manage an organization's policy management activity, including audit trails, policy creation wizard, policy training, workflow management and a best practices library. *Advantage* has a consistent way of creating policies, tracking evidence and responses from staff, test employee understanding of the policy, and track specific groups of users for real-time reporting and acceptance of policies in the organization. MetaCompliance policy management software has information assurance frameworks, such as ISO 27001. State-of-the-art cybersecurity policies are its core benefit. An important step in addressing regulatory oversight requirements is to write a compliance policy that links business practices to the compliance requirements. It should guide staff on activities and provide escalation practices for high-risk matters and/or to address questions. MetaCompliance Policy Management Software website: www.metacompliance. com/products/policy-management/.

ii. *PolicyTech* software by NAVEX Global is a secure, cloud-based entity that is easy to use and manage for policies and procedures. *PolicyTech* software is timely, works within the current standard, and provides compliance and risk management confirmation, while working within rules-based workflows and alerts to keep policies at the cutting edge to allow for updating, changing, and retiring antiquated policies. As evidence of its value, NAVEX Global has clients such as Cedars-Sinai, Toyota, Equifax, Yahoo!, Delta, Fairmont Hotels and Levi Strauss & Company to name just a few. NAVEX Global Software website: www. navexglobal.com/en-us.

iii. *PolicyManager* by PolicyMedical is an easy-to-use, reliable, web-based enterprise healthcare management software package, which is used by over 3,000 entities across North America. Equipped with advanced search, email alert, MS-Office integration and version control, users can easily connect to policies from anywhere in no time, centralizing the entire policy lifecycle across policy committees. PolicyMedical Software website: www. policymedical.com/.

iv. *PowerDMS* software by PowerDMS is crucial for organizations to update, change, manage, share and verify their most crucial content. Public and private sectors use PowerDMS to reduce risk and simplify workflows with the right amount of redundancy. The software is accessible from anywhere to learn of current policies, procedures, training, site plans, videos and photos. PowerDMS Software website: www.powerdms.com/.

v. *PolicyHub* by Mitratech has the best-in-class policy and procedure management software that handles the intricacies of managing policies and procedures through the use of robust, automated, and instinctive self-contained tools. This software creates, approves, and communicates policies and procedures to intelligent distribution, knowledge assessments and reports. It can save time, improve efficiency, and also create a secure program that demonstrates corporate responsibility and reduces the risk of noncompliance. PolicyHub Software website: http://info.mitratech.com/policy-management.html.

Let's further examine *PolicyHub* by Mitratech as we look at these five critical stages of policy management:

Stage 1: Establishing policy requirements
Mitratech requires an audit of employee understanding of the regulatory environment so that all employees are aware of the ongoing policies and practices. The data must be retained online and must always be accessible to employees.

Stage 2: Drafting policy
Mitratech procedures entail importing information from existing documents by either cutting and pasting into new documents or actually entering information into them and recording "who did it" and "when." This new information should be collaboratively drafted and edited and, of course, maintained with automatic tracking controls for records retention. Appropriate records retention should be utilized.

Stage 3: Policy deployment

Clarity and consistency of the correct policies broadly issued officially to people across the organization is how Mitratech establishes deployment. Automatic reminders are set for employees who have not signed up for policy documents, and different policies for different user groups must be specified. Individual libraries with access to up-to-date policies and procedures are required under policy deployment. Policy deployment requires the local language to be used and provided which may require some edits to accommodate local practices.

Stage 4: Testing, understanding and affirming acceptance

Mitratech requires secure confirmation by staff that they agree to the terms and conditions of the policy; in fact, they state that randomized, multiple-choice questions may be used to confirm such agreement. Automatic reaffirmation is then done periodically to represent policies to refresh and reinforce understanding. Mitratech responds to low scores on the test or nonagreement to alert policy owners.

Stage 5: Auditing and reporting

Mitratech captures and displays detailed data on effectiveness of policy deployment for auditing and reporting services to demonstrate acceptance and understanding of policies by group(s), individuals(s) and over various timetables. Audit trails must be provided, showing who received or agreed to which version, of which policy, and on what date. Mitratech also highlights ambiguous, confusing or poorly worded questions.

No new technology or software must be installed, and various policies and procedures are used as intended, are current and provide compliance evidence. Also, responsibility for managing policies and procedures is distributed among several entities (Shah, 2011).

PolicyTech by NAVEX Global is a new way for organizations to benchmark their own policy and procedure management programs. PolicyTech represents over a thousand respondents globally who influence, manage and have overall decision-making authority over their compliance program. NAVEX Global ranks maturity level of each program, labeling them as either advanced, basic or reactive. In a recent determination by NAVEX Global, less than 20% (about 14%) had "advanced" policy management programs (Figure 12.2).

Taking signals from companies with advanced policy and procedure management programs, those with basic or reactive ranks of policy management should consider the following steps to advance:

Be proactive in the review of policies – Those companies whose policy programs are ranked as advanced or maturing are more proactive and sophisticated in their approach to policy management, especially the creation and review of new policies. Those ranked as "reactive," typically only revise and review things when trouble arises, e.g., Starbucks and their restroom policy.

Leverage software tools to automate processes – Companies ranked as advanced policy management programs automate their policy management systems, which allows all employees, especially new employees, to access the latest version of policies.

Maintain a centralized repository of policies online – With the rapid pace at which policies may change, making certain that employees have the most current version of policies requires they be maintained on a company's intranet.

Require a form of employee certification – Most respondents require employees to attest to at least one policy and to do so with annual recertification.

> Policy management systems can mitigate risks by making policies known throughout the organization as part of an employee "tool kit," thereby protecting an organization by staying "ahead of the curve" relative to policy problems.

> If policy management is done well, it could add to a sustainable competitive advantage for the organization, which is all important to the strategy of the organization.

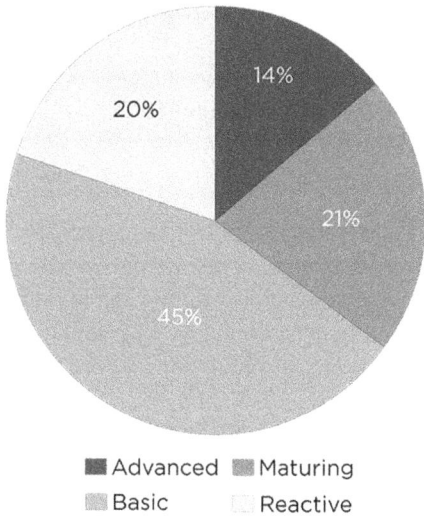

FIGURE 12.2 NAVEX Global maturity-level ranking of each program. (Source: Data from NAVEX global software website: www.navexglobal.com/en-us.)

Conceive of a policy on policies – A documented set of guidelines for how to create, review, issue, and if necessary, reconsider, and reissue policies is very important. Having a policy on policies helps ensure that they remain in the proper form, use consistent language and are easy for all employees to acknowledge.

Establish policy management as a cross-functional responsibility – Large and multi-business corporations tend to involve more than one department. As an example, an author here used to work for PANYNJ, which had a unique matrix structure with staff or functional departments (accounting, marketing, finance, engineering, etc.) intertwined with a line or divisional departments (TBT, airports, rail, World Trade Center, etc.). When the Port Authority was undergoing policy management renewal or even establishment, it was an immense undertaking.

Leave management with decision-making authority – It is the role of senior management to make decisions, manage the organization, as well as drive policy changes as appropriate. It is the role of the Board of Directors to oversee and challenge management (and policies) as appropriate. The autonomy of the Board should not be compromised so they may serve the shareholders independently. These bifurcated roles are sometimes at odds when the CEO and Board Chair role are shared, e.g., Jamie Dimon at JPMorgan Chase.

Have an escalation policy for any problems – Escalation of important policy matters needs to be assessed urgently. Whether it is escalation to the board, CEO, staff department executive management or line department executive management, a policy should be in place to elevate such matters. Mature and effective policy and procedure management programs require more robust ethical and compliance programs.

> Policies are tools to provide a demonstrated standard of conduct throughout the organization, where high achievers and early adopters follow the rules, but they also provide direction to the slower adopters who need a push.

EXAMPLE TO APPLICATION OF COMPLEX PROJECT MANAGEMENT

In the *PolicyHub* (by Mitratech) example above, for instance, one could apply the triple bottom line

> Policies also provide a defense for litigation, but when policies do not match the actions of an organization there are problems.

of project management to manage cost, schedule and scope of a project. *PolicyHub* has numerous proprietary ways in which to manage these critical project characteristics, which note differences in stages from when they're planned to be completed, much like Microsoft Project. See the numerous project management software programs available: https://en.wikipedia.org/wiki/Comparison_of_project_management_software

CLASSROOM QUESTIONS

a. How would you develop an efficient and effective online policy management program for a very large, old, traditional organization (e.g., The World Bank)?
b. How would you involve the upper echelon of a corporation (e.g., the board of directors) in your approach to develop an online policy management program for the same organization?
c. What differences would you foresee for developing an efficient and effective online policy management program for a very large, mature, traditional (1) for-profit, (2) not-for-profit, (3) governmental, (4) international, or (5) educational organization?

CONCLUSION

This chapter on policy management discusses its past, present, and future with an all-around look at the need of creating, communicating and maintaining policies and procedures. This chapter is one of the many in this "evolving toolbox for complex project management," covering such information as challenges in the 21st century, simulations, case studies, visualization, adaptability, governance and resilience.

Policy management is essential for all types of business enterprises today. Its systems can mitigate risks by making policies known throughout the organization, as described in an employee "tool kit," thereby documenting accepted practices and protecting an organization. Appropriate cases help illustrate the problems and risks in various types of business enterprises.

As discussed, policy management can be expensive and time consuming, but it is a necessary process required to keep the business enterprise striving forward without the need for corrective actions undertaken reactively. Reducing business expenses by not managing policies is shortsighted for an enterprise, and unfortunately, some business enterprises only think in the short term. If policy management is done well, as seen here, it could add to a sustainable competitive advantage for the organization, thus advancing its strategy.

This chapter discussed establishing policy requirements, drafting policy, approval process, policy deployment/communication, testing understanding and affirming acceptance, and controlling/auditing policies. Examples from organizations such as Mitratech and NAVEX Global and experiences and publically available information from companies such as Deloitte, PricewaterhouseCoopers, Ernst & Young, KPMG, Accenture, McKinsey & Company, Boston Consulting Group (BCG), Booz Allen Hamilton, Capgemini, Bain & Company, Cognizant and others specializing in policy management issues were considered.

Finally, reputation management, customer–client authority, and GDPR, which is the European rule for data protection on the Internet, were discussed. A historical perspective from when and where policy management issues began was examined, noting its introduction, growth, and now that it is here to stay, its

> Always remember that policy management is executed by people and for people, in order to make their role in management ultimately easier.

maturity. Policies are tools to provide a demonstrated standard of conduct throughout the organization, where high achievers and early adopters follow the rules, but they also provide direction to the slower adopters who need a push. Policies may also contribute to a defense for litigation, but when they do not match the actions of an organization, there are problems. Always remember that policy management is executed by people and for people, in order to make their role in management ultimately easier. People, however, are subjective according to their own biases and thought process; hence, there may be limits to the rational approach to organization management (Tversky & Kahneman, 1974).

REFERENCES

Christensen, C.M. and D. van Bever. 2014. The capitalist's dilemma. *Harvard Business Review*. June.

Deltek website: www.deltek.com.

Flyvberg, B., N. Bruzilius, and W. Rothengatter. 2003. *Megaprojects and Risk*. Cambridge, MA: Cambridge University Press.

Jensen, C. and W.H. Meckling. 1976. Theory of the firm: Managerial behavior, agency costs and ownership structure. *Journal of Financial Economics* 3(4): 305–360.

Kilpatrick, M. 2006. Complex project management: Towards a theory of cognition for ill-structured tasks. *Paper presented at PMI® research conference of the Project Management Institute*, Montréal, Québec, Canada.

MetaCompliance policy management software website: https://www.metacompliance.com/products/policy-management/.

NAVEX global software website: https://www.navexglobal.com/en-us.

Pavlak, A. 2004. Project troubleshooting: Tiger teams for reactive project risk management. *Project Management Journal*, 35(4): 5–14.

PMI. 2017. *PMBOK® Guide* – Sixth Edition. Newtown Square, PA: Project Management Institute. www.pmi.org/pmbok-guide-standards/foundational/pmbok.

PolicyHub software website: http://info.mitratech.com/policy-management.html.

PolicyMedical software website: https://www.policymedical.com/.

PowerDMS software website: https://www.powerdms.com/.

Shah, S. N. 2011. Policy management software for hospitals and clinics helps with change management. www.healthcareguy.com/2011/10/30/guest-article-policy-management-software-for-hospitals-and-clinics-helps-with-change-management/ (accessed April 26, 2013).

Shane, J., K. Strong, and D. Gransberg. 2012. Project management strategies for complex projects: Guidebook, resources, and case studies. Strategic Highway Research Program 2 Transportation Research Board of the National Academies.

Standish Group. 2001. Extreme CHAOS. www.coursehero.com/file/12691071/Standish-group-extreme-chaos-2001/ (accessed October, 2005).

Tallyfy website: https://tallyfy.com/policy-management/.

Tversky, A. and D. Kahneman. 1974. Judgment under uncertainty: Heuristics and biases. *Science* 185: 1124–1131.

Wikipedia. 2019. Comparison of project management software website: https://en.wikipedia.org/wiki/Comparison_of_project_management_software.

13 Legal Aspects Toolset

George F. Jergeas
University of Calgary

CONTENTS

CONTRACTUAL ARRANGEMENTS

There are many ways complex projects can be delivered. We selected five common contractual arrangements as examples (Moazzami, 2015; Jergeas & Lynch, 2015).

- Design–bid–build
- Design–build or Engineering, Procurement and Construction (EPC)
- Build–operate–transfer (BOT), build–own–operate–transfer (BOOT) or public–private partnership (PPP)
- Construction management
- Strategic alliances.

These contracting arrangements can be applied with modification to different projects outside the construction and engineering practice with proper consultation with legal counsel. The selected contractual arrangements are discussed below.

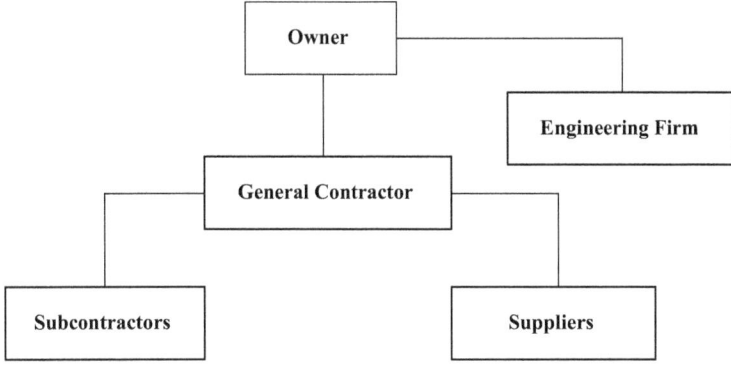

FIGURE 13.1 Design–bid–build delivery method.

Design–Bid–Build

In the design–bid–build project delivery method, the owner enters into a contract with a design firm to provide design services. The design (engineering) firm provides design documents including drawings and technical specifications required for the project. These documents will be used by the owner as the basis to make a separate contract with a contractor, which is also called a general contractor to execute the project. In design–bid–build, design and execution activities are performed by two different entities through two separate contracts. This means there is no contractual relationship between the design team and the general contractor. Figure 13.1 presents the contractual relationships in design–bid–build (Moazzami, 2015).

Design–Build or EPC

In a design–build arrangement, one single organization provides project design and construction services mainly to deliver building and infrastructure projects. On industrial projects such as oil and gas facilities, an EPC contract is an alternate term for the design–build delivery method to provide engineering, procurement, and construction services. Figure 13.2 shows the simple contractual structure when the EPC contractor has a single contract with the owner and several contracts with various subcontractors and suppliers for providing engineering procurement and construction services (Moazzami, 2015).

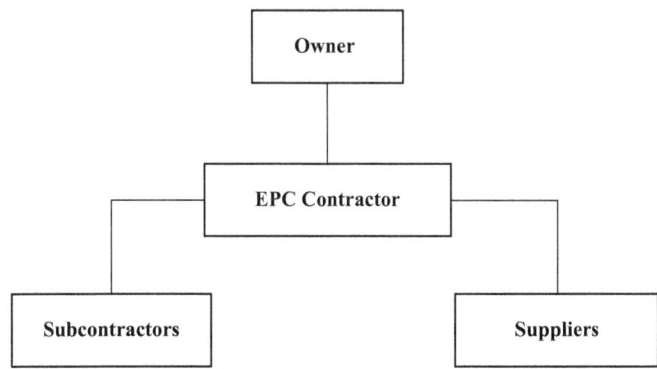

FIGURE 13.2 EPC project delivery method.

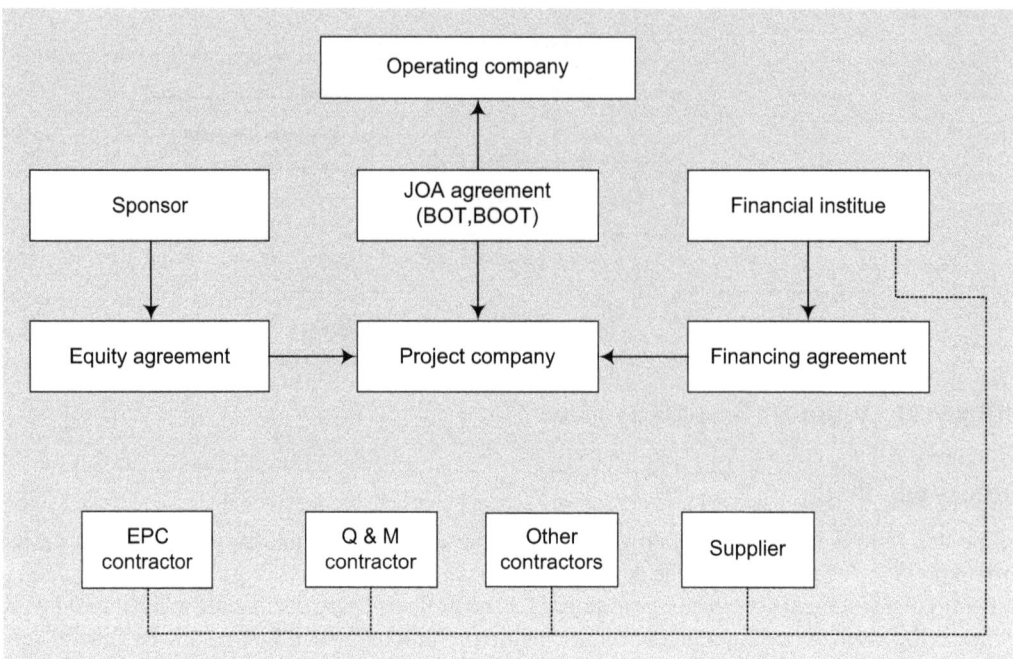

FIGURE 13.3   BOT/BOOT method.

BOT, BOOT OR PPP

On some large-scale and complex projects, the owner may elect to require external financing and operation added to the design–build services. In this situation, the owner signs a Joint Operating Agreement (JOA) with a project company. The agreement might be in forms of build–operate–transfer (BOT), build–own–operate–transfer (BOOT), or other variations. BOOT delivery method can be also called PPP. Figure 13.3 illustrates the contractual structure of a financed project involving the EPC contractor (Moazzami, 2015).

CONSTRUCTION MANAGEMENT

Construction management (CM) method is another form of project delivery in which the CM firm performs management activities on behalf of the project owner. In one version, the CM firm can act as an agent and advisor to the owner liaising between the owner and the contractor and design team.

There is a direct contract between the owner and the CM firm. There are no direct contract between the CM and engineering firm or the general contractor; however, the CM acts on the owner's behalf in managing and coordinating the engineering and general contractors. Figure 13.4 shows the contractual relationships of the CM delivery method (Moazzami, 2015).

STRATEGIC ALLIANCES

Strategic alliances have been around a long time, and in the last 20 years, the alliance profession has deeply codified the organizational architecture of best practices that create repeated success. Many of the collaborations have produced billions of dollars of value for companies like IBM, P&G, and the airline industry such as the Star Alliance, among numerous other smaller companies.

Alliances establish a formal governance structure that enables the multiple partners to work collaboratively. Alliance framework calls for a shared risk, shared reward approach to incentivizing the partners (Jergeas & Lynch, 2015).

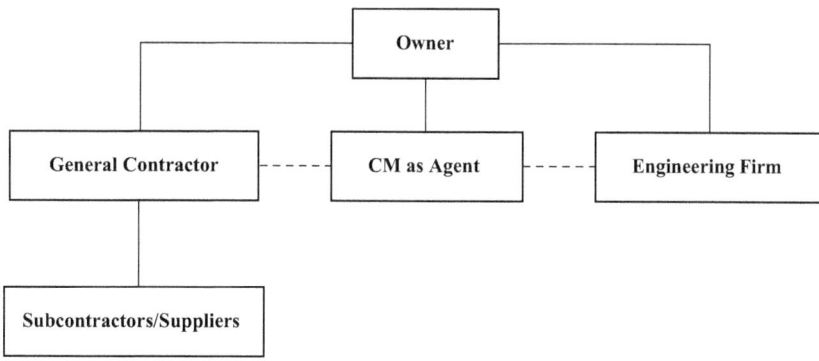

FIGURE 13.4 CM as agent project delivery method.

Because complex projects constitute a very unique environment beset by a multitude of problems, Jergeas and Lynch propose the establishment of a governance and alignment mechanism named "system integrator". The system integrator acts as a coordinator, facilitator and integrator that provide the missing pieces that none of the typical alliance participants provide. During the project formation stage, the systems integrator facilitates bringing together and alignment of the key delivery partners. Next, during the operational phase, the systems integrator provides coordination, anticipation of problems, and critical services that none of the partners have core competencies to provide but typically cause projects to fail in the end (Jergeas & Lynch, 2015).

The proposed model for effective project alliance on complex projects is shown in Figure 13.5 (Jergeas & Lynch, 2015).

ROLES AND RESPONSIBILITIES

As discussed earlier, there are many contractual arrangements for delivering projects. These contractual arrangements result in assigning different roles and responsibilities to the parties and are usually prescribed in the contract documents. In this chapter, we will focus on the role of the main parties—the contractor, the design professional (engineer) and the owner—as per the requirements of the **Design–Bid–Build** contractual arrangement.

> Contractual arrangements result in assigning different roles and responsibilities to the parties and are usually prescribed in the contract documents.

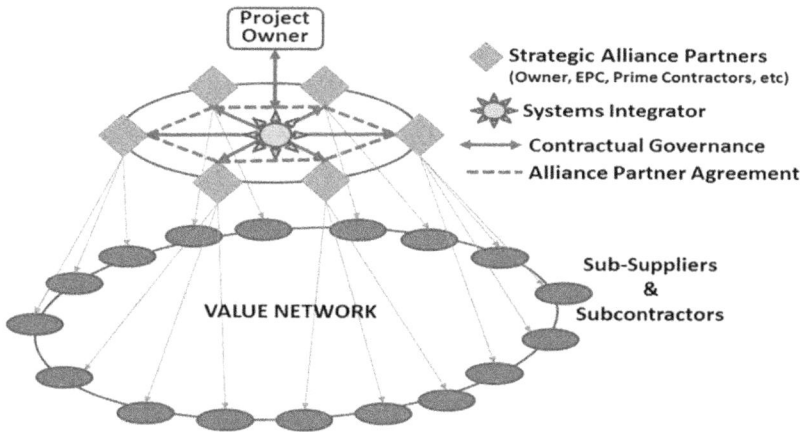

FIGURE 13.5 Strategic alliance with a system integrator.

Regardless of the contractual arrangement, each party must thoroughly review, understand and comply with the requirements of their contract. For the purpose of this chapter, a contract can be defined as a promise, or a set of promises, which one person gives in exchange for the promise, or set of promises, of another person.

THE CONTRACTOR

The contractor's role is to deliver the project in accordance with the drawings and technical specifications—within the agreed-upon timeframe and budget—complying with relevant codes, laws and regulations of the land. Noncompliance may place the contractor in a dispute with the owner or even lead to breach of contract.

> The contractor's role is to deliver the project in accordance with the drawings and technical specifications.

The contract documents with the owner outline the contractor's obligations clearly and in detail. These documents contain provisions specifically detailing the contractor's standard of performance: expected quality, timeframe, contract administration issues and compliance with statuary obligations. Below is a selection of examples of typical contract clauses, between a contractor and an owner. Contractors must read, understand and comply with these clauses:

- Notice provisions

    These clauses require the contractor to strictly comply to provide notices when certain events occur such as delays or claims. Noncompliance may result in the contractor waiving their rights to claim at a later day.

    > Any claims which the contractor may have against the Owner shall be *presented* to the Design Professional/Engineer *in writing* not later than seven (7) days after the occurrence of the delay. Failure by the contractor to present any claim within the seven (7) day period *shall be deemed to be an absolute waiver of such claim.*

- Project changes

    An existing contract can be altered by mutual agreement, by issuing a "Change Order" or "Variation Order". Change orders are the legal mechanism for amending the contract to accommodate changes, whether these be additions or deletions to the original scope of work. Change order clauses describe the scope of changes and the method for pricing and paying for a change, as well as providing a record of the total impact and consequences.

    > Change orders are the legal mechanism for amending the contract to accommodate changes, whether these be additions or deletions to the original scope of work.

    An example of a change order is shown below:

    > The Owner shall have the right, at any time, to make Changes. When a Change is proposed or required, the Owner shall provide a Contemplated Change Notice to the Contractor describing the proposed Change. The Contractor, upon receipt of a Contemplated Change Notice, shall promptly provide the Owner's Representative with a Change Quotation which shall include a method of adjustment or an amount of adjustment to the Contract Price, if any, and the adjustment in the Contract Time, if any, for the proposed Change.
    >
    > The Owner shall promptly following receipt of the Change Quotation either agree to the adjustments in the Contract Price and Contract Time or to the method to be used to determine the adjustments, or give the Contractor notice that the Change Quotation is not acceptable. If the Change Quotation is agreed to, then the Owner shall immediately issue a Change Order recording the Change, which shall be signed by the Owner and the Contractor. The value of Work performed as a result of a Change Order shall be included in invoices for payment given in accordance with the contractual terms of payment.

> If the Owner requires the Contractor to proceed with the Change before the Owner and the Contractor agree, or where the Owner and the Contractor have failed to agree upon the adjustment in Contract Price and Contract Time, the Owner shall issue a Change Directive.

The owner has a duty to issue change orders in a timely and fair manner without affecting the contractor's performance. The contractors should submit their proposal to execute a change order as soon as possible upon receiving the owner's request, and approval or rejection should follow in a timely manner.

The change order should be fair and recognize the contractor's right to include overhead and profit percentages, compensation for legitimate time delay (if any) and compensation for legitimate impact costs (if any). Below is a list of practical tips for an effective change control system to be followed by project organizations:

1. Must have a simple and written change control procedure
2. Accountability to follow procedure
3. Early communication of a probable change to all affected parties
4. Input from all stakeholders when preparing a change order
5. Develop a detailed and realistic scope for the change that may include equipment list, bill of materials, etc.
6. Conduct and include impact analysis (cost, schedule, resources, etc.)
7. Must have a defined and timely approval process that is easy for essential changes and painful for nice-to-have changes
8. All changes must be documented
9. Discipline in implementation and follow-up of the change.

The following is an example of a change order template that can be modified and used by professionals.

Example of a Change Order

PROJECT NAME:

CONTRACT NO.:   CONTRACT DATE:

CONTRACTOR NAME:

THE FOLLOWING CHANGES ARE MADE TO THE SCOPE:

REASON/JUSTIFICATION:

CHANGE TO CONTRACT PRICE

Original Contract Price: …………

Current Contract Price, as adjusted by previous change orders: …………

The Contract Price due to this Change Order will be (increased)/(decreased) by: ………

The new Contract Price due to this Change Order will be: …………

CHANGE TO CONTRACT TIME

Original Contract Completion date: …………

The Contract Time will be (increased)/(decreased) by ………… calendar days

The Contract Completion date will be extended to: …………

SIGNATURES:

- Force majeure

    Force majeure occurs when a major change in circumstances, for which neither party is responsible, makes performance impossible. Some contracts include a clause that contemplates certain events and subsequent action.

    > Either the *Owner* or the *Contractor* may claim that an *Event of Force Majeure* has taken place by giving the other party verbal notice within 24 hours of the *Event of Force Majeure* and written notice, including a proposed plan of corrective action to resolve or minimize the effect of the *Event of Force Majeure*, within 48 hours of the *Event of Force Majeure*.

    > If the *Owner* has given notice of an *Event of Force Majeure*, then the *Owner* may: cause the *Contractor* to complete the *Work*, with such adjustments to *Contract Time* as are required by the *Event of Force Majeure*; suspend the *Work* or any portion thereof; or terminate this *Contract* or any portion thereof.

    > If the *Contractor* has given notice that the *Work* or any portion thereof is affected as a result of an *Event of Force Majeure*, then, the *Contractor* shall complete the *Work*, with such adjustments to *Contract Time* as are required by the *Event of Force Majeure*.

    > If the *Owner* and the *Contractor* do not agree that the *Work* or any portion thereof is affected as a result of an *Event of Force Majeure* for which notice has been given or cannot agree on any adjustments to *Contract Time* that may be required by the *Event of Force Majeure*, then the *Contractor* shall continue to complete the *Work* in accordance with the *Work Schedule* and either party may refer the matter for resolution under Dispute Resolution Clause.

    > If an *Event of Force Majeure* exists and continues for a period in excess of 30 continuous *Work Days* and results in substantially all of the *Work* being stopped or suspended during that period, then the *Contractor* may terminate the *Contract* and the *Owner* shall pay the *Contractor* for the *Work* performed to the date of termination.

    > Any delay or failure on the part of either the *Owner* or the *Contractor* which is a result of an *Event of Force Majeure* shall not constitute default hereunder or give rise to any claim for damages.

- Project delays

    Throughout their life cycle, projects will experience many kinds of delays due to, for example:

    - Late approvals
    - Untimely decisions
    - Late design information
    - Late owner-supplied equipment and materials
    - Lack of or limited access to project site
    - Unusually severe weather conditions
    - Acts of God/force majeure events

    > **Throughout their life cycle, projects will experience many kinds of delays.**

    Project delays can be categorized into three types:

    1. Compensable delays: Delays for which the owner must assume responsibility pursuant to the terms of the contract such as scope changes. These delays entitle the contractor reimbursement of cost plus extension of time.
    2. Excusable delays: Delays that result from the so-called "neutral events", i.e., causes beyond the control of both parties such as force majeure events or unusually severe weather. Extension of time only (relief from liquidated damages).
    3. Non-compensable delays: Delays for which the contractor must assume responsibility pursuant to the terms of the contract self-inflected delays or normal weather conditions. The contractor receives no extension of time and no financial compensation or relief.

    See contractual clause below as an example.

    > If the Contractor is delayed in the performance of the Work by an act or omission of the Owner or anyone employed or engaged directly or indirectly by the Owner, contrary to the provisions of the Contract Documents, or by the Owner taking possession of or using any partially completed portion of the Work ahead of the Work Schedule, then the Contract Time shall be extended for such reasonable time as may be necessary to allow the Contractor to make up the delay.

The Contractor shall be reimbursed by the Owner for reasonable costs incurred by the Contractor as the result of such delay.

If the *Contractor* is delayed in the performance of the *Work* by an order issued by a court or other public authority having jurisdiction, and providing that such order was not issued as the result of an act or fault of the *Contractor* or any person directly or indirectly employed or engaged by the *Contractor*, then the *Contract Time* shall be extended for such reasonable time as the *Owner* may recommend in consultation with the *Contractor*. The *Contractor* shall be reimbursed by the *Owner* for reasonable costs incurred by the *Contractor* as the result of such delay.

No claim for delay and no extension of time on account of delay shall be made for delay unless notice with a *Change Quotation* is given to the *Owner* not later than 5 *Work Days* after the commencement of delay, providing, however, that in the case of a continuing cause of delay, only one notice of claim shall be necessary.

- Liquidated damages

  A liquidated damages clause specifies a predetermined amount of money that must be paid as damages for failure to perform under a contract. The amount specified in the contract must be based on a genuine/actual calculation of damages. See the example below.

  In the event of a delay to the Project Completion Date as per the Contract Schedule for which Contractor is solely responsible, Contractor shall pay Liquidated Damages to Company (owner) $10,000 per day of delay, subject to a maximum of ten percent (10%) of the Initial Contract Price.

- Dispute resolution

  Project problems are inevitable and if remain unsolved can escalate into disputes. Disputes that aren't resolved at the project level have to be expensively resolved by "Strangers to the Project" (mediators, arbitrators, and courts). Disputes may be resolved in accordance with the below contractual dispute resolution clause:

  In the event of disagreement between the parties as to the performance of the *Work* or the interpretation, application or administration of the *Contract Documents*, the *Contractor* shall perform the *Work* as directed by the *Owner's Representative*. All differences between the parties not resolved by the decision of the *Owner's Representative* and all disputes and claims of either party arising out of the *Contract* and its performance shall be settled in accordance with this Clause.

  The parties shall make all reasonable efforts to resolve all disputes and claims by negotiation and agree to provide, without prejudice, open and timely disclosure of relevant facts, information and documents to facilitate these negotiations.

  Either party shall be entitled by notice to the other to call for the appointment of a *Project Mediator*, in which case the parties shall within 10 *Work Days* thereafter jointly nominate a *Project Mediator*. If the parties do not agree on the appointment of a *Project Mediator*, then either party may request the Chair of the Alberta Arbitration and Mediation Society to appoint a *Project Mediator*, who when so appointed shall be deemed acceptable to the parties and to have been appointed by them.

  The parties shall submit in writing their dispute to the *Project Mediator*, and afford to the *Project Mediator* access to all records, documents and information the *Project Mediator* may request. The parties shall meet with the *Project Mediator* at such reasonable times as may be required and shall, through the intervention of the *Project Mediator*, negotiate in good faith to resolve their dispute. All proceedings involving a *Project Mediator* are agreed to be without prejudice, and the cost of the *Project Mediator* shall be shared equally between the parties.

  If the dispute has not been resolved within 15 days after the appointment of the *Project Mediator*, either party may by notice to the other withdraw from the mediation process.

  All disputes, claims and differences not settled as herein provided arising out of or in connection with the *Contract* or in respect of any defined legal relationship associated with it or derived from it, shall be referred to and finally resolved by arbitration in accordance with the *Arbitration Act* (name of jurisdiction/country). The arbitral tribunal shall be composed of one arbitrator where the subject of the dispute, claim or difference relates primarily to whether work required to be performed is within the scope of the *Work* or whether the *Contractor* has met the required specifications of the *Contract*, and provided that the *Work* has not yet been completed when the matter is referred to arbitration. In all other cases, the arbitral tribunal shall be composed of 3 arbitrators, one appointed by each party who shall select the third who shall act as chair.

There are many standard contract documents used in different kinds of industries and countries. The Federation Internationale Des Ingenieurs-Conseils (FIDIC), as an example, publishes a well-known contract document that is used on international civil engineering projects. In Alberta, Canada, the Construction Owners Association of Alberta (COAA) has developed a set of best practices for delivering industrial projects. COAA best practices include contract strategy, contract templates and contractor prequalification among many other practices useful as a model for application on complex projects. Professionals who are interested to learn more about COAA templates, please refer to www.coaa.ab.ca/wp-content/uploads/Best-Practices-Map.pdf.

Regardless of which standard contract or templates selected, industry professionals must refer to and consult with their legal counsel regarding the terms and conditions and also for proper modification to suit the unique circumstances of their project.

THE DESIGN PROFESSIONAL (ENGINEER)

The design professional (engineer) also has a contract with the owner. This contract typically determines the design professional's responsibilities and duties to the owner with respect to the project, including:

- Advising the owner of any technical investigations required to design the project
- Preparing the design, drawings and specifications
- Preparing the bid documents used to select and hire the contractor
- (Possibly) providing estimates for the owner's budgetary purposes
- Reviewing the contractor's bids and providing recommendations as to which contractor should be awarded the contract
- Acting as the owner's agent during execution with respect to contract administration, inspection and monitoring of the work, as well as interpreting the contract documents.

> The design professional prepares the design, drawings and specifications and acts as the owner's agent during execution.

In most contractual arrangements, there is no contract between the design professional and the contractor; however, the design professional's role in providing design and contract administration services has, in Canada, for example, resulted in a tort relationship.

The design professional plays a significant role during execution—not just ensuring that the work is done in accordance with the design but also with regard to changes/variations, monitoring progress, certifying progress and final payment, and advising the owner on how the project is progressing in terms of schedule and cost.

The design professional's duties include inspecting the work to ensure it complies with the design and issuing directives to correct deficiencies in the work. They are additionally responsible for any unanticipated situations that might arise during the project that require design changes.

The design professional also acts as the interpreter and judge of contract documents—addressing issues pertaining to the nature of the contractor's work, such as interpreting specifications and code. As a judge, the design professional may determine issues relating to delays, timeframe extensions and financial compensation.

THE DESIGN PROFESSIONAL'S RESPONSIBILITIES TO THE OWNER

The design professional's contract with the owner outlines their responsibilities regarding the scope of work, terms of agreement and financial compensation. However, the contract with the owner does not specify the design professional's obligation to abide by a specific standard of job performance, as it would for the contractor. The law sets the standard against which the design professional's performance can be judged.

The standard of care that applies to any professional (doctors, dentists and lawyers, for example) also applies to a design professional acting as a design consultant. The design professional is bound to exercise a reasonable degree of care and skill but is not expected to exercise or possess extraordinary care or skill. The design professional's performance is instead gauged against members of their industry/specialty—therefore, the expected degree of skill is that of the **average** person in their profession.

The standard of care that applies to any professional (doctors, dentists and lawyers, for example) also applies to a design professional acting as a design consultant.

The contract can, however, impose a **higher** standard of care on the design professional if they agree to perform to the standard of a *specialist* providing applicable consulting services. The design professional is then obliged to exercise extraordinary care or skill beyond the standard imposed by the law. The following is an example of contractual clause:

> The design professional/engineer acknowledges and represents that it and all employees it assigns to perform the consulting services hereunder possess the necessary qualifications, knowledge, skills, expertise and experience and the consultant agree to perform the engineering services to the standard of a specialist in the area of the consulting services being provided under this agreement.

The above example contractually modifies the standard of care expected of the design professional and imposes a **higher** standard of care. The design professional must read their contract carefully and try to negotiate eliminating such language or at least limit their liabilities to a certain manageable level as illustrated in the following example:

> The Client agrees to limit the Design professional's liability for the Client's damages to the sum of $.
> This limitation shall apply regardless of the cause of action or legal theory pled or asserted.

Contractually, the design professional has the authority to act on behalf of the owner and the owner is bound by their actions. The contractor must comply with the design professional's directives as if they were those of the owner. For example, the design professional can instruct a contractor to modify their work due to changed conditions. The contractor is not required to inquire whether the design professional has the owner's approval—this can simply be assumed. The owner is liable for any additional costs the contractor incurs by addressing the changed conditions as per the design professional's instructions.

It is recommended that the design professional clearly understand what authority they have been given by the owner to act on their behalf, both in their contract with the owner and in the owner's contract with the contractor. It is important to understand what the design professional is permitted to do on a project, in terms of directing the work as outlined in the contract, and the procedures that must be followed. Otherwise, the design professional might find themselves overstepping the boundaries of their authority and paying the contractor damages—or, if the procedures for exercising authority were not followed, paying the owner for any costs and damages incurred.

In addition to getting professional Liability Insurance and making sure they are doing a good quality job, the design professional must consider a legal protection by trying to include a clause in their contract with the owner similar to the following:

> Any representations in the tender documents were furnished merely for the *general information* of bidders and were *not in any way warranted or guaranteed* by or on behalf of the Owner or the Owner's consultants' and its sub-consultants' employees, and neither the Owner nor its consultants or its employees shall be liable for any representations, negligent or otherwise contained in the documents.

THE DESIGN PROFESSIONAL'S RESPONSIBILITIES TO THE CONTRACTOR

The design professional may be found directly liable to the contractor for losses that arise specifically as a result of the relationship between the two parties (the design professional and contractor). The contractor may be able to hold the design professional directly liable for any losses suffered as a result of (for example) the design professional's misrepresentation regarding their authority, the quality of information they've provided or their failure to disclose data or reports. Misrepresentation is a false statement of fact either innocent or fraudulent.

In a negligent misrepresentation case against a design professional, the contractor must substantiate three things:

1. The design professional owed the plaintiff a duty of care
2. The design professional breached that duty of care by his/her conduct
3. The design professional conduct caused the injury to the contractor.

The design professional is required to exercise some judgment or discretion in performing their duties under the contract. They may be expected to make decisions regarding:

- The measurement of quantities
- The contractor's right to payment
- What to be paid for "extra work"
- The degree of satisfaction for completion of the work
- The interpretation of plans and specifications
- Ordering any additional work
- What to do in the case of differing conditions
- Project delays (and justification)
- The determination of disputes.

The design professional's decisions concerning these matters may have a significant impact on the contractor's performance—and more importantly, what the contractor might be financially entitled to for doing the work.

Although the design professional is retained and paid by the owner, they must still be impartial when dealing with matters such as those listed above. The design professional must act in accordance with their own best judgment, without being influenced by their relationship with the owner—and the contractor is bound by the design professional's decisions.

Liability to both the owner and the contractor may arise if the design professional improperly exercises the discretion given to them with respect to the contract. The owner may be liable to the contractor if the design professional fails to act impartially and fulfill their duties in good faith. If, for example, the contractor is entitled to an extension of time and the design professional fails to provide it, the contractor may incur the substantial costs of accelerating work to meet the unaltered schedule. This may result in a claim by the owner against the design professional, for any additional costs resulting from an accelerated pace that could have been avoided if they had made the appropriate decision to extend the schedule.

The design professional's relationship with the contractor may also result in the design professional's liability for costs incurred by the contractor in performing their work. It is common for the design professional to provide the owner with recommendations regarding, for example, the quality/completion of the tender documents. As a result, the design professional may be liable to the contractor for negligent misrepresentations contained in the tender information. This is referred to as tort liability, which means it is not the result of any contract between the parties but is instead the result of their relationship. In a competitive bidding setting, the contractor bases their bid on the drawings, specifications and information provided by the design professional. This can result in the design professional being liable for negligent misrepresentation. The concept of negligent misrepresentation applies to any professional providing advice, including design professionals. The rules for establishing negligent misrepresentation include:

- A person with special skill and judgment makes a statement or representation of fact to a second person
- The second person relies on the aforementioned statement

- The first person knows of this reliance and is duty bound to take reasonable care
- The second person's reliance on the statement results in loss
- Therefore, the first person is responsible for the loss (unless clearly disclaimed).

THE OWNER

The owner's obligations are often not expressly set out in the contract documentation with the contractor—they arise from the inherent duty of the owner to cooperate with the contractor and to not interfere with their scope of work, method of execution and schedule of activities. The contractor has exclusive preserve regarding scope, method and schedule—unless otherwise stipulated in the contract—and the owner enters such preserve at their own risk.

> The inherent duty of the owner is to cooperate with the contractor and to not interfere with their scope of work, method of execution and schedule of activities.

Site Availability

One of the fundamental obligations of the owner is to make the project site available to the contractor. This obligation is frequently not mentioned in the contract but is a necessary implication of its express terms. If the contract contains start and completion dates, specifying a timeframe for doing the work, it follows that the site must be made available to the contractor in order for them to comply with their obligations under the contract.

However, in some instances, the owner will put out a project's call for tenders—and award a contract—before obtaining all required permits and right-of-way or access roads. These circumstances effectively make the site unavailable to the contractor. Transmission line, pipeline, railway or highway projects, in particular, require uninterrupted right-of-way in order to achieve the productivity upon which the contractor's bid price is based.

Owner-Supplied Materials

Late delivery of owner-supplied materials is an obvious default, entitling the contractor to a schedule adjustment and recovery of damages. As is the case with site availability, it is typically necessary to include a term in the contract to the effect that owner-supplied material will be provided as required by the contractor's schedule.

Design Responsibility

The owner's liability for defective design or incomplete specifications depends on the type of contract selected. The consensus in a design–bid–build arrangement is that the owners impliedly warrant their own design. The court is unlikely to impose liability on the contractor for determining whether the specified design system is adequate for the method—i.e., the contractor will not be liable for the design. Design is the responsibility of the owner or the owner's design professional.

Approvals

Late approvals issued by or on behalf of the owner are another common situation causing serious disruption to the contractor's schedule. If the design professional does not issue drawings in a timely manner, the contractor schedule may be delayed, resulting in acceleration, loss of productivity and cost overruns. When drawings are issued too late to begin work as scheduled, the contractor will have to decide whether to delay the work and seek a time extension; abandon their plan for work and re-sequence their operations; resort to alternative and possibly less economical execution methods; or proceed as best as possible based on the information in the contract drawings.

In sum, the owner is obligated to supply drawings and information in a reasonable timeframe.

Method of Execution

Any fixed price or lump sum contract implies that the contractor has control over the execution of the work. In poorly designed projects, some owners have a tendency to direct the contractor regarding how the work should be performed. Such interference with the method and timing of the work should be avoided.

Owners must remember that the contractor is solely responsible for the manner and method for completing their work under the contract and should have full power and authority to select the means, method and manner for performing such work (so long as they do not contravene applicable Occupational Health and Safety requirements).

The owner and their design professional should only be interested in the result or final product of a project, as well as conformity to the plans, specifications and contract.

Owner-Supplied Facilities

The owner is not complying with contractual obligations when they fail to provide the contractor with the facilities as promised. The owner may, for example, commit to supplying suitable and sufficient camp (bed space) for a geographically remote project. This contractual obligation can directly influence the availability and productivity of the contractor's workers. In this case, the owner's failure to provide sufficient bed space is a breach of contract.

Quantities

In some jurisdictions, there is an implied obligation that quantity estimates in a tender package will be reasonably accurate. The accuracy of estimates depends on adequate front-end planning, and the level of design performed on behalf of the owner. There are many cases of complex projects where design is rushed and inadequate, causing substantial errors to arise during execution, which may result in significant increases in quantities.

Although some owners may state that the quantities provided are estimates only, it is very important to ensure that these be more than rough guesses—project estimates must bear some relation to reality.

To address this issue, there may be a contractual provision that if the quantities vary by 15%–20%, the parties will renegotiate prices.

Change/Variation Orders

As discussed earlier, an existing contract can be altered by mutual agreement, by using the contractual "Change Order" mechanism. The owner has a duty to issue change orders in a timely manner without affecting the contractor's performance. The contractor should submit their proposal to execute a change order as soon as possible upon receiving the owner's request, and approval or rejection should follow in a timely manner.

Superior Knowledge

An owner is under no obligation to protect an inexperienced contractor from failure to make proper inquiries or from their own lack of experience. However, there is a line of authority in some jurisdictions supporting the principle that an owner with superior knowledge of project conditions, or the nature of previously installed utilities, has a duty to disclose such information. Failure to provide such essential and important information in the tender documents is negligent omission.

DISPUTE MANAGEMENT

Projects are increasingly competitive and complex, delivered in a risky and sometimes adversarial environment. Many projects are facing considerable claims and disputes, along with increasing difficulty in resolving such disputes in an effective, timely and economic manner. Valuable resources are diverted towards the preparation of claims, settlement of disputes and participation in arbitration or litigation. The net result is time wasted in a non-value-added effort.

Defining Dispute Management

A claim can be defined as a request for additional financial compensation and/or time extension. Contractors or owners can file a claim, and most claims are resolved at the project level by negotiation between the parties themselves. A claim that cannot be, or is not, resolved by negotiation at the project level becomes a dispute. Unresolved claims can be adjudicated through adversarial techniques such as arbitration or litigation or by using less adversarial methods such as mediation.

Typical Causes of Claims and Disputes

There is hesitation in industry to expend effort during the initial project phases that result in changes to design, scope and schedule during execution impacting project cost and time. Additionally, unclear communication of contract intentions has been found to cause misunderstandings regarding the risks and responsibilities assigned to contracting parties. Based on the author's experience, typical causes of claims and disputes include:

- Lack of project planning and inadequate design
- Design changes, errors/omissions and extras
- Lack of coordination between project teams
- The combination of a fixed-price contract and fast-tracking
- Insufficient bid preparation time
- Inadequate bid information
- Underestimation, mismanagement and/or bad workmanship by contractors
- Misunderstanding of contract intentions, terms and conditions
- Various kinds of project delays.

The actions of owners, design professionals and contractors can make the situation worse and serve as early warning signs that claims or disputes may occur. For owners, this includes having unrealistic project expectations regarding completion time and risks, failing to properly develop project scope and design and inappropriately shifting risks to the contractors. For design professionals, this may be in the form of passing design responsibilities on to contractors, agreeing to unrealistic expectations, having a vested interest in disputes and producing ambiguous or conflicting documents. A contractor may negatively impact the situation if they don't satisfy their contract requirements for written communications or comply with "notice provisions", cut corners to offset bid deficiencies, abuse subcontractors, fail to read the contract or assume it will not be enforced, fail to keep adequate project records or do not properly document claims (Revay, 2004, Gillan, 2014; Hudon, 2015).

The following section provides a summary of the claim management recommendations published by Navigant Construction Forum™ (Zack, 2013, 2014, 2015, 2016). If implemented, these recommendations should reduce the likelihood of disputes occurring during the execution phase of complex projects.

Planning for Disputes and Avoidance of Disputes

Communicate Roles within the Owner Organization

The owner should establish, document and circulate (in writing) the boundaries of authority and responsibility within their own project organization for the various execution phase functions.

Encourage Partnering

The owner should encourage collaborative relationships and implement partnering process on the project. Partnering is a structured project management approach whereby all parties agree to establish a trust-based relationships based on open and honest communications and a commitment to achieving mutual goals. It is a voluntary system of handling normal, everyday project issues in a cooperative and joint fashion between the parties before issues escalate into claims or disputes. In this collaborative approach, all project parties resolve that issues will be settled by employing a positive and cooperative dispute resolution approach.

> Partnering is a structured project management approach whereby all parties agree to establish a trust-based relationships.

Due to the importance of partnering to the success of projects, the following subsection provides a description of the process used by the author to build and sustain complex project teams.

Partnering Process

Partnering is a structured process for enhancing communication and team effectiveness by transforming contractual and project participants' relationships into a cooperative based on aligning the team towards common goals and objectives and procedures for evaluating their performance and resolving issues/disputes in a timely manner. The partnering process forges a common vision and direction among team members from different organizations (Rolstadas et al., 2011).

Partnering ensures that projects are designed and executed in a coordinated manner to meet the requirements of the owner. This author has applied and facilitated a partnering model on 150 complex projects in Canada. His model includes the following elements:

1. Team alignment/chartering session: This is a 1 or 2-day facilitated module combining team building and project chartering that include risk and stakeholder assessment among other things. In this session, the following documents are developed and documented through a series of workshops and discussions:
   a. Common goals and objectives statement to provide a common purpose and direction. Below is an example of goals and objectives statement:
      > We, the NE LRT Extension Team, recognize the complexity of the overall program and the individual project responsibilities, commit to a collaborative process through mutual respect and effective communication, and trust to achieve a successful opening of the LRT to Saddle Towne by September 2012.
   b. Scope of work and agreement on what is in scope and what is out of scope of work.
   c. Milestone schedule to gain an alignment on logical sequence and prerequisites. Table 13.1 presents a useful template for analyzing project milestones.
   d. Internal and external project risks and develop project risk register by asking the simple question: What could go wrong? Table 13.2 presents an example of a typical template used for identifying and dealing with project risks.

TABLE 13.1

Project Milestone Template

Milestone Analysis			
Schedule Milestone	Required Date	Likelihood Date Can Be Met (L, M, H)	Discussions of Assumptions, Constraints, Risks, Issues, Criticality, etc.

TABLE 13.2

Risk Management

No.	Description	Probability	Impact	Mitigation Action
1	Increased traffic related to game day and events	High	Low	• Detour and schedule planning
2	Utility coordination	High	High	• Early involvement • Dedicated design resources
3	Engagement participation (impact comes too late)	Med-High	High	• Complete communication plan • Clear messaging • 3D Visualization included • Various outreach tactics • Focus groups for special interest • Councilor 1-on-1's
4	Overlapping phases (rework)	Med	High	• Communicate with internal groups, councilors • Key design decisions are documented and communicated (to assist with staff changes) • Scope management • Value engineering session early • Focus meetings/communications with key design reviewers
5	Timing in relation to construction resources and other projects in the area (detour congestion)	Med	High	• Coordinate and communicate schedule with others • Staging construction
6	Timing of other projects exacerbating congestion	Low	High	• Communication with others • Provide information regularly on design details and construction timing
7	Design coordination/ completion	Low	Med	• Schedule definition – what is needed and when by Graham • Communication between groups • Regular meeting – bi-weekly coordination meetings and weekly technical meetings

 e. Stakeholders plan to identify project stakeholders and their needs, and manage their expectations in a positive and constructive manner so that the project will be a success for all. Stakeholders are individuals or organizations that may affect the project or affected by its execution or results. Table 13.3 illustrates a template used at the partnering session.

 f. Criteria that will be used to measure the success of the project. The success criteria allow the team to know if they are successful or not. The following example shows the criteria for measuring the success of a project:

  – **On time—Complete project within contract schedule dates or better**

  – **Within budget**

  – **Safety—No loss-time accidents**

  – **Quality—Work constructed to town or applicable standards**

  – **Team satisfaction**

  – **Timely resolution of any issue/disputes**

TABLE 13.3
Stakeholder Management Template

Stakeholder	Objective	Operations	Impact	Mitigation Action
Adjacent Businesses	Access to businesses Visibility	Business association where available Individual concerns	Detour plan info Schedule	1-on-1 or focus groups Lines of communication Minimum construction impact
CPR	Crossing Agreement Lease Agreement Pedastrian Crossing		Schedule	Regular meetings Follow-up
Transport Canada	Work Approval		Schedule	On-time notification Regular follow-up
Alberta Culture and Tourism	Work Approval		Schedule	Follow-up
Community Associations and Residents	Access during construction Avoid local road detours	All surrounding communities will have unique perspective/concerns		Schedule communication
SouthFoothills Medical Center	Emergency vehicle access during construction		Public safety	Staging/detour planning
City Council and Ward 7, 8 Councillors	VFM			1-on-1 meetings Schedule communication
Provincial MLA Federal MP				Schedule communication
Special Interest Groups – Pedestrians, Cyclists, Transit	Very diverse Not all same objectives Mode infrastructure built Commuters vs. recreational users	Formal structure Provide own stats on use	Schedule Conflicting input Detour plan info	Focus groups Use social media and website Designated reps from their group

    g. Critical success factors representing the prerequisites or conditions for achieving project success, i.e., without the critical success factors, success will not be achieved. The following example shows the critical success factors for an infrastructure project:
- **Timely delivery of material**
- **Good construction practices**
  - **Within tolerance**
  - **Timely decisions**
  - **Clear understanding of schedule requirements**
- **Meeting milestones**
  - **North abutment before December 25, 2009**
  - **North fill mid-January**
  - **Girder erection by March 29**
  - **Deck pour by May 19**
- **Strict compliance with safety requirements**

 – **Strict compliance with environmental requirements**
 – **Timely communication of construction activity**

h. Health check or scorecard tool to monitor and evaluate the performance and effectiveness of the team throughout the life cycle of the project. Table 13.4 shows an example of a health check template:

i. Issue or dispute resolution mechanism that encourages higher quality discussions to achieve win–win solutions to outstanding issues in a timely manner. This non-adversarial face-to-face negotiated approach focusses on resolving issues at the lowest/working managerial level with time limit for resolution. See Table 13.5 as a template.

j. Project ground rules to ensure everyone knows and agrees with what is expected of him or her. The ground rules' main objective is to have everyone agree on what is important to the team and what is appropriate behavior. See the example below:

The following Ground Rules were agreed to:

• If response is urgent, say when – if timeline is not possible, say so and establish a reasonable timeline;

• Key decisions – document and share them within 3 days (aids communication and defines future change);

TABLE 13.4
Project Team Health Check Template

Item No.	Description	Assessment		
		Communication		
1	Team communications are…	Difficult, guarded	1 2 3 4 NA	Open, up-front
2	Information flow is…	Restricted	1 2 3 4 NA	Free, open
3	Timeliness of information is…	Late	1 2 3 4 NA	On-time
4	Communications/meetings are…	Ineffective	1 2 3 4 NA	Effective
		Working Relationships		
5	Cooperation between project groups is…	Poor, detached	1 2 3 4 NA	Good, unreserved
6	Responses to issues become…	Personal, negative	1 2 3 4 NA	Project specific
7	Problems/issues are resolved by…	Senior management	1 2 3 4 NA	Site level
8	Level of trust is…	Low	1 2 3 4 NA	High
9	City support services	Late, unclear, vague	1 2 3 4 NA	Timely/relative
10	Third party utilities	Late, unclear, vague	1 2 3 4 NA	Timely/relative
		Project Requirements		
11	Safety is being…	Ignored	1 2 3 4 NA	Considered
12	Quality of work is…	Not acceptable	1 2 3 4 NA	Acceptable
13	Environmental requirements are…	Ignored	1 2 3 4 NA	Considered
14	Schedule management is…	Poor	1 2 3 4 NA	Effective
15	Cost management is…	Not acceptable	1 2 3 4 NA	Acceptable
		Stakeholder & External Issues		
16	Public complaints are…	Frequent	1 2 3 4 NA	Infrequent
17	Internal City stakeholders are…	Dissatisfied	1 2 3 4 NA	Satisfied
18	Business community is…	Dissatisfied	1 2 3 4 NA	Satisfied
		Team Satisfaction		
19	Completion of milestones are rewarded/recognized	Ineffective	1 2 3 4 NA	Effective
20	Overall satisfaction is…	Dissatisfied	1 2 3 4 NA	Satisfied

TABLE 13.5

Issue Resolution Mechanism

Level	Name	A Immediate	B 5 Days	C 5+ Days
Site	Kimberley (Graham) Mike (ISL) Kevin (City)	4 h	1 day	3 days
PM	Martin (Graham) Syed (ISL) Jeff (City)	4 h	2 days	5 days
Manager	Alisdair/Jayson (Graham) Alana (ISL) Duane (City)	8 h	2 days	5 days
Director	Tom Cole (Graham) Chris (ISL) Michael (City)	As needed	As needed	10+ days

- Meetings – every 3 weeks: agenda provided in advance (2 days); meeting minutes sent out within 1 week by Engineer;
- Open communication – include the PM in project correspondence;
- Respect decisions once made – leave disagreements at the door; trust and personal respect
- Everyone shares responsibility and celebrates success;
- Don't let issues fester – speak up.

k. Agree on the roles and responsibilities and organization structure of and lines of communication between the different organizations. See Figure 13.6 for an example.

2. Conduct health check sessions. These are regular health checks to monitor the health and performance of the team in addition to monitoring and responding to the risks and

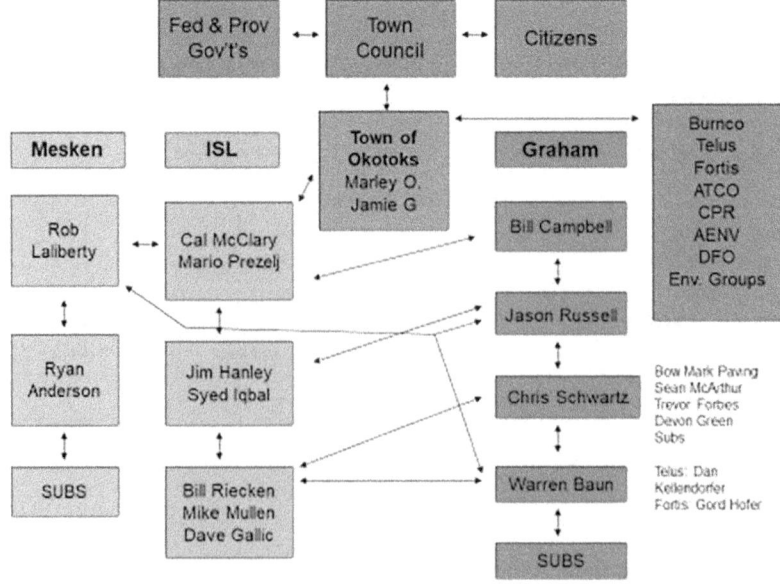

FIGURE 13.6   An example of a project organization structure and lines of communication.

stakeholders' concerns. These sessions are conducted throughout the entire duration of the project, using the health check/scorecard described above.

3. Document lessons learned sessions. At the conclusion of a project, a lessons learned session is conducted to capture good things and things to improve on future projects.

The other recommendations for planning and avoidance of disputes on complex projects as recommended by Navigant Construction Forum™ (Zack, 2013, 2014, 2015, 2016) are as follows.

Discuss Scheduling and Time Extension Requirements

One method for preventing disputes related to scheduling requirements, updates and/or time extensions is to hold a pre-execution meeting specifically for the purpose of discussing scheduling and requirements related to updates, 3-week look-ahead schedules, delay analysis and time extensions.

Review the Schedule of Values

It is common on lump sum projects for the contract to require that the contractor prepare a schedule of values to be used as a payment schedule. Owners should carefully review the proposed schedule of values to prevent the practice of front-end loading.

Designate an On-Site Owner Representative

Owners should arrange for a representative to be on site at all times when the contractor is working, since so many project decisions have to be made or approved by the owner's representative. Full-time, on-site representation helps facilitate decision-making and should result in fewer project delays, i.e., the owner should remain fully engaged in the project and be readily available to make timely decisions.

Review and Respond to All Contractor Letters

Contract documents are replete with requirements for written communications (e.g., confirmation of field agreements, submittals, change order requests, notices of delay or differing site conditions). Contractors are obligated to comply with the terms and conditions of the contract, including notice and other communication requirements. The owner is likewise obligated to respond in a timely and accurate manner and needs to review and respond to all contractor communications. Further, the owner should create a correspondence log documenting the date of receipt of the contractor communication, the issue addressed and the owner's response date (this practice also applies to e-mails).

Finally, senior or more experienced project management staff should review all owner responses to contractor communications, to ensure the information is accurate and that the response does not exacerbate a situation or leverage a routine matter into a later dispute. When done in a timely manner, this should decrease the number of claims and disputes.

Prepare a Project Management Plan

Owners should require that their design professionals and/or project managers prepare a project management plan detailing systems, policies, procedures and document logs for the following:

- Requests for information (RFIs)
- Technical submittals—shop drawings, catalogue cuts, product data, etc.
- Notices of variations/changes and proposals for extra cost
- Schedule submittals material test results
- Payment requests and payments
- Requests for deviations
- Requests for time extensions
- Claim submittals

Such standardized project management procedures should mitigate delays related to these issues and, in turn, decrease claims and disputes.

Use Standard Forms

Owners should create, or have their design professionals or project managers create, a standard set of forms for use on the project to address RFIs, change orders/variations, cost proposals, material testing, submittals, etc. Standardized forms, accompanied by standard policies and procedures, make document processing easier and help mitigate claims and disputes over issues such as late return of document submittals and disruption due to improper handling.

Agree upon Timeframe for RFI Responses and Submittal Reviews

If the contract documents do not specify timeframes for responses to RFIs and submittals, owners and contractors may, during the partnering sessions or kick-off meetings, negotiate and agree upon standard review times for various documents.

Utilize Project Photography

In the event of a claim or dispute, photographs or videos are often more easily understood than the written word. Owners should consider employing an on-site photographic monitoring system. Such systems make it easier to document delays, resolve delay claims and concurrency issues, etc. and thus prevent end-of-job disputes.

Do Not Use RFI Responses to Correct Errors or Redesign Project

Owners and their design professionals and project managers should avoid the use of RFIs to correct errors or redesign elements of the project. If the design needs to be changed during the execution to achieve the owner's needs—or if the owner simply wants a change—the owner's team should issue a notice of change and negotiate a change/variation order. Failure to do so will likely cause contractors to respond by filing change orders or constructive change claims, both of which may lead to disputes.

Do Not Object to Written Notices

Written notices of changes, delays, suspensions of work, differing site conditions, etc. are almost always required in contracts. The intent of such written notice requirements is to alert project owners to problems or issues that may have potential cost and/or impact the timeframe. The best means for an owner to avoid this situation is to not object to written notices and always remind contractors to file written notice whenever a situation (requiring written notice under the contract) arises.

Pay Promptly

Depending upon the terms of the contract, contractors may seek routine reimbursements for work successfully completed during previous periods (months). Project owners need to establish systems for the prompt handling of contractor payment requests in order to prevent (1) late payments, which may subject owners to legal challenges, or (2) slow payments, which may adversely affect a contractor's financial ability to execute the work as planned, leaving the owner exposed to claims of interference with the work. Owners must emphasize to their project staff the need for prompt handling of all payment requests and equally prompt resolution of disagreements over such payment requests.

Review and Evaluate Change Order Requests

Owners and their staff must focus on, and provide for, the timely review and evaluation of all change order requests submitted by the contractor. The evaluation of such requests—to determine entitlement under the contract—should be the first order of business for the owner staff. Ignoring change order requests, or letting them languish unanswered, is likely to lead to larger disputes.

Deal with Delay and Impact Damages

Changes and delays are nearly inevitable on projects. However, too many owners refuse to deal with delay and impact damages at the time of the delay—despite the fact that most contracts require prompt notice of delay and submittal of delay claims within a relatively short period of time after the event has passed. Once notice of delay or potential delay is filed, the owner's project management staff should meet with the contractor to find out what has happened (or failed to happen) in order to determine whether the owner can take any action to mitigate such delays.

Review All Time Extension Requests and Time Impact Analyses

"Constructive acceleration" is generally defined as compelling a contractor to complete their work on time despite legitimate, documented requests for time extensions. This type of claim arises most often when a contractor files a request for a time extension (either excusable or compensable) and the owner denies all or part of the request (or ignores the request entirely), specifically directing the contractor to complete work by the original date or threatening the imposition of liquidated damages for "failure to complete on time"—all of which forces the contractor to accelerate their efforts to complete work on time, thus incurring actual damages. Owners should ensure that their project staff deals with time extension requests promptly, objectively and in accordance with the terms and conditions of the contract.

Liquidated damages clause is a contractual provision that failure of the contractor to complete the work by a specific date will result in the contractor being required to make a specified payment to the owner for each day, week, or month that completion of project is delayed. This is not a penalty, and the amount specified should represent a genuine pre-estimate of damages.

Hold Regular Project Progress Meetings

Experience shows that communication on a project is critical to its successful completion without any outstanding disputes. Routine, face-to-face conversations on a regular basis will likely provide more—and more current—information than written communication. Routine meetings (preferably weekly) with formal written agendas should be attended by all appropriate project team members from both the owner and contractor organizations—including project managers, schedulers, quality control/quality assurance ("QC/QA") personnel, procurement managers, subcontract managers, etc. These meetings help define the project status on a contemporaneous basis, bringing problems to light at an early stage and providing an open forum for both teams to ask questions and air their concerns. If senior managers of both teams take the time to walk the jobsite together, either prior to or immediately following these meetings, each side will gain more understanding of the other's perspective, thus increasing effective project communications.

Allot Time during Progress Meetings for Contractor Needs

As part of the regular meeting agenda, the contractor should be asked to identify the specific needs of the owner, project manager and/or design team. These needs may be related to the return of certain design drawings or submittals by a certain date, responses to certain RFIs, etc. This allows contractors to advise owners of what is important (including why and by what date) for the progress of the work. In this process, contractors can work with owners to prioritize necessary items.

Require Major Subcontractor Participation in Meetings

Owners should insist that all "major" subcontractors attend and participate in all project progress and schedule review meetings. The value of having all parties in attendance is the ability to fully communicate actual project status, progress, problems, etc. The inclusion of major subcontractors enhances the communication process. Each subcontractor should be allotted meeting time to discuss their issues, in order to make the owner team aware of the actual project status. Furthermore, each subcontractor should be asked whether they are aware of any potential delays and issues affecting their work (their responses should be carefully recorded in the meeting minutes).

Promptly Circulate Meeting Minutes to All Participants

The owner staff (or designee) must keep accurate written meeting minutes. These meeting minutes serve as a "project history" establishing what the contemporaneous project priorities are; what current issues need to be resolved, by whom and when; what agreements have been reached between the project teams; etc. These minutes should be processed and sent out to all attendees within a day or so of the meeting, as well as to any designated project executives who are involved in the project but do not ordinarily attend such meetings.

Create Daily Reports from the Owner

The owner's project staff and/or their on-site representatives should keep and maintain daily reports listing: all activities worked on by the contractor and each subcontractor, by trade; daily manpower; all on-site equipment, whether used or idle; daily weather; site visitors; site inspection activities and ongoing or potential delays; etc. These reports must not include opinions, only factual comments and observations. They must be created on a daily basis and reviewed by the project manager for accuracy and completeness.

Monitor Contractor's Production

Owner staff ought to routinely monitor or request the contractor's field labor production and compare it to the planned production calculated from the baseline schedule or current schedule revision/update. This helps facilitate early identification of problems that can be corrected before they become a dispute. Alternatively, this may help defend against later delay and impact damages claims.

Prepare and Issue Deficiency Reports

Owner staff should routinely prepare interim reports of deficiencies while the work is in progress—providing them to the contractor and obtaining the contractor's commitment to immediately remedy such defects—without waiting until the end of the job. The owner staff should then follow up to ensure that appropriate corrections have actually been completed.

Carefully Review Monthly Payment Requests

Owner staff must very carefully review routine payment requests. They need to ensure that any materials and equipment claimed to be on-site were in fact delivered and properly stored. Owner staff should also perform a physical inspection of the work in progress and project status to ensure the accuracy of the payment request. Prior to authorizing payment, payment requisitions must be checked against appropriate unit prices and payment breakdown line items. Underpayment or over-payment to the contractor may be problematic for all sides and should be avoided.

Implement a Clearly Regimented Change Control Process

As discussed earlier, the owner's project management plan should establish a rigorous change management plan coordinated with the terms and conditions of the contract documents. Standard forms should be used, and a standardized procedure for the issuance of change requests, review of contractor proposals, negotiation of changes, etc. should be implemented.

Establish a Mechanism to Promptly Pay for Changes

Successfully negotiated change orders are typically included in the list of pay items, allowing contractors to proceed with the change order work and receive payment on the basis of a percentage of work completed (less retainage). Unpaid change order work can place a serious financial strain on a contractor, which in turn may impact the contractor's ability to prosecute the remainder of the work; it may also lead to a major dispute later on.

Settle All Changes in Full and Final Language

Some changes will be resolved by prospectively settled change orders, wherein the scope, time and cost of the change are agreed upon by the owner and contractor before any work is performed on the change. More frequently, however, other changes will be settled retrospectively. These are change orders where the work is completed prior to the change order being issued. In either event, all change orders, prospective or retrospective, should be settled with some sort of full and final settlement language on the face of the change order.

Advise Contractors of Significant Changes As Early As Possible

Owners and their design professionals and project managers frequently consider and scope potential changes for some period of time prior to advising the contractor about the potential change order. However, it is recommended that owners advise contractors as early as possible about changes under consideration to provide them with adequate planning time. Where possible, owners may want to involve contractors in planning and assessing potential change orders to help mitigate the impact of the change on existing project activities. The technical details of the change should remain under the purview of the design professionals, but contractors can provide input from the execution perspective to help mitigate time and cost impacts should the change move forward. This approach may also help resolve change orders more quickly and assist in obtaining agreement on prospectively priced change orders, thus avoiding later disputes over time, cost and impacts.

Minimize Design Changes during Execution

If owners and design professionals thoroughly plan and design the project, there is little need for design changes during execution. In order to minimize design changes, owners should (to the extent possible) make all critical decisions during the planning and design phases of the project. During the execution phase, owners should resist the urge to change their initial decisions or order changes to obtain betterments during execution. The chances of a dispute at the end of a project are substantially reduced if owners can minimize design changes.

Establish a Project Risk Management System

Design professionals and project managers must consider risks, develop a risk register and allocate appropriate contingencies during the planning phase. Owner staff should then implement a monthly or quarterly risk management meeting, based on the size, scope, speed and complexity of the project. Owners should recognize that as the project progresses, the elements of risk evolve from one risk to another. For example, once all underground utilities are in place and the foundation completed for a building project, underground risks drop off the risk register and are replaced by new risks, whether previously identified or newly discovered.

Consult with Legal Counsel Concerning Liquidated Damages

Owners generally view the assessment of liquidated damages as an administrative matter, wherein the owner is only responsible for withholding such damages from the contractor's payments once the work has exceeded the contract completion date. However, this is not a simple issue and is more legal than administrative. The owner should consult with legal counsel before announcing a decision to assess liquidated damages, to determine whether their case for withholding liquidated damages is warranted or valid. This may avoid disputes related to owner versus contractor delays at the end of the project.

Advise Senior Management and Legal Counsel of Potential Breaches of Contract

Breach of contract is another legal (rather than project management) issue. Such allegations will invariably give rise to legal disputes. Senior management and legal counsel should be consulted whenever the project team believes a breach has occurred or the contractor has announced their intent to breach the contract, in order to prevent such disputes. Early intervention by senior management and legal counsel may help avoid a dispute.

Submittals of Short Interval Schedules

Most standardized scheduling specifications require routine schedule updates, either on a monthly basis or when specified contract milestones are achieved. It is therefore recommended that owners work with contractors to obtain their short interval schedules. These are frequently 3-week bar charts showing what was scheduled and completed last week, what is scheduled this week and what needs to be planned for the following week. Owners should ensure these short-term schedules are consistent with the overall project schedule and determine if the contractor is executing to their own plan. Reviewing these documents will provide the owner with a clearer picture of what is happening on the project on a weekly basis, thus facilitating better communication between the project teams.

Enforce Scheduling and Time Extension Requirements

Owners should always enforce the scheduling specifications and adhere to the time extension requirements of the contract. Failure to do so may waive the requirements of the time extension and scheduling specifications to the detriment of the owner. The owner needs to keep in mind that since the contract was created by them, failure to enforce the provisions of the contract may later (during a dispute) be deemed a "waiver of the contract terms through a prior course of dealings".

Establish a Protocol for Delay and Disruption Events

Many disputes arise not as a result of the owner and contractor disagreeing about whether or not there was a delay or impact, but because the two sides cannot agree on how the claim should be documented and proven and the damages calculated. If the contract does not specify a protocol for delay analysis, the owners and contractors should (at the outset of the project) establish how time extensions and impact claims should be developed, documented, submitted and reviewed.

Document and Discuss Conflicts, Errors or Omissions in Contractor Submittals

The owner's team should review any contractor submittal (technical, administrative or otherwise) in a timely and thorough manner and in accordance with the requirements of the contract. Should the owner's team ascertain errors, omissions or conflicts in a submittal, they should document their findings and provide this information to the contractor in detail. Responses such as "Rejected, Resubmit" are virtually useless unless accompanied by a discussion of why the submittal was rejected.

Recognize the Importance of Scheduling

Owners should commit to proper schedule management practices, including:

- Thoroughly reviewing and approving or rejecting all schedule submittals and updates
- Attending all meetings where schedules are discussed, to help resolve schedule challenges and/or issues
- Insisting on strict compliance with schedule specification requirements
- Maintaining hard copy and electronic records of all documents related to schedule submittals
- Keeping accurate records of all time extension requests—when they were submitted, when/how they were responded to, and whether/when a time extension was issued.

It is possible to complete projects and resolve disputes in a timely manner without resorting to arbitration or litigation—with proper prior front-end planning, good quality design, a selection of good contractors and the application of industry best practices focused on a project team and collaborative approach.

ALTERNATIVE DISPUTE RESOLUTION

If a claim situation arises, the issue should be resolved at the project level by negotiation between involved parties. In the event the claim cannot be resolved at the project

> If a claim situation arises, the issue should be resolved at the project level by negotiation between involved parties.

level, it becomes a "dispute" that may require alternative dispute resolution (ADR), arbitration and/or litigation.

ADR can be defined as the various ways to resolve disputes without resorting to traditional, time-consuming and expensive arbitration or litigation. Arbitration is excluded from ADR, in part because:

- It is a costly and lengthy process, diverting resources away from value-added efforts
- It harms and damages the business relationships and reputation of the parties

This section identifies and discusses six nonbinding forms of ADR used in the industry when a dispute arises and the project-level staff cannot resolve the issue by negotiation (Jergeas & Lozon, 2017).

PROJECT NEUTRAL

This method involves appointing an impartial professional to offer unbiased advice and decisions. This is a third party brought on-site to listen to both parties and provide an evaluation of their positions. As a result, the parties learn at an early stage of the dispute just how strong their cases actually are in the eyes of a neutral party. The neutral evaluation may point the way towards a negotiated settlement that both parties can agree upon and which satisfies their needs.

OWNER REVIEW BOARD

In this method, owners—particularly those with larger, longer durations—establish in-house review boards to hear disputes that cannot be resolved at the project level. These boards typically consist of very senior employees (or retired senior employees) of the owner's staff. They are empaneled to review disputed issues in house in an effort to resolve issues such as personality conflicts or misinterpretation of contract requirements.

DISPUTE RESOLUTION BOARDS

A dispute resolution board (DRB) is a panel of three experienced, respected and impartial reviewers, organized before execution begins and meeting periodically at the jobsite. The owner usually selects a member for approval by the contractor, the contractor selects a member for approval by the owner, and these two individuals select the third member to be approved by both parties. The three members then select one of them to serve as chair (with the approval of the owner and contractor).

The DRB reviews the contract documents, becomes familiar with the project procedures and participants and is kept abreast of job progress and developments. They meet with owner and contractor representatives during regular site visits and encourage the resolution of disputes at the job level.

Disputes can be referred to a DRB hearing, where each party explains their position and answers questions. The DRB then considers the relevant contract documents, correspondence, other documentation and particular circumstances of the dispute.

The DRB prepares a written, nonbinding recommendation for resolution of the dispute, which includes evaluation of the facts, contract provisions and rationale for their conclusion.

MEDIATION

Mediation is a voluntary process in which the mediator assists parties in reaching an agreement in a collaborative, consensual and informed manner. In a typical mediation, the parties jointly select a mediator whose role is to assist in reaching a mutually satisfactory resolution of the dispute.

The mediator's objective is to facilitate the parties in reaching the most constructive agreement themselves. The mediator is obligated to work on behalf of each party equally and cannot render individual legal advice to any party. A mediator does not render a decision; instead, they assist the parties in assessing their respective risks and finding areas of compromise.

Any recommendations or statements made by the mediator do not constitute legal advice, and mediation is not a substitute for independent legal advice. The parties are encouraged to secure advice throughout the mediation process but are strongly advised to obtain independent legal review before signing any mediated agreement.

FACT-BASED MEDIATION

In this method, a claims consultant is jointly appointed by the parties to provide "fact-based mediation" services when resolution of the issue(s) requires some analysis before the problem can be addressed in an objective fashion. The two parties engage one entity (a claims consultant) to provide a complete analysis in an objective manner, resulting in a final report presented to both sides.

JUDICIAL DISPUTE RESOLUTION

This is similar to mediation (as discussed above), but the mediator is a judge who may have mediation training and facilitates a private, confidential and informal resolution of a dispute.

QUESTIONS FOR DISCUSSIONS

Question 1

A contracting firm entered into a contract that included the following two clauses:

- ... the contractor *shall not have any claim* for compensation for damages against the owner for any stoppage or delay *from any cause whatsoever.*
- The bidder is required to investigate and satisfy himself of everything and every condition affecting the work to be performed and the labour and material to be provided, and it is mutually agreed that submission of tender shall be conclusive evidence that the bidder has made such an investigation.

When the contractor commenced work, many delays occurred outside the control of the contractor making the project become more difficult and resulting in project delays.

The contractor sued for additional compensation. Is contractor entitled for financial compensation? Why?

Answer:

No. Because the court most likely will enforce the contract and its language. The court may rule that the contractor was paid to do the job and has assumed all the risks. The court may not modify or rewrite a valid contract by reallocating risks. This may sound unfair, but please check the legal precedence in your jurisdiction.

Question 2

A contractor entered into a contract that includes the following clause:

Any claims which the contractor may have against the Owner shall be presented to the Engineer in writing not later than seven (7) days after the occurrence of the delay. Failure by the contractor to present any claim within the seven (7) day period shall be deemed to be an absolute waiver of such claim.

The project experienced delays in receiving the owner's approval which lasted for 15 days, i.e., the delay was not caused by the contractor.

A month later, the project manager of the contractor e-mailed the engineer informing him/her of the delay and requesting $100,000 as a compensation for the damages incurred.

Should the contractor be compensated? Why?

Answer:

No. The law in Canada, for example, requires "Strict Compliance" with Notice Provision requirements. Contractors must read and understand their contract. Please check the legal precedence in your jurisdiction.

Question 3

What are the three ways design professionals protect themselves against claims from parties relying on the design professional's work but who have no contract with the design professional?

Answer:

- **Do a good quality job following best practices and codes and also checked and supervised**
- **Have a disclaimer clause included in their contract with the owner to protect them from a claim by a third party**
- **Get a Professional Liability Insurance.**

Question 4

The plaintiff in a negligence case against a design professional (the defendant) must substantiate three things: Please list them.

Answer:

1. **The defendant (the design professional) owed the plaintiff a duty of care**
2. **The defendant (the design professional) breached that duty of care by his/her conduct**
3. **The defendant's (the design professional) conduct caused the injury to the plaintiff.**

Question 5

Force majeure occurs when one of the parties is unable to perform his or her obligations under the agreement because of:

a. A deliberate action on the part of one of the parties that makes performance impossible.
b. Engineering design failure causing the structure to collapse.
c. An outside event out of the control of the parties.
d. An unforeseen event caused by the carelessness of one of the parties.

Answer: C

Question 6

What is the importance of "Change Orders?

Answer:

Change orders are the legal mechanism to change/amend the original contract. It is also the mechanism to document scope changes and impact on project cost and time.

CONCLUSION

In addition to complying with their legal and contractual obligations, owners of complex projects can employ proactive strategies, taking steps at the front end of a project and during execution to reduce the likelihood of disputes. Owners must include a plan for dispute management in their contracts, mandating a process for promptly addressing unresolved issues before they adversely impact the project. The owner's plan and strategies for avoiding disputes should include the following (Zack, 2014):

1. Choose the right design professionals/consultant and contractor.
2. Select the most suitable contractual and procurement models.

3. Involve a contractor to provide constructability input on the design.

4. Assign risks to the party best able to manage (or share) them, rather than simply trying to offload as many risks as possible.

5. Use partnering to build and sustain project teams and to address (or prevent) difficult technical, communication or relationship issues.

6. Identify disputes early by including reasonable but not impossible, timeframes for providing written notice to the owner.

7. Require the submittal of claims immediately after a disputed event is completed. Impose a deadline for fully documented claim submittals within 30–60 days of the event's completion.

8. Specify what must be included in a properly documented claim, stating that the contractor (as claimant) bears the burden of proving entitlement, causation and damages.

9. Require prompt claim review by mandating a specific timeframe for the owner representative's response to claims submittals.

10. Require that once entitlement is agreed upon, settlement negotiations shall commence within 30 days. The longer it takes to proceed to the negotiation table, the larger the claim is likely to be.

11. Mandate prompt elevation of unresolved issues to management by requiring the completion of project-level negotiations within 45 days of a finding of entitlement—otherwise, the issue will be elevated to senior management.

12. Use a dispute resolution clause that meets the project's needs in terms of timeliness, flexibility and finality—there are creative options beyond arbitration and litigation.

13. Include a DRB in the contract; the DRB hearing and recommendation should be a condition prior to any formal legal action.

14. If a DRB recommendation is rejected, require subsequent mediation. This imposes a cooling-off period and analysis of the recommendation by an experienced neutral party, which should help resolve the dispute.

Disputes on complex projects cannot be avoided, but arbitration and legal proceedings can be—with a little diligence, creativity and sound judgement. Negotiation is by far the best way to resolve a dispute; it is cost effective, and importantly, it gives parties control over the

> Disputes on complex projects cannot be avoided but arbitration and legal proceedings can be.

outcome. Prior to bidding, owners need to plan the dispute management process and embody such decisions in their contracts. Proper planning and rational decision-making concerning the disputes process should result in a professional resolution of issues by the project stakeholders and not by others.

ACKNOWLEDGMENTS

We acknowledge the contributions of several guest lecturers for the *ENCI 699: Law for Project Managers* course at the Schulich School of Engineering (University of Calgary, Alberta, Canada), including R. Eden, D. Goodfellow, J. McCartney, P. Matthews, A. McWilliam, S. Revay, R. Simpson and Dr. Mohammad Moazzami.

REFERENCES

Gillan, W.R. (2004). Facilitating the construction dispute process. *The Revay Report*, 23(1), 1–4.

Hudon, J. (2015). Quantification of construction claims. *The Revay Report*, 32(1), 1–8.

Jergeas, G. & Lozon, J. (2017). Benevolent dictatorship for major capital projects, Learnacademy.

Jergeas, G. & Lynch, R.P. (2015). *The Collaborative Construction Model: Report to Go Productivity.* Edmonton, AB: Go Productivity. AB: LEARN academy™.

Moazzami, M. (2015). A theoretical framework for implementing convertible contract in oil and gas projects. *PhD Thesis*, University of Calgary, Alberta, Canada.

Revay, S.G. (2004). Managing contract changes. *The Revay Report*, 23(3), 1–5.

Rolstadas, A., Hetland, P., Jergeas, G., & Westney, R. (2011). *Risk Navigation Strategies for Major Capital Projects: Beyond the Myth of Predictability*. Springer.

Zack, J. (2013). *Delivering Dispute Free Projects: Part 1 – Planning, Design & Bidding*. Boulder, CO: Navigant Construction Forum™.

Zack, J. (2014). *Delivering Dispute Free Projects: Part II – Construction & Claims Management*. Boulder, CO: Navigant Construction Forum™.

Zack, J. (2015). *A Crystal Ball – Early Warning Signs of Construction Claims & Disputes*. Boulder, CO: Navigant Construction Forum™.

Zack, J. (2016). *Delivering Dispute Free Projects – Does Partnering Help?* Boulder, CO: Navigant Construction Forum™.

14 Intellectual Property Toolset

Alexey Bakman

CONTENTS

BACKGROUND

The World Intellectual Property Organization[1] (WIPO) defines Intellectual Property (IP) as referring to "creations of the mind, such as inventions; literary and artistic works; designs and symbols; names and images used in commerce."[2] The World Trade Organization (WTO) refers to IP rights as "the rights given to persons over the creations of their minds."[3]

We are all intuitively familiar with a concept of personal property. A hat on our head, a wallet in our pocket, and a computer on our desk are all examples of personal property. We are also intuitively familiar with a concept of real property. A house we live in, an office we work from, and a farm down the road are all examples of real property. However, Intellectual Property is different. It concerns the often-abstract concept of "creations of the mind."

Can creations of the mind be property? Can an expression or implementation of a thought be property?

From the earliest age, we think. We come up with ideas, plans, and fantasies. We share them with others around us and ask for feedback. We learn from the wisdom of our parents, friends, and teachers. We absorb the knowledge from books, films, and the Internet. René Descartes, a French philosopher, famously stated: "I think, therefore I am." To think, to generate, to share and express, to give and take ideas is utterly human and natural.

When a project involves use of real property, such as office space, it is natural to consider the implications of such use. Do we rent or buy the real estate? What is the location and availability of such property? What are the risks and inherent costs? Similarly, when personal property such as office equipment or tools is involved, it is natural and wise for a project manager to consider the costs and availability of such resources from the start of the project. Yet, for most people, it is unnatural to visualize the abstract concept of Intellectual Property. It is unintuitive to consider the risks, costs, and burdens, associated with managing IP aspects of a new project. Intellectual property is not as obvious as a computer on a desk or a farm down the road. IP is an invisible and intangible aspect that is easy to overlook and ignore at early stages of an undertaking. It is almost as if it is not present if we do not think about it. Yet, it very much is present at every stage of a project. Every complex project generates and/or uses some kind of IP.

While it is simple to overlook the significance of Intellectual Property at initial stages of a project, it is hard to underestimate its importance in any sizeable project and in the economy in general. IP is a cornerstone of the global economy. For example, an estimated 80% of the value of US corporations lies in their IP portfolios.[4] Ignorance of IP issues can ensure failure of a project before it ever begins. Furthermore, IP liabilities can doom a project even after the project appears to be successfully completed.

Thus, it is critical that a strong, conscious, and deliberate effort is made from the very start of the project to spot and resolve all the IP issues inherent in its implementation. It is also critical to monitor the developing project for new IP issues, as they come up. At the start of every new project affirmatively ask: "What Types of Intellectual Property Does the Project Involve?" The answer may not be instantly evident, or simple, but the question will always put you on the right path of analysis and action.

This chapter arms project managers with powerful practical tools but is not intended to be used as legal advice or guidance. Like other laws, IP law varies among jurisdictions. This chapter covers some international aspects of IP law and provides tools, relevant to international IP protection. However, many of the concepts, laws, and examples presented here are based on US legal interpretations. While the laws and general concepts, governing your particular project's IP, are likely to have some similarity to those presented in this chapter, they may differ in significant details. IP laws are also in constant flux. By the time you read the contents of this textbook, some aspects of the laws may change.

It is impossible to cover all or even most of the intricacies of Intellectual Property law within constraints of a single chapter. Nor is this chapter attempting to do so.

> At the start of a new project, make a conscious effort to ask: "what types of IP does the project involve?"

IP law is a highly complex, extensive, and deeply-specialized legal field. No chapter or textbook can ever replace a consultation with a certified IP legal practitioner in your jurisdiction. The facts of each particular legal situation are unique. Proper application of law to facts is best left to highly trained legal professionals.

Instead of diving deeply into the intricacies of the law, this chapter aims to present a selection of highly practical tools and legal concepts that will be immediately useful to a project manager. It will provide managers with the tools to spot and identify the main types of Intellectual Property, present in most complex projects. Other tools will assist the reader in avoiding the most common project management IP pitfalls. Most importantly, this chapter will help you get an understanding and a feel for when and why seeking professional IP advice becomes a critically important step in project's success.

> IP law is a highly complex, specialized legal field. Seek legal advice in order to spot potential IP issues. Seek legal advice once potential IP issues are spotted.

Types of Intellectual Property

The Convention Establishing the World Intellectual Property Organization (WIPO), concluded in Stockholm on July 14, 1967 (Article 2(viii)), provides that "intellectual property shall include rights relating to:

- literary, artistic and scientific works,
- performances of performing artists, phonograms and broadcasts,
- inventions in all fields of human endeavor,
- scientific discoveries,
- industrial designs,
- trademarks, service marks and commercial names and designations,
- protection against unfair competition,

and all other rights resulting from intellectual activity in the industrial, scientific, literary or artistic fields."[5]

National Nature of IP Laws

The list above is generalized and broad. The reason for this is that WIPO is an international agency, established under the auspices of the United Nations. However, most of the Intellectual Property law is national, not international. The list above is a general recitation of the fields that are likely to be subject to IP regulations in most countries. However, a detailed list of fields of IP protection can differ considerably from country to country.

Historically, laws, concepts, and legal precedents developed independently in each jurisdiction. To this day, national IP laws vary significantly from each other. Some terms and concepts used in IP law are often unique to a particular jurisdiction. Where the same terms are used, legal interpretations of these terms may be different. Therefore, a project manager must remember that legal advice that is appropriate in one jurisdiction may not be relevant for another.

> Project management must obtain local IP legal advice in every jurisdiction that may be relevant to complex project's scope.

Four Types of IP Protection

Despite being broad, the list above is a good starting point for a project manager, attempting to gauge the extent of IP issues, involved in a project. Does your project use, generate, or in some way concern any of the above-listed concepts/fields? If you consider all aspects of a complex project, chances are that the answer is "Yes." If so, you just identified a source of a possible IP issue. The next logical step would be to determine what branch of IP the issue concerns.

Generally, IP is divided into four distinct branches: Trademarks, Copyrights, Patents and Trade Secrets.

It is far too common for an IP practitioner to hear that a client wants to "patent a trademark" or to "copyright an invention." Such phrases are dangerous misnomers, especially for someone managing a large complex project. Each of the branches of IP law generally deals with different subject matter. Each branch has its own sources of law and varying (and sometimes even contradictory) public policy aspects.

Different branches of IP law are so distinct from one another in nature that in some countries, separate government agencies are assigned to oversee each branch. For example, in the United States, Copyright Office oversees all aspects related to copyright law, while the United States Patent and Trademark Office oversees patents and trademarks. Many IP attorneys specialize in one particular branch of IP law but not in others. In fact, attorneys, practicing in certain areas, such as a patent law, may require numerous additional qualifications, not required of attorneys practicing in other fields of IP law. Therefore, even in order to obtain correct legal information or qualified professional assistance from a government agency or an attorney, a project manager must know the differences between the four branches of IP.

> The four distinct branches of IP law are Trademarks, Copyrights, Patents, and Trade Secrets. A project manager must know the differences between each of the four branches in order to spot potential IP issues, obtain proper legal information, and find properly qualified legal assistance.

Trademarks

A trademark is commonly a word, phrase, symbol, and/or design that identifies and distinguishes the source of the goods of one party from those of others.[6]

McDonald's® , Coke®, Apple®, Google®, Marlboro® and Rolex® are some examples of world-famous trademarks that we immediately associate with a particular source. Similarly, images of golden arches over a fast-food establishment, a bitten apple on the back of an electronic device, or a crown in connection with a wristwatch all help us make an instant connection of products with particular companies.

When a mark is associated with a service, rather than with a product, it may be referred to as a "service mark," rather than a trademark. However, the term "trademark" is often used in a general sense to refer to both trademarks and service marks.[7]

Trademarks can be tremendously valuable. For example, as of 2018, the values of Amazon®, Apple®, and Google® trademarks all well exceeded $100,000,000,000 US each.[8] Companies often invest great effort and financing into developing and maintaining their brands/trademarks. Customers learn to associate and value certain characteristics of a product with particular trademarks. For this reason, many diners prefer to order Pepsi® or Coke®, rather than an unknown-brand cola on the menu. The diners' choice is guided by knowledge of the consistent quality and taste characteristics of the brand-name colas. Some marks are preferred, based on customers' prior experiences of the source product, others for the unique characteristics of the products, yet others for the spirit of fame or prestige that certain brands exude, or a range of other characteristics, associated with a particular source of goods or services.

The law of Trademarks is primarily intended to protect consumers in the marketplace. When a consumer picks up that can of Coke® or a Rolex®-brand watch, the trademarks on the products and packaging immediately suggest that these products come from particular companies and possess certain consistent quality characteristics. Trademark law ensures that consumers are not

> When making trademark-related decisions, project manager must keep in mind that primary policy behind trademark law is consumer protection.

mislead by confusingly similar or infringing marks into mistakenly buying products or services, which they did not intend to buy.

The law of Trademarks is also intended to prevent unfair competition. Much of the trademark's value is in its distinctiveness and goodwill that consumers associate with the mark. It would be unfair to the trademark holder and tremendously disruptive to the market as a whole if new players could come into the marketplace and start using established trademarks on their products. Not only would the newcomer be misleading consumers as to the source of the product, but they would also be hijacking years of investment and goodwill-building that went into establishing the reputation of the mark. In a sense, the newcomers would be piggybacking on someone else's success. In the process, they would be destroying the original mark by "diluting" its distinctiveness and strength. If the newcomer's products are inferior, they would also be "tarnishing" the reputation of the original mark. Trademark laws ensure that both consumers and trademark holders are protected in the marketplace.

> Trademark law attempts to ensure that both consumers and trademark holders are protected in the marketplace.

Most complex projects, especially projects of commercial nature, involve at least some trademark issues. Think, does your project have a name? Will your project provide any goods or services that will be represented by or associated with names or designs? Will your project or any of its components be advertised or presented, using any particular symbols or catchy phrases? If so, will these names, designs, symbols, or phrases infringe on existing trademarks? Are they protectable as part of your intellectual property portfolio? Does your project involve cooperation with any third parties that use their own trademarks? How will their trademarked products or services be represented in your project? Will you be copying any aspects of another party's product or service? Will any of your project's products or services be so similar to something that already exists on the market that the source of these products and services may be confusing to consumers? These are just some sample questions that may come up in analyzing the trademark issues.

In evaluating the trademark-related risks, a project manager must remember that trademarks are not necessarily limited to words, phrases, symbols, and/or designs. The latest trends in the Trademark law have been to expand protection to other characteristics that can identify and distinguish the source of the goods and services. A look and feel of a product's packaging or a uniquely decorated service location may be protectable as a "trade dress." Furthermore, the United States Patent and Trademark Office has granted trademark protection to such characteristics as distinctive scents, sounds, and colors. Thus, a special scent in Verizon Wireless stores is protected, as are the colors Tiffany Blue, T-Mobile Magenta, and UPS Brown. The sound of the New York's Stock Exchange Bell, Tarzan's yell, Darth Vader's Breathing, and even Homer Simpson's "D'OH" are all registered sound trademarks. Thus, in a complex project, trademark issues (and opportunities) may arise from the most unexpected of places.

Patents

A patent is a state-granted property right in a useful invention. This right gives inventors or other patent holders a time-limited monopoly in the invention and allows patent holders the power to exclude others from commercially exploiting or otherwise using the invention.

Like other IP laws, patent law is highly jurisdiction-specific. However, World Intellectual Property Organization (WIPO) provides the best broad definition of a patent, suitable for international audiences:

> A patent is a document, issued, upon application, by a government office (or a regional office acting for several countries), which describes an invention and creates a legal situation in which the patented invention can normally only be exploited (manufactured, used, sold, imported) with the authorization of the owner of the patent. "Invention" means a solution to a specific problem in the field of technology. An invention may relate to a product or a process. The protection conferred by the patent is limited in time (generally 20 years).[9]

The public policy behind patent law rests on the fact that society as a whole benefits from inventive activities. We want to encourage inventors to come up with new methods and machines and to improve existing ones. We want to encourage companies to invest in innovation, while knowing that their investments will be protected from freeloaders and will provide legally defensible advantage over the competition. As society, we also want to discourage secret use of new technologies. Instead, it benefits society when new technologies are exposed, published, and shared, so that they may be analyzed and improved upon by others.

A patent is a social tool that encourages innovation and disclosure of inventions to society. In a sense, patent can be seen as a type of contract between inventor and society. In order to obtain a patent, an inventor must disclose the invention in great detail, so that others, skilled in the same art (field), would be able to understand, implement, and practice the invention.

> In order to obtain a patent, an inventor must disclose the invention in great detail, so that others would be able to understand, implement, and practice the invention.

In exchange, government/society grants the inventor a time-limited monopoly to openly control the use of technology in the marketplace. Such monopoly provides a patent holder with a marketplace advantage, allowing for the recuperation of costs and increased profits.

In the modern world, patents generally protect a broad scope of inventions. For example, patentable subject matter is defined under US law as follows: "Whoever invents or discovers any new and useful process, machine, manufacture, or composition of matter, or any new and useful improvement thereof may obtain a patent therefor, subject to the conditions and requirements of this title."[10] In the United States, this definition has been broadly interpreted to include inventions related to designs of useful articles, computer software, and even methods of doing business.

Therefore, it is critically important for a project manager to evaluate all aspects of a project for implementation of new technologies. Will the project develop any new product or process as its main purpose or as a by-product? Will the project produce any new method involving computer software or a novel business method? Are there any novel approaches in the project? If so, can they be patentable?

> At project's planning stage, and periodically thereafter, evaluate all aspects of a project for implementation of any new technologies that may be protectable by patents.

The answers to the latter questions will likely require professional evaluation. However, it is critical to know the answers at the earliest stages of the project. If new technology is to be protected with patents, applications must be filed as soon as practicable. Some jurisdictions provide limited time for a patent application to be filed once the invention is made. If public disclosure, use or sale of technology is made prior to filing for a patent, in most jurisdictions, all rights to patentability will be lost.[11] Thus, a project manager must remember that rights to patent an invention must be claimed as soon as possible or be lost forever.

> Patent rights to an invention must be claimed as soon as possible or be lost forever.

Furthermore, as important as it is to timely claim the rights to a patent, it is even more important to ensure that patent rights of others are not being infringed by your project. Think of all aspects of technology in use throughout your project. Did you legally acquire the rights to use all the technology? Does your product involve development of new technology? If so, is your new technology in any way based on or derived from technologies that may be patented and owned by someone else? If it is, you may be in violation of third-party patent holder's rights. All-out effort must be made by a project manager early on in the

> All-out effort must be made early in the project to ensure that ownership rights or licenses for use of all patented technology have been obtained.

project to ensure that ownership rights or licenses (permissions) for use of all patented technology have been obtained.

If this risk has not been cleared prior to the project's start, the undertaking may be exposed to legal liabilities that may be fatal to the entire project.

Copyright

Copyright is a legal term used to describe the rights that creators have over their literary and artistic works. In most jurisdictions, copyright law covers a broad range of works and usually includes:

- literary works such as novels, poems, plays, reference works, and newspaper articles;
- computer programs, and databases;
- films, musical compositions, and choreography;
- artistic works such as paintings, drawings, photographs, and sculpture;
- architecture;
- advertisements, maps, and technical drawings.[12]

A number of other modes of expression may be subject to copyright protection in each jurisdiction. It is hard to imagine a complex project that does not generate or utilize at least some modes of creative expression, thus becoming a subject to Copyright laws.

Just like the patent law, copyright law grants limited monopoly to authors over their works in order to promote and reward creativity. However, in the case of copyrights, protection is granted to authors for a time span that is significantly longer than the typical 20-year period of patent protection. The span of copyright protection typically extends for the entire life of the author, plus several extra decades. The US law provides for copyright term of life of author plus 70 years.[13] Unlike the relatively short lifespan of a patent, which is not extendable, copyrights may be extendable under certain circumstances.

In contrast to patent law, copyright law DOES NOT PROTECT IDEAS. This is a critical distinction. Copyright protection extends only to EXPRESSION of ideas.

In US law, this critical distinction between protection of an idea and protection of an expression of an idea was first made in the 1879 Supreme Court case of

> When deciding on IP protection mode, remember: COPYRIGHT LAW DOES NOT PROTECT IDEAS. Copyright law protects expression. Patents protect ideas embodied in inventions.

Baker v. Selden.[14] The case provides an example of this somewhat subtle, but extremely important distinction that defines the nature of copyright protection and distinguishes it from patent law. In that case, Selden copyrighted a book, describing a new system of book-keeping. Several years later, Baker published and successfully marketed a book, describing a similar system. Selden's widow filed suit for copyright infringement. Baker argued that the book-keeping system itself was not subject to copyright protection. The Supreme Court agreed with Baker, stating that "[W]hilst no one has a right to print or publish his [Selden's] book […], any person may practice and use the art itself which he has described and illustrated therein." The Court went on to say "the mere copyright of Selden's book did not confer upon him the exclusive right to make and use account-books."

The description and the expression in Selden's book were found to be protected by copyright. That is, Baker would not have a right to literally copy the content of Selden's book. But Selden did not have the right to control the idea ("useful art") by virtue of merely copyrighting the book. Protection of ideas is the domain of patent law. In US law, this principle is now codified in 17 U.S.C. § 102(b), stating: "In no case does copyright protection for an original work of authorship extend to any idea, procedure, process, system, method of operation, concept, principle, or discovery, regardless of the form in which it is described, explained, illustrated, or embodied in such work."

This important principle has been adopted in most jurisdictions and must be known and used by the project manager in considering the types of protection relevant for a project's intellectual property. Copyright protection is generally cheaper, faster and easier to obtain than patent protection. It also lasts much longer. For this reason, it is tempting to describe an idea or an invention on a piece of paper and to receive a copyright registration. However, a project manager must remember that such registration will not protect an underlying idea.

> Copyright protection is generally cheaper, faster, and easier to obtain than patent protection. But do not count on it when ideas must be protected.

The creativity protected by copyright law is creativity in the choice and arrangement of words, musical notes, colors, shapes, and so on.[15] So, if the project involves works, involving a degree of creativity, copyright protection may be a proper and effective way to safeguard project's IP rights.

> Copyright protects creativity in the choice and arrangement of words, musical notes, colors, shapes, etc. If a project creates works, involving a degree of creativity, copyright protection may be proper. If projects use such works, produced by others, the possibility of infringement must be considered.

Law grants copyright owners certain exclusive rights. These rights are of two types: economic rights and moral rights.

Moral rights give author the right to claim authorship of the work and the right to object to any modifications, such as mutilation of the work or other actions that may be prejudicial to author's reputation and honor.

Economic rights safeguard the economic value of the work. In the United States, copyright law grants author the economic rights to do and authorize any of the following:

1. to reproduce the copyrighted work in copies or phonorecords;
2. to prepare derivative works based upon the copyrighted work;
3. to distribute copies or phonorecords of the copyrighted work to the public by sale or other transfer of ownership, or by rental, lease, or lending;
4. in the case of literary, musical, dramatic, and choreographic works, pantomimes, and motion pictures and other audiovisual works, to perform the copyrighted work publicly;
5. in the case of literary, musical, dramatic, and choreographic works, pantomimes, and pictorial, graphic, or sculptural works, including the individual images of a motion picture or other audiovisual work, to display the copyrighted work publicly; and
6. in the case of sound recordings, to perform the copyrighted work publicly by means of a digital audio transmission.[16]

Does your project utilize the creative works of others? Does any stage of the project involve one or more of the six actions above that fall under the author's exclusive rights? If so, ensure that either the source works are in the public domain (freely available for use) or that licenses have been obtained from the author or a licensing agency.

> A project may obtain licenses for the use of copyrighted works from the author or a copyright licensing agency.

Trade Secrets

The World Intellectual Property Organization generally defines "Trade Secrets" as any confidential business information that provides an enterprise a competitive edge.[17] Trade secrets encompass manufacturing or industrial secrets and commercial secrets. The unauthorized use of such information by persons other than the holder may be regarded as an unfair practice and a violation of the

trade secret. Article 39.2 of The WTO Agreement on
Trade-Related Aspects of Intellectual Property Rights
(TRIPS) more concretely defines a trade secret as
information that (1) is secret, (2) has commercial value
because it is secret, and (3) has been subject to reasonable steps to keep it secret.[18]

> Project's information that is secret,
> has commercial value because it
> is secret, and has been subject to
> reasonable steps to keep it secret
> may qualify for protection as a
> trade secret.

Depending on the legal system, the protection of trade
secrets forms part of the general concept of protection
against unfair competition or is based on specific provisions or case law on the protection of confidential information.[19] Since TRIPS Agreement came into effect in 1995, its definition of trade secrets
(above) has been widely adopted into national laws.[20]

Every complex project is likely to harbor at least some trade secrets. Some projects are secret by
their very nature. For example, the development of new products, military, law enforcement, surveillance, disaster preparedness, and security projects
are among the types of projects that may have to be kept
secret in their entirety. In these cases, filing a disclosing
patent to protect technologies is not an option. In fact,
any disclosure of information may be lethal to the success of such a project. In such cases, trade secret laws
may provide the most effective, and sometimes the only,
protection a project manager can count on.

> In cases where a project is secret
> by its very nature, trade secret laws
> may provide the most effective, and
> sometimes the only, protection a
> project manager can count on.

Almost any other complex project is also likely to comprise at least some information that
provides value for the project and must be kept secret. Some examples of such information may
include a list of customers and supply sources, processes of manufacture, lists of ingredients,
expansion and advertisement strategies, consumer and employee profiles, methods of sale and
distribution, etc. According to the US Patent and Trademark Office's Trade Secret Policy, trade
secrets can include a formula, pattern, compilation, program, device, method, technique, or
process.[21]

Whether a particular kind of information constitutes a trade secret is highly fact and is
jurisdiction-dependent. However, often Trade Secret law protects such information from industrial
or commercial sabotage and breaches of confidence and contract.

The WTO Agreement on TRIPS was the first multilateral agreement to explicitly require member
countries to provide protection for "undisclosed information,"[22] that is, trade secrets. The agreement
was made between all the member nations of the World Trade Organization (WTO) and provides
the minimum standards for protection of trade secrets (and IP in general).

In the United States, trade secret protection has been strengthened, following TRIPS. US Patent
and Trademark Office's Trade Secret Policy states:

> As a member of the World Trade Organization (WTO) and a party to the Agreement on Trade
> Related Aspects of Intellectual-Property Rights (TRIPS), the United States is obligated to pro-
> vide trade secret protection. Article 39 paragraph 2 requires member nations to provide a means
> for protecting information that is secret, commercially valuable because it is secret, and subject
> to reasonable steps to keep it secret. The Defend Trade Secrets Act of 2016 created federal civil
> cause of action, strengthening U.S. trade secret protection, with a choice for the parties between
> localized disputes under state laws or disputes under federal law, heard in federal courts. While
> state laws differ, there is similarity among the laws because almost all states have adopted some
> form of the Uniform Trade Secrets Act.[23]

TRIPS definition, above, requires that in order to be protectable as a trade secret, the information
must be "subject to reasonable steps to keep it secret." Accordingly, in most countries/jurisdictions,
affirmative steps must be taken in handling and safekeeping of information in order to keep it protectable under the law. A project manager must research local laws and precedent and consult an

attorney to be aware of these requirements in his/her jurisdiction. Generally, if secrecy/confidentiality of trade secrets is breached at any stage of the project due to a secret holder's action, the protection will be lost forever. Thus, it is extremely important to take these precautionary steps from the very start of the project and to follow up with them throughout the project's existence.

DEVELOPMENT OF BASIC PRINCIPLES AND TOOLS OF IP PROTECTION

DEVELOPMENT OF MODERN IP LEGAL STRUCTURES

Records of IP laws go back to ancient times. As early as 500 BCE, the Greek state of Sybaris was granting its citizens 1-year patents for "any refinement in luxury." Craft objects from Ancient Egypt and Ancient China show unique images or symbols, imprinted or stamped by the craftsmen to identify the creator of the object. In medieval Europe, as merchant guilds grew, many used unique stamps to mark their products in order to indicate the source and assure buyer of the quality. Toward the high Middle Ages, some laws were passed in Europe, requiring silversmiths, bakers and printers to mark their products with distinctive marks. These can be seen as some of the first trademark laws.

Throughout history, states offered social and economic incentives to attract skilled artisans and new industries. These incentives ranged from free housing, debt cancellation and legal immunities, all the way to absolution from sin.[24] In 1474, Venice passed the first general patent law to regulate industrial brevets. This law codified the benefits, traditionally offered to inventors in Venice.[25]

In the English-speaking world, the roots of modern IP legislation can be traced to the Statute of Monopolies, passed by the Parliament of England in 1624, and the Statute of Anne, passed by the Parliament of Great Britain in 1710. These statutes concerned patents and copyrights, respectively, and were some of the first legislations that moved the power over IP rights from private regulators (such as guilds) to state and the courts.

As industrial revolution and capitalism picked up steam, requiring the encouragement of innovation and protection of capital investment, IP laws were gradually adopted into most national laws. These laws, for the most part, were strictly national affairs, offering little protection to foreigners. In fact, some early statutes appear to have been drafted with affirmative intent of excluding protection to foreigners.[26] Such exclusions may have promoted national interests[27] by allowing the "borrowing" of foreign patents and trade secrets.

For example, in the period between the late 18th and early 19th centuries, US patent law had strict nationality and residency restrictions. Foreigners could not apply for a patent or protect their inventions in the United States.

These restrictions may have allowed for carrying out one of the most ambitious industrial "complex projects" of the time. At the time, the British textile industry was one of the most advanced and lucrative industries in the world. Exportation of machinery or trained personnel was a criminal offense. Its patents and trade secrets were strictly protected under the British law (but not under the US law). In 1789, a trained cotton mill specialist came from Britain. This man was later referred to by Andrew Jackson as the "Father of American Manufactures." The knowledge and expertise that he brought from abroad, along with the lack of enforcement of foreign IP laws in the United States, helped him establish America's first automated textile mill. By 1815, 140 mills were operating 130,000 spindles within a 30-mile radius.[28]

At the same time, Britain itself, as well as other European nations, encouraged the "importation" of foreign technologies by granting "patents of importation." The applicant for such a patent did not have to be an actual inventor, as long as the invention was new in the country.[29]

By the late 19th century, as modes of transportation and communication improved and as global trade and manufacturing projects expanded across the world, the narrow national scope of IP laws was putting increasing burdens on the industries. Inventions, books, musical notes, designs, and brands could be copied and distributed across the world within days of creation with no rights or royalties being due to the original creators. The extent of the problem became most evident at the

International Exhibition of Inventions in Vienna, Austria, in 1873. Numerous foreign exhibitors refused to attend, fearing that their inventions would be stolen and exploited by foreign competitors.[30]

Soon thereafter, in 1883, the first major international IP law agreement was adopted. This agreement is known as **Paris Convention for the Protection of Industrial Property**. To this day, the convention is still in force and remains one of the bedrocks of modern International IP protection. The vast majority of world's countries are signatories to the convention. As of 2017, contracting members of the convention included 177 countries.[31]

Paris Convention is currently administered by WIPO. In its description of the Convention, WIPO states that it "applies to industrial property in the widest sense, including patents, trademarks, industrial designs, utility models, service marks, trade names, geographical indications and the repression of unfair competition."[32] For purposes of complex project management, the Paris Convention is most relevant in providing the guidelines for seeking international patent protection. Some of the most powerful tools, presented later in this chapter, stem from the Paris Convention.

The terms of the convention provide applicants with the right of national treatment. That is, each signatory state must grant the same rights to nationals of other contracting states that it grants to its own citizens, thus allowing for patent protection of project's IP in several foreign countries. Among other rights, most relevant to project managers, is the right of priority. The right of priority is the ability to secure a priority filing date of the patent in one country, which will then be honored by other member countries.

> Among the rights, provided under the Paris Convention, the right most relevant to project managers is the right of priority. It allows the project to secure priority filing dates of its patents in one country, which will then be honored by other member-countries.

In 1978, a **Patent Cooperation Treaty** (PCT) was adopted, allowing simplified tools and procedures for international patent filing. Relevant tools and details of filing international patent applications are discussed later in this chapter.[33]

Paris Convention was soon followed by the **Berne Convention for the Protection of Literary and Artistic Works** of 1886 (Berne Convention). Berne Convention is also still in force and forms another bedrock of International Protection, relating to Copyright law. As of 2018, contracting members of the convention included 176 countries. More information on Berne Convention can be found at www.wipo.int/treaties/en/ip/berne/summary_berne.html.

The treaty establishes the basic principles that must be followed by all contracting countries. The Berne Convention's principles are listed by WIPO as follows:

a. *Works originating in one of the Contracting States (that is, works the author of which is a national of such a State or works first published in such a State) must be given the same protection in each of the other Contracting States as the latter grants to the works of its own nationals (principle of* "national treatment").
b. *Protection must not be conditional upon compliance with any formality (principle of* "automatic" *protection).*
c. *Protection is independent of the existence of protection in the country of origin of the work (principle of* "independence" *of protection). If, however, a Contracting State provides for a longer term of protection than the minimum prescribed by the Convention and the work ceases to be protected in the country of origin, protection may be denied once protection in the country of origin ceases.*

The principles of "national treatment" and "independence of protection" are important in enabling international protection of copyrights. The principle of "automatic" protection is critically important in international as well as national laws of member countries and must be understood and remembered by a project manager. This principle ensures that most copyrightable works are subject to copyright protection at the moment of creation, without the necessity of registration process.

This principle also creates a situation where almost any book/document/record that is used in a project (even documents that are decades old) is subject to copyright protection and may expose the project to legal liability. The principle of "automatic" protection is discussed in more detail later in this chapter.

In 1891, **Madrid Agreement Concerning the International Registration of Marks** was signed. This agreement, followed by the Protocol Relating to the Madrid Agreement (1989), became the basis of international trademark protection. Madrid protocol is also currently administered by WIPO and has 103 contracting parties.

> Berne Convention's principle of "automatic protection" creates a situation where almost any book/document/record that is used in a project (even document that is decades old and unregistered with Copyright Office), may be subject to copyright protection and may expose project to legal liability.

In 1893, two secretariats were formed to administer Paris and Berne Conventions. These secretariats were later merged into United International Bureau for the Protection of Intellectual Property and later became World Intellectual Property Organization (WIPO).[34] WIPO is currently headquartered in Geneva, Switzerland, and oversees international aspects of IP.

The numerous international treaties, mentioned above, have opened the doors for large, complex transnational projects. Project managers of such undertakings now have the critically important options and tools of protecting the project's IP internationally. However, project managers must remember that

> Despite the option and ability to seek IP protection internationally, the laws governing IP protection are still primarily national laws.

despite the option and ability to seek IP protection internationally, the laws governing IP protection are primarily national laws. A project manager may use the tools provided by the Paris Convention or a Madrid Protocol to file a patent or a trademark application, seeking protection in dozens of countries. However, the final determination on whether the protection is granted and the extent of such protection will be made in each country on a national level, referencing its own laws.

Even in cases where international treaties require the countries to provide certain specific rights, the definitions and scope of such rights will be interpreted by local IP offices and judges and may significantly vary from country to country. Where separate aspects of a project may be carried out in different countries, it is entirely possible that the same actions may subject the entire project to liability in one country, but not in another. Thus, in cases where the scope of a complex project spans several countries, a project manager must obtain legal advice from local IP specialists in each jurisdiction.

> Where separate aspects of a project may be carried out in different countries, it is entirely possible that the same IP-related actions may subject the entire project to liability in one country, but not in another.

THE QUESTIONS OF LIABILITY—RELEVANCE OF INTELLECTUAL PROPERTY TO COMPLEX PROJECT MANAGEMENT

Every complex project involves IP. Generally, the more complex the project is, the more layers of IP issues are lurking below its surface.

> The more complex the project is, the more layers of IP issues exist.

Like sharks under the waves of the sea, these issues are not always evident, when present. Only a vigilant and experienced project manager will spot an occasional shadow or a fin. They are illusive and easy to ignore, while one gets their feet wet with a new project. But once the project hits deeper waters, they surface, attack, and it is too late to run for shore.

Ignorance of IP issues can easily doom an entire project before its practical realization ever begins by making it illegal and financially unfeasible.

Far too often, a legal practitioner sees cases, where tremendous time, labor, and finances have already been invested into a project only to discover that it has been legally unfeasible from the start. Much worse, sometimes lethal IP issues surface once the project appears to have been successfully completed.

Just as the beneficiaries of the project are getting ready to enjoy the fruits of their labor/investments, a third party may appear, claiming all the rights in the project and all its future profits by virtue of a patent, trademark, or some other IP right. If IP due diligence has not been carried out from the start, if risks have not been monitored and cleared, that third party may well have that right and enforce it in court. The court may even award them punitive damages. While this scenario appears as horror fiction, it is far too real and is encountered by IP attorneys on regular basis. IP issues may appear illusory to the uninitiated. However, IP laws are anything, but. Their bite is real and the jaws are wide.

> Ignorance of IP issues can easily doom an entire project before its practical realization ever begins by making it illegal and financially unfeasible.

> Lethal IP issues may surface once the project appears to have been successfully completed.

Let's assume that a project manager is tasked with administering a project that develops and implements a product (or a service, process, method, campaign, etc.) for a company. At the start of such projects, it is tempting to concentrate all effort and resources on practical ways to develop and implement the product. After all, the product is new, the market is right, the team is skilled, and equipment is in place. What could go wrong? From the IP perspective, everything.

For example, does your product or project have a name? Does it utilize distinctive logos or slogans? Will the product have distinctive packaging or look and feel to it? If so, how did you come up with this name, logo, look, etc.? Did you search the trademark registers in all jurisdictions of potential use to ensure that the same elements have not been registered as a trademark by someone else? If there is no such registration, did you investigate the marketplace to see whether similar marks are in use without registration? In some countries, such as the United States, a trademark must not necessarily be registered and may simply be used in commerce to have legal protection. Therefore, you may be infringing someone else's mark even if their mark is not registered.

If your in-depth search did not locate the same name, logo, etc., in registers or the marketplace, there are still many more questions to ask. For example, is anyone using a mark that may be confusingly similar to what you are trying to use? Will the use of the mark open up the way to accusations of unfair competition? In light of similarities with other marks, will your mark appear confusingly similar to consumers? Will your mark or its elements mislead consumers in any of the countries where the mark will be used? Can the mark be considered offensive? Can the mark be registered in every jurisdiction where it will be used, so that it can be protectable against others?

Does your project involve any literary works (including fiction and non-fiction, reference works, articles, etc.), computer programs, databases, audiovisual works, music, performance, artistic and architectural works of any type or any other matter that falls under the scope of copyright protection? Will you be using any creative works of authorship in the project? For example, will you be using pictures or designs developed by someone else as an inspiration for elements of your product? Will you be using someone else's music and images in the product's advertisement, etc.? If so, are these works copyrighted? Have you obtained licenses or other rights to use them? If your project is producing works of authorship, are these works being properly protected under copyright law? Who is the beneficiary of these rights?

If project is focused on a production of a new product, is the product truly new? Perhaps the product exists or has existed in other markets or was described somewhere. A single publication or description of your product may forever prevent the project from obtaining protection for it. Even if the product or method that you are developing is new, have you researched whether someone else has patented (i.e., invented and claimed their rights in) any of the underlying technologies or approaches, no matter how small? If they did, you may be legally prohibited from selling or using your product, unless a license is purchased from the rights' holder.

IP licensing costs may be one of the most significant expenses involved in a project and must be considered up-front.

> IP licensing costs may be one of the most significant expenses involved in a project and must be considered up-front, along with other priority expenses for the project.

What happens if you ignore IP rights of others? You may never be able to legally use or sell the fruits of the project's labors, therefore nullifying all prior efforts and expenses. Worse yet, infringement (violations) of IP rights of others exposes the violator to legal liabilities and protracted and expensive litigation. Depending on the jurisdiction and the types of intellectual property infringed, a violator may not only be liable for paying out direct damages (losses), incurred by the IP rights' owner, but may also be responsible for any profits it made, using the infringing IP, and for any legal costs, incurred by the plaintiff (the side seeking justice).[35] In some cases, the rights' owner may also be entitled to collect punitive and/or statutory damages. All of these liabilities will far exceed any benefits obtainable from a project that uses infringing technology.

Surely, you must intend to infringe someone else's rights to be subject to law suits and financial liability? This is not true. Often, infringement of IP rights occurs unintentionally.

> You can infringe someone else's IP rights without intending to do so.

It is entirely possible for a musician to hear a catchy song and then subconsciously reproduce the music in their composition. It is possible for an artist to appreciate a photograph and then, years later, to reproduce important elements of the photograph in their painting or sculpture. As long as a copyright infringer had access to a work, and sufficiently important aspects of the work were copied, the proof of actual intent to copy may not always be required. Similarly, for trademark and patent law, infringement may be found by courts without any intent. The patent or trademark holder sustains damages, regardless of whether the infringer actually knows of the right-holders registration certificates. Accordingly, it is the project manager's job (or more appropriately, a project IP attorney's job) to proactively clear any type of IP subject matter, used in a project.

One of the gravest mistakes a project manager can make is proceeding with the use of IP that even theoretically may be subject to the rights of others.

> One of the gravest mistakes a project manager can make is proceeding with the use of IP that may be subject to the rights of others, hoping to make a deal with the rights-owner later.

Sometimes, an assumption is made that a rights-holder may not notice the infringement or that a deal can be made with the rights-holder once they appear to claim their rights. Such strategy is not only illegal but may have disastrous consequences for the project. Many jurisdictions have long statutes of limitations (the span of time in which the case can be brought in court) for IP laws.

For example, in the United States, a statute of limitations for patent infringement cases is 6 years.[36] This means that, theoretically, a patent holder does not have to give warnings to the infringer and/or rush into litigation. Instead, the rights-holder may wait for 6 years, watching as his patent is exploited by the project, as investments are made in reliance on his/her technology and as profits are coming in to the infringer. Then, just as the project matures and profits increase, the rights-holder may "attack," seeking lost profits and other damages.

The equitable consideration that the patent holder waited an unreasonably long time to come out may not be a valid defense. In fact, in 2017, the Supreme Court of the United States considered such a multi-year delay in litigation as a defense in a patent infringement case.[37] The court held that "Laches [I.e. unreasonable delay] CAN NOT be invoked as a defense against a claim for damages brought within §286's 6-year limitations period."[38] The US Supreme Court made a similar finding in a 2014 case Petrella v. Metro-Goldwyn Mayer,[39] 134 S. Ct. 1962 (2014), denying the defense of laches in copyright cases.

The lesson for project managers is never to let the guard down, regarding liabilities related to IP. The fact that for years, no one made any claims of infringement, does not mean that they won't. Potential plaintiffs may be watching the project and just waiting for the right moment to strike. Project managers must remember not to put the project into a position of prey. Survey the waters before you go in.

> The fact that for years, no one made any claims of IP infringement against the project, does not mean that such claims will not be made. Statutes of limitations may be years-long, allowing for litigations to spring up years after the actual infringement occurred.

In cases where IP infringement by your project can be ruled out, there is a number of other issues to consider. For example, can you protect the innovations and other intellectual property developed in the course of the project? Is the subject matter even protectable? If not, then how can you prevent competitors from gaining advantage over you by simply copying the fruits of your labor, and avoiding the burdens of development? If you do have protectable IP, when and how must you protect it? Many jurisdictions offer a limited time window for claiming the rights in certain types of intellectual property. If these rights are not timely claimed, they may be diminished or lost forever.

Does your project involve hiring outside help? Who is working on the project? Are project workers legally considered to be employees or independent contractors? Can the project's workers/employees/contractors claim all rights to intellectual property and all profits from the project once it is complete? In many circumstances, unless precautionary steps were timely taken, they can.

Have you or anyone connected with the project described any of the project's technologies to anyone—to friends or family, to colleagues, to potential investors, or in Internet postings? If so, has the project forever lost all rights to its intellectual property by improper disclosure?

The questions above bring out just a small random sampling of a myriad of IP issues that may come up in a complex project. Each one of these issues may be determinative to the survival of the entire project. Importance of IP in project management cannot be overstated.

Each project is different. The IP questions that must be asked in your particular project will likely be different from the ones above. But in every project, a strong conscious effort must be made to find the right IP questions and get the right answers. The help of a skilled and experienced IP attorney is often indispensable in this task. Asking the right questions from the start will ensure that critical IP risks and benefits of a project are considered. Once recognized and understood, these IP liabilities and/or assets may well prove to be determinative to success of the entire complex project.

OTHER PRACTICAL TIPS AND TOOLS

REGISTRATION OF PROJECT'S PATENTS, TRADEMARKS, AND COPYRIGHTS

Register your IP. Register it as soon as possible after creation. Registering early is the only way to obtain all rights and protections, provided under the law.

> Registering IP soon after it is created is the only way to obtain all rights and protections, provided under the law.

If IP registration is not applied for in time, the rights for protection may be lost forever. We can look at US patent law as an example. The law states:

A person shall be entitled to a patent UNLESS—

1. the claimed invention was patented, described in a printed publication, or in public use, on sale, or otherwise available to the public before the effective filing date of the claimed invention; or

2. the claimed invention was described in a patent issued under section 151, or in an application for patent published or deemed published under section 122(b), in which the patent or application, as the case may be, names another inventor and was effectively filed before the effective filing date of the claimed invention.[40]

In other words, if a project generated a new invention (be it a new device, process, software-related invention, business method, etc.) and the invention becomes available to the public, the rights to patent protection may be lost forever.

> If details or embodiments of an invention become available to the public prior to patent application filing, the rights to patent protection may be lost forever.

The law lists printed publication, public use, and sale of the invention as just some of the ways that lead to the loss of patenting rights. Any leak of the invention to the public, be it from an employee bragging to a friend, a photograph posted on social media, an advertisement or a presentation to potential investors (without proper non-disclosure agreements (NDAs)) may forever banish the invention into the realm of public domain. US law provides a 1-year grace period after public disclosure, within which (in some cases) a patent application may still be filed. Most other countries/jurisdictions allow no grace period at all. Public disclosure is interpreted broadly by courts. It may occur unintentionally, without project manager ever knowing about the disclosure. Nonetheless, such disclosure may still be lethal to invention's patentability. In order to minimize the risks of inadvertent disclosure, patent application filing must not be delayed.

Likewise, if someone supersedes project's management in filing a patent application for the same or similar invention, the rights are lost. In this case not only are the rights to protecting the invention against infringement lost, but a potential competitor may actually acquire the rights to control the use of the invention. The competitor may acquire the rights by either directly applying for the patent themselves (if they invented the technology) or acquiring the rights from third parties. The competitor may then prevent the project from continuing the use of its own invention.

At one time, a few countries, including the United States, Canada and Philippines, used "first-to-invent" approach, attempting to award patents to the actual first inventor, even if someone else filed first. However, these countries have now followed the rest of the world to the "first-to-file" system of patent priority. In most jurisdictions today, whichever inventor is the first to file, receives the award of a patent.

Good ideas often come when the market has ripened for them. As a result, there are often several simultaneous competing projects, developing or exploiting the same technical idea or market opportunity. Often, the projects (especially commercial projects) are secret, with each team not knowing of the other's efforts. Such competing teams tend to come up with similar solutions at around the same time. The first team to file for a patent is usually the one that succeeds, recuperating the costs and making a profit. The project that is second to file loses all rights to patent (and often use) the invention.

Examples of such competitions for earliest patent filing are often encountered in pharmacological projects. Pharmacological companies invest billions of dollars into drug research projects. Usually, there are several simultaneous competing projects, coming up with similar solutions (drugs). Sometimes, patents to nearly identical inventions are filed by these teams within hours of each other. Whoever is the first to file, gets all rights and the ability to cover the expenses and profit. Whoever is second, has to accept the losses. Recuperation of such "second-to-file" losses is often cited by pharmaceutical companies as justification for seemingly exorbitant drug prices. For every drug that a company develops, patents, and brings to the market, there are many more drugs that a company heavily invested in and developed, but lost all rights in, as a competitor was the first to file for patent protection.

> Do not delay patent filings. If a competitor supersedes the project in filing a patent application for the same or similar invention, not only will all rights be lost, but they will be acquired by a competitor. The competitor may, in turn, prevent the project from using the invention.

Having a competitor file first and block all rights to the invention is a tremendous risk for any project. Thus, it should be a duty of every project manager to take precautions against this risk and file first. How do you know that you filed first and a competitor has not filed yet? You don't. In most jurisdictions, freshly filed patent

> In most jurisdictions, freshly filed patent applications are not immediately published and are not available for review.

applications are not immediately published and not available for review. For example, in the United States, barring special circumstances, patent applications are only published 18 months after filing.

Thus, if your project as well as your competitor's project filed for a similar invention, you may not know for a long time, which project was afforded an earlier filing date. If the competitors beat your project, acquiring earlier filing dates, and if they eventually receive a patent, such patent will be valid and backdated from the filing date. If you continued to use the invention after your competitors filed (not knowing they filed or who filed first), they may later sue your project for infringement, starting from the day they filed. It often takes years to get a utility patent. However, once patent is received, it is retroactively considered to be in force from the day the application was filed.

> Patents are generally valid and may be retroactively enforced from the day the application was filed.

In most jurisdictions, a patent applicant is limited in enforcing its rights in the invention until the patent issues. However, if and once the patent issues, a patent holder may sue retroactively, as if he/she always had a patent since the day of filing.

So, if your project's innovations are important to the project, prepare and file patent applications as soon as possible.

It is also critical for a project manager to remember: patents are not limited to complex machines and groundbreaking technologies. If your project develops a novel, non-obvious solution to a problem, it may be patentable. Whether such a solution actually constitutes patentable subject matter is highly jurisdiction-dependent. A project manager must consult qualified patent attorneys to obtain an opinion on whether a potential invention is patentable, and if so, in what jurisdictions. However, it is important to remember that in many jurisdictions, including the United States, even novel methods of doing business may be protectable.

> Patentability is not limited to complex machines. Even novel methods of doing business may be protectable.

Some IP Rights Arise Automatically by Action of Law without Registration, but Make Sure to Register Anyway

Berne Convention establishes the basic principles of copyright for all contracting countries. Most countries of the world are signatories to the Berne Convention and are thus bound by its terms. Among the basic principles,

> In most jurisdictions, registration is not required for copyright protection.

established by the Convention is the principle of "automatic" protection of copyright. It states the Protection must not be conditional upon compliance with any formality. In practice, this means that in most jurisdictions, the "formality" of registration is no longer required for the protection of copyright.

According to the US Copyright Act, "Copyright protection subsists, in accordance with this title, in original works of authorship fixed in any tangible medium of expression, now known or later developed, from which they can be perceived, reproduced, or otherwise communicated, either directly or with the aid of a machine or device."[41] Note that there is no requirement of registration. Under US law, copyrightable work acquires protection as soon as it is "fixed in any tangible medium of expression."[42] That is, as soon as a poem is typed up on a piece of paper, as soon as the last stroke of paint is applied to the surface of a canvas, and as soon as a software program is saved on a disk, these works are "fixed" and are subject to copyright protections.

Note this "fixation" requirement. It is important. In order to be protectable under copyright, a work must be "fixed in any tangible medium of expression."

> In order to be copyrightable, a work must be fixed in a tangible medium.

Simply reciting a poem without writing it down or performing a play, without recording it on video, will likely NOT create works that are protectable by Copyright law. In a way, fixation ensures that an intellectual property right is tied to an actual object,[43] be it physical or digital. However, once the work is fixed, protection attaches automatically.

If copyright protection is automatic, why does copyright registration exist? Why should a project manager bother registering copyrights? As with other types of IP, international conventions may impose certain general rights on member-states, but it is national law that provides interpretations of these rights and interjects important details. For example, as can be seen from the above quotation of the US Copyright Act, under the US law, registration of a work IS NOT required for protection. However, the owner of unregistered work cannot bring an infringement action in court. That is, the work is theoretically protected, but the rights cannot be enforced unless the work is registered. Furthermore, unless the work is registered as soon as possible, some rights may be lost forever. 17 US Code §412 sets registration as prerequisite to certain remedies in court. If a work has not been registered within 3 months of publication, the rights-holder can never ask for statutory damages or attorney fees in court. In many instances of infringement litigation, this right to statutory damages is critical to successful outcome of the case. Therefore, in most cases, prior registration is required in order to enforce project's rights over its works.

> In some jurisdictions, failure to timely register copyright, limits owner's rights.

Registration also provides proof of validity of copyright. It also ensures a public record of copyrighted work and the dates of its creation, as well as the record of ownership rights. If project's competitor ever argues first creation of a substantially similar work, earlier copyright registration is the most certain way of proving priority of rights. Likewise, if project's ownership of rights in a particular work is ever challenged, a valid registration provides proof of such rights.

> Copyright registration provides a public record of the work, of copyright's validity, and of the rights of ownership.

IP OWNERSHIP ISSUES

Challenges to IP property ownership come frequently. Sometimes, these challenges come from project competitors. But even more frequently, challenges to ownership come from within the project itself.

IP within a project is developed by project's workforce. Individuals and/or groups, working for the project, are the ones that come up with protectable subject matter, such as patentable inventions and copyrightable works. Who owns the rights in these creations? This issue has been a subject of innumerable law suits, and the answer is not as obvious as it may appear.

It is a major mistake for a project manager to assume that patent rights or copyrights in the creations belong to organization or a project simply because the work was commissioned and/or paid for by the project.

> NEVER assume that rights in the creations belong to the project, rather than to individual creators, simply because the work was commissioned and/or paid for by the project.

In fact, quite the opposite is true. For example, under the US Copyright law, all rights automatically accrue to the creator.[44] Any rights that a third party acquires stem from the creator's rights and must come by an affirmative transfer of rights, such as a contract. Simply paying someone to do the work will likely be INSUFFICIENT to ensure the acquisition of full rights to that work by the payee.

The first time that most people in the US encounter this concept is when hiring a wedding photographer. A photographer is usually paid to take and edit the pictures of the couple's most important day. Yet, when it comes to printing and sharing the pictures, the couples are often surprised to discover that these actions must be done only through the photographer's studio and at photographer's rates. Unless specifically discussed in a pre-wedding contract, even full-resolution images are often withheld from the clients to prevent them from printing high-quality images from the files. To most, this practice feels unfair and counterintuitive. However, the argument "But we paid you for these pictures!" will likely not stand in a legal setting. Unless specifically agreed to in a contract, the couple often pays a photographer for the time and effort, not for full rights in his creations. As stated above, Copyright law vests an author with numerous rights. It is up to the author to contractually transfer these rights in their entirety or piecemeal, one at a time. Thus, for example, it is possible that a wedding photographer may only grant a couple nonexclusive rights to exhibit printed versions of a photographs, but not digital versions. It is also foreseeable that he/she may retain the right to use couple's photographs for his/her own advertisement or for sale as artworks.

Imagine the same situation in the context of project management. An artist, such as a painter, a photographer, a graphic designer or a videographer is hired to create works for a project. If the project is heavily dependent on audio-visual content, such works may constitute the very core and most of the project's value. In other cases, such works may constitute the basis for later "derivative works,"[45] needed to advance the project, or they may be critical to advertisement efforts. Unless a contract exists or a special legal relationship is created between the author and the project, all rights in these creations will likely vest to the author. This means that theoretically, the author can, at any time, stop the project from using his/her work, thus derailing the entire undertaking. Alternatively, the author may leave the project and use the creations for a competitor's benefit. Therefore, it is extremely important for project management to ensure that proper legal relationship exists between the project and anyone who produces any work that can be construed as intellectual property for the project.

A legal doctrine of **Work Made for Hire** provides an important exception to the general rule that all rights in the work are owned by the author. Under this doctrine, in the United States, if a work is made for hire, an employer is considered to be the author even if an employee actually created the work.[46] Unless a detailed contract for transfer of rights exists, it is critical to ensure that any works created for the project meet all legal requirements to be considered "works made for hire." Otherwise, the project will have no rights to the work product.

> Project manager must ensure that any works created for the project meet all legal requirements for "works made for hire."

Generally, under US Copyright Act,[47] in order to qualify as "work made for hire," the work must be either:

    a. a work prepared by an employee within the scope of his or her employment
       or
    b. a work specially ordered or commissioned for use
       1.   as a contribution to a collective work,
       2.   as a part of a motion picture or other audiovisual work,
       3.   as a translation,
       4.   as a supplementary work,
       5.   as a compilation,

6.   as an instructional text,
7.   as a test,
8.   as answer material for a test, or
9.   as an atlas,

if the parties expressly agree in a written instrument signed by them that the work shall be considered a work made for hire.

While prong (a) of the statute appears relatively straight-forward, it is not. Whether a work was indeed prepared by an employee and, if so, whether such work was within the scope of his or her employment are frequently litigated issues. Just naming someone an employee of the project does not necessarily give that person a legal "employee" status.

> Whether a work was prepared by an employee and whether it was within the scope of his or her employment are frequently litigated issues. Contractually resolve them before work begins.

In determining whether creator of the work is an employee or an independent contractor, US courts look at the area of law, called Agency Law.

In US Supreme Court case *Community for Creative Non-Violence v. Reed*,[48] the court looked at a number of factors to determine whether an employer–employee relationship exists under Agency Law. Among these, the court considered the details of the status and conduct of the employer and the amount of control that the employer has over the employee and over the creation of a particular work. For example, the court may look at whether the employer can control an employee's schedule, work environment, hiring of assistants, etc. The court may look at employer's control over the details of the works and the equipment used to produce the work. It may also consider whether the employer regularly produces such works and whether it provides the employee with benefits or pays the employee's taxes. One of these factors by itself, such as control over the production of the work, will likely not be determinative as to employment status.

In general, the more control the employer has over the employee and the work, the more likely that an employer–employee relationship will be found. But that determination is rarely a given and often requires lengthy and expensive litigation to determine. Therefore, it is wise for a project manager to ensure from the beginning of the project that a legal status of each "employee" is clearly defined under the Agency Law (or its local legal equivalent).

> In general, the more control the employer has over the employee and the work, the more likely that an employer–employee relationship will be found, thus granting the rights in the work to the project.

If author of a work, commissioned by a project, does not qualify as an employee, then most likely he/she is an independent contractor. In that case, prong (b) of the statute applies. That is, in order for the work to count as "work made for hire," there must be a written agreement AND the work must fall under one of the nine enumerated categories. Both of these requirements have to be satisfied. Otherwise, the project will likely have NO rights to the works it commissioned and paid for.

> If work is created by an "independent contractor," rather than an employee, make sure that there is a "work-for-hire" contract in place AND that the work falls under one of the enumerated categories.

Legal requirements needed to ensure that a work will be considered as "made for hire" are complex and jurisdiction-specific. Determination of whether a particular work fits under the doctrine requires fact-intensive analysis that often involves application of more than one area of law. Therefore, it is important for project

> Determination of whether a work is "made for hire" requires fact-intensive analysis that involves several areas of law. Obtaining legal advice from a local attorney is imperative to ensure that works created for the project belong to the project.

governance to seek qualified legal guidance from the very start of the project to ensure that all rights automatically accrue to the project.

Due to the constraints of a single chapter, the above examples of "work made for hire" doctrine are limited to copyrights. However, project managers must be aware that similar issues may arise in the context of other types of intellectual property. For example, in cases of inventorship, unless a special relationship is created prior to the invention between the inventor and the project, full patent rights may be claimed by the inventor.

Likewise, imagine a situation where the project is developing a particular technology. An outside specialist (such as an engineer) is consulted by the project regarding one, seemingly minor detail of implementation. The engineer's suggestion is then incorporated into one of a hundred or more claims within a patent. It becomes just one detail out of hundreds. Yet, under US patent law, such engineer may have the rights of a co-inventor.[49] Absent any written agreement to the contrary, as a co-inventor, this contributor of one one-hundredth, will (by default) have equal rights in the patent to other co-inventor(s), who may have contributed 99% of the invention and the claims. He/she may then have a right license or sell rights in the entire invention without consulting other co-inventors. Therefore, it is extremely important for project management to ensure that anyone who contributes to the conception of any aspect of project's technology is legally bound to transfer his/her rights in the resulting patent to the project.

> Under US law, whoever contributes to the conception of even a single claim in an invention becomes a co-inventor with full rights to license and otherwise exploit an invention. Retain legal assistance to ensure that anyone who contributes to any technologies, developed by the project is legally bound to transfer the rights to the project.

Trade Secrets Are Not Registrable

Trade secrets are an exception to registration requirement, among other types of Intellectual Property. They are NOT REGISTRABLE.

> Trade secrets are not registrable.

Application for a patent is intended to inform the public of a useful invention. Registration of trademarks is intended to give notice to potential marketplace competitors of the use of the mark. Copyright registration serves to identify works and protect authorship. Trade secrets on the other hand must be kept . . . secret. They cannot be registered and thus revealed to the public. In fact, as soon as a trade secret is revealed, it is no longer a secret. U.S. patent and trademark office's (USPTO) statement on Trade Secret Policy states: "If a trade secret holder fails to maintain secrecy or if the information is independently discovered, becomes released or otherwise becomes generally known, protection as a trade secret is lost."[50]

> Generally, as soon as a trade secret becomes known to others, all protection is lost. Project manager must make sure that all of project's trade secrets are well guarded.

What steps must a project manager take to protect a trade secret? Again, this decision must be made by project's governance, based on the type of the secret and on relevant jurisdiction, after consultation with an IP attorney. Generally, trade secrets are only known to the absolute minimum number of people that are themselves sworn to secrecy. Some famous companies are built on trade secrets. Sometimes, one well-guarded trade secret is what gives a company its main competitive advantage. For example, Coca Cola's recipe, Mac Donald's Big Mac sauce, WD-40 and Google's search algorithm are all well-guarded trade secrets. Sometimes, companies resort to exaggerated, nearly theatrical actions to make a legal statement that their secrets are extremely valuable and closely guarded.

KFC's secret is the original recipe of herbs and spices, written down by Colonel in the 1940s. It is reported to be kept in a safe that weighs 770 pounds, encased in 2 feet of concrete. The safe is monitored by 24-hour video and a motion-detection surveillance system.[51] Likewise, WD-40 formula is kept in ultra-secure bank vaults and is only removed for moving to another vault. This secret formula was once transported for a procession to celebrate company's 50th anniversary. For this event, the formula was kept by the company's CEO, who rode on horseback, in knight's armor, in order to protect the precious secret list of ingredients.[52]

You may not need a 770-pound safe or medieval armor to protect your project's trade secrets. However, a project manager must not assume that because trade secrets do not require registration, inaction is sufficient to obtain protection. Affirmative jurisdiction-appropriate actions must be taken to protect trade secrets.

> Affirmative jurisdiction-appropriate actions, must be taken to protect project's trade secrets.

PATENTS VERSUS TRADE SECRETS

Trade secrets consist of information that is used in business and gives its holder an opportunity to obtain an economic advantage over competitors who do not know or use it.[53] Trade secret may encompass a formula, pattern, compilation, program, device, method, technique, or process.[54] Some of these areas protectable by trade secret also fall under the scope of patent protection. However, patent and trade secret protection are mutually exclusive of each other. Patents require full disclosure of the invention. Trade secrets must be kept secret at all costs. Therefore, when both types of protection are available, only one must be chosen.

> The fields of trade secret and patent protection overlap. However, these two types of IP protection are exclusive of each other. Project manager must decide which mode of protection is more appropriate.

How can project manager decide whether trade secret or patent law offers the proper mode of protection? First, it must be determined whether patenting is indeed an option. In order to be protectable by patents, subject matter of the invention must be patentable under local laws. For example, if subject matter of the invention is "business methods" and this subject matter is not protectable in your jurisdiction, then patenting is not an option. Patenting also requires a "conception" of the invention. That is, the invention must be developed beyond an abstract idea. US courts found that "conception is established when the invention is made sufficiently clear to enable one skilled in the art to reduce it to practice without the exercise of extensive experimentation or the exercise of inventive skill."[55] Trade secrets do not have this requirement. Whether an invention is patentable should be determined by project's governance in consultations with patent attorneys.

If patent protection is indeed available, is the nature of the project such that public disclosure of the technology will reduce or decimate its commercial value? If yes, then obviously trade secret protection is preferable. But on the other hand, if the nature of the technology is such that others are likely to independently discover it, then patent protection is preferable. If others (such as project's competitors) are able to discover trade-secret-protected technology independently (or otherwise, such as through information leaks), then all trade secret protections lapse. On the other hand, once patent protection is obtained, the patent holder can prevent others from using the invention, regardless of whether they obtained the invention from a patent publication or discovered it themselves.

> Patents are preferable in situations where trade secrets are likely to be discovered by competitors.

Another question to ask in deciding between patent and trade secret protection is how long the technology/secret will be in use. Patent term is generally limited to 20 years from the date of filing. Trade secrets may theoretically last forever, as long as the secret is kept. For example, if Coca Cola® KFC® or WD-40® used patents as a primary means of protection for their formulas, all protection would have lapsed a long time ago.

Trade secrets are generally appropriate for technologies/processes/information that can be beneficial, when practiced in secret. This often prevents the holder of a trade secret from fully exploiting the technology/process/information publically. Patented technologies, on the other hand, can be practiced openly, with the full weight of the legal system standing behind patent-owner's rights. Patent law protects the patent holder from a wide range of the types of infringement. Patent law also provides numerous remedies and options to the holder of the infringed patent. In comparison, trade secret "protection is very limited because a trade secret holder is only protected from unauthorized disclosure and use which is referred to as misappropriation."[56]

> In deciding between patent and trade secret protection, consider the length of protection required. Patent protection is generally limited to 20 years. Trade secret protection may go on indefinitely.

> Patent law generally provides much broader scope of protection than Trade Secret law.

In deciding between these two types of IP protection, the project manager must remember that picking one mode of protection will likely forever bar the use of the other. Obviously, content disclosed in a patent application may never be claimed as a secret. It may also be tempting to avoid the disclosure requirement and costs of patenting by keeping an invention secret. However, any public use or sale of such invention (even if details of production are kept secret) is likely to forever prevent patenting of the invention. There is also a looming danger that a third party (such as a competitor) will independently discover and patent the invention, thus prohibiting the trade secret holder from using their long-held trade secret.

A project manager should remember that public policy (and, accordingly, laws) generally encourages disclosure of useful information to society (i.e., patenting). While information withheld as a secret may be protectable in certain circumstances, do consider the risks of not obtaining a patent. In weighing of such risks and advantages of a mode of IP protection, legal consultation is often necessary and invaluable.

> While information withheld as a secret may be protectable in certain circumstances, do consider the risks of not obtaining a patent.

Non-Disclosure Agreements and IP-Related Contracts

A Non-Disclosure Agreement (NDA) is a legally enforceable contract, regulating transfer, use, or dissemination of particular confidential information. NDAs are also known as Secrecy Agreements, Confidentiality Agreements, and Confidential Disclosure Agreements, among other titles. Such agreements can be very detailed and complex, but generally state that in exchange for receiving the information, the receiving party guarantees that information will be kept confidential and/or used only for a particular purpose. Such agreement is an extremely powerful and versatile tool in the hands of a project manager.

IP laws grant a number of protections to IP holders. However, in order to receive these protections, registrations (or other specific affirmative steps in case of trade secrets) are often required. Registration of copyrights or trademarks may take many months. Patenting process often takes years. Until the registration is complete, there is often ambiguity as to whether the protection will be granted at all.

Often, information must be disclosed to third parties before its registration/patenting status is finalized. For example, information about a new product may have to be released to employees, potential investors, prototype manufacturers, etc. When such a disclosure is necessary, a project manager must ensure that information will not be stolen by others and that such a disclosure will not invalidate the project's rights in its IP.

NDAs present the IP owner with an additional layer of protection on top of the rights granted by IP laws. Such agreements allow for transcending beyond IP laws and tapping into protections offered by contract law.

> NDAs allow for transcending beyond IP laws and tapping into protections offered by contract law.

For example, a project reveals information regarding a new technology to an outside manufacturer, in order to produce several test prototypes. A NDA is signed with the manufacturer, prohibiting him from using the information in any way, other than to manufacture the prototypes for the project. Project managers have previously applied for a patent, which has not been granted yet. The manufacturer uses the information to manufacture and sell the product him/herself.

In this scenario, project's governance may wait for years for the issuance of the patent in order to bring an infringement law suit. In most cases, such a wait is not practicable, as the infringer's actions may significantly alter the market and derail the entire project. Alternatively, project's governance may immediately sue the manufacturer for the breach of contract for violating the non-disclosure agreement. As part of the breach-of-contract law suit, the project may ask, among other things for immediate injunction (a court order, ordering immediate stop to infringer's damaging actions) and for monetary damages. Even if patent never issues, plaintiffs may still prevail in this case, based solely on the principles of contract law.

IP laws are broadly drafted, as they are intended to apply to a wide variety of circumstances. Contracts, on the other hand, are extremely flexible. In a way, they can be seen as a law, agreed on by two sides to govern a particular situation. A contract can be drafted, taking into account specifics of particular IP, special circumstances of the parties and defining particular conditions of breach and remedies due to the injured parties. Thus, whenever any transfer/disclosure of sensitive information or technology occurs, it is always a good idea to have a detailed non-disclosure agreement (contract), describing the information transferred, permissible uses of such information, and consequences of misuse. Even if information/Intellectual Property being transferred is already protected by a grant of IP rights, it is often beneficial to have this additional layer of protection. If nothing else, it provides rights-holder with extra legal ammunition and the ability to elect whether to enforce the rights under IP or contract law, or both.

> Contracts are extremely flexible and adaptable. In a way, they can be seen as law, created by two sides to govern a specific situation between them. Project managers should use contracts, including NDAs, whenever possible, to ensure legal clarity in relations between a project and any third parties.

While NDAs are always a good idea, in some circumstances, they are an absolute necessity. For example, sometimes trade secrets must be disclosed. Such disclosure may be to an outside manufacturer or to a senior member of the project for safekeeping or use. In most cases, releasing a trade secret without obtaining a strong and detailed non-disclosure agreement (among other safeguards) from the receiving party, will invalidate the trade secret.

Likewise, patent laws in most jurisdictions prohibit any disclosure of the invention to the public[57] prior to applying for a patent. However, sometimes disclosures must be made, be it to a colleague for advice on a technical matter, or to a patent draftsman for preparing drawings for a patent. Unless all people to whom the invention is revealed are bound by a strict NDA (and other legal obligations), the invention may be deemed to have been disclosed to the public. Accordingly, the invention may be permanently denied patent protection.

> A project manager must resist the temptation of downloading a generic non-disclosure form, from the Internet. NDAs are legal contracts and must strictly conform to the facts of the case and laws of particular jurisdiction(s) of the project.

Numerous examples of NDAs abound on the Internet. However, a project manager must resist the temptation to download and edit a default form, substituting names and addresses. NDAs are specific legal contracts.

As such, they must strictly conform to the facts of the case and laws of a particular jurisdiction of the project.

Sometimes, a complex project may span across several jurisdictions. Alternatively, the parties to a contract may be from different countries/jurisdictions. In this case, it is always advisable to include a "choice of law" provision in a contract. Such a provision identifies the location of the courts and the laws that will be applied in interpreting the contract.

> Include a "choice of law" provision into a NDA. This provision should indicate which courts will be used and the law of which jurisdiction will be applied in interpreting the agreement.

As non-disclosure agreements are complex legal documents, project managers are strongly advised to retain legal assistance for preparing such documents.

COMING UP WITH PROTECTABLE AND STRONG TRADEMARKS

Most complex projects, especially commercial projects, involve the use of trademarks. Often, the project itself has a name that distinguishes it from other projects. Manhattan Project, Apollo Program and Operation Desert Storm are all memorable names of major projects. Products that a project produces or services it provides are also frequently given names that distinguish these products and services from others on the market. Picking a strong mark is often the first step to project's success. How can a project manager ensure that the name selected for a project or a product will be a strong mark and enjoy commercial success and a broad legal protection? Generally, the more distinctive the mark is, the more likely it is to be recognized and distinguished by the consumers. Strong, distinctive marks are generally easier to register, as well as to maintain and protect in the marketplace.

The strength of proposed marks can be gauged within a hierarchy of several categories of trademark distinctiveness. These categories are: fanciful marks, arbitrary marks, suggestive marks, descriptive marks and generic marks.[58] Whenever a project manager is coming up with a new trademark, he/she must be cognizant of where on this scale of strength the proposed trademark is likely to

> Project management must strive for highly distinctive trademarks, as such marks are generally easier to obtain and enjoy strong legal protection.

be. The more distinctive the trademark is, the better. Strong, distinctive marks are generally easier to obtain and enjoy broader legal protection. Fanciful marks are inherently distinctive.[59] Arbitrary marks are also highly distinctive.[60] Thus, a project manager must strive for a mark that falls into one of these two categories.

Fanciful Marks: These are usually the best and strongest among trademarks. A mark is fanciful when it is completely novel. Generally, fanciful mark consists of a coined term that never existed in the language before. Words like "Exxon®," "Xerox®," "Pepsi®," "Linoleum®," "Aspirin®," and "Escalator®" never existed in the English language until companies invented them and adopted as names/trademarks, identifying their goods or services.

> If the project is coming up with a new and revolutionary product, consider inventing a new and fanciful term to describe it. Such trademark will have the strongest protection in the marketplace.

As fanciful marks are unique on the marketplace, they are afforded the greatest degree of protection. If your project is coming up with a new and revolutionary product, it is well worth considering inventing a new and fanciful term to describe it.

However, a word of warning must be added, regarding the fanciful marks. When such marks are initially introduced to the market, consumers have absolutely no association between the mark and the product. A word no one has heard before may attract interest but will not offer much information, regarding the nature and use of the product. Therefore, the consuming public must be educated through market exposure and advertisement to associate the product/service with the source. However, once the association with a fanciful mark is formed in the minds of consumers, it will be strong, and they will be unlikely to confuse such mark with one from another source.

Arbitrary Marks: These marks involve the use of pre-existing words and phrases. However, such terms are, arbitrarily associated with products or services. That is, the nature of products or services is unrelated to the common meaning of the terms. For example, a bagel-shop, named "Anchor," may be considered an arbitrary trademark, as anchors have no association with bagels. The famous "Apple®" trademark is arbitrary, as fruits are generally not associated with computer equipment. However, if the term "anchor" is used for anchoring equipment or the word "apple" is used for fruit vending, such marks would NOT be arbitrary. Arbitrary marks are nearly as strong as fanciful marks, since arbitrariness in naming makes such marks highly distinctive.[61] However, just like with fanciful marks, arbitrary marks initially tell consumers little about the actual product or service.

Suggestive Marks: Unlike fanciful and arbitrary marks, suggestive marks do tell consumers something about the product. They "suggest" the nature of the product or service, without fully describing it. USPTO Trademark Manual of Examining Procedure states that suggestive marks are those that, when applied to the goods or services at issue, require imagination, thought, or perception to reach a conclusion as to the nature of those goods or services.[62] For example, trademark "WET ONES" was found suggestive, when applied in relation to the moist toilette product.[63] Likewise, "SPEEDI BAKE" for frozen dough was found to be suggestive, as it vaguely suggests a desired characteristic for dough.[64]

Suggestive marks may also be strong and are registrable (in the United States on the Principal Register) without a need for the additional proof of distinctiveness ("secondary meaning").[65]

Descriptive Marks (a.k.a Merely Descriptive Marks): These are the marks that are merely descriptive or deceptively misdescriptive of the goods or services, to which they relate. Such marks immediately tell consumers something about the goods or services, without the need for any leap of imagination. According to USPTO, a mark is considered merely descriptive if it describes an ingredient, quality, characteristic, function, feature, purpose, or use of the specified goods or services.[66] The mark need not describe all the goods and services identified, as long as it merely describes one of them.[67]

In the United States, descriptive marks are generally refused registration. If such marks are eventually registered, their scope of legal protection is generally narrow. A reason for such a refusal is that as a society, we want marks to be unique and easily-distinguishable, so as to enable consumers to easily identify products and source of the goods in the marketplace. Also, since trademarks give their owners a kind of monopoly on the use of the mark, we want to limit the extent of that monopoly. If a trademark applicant were to claim rights to a description of a product, then he/or she could prevent others, offering similar products and services, from describing their product in a marketplace.

Some marks, held as descriptive in the United States, include terms "COASTER-CARDS" to describe a coaster intended for direct mailing, "BED AND BREAKFAST REGISTRY" for lodging reservations services and "APPLE PIE" for potpourri.[68]

Ironically, descriptive marks are usually the first choice for a trademark, made by project managers unfamiliar with trademark law. It is only intuitive to pick a name relevant to a product (or service), so as to tell to consumers what the product is. Sometimes, such descriptive mark is successfully used in commerce by a project, only to discover that a trademark office will not register the mark due to its descriptive nature.

However, even in cases of descriptive marks, registration may still be possible. According to USPTO, "if a proposed mark is not inherently distinctive, it may be registered on the Principal Register only upon proof of

> Descriptive marks are usually the first choice for a trademark, made by project managers unfamiliar with trademark law. It is only intuitive to pick a name relevant to a product. However, such choice may result in weak or unregistrable trademark.

acquired distinctiveness, or 'secondary meaning,' that is, proof that it has become distinctive as applied to the applicant's goods or services in commerce."[69] Sometimes, continued use and advertisement of a descriptive mark does make it distinctive to consumers. That is consumers begin to associate the mark with a particular source of the goods or services. However, in many cases of descriptive marks, it is not easy to make such a showing.

There is often a very fine line between suggestive marks that require a leap of imagination and descriptive marks, which immediately tell something about the goods or services. Therefore, if a project manager is picking out a name that is in any way reminiscent of a project's products or services, it is very important that such name be suggestive and not descriptive. This slight distinction may be critical to the strength, ability to register, and enforceability of the mark.

> When picking out a trade name that is in any way reminiscent of a project's products or services, it is critical that such a name be suggestive (requiring a leap of imagination), and not descriptive.

Generic Marks (a.k.a. Generic Terms): These are the terms that the relevant purchasing public understands primarily as the common or class name for the goods or services. They are incapable of functioning as registrable trademarks denoting source, and are not registrable.[70] One cannot claim a monopoly over "TV" brand to sell TVs, "Pear" brand for selling pears, or "Anchor" brand for selling anchors. Nor can anyone claim rights to the "Hot Coffee" trademark for selling hot coffee. Granting monopolies over such marks would be unthinkable, as it may enable grantees to prohibit others from the customary use of these general terms in the marketplace as well as in everyday language use.

GENERICIDE WARNING

If your product is revolutionary and your commercial project becomes a success, beware of Genericide. Imagine, your project is the first to introduce a new product (or service). The trademark for the product is fanciful and/or otherwise distinctive. The product becomes an instant success, initially associated with your project. Consumers cannot get enough of the beautiful shiny "Linoleum®," miraculous "Aspirin®," life-changing "Escalator®," cheap and clean-burning "Kerosene®," or amazing "Thermos®" mug. Gradually, competitors come into the market with similar products. However, consumers are so used to your revolutionary and famous products that they continue referring to new generic products by original names. They are still referring to a lifting machine as an "Escalator®," even if it is not made by Otis Elevator Company. They are continuing to call acetylsalicylic acid tablets "Aspirin®," even when they are not manufactured by Bayer. And heat-holding mugs are still referred to as "Thermos," even though Thermos GmbH had nothing to do with the production of the generic version. Your famous marks just became victims of their own success. Because of their initial success and popularity, they lost distinctiveness in the eyes of the consumers. Consumers no longer associate a term with a particular source. Instead, the marks become generic terms for a particular type of product. They commit "Genericide." As marks lose their distinctiveness and become generic, courts strip the marks of legal protection. The famous product marks above have at least temporarily lost their protection due to their own success, in at least some jurisdictions.[71]

If your project comes out with a successful and famous product, you must watch for any signs of the mark being used as a generic term. If you spot any indications that consumers are using the mark to identify a general type of goods/services, rather than those originating from your project, immediate action must be taken to stop such use. In the United States, genericide is considered a type of abandonment of the mark. In order to prevent this type of abandonment, the owner of the mark can often take affirmative steps.

> If your project comes out with a successful and famous product, you must watch for any signs of the trademark being used as a generic term. Unless such uses are affirmatively opposed, the mark may "commit genericide" and be stripped of all legal protection.

For example, International Trademark Association (INTA) provides a list of practical tips that mark owners can take for avoiding genericide.[72] Among these is a suggestion of an aggressive advertisement campaign, educating consumers on considering the trademark as a source of the goods, rather than a common name for a type of goods. Some examples of such campaigns can be seen with Xerox® and Kleenex® marks. When these marks appeared to be at the risk of genericide, the owners of these marks ran the following advertisements: "You can't Xerox a Xerox on a Xerox. But we don't mind at all if you copy a copy on a Xerox® copier," and "'Kleenex' is a brand name...and should always be followed by an ® and the word 'Tissue.' Help us keep our identity, ours." These advertisements appear to have saved these famous brands, which are still in force today.[73]

PROPER USE OF TRADEMARKS

In order to ensure that project's marks stay strong, project's management must ensure proper use of the trademark from the very beginning.

> Ensure proper use of a project's trademarks from the first steps of the project.

Among one of the most important aspects of the proper of trademarks is providing notice of trademark ownership. Most people have seen the following marks next to a name of a product or service: ®, TM, SM. In many countries of the world, these marks indicate that the name is a claimed trademark and is someone's protected intellectual property. The ®-symbol generally indicates that the trademark has been properly registered. The TM-symbol generally indicates a mark that is not yet registered but is being used in commerce and claimed as someone's intellectual property. Often mark owners start using the TM-symbol when the product is first released and then never get to changing it to the ®-symbol, once the mark is registered. So, do not assume that the mark is not registered if it features one symbol, rather than another.

Furthermore, in some jurisdictions (such as the United States), even unregistered trademarks, used in commerce, provide owners with certain (common-law) ownership rights.

The SM-symbol is similar to the TM-symbol, except that it refers to service marks, rather than trademarks. In the United States, other notifications may indicate a registered trademark, such as the statements: "Registered in the U.S. Patent and Trademark Office" and "Reg. U.S. Pat. & Tm. Off." In Spanish-speaking countries, one may encounter "Marca Registrada" or "MR" as symbols of a registered trademark. In some French-speaking countries, one may come across abbreviations, "MD" and "MC," or statements: "Marque Déposée" and "Marque de Commerce," to indicate mark's status.

> Whenever possible, provide notice of trademark ownership and registration next to the mark.

These markings act as powerful ownership statements and warning signs on products and services. In a way, they say "This mark is owned. Do not misuse! Use properly and with caution!" If your project owns registered trademarks, use these symbols next to your mark, whenever possible. They inform consumers that the mark is a registered trademark, rather than a genus of a product (thus preventing genericide). They also warn public and competitors that the mark is protected against infringement, misappropriation, dilution, and a host of unfair business practices. Protect your marks by providing these warnings. If you are coming up with a new mark for use by the project or in any way utilizing trademarks of others (such as within the products or services you offer, or within publications or advertisements), pay close attention to these symbols and heed their warnings.

Furthermore, project manager must insure that project's marks are always used in a consistent manner.

> Always use a project's marks in consistent manner.

That is, if you capitalize a particular letter in your trademark, always capitalize that particular letter, wherever you use the mark. For example, if the first or the last or the middle letter is capitalized, or the entire mark is in caps or in lower font, keep the styling consistent everywhere. Likewise, if you use a particular font or color to present your trademark, use this styling selection, whenever possible. Such consistency indicates that the mark is not a mere word or a reference to a product, but a valid indicator of a particular source of goods and services and a legally protected aspect of intellectual property.[74]

If your project's trademarks are adjectives, keep them as adjectives. A mark should properly refer to the source of the goods or services, NOT to the type of product or service. The mark, as an adjective, should qualify a generic noun that defines the product.[75] Xerox Corporation's ad, presented above, offers a perfect example and guidance on this point: "You can't Xerox a Xerox on a Xerox. But we don't mind at all if you copy a copy on a Xerox® copier." Never misuse your project's marks by using them as verbs (unless they are verbs to begin with), or by otherwise changing the part of speech that a mark originally belongs to. Avoid using a mark as a plural or an abbreviation, unless it is initially a plural or an abbreviation. Educate consumers, retailers, dealers, advertisers, and anyone who comes in contact with the mark as to the proper use of the mark. Police the use of the mark to ensure that others do not misuse the mark in this way.

> If a mark is an adjective, it must be kept an adjective. A mark should refer consumers to the source of goods, NOT to the type of goods.

> Educate anyone who comes in contact with the mark as to the proper use of the mark.

> Make your marks stand out. Distinguish them from other words and symbols.

When possible, remember to distinguish your marks from other words in a text or audio/visual material.

In text, it is often helpful to capitalize a word mark and to provide a registration symbol next to it, so that the mark stands out from other words on the page.[76] In some circumstances, it makes sense to make extra space

> Unlike other types of intellectual property, trademarks do not expire and can theoretically exist in perpetuity.

around the mark or to use a different font color to emphasize it. It is often helpful to state the word "brand" and the type of goods next to the mark, at least once, when referring to the projects goods or services. When giving an interview, regarding the project's new product or service, the manager should refer to [Insert Trademark]-brand [generic product], such as "Anchor®-brand bagels." Such references will ensure that the mark is presented and treated as a proper trademark.

Generally, unlike other types of intellectual property, trademarks can exist forever.

As long as they are timely renewed, used in commerce, and properly managed, they will not expire. Some of the trademarks have been in existence for hundreds of years. However, proper management is the key. If trademarks are important, project management must closely monitor the use of the marks in the marketplace by both the project and third parties. It is also the

> While governments register and grant IP rights, they rarely proactively enforce them. Project's management must act as project's IP police, spotting potential violations and raising alarms.

responsibility of the project management to police the marketplace against any misuse or infringement of the trademarks and other intellectual property.

POLICING FOR IP INFRINGEMENT

While governments register trademarks and copyrights and grant patents, they rarely proactively enforce IP rights. Some jurisdictions do have criminal laws on the books, intended to prevent large-scale infringement. However, in order to recover damages or receive an injunction, preventing illegal use of IP, an injured party must bring an action in civil court. Therefore, project's management (or attorneys hired by the project) must act as IP police, spotting potential violations and raising alarms.

In some areas of IP law, investigating and stopping improper uses is the IP-holder's affirmative duty. Policing is especially important for trademarks. For example, in the United States, an owner of a trademark has affirmative duty to police the use of the mark.[77] Usually, such policing involves regularly monitoring the databases of registered and applied-for marks and searching the Internet for uses of the same or similar marks. A number of companies exist that offer the service of monitoring registered trademarks for possible infringement. A project manager must respond and follow up on all consumer complaints that report confusingly similar names.[78] Depending on the nature of the mark and the jurisdiction, a number of other affirmative steps may be taken in policing the mark.

> Among other affirmative policing actions, a project manager must regularly check the Internet and registration databases for similar marks, as well as to respond and follow up on all consumer complaints that report confusingly similar names in commercial use.

Once project management finds a potential misuse of the mark, affirmative steps must be taken to immediately stop such misuse.

Sometimes, reaching out to the infringing party and issuing a warning will be enough to stop undesirable actions. Sometimes, legal action must be brought and the courts and/or law enforcement involved. In either case, local IP attorneys should be involved in resolving any misuse.

> If any misuse of project's marks is discovered, immediate affirmative steps must be taken to stop misuse.

Failure to police and legally enforce the mark may result, among other consequences, in weakening of the mark's legal enforcement strength. Mark's distinctiveness may be blurred by the use of similar marks, or its reputation tarnished due to bad publicity or confusion with products or services from a lower-quality source. Sometimes, inaction may lead to loss of all rights in the mark.

> Failure to police project's trademarks may result in the loss of all rights in the marks.

CUSTOMS AND BORDER PROTECTION AGENCIES – A TOOL AGAINST INTERNATIONAL IP INFRINGEMENT

If your project successfully comes up with a popular product or service, chances are, it will be instantly copied. In today's interconnected world, infringement and theft of intellectual property is truly a problem of global proportions. Markets exist in most major cities of the world, where counterfeit goods, bearing some of the most famous trademarks, are sold. It takes mere days from the presentation of new fashions in Milan to counterfeit copies of these fashions, flooding the markets around the world.

If you diligently protected your project's IP in the country of development and/or manufacture, as well as in other countries, you have a remedy. In all countries, where project's IP is registered, authorities may be notified and/or a law suit may be brought against the infringers. A rights-holder may ask the courts not only for damages (i.e., financial reimbursements) but for an injunction, requiring an immediate cessation of infringing activity. Depending on the type of IP infringed and on the jurisdiction, the definition of "infringers" that may be held liable is very broad. IP infringement may be direct or contributory. In many cases, the rights-holder may bring a law suit not only against the manufacturers/providers of infringing goods but also against the importers, distributors, retailers, and others that profit from infringing activity.

> In many cases, the rights-holder may bring a law suit not only against the manufacturers/providers of infringing goods, but also against the importers, distributors, retailers, and others that profit from infringing activity.

However, the manufacturer of infringing goods is likely to be located abroad, outside of the country where project's rights are protected. Most often, the production of infringing goods will occur in a locale where the IP rights have not been registered or in a jurisdiction that does not offer strict enforcement of IP rights. How can project's management prevent such an infringer from sending counterfeit goods into the countries where IP is protected? For this purpose, border protection agencies of many countries offer a powerful tool.

Each registrant of a US Trademark receives a letter, attached to the registration certificate. The letter states: "Please note that U.S. Customs & Border Protection (CBP), a bureau of the Department of Homeland Security, maintains a trademark recordation system for marks registered at the United States Patent and Trademark Office. Parties who register their marks on the Principal Register may record these marks with CBP, to assist CBP in its efforts to prevent the importation of goods that infringe registered marks. The recordation database includes information regarding all recorded marks, including images of these marks. CBP officers monitor imports to prevent the importation of goods bearing infringing marks, and can access the recordation database at each of the 317 ports of entry."

That is, as long as the mark is registered, the owner of the mark may put US Customs on the lookout for the infringing goods.[79] The form, filed with the Agency, requests applicants to provide detailed information regarding the likely nature of the infringing products, as well as where they are likely to arrive from and when. As long as the mark owner provides CBP with detailed information on what to look for, protection against infringing goods crossing the border is likely to be highly efficient. Registration with CBP is also a relatively inexpensive process with government fees at the time of this writing, being under $200 per recordation.

US law also allows for similar protection of registered copyrights.[80] US law states: "Claims to copyright which have been registered in accordance with the Copyright Act of July 30, 1947, as amended, or the Copyright Act of 1976, as amended, may be recorded with Customs for import protection."[81]

> As long as your project's trademarks and copyrights are properly registered, they may be recorded with customs, to prevent the importation of infringing goods.

Information and guidance, concerning the recordation of trademarks and copyrights with CBP, is available on their website: www.cbp.gov. CBP recordation page is available at https://iprr.cbp.gov/.

While patents and trade secrets may not be directly registered with CPB, there is a mechanism for border interceptions of products in violation of these IP laws as well. In the United States, in order to stop such cross-border violations, project management may file a complaint with the US International Trade Commission. In response to the complaint, the Commission may open what is referred to as "Section 337 investigation." According to the commission, such investigations "most often involve claims regarding intellectual property rights, including allegations of patent infringement and trademark infringement by imported goods. Both utility and design patents, as well as registered and common-law trademarks, may be asserted in these investigations. Other forms of unfair competition involving imported products, such as infringement of registered copyrights, mask works or boat hull designs, misappropriation of trade secrets or trade dress, passing off, and false advertising, may also be asserted. Additionally, antitrust claims relating to imported goods may be asserted. The primary remedy available in Section 337 investigations

> In the United States, in cases of unfair competition from abroad, project's management can file a complaint with US International Trade Commission. The Commission is empowered to consider and rule on a broad range of complaints, involving IP violations and unfair competition.

is an exclusion order that directs Customs to stop infringing imports from entering the United States."[82] A number of additional remedies to the IP owner are available within the scope of such investigations as well.

Additional information on Section 337 investigations may be obtained on US International Trade Commission's website: www.usitc.gov/. A PDF publication, concerning Section 337 investigations, is available at www.usitc.gov/intellectual_property/documents/337_faqs.pdf.

A growing number of countries are providing IP owners with tools to prevent cross-border infringement. Many countries have introduced a notification system for alerting customs and intercepting infringing goods. Among countries with such customs notification systems are: Argentina, Australia, China, European Union (EU), Hong Kong, India, Japan, Malaysia, Mexico, New Zealand, Russia, Singapore, South Korea, Taiwan, Thailand, Turkey, Ukraine, United States, Vietnam, and many others.[83]

> Numerous countries allow for protection against foreign infringement through customs notification systems. If your project manufactures a product that may be infringed in this manner, set up this protection.

Interception and destruction of goods at the border not only prevents the entrance of such goods onto the market, but also causes financial damage to the infringer. It sends a strong message of discouragement to future infringement. It is a powerful tool of IP protection. But this tool is generally available in each country only for IP that is properly registered in that particular country. This once again emphasizes the need for a project to properly and timely (before any problems arise) register IP in every country that may have a legal or business connection to the project or its products or services. Once IP is registered, this powerful tool is often just a single customs-registration form away.

TRADEMARK SEARCH

It is an absolute duty of a project manager to ensure that any term (word, phrase, symbol, design, etc.) that the project publicly uses does not infringe IP rights of others. If it does, such term will not only be unregistrable and unprotectable by the project, but may subject the project (and any organizations behind the project) to legal liability. As stated above, infringement of someone's trademark does not require intent. Simple failure to do proper IP due diligence may land a project in significant legal and financial trouble.

> It is an absolute duty of a project manager to ensure that any term that the project publicly uses does not infringe IP rights of others.

Therefore, before any new name is adopted for use by the project, a thorough search must be conducted to determine whether the mark may be used. The search must be as broad as possible and include, among other sources, national trademark registration databases and a search of unregistered uses of the mark (as such uses may also confer legal rights). A search must be done in all jurisdictions, where the mark will be used or where future use is anticipated.

> Before any new name is adopted for use by a project, a thorough search must be conducted to determine whether the mark may be used.

It is highly recommended that such search be conducted by an IP attorney in every jurisdiction. In deciding what marks to consider for adoption, project managers may do a basic search of registered trademarks themselves. Usually, searchable trademark databases are maintained by intellectual property offices in each jurisdiction and are available on their websites. In the United States, such database is maintained by the US Patent and Trademark Office and is accessible on Office's website: www.uspto.gov/trademarks-application-process/search-trademark-database.

A database of existing registered marks is an important tool for project management in picking candidates for potential trademarks. However, for final clearance for registration and use, qualified advice of an IP attorney should be obtained.

WIPO Global Brand Database Tool—This extremely powerful tool for global trademark searching is hosted and maintained by the World Intellectual Property Organization. It is accessible at www.wipo.int/branddb/en/.

> WIPO Global Brand Database Tool allows for simultaneous searching of trademarks across multiple national and international sources.

The database allows for simultaneous searching of trademarks across multiple national and international sources. In addition to trademarks, this collection of data includes emblems, appellations of origin and geographic indications.

In cases where a complex project spans more than one country or if there is any likelihood that the project's marks will be used internationally, this database must be searched, along with national databases. If a search of this database reveals similar marks in other countries, project's management must be cautious in proceeding with the use of the marks.

In case of a project's success, such marks will predetermine limitations on potential global expansion of the brand and therefore put a limit on potential growth of project's products or services. It is always best to start a new project from scratch by using marks that bear as few limitations of use as possible. This includes limitations on international expansion. Therefore, even if a particular mark is registrable and protectable in the project's home-country, a project manager must pay close attention to the results of the international trademark search.

> If international trademark search reveals similar marks in other countries, project's use of such marks will set limits on brand's global expansion. Project's management should invent trademarks that are unique and distinctive enough to allow for geographically unlimited expansion of project's goods and services.

One unique and important feature of the WIPO Global Brand Database Tool is the image search. It allows the user to upload an image and search for visually similar trademarks around the world. This feature is particularly useful for searching graphical marks.

> WIPO Global Brand Database Tool offers highly useful image search feature, which allows the user to find visually similar marks across the world.

In conducting a trademark search or analyzing the results of such search, a project manager must remember that failure to find the exact mark among existing registrations does not clear the mark for use. Trademarks often have a scope of protection that goes far beyond their exact representation. Even though a proposed trademark may have a different spelling or look somewhat different from registered marks, it may still infringe existing marks due to the similarity of the marks in their entireties as to appearance, sound, connotation, and/or commercial impression.

> Your project's trademark need not be an exact copy of another in order to infringe. Even though a proposed trademark may be different from existing marks, it may still infringe existing marks due to the similarity of the marks in their entireties as to appearance, sound, connotation and commercial impression.

An important question to ask prior to registration or any use of a trademark is whether consumers are likely to be confused or misled, by the use of the proposed mark. Various jurisdictions may approach this issue differently. In the United States, for example, the points to consider (other than similarity of the marks) would be the relatedness of the goods or services, sophistication of potential consumers, similarity of trade channels in which the goods or services will be offered, other similar marks on the market, and any agreement between the holders of the marks, among a multitude of other factors.[84]

Whether a mark can (and should) be used in the marketplace is a complex legal (and marketing) issue that often determines the success of the project. As such, it must be made, based on the entirety of available information. Proper, professionally conducted, trademark search and opinion uncovers the most valuable bits of such information.

PATENT SEARCHING

In most jurisdictions, in order to qualify for a patent, an inventor must show that the invention is novel. In the United States, the invention will not be granted if it was already "[…]patented, described in a printed publication, or in public use, on sale, or otherwise available to the public before the effective filing date of the claimed invention."[85] This is perfectly logical from a public policy perspective. Why would society reward anyone with a monopoly for disclosing information that is already in the public domain? Likewise, the invention must be non-obvious. In the United States, the invention is not patentable "[…]if the differences between the claimed invention and the prior art are such that the claimed invention as a whole would have been obvious before the effective filing date of the claimed invention […]."[86]

If a project is producing new technologies, a search may indicate the likelihood that a patent may be obtained and thus the likelihood of protection against competitors. If similar inventions have been described in the past, or if pre-existing publications make the invention obvious, obtaining patent protection may be difficult or impossible. Alternatively, if patenting is possible, the scope of protection may be narrow. In many instances, if the technology cannot be adequately protected by patents, then financial investments into such technology may be unwarranted. In some cases, trade

> Patentability search, or "prior art search," is a tool that provides project manager with information, regarding the "state of the art.

> A timely patentability search may save the project significant futile investments into unprotectable technologies.

secret protection must be considered and meticulously implemented from the start. Thus, a timely patentability search may provide guidance to the right mode of protection. It may also save the project from significant expenses, related to patenting and from futile investments into unprotectable technologies.

Even more importantly, a search of existing patents indicates whether or not a technology may be freely used. It is critical for project management to know whether someone has a patent, covering an aspect of technology in use by the project. If they do, all use of such technology must be stopped or avoided, unless and until a license is obtained from the rights-holder.

> A search may reveal third-party patents, covering some technologies, in use by the project. All use of such technology must be stopped or avoided, unless and until a license is obtained from the rights-holder.

Failure to do so jeopardizes all of the profits and benefits of the project, which may later have to be disgorged to the owner of the patent.

Often, project managers turn to patent attorneys, asking to patent a particular device or method, so that they may have a right to use the invention in commerce. Such a request reveals a significant misconception that the general public holds about patents. There is a frequently encountered belief that a patent is required in order to practice a certain invention. However, a patent DOES NOT give the right to the inventor or the owner of the patent to make, use, or sell anything.[87]

> A patent DOES NOT give the right to the inventor to make, use, or sell anything.

For example, an inventor may receive a patent for a new type of nuclear weapon or a dangerous psychedelic drug. However, the issuance of a patent does not give an automatic right to test such a weapon or sell such a drug. If use of the invention is generally prohibited by any laws, prohibition extends to the patent holder as well. According to WIPO Intellectual Property Handbook, the effects of the grant of a patent are that the patented invention may not be exploited in the country by persons other than the owner of the patent unless the owner agrees to such exploitation. Thus, while the owner is not given a statutory right to practice his invention, he is given a statutory right to prevent others from commercially exploiting his or her invention.[88]

In the example above, while the inventor does not have the right to test the weapon or administer the drug, he/she does have a right to prevent others (that would otherwise be allowed to test or administer) from doing so. If a government defense agency is interested in testing the new weapon or a pharmaceutical company is interested in distributing the drug, they must obtain permission (i.e., license) from the owner.

This right to prevent others is what allows the patent-owner to derive material benefits from the invention. The benefits are derived either by enjoying a monopoly in practicing the invention alone in the marketplace or by selling licenses to the invention.

> This right to prevent others from practicing the invention is what allows the patent-owner to derive material benefits from the patent by enjoying a monopoly or selling licenses.

Few patents are granted on revolutionary, ground-breaking machines or processes, the likes of which never existed. Rather, most inventions are improvements of pre-existing technologies. Sometimes, a patent application for a novel and non-obvious improvement may be filed without sufficient preliminary research. Sometimes, such unresearched applications even issue into patents, if some of the invention's aspects are unique and non-obvious. However, if someone else has patent-protected rights to the underlying technology, the holder of the improvement-patent, will not have full rights to manufacture and sell their patented invention. Just like in the examples above, if the law (in this case, patent law or patent-protected superior rights of others) prohibits the inventor from using the underlying technology, necessary for their invention, the inventor (or the project) may be prohibited from using their own patented invention.

> If the law or pre-existing IP rights of others prohibit the use of underlying technology, the inventor may be prohibited from using their own patented invention.

Of course, the invention may be used and practiced once (and if) a license is obtained from the holder of the underlying patent. But this often means that some of the profits and benefits of the invention have to be shared with the senior rights-holder.

An in-depth patent search can reveal pre-existing rights to underlying technologies that can prohibit the use of project's technologies, even if a patent for such technologies is eventually secured. Awareness of such pre-existing rights can be a critical bit of knowledge that may radically influence planning and financial calculations for the entire project.

Like with trademark searches, most countries' intellectual property offices provide access to searchable databases of patents and patent applications. In the United States, such database is maintained by the US Patent and Trademark Office. Detailed advice on proper searching and analysis techniques as well as links to the databases are available at www.uspto.gov/patents-application-process/search-patents.

> US Patent and Trademark Office provides a full searchable database of US patents and patent applications and provides extensive materials on patent searching and analysis.

PATENTSCOPE is a powerful searching tool, provided by the World Intellectual Property Organization. It is a database that can be used to search International Patent Cooperation Treaty (PCT) applications, as well as patent documents from some national and regional patent offices. This tool can be accessed at https://patentscope.wipo.int.

> PATENTSCOPE is a highly useful tool for searching International Patent Cooperation Treaty (PCT) applications.

Of course, not all complex projects invent and patent their own technology. Often, a project will negotiate licenses with patent-rights holders in order to obtain rights to use and implement protected inventions. However, a project cannot buy more rights than a patent-owner holds.

> A project can not buy or license more rights than a patent-owner holds. Always research the scope of patent-holder's rights.

Often, a patent will describe the invention in its entirety on its face. However, the project manager must always ask whether the patent truly removes all obstacles to practicing the invention. If a patent licensed by a project is an improvement on an underlying, protected invention, it can only confer limited rights onto the licensee.

When someone buys real property, such as a house, a title search is commonly executed to investigate whether the seller has full rights to the property. Similarly, when buying or licensing intellectual property, such as a patent, a thorough patent search must be professionally executed and analyzed to ensure that the licensor or seller indeed has full rights to the invention.

COPYRIGHT REGISTRATION

In the United States, registrations of copyrights are processed by the US Copyright Office. Copyright Office is a separate federal department within the Library of Congress.

In order to register a work, an applicant must submit a completed application form. Along with the form, nonreturnable copy or copies of the work to be registered must be submitted. Copyright applications, as well as digital copies of the work, may be submitted online at www.copyright.gov/registration/. Depending on the nature of the work and on work's publication status, sometimes physical copies of the work must be mailed to the Office.[89]

The project manager must be aware that all applications submitted to the Copyright Office become public records. All information provided on the application as well as copies of the work become accessible to the public.[90] Some or all of this information may be available over the Internet. Therefore, great care must be taken not to reveal any information that may be confidential or may constitute project's trade secret.

> All information provided on and with the US copyright application, becomes accessible to the public. In filing for copyright protection, project manager must be cautious not to reveal any information that may be confidential or may constitute a trade secret.

US Copyright Office publishes a series of extremely helpful and detailed circulars, intended to provide information to the general public. The circulars are well written and well-organized, and cover a broad range of topics, related not only to registration but also to maintenance, work-for-hire issues, licensing, and a number of other important copyright-related issues. As these circulars are highly authoritative, they constitute an important tool for a project manager, who encounters copyright-related issues. The circulars are available for download online at www.copyright.gov/circs/.

> US Copyright Office publishes a series of circulars that explain all aspects of Copyright registration and law. These circulars are a powerful tool for a project manager, encountering copyright-related issues.

While details of copyright registration may differ from country to country, general procedures are usually similar to the ones described above.

INTERNATIONAL ASPECTS OF COPYRIGHT REGISTRATION

There is no such thing as an "international copyright" that will automatically protect the author's works throughout the world.[91] Protection against unauthorized use in a particular country depends on the national laws of that country.[92] However, as described earlier in this chapter, broadly adopted international copyright treaties have allowed and greatly simplified registration of copyright in multiple countries. If copyrightable works constitute an important component of project's IP, a project manager must retain legal representation and seek registration of copyrights in each individual country that may be relevant to the project. The project manager must remember that even though certain copyright protections may attach without the need for formal registration, in most jurisdictions, works must be registered early and properly in order to obtain full protection.

> There is no such thing as an "international copyright." Protection depends on national laws. Thus, legal representation must be obtained and registrations applied for in each individual country that may be relevant to project's success.

TRADEMARK REGISTRATION

Just like other IP laws, trademark laws are highly country-specific. One of the major distinctions in trademark laws of various countries is what occurrence gives right to trademark claim and ownership. In some countries, rights to a mark primarily arise from actual use of the mark in the marketplace. Such countries are referred to as "first-to-use" trademark jurisdictions. In other countries, the rights arise from actual registration of the mark. Such countries are referred to as "first to file."[93] Some countries apply a combination of these approaches.

> In some countries, the rights and priority of use in the mark arise from first use in commerce, in others, from first registration of the mark.

In first-to-file jurisdictions, the first party to register a trademark generally receives all rights of priority. In first-to-use countries, a project must actually use the mark in commerce, prior to obtaining registration and any benefits, associated with registration. "First to use" versus "first to file" is an important legal distinction that determines the proper first steps that a project manager must take with project's marks.

A project manager must research and know which of these legal principles applies in project's jurisdiction(s). In first-to-use jurisdictions, expediting use of the mark in commerce may be required to ensure priority of rights. In first-to-file jurisdictions, early filing may be critical.

> In first-to-use jurisdictions, expediting use of the mark in commerce may be required to ensure priority of rights. In first-to-file jurisdictions, early filing may be critical.

United States is the first-to-use jurisdiction. Generally, a trademark cannot be registered unless the applicant files a "statement of use," along with a specimen (proof) showing the actual use of the mark in commerce in the United States (some exceptions exist for foreign applications).[94] It is possible to apply for trademark registration in the United States prior to using the mark, as long as the applicant has a good-faith intent of using the mark in commerce. Such an application will be termed "intent to use application." The application will be examined for registrability. Any issues that arise with such application may be resolved with the examiner in the regular course of business. The Office may even issue a "notice allowance," approving the registration of the mark, based on the "intent to use" application. However, the registration certificate will not be issued and the mark will not be registered until the applicant provides "statement of use" and proves the use of the mark in commerce.

As registration provides multiple rights and advantages even in first-to-use countries, project management should not wait for the actual use in order to file the application. "Intent to use" application may be filed for all marks intended for projects' use, while such marks are being prepared for commercial exploitation. Examination of trademarks takes months. This means that for months, the applicant cannot be certain whether the mark will be approved for registration. Therefore, it often makes sense to file as early as possible, so that the mark is past the examination stage by the time it's used in commerce.

Early filing also allows the applicant to gauge early-on whether there are issues that make the mark unregistrable. This way, if registration is denied, there may still be time to register an alternative mark.

If a mark has been in use for a considerable amount of time prior to registration, it is likely that consumer associations have already formed and financial investments have been channeled into the mark. In these cases, it is much more difficult to abandon the mark and switch it

> Early trademark registration filing allows project management to gauge early-on whether there are issues that make project's marks unregistrable. This way, if registration is denied, there may still be time to register alternative marks.

for another, in case of registration failure. Therefore, even in first-to-use jurisdictions, project managers must ensure that the mark is registrable by filing for registration as early as practicable.

In most countries, trademark registration applications may be filed online on the website of the national Intellectual Property Office. In the United States, the application may be electronically filed at www.uspto.gov/trademarks-application-process/filing-online. The site contains all necessary forms and information on filing. A fee must be submitted, along with the application.

Once the application is received, it is reviewed to ensure that minimum filing requirements have been met and forwarded to the trademark examiner.[95] This stage may take several months. The examiner will then examine all parts of the application, including the drawings and specimens for compliance with technical requirements. The mark will then be examined for the purpose of determining whether it can serve as a trademark or a service mark. A search will be conducted to identify any conflicting marks that may prevent registration. If examiner identifies any issues with the application or the mark that prevent registration, they will issue an office action (a letter), identifying the problems and requesting a response. Frequently, genericity of the mark, insufficiency of specimen, and existence of confusingly similar marks will be among the reasons for the rejection. The applicant may respond to the office action by citing arguments in response. If the arguments are found persuasive, the mark will be allowed for further processing toward registration. If the arguments are not persuasive, a final office action rejecting the registration, will issue.[96]

In the United States, if the mark is approved by the examiner, it moves into the publication stage. The mark will be published in the "Official Gazette," which is a publication issued by the USPTO. The publication informs the public of USPTO's intention to register the mark. It opens up a 30-day period in which a third party may file an opposition to the registration. If no opposition is filed, the mark will register and the certificate of registration will be mailed to the applicant in due time.

The trademark application process in most countries is similar to the US process described above. However, it should be noted that in many countries, the opposition period is not part of the registration process. Rather, the period to oppose opens up after the mark is already registered and may be significantly longer than in the United States.

Maintaining National Trademark Registration

Once a trademark is registered, it must be maintained and periodically renewed. Maintenance requirements and deadlines differ from country to country and are listed on national IP offices' sites. Most frequently, maintenance requires payment of fees and forms, confirming continuous use of the mark.

For example, in the United States, "Combined Declaration of Use and Incontestability under Sections 8 & 15," along with the fees is filed between the 5th & 6th year after registration. "Combined Declaration of Use and/or Excusable Nonuse/Application for Renewal under Sections 8 & 9," along with the required fees is then filed every 10 years after registration.[97] US registration maintenance and renewal forms are available on USPTO's website: www.uspto.gov/trademarks-application-process/filing-online/registration-maintenancerenewalcorrection-forms.

Unlike patents or copyrights, trademarks do not have a set expiration date and may theoretically exist forever. However, failure to file the required maintenance forms will result in cancellation and/or expiration of the registration. Far too many project managers recognize the importance of registering the trademarks, but forget to maintain them. Once the mark is canceled or expires, it is frequently very difficult or impossible to revive. All priority dates may be lost, and in some cases, status of foreign registrations may be effected. The mark may have to be refiled anew and is not guaranteed to register. Therefore, it is extremely important for project management to monitor and timely file all the renewal documents and fees. It is strongly recommended that IP management firm is retained by the project to ensure timely IP maintenance.

International Trademark Filing

MADRID SYSTEM, also known as Madrid System for the International Registration of Marks, is a powerful and highly efficient tool for global management of project's trademark portfolio. The

System legally stems from Madrid Agreement Concerning the International Registration of Marks (1891), as well as the Protocol Relating to the Madrid Agreement (1989), described earlier in this chapter.

Madrid System provides a cost-effective and efficient way for trademark holders to ensure protection for their marks in multiple countries through the filing of one application with a single office, in one language, with one set of fees, in one currency.[98] A single application with the System allows the applicant to apply for protection in up to 119 countries. Moreover, as project's IP needs evolve, this dynamic tool allows project management to modify, renew or expand project's global trademark portfolio through one centralized system.[99]

> Madrid System provides a cost-effective and efficient way for trademark holders to ensure protection for their marks in multiple countries through the filing of one application with a single office, in one language, with one set of fees, in one currency.

However, it must be reiterated that IP laws, and trademark laws in particular, are national laws. Madrid System allows for highly efficient and expedient transmittal of trademark applications to multiple national offices. However, application through the System DOES NOT guarantee registration in all countries selected by the applicant. It remains the right of each country or contracting party designated for protection to determine whether or not protection for a mark may be granted.[100] Once the trademark office in a designated country

> Application through Madrid System DOES NOT guarantee registration in all countries selected by the applicant. It remains the right of each country to determine whether or not protection for a mark may be granted.

grants protection, the mark is protected in that country in the same way as if that office had registered it.[101] That is, the registration and maintenance in each country remains subject to local laws and interpretations.

General overview of the Madrid System, along with legal and educational resources, is available on WIPO's website: www.wipo.int/madrid/en/. Forms, required for the Madrid System filings, are available at www.wipo.int/madrid/en/forms/. National IP offices of contracting countries also provide a wealth of jurisdiction-specific resources, relating to the System. For example, USPTO provides a wide collection of links and resources on its Madrid Protocol webpage, available at www.uspto.gov/trademark/laws-regulations/madrid-protocol.

PATENT APPLICATION PREPARATION AND FILING

Patent laws are, perhaps, the most technical and complex of all Intellectual Property laws. Preparing and drafting a patent application is not a task for the uninitiated. In most countries, it takes years of education and professional legal experience in order to properly draft specification and claims for a utility patent application. In some countries, a plurality of graduate degrees are required to take patent attorney certification tests.

Unlike with Trademark and Copyright registration filings, patent applications require much more than filled-out forms. Utility patent applications may be hundreds of pages long and may include numerous drawings and graphs. Most application parts must be drafted from scratch. Much of the language used in the applications "comprises" specific legally mandated jargon that often appears nonsensical and grammatically incorrect to people unfamiliar with patent drafting. Utility patent is a highly complex legal document that, in great detail, conveys and claims the scope of inventor's rights. Therefore, more than in any other IP field, it is critical that project management retains a law

> Unlike with Trademark and Copyright registration filings, patent applications are often drafted from scratch and may be hundreds of pages long.

firm, staffed with nationally registered/certified patent agents and attorneys in order to prepare patent applications and manage patent-related issues.

From the first steps of the project, patent practitioners must be involved, in order to identify what aspects of the project's IP may be eligible for patent protection and whether such protection is advisable. Often, great ideas may not be patentable, for a number of reasons. Often an idea is not sufficiently conceived or reduced to practice to become a patentable invention.[102] In other cases, the invention may be outside the scope of patent law. For example, in the United States, abstract ideas, laws of nature, and natural phenomenon cannot be patented.[103] If patent protection is determined to be appropriate for a project's invention, a decision must be made as to the type of patent protection.

> In the United States, abstract ideas, laws of nature, and natural phenomenon can not be patented.

In most jurisdictions, several types of patent applications exist. For example, the US Patent and Trademark Office issues three primary types of patents, each of which involves differing application and maintenance procedures. These types of patents are as follows:

Utility patent is the most common type of patent, and it is what most people associate with the term "patent." As the name suggests, such patent describes and claims (usually in great detail) utilitarian aspects of the invention. Such patents are also referred to as "Patents of Invention" They may be applied for and granted to anyone who invents or discovers any new and useful process, machine, article of manufacture, or composition of matter, or any new and useful improvement thereof.[104] Such patents can offer the broadest scope of protection and have a term of 20 years. Utility patent applications generally take the most time and effort to prepare and file. Accordingly, preparation, filing, and prosecution of such applications often require significant legal expenditures by the project.

Design patents may be granted to anyone who invents a new, original, and ornamental design embodied in or applied to an article of manufacture.[105] US design patents offer a 15-year term of protection. Design patent applications primarily consist of the drawings, illustrating the ornamental designs. Often, there is a very fine line between the applicability of design patents and copyrights. In most cases, one, but not the other is applicable. In rare cases, both may have relevance.[106] A decision on which mode of protection applies to a particular project's designs should be made after consultation with a patent attorney.

Plant patents may be granted to anyone who invents or discovers and asexually reproduces any distinct and new variety of plant.[107] Such patents generally provide a 20-year term of protection.

Once it has been determined what types of patent applications are appropriate, project management should direct project's legal representatives to prepare and file these types of patent applications as promptly as possible. Early filing will provide a project with earlier priority filing date. As previously mentioned in this chapter, an earlier priority filing date can make a critical difference in a project's ability to obtain patent protection. There are many strict legal deadlines for filing of applications. But even if deadlines are adhered to, an article by anyone, appearing in any media (including the Internet) between the time of invention and patenting may have an effect on an invention's novelty or obviousness, making it unpatentable. Alternatively, a competitor may beat the project to filing for the same of similar invention, thus intercepting all of the project's rights to patent (or even use) the invention. Accordingly, patent filings must be executed as soon as possible, and the earliest possible priority filing dates must be secured.

> Once it has been determined what type of patent applications are appropriate, project management should order proper filing, to provide a project with earliest possible priority filing date.

Yet, sometimes, it can take weeks or even months to properly draft and file a complete utility patent application. Utility applications also require significant up-front financial investments. In cases, where priority dates for multiple inventions must be obtained as soon as possible, time and financial pressures may be considerable. Under these circumstances, project management may consider a utility patent filing tool known (in the United States and some other countries) as a *Provisional Patent Application.*

Provisional patent application is a national utility application that has lower costs and simplified procedural filing requirements, in comparison with the regular utility application. Simplified procedural requirements allow for more expeditious application drafting and filing, thus resulting in earlier priority dates. Unlike a full non-provisional utility application, a provisional application is not required to have a formal patent claim, information disclosure statement or an oath or declaration.[108]

In the United States, provisional patent applications DO NOT result in the issuance of a patent. In fact, provisional patent applications are not even examined for their substance. Instead, they act as place-holders for filing of complete non-provisional applications.

Upon filing of a provisional application, project management will have up to 12 months to prepare and file a non-provisional application. The later-filed non-provisional application may refer to the earlier filing and claim its priority date. This way, when a complete application is filed, up to a year later, it will be treated as if it was filed at the time of the provisional application. Provisional application expires 12 months after filing (unless it is timely converted to a non-provisional application). This 12-month period cannot be extended.[109]

In the United States, pending Provisional Application allows for the term "Patent Pending" to be applied in connection with the description of the invention.[110]

An extra year of waiting gives the applicant/project an opportunity to properly draft and file full non-provisional application, as well as to resolve any financial aspects of filing. In some cases, a year provides sufficient time for the project to test the market and to decide whether further investments into patenting of a particular technology make commercial sense.

Project management must remember that while many aspects of the provisional application are simplified in comparison with the non-provisional application, enablement requirement is not changed. That is, the description must still be detailed enough to enable one, skilled in the art, to practice the invention without undue experimentation.[111] Therefore, filing brief summaries of the invention that leave much to reader's imagination will not be legally sufficient for obtaining a priority filing date.

Project management may also consider provisional application filings for another advantage that such

---

Whenever a project is faced with significant up-front costs and time pressures, related to filings of multiple utility patent applications, project management should consider the filing tool, known as a Provisional Patent application.

Provisional patent applications have simplified procedural requirements. This provides for expeditious application drafting and filing, which allows the project to secure earlier priority dates for its patent applications.

In the United States, provisional patent application DOES NOT result in the issuance of a patent. Instead, for 12 months, it acts as a place-holder for a complete non-provisional application.

In the United States, a filed provisional application enables a project to use the term "Patent Pending" in reference to its inventions, even though full non-provisional application has not been filed yet.

An extra year of waiting, gives applicant an opportunity to resolve any drafting and financial issues, related to non-provisional filing.

Provisional application's description must be detailed enough to enable practice of the invention without undue experimentation.

applications offers. Typically, the term of a utility application is limited to 20 years. However, when a non-provisional application claims and receives priority of an earlier-filed provisional application, the term of protection covers the time of the provisional appli-

> Provisional filing can allow a project to obtain up to an extra year of patent protection.

cation's pendency, plus the full 20-year term of a later-filed application. This way the provisional filing can allow a project to obtain up to an extra year of patent protection. In some cases, an extra year of patent protection can bring in significant extra income. For example, an extra year of protection for a popular patented drug can be worth billions to a pharmacological project. Therefore, project management should carefully consider and discuss an option of filing a provisional patent application, with their attorneys, even in cases where a full non-provisional one can be timely filed.

The option of provisional patent application does not exist in all countries. Therefore, project managers must check local jurisdiction, regarding the availability of this tool. However, in some countries where it exists, provisional filing can offer even more benefits than in the United States. For example, in Australia, an applicant may request examination of provisional patent applications. This option can allow project's management to gauge invention's patentability prior to filing a finalized patent application.

In some countries, where provisional patent applications technically do not exist, some way of obtaining advantages of a provisional filing may still be available. For example, in Canada, applicants may file imperfect applications and are then provided time to amend them. This provision allows for prompt filing to obtain a priority date, when timing is of the essence. Canadian Intellectual Property Office advises: "Since the formal requirements for receiving a filing date are not as stringent as before, you may now file a description of your invention without everything being in perfect order. You then have 15 months to complete the application with no additional fee."[112]

In most countries, patent applications may be filed online on national Intellectual Property Office's website. In the United States, patent filings may be made through the EFS-Web System, available on the USPTO's website. Access to the EFS-Web System, as well as a wealth of resources and tools related to patent filings, is available at www.uspto.gov/patents-application-process/applying-online/efs-web-guidance-and-resources. Additional information on US provisional patent application filing may be obtained at www.uspto.gov/patents-getting-started/patent-basics/types-patent-applications/provisional-application-patent.

PATENT MAINTENANCE

In most countries, patents must be maintained, in order to remain in force. In the United States, regular maintenance fees must be paid for maintenance of utility patents. The payments are generally required three times throughout the "life" of the utility patent. USPTO requires patent maintenance payments at 3 to 3.5 years, 7 to 7.5 years, and 11 to 11.5 years after the date of issue.[113] In the United States, maintenance fees are not required for design or plant patents.[114] More information on US patent maintenance fees is available at www.uspto.gov/patents-maintaining-patent/maintain-your-patent.

As is the case with trademarks, patent maintenance requirements and deadlines differ from country to country and are listed on national IP offices' sites.

INTERNATIONAL PATENT PROTECTION

When it comes to protecting inventions internationally, project management has several options. If countries, where patent protection is required, are identified at the time of the initial filing, a patent may be filed in each country, where protection is sought, individually. Such a "manual" approach may be necessary, if some of the countries, where project's filings will be made, are not members of international IP conventions.

If all countries, where filings will be made, are members of Paris Convention, then application may first be filed in one of the countries. Then, as long as filings in other countries are made within 12 months, these filings may claim priority of the initial application. This approach is useful where applications will only be filed in several countries and a project manager knows the exact countries from the start of the patenting process.

However, in most cases, it is not clear from the start where and whether foreign filings will be made. Patent prosecution in some countries can cost tens of thousands of dollars and beyond. Sometimes, it may be preferable to wait and gauge invention's success and patenting progress in one country prior to deciding to embark on a patenting spree around the world. In other cases, invention may be so promising and competition so intense that project management may want to file for protection in as many countries as possible. In both of these situations **Patent Cooperation Treaty,** commonly known as the **PCT,** is an indispensable tool for managing global patent filings.

The **PCT** is an international treaty with more than 150 contracting states. The PCT allows project management to seek patent protection for an invention in a large number of countries simultaneously by filing a single "international" patent application instead of filing several separate national or regional patent applications.[115]

> The PCT is an indispensable tool for managing global patent filings. The PCT allows project management to seek patent protection for an invention simultaneously in a large number of countries by filing a single "international" patent application.

PCT greatly simplifies the process and reduces the costs of international filing, as the applicant can designate any or all of the contracting entities/states on one application and have copies of patent application forwarded to these states. If not for this tool, just for initial filing, project management would have to find and retain patent practitioners in each country, which itself would be a lengthy and expensive proposition. The patent application would then have to be translated into local languages and filed, in accordance with local regulations in each jurisdiction. All of this would have to be done promptly and simultaneously, so as to retain application's priority. PCT lets applicants avoid these burdensome first steps.

> One PCT feature that is particularly useful to project management, is the ability to temporarily delay selection of the countries, where patent application will be filed, without losing the priority of filing.

One PCT feature that is particularly useful to project management is the ability to temporarily delay selection of the countries, where patent application will be filed, without losing the priority of filing.

When PCT application is filed, all member-states are designated by default. However, this does not mean that application must be prosecuted in all of the designated countries. Processing of a PCT application is a lengthy multi-step process. In some cases, the applicant may be allowed a time period as long as 30 months from the priority date in order to eventually decide in which of the designated countries/jurisdictions to proceed with obtaining a patent.

> The PCT application does not directly result in the issuance of a patent in any country. Rather, it is part of a process that eventually leads to filing of national patent applications.

A project manager must remember that just as there is no such thing as international trademark law, there is no such thing as international patent law. There is also no such thing as "international patent." The granting of patents remains under the control of the national or regional patent offices.[116] The PCT application itself does not directly result in the issuance of a patent in any country. Rather, a PCT application is part of a process that eventually leads to a number of applications, filed with national/regional patent offices in what is called the "national phase." These applications will, in turn, be examined by these offices in accordance with local laws and may be granted in some jurisdictions and rejected in others.

PCT does not completely shield the applicant from the costs and complexities of eventually dealing with local patenting authorities in each jurisdiction. However, it simplifies the process of entering the national stage in each jurisdiction. The delay of over 2 years for entering the national stage lets project managers carefully consider the plan of action. In many cases, by the time the PCT national stage selections are due, the project will have had time to test the invention in the marketplace. By then, it may be clear whether the invention is worth the protection and further investment.

By the 30-month mark, the perspectives of obtaining a strong patent on the invention (at least in the applicant's home-country) should also be apparent. Opinions on the likelihood of protection in foreign countries will also have been provided as part of the PCT process, giving project manager solid basis for deciding on the next steps in IP protection.

> The delay of over 2 years for entering the national stage that PCT provides lets project managers' ample time to plan for international patent protection.

PCT filing provides all of these great benefits for the effort of filing a single application and a fee, which at the time of writing, amounted to only several thousand US dollars. PCT is one of the most powerful tools in project manager's IP toolkit, as it grants project management ample time and information to make a well-substantiated decision, regarding international patent protection.

A list of PCT contracting states is available at www.wipo.int/pct/en/pct_contracting_states.html.

Information on the PCT process, including links to fees, forms, and filing tools, is available on WIPO's website: www.wipo.int/pct/en/.

A helpful list of commonly asked questions, regarding the PCT, is available at www.wipo.int/pct/en/faqs/faqs.html.

Much PCT-related information is also available on IP Office websites of PCT member-states. Valuable information and links, regarding all stages of the PCT process, are provided on USPTO's website: www.uspto.gov/patents-getting-started/international-protection/patent-cooperation-treaty.

Remember that while there are tools for international IP management, there is no such thing as international IP law. Local Intellectual Property Offices provide a wealth of information on regional aspects of IP protection and registration. A list of national Intellectual Property Offices throughout the world is available at www.wipo.int/directory/en/urls.jsp.

QUESTIONS

You are a project manager of a top-secret project, established by a major international cosmetics manufacturer. The goal of the project is development of a new and unique all-natural sunscreen lotion with a previously unheard-of ultra-high sun protection factor (SPF) of 10,000. Along with the new lotion, the project must develop marketing strategies and advertising materials for this new product.

A foreign-based research team was hired to perform scientific research and testing for your project. The team has determined that in order to achieve the high protection characteristics, the lotion must contain a unique combination of rare-earth metals, finely ground fermented frog livers, bird feathers, and fish scales.

Although the researchers have achieved some success, the optimal ratios of scales to feathers and fermentation times are yet to be determined. You recently learned that some of the members of the research team have been anonymously posting questions to their colleagues in an online forums, asking for suggestions on frog-liver fermentation.

You have learned from one of the project's employees, who used to work for a competing cosmetics manufacturer and maintains friendships with old co-workers that competitors have set up a similar project. They are also working on a lotion with similar protective characteristics. Competitor's employees were observed visiting local poultry and fish markets. Reliable reports suggest that a fish tail, sticking out of the shopping bag of one of the competitors' employees, was unscaled.

Your project's product will have a unique look and feel. Rare-earth metals provide a distinctive pea-green color that glows in the dark. Fermented livers contribute to the unique exaggeratedly organic aromas of the product. Your marketing team is developing an unusually shaped container. The container will be bird-shaped with fish-scale outside texture.

The marketing team has also just presented you with the first version of a music video commercial. The commercial is set to a tune of a famous catchy song about sunshine. In the video, an actor, costumed to look like a famous dead rock star, is dancing by a shelf, with multiple sunscreen lotions positioned on the shelf. He picks up lotions of famous brands, pronounces their names and low SPFs, and then dumps them into garbage one by one. He then picks up the project's brand. He smiles, hugs the container, and lets out a loud Tarzan scream.

1. As a project manager, should you be concerned about any IP issues in your project?
2. Identify protectable aspects of project's IP. What steps would you take to protect project's IP? Why would you take these steps and not the alternatives?
3. Come up with three great name options for the product. What are the strengths and weaknesses of each name? Which of the three is the best option? Why?

CONCLUSION

This chapter provides a basic overview of some of the IP issues that are extremely relevant in complex project management. It aims to arm project managers with powerful practical tools to recognize and tackle these issues. IP law is a broad and complex field. Constraints of a single chapter allowed us to explore just the tip of the iceberg of a range of IP issues that may come up in a complex project and to present just some of the multitude of available tools. Many more tools are waiting to be discovered, when you follow the links, suggested in this chapter.

This chapter intends to give project managers a general feel of the extensively pervasive nature of IP issues. The intent of this chapter is to make him/her understand that IP issues stealthily permeate all aspects of complex project management. IP issues are easy to overlook and hard to remedy.

Most of the IP management tools and tips, presented in this chapter, are highly practical. Their purpose and importance may not be obvious at first glance. However, once a project manager starts applying some of them to the actual projects, he/she will be able to spot the critical threats, major expenditures and hidden opportunities that would otherwise lurk unnoticed. Through practicing the use of these tools, many of the threats will be neutralized and opportunities realized. Once these tools are regularly applied to ongoing projects, their use will soon become second nature. The project manager will then be holding the reins of control over powerful legal forces. Project's competitiveness and chances of success will significantly increase.

These tools will not magically turn a project manager into a seasoned IP practitioner. However, this is not necessary. Luckily, there are seasoned IP practitioners available for hire in every jurisdiction. Project manager's most powerful IP skill is a realization that IP issues exist within a project. The second most powerful skill is the identification of these issues. This chapter's tools hone and perfect these skills. The skills, in turn, put reader on the path of finding, choosing and competently interacting with a project manager's most powerful IP tool—a locally licensed and experienced IP attorney.

> Project manager's most powerful IP skill is a realization that IP issues may exist. Project manager's most powerful tool is an experienced IP attorney.

REFERENCES

1   A specialized agency of the United Nations (UN) system of organizations, tasked with administering the international aspects of intellectual property law.

2   www.wipo.int/about-ip/en/, About IP, Accessed December 14, 2018.

3   www.wto.org/english/tratop_e/trips_e/intel1_e.htm., World Trade Organization, What Are IPRS, Accessed December 15, 2018.

4   Dan Breznitz, Michael Murphree, What the U.S. Should Be Doing to Protect Intellectual Property, Harvard Business Review, January 27, 2016, https://hbr.org/2016/01/what-the-u-s-should-be-doing-to-protect-intellectual-property.

5   World Intellectual Property Organization (WIPO), Convention Establishing the World Intellectual Property Organization (as amended on September 28, 1979), Article 2 (viii), https://wipolex.wipo.int/en/text/283833, Accessed January 23, 2019.

6   USPTO, Trademark, Patent, or Copyright? www.uspto.gov/trademarks-getting-started/trademark-basics/trademark-patent-or-copyright, Accessed December 24, 2018.

7   Id.

8   Jessica Tyler, The 10 Most Valuable Brands in 2018, /www.inc.com/business-insider/amazon-google-most-valuable-brands-brand-finance-2018.html, Accessed January 23, 2019.

9   WIPO Intellectual Property Handbook: Policy, Law and Use. Chapter 2: Fields of Intellectual Property Protection, www.wipo.int/export/sites/www/about-ip/en/iprm/pdf/ch2.pdf, Accessed December 23, 2018.

10 35 U.S.C. 101.

11 See, for example, 35 U.S. Code § 102(a)1 "A person shall be entitled to a patent unless the claimed invention was patented, described in a printed publication, or in public use, on sale, or otherwise available to the public before the effective filing date of the claimed invention."

12 www.wipo.int/copyright/en/, Accessed December 23, 2018.

13 www.copyright.gov/circs/circ15a.pdf, US Copyright Office Circular 15A—Duration of Copyright, Accessed December 23, 2018.

14 101 U.S. 99 (1879).

15 www.wipo.int/export/sites/www/about-ip/en/iprm/pdf/ch2.pdf. See also BASIC NOTIONS OF COPYRIGHT AND RELATED RIGHTS, Document prepared by the International Bureau of WIPO, page 3, www.wipo.int/export/sites/www/copyright/en/activities/pdf/basic_notions.pdf, Accessed December 24, 2018.

16 17 U.S. Code § 106—Exclusive rights in copyrighted works.

17 WIPO, What is a Trade Secret? www.wipo.int/sme/en/ip_business/trade_secrets/trade_secrets.htm, Accessed December 23, 2018.

18 See TRIPS, Article 39.

19 WIPO, What is a Trade Secret? www.wipo.int/sme/en/ip_business/trade_secrets/trade_secrets.htm, Accessed December 23, 2018.

20 http://e15initiative.org/wp-content/uploads/2015/09/E15-Innovation-LippoldtSchultz-FINAL.pdf—Trade Secrets, Innovation and the WTO—Douglas C. Lippoldt and Mark F. Schultz—August 2014.

21 www.uspto.gov/patents-getting-started/international-protection/trade-secret-policy, Accessed December 23, 2018.

22 http://e15initiative.org/wp-content/uploads/2015/09/E15-Innovation-LippoldtSchultz-FINAL.pdf—Trade Secrets, Innovation and the WTO—Douglas C. Lippoldt and Mark F. Schultz—August 2014, Accessed December 23, 2018.

23 USPTO Trade Secret Policy, www.uspto.gov/patents-getting-started/international-protection/trade-secrets-policy, Accessed December 24, 2018.

24 Kostylo, J. (2008) 'Commentary on the Venetian Statute on Industrial Brevets (1474)', in Primary Sources on Copyright (1450–1900), eds L. Bently & M. Kretschmer, www.copyrighthistory.org.

25 Id.

26 www.patentsavers.com/2016/04/early-u-s-patent-law-system/, Accessed December 20, 2018.

27 See, for example, www.essaysinhistory.com/thomas-jefferson-and-the-patent-act-of-1793/., Accessed December 12, 2018.

28 James Pooley, Was America's Industrial Revolution Based on Trade Secret Theft? www.ipwatchdog.com/2017/07/05/americas-industrial-revolution-based-trade-secret-theft/id=85377/.

29 Id.

30 www.wipo.int/about-wipo/en/history.html.

31 www.wipo.int/treaties/en/ShowResults.jsp?treaty_id=2—WIPO-Administered Treaties.

32 WIPO—Paris Convention for the Protection of Industrial Property—www.wipo.int/treaties/en/ip/paris/.

33 For more information on Paris Convention and its terms, see www.wipo.int/treaties/en/ip/paris/summary_paris.html.

34 www.wipo.int/treaties/en/convention/.

35 See, for example, 15 U.S.C. § 1117, or 35 U.S.C. § 284, US laws establishing recovery in cases of Trademark and Patent infringement, respectively.

36 35 U.S.C. § 286.

37 SCA Hygiene Products Aktiebolag v. First Quality Baby Products, LLC 580 U.S.(2017).

38 https://supreme.justia.com/cases/federal/us/580/15-927/.

39 Petrella v. Metro-Goldwyn Mayer, 134 S. Ct. 1962 (2014).

40 35 U.S.C. 102 (a) (1).

41 17 U.S.C. §102(a).

42 *Id.*

43 FIXED PERSPECTIVES: THE EVOLVING CONTOURS OF THE FIXATION REQUIREMENT IN COPYRIGHT LAW, Evan Brown, Cite as: 10 Wash. J.L. Tech. & Arts 17 (2014), http://digital.lib.washington.edu/dspace-law/handle/1773.1/1388.

44 See Copyright Office Circular 9—Works Made for Hire, www.copyright.gov/circs/circ09.pdf, Accessed January 23, 2019, "From the moment it is set in a print or electronic manuscript, a sound recording, a computer software program, or other such concrete medium, the copyright becomes the property of the author who created it."

45 Define Derivative Work.

46 See Copyright Office Circular 9—Works Made for Hire, www.copyright.gov/circs/circ09.pdf, Accessed January 23, 2019.

47 17 U.S.C. §101.

48 *Community for Creative Non-Violence v. Reed.* 490 U.S. 730 (1989).

49 See, for example, Ethicon Inc. v. U.S. Surgical Corp., 135 F.3d 1456.

50 USPTO Trade Secret Policy www.uspto.gov/patents-getting-started/international-protection/trade-secret-policy, Accessed January 12, 2019.

51 R. Mark Halligan, David A. Haas, The Secret Of Trade Secret Success, FORBES, February 19, 2010.

52 *Id.*

53 USPTO Trade Secret Policy, www.uspto.gov/patents-getting-started/international-protection/trade-secret-policy, Accessed January 12, 2019.

54 *Id.*

55 Hiatt v. Ziegler, 179 USPQ 757, 763 (Bd. Pat. Inter. 1973).

56 USPTO Trade Secret Policy, www.uspto.gov/patents-getting-started/international-protection/trade-secret-policy, Accessed January 12, 2019.

57 See 35 U.S. Code § 102(a)1.

58 See Trademark Manual of Examining Procedure, October 2018. Section 1209—Refusal on Basis of Descriptiveness, https://tmep.uspto.gov/RDMS/TMEP/Oct2012#/Oct2012/TMEP-1200d1e6980.html, Accessed January 14, 2019.

59 See, for example, Trademark Manual of Examining Procedure, October 2018. Section 1209.01—Distinctiveness/Descriptiveness Continuum, https://tmep.uspto.gov/RDMS/TMEP/Oct2012#/Oct2012/TMEP-1200d1e6980.html, Accessed January 14, 2019.

60 *Id.*

61 See generally Trademark Manual of Examining Procedure, October 2018. Section 1209—Refusal on Basis of Descriptiveness, https://tmep.uspto.gov/RDMS/TMEP/Oct2012#/Oct2012/TMEP-1200d1e6980.html, Accessed January 14, 2019.

62 Trademark Manual of Examining Procedure, October 2018. 1209.01(a) Fanciful, Arbitrary, and Suggestive Marks, https://tmep.uspto.gov/RDMS/TMEP/Oct2012#/Oct2012/TMEP-1200d1e7036.html, Accessed January 14, 2019.

63 Playtex Products v. Georgia-Pacific Corp, 390 F.3d 158.

64 See In re George Weston Ltd., 228 USPQ 57 (TTAB 1985).

65 See, for example, Trademark Manual of Examining Procedure, October 2018. Section 1209.01—Distinctiveness/Descriptiveness Continuum, https://tmep.uspto.gov/RDMS/TMEP/Oct2012#/Oct2012/TMEP-1200d1e6980.html, Accessed January 14, 2019. "Fanciful, arbitrary, and suggestive marks, often referred to as "inherently distinctive" marks, are registrable on the Principal Register without proof of acquired distinctiveness."

66   Trademark Manual of Examining Procedure, October 2018. 1209.01(b) Merely Descriptive Marks, https://tmep.uspto.gov/RDMS/TMEP/Oct2012#/Oct2012/TMEP-1200d1e7074.html, Accessed January 14, 2019.

67   *Id.*

68   *Id.*

69   Trademark Manual of Examining Procedure, October 2018. 1212 Acquired Distinctiveness or Secondary Meaning, https://tmep.uspto.gov/RDMS/TMEP/Oct2012#/Oct2012/TMEP-1200d1e10316. html, Accessed January 14, 2019.

70   Trademark Manual of Examining Procedure, October 2018. 1209.01(c) Generic Terms, https://tmep. uspto.gov/RDMS/TMEP/Oct2012#/Oct2012/TMEP-1200d1e7132.html, Accessed January 14, 2019.

71   It should be noted that at least some of the referenced marks in "Genericide" examples remain as registered and protected trademarks in some jurisdictions.

72   Gary H. Fechter, Elina Slavin, Practical Tips on Avoiding Genericide, INTA Bulletin, November 15, 2011 Vol. 66 No. 20, www.inta.org/INTABulletin/Pages/PracticalTipsonAvoidingGenericide.aspx, Accessed January 15, 2019.

73   *Id.*

74   See generally INTA—A Guide To Proper Trademark Use For Media, Internet and Publishing Professionals http://inta.org/Media/Documents/2012_TMUseMediaInternetPublishing.pdf, Accessed January 24, 2019.

75   *Id.*

76   *Id.*

77   See, for example, Ben Wagner, A Primer on Policing Your Trademark, www.mintz.com/insights-center/ viewpoints/2251/2013-04-primer-policing-your-trademark, Accessed January 24, 2019.

78   *Id.*

79   Such registration is granted in US law under 19 CFR 133.1.

80   19 CFR 133.31(a).

81   *Id.*

82   See United States International Trade Commission, About Section 337, www.usitc.gov/intellectual_ property/about_section_337.htm, Accessed January 19, 2019.

83   The IP Exporter, June 12, 2013, Understanding Foreign IP Customs Notification Registration Procedures, https://theipexporter.com/2013/06/12/understanding-foreign-ip-customs-notification-registration-procedures/, Accessed January 19, 2019.

84   See, for example, Trademark Manual of Examining Procedure, October 2018, TMEP §1207.01, Likelihood of Confusion, https://tmep.uspto.gov/RDMS/TMEP/Oct2013#/Oct2013/TMEP-1200d1e5044.html, Accessed January 24, 2019.

85   17 U.S.C. §102(a).

86   17 U.S.C. §103.

87   WIPO Intellectual Property Handbook: Policy, Law and Use. Chapter 2.

88   *Id.*

89   See, generally, US Copyright Office Circular 2, Copyright Registration, www.copyright.gov/circs/ circ38a.pdf, Accessed January 18, 2019.

90   *Id.*

91   US Copyright Office Circular 38A, International Copyright Relations of the United States, www. copyright.gov/circs/circ38a.pdf, Accessed January 18, 2019.

92   *Id.*

93   See, for example, Lanning G. Bryer, International Trademark Protection, Acquisition of Trademark Rights, page 16, www.inta.org/trademarkadministration/Documents/Bryer_TMs_in_Business_ Transactions.pdf, Accessed January 20, 2019.

94   For general steps to obtaining US trademark, see USPTO—Trademark process—An overview of a trademark application and maintenance process, www.uspto.gov/trademarks-getting-started/trademark-process#step1, Accessed January 20, 2019.

95   *Id.*

96   *Id.*

97   See USPTO—Registration Maintenance/Renewal/Correction Forms, www.uspto.gov/trademarks-application-process/filing-online/registration-maintenancerenewalcorrection-forms, Accessed January 24, 2019.

98   USPTO, Madrid Protocol, www.uspto.gov/trademark/laws-regulations/madrid-protocol, Accessed January 20, 2019.

99   WIPO—Madrid—The International Trademark System, www.wipo.int/madrid/en/.

100   USPTO, Madrid Protocol, www.uspto.gov/trademark/laws-regulations/madrid-protocol, Accessed January 20, 2019.

101   *Id.*

102   See, for example, Solvay S.A. v. Honeywell International, 742 F.3d 998, 1000 (2014), "Making the invention requires conception and reduction to practice. While conception is the 'formation, in the mind of the inventor, of a definite and permanent idea of a complete and operative invention,' reduction to practice 'requires that the claimed invention work for its intended purpose.'"

103   SEE USPTO—MPEP § 2106—Patent Subject Matter Eligibility, www.uspto.gov/web/offices/pac/mpep/s2106.html, Accessed January 21, 2019.

104   See USPTO—General information concerning patents, www.uspto.gov/patents-getting-started/general-information-concerning-patents, Accessed January 21, 2019.

105   *Id.*

106   Choosing between Design Patent and Copyright Protection? Sometimes You Don't Have To, April 20, 2016, https://mabr.com/choosing-between-design-patent-and-copyright-protection-sometimes-you-dont-have-to/.

107   See USPTO—General information concerning patents, www.uspto.gov/patents-getting-started/general-information-concerning-patents, Accessed January 21, 2019.

108   USPTO—Provisional Application for Patent—www.uspto.gov/patents-getting-started/patent-basics/types-patent-applications/provisional-application-patent, Accessed January 21, 2019.

109   *Id.*

110   *Id.*

111   See, for example, Storer v. Clark, 860 F.3d 1340 (Fed. Cir. 2017).

112   Government of Canada, Canadian Intellectual Property Office, Requirements For Filing Patent Applications In Canada, www.ic.gc.ca/eic/site/cipointernet-internetopic.nsf/eng/wr00315.html, Accessed January 21, 2019.

113   USPTO—Maintain Your Patent, www.uspto.gov/patents-maintaining-patent/maintain-your-patent, Accessed January 21, 2019.

114   *Id.*

115   WIPO—Protecting your Inventions Abroad: Frequently Asked Questions About the Patent Cooperation Treaty (PCT), www.wipo.int/pct/en/faqs/faqs.html, Accessed January 21, 2019.

116   *Id.*

15 Systems Thinking Toolset

Michael Emes
University College London

David Cole
Independent Consultant

CONTENTS

BACKGROUND

INTRODUCTION

The vast majority of projects fail when judged by their ability to deliver their specified objectives on time and to cost, especially when there are significant organisational change, software or information technology components (The Standish Group 2014). This continues to be the case, despite increases in computing power and knowledge. Various factors contribute to the problems in these projects including excessive ambition, over-optimistic planning and failure to manage risk (Flyvbjerg 2003, Flyvbjerg, Garbuio, and Lovallo 2013, Kahneman 2011). At the heart of the problems of many modern complex projects, however, is a failure of understanding and, in particular, a failure to appreciate how changes in one part of a system (product, service or project) may have subtle knock-on effects elsewhere in the system, often not immediately, but developing over time.

> At the heart of the problems of many modern complex projects, however, is a failure of understanding, and in particular a failure to appreciate how changes in one part of a system (product, service or project) may have subtle knock-on effects elsewhere in the system, often not immediately, but developing over time.

Research into conditions for project success (APM 2015) found that only 22% of projects were wholly successful, with 12% not meeting their budget and 17% not delivering on time (although 90% of respondents considered their projects successful to some degree). Respondents identified the top three success factors as (1) clear goals and objectives, (2) project planning and review, and (3) effective governance. Systems thinking not only supports the definition of projects but also their effective execution, by exploring both the complexity of the problem and how the proposed solution will be implemented. From this, a project can be defined that can be delivered according to cost, schedule and performance expectations.

THE IMPORTANCE OF SYSTEMS THINKING

Systems thinking provides a framework to address complex, uncertain and interconnected problems which do not have simple solutions. It recognises that the world is made up of organised groups of entities which have 'emergent properties' (Aziz-Alaoui and Bertelle 2009), or properties or behaviour that cannot be predicted by analysing the parts in isolation or through simple aggregation. While simple projects may contain relatively few interactions, complex projects involve significant interconnection in the problems they address or in the dynamic nature of the environments within which their solutions operate. Here, the application of systems thinking can improve project success rates and reduce delays and budget over-runs by:

- Enabling understanding of what created the problem, its environment and why the system behaves as it does. Dynamic complexity and emergent behaviour are easier to understand when component relationships and the cause and effect cycles are understood, including the human aspects.
- Enabling the definition of effective solutions and implementation strategies, together with clear objectives for their realisation. By revealing assumptions and making them explicit, risks can also be more comprehensively identified, assessed and managed.

- Enabling understanding of how the proposed solution fits into the broader environment so that unintended consequences can be addressed before implementation.
- Facilitating contribution and buy-in from stakeholders as a working understanding of systems thinking can be taught relatively quickly. For example, drawing causal loop diagrams helps teams develop a shared understanding of the problem and their place in it (and possible contribution to it), without apportioning blame. For broader audiences, the diagrams used by systems thinking are effective for communicating the problem, its causes and the proposed solution.
- Enabling dynamically complex problems to be addressed, characterised by non-linearity and interdependencies. The traditional top-down 'divide-and-conquer' approach is often unsuccessful for these problem types as it is neither clear where the 'top' is nor how the problem should be divided. Systems thinking enables these problem types to be addressed holistically to identify the key leverage points where interventions will have the greatest effectiveness.

OVERVIEW OF SYSTEMS THINKING

HARD SYSTEMS THINKING

Although aspects of systems thinking can be recognised in engineering projects as far back as for the building of the pyramids, the emergence of systems thinking in a project context is usually linked with the management of technology projects during and after the Second World War. Application of organised, systematic or hierarchical approaches to planning projects was evident in the development of Gantt charts (Clark 1922) and Network charts such as those used in the programme evaluation and review technique (PERT) for critical path analysis (US Department of the Navy 1958).

Books referring to 'systems engineering' by name appeared in the 1950s and 1960s, such as Goode and Machol (1957), and systems engineering was developed as an approach to engineer functional systems that spanned disciplines. Early projects were mainly military and space based (Westerman 2001), and in this context, systems engineering was established as an approach to optimise complex systems with major technological challenges and significant schedule pressure but much less concern over cost.

According to Hitchins, 'The modern philosophy – the "why" and the "how" of today's systems engineering developed mostly at NASA in the 1960s and 1970s' (Hitchins 2003, 76). The early military and aerospace approach to systems engineering emphasised the activities involved rather than the holistic principles. The Defense Systems Management College produced a Systems Engineering Management Guide (DSMC, 1983) which explained the steps in systems engineering, starting with requirements analysis and ending in the synthesis of alternative solutions.

SOFT SYSTEMS THINKING

Researchers at Lancaster University in the 1960s had great expectations of the impact of systems engineering, stating that 'it is not unreasonable to claim that a new industrial revolution is now on its way with the advent of systems engineering, a revolution which is going to exert a major influence on how industry can be organised so as to integrate properly the potentialities of people and the possibilities of technology' (Jenkins and Youle 1971). The Lancaster team recognised that although the systems approach was originally inspired by the need to solve technological challenges with relatively little consideration of human interactions with the system, some of the same thinking could be adapted to apply to so-called 'human activity systems' such as business problems. Soft Systems Methodology (SSM) was developed as an attempt to address questions that were outside the scope of hard systems engineering as it was defined at the time (Checkland 1981). Since the 1970s, systems thinking approaches have developed significantly, and a broad range of project complexity can now be tackled with systems thinking (Jackson 2003).

Traditionally, problems are analysed and solutions developed using the approach of top-down functional decomposition (NASA 2007), where the primary purpose or function of the solution is defined and agreed, with all further definition flowing from this. Decomposition continues until buildable components are identified. These components are then built and integrated at increasing levels until the whole solution is obtained. This approach works well for problems with a single agreed cause or where there is no pre-existing solution that the new solution must operate within.

There is an increasing need to address problems where there is no agreement on the boundary for the solution, however, or where there is a need to take into account the dynamic complexity of the environment within which the solution will operate. The holistic approach provided by systems thinking is needed.

Defining Systems Thinking

Systems thinking can be thought of in terms of its purpose and the means of its practical application.

Purpose of Systems Thinking

We can consider the purpose of systems thinking as a discipline for seeing wholes rather than parts, for seeing patterns of change rather than static snapshots, and for understanding the subtle interconnectedness that gives the systems their unique character (Senge 1990).

Application of Systems Thinking

Systems thinking uses approaches, tools and methods to help explore and map dynamic complexity:

- Approaches to think about and frame dynamically complex problems and their solutions
- Diagrams and other tools to visualise and understand dynamic complexity
- A vocabulary with which to express understanding of dynamic complexity
- Methods to coherently apply the approaches and tools.

Application of systems thinking requires a conceptual move from considering the individual system components to considering the structure of the system, its interconnections and the interactions between its components and its environment. Systems thinking does not replace traditional top-down thinking but complements it to enable problems that are intractable using a top-down approach to be addressed successfully.

Systems Thinking Principles

Various authors have suggested a broad range of principles that should be adopted as we encounter systems (Meadows 2008, SEBoK authors 2017) or as we seek to develop them (Emes, Smith, et al. 2012). Amongst the most important general principles are the following:

- Understand the bigger picture by identifying the patterns and trends generated as inputs, outputs and the environment change over time. Analyse external events and their internal responses. This enables the problem to be framed in terms of patterns of behaviour over time rather than focusing on particular events.
- Recognise that a system's structure generates its behaviour, that the context of relationships must be understood and identify the (often circular) cause and effect relationships, rather than focusing on details. Appreciate that internal actors managing policies and system operation are responsible for system behaviour rather than assuming that behaviour is driven by external forces.
- Make assumptions explicit and test them. Recognise that all models are working hypotheses with limited applicability (Box 1979) rather than seeking to prove models by validating

them with existing data or new measured data. While things cannot always be measured, they can always be quantified.

• Change perspective to increase understanding. See causality as uncertain and ongoing (with feedback influencing causes, and causes driving each other), rather than viewing causality as running only one way, with causes being independent.

These systems thinking principles are embodied in the recognition that events are the 'tip of an iceberg' of a deeper structure, as shown in Figure 15.1 (Senge 1990).

Events

The Events level is the level at which we experience the world and experience the symptoms of problems. While some problems observed as events can be addressed directly, not all can be solved by treating the symptoms and a deeper analysis is needed.

Patterns of Behaviour

Looking below, the Events level often identifies patterns of behaviour to enable events to be forecast and hypotheses for their causes developed and tested.

System Structure

Consideration of System Structure (including consideration of the broader environment within which the system operates) identifies the causes of the pattern(s) observed.

Mental Models/Mindset

Mental models (or mindset) are the (often subconscious) assumptions, beliefs and expectations that cause systems to function as they do and are used to drive problem analysis and solution definition. As Senge (1990, 53) points out 'The deepest insight comes when they realise that their problems and their hopes for improvement are inextricably tied to how they think'.

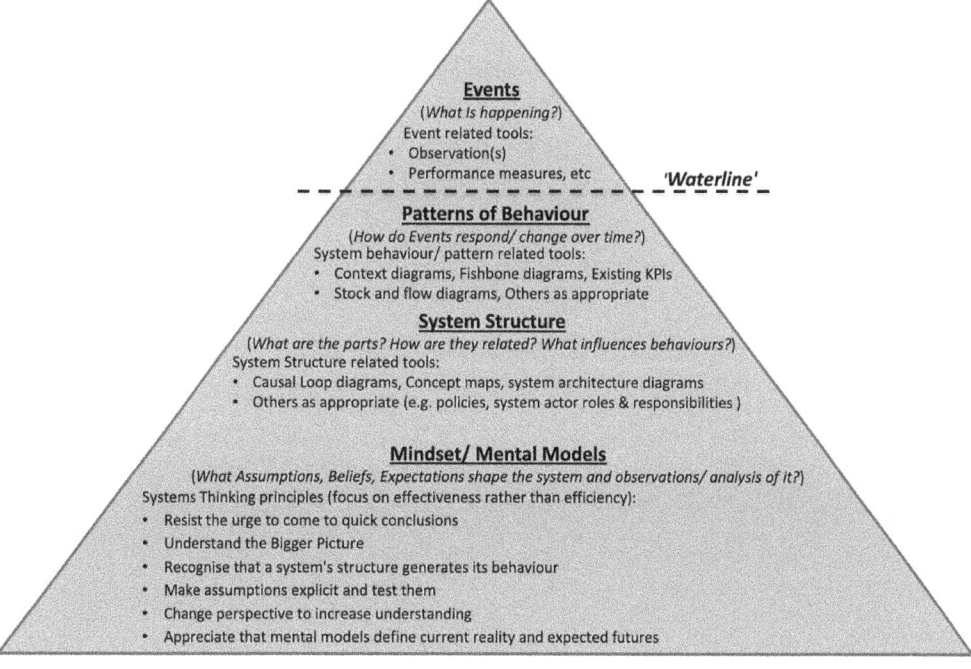

FIGURE 15.1 Systems thinking iceberg, based on Senge (1990).

Both for the solution to the problem and for the management of its development, thinking should focus on what is required for an *effective* solution as opposed to an *efficient* solution. Efficiency is important so that resources are not wasted, but consideration of effectiveness – the extent to which higher-level goals are achieved – should come first. Consideration of effectiveness should be in terms of the recipients or users of the system outputs.

Both for the solution to the problem and for the management of its development, thinking should focus on what is required for an *effective* solution as opposed to an *efficient* solution.

SYSTEMS THINKING APPROACH

The application of systems thinking comprises the three main generic steps described below and shown in Figure 15.2. These steps are derived as a combination of the methods from Seddon's Check-Plan-Do cycle (Seddon 2008), SSM (Checkland 1981) and Senge's 'The Fifth Discipline' (Senge 1990). Note that here, the term 'solution' is perhaps a little idealistic. Complex problems cannot be simply and permanently solved through the design of the perfect system or project. Rather, the solution represents an intervention that makes a positive step towards addressing the problem. Addressing complex problems will necessarily be an ongoing, iterative endeavour. With this caveat in mind, systems thinking application typically follows the steps below. The main diagram tools given are an initial suggestion and are described later in this chapter. Other tools should be used where they can provide insight into the problem.

Define the As-Is Situation to Understand the Problem and Its Causes

There is not just one way to define a system; the system or a problem can be defined in different ways. A system is an imaginary construct (Emes, Bryant, et al. 2012, Martin 2008), and the way we conceive of and architect a system can be done in many different ways depending on our point of view or purpose. Similarly, a system or a situation may be seen as pathological by some stakeholders and normal or even desirable by others.

Systems thinking helps by encouraging a broad range of stakeholders to participate, such as through the development of rich pictures. This involvement provides a

A system is an imaginary construct (Emes et al. 2012, Martin 2008), and the way we conceive of and architect a system can be done in many different ways depending on our point of view or purpose.

A system or a situation may be seen as pathological by some stakeholders and normal or even desirable by others.

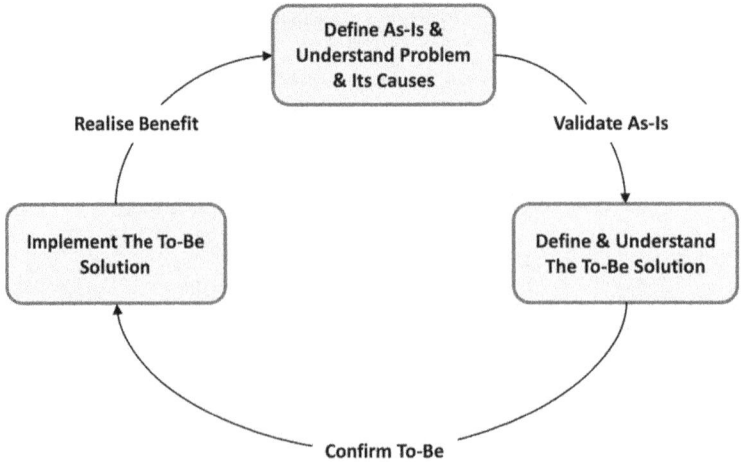

FIGURE 15.2 Generic systems thinking process.

better view of the full breadth of the scope and helps to communicate why there is a problem that needs to be addressed. This should also help later, as the stakeholders will be better engaged and more likely to buy into the proposed solution.

Understanding the problem also critically involves risk identification; risk management should start at the very beginning of the project. Assumptions must be made explicit and tested. As a minimum, each assumption should be stated and the risk that the assumption is incorrect should be evaluated. Mental models play an important role here. They define both the current perception of reality and also frame the expected future. The full scope of the problem should be documented in a project brief or as a separate problem scoping document, which may be added to as the analysis of the detail of the problem is developed further. This forms the basis for development of the business case and governs the later application of change control.

The performance of the system over time should also be examined, using techniques such as causal loop diagrams to highlight important variables and stock and flow diagrams to model how key variables may change over time. The latter can be used to investigate how progress in projects is impacted by various factors. Stock and flow diagrams have been successfully used during litigation, for example, to explore the impact on cost and schedule of changes to customer requirements (Cooper 1993).

Main activities
- Describe and understand the bigger picture and define the system boundary
- Identify and describe cause-and-effect relationships between system elements and the environment
- Make assumptions explicit and test these assumptions
- Develop a conceptual model of the problem and test this on the causes of the problem
- Resist the urge to come to quick conclusions: consider issues fully
- Change perspectives to increase understanding
- Iterate as necessary.

Thinking
- Identify, describe and understand patterns of behaviour over time
- Identify and consider the actors (internal and external) responsible for behaviours and the policy environment(s) within which they operate
- Understand the context(s) of behaviours, actors, policies and relationships identified.

Core Tools
- Fishbone diagrams, context diagrams, actor maps, trend maps, causal loop diagrams, stock and flow diagrams, and checklists.

Define and Understand the To-Be Solution That Will Address the Problem

A system is a body consisting of interconnected parts. We must understand that seeing the world through a systems lens is to recognise that a whole consists of parts, and these parts must be designed to work together and to deliver value to a higher-level system (or a supersystem). There will be constraints placed upon the system by the supersystem (Emes, Smith, et al. 2012).

In the past, many major technological advances were made by lone innovators driven largely by individual inspiration and effort (Cadbury 2004). In the globally connected information age (Castells 2010) in which we now live, though, even where the vision for a new concept comes largely from an individual, the project that develops the product into a commercial reality almost always transcends discipline, enterprise and national boundaries.

The natural approach to solving large or difficult problems is to start by breaking the problem down into manageable parts – the reductionist or Cartesian philosophy (Hitchins 2003). This top-down or 'divide-and-conquer' approach works for straightforward 'painting by numbers' (Obeng

1995) projects with clear objectives and a well-understood process of how to deliver them. Here, tasks are easily defined and effort is efficiently shared (like building a long wall). For more complicated or intricate projects with a large number of interfaces, specialist skills are required (such as for most modern projects). Here, a divide and conquer approach can still be effective, as long as careful consideration is given to how the system is partitioned to facilitate the design and build process. Modern supply chains are often internationally distributed, requiring careful mapping of activities to organisations and clear definition of responsibilities and interfaces. Projects today are rarely straightforward exercises in which a known recipe is applied to achieve a clearly defined goal, however. Instead, there is uncertainty, either in terms of the best way of reaching the goal or in terms of the goal itself. Objectives or requirements are particularly contentious for projects with a social dimension such as schools, transport systems or health services. Unfortunately, time pressures and traditional approaches to problem-solving often favour 'quick fixes', which may provide some positive results in the short term but ultimately prove ineffective in addressing the higher-level purpose, sometimes even making the problem worse rather than better (Meadows 2008, Senge 1990, Zokaei et al. 2010).

Systems must operate in situations of uncertainty and ambiguity. Many situations cannot be completely described fully; the data necessary to understand a system fully is not available. This aspect of living in an uncertain and unpredictable world is natural for a system. Studies have shown risk management to be a critical success factor for project management in a number of sectors (Tsiga, Emes, and Smith 2016a,b, 2017).

Main activities
- Construct and test hypotheses for solutions to the problem
- Define the 'To-Be' that addresses the problem. This may suggest whether a portfolio, programme or project is required. The formal definition of this is done under the next step
- Develop the conceptual model of the new system
- Make assumptions explicit and test these assumptions
- Resist the urge to come to quick conclusions: consider issues fully
- Change perspectives to increase understanding
- Iterate as necessary.

Thinking
- Identify, describe and understand causality and understand how behaviour is generated
- Identify and describe the feedback (reinforcing and balancing) that influences causes and the relationships between the causes
- Consider how suitable data to enable measurement can be gathered, although this data may not necessarily be quantified.

Core tools
- Context diagrams, concept maps, actor maps, trend maps, causal loop diagrams, stock and flow diagrams.

Implement the To-Be Solution

Having considered the problem and concluded what sort of solution or intervention system is required, systems must be able to respond to external change. Doing this in an effective or constructive way is an important feature of a system.

Systems themselves will in general change over time. In the language of James Martin's Seven Samurai model, this To-Be solution can be thought of as an 'intervention system', which must be sensitive to the needs of its stakeholders but crucially may need to interact with pre-existing elements ('collaborating systems') and will need further systems to sustain its effectiveness (Martin 2004). A system may go through a series of life cycle stages, and the demands and expectations of the system will vary at different stages in its evolution. One of the key decisions when managing an enterprise or a project is to make a judgement on the relative importance of short-term performance relative to the long-term performance. In finance and value management, the concept of a discount rate (Brealey, Myers, and Allen 2005) is introduced to quantify the extent to which we prefer to consume benefits today rather than in a year's time. This discount rate can be compounded to make asset management decisions relating to the performance and maintenance of a system. Systems engineering wisdom encourages early investment in projects to uncover potential problems before too many design decisions have been made (so-called 'left shift' of the project's effort profile). Financial management usually pushes us to delay expenditure in projects as much as possible, however, and the sweet spot between these competing forces is sometimes hard to find (Emes, Smith, and James 2007).

No matter how thorough our planning, change is inevitable in large projects and it's typical for goals to evolve during the life span of complex projects (Aramo-Immomen and Vanharanta 2009). When applying systems thinking, we plan for evolution. Implementing an intervention is just a step in the ongoing process to addressing the problem. Hopefully this intervention will allow the realisation of some benefits, but as Figure 15.2 demonstrates, the process of implementing the solution is iterative.

In the face of these challenges, we use systems thinking as a lens to identify where interventions will have the greatest impact, ensuring that the whole system performance is optimised, accounting for all relevant factors when implementing change (including the interests of the full range of relevant stakeholders and interactions over time). For almost all modern projects, some form of systems thinking will be helpful, but it is particularly important in the face of complexity, where projects are not clearly defined or understood.

Main activities
- Compare the As-Is and To-Be conceptual models to define the changes required
- Define the changes (are both desirable and feasible) that will be implemented by the portfolio, programme or project to be established to implement the solution
- Define the portfolio, programme or project to implement the changes. This includes development of appropriate documentation as required by the organisation: e.g. Blueprint, Project Definition
- Implement the changes by implementing the portfolio, programme or project
- Communicate the changes being implemented
- Transition the solution into operation
- Monitor/review the effectiveness of the changes
- Evaluate the residual As-Is problem.

Thinking
- Recognise that all models are working hypotheses with limited applicability.

Core Tools
- Fishbone diagrams, context diagrams, concept maps, actor maps, trend maps and causal loop diagrams. Portfolio, programme and project documentation associated with the organisation(s) involved.

RELEVANCE OF SYSTEMS THINKING FOR MANAGEMENT OF COMPLEX PROJECTS

Systems thinking can be applied in projects as described above, but the value of systems thinking is increased when the projects have a high level of complexity.

SYSTEMS AND COMPLEXITY

There are various ways of interpreting complexity in projects, such as based on project value or scale, number of organisations, components or interactions, intricacy and technological challenge. Complexity can be divided into the categories of structural complexity, technical complexity, directional complexity and temporal complexity (Remington and Pollack 2007). The clarity of goals and the clarity of method of process can be used to understand the value of various systems thinking approaches for a specific project. Sheffield, Sankaran, and Haslett (2012) define complex projects as those that have a large number of interactions and many components, as shown in Figure 15.3. In this context, systems thinking is seen as the natural response to managing complexity.

Other authors see complex projects as those that have a high level of unpredictability, in particular involving a lack of clarity in the goals. Snowden and Boone (2007) distinguish between complicated and complex projects. Complicated projects can be characterised as having high intricacy (usually with both a large number of components and a large number of interactions) and require a high level of specialist knowledge to complete. Complex projects, on the other hand, can be seen as having an additional characteristic of a high level of uncertainty in the organisational domain – with unclear or incompatible stakeholder needs (Loch, de Meyer, and Pich 2006). These complex projects are likely to include a large number of unknown unknowns (Snowden and Boone 2007).

Building on Turner and Cochrane's (1993) framework, Obeng (1995) describes a simple but effective way of characterising projects according to (1) the level of clarity of the goals and (2) understanding of the process or method for achieving them (Table 15.1).

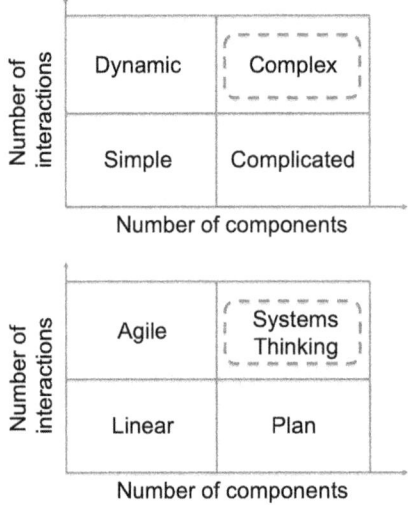

FIGURE 15.3   Types of systems and projects and types of project management methods (Sheffield, et al. 2012).

TABLE 15.1
Project Characterisation

Clarity of Goals	Clarity of Methods/Process	Description (Obeng 1995)
High	High	Painting by numbers
High	Low	Quest
Low	High	Making a movie
Low	Low	Fog

A slight simplification of the Snowden and Boone (2007) model that the authors have found to be effective is to characterise projects into one of just three levels:

- 'Simple', where elements are relatively independent. The solution and approach are agreed and unambiguous, even if the solution stretches technologies. It should be noted that Simple projects are not necessarily easy.
- 'Complicated', where the solution requires different components and interfaces, together with a range of skills for their integration. Careful consideration of how the solution must be partitioned is needed to design and build it. However, the objective and approach are both clear, even if these stretch solution technologies.
- 'Complex', where there is a range of interpretations of the problem. This may be due to differing perspectives on the problem and of the dynamic ways in which it may evolve. There may also be many views on what the scope of the solution should be and the approach to be used.

Applying systems thinking will not give major benefit for simple projects due to their lack of connectedness (although it could help transition-to-operations planning where the 'no-connectedness' aspect may not hold). The value proposition for applying systems thinking is much clearer for complicated and complex projects.

> The value proposition for applying systems thinking is much clearer for complicated and complex projects (than for simple projects).

In the context of the development of large technological projects with high clarity of goals and high clarity of methods, systems thinking typically involves application of the systems engineering approach (Stevens et al. 1998). Where there is clarity of goals but not clarity of methods such as in the product development context, similar structured approaches tend to be applied, but with the addition of a level of experimentation in the search for a solution, with a recognition that selectionist or iterate and learn approaches to risk management are likely to be needed (Loch, de Meyer, and Pich 2006, de Meyer, Loch, and Pich 2002).

Even when the value proposition for a product is well understood, however, the range of stakeholders and viewpoints involved in most modern projects means that there is rarely a common vision of the best way to tackle the project (the clarity of goals is low). Even apparently straightforward projects may have elements of ambiguity and complexity. In this situation, the application of systems thinking is essential to manage complexity by engaging with stakeholders and confronting and challenging assumptions at an early stage in the project's life cycle.

EVIDENCE OF APPLICATION OF SYSTEMS THINKING IN PROJECTS

According to Sheffield et al. (2012), relatively few project managers (PMs) employ systems thinking to manage complex projects, even though a few simple tools could bring unique benefits to problem solving for these projects.

More recent research funded by the Association for Project Management finds that the majority of PMs use some form of systems thinking at least half of the time, but many recognised systems thinking tools like rich pictures and SSM are not widely used. This research also finds that the use of systems thinking tools is slightly greater for more experienced PMs and for those working in sectors that deliver complex technological products such as defence and aerospace (Emes and Griffiths 2018).

DEVELOPING AN ORGANISATIONAL SYSTEMS THINKING CAPABILITY

Systems thinking complements the other skills staff already have rather than being a standalone discipline. A systems thinking capability in an organisation cannot be bought in from outside but must be developed by staff at all levels. A manager in the systems thinking applications study by Senge (1990) points out that 'Systems thinking is only truly learned by doing, by action learning: it is only by doing that managers can unlearn, can find out for themselves where their current beliefs about the design and management of work are flawed, in order to put into place something that works systematically better, and can systemically be further improved'. A typical approach to develop systems thinking capability uses the following steps:

- As with all major organisational change activities, senior management support is vital.
- Mobilise a small team of staff who collectively undertake the end-to-end activities to which systems thinking is to be applied. This team should be led by a manager who believes systems thinking can give significant benefit but may not be a systems thinking practitioner. The team will diagnose the problem and design and implement the changes identified by the systems thinking application. The members of this team need to be freed up from their day-to-day responsibilities and given the scope and time to experiment with different approaches.
- Teach the team the principles and practice of systems thinking. A working knowledge can be taught relatively quickly and be provided by an external consultant if necessary. This consultant should be available for a further period of time to provide systems thinking guidance and support as requested by the team. Using the team to collectively understand the problem and jointly develop the solution enables broad participation to improve the As-Is and To-Be analysis, improve job satisfaction for the people involved and gain their buy-in to the need for change and to the solution developed.
- The team should establish a set of principles to drive solution definition development. These principles should relate to what an effective solution looks like for the 'customer' of the solution rather than simply an efficient one for the organisation using or providing this solution. Facilitation and advice from an external consultant can be useful for this.
- Develop and implement the solution. Although this may be undertaken by others, the team should be available to clarify and resolve any issues with the interpretation or implementation of the solution.
- Transition the solution into operation. This gives the organisation an initial systems thinking capability.
- To expand the systems thinking, a 'roll-in' approach can be used whereby the initial team rolls in further staff into the application of systems thinking as they apply it elsewhere in the organisation.

REVIEW OF SELECTED PRACTICAL TOOLS FOR SYSTEMS THINKING

Systems thinking uses various diagrammatic and visualisation tools to support the approaches described above. Development of the diagrams should involve individuals working in the parts of the organisations covered, with these individuals either being part of the project team or through workshops. Joint development of the diagrams facilitates the processes. Frequently used diagram types are described below. It is not an exhaustive list, and where others provide insight, they should be used. Examples are given at the end of the chapter.

RICH PICTURE

Context diagrams enable definition of the system in its environment. Checklists (e.g. PESTLE: political, economic, social, technological, legal, environmental) can be used to identify the influence of relevant factors and captured to come to a shared understanding of the problem. A rich picture is a graphical representation of the stakeholders interested in a project. Rich pictures encourage a deep consideration of a situation from the perspective of multiple stakeholders, uncovering sympathies and tensions between the various actors. This can form an excellent

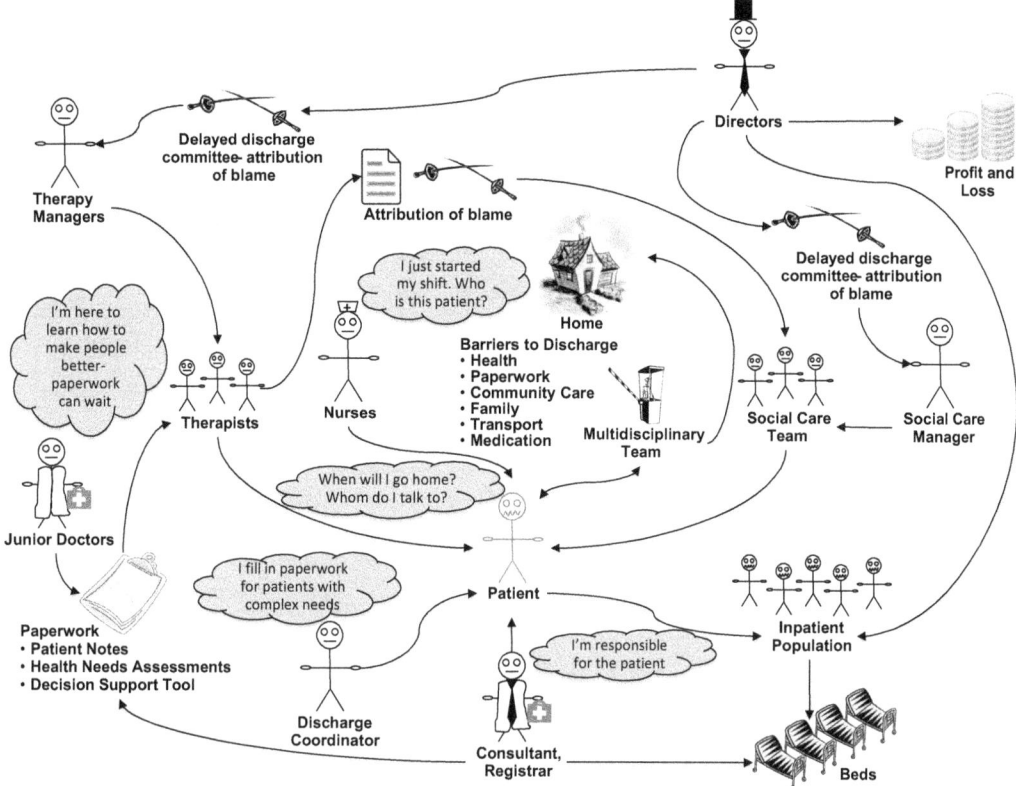

FIGURE 15.4 Rich picture demonstrating the problem of discharging patients from acute hospitals (Emes et al. 2017).

foundation for the requirements management process (Niu, Lopez, and Cheng 2011). An example of a rich picture used to understand the challenges with managing the discharge of patients from an acute hospital is given in Figure 15.4. A variation is the rich picture defined as part of SSM (Checkland 1981).

Soft Systems Methodology

SSM builds on rich pictures to explore relationships within projects to enable better decisions to be made (Jackson 2003). From the expression of a problem situation in a rich picture, SSM goes on to develop 'root definitions' of relevant systems, enriched using the mnemonic CATWOE (Customers, Actors, Transformation, Worldview, Owners and Environmental constraints) and conceptual models, to reveal different logical ways of interpreting and managing the problem situation. Comparing these conceptual models with reality gives ideas for interventions that may be effective in addressing the problem.

Concept Map

Concept maps and mind maps represent (often hierarchical) ways of structuring the knowledge concepts of a topic, starting with the main concept and breaking this down to show its subtopics. The relationships between these are articulated on the links (e.g. 'causes', 'contributes to'). An example is given in Figure 15.5.

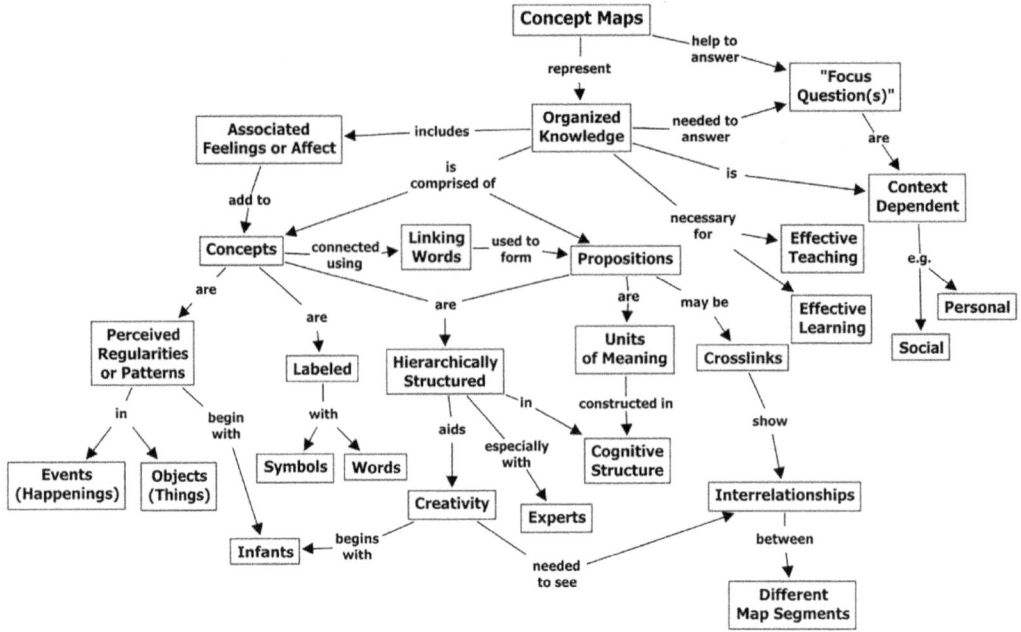

FIGURE 15.5   Concept map example (Novak and Canas 2006).

ACTOR MAP

Actor maps depict the key organisations and roles that make them up and are affected by the system (many, if not all, of whom are likely to be stakeholders in the solution). An example is given in Figure 15.6. An extension is the policy structure diagram, which focuses on how an organisation weights factors at various decision points (and the roles in the organisation that define these weights).

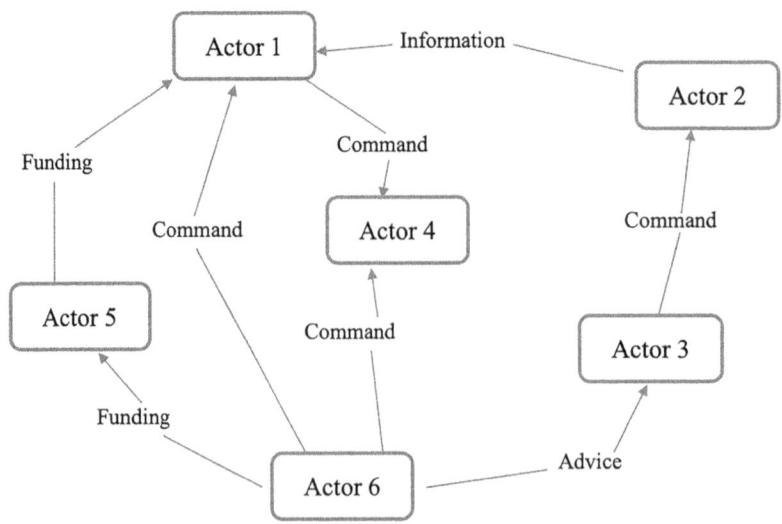

FIGURE 15.6   Actor map example.

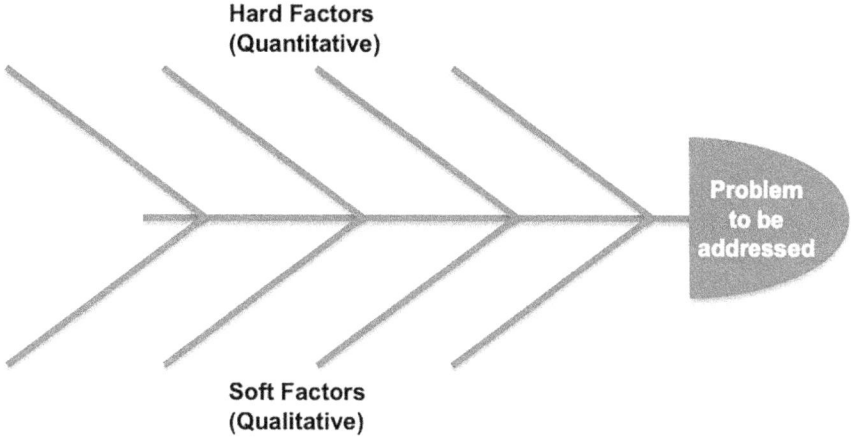

FIGURE 15.7   Fishbone diagram.

FISHBONE DIAGRAM

Fishbone diagrams are a variation of the standard fishbone or Ishikawa diagram to structure thoughts and distinguish hard (at the top of the diagram) and soft (at the bottom of the diagram) variables that affect the problem of interest. An example is given in Figure 15.7.

TREND MAPS

Trend maps depict the trends influencing the system. Trend maps should be developed using the knowledge of people familiar with the system and its context. A trend map over time is useful to visualise important activities, changes and other events to identify contextual factors (e.g. social, economic and political). An example is given in Figure 15.8. A variation on the trend map is the cumulative sum (CUSUM) chart from the total quality management (TQM) world that can be used to monitor a process based on averaging samples at a given time.

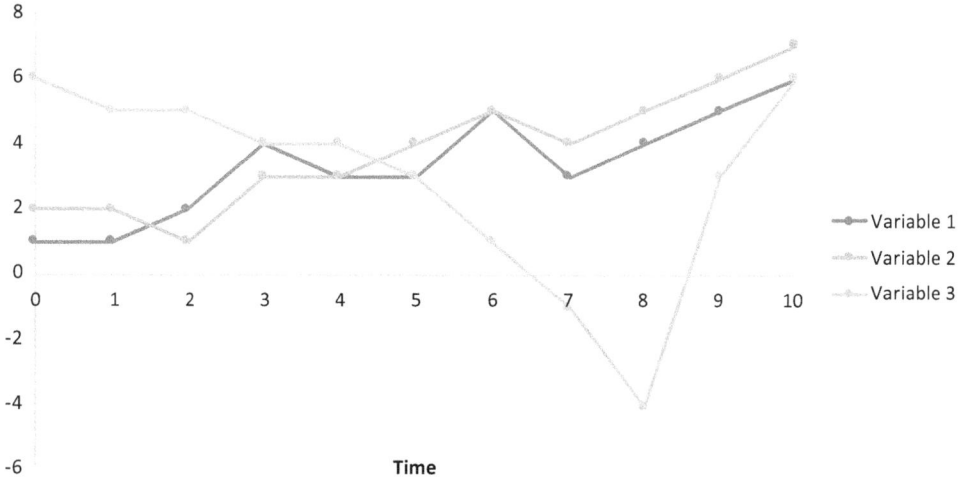

FIGURE 15.8   Trend map example.

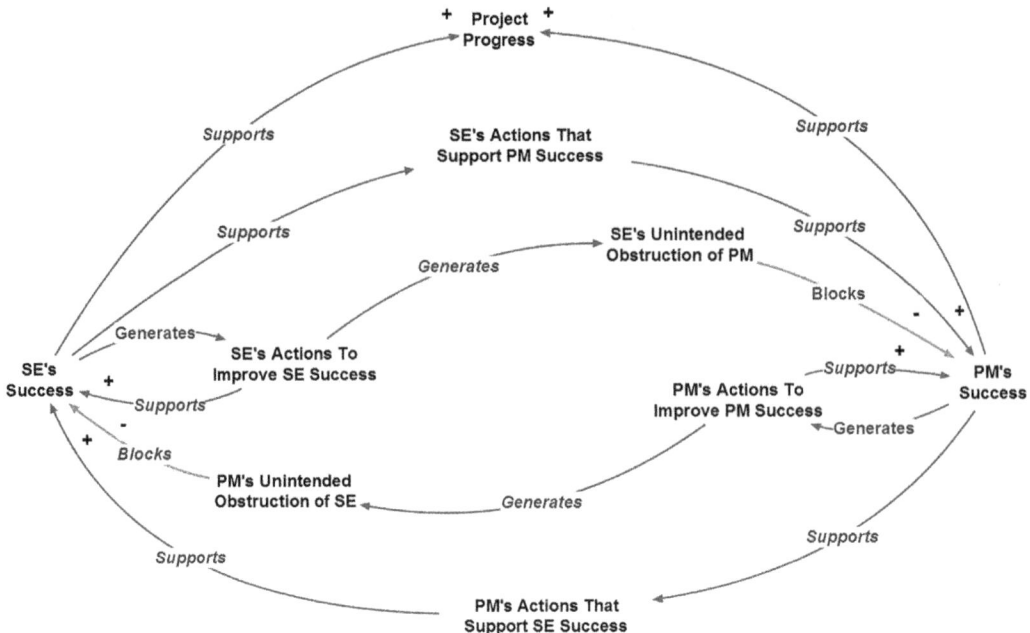

FIGURE 15.9 Causal loop diagram example.

Causal Loop Modelling

Causal loop diagrams represent the causal relationships between system elements and identify reinforcing and balancing processes (positive and negative feedback). An example is given in Figure 15.9, showing the accidental adversaries archetype, used here to model unproductive tensions between systems engineer (SE) and PM (Cole and Emes 2017). In addition, common situations are addressed by the 'system archetype' causal loop diagrams (Senge 1990).

Stock and Flow Diagrams

Stock and flow diagrams build on the cause and effect logic of causal loop diagrams but add quantification of key variables, using stocks (things that build up over time like the level of water in a bath) and flows (the rate of change of levels or stocks), as shown in Figure 15.10.

PERT Chart and Gantt Chart

While not really considered systems thinking tools, PERT charts and Gantt charts require the project activities that will be undertaken to develop and deploy the solution to have been decomposed. They can be regarded as one of the end products from the application of systems thinking and its interface into the project management processes.

PERT charts show the interconnections of the activities to give the 'build logic' for the development and deployment of the solution. In presenting the interconnections between activities, PERT charts generally do not provide the detailed schedule for these activities.

Gantt charts show the time-based relationship of the activities to give the schedule for the development and deployment of the solution. In presenting the schedule for each of the activities, Gantt charts generally do not provide the detailed logical relationships between these activities.

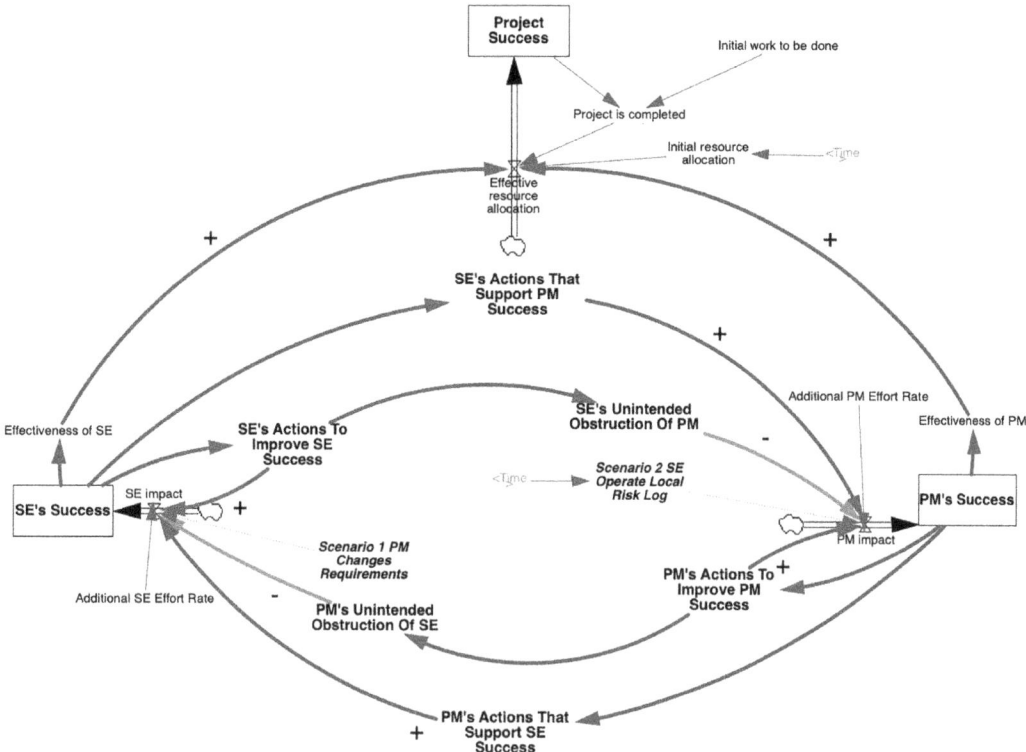

FIGURE 15.10 Stock and flow diagram example.

Mapping Systems Thinking Principles and Tools across the Project Life Cycle

The Association for Project Management undertook a survey (APM 2015) to identify the critical factors for project success. Table 15.2 maps these and the systems thinking principles and tools across a generic project life cycle.

EXAMPLE OF APPLICATION OF TOOLS TO COMPLEX PROJECT MANAGEMENT

Background

This example is adapted from a report from the Wales Audit Office into systems thinking case studies (Zokaei et al. 2010) and examines the case of housing management service in a UK council.

In 2010, the Portsmouth Council Housing Management Service was responsible for the maintenance of approximately 17,000 houses (18% of rental properties in the city), operated with a staff of approximately 600 and a budget of £80M. In July 2006, the Housing Service was rated as three out of four on the council's Comprehensive Performance Assessment scorecard and achieved Beacon status. The repairs and maintenance service was rated as a 'good 2 star service, with promising prospects for improvement'. The Audit Commission rated the council as doing very well. In addition, the council's key performance indicators (KPIs) showed repairs were being carried out within budget and within the time targets as specified centrally and indicated that 98% of tenants were happy with the service.

TABLE 15.2

Mapping Systems Thinking Principles and Tools across the Project Life Cycle (APM 2015)

	Identify the Right Problem	Define the Right Solution	Initiate the Project Right	Execute the Project Right (Inc Transition)
Project success factors	• Clearly identified goals and objectives • Competent project teams • Proven methods and tools • Commitment to project success • Capable and active sponsors • End users and operators engaged	• Clearly identified goals and objectives • Competent project teams • Proven methods and tools • Appropriate standards • End users and operators engaged	• Clearly identified goals and objectives • Thorough pre-project planning and progress monitored and reviewed • Competent project teams • Proven methods and tools • Effective governance • Secure funding • Supportive organisations • End users and operators engaged • Appropriate standards	• Clearly identified goals and objectives • Effective governance • End users and operators engaged • Competent project teams • Aligned supply chains • Proven methods and tools • Secure funding • Supportive organisations • Appropriate standards
Systems thinking principles applied	• Understand the bigger picture • Recognise that a system's structure generates its behaviour • Make assumptions explicit and test them • Change perspective to increase understanding • Appreciate that mental models define current reality and expected futures • Iceberg model	• Understand the bigger picture • Recognise that a system's structure generates its behaviour • Make assumptions explicit and test them • Change perspective to increase understanding • Appreciate that mental models define current reality and expected futures • Iceberg model	• Understand the bigger picture • Make assumptions explicit and test them • Change perspective to increase understanding • Appreciate that mental models define current reality and expected futures • Iceberg model	• Understand the bigger picture • Make assumptions explicit and test them • Change perspective to increase understanding • Appreciate that mental models define current reality and expected futures
Systems thinking tools applied	• Rich picture (As-Is) • Fishbone diagram (As-Is) • Actor map (As-Is) • Concept map (As-Is) • Trend map (As-Is) • Causal loop diagram (As-Is) • Stock and flow modelling (As-Is)	• Rich picture (To-Be) • Actor map (To-Be) • Concept map (To-Be) • Stock and flow modelling (To-Be)	• Rich picture (To-Be) • Concept map (To-Be) • Project definition documentation (Inc Business Case)	• Rich picture (To-Be) • Concept map (To-Be) • Actor map (To-Be) • Life cycle definition (e.g. Waterfall, Agile) • Project execution documentation (e.g. Gantt chart, PERT chart)

At the same time, however, local councillors' surgeries were full of tenants complaining about waiting for repairs. The Head of Housing Management decided to investigate further to make sense of the situation – everything suggested their service was good, except the recipients (i.e. the tenants).

Through discussions with colleagues, he discovered systems thinking and decided to find out more by attending a course and to apply systems thinking to the problem. A major enabler for this was that Portsmouth Council retains direct ownership and management of its housing stock.

PROJECT APPROACH

A small team was established to investigate the situation comprising council operational staff supported by a consultant to teach them systems thinking and to facilitate discussions.

The project followed the generic life cycle as described above: 'Check' (define the As-Is to understand the problem and its causes), 'Plan' (define and understand the To-Be solution to resolve the problems), 'Do' (implement the solution) life cycle (Seddon 2008). The description below uses the elements of the systems thinking iceberg to describe how the Portsmouth team undertook the project.

Check (Define the As-Is to Understand the Problem and Its Causes)

System purpose: The council's stated purpose for the housing maintenance and repair service was to 'manage all activity in order to meet the targets and keep down costs'. It was noted that this does not mention anything about maintaining or repairing houses and is primarily related to 'busyness' rather than delivery.

The team started by deciding to consider the problem from the perspective of the tenants as the users of the system.

Events

The team gained insight into tenants' requirements and the types of demand coming into the system by listening to on-phone demands in the council call centre and observing and talking with tenants at council reception counters.

Demands received from tenants as either calls or visits were categorised as either 'Value' demands (e.g. 'I need something fixed') or 'Failure' demands (e.g. 'You've been to repair, but it's not finished'). The teams calculated that Failure demands were 60% of all demands. They also noted that more than half of calls to the call centre were essentially self-generated as they were to ask when repairs would be completed or to tell the council that they hadn't done a good.

The team also looked at the council's satisfaction survey results that showed 98% of tenants were happy with the service they received. They found that this survey did not measure the effectiveness of repairs but only whether workmen had been friendly and cleaned up after themselves.

Patterns of Behaviour

When the team analysed the demand for repairs (through both the call centre and though visits to the reception counters) by developing trend maps, they found that demand for repairs was very predictable by time of year, month and even day rather than being unpredictable as was initially assumed.

The team also found that trades-people's behaviour was driven by each 'patch' having a set monthly spend as released by the council's financial system. As a result, repairs were only undertaken on the problem reported, not any others found. This behaviour was being driven by the need to keep to the monthly budget and the assumed unpredictability of demand, i.e. 'I need to spend as little as possible of the budget now as I don't know what repair work is likely to be coming up'.

Repairs were undertaken on the basis of what the tenant had reported as the problem, with no expert diagnosis by skilled staff prior to sending a trades-person to undertake the repair. Skilled staff were regarded as a scarce resource that had to be guarded rather than used immediately following a demand for a repair.

The monthly budget and assumed unpredictability of demand also meant that trades-people's vans did not carry a full range of spare parts, and repairs were often delayed by to the need to go and get the correct spare parts.

Analysis showed that the approach of only undertaking repairs to problems reported was a false economy, with many repair types (e.g. tap washers) being undertaken multiple times when it would have been cheaper to replace the whole tap. In other cases, the repair was to the symptom of the problem rather than the problem itself, with the same problem recurring (often many times).

System Structure

The team developed process diagrams to show the flow of work for different repair types from initial request to completion of the task.

Statistical process charts developed from these demonstrated the capacity and performance of the system. Predictably and reliably, it took a maximum of 98 days to complete a repair, with a mean repair time of 24 days.

Analysis of the system's capability to deliver showed that none of the measures actually related to its purpose but its activities, budgets or performance against centrally specified measures. It is worth noting that as the initial definition of the purpose of the system was about 'busyness', it is not surprising that measures are also about 'busyness' rather than delivery. In addition, the KPIs being produced didn't provide anything useful to show how the system was performing and understand how it should be managed (which is the purpose of KPIs).

Mental Models

The team analysed the information they had collected and generated about event, patterns of behaviour and system structure to identify the mental models that underpinned the system. They concluded:

The purpose of the system, including its policies, processes, people and technologies, was to 'Manage all activity in order to meet the targets and keep down costs'. This drove the system in the wrong direction, however, focusing on minimising short-term spend without considering long-term effects. The event data on repeated repairs of the same problem showed that this was not the best way to reduce overall spend and also did not meet the tenants' needs.

A basic assumption underpinning the system was that repair demand was unpredictable. However, the trend analysis showed that demand was actually very predictable by time of year, month and day. The assumption about the predictability of demand should still be one of the basic assumptions for the system, but the assumption should be the opposite of the original.

Rather than regarding skilled staff as a scarce resource to be guarded, their early intervention would reduce both costs and further calls on the service by diagnosing problems correctly and thus reducing waste. Early intervention by skilled staff would ensure problems are diagnosed correctly and the right repairs done. This would not only reduce repair times but also reduce Failure demand.

As the team considered better approaches, it became clear that the current processes made the implicit assumption that trades-people were not trustworthy or honest. In addition to the budget considerations, the reason why vans did not carry a broad range of spares was because the council did not trust its trades-people.

A systems thinking iceberg summary of the As-Is situation from the analysis undertaken during the Check phase that includes linkages between aspects is given in Figure 15.11.

Plan (Define and Understand the To-Be to Redesign the System to Resolve the Problems)

The team identified the steps in the process that added 'value' from the tenant's perspective. These were (1) ensuring access to the property (when the tenant would be there. Previously, trades-people

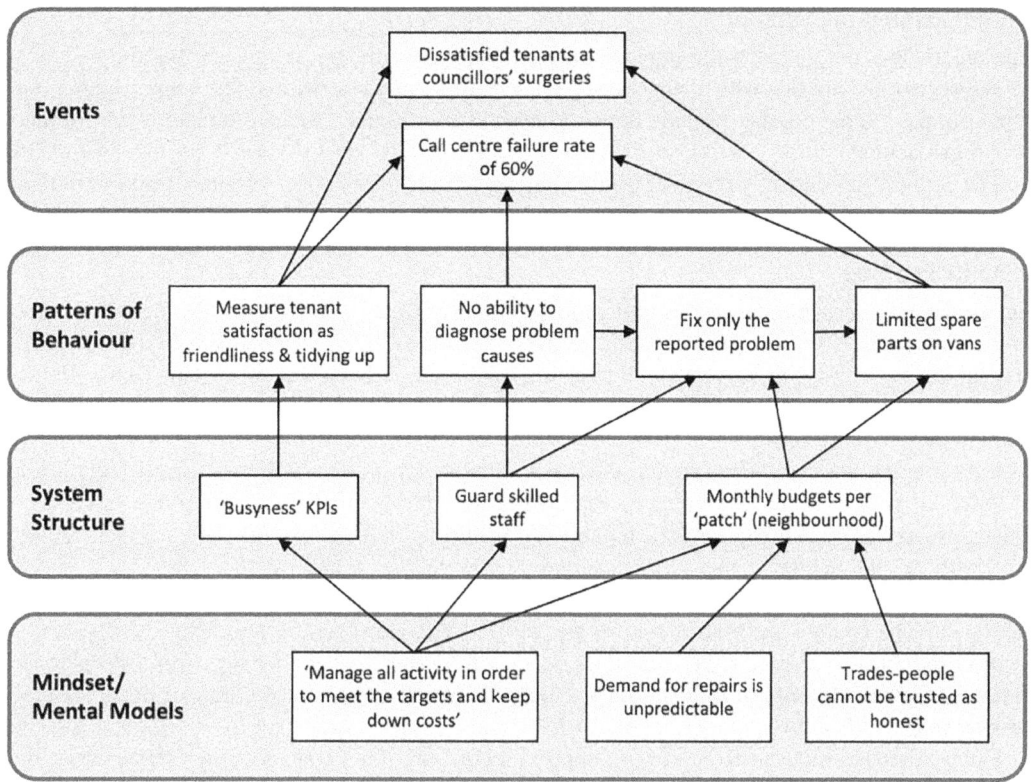

FIGURE 15.11 Systems thinking iceberg and relationships for As-Is situation.

would turn up driven by their availability and the tenant might not be in. This added to delays), (2) correct diagnosis of the problem and (3) completion of repairs.

The team considered new ways of delivering repairs, and further ways to improve were identified (e.g. fixing all the problems enables the council to not only satisfy its tenants but also proactively look after its housing stock to prevent future maintenance issues). This is a Win-Win for both the tenants and the council.

Mental Model Changes

The main changes were to the Mindset/Mental models that underpin the council's housing maintenance and repair system. The main conclusions were as follows:

- The purpose of the system should be recast from 'Manage all activity in order to meet the targets and keep down costs' to 'Do the right repair at the right time'. The new purpose aligns with tenants' needs and also reduces the overall cost of maintaining the council's properties.
- Proactively maintaining properties by fixing all problems at once reduces council costs.
- The team realised it is better to fix everything which may need fixing at the same time. To do this, they must trust their trades-people, which the old system did not. Other mechanisms can be used where trades-people don't operate honestly.
- Skilled staff are a scarce resource, so use them wisely. Rather than guarding them, use them at the point(s) in the process where they can add the most value (to the tenants and to the council).

Do (Implement the Solution)

The report does not describe the implementation approach, although it appears that the new processes were rolled out first with changes to IT systems following afterwards. The council's intervention also led contractors that provide housing services to also acknowledge problems within their own organisations.

The council subsequently required all its housing contractors to apply systems thinking to their organisations. One did not and as a consequence did not have its contract renewed.

Benefits Realised

Repair times: Average repair times reduced from an average of 24 days to an average of 6.9 days to fix the originally reported repair, with the average time to fix all repairs identified at the property taking 11.2 days. This gave a serial total of 18.1 days, but this is not comparing like with like as new average for the new approach is for all repairs. The council was able to take less time to do more.

Customer satisfaction: The original customer satisfaction measures showed a satisfaction rate of 98%, but what was actually being measured was whether council staff were polite rather than whether work had been completed and tenants were happy with the work done. The true satisfaction level with the old service, as measured during the redesign, was actually 60%.

Costs: Reactive repair costs initially increased due to latent demand in houses that had become run down. The council reported that costs per job fell 7%. However, by removing the distinction between planned and reactive repairs, the council discovered that the savings from the planned maintenance budget more than funded the increase in reactive repair costs. Repair costs also fell year by year.

Failure demand reduced from 60% to approximately 14% (and was reducing further when the report was written). This freed up almost half of the call centre capacity.

Supply chain: Dramatic results were achieved by supply chain contractors who adopted systems thinking. One contractor discovered they previously had no measures to help them learn and improve their business.

Organisational performance: The council's aim was to become an end-to-end repairs business. Using the same staff, they increased capacity from a mean of 85 to 225 jobs per day (a factor of just over 2.5). This allowed additional work to be taken on from other council contractors.

Additional benefit: After the initial application, the council applied systems thinking to their 'Green and Clean' service, which keeps neighbourhoods clean and tidy, with particular focus on removing bulk refuse quickly. Subsequently, Hampshire Fire and Rescue wrote to the council saying they had noticed a dramatic reduction in fires on estates. They didn't know what the council had done, but whatever it was, it was working.

QUESTIONS FOR THE CLASSROOM

1. How would you recognise a complex project? What should you look for?
2. In the example, what were the key 'emergent properties' of the system that could not have been predicted by simply looking at the individual parts of the system?
3. What was the main aspect that enabled Portsmouth Council to successfully apply systems thinking in its Housing Maintenance and Repair department? Why is this important?
4. The original purpose of the housing maintenance and repair system was to 'manage all activity in order to meet the targets and keep down costs'. How did this influence the patterns of behaviour and the structure of the original housing maintenance and repair system?
5. Consider how the investigation of the discrepancy between council KPIs, tenant satisfaction survey results and what local councillors were reporting from their surgeries would

have been different if the assessment had been undertaken from the perspective of the council rather than the tenants.

CONCLUSION

Systems thinking is not new; aspects of it go back to the building of the pyramids. Its value has never been greater, however, with modern, complex projects increasingly characterised by interconnectedness and increasingly plagued with ambiguity over their objectives.

Application of systems thinking requires a conceptual move from considering individual system components to considering the structure of the system, its interconnections and the interactions between its components and its environment. Systems thinking complements top-down thinking to enable problems to be addressed successfully that would be intractable using only traditional or reductionist approaches. Systems thinking is a lens or mindset that can be used alongside other skills that staff already have and can reveal new and different insights.

> Application of systems thinking requires a conceptual move from considering individual system components to considering the structure of the system, its interconnections and the interactions between its components and its environment.

> Systems thinking is a lens or mindset that can be used alongside other skills staff already have, and can reveal new and different insights.

The key aspects for successful application of systems thinking are to:

- Move from a view of the world that considers only causes and effects between components to consider systems in terms of their elements, structures, environment and the interactions and relationships between these.
- Appreciate that complex problems cannot be understood from a single perspective and context, and different viewpoints and contexts are necessary to properly understand these types of problem.
- Apply systems thinking without jumping to quick conclusions: understand the bigger picture, recognise that a system's structure generates its behaviour, make assumptions explicit and test them, and change perspective to increase understanding.

The main benefits from applying systems thinking are:

- Providing solutions to complex problems that traditional top-down decomposition is not well suited to tackle.
- Identification of the right project to address the right problem and undertaking this project in the right way.
- Improved project outcomes, including better identification, and avoidance or mitigation of unintended consequences.
- Reduced project delays and cost overruns.
- Increased organisational learning for current and future projects.

REFERENCES

APM. 2015. Conditions for project success. accessed 14/01/2019. www.apm.org.uk/media/1621/conditions-for-project-success_web_final_0.pdf.

Aramo-Immomen, H., and H. Vanharanta. 2009. Project management: The task of holistic systems thinking. *Human Factors and Ergonomics in Manufacturing* 19 (6):582–600.

Aziz-Alaoui, M., and C. Bertelle. 2009. *From System Complexity to Emergent Properties*. Berlin: Springer.

Box, G. E. P. 1979. Robustness in the strategy of scientific model building. In *Robustness in Statistics*, edited by R. L. Launer and G. N. Wilkinson. New York: Academic Press, 201–236.

Brealey, R. A., S. C. Myers, and F. Allen. 2005. *Corporate Finance*. 8th Edition (International Edition) ed. New York: McGraw Hill.

Cadbury, C. 2004. *Seven Wonders of the Industrial World*. London: Harper Perennial.

Castells, M. 2010. *The Rise of the Network Society*. Chichester: Wiley-Blackwell.

Checkland, P. B. 1981. *Systems Thinking, Systems Practice*. Chichester: Wiley.

Clark, W. 1922. *The Gantt Chart: A Working Tool of Management*. New York: Ronald Press.

Cole, D., and M. R. Emes. 2017. Common aspiration, different perspectives': SE & PM as accidental adversaries. *INCOSE UK Annual Systems Engineering Conference*, Warwick, 21–22 November.

Cooper, K. G. 1993. The rework cycle: benchmarks for the project manager. *Project Management Journal* 24 (1):17–21.

de Meyer, A., C. H. Loch, and M. T. Pich. 2002. Managing project uncertainty: From variation to chaos. *MIT Sloan Management Review* Winter.

DSMC. 1983. *Systems Engineering Management Guide*. Fort Belvoir: Defence Systems Management College.

Emes, M. R., P. A. Bryant, M. K. Wilkinson, P. King, A. M. James, and S. Arnold. 2012. Interpreting 'systems architecting'. *Systems Engineering* 15 (4):369–395.

Emes, M. R., and W. Griffiths. 2018. Systems thinking: How is it used in project management. In *Research Fund Series*. Princes Risborough: APM.

Emes, M. R., A. Smith, and A. M. James. 2007. Left-shift vs the time value of money: Unravelling the business case for systems engineering. *International Council on Systems Engineering UK Spring Conference*, Swindon, 16–18 April.

Emes, M. R., A. Smith, A. M. James, M. W. Whyndham, R. Leal, and S. C. Jackson. 2012. Principles of systems engineering management: Reflections from 45 years of spacecraft technology research and development at the Mullard Space Science Laboratory. *INCOSE International Symposium*, Rome.

Emes, M. R., S. Smith, S. Ward, A. Smith, and T. Ming. 2017. Care and flow: Applying soft systems methodology to understand the patient discharge process. *Health Systems* 6 (3): 260–278.

Flyvbjerg, B. 2003. *Megaprojects and Risk: An Anatomy of Ambition*. Cambridge: Cambridge University Press.

Flyvbjerg, B., M. Garbuio, and D. Lovallo. 2013. Delusion and deception in large infrastructure projects: Two models for explaining and preventing executive disaster. *California Management Review* 51 (2):170–193.

Goode, H. H., and R. E. Machol. 1957. *System Engineering. An Introduction to the Design of Large-Scale Systems*. New York: McGraw Hill.

Hitchins, D. K. 2003. *Advanced Systems Thinking, Engineering and Management*. Norwood, MA: Artech House.

Jackson, M. 2003. *Systems Thinking: Creative Holism for Managers*. Chichester: Wiley.

Jenkins, G. M., and P. V. Youle. 1971. *Systems Engineering. A Unifying Approach in Industry and Society*. London: C.A. Watts & Co. Ltd.

Kahneman, D. 2011. *Thinking, Fast and Slow*. New York: Farrar, Straus and Giroux.

Loch, C. H., A. de Meyer, and M. T. Pich. 2006. *Managing the Unknown: A New Approach to Managing High Uncertainty and Risk in Projects*. Chichester: Wiley.

Martin, J. N. 2004. The seven samurai of systems engineering: Dealing with the complexity of 7 interrelated systems. *International Council on Systems Engineering Annual Symposium*, Toulouse, France, 20–24 June.

Martin, J. N. 2008. Using the PICARD theory of systems to facilitate better systems thinking. *INCOSE Insight* 11 (1):37–41.

Meadows, D. H. 2008. *Thinking in Systems*. White River Junction, VT: Chelsea Green.

NASA. 2007. *NASA Systems Engineering Handbook*. Washington DC: NASA.

Niu, N., A. Y. Lopez, and J.-R. C. Cheng. 2011. Using soft systems methodology to improve requirements practices: An exploratory case study. *IET Software* 5 (6):487–495.

Novak, J. D., and A. J. Canas. 2006. The theory underlying concept maps and how to construct and use them. Institute for Human and Machine Cognition. www.academia.edu/688215/The_theory_underlying_ concept_maps_and_how_to_construct_and_use_them

Obeng, E. 1995. *All Change!: The Project Leader's Secret Handbook*. London: Financial Times/Prentice Hall.

Remington, K., and J. Pollack. 2007. *Tools for Complex Projects*. Abingdon: Gower Publishing.

SEBoK authors. 2017. Principles of systems thinking. In *The Guide to the Systems Engineering Body of Knowledge (SEBoK)*, edited by BKCASE Editorial Board. www.sebokwiki.org/wiki/Guide_to_the_ Systems_Engineering_Body_of_Knowledge_(SEBoK)

Seddon, J. 2008. *Systems Thinking in the Public Sector.* Axminster: Triarchy Press.

Senge, P. M. 1990. *The Fifth Discipline: Art and Practice of the Learning Organization.* Revised Edition ed. Chatham: Random House.

Sheffield, J., S. Sankaran, and T. Haslett. 2012. Systems thinking: Taming complexity in project management. *On the Horizon* 20 (2):126–136. doi:10.1108/10748121211235787.

Snowden, D. J., and M. E. Boone. 2007. A leader's framework for decision making. *Harvard Business Review* 85 (11).

Stevens, R., P. Brook, K. Jackson, and S. Arnold. 1998. *Systems Engineering: Coping with Complexity.* London: Prentice Hall.

The Standish Group. 2014. The chaos report. accessed 5th December. www.projectsmart.co.uk/docs/chaos-report.pdf.

Tsiga, Z. D., M. R. Emes, and A. Smith. 2016a. Critical success factors for projects in the space sector. *Journal of Modern Project Management* (January–April) 3:56–63.

Tsiga, Z. D., M. R. Emes, and A. Smith. 2016b. Critical success factors for the construction industry. *PM World Journal* 5 (8).

Tsiga, Z. D., M. R. Emes, and A. Smith. 2017. Critical success factors for projects in the petroleum industry. *ProjMAN - International Conference on Project Management*, Barcelona, 8–10 November.

Turner, J. R., and R. A. Cochrane. 1993. Goals-and-methods matrix: Coping with projects with ill defined goals and/or methods of achieving them. *International Journal of Project Management* 11 (2).

US Department of the Navy. 1958. Program evaluation research task: Summary report. Washington DC: Government Printing Office.

Westerman, H. R. 2001. *Systems Engineering. Principles and Practice.* Norwood, MA: Artech House.

Zokaei, K., S. Elias, B. O'Donovan, D. Samuel, B. Evans, and J. Goodfellow. 2010. *Lean and Systems Thinking in the Public Sector in Wales.* Lean Enterprise Research Centre. www.lgcplus.com/download?ac=1213980

16 Test and Evaluation Toolset

Keith Joiner and Mahmoud Efatmaneshnik
University of New South Wales – Canberra at the
Australian Defence Force Academy

Malcolm Tutty
Air Power Development Centre

CONTENTS

INTRODUCTION

Test and evaluation (T&E) is a powerful and evolving discipline designed to inform decision makers to successfully deal with the development and operation of modern systems complexity. T&E is the process of assessing validity or reliability of a system or component at various stages of its life cycle against a set of requirements and specifications through testing. It is one of the key Systems Engineering (SE) management functions that ensures a coordinated and consistent testing effort is applied to the system for the entire system life cycle. T&E is also fundamental to Project Management (PM), irrespective of the adherence to SE principles or whether a project has significant developmental aspects. Primarily what T&E delivers is the evidence for informed decision-making, which in the simplest of PM constructs is the rigour regarding capability as shown in Figure 16.1.

T&E applies throughout the entire system life cycle (Figure 16.2) and involves both the customer and the contractor. Basically, a test has the following major elements: test planning, a (set of) test scenario(s) and a (set of) evaluation measure(s) into a method of test, a data collection plan, test execution, analysis and most importantly test reporting (Tutty, 2016). The subject of testing may be a single component, a product, a system or a system-of-systems. Different types of systems T&E are common practice in Defence context such as:

- **Developmental T&E (DT&E).** DT&E refers to the test and evaluation activities undertaken during the Acquisition Phase of the system life cycle to support the design and development effort.

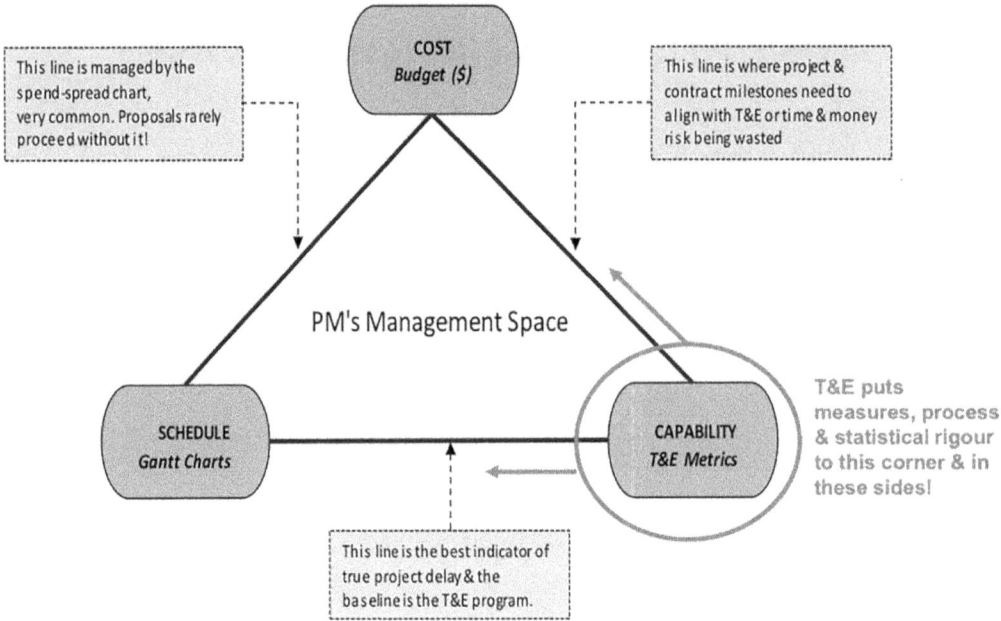

FIGURE 16.1   The T&E contribution to the basic PM iron triangle.

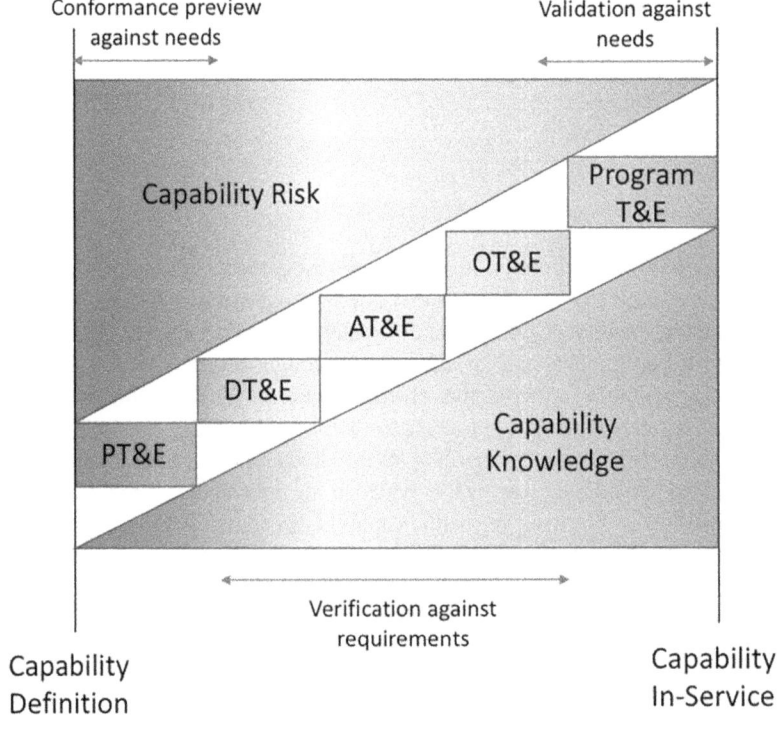

FIGURE 16.2   A typical PMO-based T&E practice from capability definition phase to deployment. The gradual and continuous application of different T&E practices throughout various stages leads to a gradual reduction in all sorts of risk to capability validity.

- **Preview T&E (PT&E).** PT&E is that DT&E conducted to evaluate the feasibility and performances of alternative capability options and identify risk areas prior to a final decision to acquire. PT&E is a relatively new T&E developed largely to account for difficulties in developing requirements for revolutionary new capabilities or choosing solutions during solicitation that are believed to be off the shelf or have largely mature subsystems, and technical risk analysis has determined that customer- and risk-focused, try-before-buy preview is required to demonstrate or trial prototypes independently of potential contractors.
- **Acceptance T&E (AT&E).** As DT&E completion approaches, AT&E activities become increasingly relevant. AT&E is that DT&E focused on the formal acceptance testing conducted on the system to enable the customer to accept the system from the contractor. AT&E effectively forms the boundary or transition between the Acquisition Phase and the commencement of the Utilization Phase.
- **Operational T&E (OT&E).** OT&E is the term associated with the T&E effort that is focused on the functional or operational testing of the system and its components, conducted under realistic operational conditions by representative operational personnel. OT&E is normally conducted for a period of time following design acceptance of the system by the customer leading to, and supporting, acceptance into operational service of the final operating capability.

> T&E applies throughout the entire system life cycle and involves both the customer and the contractor, and the challenge in T&E has been to conduct T&E early enough to optimally influence design and reduce project risk rather than be fait accompli at the end.

This chapter provides an overview of tools and methods and their evolution for better management of T&E as a significant part of system design and development projects. Examples of how each tool may be applied are provided in the relevant section. The chapter is organised in five sections as follows. The "T&E Fundamental Characteristics and Their Evolution" section covers foundational developments in the principal characteristics of T&E for decision-making such as timeliness, rigour, efficiency, independence and user acceptance. The "Testing Complicated & Complex Systems" section builds on the first in order to show how these T&E evolutions are being used to deal with complexity in systems and systems-of-systems, starting with interconnectedness, then malicious cyberthreats and finishing with autonomy and artificial intelligence (AI). The "Complex Systems Governance as an Assurance Strategy Alternative to T&E" section examines a foundational re-examination of the governance of complex systems that is not exclusive to either PM or SE but instead looks simply at what must be monitored, by who, when and how. This final section looks at where T&E would fit in this refocused governance approach and its prospect for use. The chapter synthesises these dimensions to conclude with a review of the evolution of T&E in support of project complexity. Table 16.1 presents a summary of these tools, including a brief description of each tool, where to find it (if applicable) and a reference to the section where the tool is discussed in depth. The first three tools apply to the general management of T&E in any project, and the rest are specifically designed for T&E management of complex systems.

T&E FUNDAMENTAL CHARACTERISTICS AND THEIR EVOLUTION

In this section, fundamental T&E characteristics, such as timeliness, rigour, efficiency, independence, and verifiability of their practices and process, are discussed. The evolutions of these characteristics are discussed in detail and further clarified with a few practical and real-world examples.

TABLE 16.1

Summary of Managerial Tools for T&E Management of Complex Systems Projects Presented in This Chapter

No	Tool	Description	Address	Section
1	T&E in project scoping for previewing (trialling)	Your organisation needs skills in project scoping and relating project risks into PT&E (trialling) that can be done pre-contract as part of tendering and offer-definition activities. Such activities should be done at the level of claimed technological readiness; that is, if necessary aspects will be modelling and simulation, structured demonstrations and/or prototypes. Such preview ensures you get into the right projects the right way!	T&E planning in project scoping and PT&E are taught in the University of New South Wales (UNSW) subject ZEIT 8231.	"Timeliness"
2	Combinatorial test design methods	Your organisation needs skills in combinatorial test design and analysis (HTT) from the design-for-six-sigma stable, as these provide rigour and efficiency in evaluating highly complicated (mission) systems where the permutations (possibilities) are very high and such testing can be conducted safely.	Combinatorial test design and analysis techniques are taught in the UNSW subject ZEIT 8034 (https://youtu.be/ZK8dmC10AVo) and by specialist training providers like Air Academy Associates (www.airacad.com). Example underpinning software tools are Quantum XL (Six Sigma Products Group – www.sixsigmaproductsgroup.com/) and rdExpert(lite) (Phadke Associates Ltd – https://paportal.phadkeassociates.net/OnlineStore.aspx).	"Rigour & Efficiency"
3	Independent T&E oversight	Your organisation needs independent T&E representation in the PMO and at each project gate review to ensure T&E evidence is credible and understood in the context of all the key decision makers.		"Independence and User Acceptance"
4	ICT and software T&E types	If your projects involve ICT or software-intensive functionality, your organisation needs skills in the four types of T&E key to ICT projects: usability T&E, integration, performance and cybersecurity. The project governance and associated gate reviews need to see evidence of progress in all four throughout the life cycle even if just from V&V modelling and simulation and desktop experimentation.	Joiner et al. (2018).	"Independence and User Acceptance"

(Continued)

TABLE 16.1 (*Continued*)
Summary of Managerial Tools for T&E Management of Complex Systems Projects Presented in This Chapter

No	Tool	Description	Address	Section
5	I3 assurance	If your projects involve delivering, or delivering into, highly interconnected system-of-systems and/or require assured cyber-resilience through-life, then you will need to invest in I3 assurance strategies and skills, such as: • Creating more representative environments through federating SILs and HWILs to enable regular testing against new threats, new roles, new environments or developing capabilities. • Creating regular experimentation exercises with a rhythm of testing for I3 assurance using the infrastructure above. • Adopting use of six-sigma test design and analysis methodologies (see Tool 2) so as to build operational models and efficiently handle permutations. • Adopting cybersecurity T&E in acquisition policy, infrastructure and staff competency targets (see next section). • Assigning dedicated acquisition staff, closely aligned with P3O staff, focused on the infrastructure, competencies and other wherewithal to give I3 assurance.		"Interconnectedness and System-of-Systems"
6	Cybersecurity	If your projects require assured cyber-resilience through-life, then you will need to invest to varying degrees in staff and infrastructure for the following: • Cybersecurity tabletop facilitators for early and repeated cyber-risk management and cyberthreat deconstruction. Such facilitator courses are so far only offered in the United States. • Cooperative vulnerability and penetration assessment (CVPA), otherwise known as cyber defenders (blue team) and cyber attackers (red team). • Combinatorial test design and analysis (Tool 2) but up to six-way rigour with the associated software to do orthogonal test design to this extent (such as the full rdExpert).		"Cybersecurity T&E"
7	Usability T&E for autonomy and/or AI	If your projects involve delivering autonomous systems then you will need to invest in strategies and skills to establish and maintain positive, active and meaningful human control through techniques like usability T&E run by human factor specialists working with representative operational users and managers (commanders). For AI, there is also the need for significant investment in through-life test infrastructure for through-life monitoring actual life to the projected accelerated life learning.		"Autonomy and AI"
8	CSG entry-level assessment	If you manage a program or portfolio of complex projects, then your governance structures could be improved by an assessment survey in CSG. The assessment is on the scales shown in Figure 16.10, and extant pathologies can plot a path to better meet project demand by moving resources across dimensions as necessary.	The CSG assessments are administered through Professor Charles Keating at Old Dominion University in Norfolk Virginia or any of his international collaborators.	"Complex Systems Governance as an Assurance Strategy Alternative to T&E"

Timeliness

One of the tragedies of T&E is the number of project staff who conceive, plan and perpetuate the misconception that T&E is something done at the end, once a system is built and the operators and maintainers have been trained. In fact, successful programs show that experimentation and T&E should be done to inform decision-making, and if it is done only at the end of projects, it can only inform final operational acceptance or project failure: a poor binary choice. Fortunately, SE lays a foundation for T&E to inform iterative design (Reynolds, 1996; McShea, 2010; Fox et al., 2004) and helps deliver timely T&E in engineering fields like Defence (Murphy et al., 2015), transport (Gray et al., 2017), mining (Azevedo & Olsen, 2017) and increasingly service sectors (Antony, 2014) and Information and Communications Technology (ICT) (Joiner et al., 2018). Furthermore, the advent of Program Management Offices (PMOs) with structured phasing, gating and review of projects has increasingly demanded evidence-based phasing, with an associated pathway of T&E as illustrated in Figure 16.2 (Joiner, 2015a; Smith et al., 2016; Gray et al., 2017).

Notwithstanding such progress in systematic earlier use of T&E, there is a significant roadblock in experimentation and T&E going any earlier than about halfway through the main developmental contract. Outsourcing of development, production and support of complex systems-of-systems to prime contractors means a heavy reliance on these primes for any technical wherewithal to do early PT&E; that is, until a contract is in place. Strategies to do early pre-contractual PT&E and prototyping as part of project scoping are really important to portfolios getting into the *right projects* with the *right prime contractors* and reasonable awareness acquisition risks (Joiner, 2015b; Copeland et al., 2015). These PT&E planning strategies can leverage the tendering process to do PT&E through *offer-definition* activities. Such previews target high-to-medium risk across as many, or as few, of the tendering companies as is sensible to do so. Preview testing is as common sense as the test drive of a new or used car and yet is often fiercely avoided in project scoping, ironically on the basis of most projects involving significant proportions of mature or *off-the-shelf* systems and applications. Such avoidance is part of the phenomenon of *project over-optimism* noted by researchers (Smith et al., 2016; Ghildyal, Chang, & Joiner, 2018) and auditors in both the United States and Australia (Australian Senate, 2012; Shergold, 2015). To overcome this over-optimism, U.S. Department of Defence (DoD) and others with a serious track record of system development projects have dedicated project life cycle phases devoted to technological maturation and risk reduction (Murphy et al., 2015). This maturation phase came initially from major reviews of the knowledge points of acquisition projects in 1998–1999 (GAO, 1998, 1999; Orlowski et al., 2017) and now is linked to the well-researched use of Technological Readiness Levels (TRLs) (GAO, 2016) initially practiced by NASA (Mankins, 1995). Such approaches are to avoid getting into acquisitions that develop substantial cost and schedule overruns based on undisclosed technological focused immaturities, integration risks and developmental risk (Australian Senate, 2012). Complexity is meant to be confined to non-linear and emergent behaviours that could not have been foreseen (Javorsek, 2016), and therefore if projects are to avoid surprises in the presence of complexity challenge, then the risk-based, pre-contractual T&E or trials are the key practices to consider. Considerations and key points that should be built into project scoping checklists are as follows:

- Mature or *off-the-shelf* technology can be readily demonstrated.
- Developmental technology can be modelled and simulated, followed by mixing with any *off-the-shelf* elements and then successfully demonstrated.
- Risks in contractual partnering, release issues or configuration changes in a tendering prime's bid, can be disclosed with PT&E before projects are in contract.
- The attendance of representative users at preview demonstrations could disclose risks associated with undocumented, miswritten and misunderstood *concept-of-operations*, mission scenarios or user requirements by a tenderer.
- Seeking to perform preview demonstrations *in-country* (locally) could resolve risks associated with potential importation difficulties and regulatory obstacles.

A major U.S. study into DoD projects (Copeland et al., 2015) found significant schedule savings in projects that had undertaken a prototype demonstration when compared to those projects that did not, as well as evidence of less costly design and *through-life* configuration changes. A major multi-departmental review of testing in support of ICT governance (Joiner et al., 2018) found that software continues to be developed late in programs, and the fact that it can be modelled and simulated before it is coded is not widely appreciated. As a strategy for system adaptability and changeability, most functions are designed in the software not hardware (Cofer, 2015),

As a strategy for system adaptability and changeability, most functions are designed in the software not hardware, and early usability test demonstration of software functionality, modelled if necessary, is key to ensuring both the project sponsor and potential contractors understand the right user requirements before contract.

yet such reviews find there is still a critical need to emphasise the software design cycle (Kramer Sahinoglu, & Ang, 2015) and fundamentals like early usability testing, especially in demonstrations pre-developmental contract. Early usability test demonstration of software functionality, modelled if necessary, is key to ensuring both the project sponsor, and potential contractors understand the right user requirements before contract. The more mature the applications, the easier such testing ought to be. Further, the more developmental the application, then the more important is prototyping to ensure suitability.

Nearly all modern DT&E relies on a degree of modelling and simulation (Elele et al., 2016; Joiner & Tutty, 2018), and so experienced systems engineers, experimentation and T&E professionals there can merge Live, Virtual and Constructive (LVC) simulations to create as much realism as possible for early insights into Integration, Interoperability and Information (I3) assurance as possible. Project scoping in less advanced acquisition contexts is more likely to be based on studies without formal SE or T&E oversight. Sometimes these studies or prior works risk being implied from tenderers (the primes) and foreign agencies or internal 'studies' (Australian Senate, 2012). Often if such studies are scrutinised, they are to varying degrees at risk of being unverified, not accredited, commercially not releasable (until contract or not at all due third-party) or withheld by a foreign government. Furthermore, if such modelling, simulation or prior testing is reviewed by system engineers or T&E professionals with some preview, the configurations, roles (missions) and environments (CRE) may lack analogy in ways not normally disclosed until much later (Australian Senate, 2012; Joiner, 2015b) resulting in Type I or II errors as shown in Figure 16.3.

In summary then, to improve the timeliness of T&E in complex projects, the most promising measure is to focus the PMO on checklists and skills training to elicit risk-based pre-contractual PT&E. Such preview offers the greatest risk illumination, particularly in software functionality with a focus on usability; if necessary, using simulated coding and middleware. Given the propensity for *project over-optimisation* and the enthusiasm of investors at early stages, some development of board-level project approvers in preview thinking may be warranted. An excellent board-level axiom is that *the extent of pre-contractual preview T&E should be at the level of claimed (tender) technological readiness*. Further, a good follow-up rule is that such preview should be overseen by

| | | True State of Nature | |
		H_0	H_1
Conclusion Drawn	H_0	Conclusion is correct	Conclusion results in a Type II error
	H_1	Conclusion results in a Type I error	Conclusion is correct

FIGURE 16.3 Capability and the experimental method.

qualified T&E or system engineering professionals with some independence from the project scoping. A good further reading on project scoping and the need for preview trialling is in the Shergold report (2015). Shergold (2015) found a systemic lack of public policy de-risking in the Australian public service, also reported in the study by Banks (2014). This suggested reading is not about the usual high-technology risks but rather other political and public complexities which can be more universal to scoping projects in many public service departments and commercial contexts.

Timeliness Example. Examples of the potential benefits of preview testing are numerous, and they are usually done in hindsight by auditors. Large Defence projects are an excellent source of such lessons learnt because they attract rigorous independent auditing and scrutiny. Such sources include the Sea Sprite helicopter project (Australian Senate, 2012), Army heavy truck project and the Landing Helicopter Dock (LHD) ship project (both covered in ANAO (2015), Australian Parliament (2016)). In each of these instances, auditors found preview testing prior to contract would have disclosed technical and operational risks around the immaturity of substantive system elements and the implications of differences in CRE. Unfortunately, few of the lessons-learnt examples include significant software examples, although the Sea Sprite project did claim high maturity and low risk based upon hardware without due regard for safety-critical helicopter software. The bias of lessons learnt to hardware is not likely to last long, as many complex software-intensive projects are underway and are experiencing significant delays due to that software complexity. A good example is the Joint Strike Fighter aircraft (U.S. DoD, 2013–2017), while other Australian projects include the combined Defence and Civil Aviation Safety project to replace the air traffic control systems and Defence electronic health records. Such software-intensive projects usually claim low-risk adaptations of extant applications and operating systems into business contexts where operators are being replaced or augmented with digitisation and automation. Clearly risks exist in adequately capturing user requirements and in integrating the always unique architecture of such system-of-systems, which if not disclosed before contract, can lead to project overruns. PT&E is about documenting these risks and creating credible *offer-definition* activities to explore, document and account for such risks before a development contract and certainly before a production decision.

In the case of a proposal for a military electronic health record, one author was involved in assessing risks and proposing PT&E. The most uncertain area of the requirement was in deployed use of electronic health records, such as on-board deployed ships, aircraft detachments or counter-insurgency land bases. Such records were for the benefit of deployed personnel (i.e. access to records) but also for acquired personnel such as detainees, immigrants, prisoners of war or humanitarian assistance. The other major risk was in integrating a hitherto civilian hospital architecture to military systems with their additional security and deployment. The confluence of these two main risks is in a deployable architecture serving medical assistants in war-like circumstances, where the architecture may be insecure or provide functionality not required and often labelled *creeping featurism* (Wickens et al., 2014). Clearly such a mission would be the raison d'être of Defence medical support. In this example, the greatest technical risk (integration) and operational risk (usability) coincide with the most critical of missions for the business capability (at least the most dangerous if not most likely). The PT&E proposed was to fund a prototype software architecture to be trialled with deployed medics on realistic operational exercises. The all-up cost of such an activity came close to 1% of the estimated acquisition cost and 6–9 months of activity. Such upfront expenditure is usually apropos when military projects on average experience around 30% overrun in schedule and cost (DMO & ANAO, 2014) and ICT projects much worse (Flyvbjerg & Budzier, 2011; Downes, 2015; Hecht, 2015).

RIGOUR AND EFFICIENCY

From around the turn of the millennium, the development of test design and test analysis techniques and their spread into industry has been phenomenal, leading to very significant improvements in the rigour and efficiency of testing (Johnson et al., 2012; Lednicky & Silvestrini, 2013; Chu, 2016;

Kuhn et al., 2016; Joiner, 2018). The reform has been enabled by the development of software packages using scientific methods for multifactor, multi-response test design, hitherto only possible in some university fields. The methods are further enabled by a management reform methodology known as six sigma (Antony, 2014).

> From around the turn of the millennium, the development of combinatorial test design and multifactor, multi-response test design techniques and their spread into industry has been phenomenal, leading to very significant improvements in the rigour and efficiency of testing.

The U.S. DoD has illustrated that the six-sigma test methods can be decoupled from management practice for just their test benefit. The U.S. DoD has adopted these six-sigma test design methods and competencies for all their acquisition and test practitioners since 2009, specifically for the additional rigour and efficiency they offer (Murphy et al., 2015; Chu, 2016). The U.S. DoD is a somewhat rare example of public sector use of the test design and test analysis aspects of six sigma, and there is only limited adoption of the broader six-sigma aspects in some governments (Rodgers, Antony, & Kregel, 2018). Commercial imperatives such as *complex design before build* and *efficient production with acceptable variation* have driven the use of the methods well ahead of the commercial and public sectors (Antony, 2014).

The six-sigma test methods are inherently focused on preview of models and simulations before decisions to commit to full development and full production. Where the test design methods have been adopted, the skills and techniques have enabled test practitioners to adopt even more efficient techniques for screening in high-factor complex systems, such as the combinatorial high-throughput testing (HTT) (Tatsumi, 2013; Kuhn et al., 2016). Such test design techniques use combinatorial mathematics to test all *two-way* combinations of factors in the least possible number of test runs. For example, if a system has 10 factors all with multiple levels that need evaluation, varying from three to five levels each, one needs over 2,600 test points using traditional approaches; however, using HTT methods makes possible the screening of every two-way combination in as few as 25 test runs. These methods can be very efficient in validating that all factors work together at all levels or initially screening which factors are significant and which are not. Screening in testing is critical to ensure that when detailed rigorous testing does occur, say to verify a model, then the testing can focus only on the key factors. Unfortunately, classical test design methods, many entrenched in standards or passed between subject-matter experts, have made simplifying assumptions about which factors are historically significant and which can be assumed insignificant. Yet, modern conformal structural designs and complex highly connected systems often do not model or simulate well to such historical assumptions (Joiner, 2018). Also, these historical test design methods often make the simplifying assumptions so as to test *One-Factor-At-a-Time* (OFAT). Such OFAT methods are inherently inefficient even if they can be made to adequately characterise systems under test (Antony, 2014). Often in industry and the U.S. DoD, the use of six-sigma multifactor test design techniques has not only been shown to be far more efficient but to find superior optima that give business competitive edges or for a Defence force, overmatch against near-peer adversaries (Lednicky & Silvestrini, 2013; Antony, 2014; Murphy et al., 2015; Mackertich et al., 2017). This usually occurs when screening discloses factors, or interactions between factors, which are no longer overlooked and which can, when included in modelling, give a superior solution.

Such advantage can be used in DT&E to improve designs between iterations or in OT&E to optimise the use of systems in different CRE (Murphy et al., 2015) as shown in Figure 16.4. Also, when done properly, such test design and its test analysis techniques enable efficient sharing of DT&E and OT&E (Lednicky & Silvestrini, 2013). This sharing is precisely because the same tests can efficiently have more factors and output metrics and the test designs keep the effects rigorously independent (orthogonal). Using these techniques requires the subject-matter experts to *let go* of seeing the solution or optimum evolve with each test and instead collect all the test points before then mathematically sorting the effects of each factor and a collective probabilistic model. For some qualification engineering standards, the new methods would change the approach fairly

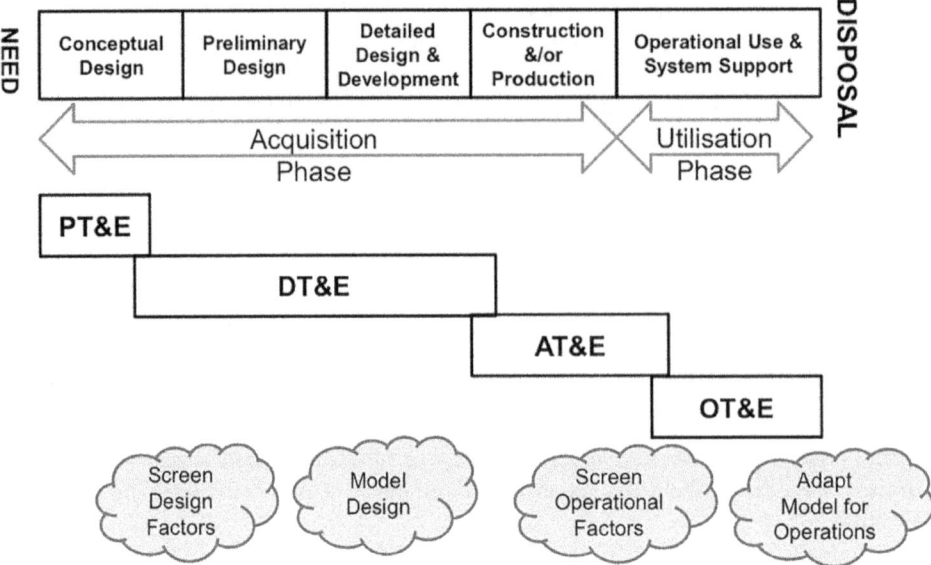

FIGURE 16.4 Showing the use of screening and modelling test methods across the life cycle to build progressively better operational capability.

significantly (i.e. Henry & Joiner, 2017). Many of the standards test *to the centre*; that is, they test the system where it is most often used in the centre of the anticipated operational envelope and more scarcely outside of that. For many operational testers, this is why when weather is bad, or missions less anticipated, modelling accuracy or advice on a system's performance 'off-design' deteriorates chronically. The six-sigma test design techniques usually start wide on all factors (i.e. each set high and set low) and focus down the test envelope only where proven to do so through screening. Also, the six-sigma system test approaches characterise far more evenly than *testing-the-centre* approaches. Note that aircraft stores compatibility is one of the disciplines that is mandated in the 'over testing' of the entire carriage, employment and jettison operating envelopes (speed, altitude, G, etc.) for all aircraft stores' configurations and downloads (MIL-STD/HDBK-1763, 1984). This enabled the mission reliability and effectiveness of aircraft stores and electronic warfare (EW) pods, for example, to go from less than 50% in the Vietnam era to over 95% in Gulf War I (Tutty, 2017). Such rigorous methods often disclose hitherto unknown interactions that can be exploited and deliver better modelling and simulations and due to the mathematical efficiency does so repeatedly with less, and not more, testing (Grafton et al., 2018). Further because the test methods are founded in statistics (probabilistic), the output are not just averages but distributions with confidence limits.

> Confidence limits fundamentally improve the board-level decision-making (Rucker, 2014) because they inform all stakeholders of risks in proceeding at each project gate with the sufficiency or adequacy of testing so far.

Rigour and Efficiency Examples. A well-documented example of applying six-sigma test methods to a development for improved rigour, efficiency and cyber-resilience is by Mackertich et al. (2017). This example won an Institute of Electrical and Electronic Engineers (IEEE) and Software Engineering Institute (SEI) award. An example of applying the new test methods progressively to a Defence unit is given by Grafton et al. (2018), where they emphasise that the improved test efficiency gave additional rigour in operational models and advice. This adoption of unit-wide use of the six-sigma test methods was underpinned, similar to the U.S. DoD beforehand, by a rigorous competency program for staff (Joiner, 2018). What was particularly beneficial for the staff in the Grafton et al. (2018) example was that the six-sigma methods were found to be empowering for the test practitioners. The multifactor test design process of screening enabled staff to investigate the *what-if* of factors not looked at before and the *exactly why do we* of factors assumed by subject-matter experts. This new exploration is enabled within the existing resource allocations by the new efficiency. Furthermore, screening led to candidate solutions and enquiry that in some instances came from deliberately starting with input factors set high and set low and which were not focus areas beforehand. Without disclosing the specific outcomes, the testing led to more operational alternatives for countering potential adversaries – all without additional test expenditure. For the staff involved, such ingenuity is inherently rewarding, especially in dealing with the more complex systems and potentially in the future with greater autonomy and cyberthreats.

Another example of the six-sigma methods used in a Defence context was exploring the means to validate a complex operational modelling and simulation capability. The capability was one where the weaponry is primarily a deterrence and almost never subject to end-to-end testing. The model and simulation were thus an operational capability in its own right – supporting operational training and decision-making. The validation test plan was developed by the foreign DoD and upon close inspection was using representative sampling in such a way as to retain multifactor orthogonality (independence) of the factor effects. Put simply, the very few actual validation runs were carefully chosen to build or reinforce the (probabilistic) accuracy and limit any errors therein. Compared to prior practices of choosing validation points from representative population metrics, where the efforts have limited residual benefit beyond the confidence to accept or reject the model, this methodology is far superior. This analysis illustrated that something as complex as nearly orthogonal Latin hypercube (test) design (NOLHD) can be simple to use and have enormous rigour and efficiency gains.

INDEPENDENCE AND USER ACCEPTANCE

The importance of independence in T&E is well established (Reynolds, 1996) and is reflected in the construct of the *Office of Operational T&E* in the U.S. DoD and the Title 10 congressional laws under which it operates. However, in the literature of Portfolio/Program/Project Management Offices (P3O), the concept appears to be rarely stated (Joiner, 2015a). This may be because impartiality is an underpinning principle in modern project conceptions, reflected in such tasks as professionally balancing tension fields like cost, schedule and capability, such that independence is implied. In military contexts and some commercial settings, there is the notion of being too close to the initial and ongoing funding of projects to be as rational, or at least risk-minded, as projects ought to be. This potential bias, coupled with unhealthy bureaucracy and interservice rivalry, is perhaps best portrayed in the movie *The Pentagon Wars* (1998), based on the real-life book by Colonel Burton (1993). *Pentagon Wars* shows the early attempt to bring independence into the T&E of military acquisitions by having officers of a difference service oversee the T&E of the service seeking the capability. In the case of the book and movie, it is an Air Force officer trying to independently certify T&E for the U.S. Army on the Bradley Fighting Vehicle.

Good project scoping will involve *contestability* or, put differently, provide an opposing view to that of the proceeding with a project and often options on how to proceed. The aim of such *contestability* is to ensure a robust examination of alternatives and risks (Peever et al., 2015; Shergold, 2015; Joiner, 2015a; Ghildyal, Chang, & Joiner, 2018). PT&E is a key element of exposing and dealing with alternatives and risks; however, it is often surprisingly hard to convince project champions of the merits of PT&E. Such champions often perceive such early T&E as a threat to project approval by placing demands on schedule and cost early on when new funding is often the most constrained. Project over-optimism is often exacerbated by operational executives with limited time seeing demonstrations or presentations by sales persons, often overseas, where the cost, time and solicitation processes to properly examine the demonstrations are often beyond the limited budgets given to project scoping. Unfortunately, a common problem is larger bureaucracies incentivising project managers to get on contract (Smith et al., 2016) in such a way as to reward *paper-based* and *siloed* risk estimation instead of *activity-based preview T&E* (Joiner, 2015b). Much of the language of project scoping is about *evidence-based* and words like comparative analysis, or study; however,

> A flexible P3O gating process should simply fund a short examination phase as part of the conclusion of tendering which involves *offer-definition* activities and PT&E.

closer examination from author experience often reveals there is a very limited process of asking questions. For example, a *request-for-tender* is followed by an answer (tender) and then some evaluation of read across, but scope for clarification with preferred tenders and interactive negotiations are often heavily constrained and lacking practical informative early T&E (if necessary through modelling and simulation). Even in later design certification (acceptance), there are usually claims of 'evidence-based' decision-making that uses terms like analysis, demonstration, inspection and test, but the actual activities and parties undertaking them are often well removed from the acceptance decision, difficult to review and lacking aggregation until the very end, especially in software.

Figure 16.5 was produced as an attempt to improve understanding of independence in T&E. In the illustration, test results are passed unfettered to the key stakeholders who are simplified to the technical and operational authority and the investor. While simple in its construct and principle, if the project has not been scoped well or contracted well for T&E, then often such results are untimely (i.e. too late), lacking rigour (i.e. an average performance stated without confidence) and without the funds being planned and resources to do anything (i.e. fait accompli).

An amalgamation of common late test experiences might help illustrate the importance of independence of testing. Consider the unhelpful scenario where a test result on say a software load is completed after the following:

- The bulk of the project funds have been spent
- The hardware is in production
- There was no scheduled software design iteration (using 'The Big Bang Theory' or 'one-shot wonder' strategy of hope)
- The software coders have dispersed to other tasks
- No representative users were involved (i.e. usability T&E)
- Cybersecurity vulnerability and penetration testing are yet to be performed.

Now consider that independence is meaningless if the technical authority can see only process conformance, the operational authority is obliged to transfer the risk to operations and the investor's funds and schedule are depleted.

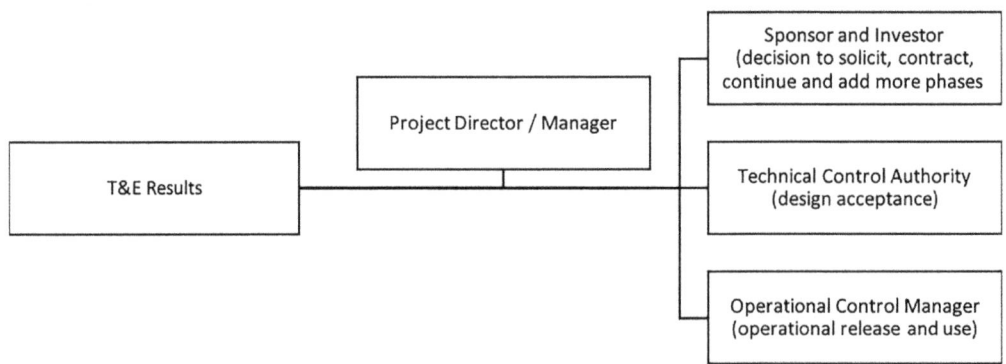

FIGURE 16.5   Illustration of independence in T&E. (Adapted from Joiner, Atkinson and Sitnikova (2017).)

One very positive development for independence that is occurring, even somewhat surprisingly in some public service departments, is the development of strong PMOs (Joiner, 2015a). PMOs are strongly vested in *on-going* or *through-life* management of capability, often at the level of highly interconnected system-of-systems. As such, considering again the example of early transfer of underdeveloped and underfunded capabilities into service, a major and disproportionate impact of such early transfer is on sustainment costs, the complexity of sustainment and a decline from the outset in the cyber-resilience of the cyberattack surface of the system-of-systems. Many of the incentive problems noted by Smith et al. (2016) can be reduced by the PMO construct, including project manager continuity, if project managers are appropriately constrained to work within their program.

> PMOs are incentivised in gate reviews to test early and often so as not to inherit an operational burden.

Independence in T&E can also be constructed in such a way so as to act on behalf of the user, usually by insisting on knowledgeable representative users being involved in requirements definition and in earlier usability T&E. The stakeholders shown in Figure 16.5 ought to essentially act for the users as well. For example, the operational authority should champion the operational users, and the technical authority should be championing safety, reliability, maintainability and even elegance by design (Efatmaneshnik & Ryan, 2018). A possible fix in the diagram would be to show users as a separate stakeholder but rarely do they have such input or indeed scope. Operational test authorities, as evidenced in *Pentagon Wars* and the subsequent congressional charter of U.S. DoD office of T&E, are chartered to work for the safety and mission effectiveness (i.e. the lives) of sailors, soldiers and airmen. User acceptance, like design acceptance, must not be left until a fait accompli, where there is no time or budget for anything other than operational workarounds with their associated compromise to safety or mission (as shown at Figure 16.3 with Type I Error – accepting bad capability as good or even a Type II error – failing to accept good capability). Pressure on acquisitions to reduce schedule, outsource and, in particular, reduce travel budgets, has meant that investments like resident project offices for the design development and production facilities are at risk. As such, these resident project staff may not be there to ensure that representative users are arranged for key test activities, and in their absence, the contract developers are unlikely to be incentivised or powerful enough to make such arrangements. The lack of a resident project office and access of the crew to the Australian LHD ship during assembly and test was raised in audit testimony to Parliament (Australian Parliament, 2016). In this project, this limitation led to a high transference of risk to Navy for completion of the project during operational test. Wernas and Joiner (2018) have shown the effect of, and importance of, such representative users in maintainability of these ships, along the lines first proposed by Blanchard in 1975 (6 Ed., 2013) as *front-end* logistics.

The failure to provide independent T&E for users is perhaps no more disappointing than in software-intensive or ICT projects. Early in the information age, the efficacy of structured computer-based usability T&E was researched and documented, along with sound computer-interface metrics (e.g. Wickens et al., 2014). However, significant failures in about one-sixth of ICT projects (Flyvbjerg & Budzier, 2011) can be attributed to failures to conduct usability T&E early enough or in most instances not at all. One of the authors has conducted OT&E on over 20 computer applications, each costing multimillion dollars and many years of project effort, where unfortunately users were not involved in the development of the functionality between the initial requirement writing and the operational fielding. In one instance, the application, important though it was, involved just three users who first saw the application in operational test.

Similar to earlier, PMOs are a positive development for more user-centric metrics in judging the progress of ICT projects. As documented by Joiner et al. (2018) and illustrated in Figure 16.6, usability T&E is at the heart of the Australian Department of Human Services (DHS) project process. This process was developed by the new PMO to guide and check their mostly ICT projects from

> All chances of the ideal three to five iterations for optimum efficiency, as documented by Wickens et al. (2014), are lost when users are not formally involved.

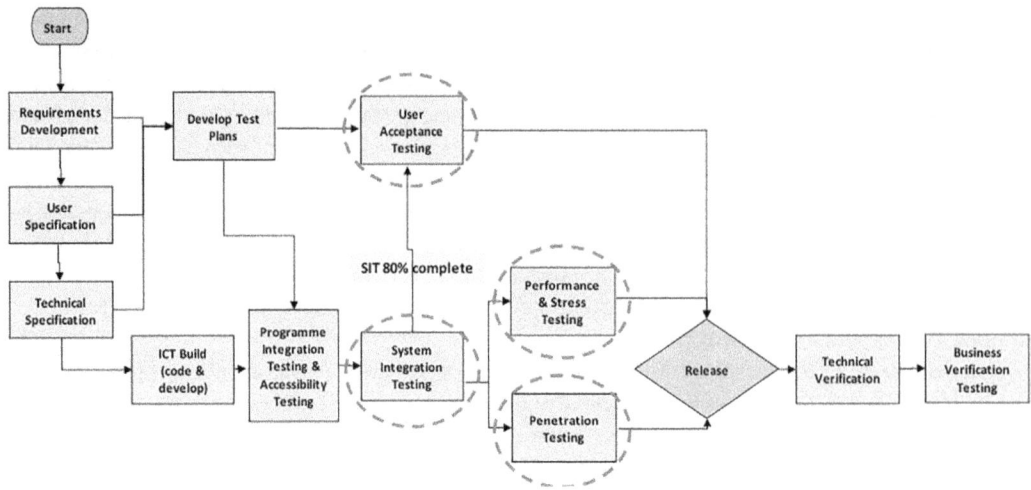

FIGURE 16.6   Example of test-led ICT Project Governance model by a PMO. (Adapted from Terrell (2016).)

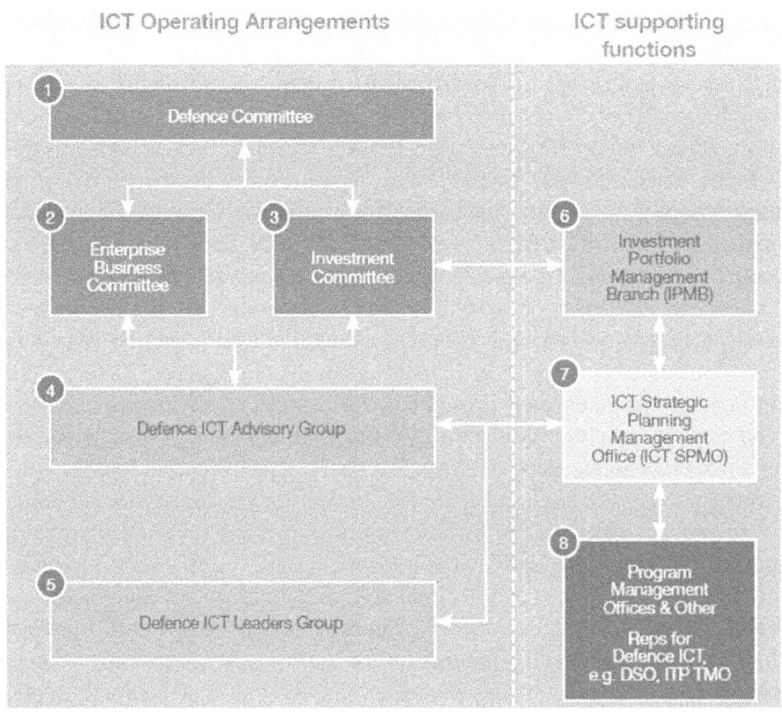

FIGURE 16.7   ICT Governance Structure in Australian Defence (CIOG, 2016, p. 7).

debuting without adequate user and other testing (Terrell, 2016). The DHS project process diagram has four dominant T&E boxes. When this project process is compared to others like the DoD (Figure 16.7), the DHS one has relative simplicity, is evidence-based and test-led. Joiner et al. (2018) find from this work and others that there are four key T&E types for good ICT governance to attain and maintain cyber-resilient systems. These findings can be extrapolated to some extent to any software-intensive project. The four test types are mutually reinforcing for governance and cyber-resilience, since they collectively enable trade-off decisions. Such trade-offs have at their heart usability and a need for independence.

Independence and Usability Example. A project tested by an author for digitisation of battlefield management adopted an *off-the-shelf* system from a foreign country. Operational testing was the first structured usability T&E, and it determined some fairly major cultural and structural limitations in the *off-the-shelf* user interface. Limitations continued to be found by users in trials from different areas (e.g. logistics, ordnance) and in connectivity with other systems. Moreover, cybersecurity of the *off-the-shelf* system was largely frozen with little capacity for system modification. A decision was made to have a second phase of the acquisition project to redevelop the battlefield management system and to include three funded design iterations with a stable representative user membership engaged to formally trial usability between each iteration. The process delivered a far better product with scope for indigenous and ongoing integration to other systems and for cybersecurity. This example illustrated that complex interconnected architectures like Defence command, control, communications, computers, intelligence, surveillance and reconnaissance (C4ISR) capabilities need standard test-led governance arrangements like that used by DHS. Further, all such capabilities need some degree of development, integration and usability so as to work as a cyber-resilient system-of-systems with regular testing against new cyberthreats and for new additions of systems.

> Very representative system and software integration laboratories are quintessential to maintaining complex systems with resilience.

TESTING COMPLICATED AND COMPLEX SYSTEMS

Much has been written about the correct use of the term complex (Keating & Bradley, 2015; Pyke, 2016; Keating, 2017). The concern of many of these references is that complexity could excuse examination of projects that ought to have been scoped better and/or governed better (Keating et al., 2018). Another line in examining complexity in projects has been that complexity usually does not simultaneously exist in all disciplines, such as political, financial and technical (Pyke, 2016). The argument in these instances is that more careful use of appropriate processes in each discipline might better compartmentalise (thus harness) complexity within a project. A third concern in the overly quick use of the label complexity is with regard to modelling and simulation (Joiner & Tutty, 2018).

True complexity is more about emergent properties and nonlinearity in outputs and feedback that cannot be predicted, whereas complicated systems (ships, planes, tanks, ambulances, modern buildings, etc.) are those that are difficult to model and develop, but that with deconstruction and exacting process can be predicted (Keating & Bradley, 2015; Tutty, 2016). Careful examination of many projects, such as Australia's Collins Class submarine, shows that poor scoping, poor project processes and lack of contemporary test practices (even allowing for the times) made prediction unlikely (RAND, 2011; ANAO, 2002; Joiner & Atkinson, 2016).

> Many allegedly complex project acquisitions have mixed maturity of systems (GAO, 2016) and deliver into complex interconnected legacy system-of-systems and yet do so without modelling and simulation for the new systems or a representative test environment for the interconnected legacy system-of-systems. That is poor planning not true complexity.

Without diverging too much, there is a strong argument that much of the complexity, even including political uncertainty, was of the project's poor tactical, operational and strategic levels of planning. The use of land-based test sites (LBTSs) to representatively qualify and do usability T&E prior to build was extensively developed by the U.S. Navy in the period from 1975 to 1980 (Asher, 1978; Guido & Light, 1978; Stark & Stembel, 1981) and by the time of the Collins scoping (1989) was contemporary best practice (Mcguigan & Boylan, 1986; Cairns, 2011). Indeed, early plans for the Collins project were for extensive LBTS to be developed and continued in sustainment support *through-life*; however, budget pressures around the cost of spreading the build and a need to show success with production are understood to have led to a curtailing of LBTSs to be almost non-existent when compared to contemporary practice (Coles et al., 2011; Coles, Greenfield, & Fisher, 2012).

Technical and operational faults were found and publicly questioned only when four boats were in the water (ANAO, 2002; RAND, 2011). There is a risk that without appropriate governance (Bradley et al., 2017) such a budget dilemma may befall Australia's newest future submarine, as early as next year when the true cost of the current design and proposed manufacture is expected to be known. The time to build LBTSs, qualify systems and do usability T&E before submarine manufacture is most of the 2020 decade (Joiner & Atkinson, 2016; Joiner, Atkinson, & Sitnikova, 2017). It may be tempting to curtail this qualifying work as added cost and a schedule risk to continued project support. The point of this example is to help posit complicated versus complex, and how much the latter's emergence can be based on not being able to predict because T&E with modelling, simulation and prototype software infrastructure has not been planned or followed through.

Probably the most insightful examination of the factors causing surprises in major Defence projects, at least for Australia, was the bipartisan Senate Inquiry into Defence Procurement (Australian Senate, 2012). In short, two key findings were that project over-optimism led repeatedly to a belief that capabilities were more mature than they turned out to be and that pre-contractual PT&E was fundamental to de-risking. A second major element in Defence projects in Australia was disclosed in the Government response to an audit of Defence T&E in 2002. The Government and Defence at the time claimed that Australian Defence does not do developmental projects (ANAO, 2002), which if true would significantly reduce validity of excuses around project and technical complexity. The point here is one of early misconceptions and assumptions relative to all project scoping. One consequence of the Australian Defence (mis)conception is that the project life cycle of Defence to this day, when compared to the U.S. DoD, can be argued to (still) be missing the key phase of 'Technological Maturation and Risk Reduction (TMRR)' and the key milestone around a production decision (Milestone C) separate from the decision for development acquisition contract. Again, the point here is not to understand Defence projects, nor even an Australian perspective, but to appreciate the following principles:

- PT&E should always practically check the claimed level of maturity to some extent before developmental contract, even if that is based on experiments and modelling and simulation.
- All PT&E should include some usability T&E, again even if it is based on modelling and simulation if necessary, in order to ensure that the user requirements envisaged are indeed as correct as they can be at the time of contract.
- Any capability being acquired to operate in a system-of-systems will need a representative test environment (even if key systems and elements are claimed to be 'commercial of the shelf') to qualify the design, do further usability T&E, check interconnectedness and synergy with legacy systems (the most often overlooked aspects – looking across the system-of-systems not just within the project) and to support the capability *through-life*.
- Production decisions always need representative T&E and stakeholder consensus around those decisions.

> The most insightful examination of the factors causing surprises in major Defence projects was project over-optimism leading repeatedly to a belief that capabilities were more mature than they turned out to be and that pre-contractual preview T&E was fundamental to de-risking.

Not following such principles invites 'apparent complexity' and does not set up acquisition processes and staffs at all well for dealing with true complexity.

This introduction to the second part of this chapter is not all reflective of past practices. Looking forward, it is often said that systems will be more complex, and the main reasons are autonomy, interconnectivity and, thus, multigenerational interdependence and AI (Nielsen, 2017; Keating, 2017; Roedler, 2018). The prediction is certainly supported by the weight of T&E literature (ITEA Journal, 2011–2017); however, that literature is by no means pessimistic. Taking probably the most threatening of these for time-dependent complexity – AI or sometimes referred to as complex adaptive systems or cognitive systems – it is possible to represent learning in AI to a test-representative environment to predict development within some bounds as an extension to autonomy (Sustersic, 2017). Furthermore, it is possible to provide sophisticated continuous monitoring of AI as complex adaptive systems – testing development *through-life* with immersive technologies (Panei, 2017). The fundamental metrics of such T&E of intelligent systems are about trust (Madhavan, 2017) to achieve meaningful human control (Roff & Moyes, 2016). Such approaches are in many ways analogous to structural fatigue or *accelerated-life* testing, where a representative article is put through a carefully designed compendium of possible life events, and some further articles are instrumented in service so as to check *through-life* the validity of the anticipated or *accelerated-life* to the *real-life*. Certain inspections and other checks augment the *through-life* management, and in some instances, where configuration, role or environment differ from the anticipated, then fatigue testing may need to be done again differently. Already, highly autonomous systems with cyberthreat risks are having smart-monitoring fitments known generically as 'shadowed systems' or *sidecars*, to check for anomalous behaviours that might predict some form of compromise (U.S. DoD DSB, 2016, p. 93). Such *sidecar* fitments are a necessity of numerous legacy systems being in service that are now receiving additional connectivity and being subjected to ever-evolving malicious threats for which they were not designed or tested (Joiner et al., 2018).

This part of the chapter will now examine in greater depth evolutions in T&E for three complexity or complicating factors:

- Interconnectedness (system-of-systems)
- Malicious use of cyberspace and therefore the need for cyber-resilience
- Autonomy and AI.

T&E evolutions are kept as the common theme in these subsections, drawing on the fundamentals addressed in the previous section.

INTERCONNECTEDNESS AND SYSTEM-OF-SYSTEMS

A major review of the approach of the U.S. DoD to a growing complexity in its systems by Joiner and Tutty (2018) looked specifically at the initiatives used to give assurance to I3. The review contrasts the Australian DoD with the U.S. DoD to show the efficacy of the U.S. DoD initiatives, especially their collective synergy, but also to show the challenge to an alliance when one partner embraces significant shift in rigour and threat posture. While the context is Defence, the principles should readily apply to business transformation in digitisation, internet usage and cybersecurity. The types of initiatives examined are also universal to today's complexity challenges:

- Creating more representative environments through federating SILs and HWILs to enable regular testing against new threats, new roles, new environments or developing capabilities.
- Creating annualised experimentation exercises with a battle rhythm of testing for I3 assurance using the infrastructure above. Such exercises are at the portfolio or program level, and therefore, this unburdens projects bringing new capability from needing to organise access to much of the legacy system infrastructure, the latest cyberthreats and representative operators, while also compelling certain fundamental T&E assurances.
- Adopting use of six-sigma test design and analysis methodologies in acquisition policy, project milestone approval gates and acquisition staff competencies, so as to achieve the additional rigour and efficiency of these methods, especially in more complicated high-factor, highly connected (interdependent) environments.
- Adopting cybersecurity T&E in acquisition policy, infrastructure and staff competency targets (see the next section).

- Assigning dedicated acquisition staff, closely aligned with P3O staff, focused on the infrastructure, competencies and other wherewithal to give I3 assurance. In some instances, this requires augmentation of the current design or technical oversights to include new certifications, especially around key protocols for interconnectivity, even where this includes new skilled personnel, necessary to fuse functions.

The threat of greater interconnectivity and information flow is countered by greater interconnectivity of the representative test systems. Similarly, the cyberthreat activity is countered or met with greater cyber-resilience activity in new activities and strict battle rhythms for

> These initiatives can at their simplest be described as turning threats into opportunities.

these to occur. The sophistication of complicated systems and of cyberthreat adversaries is countered with new *higher-order* competencies and education for acquisition and in-service management staff. Ever has business sought to turn threat to advantage – what is remarkable is when public service bureaucracies like the U.S. DoD exhibit adroitness with such persistent initiatives. In that context, the fact that other DoDs have been slow to respond is perhaps not so surprising. A really encouraging sign of the U.S. DoD initiatives has been its effect on other U.S. departments like Homeland Security and on Defence Industry.

Simply setting off in the direction of an exemplar like the U.S. DoD, even adjusting for scale, can be daunting. In an effort to add quantitative modelling to the qualitative work of Joiner and Tutty (2018), research by Joiner, Efatmaneshnik, and Tutty (2018) created two approximated project acquisition models for the system-of-systems acquisition approaches of the U.S. and Australian DoDs. This work found the 90% confidence limits for project achievement were about the same for both DoDs for the best instances, that is about 9 years but for the worst instances were substantially better for the U.S. DoD compared to the Australian DoD, that is, 18 years compared to 26 years (more likely cancelled or with costly reincarnation). This modelling illustrates the advantage of the U.S. DoD's additional and earlier systematic integration T&E. The real prospect of such modelling of system-of-system acquisition test approaches is in being able to tailor the extent of I3 assurance wanted in different acquisition programs, that is, explore different alternatives before undertaking multi-year reform by a P3O. An illustration of the work is at Figure 16.8.

Both the qualitative and quantitative research on I3 assurance testing for system-of-systems also used Tutty's preferred terminology of *families-of-system-of-systems* (Tutty, 2016). The family reference adds another layer to system-of-systems for say consideration of ships operating independently but interconnected and synergistically with helicopters and satellites. The terminology acknowledges each of these can be a system-of-systems to design, produce and maintain in their own right but then must operate as a family in-service and joint task forces when deployed. Moreover, the family reference implies multigenerational development, adaptability in missions and with some elements integrated by design and others interoperating through generic interfaces. Using the family analogy, the I3 assurance testing approaches is analogous to the type of *growing up and schooling* new capabilities receive with, and for, the family.

In summary, the best response to interconnectedness and system-of-systems is to be prepared through federation of representative systems to test early, for example, using LVC simulation, and then to test often and *through-life*. The wherewithal includes test infrastructure and competencies. There are many good examples that can be leveraged, especially in the U.S. DoD.

Cybersecurity T&E

The U.S. DoD is exemplary in cybersecurity T&E and surprisingly has again be adroit (since about 2009) in dealing with cyberthreats using new cyber-resilience initiatives, especially relative to Australia (Joiner, 2017). The U.S. DoD began by exposing its new capabilities to the new cyberthreats at the *right side* of the acquisition life cycle (Joiner, 2017), that is, as capabilities go

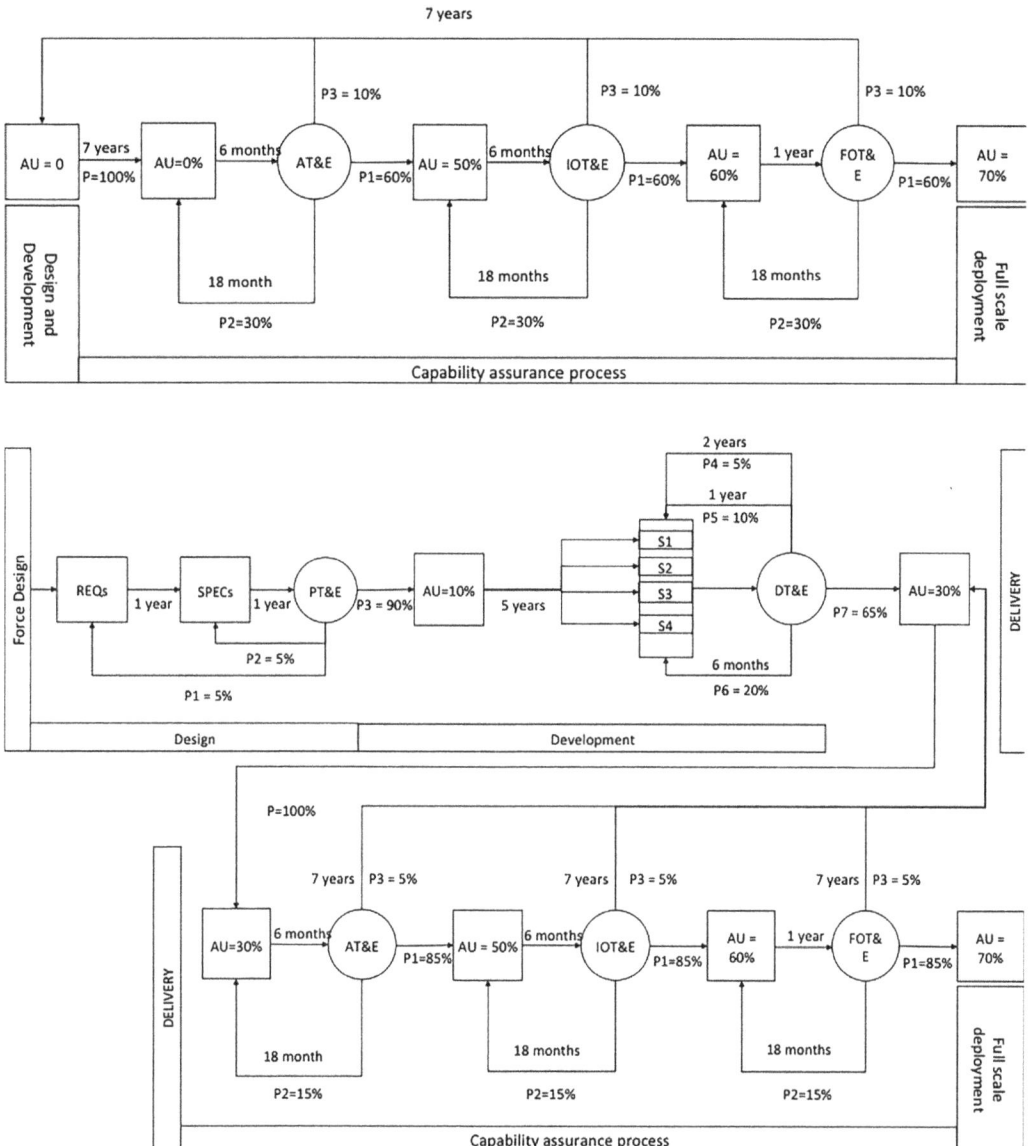

FIGURE 16.8 I3 assurance test model for the Australian DoD (top) and for the U.S. DoD that has pre-delivery and post-delivery assurance processes (bottom). The process probabilities and assured utility (AU) values were assigned subjectively (Joiner et al., 2018).

into initial service. This was remarkable since acquisition thinking would say such testing against threats that were not a requirement in the design would be *unfair*, yet any operational analysis would see the threat as contemporary and demand to quantify the risk on operational systems. Several years of arguing for Australia to follow the U.S. DoD and start by testing operational systems have continued to meet the *unfair* label by senior professional project engineers, while the operators therefore remain comparatively unaware of the new risks. The U.S. DoD approach led to their service chiefs getting candid cybersecurity T&E results on newly fielded systems, such that these chiefs invested in a higher level of measures for cyber-resilience than has occurred in other countries like Australia. This investment is orders of magnitude different, even when adjusting for the comparative size of the two DoDs' total budgets. Since 2015, the U.S. DoD acquisition processes have been fully revised to include cybersecurity processes and T&E throughout, as illustrated by the six-step process (Brown et al., 2015). The U.S. industry, led by International Council on Systems Engineering, has mapped the U.S. DoD cybersecurity processes to normal SE processes and found the quintessential 53 or so additions to normal practice in order to be cybersecurity compliant and consistent (Nejib, Beyer, & Yakabovicz, 2017). This work is not only about the *what* but also the *when*. These processes are percolating out into wider departmental and industry use in the United States but are arguably at risk of a *not-ready* or *not fair* outside of the United States. The risks in not adopting cybersecurity reforms are simply not believed in some countries and industries or, if spoken about, are not followed through with infrastructure and training investments anywhere near the scale necessary to pace the United States as a trusted military or business ally. The forefront of achieving cyber-resilience of systems is of course in ICT projects or software-intensive systems (Joiner et al., 2018) although not exclusively so, given capabilities for electromagnetic probing of systems hitherto perceived to be *stand-alone* (Joiner, 2017).

The work by Joiner et al. (2018) into ICT governance for cyber-resilient systems takes a test-led approach, premised around a PMO, to build better *in-house* capacity and thus governance for cybersecurity T&E. Such ICT governance approaches enable trade-offs between usability, system integration, system performance and cyber-resilience. This work also recommends the U.S. DoD approach of federating representative test infrastructure in order to increase the pace and realism with which potential and real cyberthreats can be anticipated and defended, both for developed (fielded) and *developing* systems. Fundamental to understanding cybersecurity T&E is to understand that the *cyber attack surface* includes every exploitable connection that is cyber-enabled (Brown et al., 2015). As such, representative cybersecurity tests cannot be performed without the representative presence of the full system-of-system or *family-of-systems-of-systems* in which a system will operate. These criteria are why there is a *high bar* or impediment to many industries and departments adopting cybersecurity T&E for cyber-resilient systems. As explained early, many departments and industries have harvested savings to outsource their computer applications, software support and operating systems, such that they no longer have design authority, *in-house* representative systems, intellectual property rights, trained T&E staffs (ICT testing generally) and so forth (Joiner et al., 2018). From a government perspective, the impediments to cybersecurity testing are even higher, since most outsourced technical support will need to amend the support contract to cover the new requirements; however, with cyber, releasing the threats to industry is unlikely to be a sensible security option. Provisions are needed for *multi-security* T&E where after cybersecurity testing, most commercial support contracts will need to be provided only architectural modification requirements without full disclosure of the cyberthreats used (Christensen, 2017).

> Much of cybersecurity requires heavy reinvestment in *in-house* test capabilities, much of which has to necessarily be federated.

No treatment of cybersecurity T&E is complete without an overview of the problem of attribution. Put simply, attribution in cyberspace is prohibitively difficult, and there is an absence of international and domestic laws with which to prosecute what ought to be, in most instances, criminal or even war-like behaviour (espionage) (Heinl, 2016). As such, this is a further impediment to any outsourcing in contracts in the usual means of risk transference, via punitive measures such as liquidated damages. Cyber operatives have the time and often the resources to systematically look for the weakest link, or even the confluence of minor weaknesses, to determine an exploitation strategy. Worse still, such strategies need not be used until a time of choosing in some form of coordinated strike. Hence, one of the best Defences for cybersecurity T&E is combinatorial testing techniques from the six-sigma stable, since these are highly systematic and rigorous but still efficient (Joiner et al., 2018). A common cybersecurity technique being used in the absence of combinatorial techniques is random or *fuzz* testing, which by randomness has weaknesses and can be inefficient (Kuhn, Kacker & Lei, 2010; Kuhn et al., 2016). The likely reason for widespread use of *fuzz* testing is that it requires less training to operate and is easier to give apparent rigour through somewhat misleading measures like *weeks of testing*.

Works like (Christensen, 2017) find after years of practice that well run *tabletop* exercises are key to systematically understanding what cyber risks are realistic threats, how they might manifest and what defensive measures could detect and mitigate such threats. As with all such *play*, diversity of experiences and ideas is crucial, and as such, these *tabletop* exercises are now being run by trained cybersecurity facilitators who are increasingly accredited.

Following systematic *table topping*, equally systematic test methods are needed to quickly screen which permutations of threat, defence and architectural variances will be problematic: what is now called *heat mapping* (Troester, 2015). Because cybersecurity T&E was rapidly developed in the period from 2009 to 2015 by the U.S. DoD when six-sigma combinatorial HTT methods were also being widely disseminated and used by U.S. DoD acquisition and test staff, these efficient and rigorous methods are primarily used by U.S. cybersecurity practitioners to create the *heat map* screening. Unfortunately, outside of the United States, such test design skills and packages are not in common cybersecurity use. The work by Joiner (2018) has shown that the competencies for such test design can be readily taught and applied.

In summary, the evolutions in cybersecurity T&E have been prodigious in the United States, but there are significant infrastructure, contractual, educational and experiential impediments to their use in wider countries and industries. Work on key governance requirements for cyber-resilient systems (Joiner et al., 2018) and on key competencies (Joiner, 2018) are readily available to begin transformation; however, infrastructure and other costs will take most departments and businesses years of sustained effort. The synthesising of computer-based systems into every aspect of life and all facets of operations means that the T&E evolution necessary to catch up for each department and business needs to begin soon and be sustained. The directions given here, and in the supporting literature, are unequivocal.

Autonomy and AI

Autonomy exists in tangible systems like the driverless mining trucks and trains, to the signalling in public transport systems, autopilots in planes and highly autonomous unmanned aerial vehicles and satellites (Nielsen, 2017; Keating, 2017; Roedler, 2018). Testing of autonomous systems requires establishing, like all other capabilities, assurance of functionality (Cofer, 2015; Panei, 2017). The difficulty with autonomy is that it essentially automates an extent of human operator control from what is known as *in-the-loop* (direct say) to *on-the-loop* (monitoring) to *out-of-the-loop* (prior consent) (IEEE, 2018). Sometimes such automation is for speed (e.g. supersonic missile defence) and sometimes for consistency (e.g. auto pilot). In many instances, modern automation is requiring *higher-order human-like functions such as strategies and decision-making, not simply control* (Joiner and Tutty, 2018). Testing such software becomes a matter of trust (Sustersic, 2017), and well-tested autonomous systems are often referred to as *trusted-systems* (U.S. DoD DSB, 2016). Trust is a human dimension, and thus, usability T&E with human factor specialists are essential to establishing trusted autonomy. When software performs *higher-order* functions, it does so for, or on behalf of, a responsible human at some level, and thus, these persons in a representative sense must be part of the usability T&E to establish trust, starting right with the early modelling and simulation of functional requirements that set the functional trust boundaries (Madhavan, 2017). Such ethically based design considerations are being codified by the IEEE (2017) and for weaponry through proposed UN conventions (Roff & Moyes, 2016). The latter proposes that meaningful human control in sociotechnical systems derives from the following characteristics that pervade the T&E of such autonomous systems and change some T&E from its usual quantitative-only measures (e.g. reliability) to one that includes perceived trust by the responsible human.

> It may be counter-intuitive to some project managers that the most difficult of usability T&E will be for the most autonomous of systems that have less and less users.

- Technology is predictable
- Technology is reliable

- Technology is transparent
- The user has accurate information (to include contextual awareness)
- There is timely human action and a capacity for timely intervention
- There is accountability to a certain standard.

An example of trust measures that requires responsible human interpretation from a representative population is the term *timely* in the context of the above measures. One aspect of increasing concern regarding the reliability of autonomous systems is the extent of possible interference that must be considered in the design and T&E. Hitherto, it was generally sufficient to consider only *use*, *abuse* and *misuse* and to constrain the latter with likelihood assessments. The now *malicious* use of cyberspace has added malicious intent to the design and operational cases that many systems ought to be reasonably resilient to, particularly autonomous systems (Nielsen, 2017). This poses another reasonableness test for early usability T&E around functional requirements that ought to also involve cybersecurity *tabletop* evaluation, without which the malicious cases risk being nonexistent, unrealistic or speculative (Christensen, 2017).

Moving now from autonomous to AI in systems, the primary lens with which to view the change is the ability for an intelligent system to learn and adapt using what are known as reward functions. Earlier, the concepts of *accelerated-life* testing and *through-life* test monitoring were offered as conceptual (epistemological) bridges for strategies to adequately build tests and evaluation for AI. If a representative sample of humans can set the reasonable bounds of learning experiences for an artificially intelligent system, expose such systems to those diverse experiences and follow the learning and responses, then reasonably, there is a case for trust and meaningful human control (Sustersic, 2017). There is of course one key caveat that is necessarily based on the analogy to *accelerated-life* testing and that is to monitor *through-life* the actual experiences and ensure they remain within the bounds of the *accelerated life* construct. Such *through-life* monitoring is *through-life* continuous testing and needs to be managed as such by competent test teams with time-dependent test competencies and of course understanding of the learning of artificially intelligent systems (Normann, 2015; Panei, 2017).

A final caution for modelling and verifying complex adaptive systems such as AI comes from both Taleb (2007) and Javorsek (2016). The latter counsels support of the former in that *human agency in complex adaptive systems can be better modelled with heavy-tailed distributions (i.e. non-Gaussian) where extremes are not exponentially bounded* (Joiner & Tutty, 2018). In essence, Gaussian distributions are more often associated with, and achieved with, real humanity; whereas artificially intelligent systems, which when they go wrong are not as constrained to such bounds of reasonableness. Consequently, *heavy-tailed* distributions when aggregated show a greater risk of *off-design* outcomes that appropriately lead to more checks and balances in the design and T&E. An interpretation of Javorsek's (2016) caution is that risk profiling needs to be *heavy-tailed*, so verification includes wider extremes. In essence, this is a caution against anthropomorphising complex adaptive systems with historically human distributions because it inappropriately limits the verification testing.

So, autonomous and artificially intelligent systems are increasing and are not beyond evolved T&E processes to capably deal with. That said, like many of the complexity issues dealt with earlier, acquisition and T&E staffs need new competencies, new T&E methods and the test infrastructure that supports accelerated learning and *through-life* continuous test monitoring and timely test reporting.

> Contrary to perceptions, meaningful human control of such systems requires early and more difficult usability T&E with representative responsible humans, structured around measures that are essentially ones of trust.

An example of an advanced autonomous system being acquired by Australia is the Triton MQ4 unmanned aerial vehicle. This capability is to operate long-range autonomous surveillance of maritime approaches, such as can be found between Perth and Darwin. The capability predominantly relies on satellite links to operate both the flying and sensors from a central control facility. The geostrategic significance of such distances is impressive and in most other countries would be considered a global projection but for Australia's size is finally a capability apropos to the limited population and geographical extent. Australia appears to have chosen to acquire the capability as promissory *off-the-shelf* and not cooperatively join the development of the capability with test aircraft. Early usability T&E around the meaningful human control and trust risks being abrogated to the U.S. DoD if such autonomous systems are adopted rather than co-developed. The use of such a complex and autonomous system, within a maritime, air and space surveillance and response *family-of-system-of-systems* also reasonably requires a federated battle laboratory of representative SILs and HWILs that go well beyond service and Defence-department boundaries into border protection and customs. Many of the system-of-systems that the Triton aircraft must network with are not of U.S. origin and the role therefore does not directly translate, nor the *cyber attack surface*, and certainly not the environment, where Australia's top third of the continent is 'hot-humid' not found within the United States. Moreover, the sovereignty of *rules-of-engagement* warrants a close examination of meaningful positive active human control and trust (Tutty & White, 2018). Not procuring early modelling, simulation and test aircraft risks not adequately doing the necessary T&E to understand, leverage and even optimise such a capability until many years into operating the real capability. It is almost impossible to fully represent the mixed *cyber attack surface* of the mixed European, Australian and U.S. *family-of-system-of-systems* without a representative Triton SIL and HWIL that is federated to the other system-of-system SILs and HWILs. Bargaining for *off-the-shelf* capabilities can be argued to be a very short-term gain and, in principle, means far less informed decision-making at every stage into and *through-life*. The point of this example is not about the Triton capability but the need to prepare for such complex systems with a deeper foundational understanding and commitment to be involved and informed through T&E and the necessary T&E infrastructure and competencies, especially to cyberthreats and to build trust in your specific configuration, roles and environment with meaningful and documented human control.

COMPLEX SYSTEMS GOVERNANCE AS AN ASSURANCE STRATEGY ALTERNATIVE TO T&E

One promising theoretical development in governance of complex systems that could foundationally improve the use of T&E for improved decision-making is known as Complex Systems Governance (CSG) theory (Keating & Bradley, 2015; Keating et al., 2018). The CSG theory took *first-principle* elements of systems theory, management cybernetics and governance to produce a new model of how to do governance for complex system. CSG is shown at its simplest meta-cognitive level in Figure 16.9.

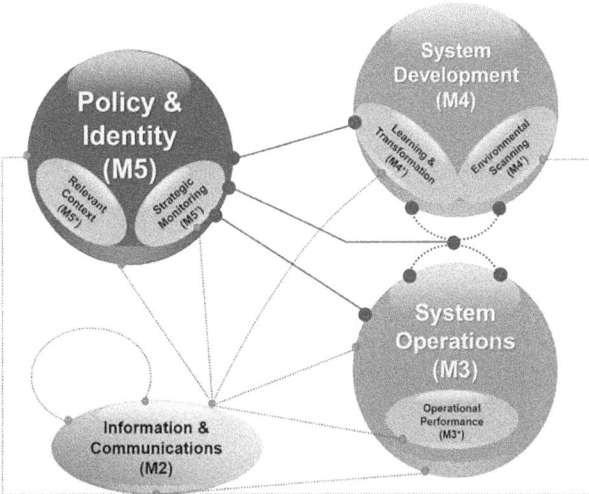

FIGURE 16.9 CSG reference model. (Reproduced with permission from Keating and Bradley (2015), Keating et al. (2018).)

The CSG model is not constrained to an alliance between fields like PM or SE where vital clues and information may not be connected. Further, the model has a key node of *policy and identity* alongside more recognisable elements of a 'system development' and 'system operations.' The *policy and identity* node is to account for the necessary existential evolution of complex systems based on emergent properties, be they technical, social or political. There is a high emphasis, as you would expect, on strategic monitoring, particularly on the nexus between systems development and systems operations, where some of the T&E evolutions discussed in this chapter are crucial. For example, a federation of SILs and HWILs that adequately represents the operational architecture that is augmented to routinely test developing new systems in the system-of-systems and further can collectively be routinely tested against evolving new threats like cyber threats, clearly is an evolutionary advantage to complex system-of-systems.

The path to adopting CSG theory into practice has been researched, including the difficulties of substantial and somewhat competing cultures like PM and SE (Keating et al., 2018). Wholesale adoption of *clean-sheet* governance is almost never possible, and so instead a method has been developed with careful alignment to the theory principles to move existing governance structures closer to CSG theory by accretion. This method uses some 53 pathologies to identify any dysfunction in CSG against the CSG theory and is known as the Metasystem Pathology Method (M-Path Method) (Katina, 2016a,b). The *entry-level* assessment of an existing governance structure is not onerous and is performed by surveying perceptions of the extant governance by the milieu inhabitants. The surveys are standardised questionnaires that measure perceptions of how well governance metafunctions are being performed, by comparing 'governance capacity' versus 'governance demand' on seven dimensions shown in the spider diagram in Figure 16.10.

The new CSG theory therefore offers opportunity for better governance of complex systems especially through better use and valuing of T&E in each of the governance characteristics.

The fit of *capacity* versus *demand* avoids striving for utopia, which is a risk in many P3O maturity or other benchmarking models. Moreover, such an approach is inherently tailoring to the unique complexity, flexibility, change, uncertainty and other environmental characteristics of each governance of *families-of-system-of-systems*

> T&E is a powerful agent to any governance, mixing well with the processes of SE and the balance of PM, and yet arguably never quite being recognised for its true synergy and power by either field.

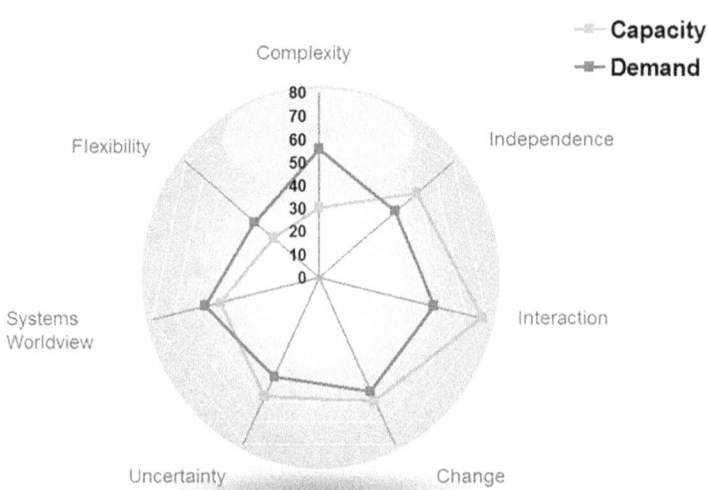

FIGURE 16.10 Complexity – systems capacity versus environmental demand. (Reproduced with permission from Keating et al. (2018).)

(see Tutty, 2016 Annex B). In part, this may be because projects without the P3O overlays are somewhat ephemeral, meaning they miss the complexity of interconnected *through-life* systems. Second, SE is design-centric and can confine T&E to being a means to limited ends or a low-level method. The CSG construct helps explain why, in entities like the U.S. DoD, T&E has emerged as a critical governance element in its own right, looking into and supporting both PM and SE, but omnipresent at the decision table (Joiner, 2015a; Chu, 2016). In the CSG theory, this is because T&E can independently inform with evidence on each of the seven scales of assessing governance adequacy (Figure 16.10), especially:

- *Complexity* through representativeness of configuration, role and environment and in the predictability of modelling and simulation
- *Independence* from the assigning project and developing contractor as outlined earlier
- *Interaction* through the necessity of collaboration to get test events to happen and the power of such test events to build cohesion around deficiencies and opportunities
- *Change* through the power of particularly PT&E to showcase constructively new technologies and operating concepts
- *Uncertainty* through the power of test to disclose technical and operational problems, uncover stakeholder concerns, especially in interconnectedness
- *Systems worldview* through the power of test to unearth limitations and assumptions
- *Flexibility* where it exists to get testing done, especially at the functional level through innovative modelling and simulation.

QUESTIONS FOR DISCUSSION

(1) What combination of T&E tools are appropriate for any given complex project?

This chapter prescribes eight T&E tools for best practice in complex projects. Of those eight tools, the first on structured preview testing of risk prior to developmental contract (No. 1) and the second on combinatorial test design techniques (No. 2) are the tools always appropriate. These two tools will fundamentally reduce the discovery on projects so that such discovery is only for the truly emergent elements and will give modern efficiency and rigour to testing from the outset. Of the remaining T&E tools listed, the essential skills for ICT and software testing (No. 4) need only be applied when the software functionality is predominant, while the remaining five are only essential when projects or programs are at scale or truly complex in each of those dimensions; such as:

(No. 3) – Essential if there is ongoing stakeholder uncertainty or political interference possible
(No. 5) – Essential if there is high degree of multigenerational interconnectedness and operational dependency in the operational collection of system-of-systems
(No. 6) – Essential if there is an envisioned cyberthreat or critical continuous operations (cannot be shutdown temporarily) and software-based and/or internet-based control elements
(No. 7) – Essential if there is autonomy or AI contemplated for the project
(No. 8) – Highly desirable if the project or program operates within a portfolio with other complex projects such that it has both internal and external governance factors extending for periods longer than 3–5 years.

(2) How can PT&E and model-based T&E benefit from the current tools?

PT&E, if necessary involving virtual modelling from functional model-based systems engineering (MBSE), is one of the listed T&E tools in its own right, essential to preview risks prior to development contract and desirably before final solicitation. However, of the other T&E tools, combinatorial test design (No. 2) gives rigour and efficiency to PT&E because of the power to screen significant factors, while many of the other tools listed also bring synergy to PT&E. For example, early software usability T&E (No. 4), early I3 assurance (No. 5) and early cybersecurity table-topping (No. 6) validate, enrich and inform the right requirements before developmental contract.

(3) How can we evolve the current tools so that they can be used with highly complex projects?

The greatest way to evolve current T&E tools is to develop wider use of CSG (No. 8) and refine its pathological methods, so that program directors and project managers are better at isolating what is truly complex from that which is just complicated. There is also a fundamental lack of textbooks and dedicated courses on all the listed T&E tools.

> When complexity can be isolated, then model-based system development and testing can be used for the complicated elements, while the emergent complex elements can be subjected to more frequent test regimes to determine when emergence occurs and quickly and appropriately test compensations as necessary.

CONCLUSION AND FUTURE DEVELOPMENTS

What has emerged in the fundamental characteristics of T&E is a mixed scorecard of evolution. This summary will start with the more stagnant of fundamentals and move to the most evolving. The PM movement, breaking capability acquisition into unique tasks as it does, has been found to struggle with *project over-optimism* early, testing too late and failing to realise benefits. However, the P3O movement appears to scaffold projects into programs with benefits realisation and sustainment in their governance of projects. Those PMOs with strong authority to critically invest in, and divest of, projects are looking to T&E early as the evidence-based assurance to de-risk projects – if you like to get into the *right projects* in the *right way*. As such, PMOs are likely to be a welcome governance force for more timely and credible use of T&E. Similarly, PMOs appear to offer governance that is more inclined to conduct T&E independently and with earlier and more frequent user acceptance in the T&E methods. Software testing has somewhat ironically tended to come after hardware, despite the ability of software to more easily model and simulate code around functionality. Now that software-based functionality predominates and there is pervasive cyberthreats to that front, PMOs need to invest far more in software/system integration laboratories that are more representative of fielded and potential future architectures. Unfortunately, many public service departments and business have outsourced software support to a *shopping mall* of applications, operating systems and dedicated processing. In essence, they have harvested efficiencies that make the move to more cyber-secure, user-focused and *in-house* SILs very hard. Whether senior executives see these risks and move to representative system-of-system SILs, will dictate whether their T&E has the independence and usability to wisely govern the efficacy and security of both their fielded and developing systems.

The most evolving of the T&E fundamentals has been in the rigour and efficiency of test design, led by the innovative test design methods of the six-sigma movement. The success of United States, Japanese and Korean industries and U.S. Defence in using six-sigma test methods for better rigour and efficiency are remarkable. Where these methods have been employed with policy support, test practitioners report them as empowering. Put another way, these are test design and analysis methods that help practitioners readily deal with multifactor interconnected complicated systems. Importantly, these methods can be as readily used at the level of service industries and customer satisfaction as they can at the heart of a system's processing technology. Where the methods have evolved to include combinatorial (high-throughput) testing, this has enabled high rigour in cybersecurity testing and highly efficient screening of very large numbers of factors typical of system-of-systems. The only frustration with such rapid evolution is that it is occurring in isolated pockets, especially outside the United States, and as such, each area has to sceptically learn the benefits.

Hence, T&E characteristics, born as they are to inform decision-making, are proving to be powerful agents in governing more complicated and interconnected systems, especially for PMOs.

Testing of complicated and complex systems is evolving rapidly in response to technology and demand. Much is labelled as complex that ought to be predictable or at least manageable if the modelling, simulation and testing were set up right for timely and informed decisions. An identified theme was the need to meet increasing interconnectedness with better interconnected (federated) system integration laboratories for more timely, representative and pervasive T&E. Second is the need to overcome the rapidity and sophistication of cyberthreats with more frequent and sophisticated testing for cyber-resilient systems, starting with the

In the case of autonomous and artificially intelligent systems, T&E has evolved considerably using analogous themes to earlier T&E approaches: namely *accelerated-testing* and continuous *through-life* test monitoring; the fundamental metrics of such T&E of intelligent systems are about trust to achieve meaningful human control.

contemporary threat to legacy systems. Both of these themes require a move back to more *in-house* testing or at least partitioned *multi-security* testing on a federation of SILs to represent the system-of-system being operated. In the case of autonomous and artificially intelligent systems, T&E has evolved considerably using analogous themes to earlier T&E approaches: namely *accelerated-lifing* and continuous *through-life* test monitoring. Counter-intuitively to some, meaningful human control of autonomous and intelligent systems requires more complex usability T&E focused on representative responsible humans so as to build the criteria and measures for trust. These evolutions build on extant acquisition and T&E competencies but place greater test demand, especially for T&E infrastructure, T&E networking and structured experimentation exercises. Departments and industries need to invest in the extended competencies of their acquisition and test staffs for these complicated factors, build test capacity and the infrastructure and processes to manage and report in a timely manner on such testing. Anything less, risks ill-informed importing without the necessary I3 assurance – or in the case of complex adaptive systems, without the trust for positive, active meaningful human control. T&E is rapidly evolving both through better methods, like the six-sigma test design and analysis, and by demand, through complex system-of-systems born mostly from greater interconnectedness and software functionality. The six-sigma methods provide substantially improved efficiency and rigour in testing and they enable practitioners to handle large multifactor complex systems testing with a degree of empowerment. The interconnectedness and growing complexity of system-of-systems is requiring those who develop such system-of-systems to have more representative and federated test sites and to test more often, particularly to meet cyberthreats with greater system resilience. The new T&E methods and the necessary T&E infrastructure for complex adaptive systems are both substantial challenges to self-assessing competency models and to acquisition outsourcing, since practitioners are likely to be ignorant of new methods and there are bound to be substantial impediments to adequately federating outsourced systems, particularly for cybersecurity T&E.

Areas of T&E that are not as evolved include T&E timeliness, independence and usability, as well as test methods for autonomy and artificially intelligent systems. The development and use of program-level management (i.e. P3O) shows prospects for improving T&E timeliness, independence and usability since these levels of management are more motivated to long-term objectives and earlier risk disclosure – setting the *right projects* for the *right reasons* and *adjusting often*. Similarly, P3O management structures are more likely to be motivated to invest in the T&E infrastructure to competently govern complex adaptive system-of-systems *through-life*. A new *clean-sheet* governance theory known as CSG offers substantial prospects for improved T&E use and rigour, as it focuses on strategic monitoring at the nexus between systems development and systems operations with a recognition of evolving system identity. There are now robust *entry-level* methods to readily be assessed against this new theory in order to diagnose improvements through sound pathologies, such that existing governance can improve incrementally only where necessary. This new governance model and its pathologies can be closely aligned to improving the timeliness, rigour and other fundamental characteristics of T&E for better-informed decision-making.

A total of eight T&E tools have been highlighted in this chapter. Complex projects by their nature have evolving characteristics and need good governance. T&E can provide that governance the evidence needs to make timely and informed decisions as characteristics emerge, involving trade-off such as those highlighted for ICT governance between usability, integration, performance and cybersecurity. However, as the tools featured through the chapter show, system complexity necessitates iterative testing, beginning with modelling and simulation (if necessary) and continuing through life, and that testing has increasing interconnectedness, more threat permutations, more configurations and more complex roles. The evidence from the industry and Defence is that combinatorial test design skills are quintessential to all types of T&E to achieve rigour and efficiency. As such, Tool 2 is the most enabling for the other tools, and consequently, more exposure to such methods (and understand the suitability measures and safety principles) even during undergraduate studies is strongly supported. That said though, without the necessary governance structures to set up and to run T&E, such investments like combinatorial test design would just deliver results too late to influence positive project outcomes.

Despite tremendous development in the field of T&E management, there remain critical questions that will enable its future evolution, most importantly, the question of tighter integration of sponsor/capability developers/operational users, PM and T&E. There have been recent serious attempts at integrating PM and T&E, for example, in the study by Rebentisch (2017). Despite all the difficulties and challenges, there are benefits in the integration for both operational use, PM and T&E. There is also a lack of any tangible tool for such integration. New models of SE and T&E that allow evaluation of PM-related metrics might provide a good avenue for this development (Efatmaneshnik et al., 2018). Other T&E areas of interest for further development are as follows:

- development of tools that facilitate distributed T&E
- T&E of human-system interfaces for the positive active control of autonomous and intelligent systems particularly those with swarm capabilities
- T&E of non-functional requirements
- more real-time T&E tools and reporting methods that are useful to key decision makers
- most importantly T&E of resiliency issues seamlessly embedded in operational families-of-system-of-systems with useful AI assistants meaningfully aiding timely human decision-making, governance and agency.

ACKNOWLEDGEMENTS

This work was heavily influenced by the mentoring and research with numerous outstanding testers who research passionately what they do, particularly Associate Professor Simon Atkinson, Dr Leanne Rees, Dr Elena Sitnikova, Dr Craig Benson, Associate Professor Michael Ryan, Pete Christensen, Christopher Skinner, Amit Ghildyal, Narelle Devine, Pete Nikoloff, Kate Yaxley and many others. This work would not be possible without the dedication and support of the International T&E Association and their journal run professionally by Dr Flash Gordon. We also acknowledge the significant influence of the Australian Senate's Inquiry into Defence Procurement (2012), particularly Chapters 2 and 12.

REFERENCES

Antony, J. 2014. *Design of Experiments for Engineers and Scientists*. London: Elsevier.

Asher, M. W. 1978. Land based test center: A tool for design and construction of FFG 7 class frigates. *Naval Engineers Journal* 90(4): 79–86.

Australian National Audit Office (ANAO). 2002. Audit Report No. 30: 2001–02 Test and evaluation of major defence equipment acquisitions. Canberra: Department of Defence. www.anao.gov.au/sites/g/files/net616/f/anao_report_2001-2002_30.pdf (accessed 7 April 2018).

Australian National Audit Office (ANAO). 2015. Report No. 9 2015–16: Test and evaluation of major defence equipment acquisitions. Canberra: Department of Defence.

Australian National Audit Office (ANAO). 2018. Report No. 6 2018–19: Army's protected mobility vehicle – Light. Canberra: Department of Defence.

Australian Parliament. 2016. Joint Parliamentary Committee for Accounts and Audit (JCPAA) hearing with Defence and the Australian National Audit Office. www.aph.gov.au/Parliamentary_Business/Committees/Joint/Public_Accounts_and_Audit/Reports_Nos_52_3_and_9 (accessed on 3 March 2016).

Australian Senate. 2012. Chapter 2 & 12. *Senate Inquiry into Defence Procurement*. Canberra: Australian Parliament House.

Azevedo, K. & Olsen, D. 2017. System engineering analysis of construction equipment operation in the Latin America. *27th Annual INCOSE International Symposium (IS 2017)* Adelaide, Australia, July 15–20.

Banks, G. 2014. Restoring trust in public policy: What role for the public service? *Australian Journal of Public Administration* 73(1): 1–13.

Blanchard, B. S. 2013. *Logistics Engineering & Management*, New International Edition. Harlow: Pearson Education.

Bradley, J. M., Joiner, K. F., Efatmaneshnik, M., & Keating, C. B. 2017. Evaluating Australia's most complex system-of-systems, the future submarine: A case for using new complex systems governance. *Proceedings of the 27th Annual INCOSE International Symposium (IS 2017)*, Adelaide, Australia, 15–20 July.

Brown, C., Christensen, P., McNeil, J., & Messerschmidt, L. 2015. Using the developmental evaluation framework to right size cyber T&E test data and infrastructure requirements. *The ITEA Journal* 36(1): 26–34.

Burton, J. 1993. *The Pentagon Wars: Reformers Challenge the Old Guard*. Annapolis, MD: Naval Institute Press.

Cairns, J. A. 2011. DDG51 Class land based engineering site – The vision and the value. *Naval Engineering Journal* 123(2): 73–83.

Chief Information Officers Group (CIOG). 2016. Defence Information and Communications Technology (ICT) Strategic direction 2016–2020. Department of Defence. www.defence.gov.au/CIOG/_Master/docs/Defence-ICT-Strategic-Direction-2016-2020.pdf (accessed 7 April 2018).

Christensen, P. 2017. Cybersecurity test and evaluation: A look back, some lessons learned, and a look forward. *ITEA Journal* 38(3): 221–228.

Chu, D. S. C. 2016. Statistics in defense: A guardian at the gate. *ITEA Journal*. 37: 284–285.

Cofer, D. 2015. Taming the complexity beast. *ITEA Journal* 36: 313–318.

Coles, J., Greenfield, P., & Fisher, A. 2012. *Study into the Business of Sustaining Australia's Strategic Collins Class Submarine Capability*. Canberra: Department of Defence.

Coles, J., Scourse, F., Greenfield, P., & Fisher, A. 2011. Collins class sustainment review phase 1 report. Canberra: Department of Defence.

Copeland, E., Holzer, T., Eveleigh, T., & Sarkani, S. 2015. The effects of system prototype demonstrations on weapon systems. *DefenseAR Journal*. 22(1): 106–134.

Defence Materiel Organisation (DMO) and Australian National Audit Office (ANAO). 2014. Report No. 14 2014–15: 2013–14 Major projects report. Canberra: ANAO.

Downes, L. 2015. What is 5G and why should lawmakers care? *Washington Post 26th of October 2015*. www.washingtonpost.com/news/innovations/wp/2015/10/26/what-is-5g-and-why-should-lawmakers-care/?noredirect=on&utm_term=.7daded5274be.

Efatmaneshnik, M. & Ryan, M. 2018. On the definitions of sufficiency and elegance in systems design. *IEEE Systems Engineering Journal*. doi:10.1109/JSYST.2018.2875152 "early view".

Efatmaneshnik, M., Ryan, M. J., & Shoval, S. 2018. A framework for testability analysis from system architecture perspective. *Proceedings of INCOSE International Symposium* 28, pp. 819–834. doi:10.1002/j.2334-5837.2018.00518.x.

Elele, J. N., Hall, D. H., Davis, M. E., Turner, D., Faird, A., & Madry, J. 2016. M&S requirements and V&V requirements: What's the relationship? *ITEA Journal* 37: 333–341.

Flyvbjerg, B. & Budzier, A. 2011. Why your IT project might be riskier than you think. *Harvard Business Review* 9: 24–27.

Fox, B., Bioto, M., Graser, C., & Younossi, O. 2004. *Test and Evaluation Trends and Costs for Aircraft and Guided Weapons*. Santa Monica, CA: RAND Corporation for United States Air Force.

GAO (U.S. Government Accounting Office). 1998. *Best Practices: Successful Application to Weapon Acquisition Requires Changes in DoD's Environment*. GAO/NSIAD-98-56. Washington, DC: GAO.

General Accounting Office (GAO). 1999. *Best Practices: Better Management of Technology Can Improve Weapon System Outcomes (GAO/NSIAD-99-162).* Canberra: Department of Defence, General Accounting Office.

General Accounting Office (GAO). 2016. *Technology Readiness Assessment Guide: Best Practices for Evaluating the Readiness of Technology for Use in Acquisition Programs and Projects (GAO-16-410G).* Canberra: Department of Defence, General Accounting Office.

Ghildyal, A., Chang, E., & Joiner, K. F. 2018. A survey of benefit approaches and innovation loop required for IT governance. *Archives of Business Research* 6: 10.

Grafton, R., Brown, G., Novakovic, Z., & Joiner, K. F. 2018. Developments in test design and analysis techniques for aircraft survivability assessments. *Systems Engineering, Test and Evaluation Conference,* Sydney, Australia, 30 April–2 May.

Gray, A., James, A., Nasser, H., Richardson, K., & Rooke, K. 2017. Foundations for improved integration – Using systems engineering in programme and project management. *27th Annual INCOSE International Symposium (IS 2017)* Adelaide, Australia, 15–20 July.

Guido, A. J. & Light, S. P. 1978. Ship design for maintainability: Experience on FFG 7 class. *Naval Engineers Journal* 90(2): 75–84.

Hecht, M. 2015. Verification of software intensive system reliability and availability through testing and modelling. *ITEA Journal* 36: 304–312.

Heinl, C. H. 2016. The potential military impact of emerging technologies in the Asia-Pacific region: A focus on cyber capabilities. *Emerging Critical Technologies and Security in the Asia-Pacific,* Ed. R. A. Bitzinger. 123–137. Hampshire: Palgrave Macmillan,

Henry, R. & Joiner, K. F. 2017. Improving small arms ammunition qualification with statistical test techniques from U.S. defense. *PARARI 2017- Australian Explosive Ordnance Safety Symposium,* UNSW Canberra, 21–23 November.

IEEE 2017. Aligned design: A vision for prioritizing human well-being with autonomous and intelligent systems, version 2. *IEEE Global Initiative on Ethics of Autonomous and Intelligent Systems,* ISBN 978-0-7381-xxxx-x, New York, USA, December 2017.

IEEE Standards Association P7009. 2018. Standard for fail-safe design of autonomous and semi-autonomous systems. Available at http://standards.ieee.org/develop/indconn/ec/autonomous_systems.html (accessed on 8 February 2018).

Javorsek, D. 2016. Modernizing flight test safety to address human agency. *ITEA Journal* 37: 325–332.

Johnson, R. T., Hutto, G. T., Simpson, J. R., & Montgomery, D. C. 2012. Designed experiments for the defense community. *Quality Engineering* 24(1): 60–79.

Joiner, K. F. 2015a. Implementing the defence first principles review: Two key opportunities to achieve best practice in capability development, Canberra: Australian Strategic Policy Institute, *Strategic Insights* No. 102, at www.aspi.org.au.

Joiner, K. F. 2015b. How new test and evaluation policy is being used to de-risk project approvals through preview T&E. *ITEA Journal* 36: 288–297.

Joiner, K. F. 2017. How Australia can catch up to U.S. cyber resilience by understanding that cyber survivability test and evaluation drives defense investment. *Information Security Journal* 26: 74–84. doi:10.1080/19393555.2017.1293198.

Joiner, K. F. 2018. Six-Sigma reform and education in Australian defence: Lessons-learned give rigour and efficiency to ordnance, aircraft and ship testing. *Conference: Proceedings of the 7th International Conference on Lean Six Sigma,* Dubai, UAE, 7–8 May, Eds. J. Antony, C. Laux, & B. Rodgers, pp. 87–92, ISBN:978-0-947997-05-2.

Joiner, K. F. & Atkinson, S. R. 2016. Australia's future submarine: Shaping early adaptive designs through test and evaluation. *Australian Journal of Multi-Disciplinary Engineering* 1: 1–23.

Joiner, K. F., Atkinson, S. R., Christensen, P. H., & Sitnikova, E. 2018. Cybersecurity for allied future submarines. *World Journal of Engineering and Technology* 6: 696–712. doi:10.4236/wjet.2018.64045.

Joiner, K. F., Atkinson, S. R., & Sitnikova, E. 2017. Cybersecurity challenges and processes for Australia's future submarine. *Proceedings of the 4th Submarine Science, Technology and Engineering Conference,* Adelaide, Australia, pp. 166–174.

Joiner, K. F., Efatmaneshnik, M., & Tutty, M. 2018. Modelling the efficacy of assurance strategies for better integration, interoperability and information assurance in family-of-system-of-systems portfolios. *Complex Systems Design & Management (CSD&M) Conference,* Singapore, 6–7 December 2018.

Joiner, K. F., Ghildyal, A., Devine, N., Laing, A., Coull, A., & Sitnikova, E. 2018. Four testing types core to informed ICT governance for cyber-resilient systems, *International Journal of Advances in Security* 11: 3–4.

Joiner, K. F. & Tutty, M. G. 2018. A tale of two allied defence departments: New assurance initiatives for managing increasing system complexity, interconnectedness, and vulnerability. *Australian Journal of Multidisciplinary Engineering* 1: 1–22.

Katina, P. F. 2016a. Metasystem pathologies (M-Path) method: Phases and procedures. *Journal of Management Development* 35(10): 1287–1301.

Katina, P. F. 2016b. Systems theory as a foundation for discovery of pathologies for complex system problem formulation. *Applications of Systems Thinking and Soft Operations Research in Managing Complexity*, Ed. A. J. Masys. 227–267. Geneva: Springer International Publishing.

Keating, C. B. 2017. Complex systems problem domain: Landscape of a modern project management practitioner. *Keynote presentation to the Project Governance and Controls Symposium*, University of New South Wales, Australian Defence Force Academy campus, Canberra, 4 May.

Keating, C. B. & Bradley, J. M. 2015. Complex system governance reference model. *International Journal of System of Systems Engineering.* 6(1): 33–52.

Keating, C. B., Katina, P. F., Joiner, K. F., Bradley, J.M., & Jaradat, R. 2018. A method for identification, representation, and assessment of complex system pathologies in acquisition programs. Presented at *15th Annual Acquisition Research Symposium, Graduate School of Business & Public Policy at the Naval Postgraduate School*, Monterey, California, 9–10 May.

Kramer, W. F., Sahinoglu, M. & Ang, D. 2015. Increase return on investment of software development life cycle by managing the risk—A case study. *Defense AR Journal* 22(2): 174–191.

Kuhn, D., Kacker, R., & Lei, Y. 2010. Practical combinatorial testing. NIST Special Publication 800-142, October.

Kuhn, D. R., Kacker, R. N., Feldman, L., & White, G. 2016. Combinatorial testing for cybersecurity and reliability, information technology bulletin for May 2016. Computer Security Division, Information Technology Laboratory, National Institute of Standards and Technology, U.S. Department of Commerce. Available at www.nist.gov/publications/combinatorial-testing-cybersecurity-and-reliability.

Lednicky, E. J. & Silvestrini, R. T. 2013. Quantifying gains using the capability-based test and evaluation method. *Quality Reliability Engineering International* 29(1): 139–156.

Mackertich, N., Kraus, P., Mittlestaedt, K., Foley, B., Bardsley, D., Grimes, K., & Nolan, M. 2017. IEEE/SEI software process achievement award 2016 technical report, Raytheon Integrated Defense Systems, Design for Six Sigma Team.

Madhavan, P. 2017. From automation to autonomous systems: The story of human trust. *ITEA Annual Symposium*, Washington, October 2017.

Mankins, J. C. 1995. Technology readiness levels: A white paper. Advanced Concepts Office, Office of Space Access and Technology, NASA, April 6, 1995.

Mcguigan, D. B. & Boylan, W. J. 1986. Ship systems test process – concept and application. *Naval Engineers Journal* 98(3): 157–170

McShea, R. 2010. *Test and Evaluation of Aircraft Avionics and Weapon Systems*. Raleigh, NC: SciTech Publishing Inc.

MIL-STD/HDBK-1763. 1984. Aircraft stores compatibility: Systems engineering data requirements and test procedures. US Department of Defence Standard/Handbook, Revision 15 June 1998, USA.

Murphy, T., Leiby, L. D., Glaeser, K., & Freeman, L. 2015. How scientific test and analysis techniques can assist the Chief Developmental Tester. *ITEA Journal* 36: 96–101.

Nejib, P., Beyer, D., & Yakabovicz, E. 2017. Systems security engineering: What every system engineer needs to know. *27th Annual INCOSE International. Symposium*, Adelaide, July.

Nielsen, P. 2017. Keynote speaking topic: Systems engineering and autonomy: Opportunities and challenges, *27th Annual INCOSE International Symposium (IS 2017)*, Adelaide, Australia, July 15–20; given as Director and CEO, Software Engineering Institute, Carnegie Mellon University.

Normann, B. 2015. Continuous system monitoring as a test tool for complex systems of systems. *ITEA Journal* 36: 298–303.

Orlowski, C. T., Blessner, P., Blackburn, T. & Olson, B. A. 2017. Systems engineering measurement as a leading indicator for project performance. *ITEA Journal* 38: 35–47.

Panei, V. 2017. Unmanned and autonomous systems test (UAST). *ITEA Test and Evaluation Conference*, Washington, October.

Peever, D., Hill, R., Leahy, P., McDowell, J., & Tanner, L. 2015. First principles review: Creating one defence. Canberra: DOD.

Pyke, A. 2016. How rough is your project, *Project Governance and Controls Symposium*. University of NSW, Canberra, 9 May.

RAND Corporation. 2011. Learning from experience, Volume IV – Lessons from Australia's collins class submarine program. Santa Monica, CA: RAND Corporation on Behalf of Australian Department of Defence. www.dtic.mil/dtic/tr/fulltext/u2/ a552686.pdf.

Rebentisch, E. 2017. *Integrating Program Management and Systems Engineering: Methods, Tools, and Organizational Systems for Improving Performance*. New York: John Wiley & Sons.

Reynolds, M. T. 1996. *Test and Evaluation of Complex Systems*, Chichester: John Wiley & Sons.

Rodgers, B., Antony, J., & Kregel, I. 2018. The role of government in leadership for Lean Six Sigma in the public sector, *Conference: Proceedings of the 7th International Conference on Lean Six Sigma*, Dubai, UAE, 7–8 May, Eds. J. Antony, C. Laux, & B. Rodgers, pp. 190–198, ISBN:978-0-947997-05-2.

Roedler, G. 2018. Keynote presentation at systems engineering. *Test and Evaluation Conference*, May, Sydney, Australia; given as President INCOSE.

Roff, H. M. & Moyes, R. 2016. Meaningful human control, artificial intelligence and autonomous weapons. Briefing paper prepared for the *Informal Meeting of Experts on Lethal Autonomous Weapons Systems*, UN Convention on Certain Conventional Weapons, April.

Rucker, A. 2014. Improving statistical rigor in Defense T&E: Use of tolerance intervals in designed experiments. *Defense AR Journal* 21(4): 804–824.

Shergold, P. 2015. Learning from failure: Why large Government policy initiatives have gone so badly wrong in the past and how the chances of success in the future can be improved, Australian Public Service Commission, Canberra, Australia.

Smith, N, White, E., Ritschel, J., & Thal, A. 2016. Counteracting harmful incentives in DoD acquisition through test and evaluation and oversight. *ITEA Journal* 37: 218–226.

Stark, R. E. & Stembel, D. M. 1981. Detail design-FFG 7 class. *Naval Engineers Journal* 93(2): 109–119.

Sustersic, J. 2017. Constrained learning for assured autonomy. *ITEA Annual Symposium 2017*, Washington October; given as Autonomous and Intelligent Systems Division Applied Research Lab–Penn State University.

Taleb, N. N. 2007. *The Black Swan - The impact of the Highly Improbable*. London: Penguin Books.

Tatsumi, K. 2013. Combinatorial testing in Japan, *ICECCS 2013*, 16 July, Singapore, Association of Software Test Engineering (ASTER) & Fujitsu Ltd.

Terrell, K. 2016. Going the extra mile, keynote presentation at project. *Governance & Controls Symposium*, 11th May, University of NSW, Canberra, Australia.

Troester, T. 2015. National cyber range overview. Presentation to ITEA Cybersecurity Workshop: Test and Evaluation to meet the Persistent Threat, Belcamp MD, February.

Tutty, M. G. 2016. The profession of arms in the information age: operational joint fires capability preparedness in a small-world, PhD Dissertation, University of South Australia, 1 January.

Tutty, M. G. 2017. The profession of arms in the information age. V2.2, Air Power Development Centre, Defence Estate Fairbairn, ACT, June 1.

Tutty, M. G. & White, T. 2018. Unlocking the future: Decision-making in complex military and safety critical systems. *Systems Engineering Test and Evaluation Symposium*, Sydney, NSW, 30 April–2 May 2018.

U.S. DoD. 2017. F35 Joint strike fighter: Financial year 2012/2013/2014/2015/2016 DoD programs. Annual Director Operational Test & Evaluation (DOT&E) Report(s) to Congress, January. Available at www.dote.osd.mil/annual-report/index.html.

U.S. DoD Defense Science Board (DSB). 2016. Summer study on autonomy, June, 28–30, www.hsdl.org/?view&did=794641 (accessed on 24 Aug 2017).

Wernas, M. & Joiner, K. F. 2018. Screening the important factors in supportability test and evaluation activities for ships. *Systems Engineering, Test and Evaluation Conference*, Sydney, Australia, 30 April–2 May.

Wickens, C., Lee, J, Liu, Y., & Becker, S. 2014. *An Introduction to Human Factors Engineering*, 2nd Ed. New York: Pearson Prentice Hall.

17 Advanced Visualization Toolset

Thomas A. McDermott
Stevens Institute of Technology

CONTENTS

BACKGROUND

In January 2001, after 9 years in development, a new fighter aircraft, the F-22 Raptor, completed the maiden flight of the most advanced and complex aircraft avionics system of its time. At the time, I was the project manager of the avionics team, a complex virtual organization of over 4,000 employees across 14 companies. The successful flight was a critical project management milestone that allowed the first production contract award for the F-22 by the United States Congress.

The complexity of the F-22 avionics product was immense, with roughly 40 major hardware and software components to develop and integrate, over 1 million lines of software code, and nearly every function in the system newly created. It was innovative – the first aircraft avionics system to fully integrate all flight, navigation, and tactical capabilities into a single pilot information display. The complexity of the enterprise was also immense, with 3 prime contractors, 11 major subcontractors, and development facilities across more than 10 states. The project had a long life cycle, in not just schedule time but also in the number of team leadership changes. External complexity was a challenge – the United States Congress had mandated the schedule for this first flight, accelerating the planned milestone by 4 months. This effort had all the dimensions that separated complex project management from traditional project management.

We managed this effort as a team spanning multiple organizations with multiple disciplines, experience, and cultures. At the center of the effort was a book, the *Avionics Program Plan*, used to manage the complexity of the team and system integration. On the day of that first flight, the team was on version 24 of that book – 24 times the complex plan had "failed" and had been recovered by program leadership. The *Avionics Program Plan* was a book of data, including items like all the major equipment delivery dates and capabilities. It was a book of information, capturing all the interdependencies of the complex system integration and flight test plans. It was a book of knowledge, documenting the shared information that this team of over 4,000 people used to manage the effort on a daily basis. The continuous communication of shared knowledge is the core of complex project management.

Sometime after the successful first flight, senior program leadership decreed there would be no more iterations of the book. Every new iteration documented further program schedule (and often cost) changes, mostly negative. It was like the book was causing the shifts in the plan, not the complexity of the project itself. This is traditional project management at its worst: decreeing that a program be managed to a rigid plan, independent of complexity. We learned on this program that is the ability of a talented and dedicated team to visualize and revisualize the complexity of the plan that leads to success, not the plan itself. And it is the shared visualization of data, information, and knowledge across key individuals of the team at every level that allows the complex product to come together successfully.

INTRODUCTION

In project management, visualization tools allow the mapping of large amounts of data to visual attributes that aid human information processing. Project management applications are concerned with visualization of data, information, and knowledge, all used to improve the decision-making of project leadership. A key aspect of project leadership is understanding what types of visualization approaches to use and how to use them in the context of dynamically changing project performance, particularly in highly complex projects. Although there is a large body of literature on managing complex projects, very little of it addresses the role of visualization in project execution.

There are thousands of project management tools and associated visualization forms that support decision-making in project management. Current project managers have access to and must manage large quantities of project data and associated visualization forms for project reporting, project collaboration, and facilitation of team decision-making. Examples from our *"Avionics Program Plan"* include work breakdown structures, organization charts, Gantt schedule charts, program evaluation and review technique (PERT) charts for integration planning, team collaboration calendars, staffing curves, tables of key data, progress histograms and aggregate trend charts, pictorial diagrams of key system components, wiring diagrams of interfaces, conceptual diagrams of process flows, value stream maps, risk categorization and prioritization, and accompanying narrative explanations. As stand-alone artifacts, these represent data and information. However, in book form, they tell a story reflecting the collective knowledge of the team. Project complexity and pace drive the need for integration of project information with project team knowledge, in forms that support rapid communication to a wide set of stakeholders.

Challenges of complexity in project management have four dimensions: the scale of the project and supporting enterprise; the variety of organizational disciplines, processes, and tools that might be used to execute the project; the external context the project is surrounded by, including social and political factors and market dynamics; and the uncertainties created by newness, originality, and innovation. Project complexity is influenced by these dimensions across a larger enterprise on both the provider and user sides, and project decisions must consider program trades in conjunction with enterprise business goals, user needs, enterprise standards and practices, and rapid market or other external context changes. Project views used to manage the effort must consider the project as a boundary, but also the larger enterprise as a boundary, in a combined project/enterprise decision space.

Project complexity will not be "solved" by a new tool or visualization method, only by how project leadership learns to use tools and appropriate visualizations to gain insight into and solve problems within an evolving project. The primary project data visualization challenge is to support the combination of qualitative or heuristic decisions that must be made in conjunction with quantitative data-driven decisions. These might be categorized

Project complexity will not be "solved" by a new tool or visualization method, only by how project leadership learns to use tools and appropriate visualizations to gain insight.

as knowledge-driven versus information-driven decisions. In complex projects, exchange of knowledge and creation of appropriate social networks for exchange are often more critical than information. How the network of decision makers use data, information, and shared knowledge is paramount.

A paper "book" was the norm in 2001, but today and in the future, digital visualization tools and shared digital "truth" data should be the standard. Modern data collection, data warehousing, data analysis, and visualization tools are integrating traditionally stand-alone project management processes into tools that allow interactive visualization and data management. These tools include dashboards that support people, team, project, and enterprise-level views of the project. They place the data in a model-based framework for sensemaking that relates performance data to decision or process models that are more conceptual and in line with management heuristics. Computer visualization will improve traditional book-form or document-based project complexity management but only if the tools allow and create a similar book-form "story," which can be shared across the leadership team. When decisions remain obvious, current computer-based data management and visualization tools work very well. In complex and uncertain situations, the flexibility of these tools to create the story falls short and decision makers must explore data in more qualitative frameworks. The challenge for project leadership is to situationally master the combination of tools and visualization forms that visualize data and information to collect, transfer, and communicate shared knowledge.

The central thesis of this chapter is that as complexity increases, the role of knowledge transfer in social networks becomes more critical and the ability to visualize knowledge (as opposed to information) becomes paramount to project success. The Project Management Institute (PMI), in their Pulse of the Profession 2015 report, *Capturing the Value of Project Management Through Knowledge Transfer*, found that organizations that are effective at knowledge transfer improve their project outcomes by 35% and also that 34% of unsuccessful projects are adversely affected by lack of timely or accurate knowledge transfer (PMI 2015). Complex project management is a continuous and ever-changing process, characterized by human social learning cycles followed by (hopefully much longer) execution cycles. There is a growing set of project management literature describing project complexity but much less on how the role of human management and decision-making drives a complex project from inception to completion. The relationship between knowledge transfer and project success, combined with dimensions of project complexity, has not been fully explored.

This chapter first provides a framework that relates project complexity, team knowledge transfer, and leadership behaviors. This framework is derived from the domains of systems thinking and complexity science. Following that framework, the chapter discusses the role of data, information, and knowledge visualization to the application of the framework. The chapter concludes by highlighting the key leadership competencies needed to visualize and communicate complex project strategy and decisions.

PROJECT MANAGEMENT COMPLEXITY

It is useful to summarize project complexity around the patterns described by Geraldi and Albrecht in their paper "On Faith, Fact, and Interaction in Projects" (Geraldi and Adlbrecht 2007). They describe complexity in project management as three dimensions they call "Complexity of Fact," "Complexity of Faith," and "Complexity of Interaction." *Complexity of Fact* is a measure of the number of entities and their interdependence as an issue of interdependent information. Given large complexity of fact, project managers find it difficult to obtain and use information rapidly enough to support their decision-making. They search for available higher-level abstractions or simplifications to base their decisions. In this dimension, there exists a fundamental problem of *abstraction* – one needs to visualize the whole of the project and represent information in patterns that are embedded in the whole. Visualization frameworks that maintain the holistic perspective while allowing access to detailed information are necessary. The project visualization challenge is to create situational trust and transparency – does one trust the visualized patterns and does the source of the data remain transparent to the decision maker so they may look deeper as needed.

Complexity of Faith relates to project *uncertainty*, often associated with the newness of the problem being solved. Factors related to the maturity of the product or process and external stakeholders, as well as the dynamics of multiple changes and feedback, drive the complexity of the project. Complexity of Faith implies a need for learning, a factor that has driven the widespread adoption of iterative software development models. Complex projects begin with "Faith" and work their way toward "Fact." Donald Rumsfeld famously described this process when he stated "as we know, there are known knowns; there are things we know we know. We also know there are known unknowns; that is to say we know there are some things we do not know. But there are also unknown unknowns – the ones we don't know we don't know" (Rumsfeld 2011). One way to view the execution of a complex project is a "burndown of uncertainties," shown in Figure 17.1, where unknown-unknowns and known-unknowns are eliminated as the project evolves. In periods of high uncertainty, project decisions rely more on shared knowledge of the project team than on the availability of project-related information. Project visualization methods and tools in these periods must encourage facilitation, knowledge transfer, and more abstract conceptualization of decision alternatives.

Complexity of Interactions exist at interfaces between systems. These include people, disciplines, locations, external stakeholders, and social and political factors. For all complex projects, it is important to frame the project interactions in a larger enterprise framework. Enterprises are complex adaptive systems, in which both the human/social participation and the project itself co-adapt over time. To address Complexity of Interactions, a project must develop social constructs such as integrated product teams and associated knowledge-sharing mechanisms. As shown in Figure 17.2, these mechanisms are not just related to project leadership and employee skills but also organizational purpose, employee motivation and culture, and external influences.

Understanding interactions drives the need for facilitation and conceptualization methods and tools. In this case, the visualization challenge is to represent the human interactions internal to and external to the project. Moldoveanu and Leclerc summarize this challenge as mapping of decision agents and incentives, mapping of evolutionary strategies, and mapping of relationship networks (Moldoveanu and Leclerc 2015). The importance of social networks, decision agents, and evolution of strategy is a consistent theme in complex project management and should remain front and center in the mind of leadership in these situations.

> The importance of social networks, decision agents, and evolution of strategy is a consistent theme in complex project management.

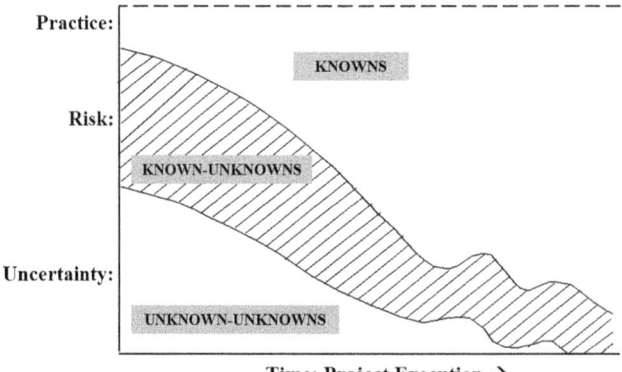

Time: Project Execution →

FIGURE 17.1   A project represented as a burndown of uncertainties.

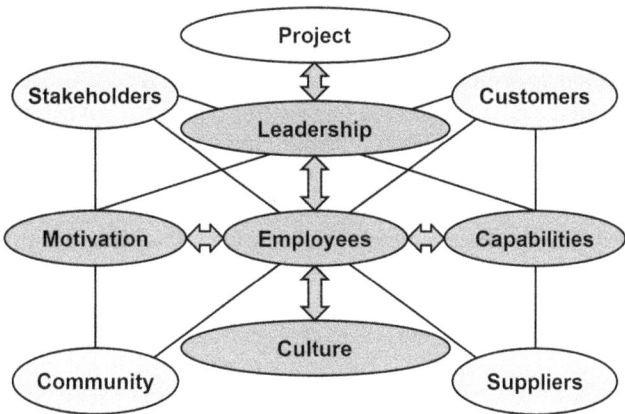

FIGURE 17.2   The project organization as a complex adaptive system.

To manage the relation between faith, fact, and interaction, project management must develop strategies and tools that support (1) exchange and visualization of information across social networks, (2) exchange and visualization of knowledge between decision agents, and (3) evolutionary planning that includes cycles of learning. The presence of all three of these strategies implies that project learning be data-driven, so visualization methods that support access to both information and knowledge must be used. Methods for data visualization in complex projects must include tools for information visualization and tools for knowledge visualization in processes that support effective organizational learning.

These methods are often discussed on literature related to systems thinking and the complexity sciences. These two disciplinary domains and their relationship to management of project complexity are discussed next.

COMPLEX PROJECT MANAGEMENT AND SYSTEMS THINKING

Moving from traditional project management approaches to complex project management requires a mindset that can deal with complexity and uncertainty. Dombkins (2008) argues that complex project management is a job specialization requiring skills and a mindset closely associated with systems thinking. He suggests complex project management requires a unique class of project management certification with competencies and training related to management of uncertainty. McDermott and Freeman (2016) further describe three systems thinking characteristics and competencies that are essential to complex project management. These are sensemaking, adaptive and computational thinking, and a design mindset. Each of these competencies uses data-driven activities that encourage visualizing the relationships between project uncertainties and the underlying project activities.

Sensemaking is a collaborative process that involves collection of knowledge, visually describing or modeling a problem or solution in the wider context, and learning by doing. It results in artifacts like the book-form program plan mentioned earlier – such artifacts are "maps" of program knowledge. Sensemaking is associated with decision-making not as a direct process of applying theory or practice but by developing an understanding of the larger context in order to create a framework for selecting appropriate theories or practice. In management of complex projects, sensemaking is used to diagnose project execution issues (problem definition) and to develop strategy options for execution in the future (problem solution). This allows the right methods, processes, tools, and skills to be applied situationally as a project evolves.

Literature on sensemaking bridges the disciplines of psychology, leadership, and information science. Klein, Moon, and Hoffman (2006) described a psychological view of sensemaking as the process of understanding connections between people, processes, and events in order to anticipate future trajectories and act effectively – a project *leadership* characteristic. Dervin (2005) described sensemaking as a methodology of "systematic and reliable dialogue," which verbalizes the situation, context, and potential outcomes focused on bridging gaps between peoples' understandings – a project *team* characteristic. Russell, Stefik, Pirolli, and Card (1993) describe sensemaking as a learning process of forming and reforming models that can be used as a hypothesis to test specific representations of information – a project *visualization* characteristic. This is where project leadership, team interactions, and visualization tools come together. In uncertain situations, sensemaking is the ability to frame issues and gain situational insight collectively from people, data, and associated knowledge. Sensemaking supports a leader's abilities to continually refine or redefine the project execution strategies that allow success to emerge from complexity.

Adaptive and computational thinking is the ability to situationally adjust a team's thinking and related activities by employing analytics and simulation methods that make sense of large amounts of data (or to understand when data is lacking). This includes the ability to situationally discriminate and filter information to maximize cognitive function for decision-making – the primary role of visualization. Courtney, Lovallo, and Clarke (2013) proposed a model for data-driven decision-making that can be used to select the right data analysis and visualization tools for a given situation. This is shown in Figure 17.3. The model suggests that project leadership base their decision strategies on their level of causal understanding of the situation and how broad the range of possible outcomes might exist. If the causal relationships in the situation are understood and the outcome of the decision well understood, conventional analytical and visualization tools can be applied. If the causal relationships are understood but there is a range of possible outcomes, running multiple quantitative scenarios using broader data sets is advised. If the causal relationships are understood but the outcomes cannot be predicted, qualitative scenarios or historical case studies should be pursued, and data should be tested (Courtney, Lovallo, and Clarke 2013). Finding unique causal relationships in complex situations is a scenario-driven activity, and the complex project manager needs specialized skills in leading scenario-driven analysis using hypothetical or historical case-driven scenarios. Scenario-driven analysis is a team facilitation process and requires leadership with strong facilitation skills.

Evolution of strategy and planning in complex project management is a process of iterative design. Building a *design mindset* in complex project leadership moves the entrained thinking of the project team away from continued use of available data, methods, and tools to a participatory team process of understanding and selecting new data, methods, and tools to be used in an iterative process of design for the new situation. Each "cycle" in the dynamics of a complex project becomes a new opportunity to design the project for success.

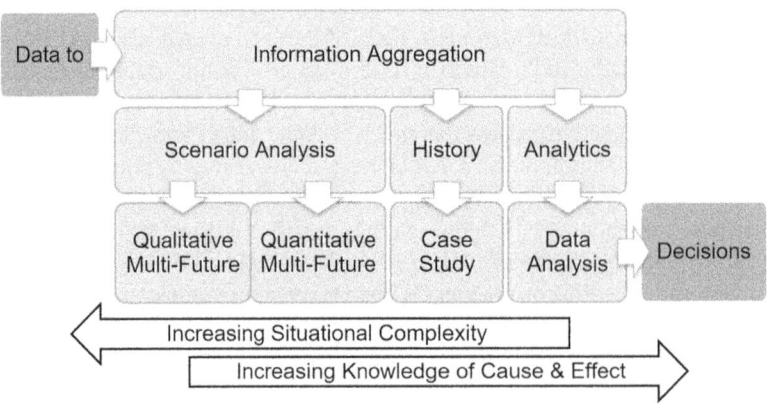

FIGURE 17.3   Data-oriented approach to decision-making.

In design, decisions require "zooming out and in" on multiple stakeholder perspectives of value. In uncertain situations, leadership must "zoom out" the decision process moves from established data-driven methods to higher levels of abstraction, then "zoom in" to select new data-driven methods and measures that have more relevance to the situation at hand. Collins and Hansen (2011) provided the "zoom out, then in" framing in their book *Great by Choice*, describing such skills as the ability to sense and analyze change in the larger system, combined with the ability to focus in on plans and objectives for change. This mindset is essential to understanding the measurement constructs of a complex project and the process to build a data-driven visualization framework for decision-making. Figure 17.4 provides a process cycle for dealing with decisions in uncertainty (McDermott and Freeman 2016). The sensemaking process frames the project in terms of fact (entities and attributes), faith (emergence), and interaction (stakeholder perspectives). It builds a new set of conceptual models and strategies for project execution. Computational thinking then relates existing project/enterprise behavioral constructs (generally performance measures) to new constructs selected to guide future project execution. This can be thought of as new project success measures selected to encourage short- and long-term changes. Finally, implementation of these strategies is a design process where new product, process, and organizational constructs are structured, specified, and tested for improved execution.

As complex projects can be characterized by a "burndown of uncertainty," the specialized skills and tools of systems thinking provide leadership framework to iteratively evaluate, analyze, redesign, and test project execution strategies as the situational context of the project evolves. Visualization tools in systems thinking support an iterative process for exploring the larger context of the project and for creation of a project learning environment. These tools will be discussed in the "Visualization in Complex Project Management" section.

> Visualization tools should support an iterative process for exploring the larger context of the project and for creation of a shared learning environment.

Two additional specialized leadership skills in systems thinking are "seeing project context" and using those views to execute a "process of learning." These characteristics define an experienced project leader's ability to continually "zoom out, then in" on project problem and solution sets in order to make effective decisions in complex and uncertain situations.

Seeing Project Context: In practice, all projects exist as part of a larger environment, and the behavior of the project will be influenced by the factors of the environment they are in. This is the core to the "Complexity of Interaction" dimension for any complex project. Many project interactions are decoupled from the project execution but still influence project behaviors. These can range from political and enterprise leadership exerting influence all the way down to events that affect employee morale. In addition, the context of the project exists separately from the context of the product that is being developed in the project. External market or technology shifts can play a major role. Systems thinking encourages one to consider the behaviors of the project itself with an understanding of the project's external context. This is in effect modeling or mapping the project

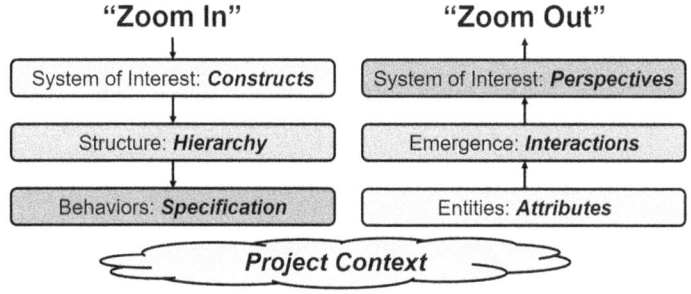

FIGURE 17.4   Zooming out and in as a sensemaking process for project learning.

and external context together in order to visualize the internal and external interactions that drive project execution. It is unlikely that internal project data is reflecting these external drivers, creating a situation where new sensemaking is required.

Process of Learning: Finally, system thinking recognizes the process of change in systems is a process of learning. It begins with high levels of uncertainty and progresses toward an end state which results in a defined system structure and performance attributes which achieve the intended long-term goals. Understanding of the system needs to consider both what is known and what is unknown to effectively plan for shared outcomes. Planning for change and actions to produce change is informed by models of the system that envision a desired future outcome or set of effects. In this area, tools that capture theories of change are useful.

These five practices form the core "sensemaking loop" that supports the process of complex project management. This loop is shown in Figure 17.5.

The foundation of the sensemaking loop is visualization. Visual modeling is used to frame the project within the enterprise system that is managing it. Projects have goals, boundaries, a structure of (internal and external) entities and interfaces, and interactions between those entities. Complex projects also exhibit feedback and nonlinearity, and emergent behaviors and adaptation are important characteristics of the complex project life cycle. Systems thinking is a goal-setting activity, and its visual modeling tools are focused on understanding the underlying complexity of the project in order to adjust and refine goals. Systems thinking generally begins with "soft systems" methods and tools that conceptually model and visualize the project/enterprise structure and dynamics, then moves to "hard systems" methods that use data and computation to understand project/enterprise feedback paths for predicting results of change strategies. Basic visual modeling tools can be quite simple (mind maps for organizing information, PERT charts for relating events, etc.). It is the specialized skills of the systems thinker that determine when and what tool to use to address the situation at hand.

Systems thinking, however, does not fully represent the challenges of complex project management because its theory and tools assume that the system can be structured in advance to produce solutions that achieve those goals, and that performance can be predictively modeled. In truly complex situations, the solutions cannot be known ahead of time, and one must act in the system in order

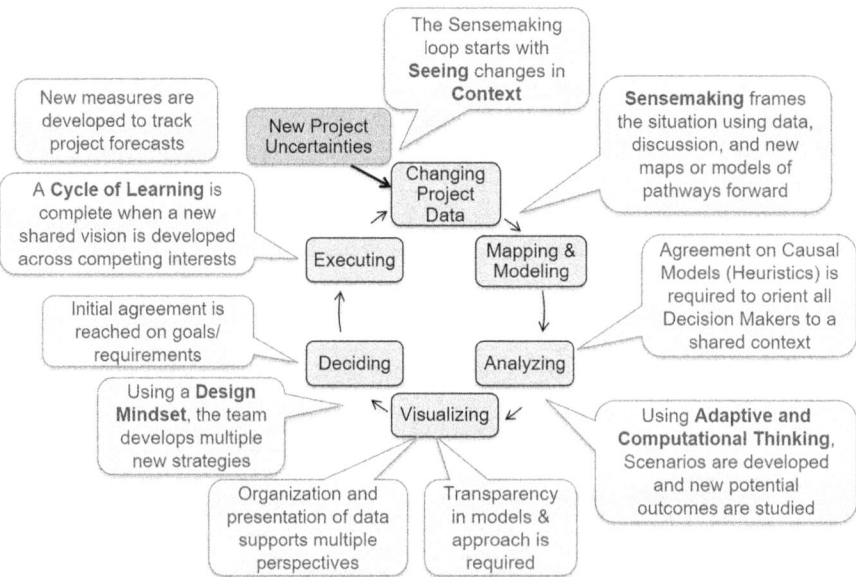

FIGURE 17.5 The full sensemaking loop.

to discern likely outcomes. In the domain of project complexity, additional strategies and tools are necessary that have their roots in complexity science.

PROJECT MANAGEMENT AND COMPLEXITY THINKING

As mentioned previously, social networks, decision agents, and evolution of strategy are keys to complex project management. Axelrod and Cohen (2008), in *Harnessing Complexity*, describe an execution framework for addressing complexity in the type of complex adaptive systems that define management of complex projects. The framework centers around the decision agents in the project, artifacts they use, and their strategies to influence behaviors and outcomes. To understand complexity, the following questions are asked:

- What are the strategies, agents, and other artifacts of the system? How are they connected? What are their heuristics? What are their processes and tools?
- Who are the primary decision agents, and in particular who can copy strategies from whom?
- What can one observe about the agents that are driving the system?
- How might one classify the agents and strategies? Who and which will be most useful?

Knowledge transfer and social learning activities are only successful if an appropriate variety of agents and their strategies are encouraged and harnessed. In complex systems, the law of requisite variety states that if a system is to be stable, the number of states of its control mechanism must be greater than or equal to the number of states in the system being controlled (Ashby 1958). The variety of the system is the number of distinct states it can be in. Complex projects experience continual state changes that make traditional project management tools ineffective. It is the selection and attainment of an appropriate variety of responses to changing situations that make complex projects successful. Beer (1985) further states the measure of variety represents the minimum number of choices needed to resolve uncertainty, which can be related to project management strategies and resources. Building this variety of agents and responses is a knowledge transfer process and is the core complex project management specialization. In simpler terms, complex project management is about bringing together social networks involving a variety of agents (people or groups) from different backgrounds and experiences, having them select and copy new strategies that might be more successful, and then communicating evolutionary change strategies through selected change agents. To manage complexity, Axelrod and Cohen recommend considering the following questions:

- What are the processes to copy strategies that create or destroy variety? What interventions could create or destroy variety? Is there a balance?
- What are the patterns of interaction and how might they be changed?
- What success criteria does the system use to select patterns? Is the selection dependent upon agents or strategies?
- How should selection of agents or strategies be used to promote change?

These questions suggest the types of actions that should be taken in complex situations, as success will be dependent on how one selects decision agents and what strategies they bring with them and collectively agree upon. Specialized activities for complex project managers include building the networks of interaction for decision agents, developing execution strategies that balance exploration and the routine, and use of social networks to select strategies for success. Boulton, Allen, and Bowman describe this as building a "learning multi-agent model" – networks of individual agents who act according to their experience and their beliefs. Inducing evolution in complex systems requires self-organization around new shared beliefs (Boulton, Allen, and Bowman 2015).

Creation of new shared beliefs is a knowledge transfer activity where new strategies are copied from one agent to another.

Snowden and Kurtz (2003) recognized that the project team is a knowledge exchange group that creates and executes these types of actions. They are the "learning multi-agent model." Snowden suggested that managing complexity requires growth in concept and in practice from managing knowledge as a thing to knowledge as both a dynamic flow and a thing (Snowden 2003). While traditional project management might succeed with managing knowledge artifacts, complex project management focuses on flows of knowledge in response to uncertainty and changing situations. Knowledge exchange requires both content and a context. The content can only be exchanged if the context can be shared (language, education, experience, culture, etc.). Complex project management requires development of a common context for knowledge exchange between project teams and a process for creating that exchange. For example, the F-22 project's continuously iterated "book-form" plan was maintained by a team of decision agents using regular week-long knowledge-sharing events where the agents negotiated the new plan on behalf of their organizations. No attendee could leave until the new "cycle of learning" was complete.

> In complex project management, one must move from managing individual knowledge artifacts to managing flows of knowledge.

Traditional project management tools and their visualization strategies are part of that exchange but often work against it – they represent project execution only in the context of the tool and force information exchange over knowledge transfer. Visualizing a future path of execution, in complexity and uncertainty, requires representation of historical information in a framework that links varieties of knowledge to tell a story. Referring to the F-22 example, each version of that book told the revised story.

Figure 17.6 depicts Snowden's original version of the Cynefin framework, a model for common sensemaking, which was focused on the exchange of information and knowledge. The framework depicts four domains of knowledge, loosely separated by context (Snowden 2003).

1. Bureaucratic/structured domain of knowledge, which can be transferred by teaching, focused on low levels of abstraction and mature practices.
2. Professional/logical domain of knowledge, which can also be transferred by teaching, focused on high levels of abstraction and expert knowledge.
3. Informal/interdependent domain of knowledge, which can be developed by shared learning, focused on high levels of abstraction and iterative learning.
4. Uncharted/innovative domain of knowledge, also which can be developed by shared learning but at low levels of abstraction and focused on experimentation.

FIGURE 17.6 Early depiction of Snowden's Cynefin framework identifying knowledge domains (Snowden 2003).

Snowden and Kurtz' later iterations of the Cynefin framework became more focused on system ontologies in general, arriving at the corresponding domains of (1) obvious, (2) complicated, (3) complex, and (4) chaotic as shown in Figure 17.7. In complex project management, these domains help project leadership to make sense of where things are and to arrive at the types of actions to be taken.

In the evolution of Cynefin, best practices only exist in the known space, what evolved to be called simple and now is most often called obvious (Snowden 2012). Most traditional project management tools and visualization strategies target this area. The complicated domain, previously called knowable, is the domain of good practice where agents are experts. With respect to complex project management, a major issue in the complicated domain is entrainment of thinking. Project management tools and visualization strategies are often driven by the codification of expert language taught by the experts, which may or may not be relevant to the real situation at hand.

The complex domain is characterized by management of patterns. Patterns are emergent properties of a complex system. Snowden states that "by increasing information flow, variety, and connectedness either singly or in combination, we can break down existing patterns and create the conditions under which new patterns emerge" (Snowden 2012). The domain of chaos is the place where new patterns of thinking and doing can be encouraged. For F-22, the team gathered and revised the book when internal or external situations "broke" the existing plan. In complex project management, it is only when the program plan is irretrievably broken will new success strategies come into place and then only based on the transfer of new knowledge between decision agents. Snowden also states, "in this domain, leadership cannot be imposed, it is emergent" (Snowden 2012). Informal leadership networks, aided by visual knowledge exchange approaches, emerge to rebuild the program plan.

A key aspect of this model is the role of informal networks in the complex domain, where decisions are more the domain of shared experiences, culture, and heuristics than explicit information frameworks. The flow of knowledge between obvious and complicated is clear, it is where expert knowledge sets like the Project Management Body of Knowledge (PMBoK) excel. But knowledge also flows from obvious to chaotic to complex, which is the cycle of learning often necessary in management of complex projects. As a sensemaking tool, Cynefin is a framework to relate the certainty of our decision-making versus the certainty of our assessment of the situation. In areas

FIGURE 17.7 Current version of the Cynefin framework (Snowden 2017).

where both the situation and the appropriate decisions are uncertain, informal networks of decision agents and their shared knowledge and beliefs should offer better program management strategies (Snowden 2012).

The complexity thinking domain gives us frameworks to evaluate two types of visualization, those that focus on data and information transfer and associated project execution and those that focus knowledge transfer and associated strategy development. Both have application in complex project management. Having an organization basis for and learning when to apply each type is the key to successfully managing complex projects.

VISUALIZATION IN COMPLEX PROJECT MANAGEMENT

Concepts of visual project management argue that managing the complexity of the data produced requires visualization of patterns that can be evaluated with speed and by multiple stakeholders. Visualization approaches common across the project management domain work well for simpler projects but become overwhelmed as the complexity increases. In addition, few of these approaches support visualization of knowledge.

Most project management visualizations take the form of system diagrams oriented to brainstorming, organization and ranking, and implementation and process. The building blocks of all such tools are the models described as the "Seven Management and Planning Tools" documented by the Union of Japanese Scientists and Engineers in 1976. For a detailed description on the use of these models, refer to the study by Brassard (1996). The seven are the affinity diagram, interrelationship digraph, tree diagram, activity diagram, matrix diagram, prioritization matrix, and Process Decision Program Chart (PDPC).

- Affinity diagram – used to gather and then group large amounts of information, most often in interviews or brainstorming sessions. Examples include mind maps; checklists; calendars; and strengths, weaknesses, opportunities, and threats (SWOT) diagrams.
- Interrelationship diagram – used to show cause and effect relationships between information items, usually as logical relationships related to a central idea. Examples are PERT charts and the cause and effect diagrams.
- Tree diagram – used to take items of information from the Affinity and Interrelationship diagrams and map them down to the lowest level of detail. Examples are work breakdown structures and organizational charts. The PDPC is a special form of tree diagram used to organize chains of events using both information and knowledge, such as a failure modes and effects diagram.
- Activity diagram – used to plan a sequence or schedule of activities associated with implementation plans. Examples include Gantt or swim lane charts.
- Matrix diagram – used to collect large sets of information for comparison purposes. Example matrix diagrams include N2 diagrams and Design-Structure Matrices. A prioritization matrix is another special form used to prioritize information and knowledge together using weighting criteria. Examples include the house of quality and cost-benefit matrices.

Figure 17.8 shows a common set of project management data and information models using these five common forms: an affinity diagram for the statement of work (SOW), a tree diagram for the work breakdown structure (WBS), an interrelationship diagram creating an Activity Network, a matrix diagram used as a design structure matrix (DSM), and an activity diagram for the Project Schedule. These tools create a resourced project schedule shown feeding into a common data visualization pattern, a Funds Expenditure Plan.

Lengler and Eppler's Periodic Table of Visualization (not reproduced here) identifies and categorizes 104 different types of visualization, with links to examples (Lengler and Eppler 2007).

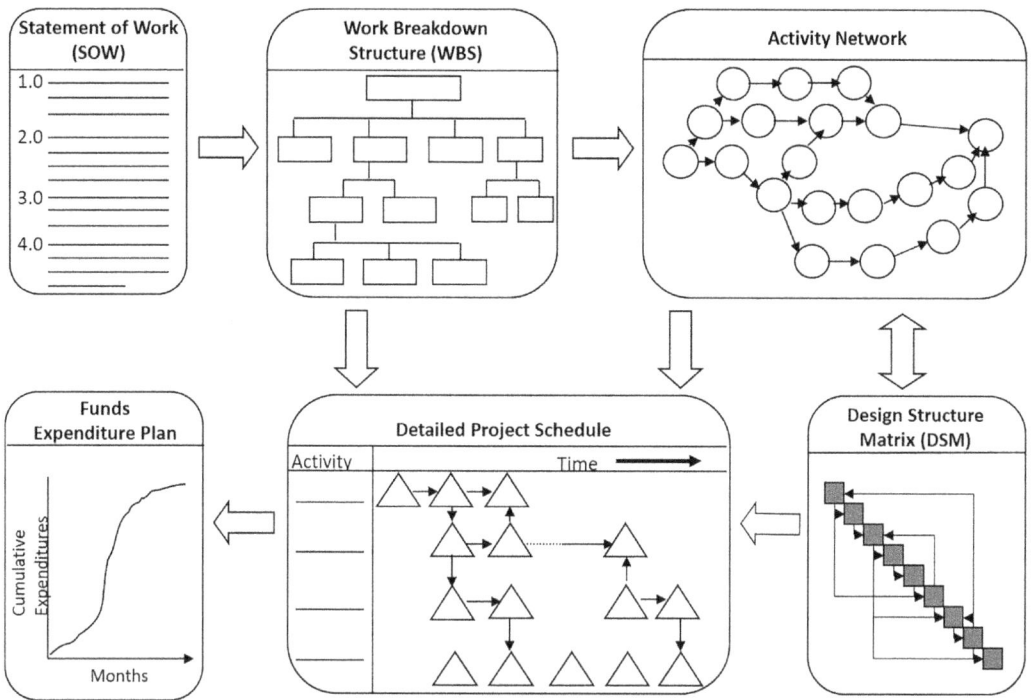

FIGURE 17.8 Common project management visualization models and patterns.

Many of these take the five basic forms above and add patterns to them, such as taking a matrix and visualizing the data in a line chart or pie chart. Lengler and Eppler categorized visualization types into a progression of forms that are useful to represent data, information, and higher-level forms that support conceptual knowledge:

- Data visualization – visualizing quantitative data. Examples are matrices, line charts, and scatter plots.
- Information visualization – visualizing representations of data in forms to amplify cognition. Data is transformed into images, often interactive. Examples are spider charts and flow charts.
- Concept visualization – visualizing and elaborating qualitative concepts, ideas, plans, and analyses. Examples are mind maps, causal chains, PERT and Gantt charts, and swim lane diagrams.
- Metaphor visualization – positioning information graphically to organize and structure information, using metaphor to provide insight about the displayed information. Examples are metro or tube maps, bridge maps, and funnels.
- Strategy visualization – visualizing the analysis, development, formulation, communication, and implementation of strategy. Examples are organizational charts, house of quality, business model or strategy canvas, stakeholder maps, and fishbone diagrams.
- Compound visualization – complementary use of multiple different formats in one schema or frame, often to represent stories. Examples are cartoons, rich pictures, and knowledge maps.

In addition, the periodic table is categorized to show which visualization methods are most often used to visualize structure or process, highly abstracted versus detailed data or information, and to produce divergent or convergent thinking. For example, a mind map supports depiction of structure, in both high-level abstractions and detail, to capture products of divergent thinking. As such, it is often used in brainstorming activities. A timeline (PERT chart or roadmap) depicts process, at high level of abstraction, to capture convergent information. It is primarily used to inform strategy. Selecting the right type of visualization to maximize communication and both information flow and knowledge is an important skill in complex project management.

The core of project visualization then takes the general form of the five basic models and adds useful patterns on top of them to communicate the underlying data, information, and knowledge. Communication with these tools is aided by the use of color, shape, and relative position and also by the placement of the visualization form into a broader context that helps tell the story. These additional means of communication add qualitative information to the patterns, such as priority, focus and attention, and risk. The qualitative information often captures the presenter's knowledge in addition to the information.

KNOWLEDGE VISUALIZATION

While data and information visualization are targeted at displaying fact, knowledge visualization focuses on sharing insights. In exchange of knowledge, this could be personal experiences, perspectives, values, goals, expectations, etc. Eppler and Burkhard (2004) classify certain visualization tools as unique to knowledge visualization based on their content and form. Examples are concept maps, rich pictures, visual metaphors, and other conceptual diagrams. Data visualization tends to be limited by entities and relationships that describe the underlying data, while information visualization abstracts the data to visual patterns that improve cognitive interpretation, and knowledge visualization forms represent more abstract concepts like perspectives, circumstances, and phenomena that are distinct from but can be associated with the data (Eppler and Burkhard 2004).

Knowledge visualization will often use text-based content in a form that emphasizes relationships or patterns. While information visualization is typically used to explore large amounts of abstract data, knowledge visualization is more used to aid in communication of abstract knowledge. The visual models provide the conceptual language for shared context that is required for knowledge flow. Knowledge visualization tools tend to support the sensemaking process, helping the observer to fill in additional insights based on patterns in the underlying information and data.

> Visual models provide the conceptual language for knowledge flow. The key to complex project management is to bring people together to exchange their knowledge in a visual form.

We use information visualization to increase the speed and accuracy with which individuals can process data. We use knowledge visualization to increase the speed and quality of knowledge exchange between individuals. Whitla (2017) introduces the idea of a "meaning curve" that plots on one axis "communication of shared meaning" and on the other "data accuracy." To communicate shared meaning in a set of data to a diverse group of stakeholders, one must generalize the information or at least select representative components. The accuracy present in highly detailed data visualization is only meaningful to people with deep shared understanding of the basis for that data, while transferring insight to a large group requires much more highly abstract forms. Selection of partial data often leads to bias and entrained thinking as the meaning of the data becomes the opinion of the presenter.

Take, for example, project earned value management (EVM) data, relating cost and schedule performance to actual versus planned completion of work. An EVM table organizes the underlying data in a visual form where the content is easily organized for those that understand the meaning of EVM data. The data itself also only represents history (either a historical plan or actual history to date). Figure 17.9 provides an example tabular format (a matrix diagram) of EVM data feeding a typical EVM line chart (information visualization) showing the EVM cost metrics over time.

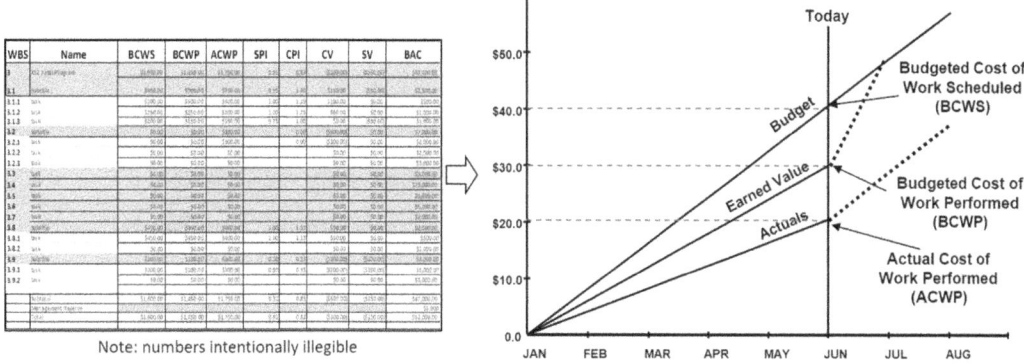

Note: numbers intentionally illegible

FIGURE 17.9   EVM data and information visualization.

The table can calculate data that predicts possible future performance, but the cognitive analysis required to gain insight on the certainty or uncertainty of those predictions is lost in the calculation. In order to visually abstract the information from the data, it is placed into the time-based graph that allows rapid visual processing of the data by larger groups. Historical trends in the data can be extrapolated into the future by projecting the future data as a continuation or change in trend. This is the purpose of information visualization. However, Figure 17.9 illustrates a typical limitation of information visualization under uncertainty. In this example, budgeted cost of work scheduled (BCWS) is less than the earned value of budgeted cost of work performed (BCWP), which indicates a behind schedule condition. Actual cost of work performed (ACWP) is also less than the earned value (BCWP), which reflects less cost to do the work than planned. Does the projected schedule recovery back to plan mean the project will continue to underspend, performed back to planned spend, or overspend even more to recover the schedule? There is no way to tell from the trends in this figure, as the condition of being behind schedule while also at less cost than planned reflects uncertainty in the plan. This is a simple example representing complexity of faith.

If the project is highly complex in all dimensions, the data can still be abstracted into information portraying highly abstract schedule and cost trends, but the usefulness in predicting future outcomes from the information is limited because of the uncertainty in the underlying data. At this point, traditional project management will often rely on simple heuristics or formulaic predictions, and decision-making often falls back to intuition. In these situations, knowledge visualization forms can be helpful. That knowledge is normally created by a project team that conducts a detailed "bottoms-up" analysis to create a better predictive estimate and a new recommended plan. The bottoms-up analysis produces new data and information using team knowledge but does not necessarily produce new shared knowledge. This is where knowledge visualization can be employed to help the team improve its analysis. Figure 17.10 shows an example knowledge transfer activity during an airline merger project (New York Times 2011). The key to dealing with complexity is to bring all the representative agents and their disciplines together to exchange their knowledge in a visual form.

Eppler and Burkhard (2004) argue that many knowledge transfer forms do not fully exploit the capabilities of human visual processing. Knowledge visualization is most useful in interdisciplinary situations, where different stakeholders with different experience and backgrounds come together for combined problem solving. In these settings, visualization tools like Figure 17.10 make tacit information and knowledge such as relationships between stakeholders or processes visible for communication. They provide a common context (language, formats, etc.) for knowledge to be communicated.

Knowledge visualization is also useful in situations where creation of new knowledge is needed, such as innovation or strategic change. Conceptual and metaphoric forms can be quickly developed and changed by the collective groups (Eppler and Burkhard 2004). Many problem definition tools

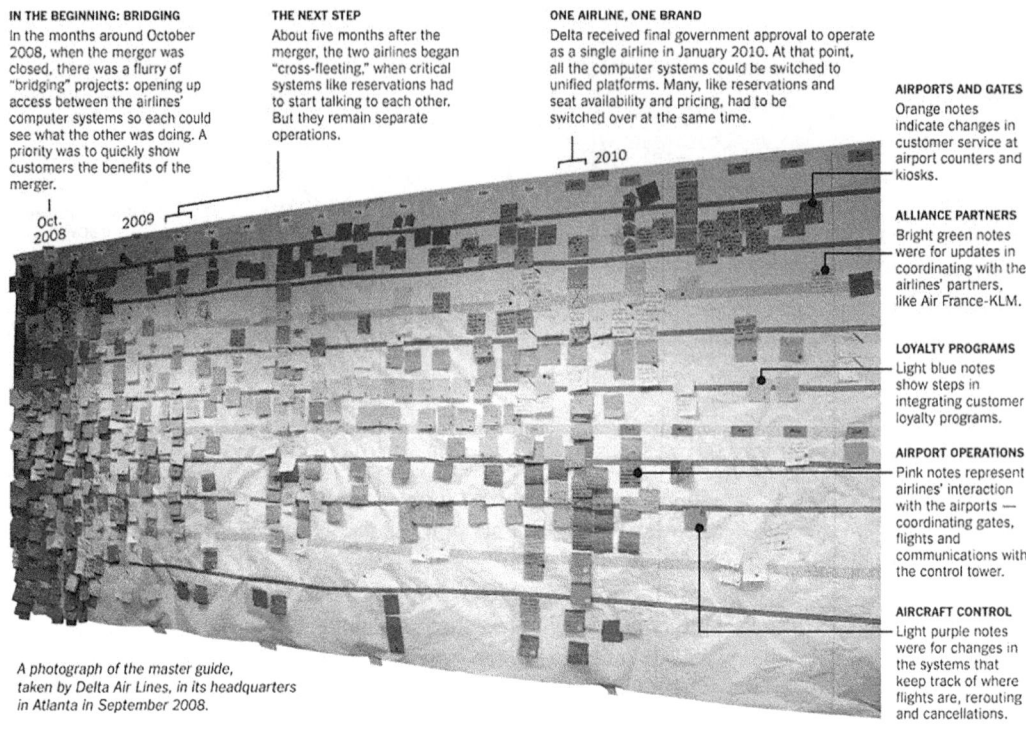

IN THE BEGINNING: BRIDGING
In the months around October 2008, when the merger was closed, there was a flurry of "bridging" projects: opening up access between the airlines' computer systems so each could see what the other was doing. A priority was to quickly show customers the benefits of the merger.

Oct. 2008   2009

THE NEXT STEP
About five months after the merger, the two airlines began "cross-fleeting," when critical systems like reservations had to start talking to each other. But they remain separate operations.

ONE AIRLINE, ONE BRAND
Delta received final government approval to operate as a single airline in January 2010. At that point, all the computer systems could be switched to unified platforms. Many, like reservations and seat availability and pricing, had to be switched over at the same time.

2010

AIRPORTS AND GATES
Orange notes indicate changes in customer service at airport counters and kiosks.

ALLIANCE PARTNERS
Bright green notes were for updates in coordinating with the airlines' partners, like Air France-KLM.

LOYALTY PROGRAMS
Light blue notes show steps in integrating customer loyalty programs.

AIRPORT OPERATIONS
Pink notes represent airlines' interaction with the airports — coordinating gates, flights and communications with the control tower.

AIRCRAFT CONTROL
Light purple notes were for changes in the systems that keep track of where flights are, rerouting and cancellations.

A photograph of the master guide, taken by Delta Air Lines, in its headquarters in Atlanta in September 2008.

FIGURE 17.10   Example knowledge visualization exercise (NYT 2011).

rely on these metaphoric forms such as stakeholder personas, force field analysis, or the six thinking hats or on conceptual forms such as rich pictures, value stream maps, or stakeholder influence maps. Figure 17.11 shows two stakeholder visualization forms that support visualization of knowledge about stakeholders known to team. The first, a knowledge wheel, combines stakeholder types around the wheel with simple narrative statements of their values to be derived from the project on the spokes. The second, a stakeholder influence map, relates relative interest in the project versus influence on the project. Key to the visualization in these are the uses of combined structure and narrative and relative spatial placement.

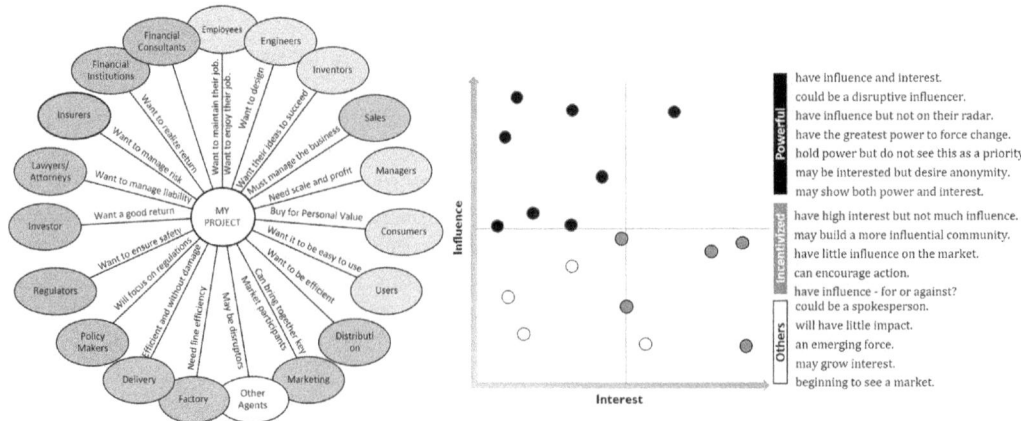

FIGURE 17.11   Two forms of stakeholder knowledge map.

Eppler and Burkhard (2004) also mention a third use of knowledge visualization, to counter information overload. In this case, the skills of the visualization user can select and represent the appropriate abstractions to convey meaning to broader stakeholder groups. This is where the disciplines of systems thinking and complexity thinking are most useful. The two diagrams of Figure 17.11 are examples that balance high levels of abstraction while retaining meaning.

SENSEMAKING AND KNOWLEDGE VISUALIZATION

Knowledge visualization is used in the sensemaking process when uncertainty prevails and solutions are unclear. What separates the complex project manager from a traditional project manager is the ability to recognize these situations and facilitate team social interactions that result in knowledge exchange and selection of new project models and patterns. Sensemaking frameworks are conceptual modeling tools that support collection and organization of team knowledge and associated information. They generally begin with qualitative analysis across groups, often called soft system methods, and then proceed to mapping or conceptual modeling activities. Conceptual modeling forms move the visual communication process away from sets of things toward sets of relationships between things. Recall that the sensemaking process "zooms out" from understanding of entities and their attributes, to understanding of the interactions that create emergence, to the exploration of different stakeholder perspectives. Part of this process is the formation of narratives (generally scenarios) that express the realized or desired emergence in the project situation in the perspective of the project participants. Thus, soft systems methods and accompanying narrative and conceptual model development are essential to diagnosing problems in complex projects and then defining strategies for correction and completion.

Soft systems methods: Complex project management relies on both "soft systems methods" and "hard systems methods" for decision-making. Soft systems methods are focused on facilitating group insights (faith and interaction), whereas hard systems methods focus on analyzing data (fact). A soft systems method takes an observed narrative or view of the real-world situation, re-expresses it as a conceptual model that abstracts the key issues or questions from the real world, creates new conceptual models of the desired world, and then compares the new models against similar or evaluated real-world changes (Checkland 1981). The conceptual model reflects participant mental views or mental models of the system, these are then converted to new actionable models in a facilitation process using language, diagrams, and pictures. Soft systems methods utilize narratives designed to describe system purpose and goals, and visual methods to portray the systemic aspects (structures, relationships, etc.) defined by the narratives. Figure 17.12 is an example of "rich pictures" model combining narrative (pictures and text) with structure (boxes and links). This is an artifact from value stream mapping using a "brown paper" model to map process flow using pictures of process artifacts.

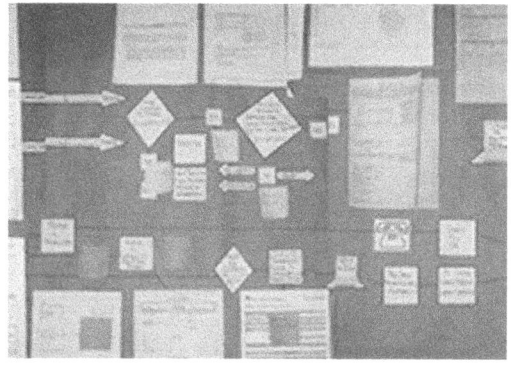

FIGURE 17.12   Knowledge visualization using a value stream map.

The purpose is to properly conceptualize, structure, relate, and validate the relationships in the complex projects from factual data and information to the higher levels of abstraction or aggregation needed to relate meaning and knowledge. This is key to the abstraction of "Fact" to "Faith." Narrative and conceptual forms of visualization are necessary. Eppler and Burkhard (2004) list heuristic sketches, visual metaphors, knowledge animations, conceptual diagrams, knowledge maps, and scientific charts as appropriate knowledge visualization approaches. McDermott and Freeman (2016) relate these forms to specialized leadership skills in narrative formation, conceptual modeling, and contextual data analysis.

Narrative formation: Narrative forms are used to express emergence, relating to evolving situations in the project. The narrative serves as a mental expression of different events, phenomena, or observations as episodes that have meaning (Polkinghorne 1988). The importance of the narrative is a description of the activities that produce change in the project. These activity descriptions also capture the decision agents that are involved in the activities and their relationships. If the situation is complex, a richer narrative becomes quite useful – in particular, a narrative that could be described as a story. This richer narrative serves to capture the activities along with the system structure, stakeholder perspectives, organizations and governance, and even ideas that would provide a greater communication and understanding of the holistic view of the system. Managers experienced with complex projects often cite "storytelling" as a primary means to articulate change strategies in complex projects but have difficulty explaining how those stories are conveyed. In the domain of knowledge visualization, effective narrative tools are visual metaphors and knowledge animations, along with soft systems methods that blend narrative and conceptual diagramming. Figure 17.13 shows an urban transportation project planned in a systems thinking tool, called a "Systemigram." The Systemigram explicitly blends a narrative story with a diagram representing the phrases in the story (Boardman 1994). The combination of textual and visual form is a very powerful knowledge-sharing tool.

Conceptual modeling: The systemigram diagram in Figure 17.13 is an example of a conceptual model known as a "concept map." The concept map is useful for representing knowledge because it allows explicit modeling of contextual information as relationships between concepts. Conceptual modeling tools are used to capture higher-level textual or descriptive models of the project that can then be used to decompose the project into lower sets of measures that can be

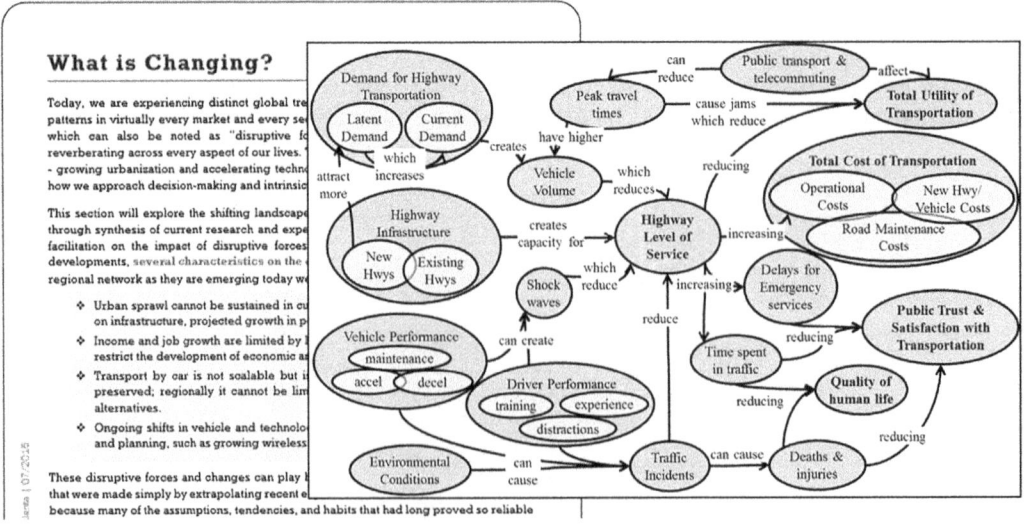

FIGURE 17.13   A "Systemigram" of an urban transportation project.

assessed analytically. Humans have a unique ability to move between analytical and conceptual models of almost anything in our experience base. The challenge is one of visualization – can we craft the appropriate linkage between high-level conceptual representations and low-level analytics?

> The challenge of knowledge visualization is linking high-level conceptual representations and low-level analytics.

Types of conceptual models can generally be grouped as network diagrams, structural diagrams, behavioral or process diagrams, causal diagrams, and concept or knowledge maps. Of these, concept maps and network diagrams are much less integrated into most project management toolsets and represent opportunities for improvement in the discipline. In addition, the process of conceptual modeling is seldom standardized. A formal conceptual modeling workflow creates an artifact like the "Avionics Program Plan" which is most often a book-form document that captures the story in its organization of artifacts and narrative along with the data and information collected into various conceptual diagrams.

Contextual data analysis: Exploring project uncertainties is essentially a data collection activity. The soft systems viewpoint helps us see the system from a variety of perspectives both internal and external to the project. This is not just collection of project data, but it relies on data collection from project team's other experience, disciplinary backgrounds, and heuristic knowledge as well. A diversity of individual backgrounds is necessary, and interviews, discussion, reading, and experimenting with others' opinions are required to collect the data. A "mind map" is one type of visualization tool that is useful to collect and organize the data. This wider data set can later be pruned as the central problems or questions of interest are determined from exploration of a problem or solution space. The various knowledge visualization and conceptual modeling forms are used to make sense of the data and show new relationships between components. The relationships provide insight, but for new strategy formation, program leadership must analyze causal relationships that would drive effective change. The tie between sensemaking and decision-making is a combination of data and causal knowledge.

Selection of the right visualization pattern for the situation at hand is a specialized skill of the complex project manager. In complex situations, one must choose patterns that maximize communication of insights and avoid propagation of entrained thinking. Croll and Yoskovitz (2013) advise when in domains of uncertainty to keep the following contextual data analysis patterns in mind:

- Qualitative/quantitative: qualitative data is unstructured, anecdotal, and hard to aggregate; quantitative data provides hard numbers but less insight. Correctly choosing the type of data based on current context is important for meaningful communication and is a specialized skill.
- Prescribed/actionable: existing prescribed metrics might make you feel good or bad, but they don't change how you act. Actionable metrics change your behavior by helping you pick a course of action. Make sure the data collected supports development of actionable strategies.
- Exploring/reporting: exploratory metrics are speculative and try to find unknown insights, while reporting metrics keep you abreast of normal, managerial, day-to-day operations. Most traditional project management tools focus on reporting metrics. Exploratory metrics are identified through "questions of interest" that attempt to define project issues and uncertainties, and data is collected to help answer those questions.
- Leading/lagging: leading metrics give you a predictive understanding of the future; lagging metrics explain the past. Data that reflects and can be tracked to confirm predictions of change must be used to inform change strategies. Contextual models that project historical data trends into the future must be developed for the project situation.

- Correlated/causal: project measures that change together are correlated, a measure that causes another measure to change is causal. Effective change strategies for dealing with situations in complex project management must be careful to find causal bases for change.

> Selection of the right visualization pattern for the situation at hand is a specialized skill of the complex project manager.

Contextual data analysis is not about selecting the right project management tool and visualization approach, it is actually about building new models for placing the data in a strategy framework that can be used to gain insight on project situational uncertainties. Croll and Yoskovitz recommend finding and developing contextual data analytics that focus on changing the project behaviors. If the visualized measure does not inform program management on what to do differently, then it may not be of much use. The specialized role of complex project managers is to continually ask new questions, search for new data and models, and not lock their thinking into existing project performance measures.

INTERACTIVE DATA AND KNOWLEDGE VISUALIZATION

Many implementations center on integration of typical project views into dashboards. This is ineffective to counter most forms of project complexity, unless the data and information visualization forms can be integrated into a meaningful story. Many project management dashboard tools have very prescriptive views and limited ability to contextually arrange informational views. This is why book-form documents remain popular. There are several trends in data and information management that may change this. Williams (2015) recommends a move from "push" driven to "pull" driven project communication. In traditional project management, the leads make decisions on project information flow and "push" it up the chain. The consumer of the flow does not get to choose the content or format. This leads to regular standardized project information transfer but is not useful to manage uncertainties. In "pull"-based communication, the data and information are posted to a common data store or warehouse. The consumer of the information can decide what they want to see and in what form. This provides an opportunity for the project leadership to have more interaction with the project team about the data and information that are most important to them, particularly where uncertainties or anomalies can be explored. Visual data and information artifacts can then be placed into more qualitative and knowledge-driven visual frameworks (Williams 2015). The value stream mapping process described previously in Figure 17.12 is a convenient example. A value stream map is a conceptual diagram representing flows of value in an organizational process. It is created by a project team using their knowledge and example data artifacts from current processes which are then linked together as a node and link diagram to produce a more efficient flow. The facilitated visual diagramming and associated data-driven knowledge accumulation allows the team to share their knowledge and create new shared knowledge of the entire process. In the spirit of Cynefin, the "best practice" artifacts are brought into the domain of informal/interdependent learning to discover better practices. Having decision agents create new value stream maps that realigned cross-team processes was a core facilitation practice used in updating the F-22 book.

Future interactive project management tools should provide centralized project "truth data" that is linked and accessible to any authorized stakeholder and modeled to support rapid development of new models, analytics, and integrated visualization dashboards. The speed of information flow in current software agile processes and the move toward more fully digital engineering and project development environments (Industry 4.0) will produce the infrastructure to support more complex project visualization. However, no single set of computer visualizations will replace the specialized skills of the project manager in team facilitation and selection of modeling and visualization tool.

The shortfall of these pull-based methods is that (1) they still do not scale beyond well-bounded components of the actual process (example: a process), and (2) as strategy-driven methods, they do not support complex project uncertainties where there is little or no experience. This is the Cynefin uncharted/ innovative learning domain. In these uncertain situations additional conceptualization, tools are needed.

IMPLICATIONS FOR PROJECT LEADERSHIP SKILL DEVELOPMENT

The International Center for Complex Project Management (ICCPM) Complex Project Manager Competency Standards Version 4.1 provides an existing basis for the specialized skills necessary in complex project management. They define the significant role of underpinning knowledge in competency development and assessment of executive and complex project managers. These include breadth of underpinning knowledge in the project domain or market and depth of knowledge in areas related to the architecture of the project and product. Architectural experience is important because it combines the technical and business decisions of the project. The competency standards identify the following characteristics of complex project managers in relation to the use of data, information, and knowledge (ICCPM 2012):

Item 3.2.5: they establish a collaborative environment, data collection, data warehouse, and performance assessment system that support life cycle management, performance assessment, and timely feedback.

Item 3.2.6: they conduct ongoing tracking, analysis, relevance and review of key project performance indicators using that data.

Item 3.2.12: they focus on performance measure collection, analysis, and reporting in order to communicate meaningful information to different types of stakeholders and ensure the presentation of project performance data is clear and understandable.

Item 7.6.2: they are able to establish effective environments where shared meanings among stakeholders about situations, the project, and organizational processes are achieved.

Item 7.6.3: they use storytelling to create a positive and engaging environment for staff and external stakeholders to share knowledge and ideas.

The background and examples in this chapter provide a set of theories, methods, and tools that back up these five core competencies.

DISCUSSION QUESTIONS

The following questions are provided to encourage further reflection and learning:

1. Can you "map" a complex project in your experience across the three dimensions of "fact," "faith," and "interaction" in order to gain better insight into its eventual performance? Try to build a visual model (a diagram) of this.
2. How well did team activities you participated in complete a full sensemaking loop to produce a real cycle of learning? Can you explain success or failure based on components of that loop? What visual models and visualizations were used in the process, and did they encourage team knowledge transfer and learning?
3. The next time you use a project management visualization form, try to list its type (data, information, concept, metaphor, or strategy) and use to communicate (data, information, or knowledge). Does this help you select the right visualization tool?
4. Can you recall a project leader who effectively used data and information visualization, knowledge visualization, and team facilitation to address project complexity and uncertainty? What were the tools used, when were they used, and how were they facilitated? Can you model this yourself?

SUMMARY

As project complexity increases, the role of knowledge transfer in social networks becomes more critical, and the ability to visualize knowledge (as opposed to information) becomes paramount to project success. The complex project manager must learn to adapt their communications and management facilitation based on changing dynamics in the project. Cycles of learning produce new knowledge for decision-making and require the use of knowledge visualization practices and forms. This is a specialized skill most closely related to systems thinking and complexity thinking.

REFERENCES

Ashby, W. R. 1958. Requisite variety and its implications for the control of complex systems. *Cybernetica* (Namur) 1(No 2), 83–99.

Axelrod, R. and Cohen, M. 2008. *Harnessing Complexity*. New York: Basic Books.

Beer, S. 1985. *Diagnosing the System for Organisations*. London and New York: John Wiley & Sons.

Boardman, J. T. 1994. Process model for unifying systems engineering and project management. *Engineering Management* 4(1), 25–35.

Boulton, J., Allen, P., and Bowman, C. 2015. *Embracing Complexity: Strategic Perspectives for an Age of Turbulence*. Oxford: Oxford University Press.

Brassard, M. 1996. *The Memory Jogger Plus+, Featuring the Seven Management and Planning Tools*. Salem, NH: GOAL/QPC.

Checkland, P. B. 1981. *Systems Thinking, Systems Practice*. Chichester: John Wiley & Sons.

Collins, J. and Hansen, M. 2011. *Great by Choice: Uncertainty, Chaos, and Luck - Why Some Thrive Despite Them All*. New York: HarperCollins Publishers.

Courtney, H., Lovallo, D., and Clarke, C. 2013. Deciding how to decide. *Harvard Business Review* 91(11), 62–70.

Croll, A. and Yoskovitz, B. 2013. *Lean Analytics: Use Data to Build a Better Startup Faster*. Sebastopol, CA: O'Reilly Media.

Dervin, B. 2005. Chapter 2: What methodology does to theory: Sense-making methodology as exemplar, in K. E. Fisher, S. Erdelez, and L. McKechnie (Eds.). *Theories of Information Behavior*. ASIS&T Monograph Series. Medford, NJ: Information Today, 25–30.

Dombkins, D. H. 2008. The integration of project management and systems thinking, in *Project Perspectives*. Nijkerk, Netherlands: International Project Management Association, 20, 16–21.

Eppler, M. and Burkhard, R. 2004. Knowledge visualisation: Towards a new discipline and its fields of application, in D. G. Schwartz (Ed.). *Encyclopedia of Knowledge Management*. Hershey, PA: Idea Group Reference, 2005, S. 551–S. 560.

Geraldi, J. G. and Adlbrecht, G. 2007. On faith, fact and interaction in projects. *Project Management Journal* 38(1), 32–43.

International Center for Complex Project Management (ICCPM). 2012. Complex project manager competency standards version 4.1 (August 2012).

Klein, G., Moon, B., and Hoffman, R. R. 2006. Making sense of sensemaking 1: Alternative perspectives, *Intelligent Systems* 21(4), 70–73.

Lengler, R. and Eppler, M. 2007. Towards a periodic table of visualization methods for management. *IASTED Proceedings of the Conference on Graphics and Visualization in Engineering (GVE 2007)*, Clearwater, FL.

McDermott, T. and Freeman, D. 2016. Systems thinking in the systems engineering process, new methods and tools, in M. Frank, S. Kordova, and H. Shaked (Eds.). *Systems Thinking: Foundation, Uses and Challenges*. Hauppauge, NY: Nova Science Publishers.

Moldoveanu, M. and Leclerc, O. 2015. *The Design of Insight: How to Solve any Business Problem*. Stanford, CA: Stanford University Press.

The New York Times (NYT). 2011. How to merge two airlines. Accessed from: https://archive.nytimes.com/www.nytimes.com/interactive/2011/05/18/business/delta-northwest-merger-graphic.html. Retrieved November 2018.

Polkinghorne, D. E. 1988. *Narrative Knowing and the Human Sciences*. Albany, NY: State University of New York Press.

Project Management Institute. 2015. Pulse of the profession® in-depth report: Capturing the value of project management through knowledge transfer. Newtown Square, PA.

Rumsfeld, D. 2011. *Known and Unknown: A Memoir.* New York: Sentinel.

Russell, D. M., Stefik, M. J., Pirolli, P., and Card, S. K. 1993. The cost structure of sensemaking. *Proceedings of the INTERACT'93 and CHI'93 Conference on Human Factors in Computing Systems*, Amsterdam, Netherlands, 269–276.

Snowden, D. 2003. Complex acts of knowing: Paradox and descriptive self-awareness. *Journal of Knowledge Management* 6(2), 100–111. May 2002.

Snowden, D. 2012. The origins of Cynefin, parts 1-7. Accessed from: http://cognitive-edge.com/blog/entry/3505/part-one-origins-of-cynefin. Retrieved November 2018.

Snowden, D. 2017. Liminal Cynefin – image release. Accessed from: http://cognitive-edge.com/blog/liminal-cynefin-release-image. Retrieved November 2018.

Snowden, D. and Kurtz, C. 2003. The new dynamics of strategy: Sense-making in a complex and complicated world. *IBM Systems Journal* 42, 462–483.

Whitla, S. 2017. The meaning curve: Where do you need to be? Visual Meaning Ltd. Accessed from: http://meaning.guide/index.php/2017/11/13/meaning-curve-need/. Retrieved November 2018.

Williams, P. R. 2015. Visual project management. PMI® Global Congress 2015, EMEA, London, England. Newtown Square, PA: Project Management Institute.

18 Social Network Analysis Toolset

Kon Shing Kenneth Chung
The University of Sydney

CONTENTS

INTRODUCTION AND BACKGROUND

There has been a significant rise in the study and application of social network (SN). They are widespread in the areas of anthropology, social psychology, human behaviour, computational and data science, and business and management studies. In project management (PM) literature, however, and more so in practice, the study of SNs and its application in the management of projects is still relatively scant, if not in its infancy, but growing. Figure 18.1 shows the number of articles found with the term "social network" in its title and abstract in the *International Journal of Project Management*, a leading journal in the field. It can be seen in the figure that from 2010 to 2017, the total number of articles is 108.

So, what is a SN? What does it mean by the analysis of such networks? Is it the study of social media such as Facebook and so on? Interestingly, when I mention to others about my interest in the study of SNs, I am often met with such responses – *Right! So you are into Facebook and Snapchat, are you? or Wow, interesting, so you are the network guru of Tinder? I'll know who to refer to then when it comes to social networking!*

Firstly, there is a difference between the study of SNs and social media. While the study of social media is about the use of (generally online) social connectivity tools such as Facebook, Snapchat,

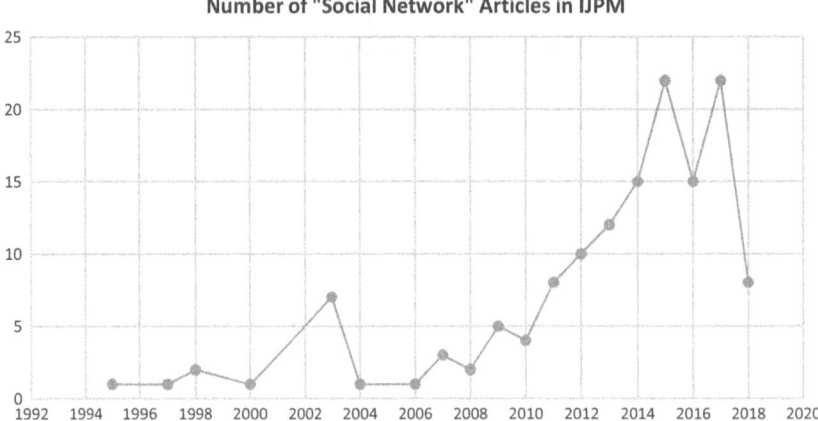

FIGURE 18.1   Number of "social network" articles in *International Journal of Project Management* between 2010 and 2017.

Instagram and Tinder, the study of SNs is about the study of structure, connections, where one is located in a network, and the like.

More formally, the study of SN concerns itself with "The disciplined inquiry into the patterning of relations among social actors, as well as the patterning of relationships among actors at different levels of analysis (such as persons and groups)" (Breiger 2004).

In the most general form of a network, a network is defined by a node and a connection to that node. In a SN, a node refers to an actor – simply put, a person, and the connection refers to a particular relationship such as friendship or advice-giving. In a team network, the node refers to the team and the connection may represent a relationship such as who refers to whom for advice, or who works with whom. Likewise, an organisational network can be represented as a supply-chain network showing which organisation supplies materials and resources to a particular organisation. When the same network thinking is applied to spread of disease (who spreads disease to whom), we have an epidemiology network. Likewise, network thinking can be applied to traffic networks, computer networks, website networks, food chain network and so on. In such instances, the study is no longer about SNs (i.e. human networks) but falls under the paradigm of "Network Science".

What intrigues a number of scholars and practitioners is that the study of SNs encapsulates both theory and methodology. For instance, you could apply analytics of SN methodology to both a project team network (showing who seeks advice from whom) for understanding project success (Toomey 2012) or a project dependency network within a project portfolio (showing which project depends on which within the portfolio) (Killen and Kjaer 2012). Furthermore, theories and concepts in SNs may also apply in both situations (for instance, the notion of brokerage would apply in both scenarios, but the interpretation of its implications would only differ in context).

As you read through the following sections, your understanding of SNs concepts and methodology will grow, and you will begin to appreciate the power of SNs that is usually overlooked because of its invisible nature.

BRIEF HISTORY OF SOCIAL NETWORK ANALYSIS

The origins of social network analysis (SNA) can be traced back to as early as the 1930s–1950s (Scott 2000). Two pioneering schools of thought in this area were the Harvard scholars such as George Homans and his student Harrison White, and the Manchester anthropologist such as John Barnes, who coined the term "social network" in 1954. At the same time, scholars such as Wolfgang Köhler, Jacob Moreno and Kurt Lewin were advanced in the field of sociometric analysis, which were hugely influential for operationalising SNs. A chart depicting pioneers who contributed to this field is shown in Figure 18.2.

Since Putnam's work on social capital (1993, 1995), there has been a huge surge in the popularity of SNA. Researchers such as Ron Burt (1997, 2000, 2005) and Nan Lin (2000) took a structural perspective of social capital where social capital was operationalised in terms of structure and brokerage capacity. Later, Migram's (1967) famous "small-world" studies helped popularize the notion of SNs and its analysis further. These small-world studies showed that, in general, people are connected to one another on an average of six degrees. With the rise of social media from early 2000, Facebook now confirms that the average degrees of separation based on its own data is 3.5 (Bhagat et al. 2016).

With the growing popularity of SNA and its application, SNA studies can now be broadly categorised into two schools of thought: the heterophily theory and structural role theory perspectives (Kilduff and Tsai 2003). SNA studies under the heterophily theory are based on the premise that new information and resource benefits accrue to those either broker groups that are otherwise not connected or because of a person who joins group and who is an effective broker. Studies based on the heterophily theory are generally motivated by theories such as the strength of weak ties (Granovetter 1973) and/or structural holes theory (Burt 1992). These theories are discussed in the following section. On the other hand, studies that are based on the structural role theory seek to understand the psychological aspect of the role setting of actors within the network. In other words, they seek to understand the extent to which the networks are cohesive or the extent to which networks demonstrate structural equivalence (Kilduff and Tsai 2003).

With technological advancements in computation power over the past two decades, there has also been significant advancements in terms of SN methodology, particularly with respect to statistical models. Since the 1990s, there has been an increase in the development of exponential random graph models (ERGMs). These are models that treat the SN as the dependent variable, and the researchers seek to explain the network structure by accounting for social processes that are stochastic in nature that led to the formation of the observed network.

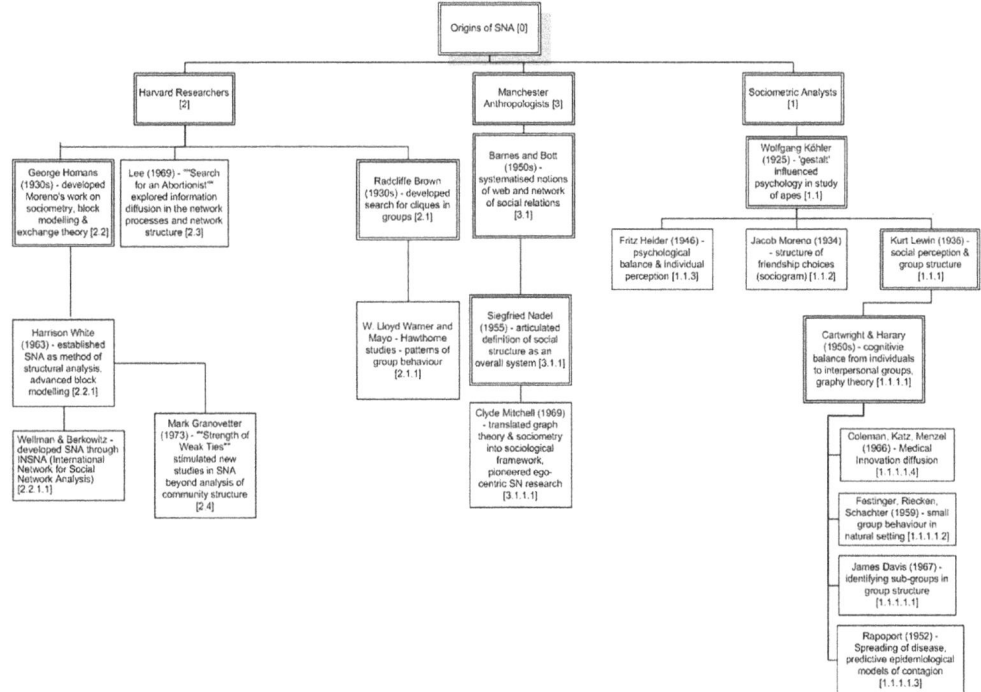

FIGURE 18.2 Pioneers from sociology, anthropology and sociometry who contributed to social networks.

ERGMs use the number of nodes of the observed network (i.e. the network data that is collected) to determine a probability distribution of all possible network configurations possible. The observed network is then examined as part of this probability distribution. In other words, it helps to explain the extent to which the observed network deviates significantly from chance. An example provided by Robins et al. (2007) *is whether the observed network shows a strong tendency for reciprocity, over and above the chance appearance of a number of reciprocated ties if relationships occurred completely at random. In other words, do actors in the observed network tend to reciprocate relationship choices?* In the context of project managers, for instance, it allows the question, do project team members only informally communicate with members who are in the same department or different department.

Most famous among these are the p* models developed by Frank and Strauss (1986), Wasserman and Pattison (1996) and Robins et al. (2007).

As it is beyond the scope of this chapter to provide an overall history of SNA, the book entitled *The Development of Social Network Analysis* by Linton Freeman (2004) offers a comprehensive account of the development of SNA.

RELEVANCE OF SNA TO COMPLEX PM

SNA can be applied from a PM organisational context at the micro, meso and macro levels. At the micro level, typical examples of SNA include analysis of advice-seeking or information-sharing networks of project team members or other internal stakeholders (e.g. suppliers, contractors, customers) (Anichenko, Chung, and Crawford 2016, Chung and Crawford 2015, Chung, Du, and Crawford 2017). At the meso level, an example could be a SNA of intra-organisational relationships within the project organisation (Li et al. 2011, Pryke 2017, Turner and Müller 2003, Chinowsky, Diekmann, and O'Brien 2010). At the macro level, examples include SNA of supply-chain network of a megaproject, where the network shows who supplies information and/or resources to whom or SNA of a project portfolio showing the dependence of projects within the portfolio – that is, which projects must be preceded before another one begins (Killen and Kjaer 2012, Borgatti and Li 2009).

For complex projects, in general, at all the levels, there are several notions and concepts of SNA that applies to complex PM. These may be conceived at the network levels, actor levels and the relational levels. It is useful at this juncture to point to key relevant concepts the body of SN literature that relates to PM. This can be understood from three levels: the *network level*, the *actor* or *individual level* and the *relational level*.

It is to be noted that the mathematical expressions of the following concepts are purposely avoided to make the content generally accessible. For those interested in the mathematics and graph theoretical expression of network measures, please refer to Wasserman and Faust (1994) and Scott (2000).

Network level: A key concept at the network level is the notion of network closure, network density and centralisation.

Network closure is used to represent the extent of clustering that occurs among actors. If we take a network of three actors – say A, B, and C, and A trusts B, and A trusts C, then it is very likely that over time, B will also trust C or vice versa. This in effect means that closure has occurred within the triad network (Burt 2000). Thus, from a networks' perspective, closure is useful to represent the extent to which there is some form of trust building occurring within the network. This is also referred to as bonding social capital. On the negative end, networks with high closure can also exhibit constraint where the emergence of group norms results in the restriction of free-flowing ideas within the clique (Tortoriello and Krackhardt 2010).

Network density measures the ratio of existing ties to all possible ties in the network. In other words, it measures the cohesiveness of the network, with values of 1 meaning high connectivity and 0 meaning low connectivity. Networks where everyone is connected to everyone are "maximally connected cliques". Such networks connote a sense of homophily (McPherson and Smith-Lovin 1987) – characterised by how tightly bound individuals are – and suggest shared values such as membership, interests and belonging (Reagans and McEvily 2003). Thus, a highly connected network of stakeholders here can mean that they collaborate or share information closely or that they could be a coalition that could be influential for advocating or resisting change, for instance.

Related to the notion of network density is the idea of network centralisation. This represents the extent to which the density of the network is focused around a certain actor. A network with high centralisation is reminiscent of an organisational structure where the leader is at the top of the hierarchy with subordinates below him or her. In a series of experiments dubbed, the "MIT experiments", Bavelas (1950) led a seminal study that demonstrated the importance of the structure of communication networks and its impact on communication flow and performance. He demonstrated that structures that were highly centralised were conducive to solving tasks that were simple and straightforward. For more complex tasks, however, a decentralized structure was better suited.

Actor level: At the actor level, it is the location of actors with respect to others within the network that is of interest. The most common and perhaps most useful SN concept here is the idea of centrality. Linton Freeman (1978) is often credited with his work on the centrality concept. According to him, the three most common forms of centrality are as follows:

- Degree centrality: Measured as the number of ties to or from a particular actor, it indicates an actor's communication activity.
- Closeness centrality: Measured as the extent to which an actor is close to all others within the network, it indicates the independence of the actor in terms of his/her ability to reach all others within the network. It is also a proxy for minimum cost of time and efficiency for communicating with others within the network.
- Betweenness centrality: Measured as the extent to which an actor lies in the shortest path to all others within the network, it also indicates the actor's potential to control communication.

Burt (1992) builds further on the notion of betweenness centrality and argues that those having high betweenness centrality also play a crucial role in brokering information and control benefits particularly where there are two or more groups (which are internally closely connected) that are not connected. Referring to this as a hole in the network structure, Burt suggests that those occupying this 'structural hole' (i.e. providing the brokering connection to the disconnected groups) stand to gain from information benefits and others including creativity, good ideas, job promotions and so on. Thus, for project professionals, "these kinds of centrality imply three competing 'theories' of how centrality might affect group processes" (Freeman 1978).

Tie level: At the tie level, the most relevant theory is the "strength of weak tie" theory postulated by Mark Granovetter (1973) whose work here was seminal in terms of its application to information theory, communication studies, economics and management. Granovetter reasoned that as networks become denser, the rate of novelty of information being diffused becomes lower. This is because information becomes quickly redundant such that everyone in the network knows that the others know. New or novel information must hence come from weak ties (to those or groups outside the closely knit group). This is crucial in terms of innovation and generating new ideas, and therefore, the "strength of weak ties". More recently, it has been demonstrated that while weak ties are important for facilitating new and novel information transfer, the strong ties are instrumental for solving complex problems (Montjoye et al. 2014, Pentland 2012). For project professionals, this means that the strength of relations or ties need to be accounted for in terms of understanding the potential for information transfer, innovation, and collaborative complex problem solving.

It can, thus, be inferred that at the network, actor and tie levels, there are relevant theories and concepts in SNs, which may be pragmatically applied in PM. To illustrate, the example used below is in the context of stakeholder identification, analysis, intervention and management (see Table 18.1).

To illustrate the usefulness of such a perspective in project conflict management, for instance, network thinking as described at all levels in Table 18.1 can be applied either in a *proactive* manner or a *reactive* manner. I will discuss the latter first, as most SNs are generally a snapshot of social structure in retrospect.

Because the nature of SN data depends on ties that have developed over time (e.g. collaboration or social relationships), one retrospectively investigates into the structure of relationships for the purpose of understanding sources of conflict; the dynamics of power and the identification of bottle-necks, brokers and opinion leaders. In this sense, one is operating in a *reactive* manner to the conflict at hand and is using SNA to understand where conflict lies and how network structure is conducive to both conflict and collaboration.

To illustrate, Robins et al. (2011) analysed how various stakeholders (comprising organisations from private and public sectors) with seemingly significant interest and involvement in the governance of the Swan River environmental project (in Perth, Australia) as a result of changed legislation had con-flicting interests and contested to put others' policies down in order to push forward their own policy. Using a qualitative interview approach, they revealed that while one stakeholder organisation would consider a relationship as crucial with another for the governance of the project, the same organisation would also indicate difficulties in working with the other stakeholder organisation (i.e. ease of working with the other organisation was indicated by a negative tie). The research utilised the blockmodelling approach (which determines the set of organisations that are subject to similar structural opportuni-ties and constraints via the incoming and outgoing ties) and ERGM techniques to reveal four primary network groups that showed relatively little reciprocity, coordination and interactions in inter-group ties. In summary, the results of the study showed that SN methods are useful for analysing a networked governance system for understanding cooperation and conflict. In particular, they concluded that "simple exploratory network techniques can go a long way to understanding the major effects within an organisational system, but a more complete analysis using network statistical models can delineate and separate the impact of the different networks in describing the underlying structural logic".

As the old adage goes, "prevention is better than cure", it is useful to apply SN thinking in ways that proactively minimise opportunities for conflict. In Pentland's study of collaboration within proj-ect teams where "far-flung and mixed-language teams often struggled to gel", he found that by con-sistently and unobtrusively allowing the project team members to view their team communication patterns and giving simple guidance as to what constituted good team collaboration, members who failed to engage with high energy were eventually able to do so within a week's time (Pentland 2012).

TABLE 18.1

Summary of Relevant Theory and Concepts Useful for Stakeholder Management

Levels in Network Thinking	Relevant SN Theory or Concept	SN Construct	Relevance to Stakeholder Management
Network level	Bavelas' MIT experiment	• Centralization • Network density	Allows assessment of how closely knit or sparse the connections of stakeholders are and the degree to which these connections focus around a central stakeholder
Actor level	Freeman's centrality concept Burt's structural hole theory	• Degree • Betweenness • Closeness • Constraint	Allows identification of stakeholders who have high information flow, brokerage potential, independence and lack of reach, respectively.
Tie level	Granovetter's strength of weak tie theory	• Tie strength	Allows assessment of how influential or close or how strong or weak a connection is of one stakeholder to others.

Furthermore, the concentration of engagement that was initially highly focused within members of the same culture and background had by a week spread out more evenly across the entire team.

In this research, Pentland made use of sociometric tags that were worn as badges by the project team members. These badges had wireless capabilities that captured body movements (such as the movement of one's hands when one speaks), vocal attributes (volume, tone and pitch) as well as who were interacting with whom. So, if a team member is quiet and disengaged, then the member is said to demonstrate low energy and engagement. On the other hand, if a team member is highly interactive with others beyond the project team, excited in an idea and is enthusiastically sharing it with others, then the person is said to demonstrate high energy and engagement.

Using this sociometric data, Pentland was able to visually display the communication patterns to the project team on a daily basis. The change after a week in the pattern of communication was remarkable. Teams became more engaged with one another, despite difference in nuances of language, culture, background and geographic location. Furthermore, whereas there was an initial development of "teams within a team", which could easily and potentially give rise to conflict and disputation, by the end of a week, the energy and engagement had spread out more evenly to those who were considered "low energy" at the first week. Ultimately, there was a significant change in that almost all team members demonstrated higher energy and engagement by the end of the week. This approach towards conflict management is thus a *proactive* one and does not require any sort of direct intervention by senior executives of the project.

DATA COLLECTION FOR SNA

There are two main approaches to SN data collection – whole network and egocentric network approaches. Below is a brief description of these approaches.

WHOLE NETWORK APPROACH

Also called the "sociocentric network" approach, the focus of a whole network analysis is on measuring the structural patterns of those interactions and how those patterns explain outcomes, like the concentration of power or other resources, within the group. The underlying assumption is that members of a group interact more than would a randomly selected group of similar size. Sociocentric network analysts are interested in identifying structural patterns in cases that can be generalised (Wellman 1926, Garton, Haythornthwaite, and Wellman 1997).

In a whole network study, the actors of the network are usually known or easily determined. This is because a sociocentric network study usually focuses on "closed" networks implying that the boundaries of a whole network are a priori defined. In many cases, this approach remains the gold standard because of its ability to gather data for the entire network. The network represents the saturation sample of interests, and the analysis allows for the results to be generalised to the population.

Data collection using a whole network approach usually involves listing the names of the actors in the form of an adjacency matrix. When respondents are administered the network survey consisting of the roster of names, they usually check off the names of people whom they know depending on the name-generator question asked. For example, in a project team that consists of 15 members, a whole network study may be conducted in order to understand the advice-sharing network of the team. A roster of the names of all the members (excluding the respondent's) will be presented to each of the team members.

A simple name-generation question such as "In the past 2 months, who have you sought advice from to carry out your work in project X?" Obviously, a more detailed specification of what the work (e.g. the development of the feasibility study) may be is provided here.

An example of a well-applied whole network approach was in the context of a large Australian bank (part of the big four) where, at the time of the study, an important phase of a large transformational project had just been carried out (Monaghan and Chung 2016). In this project, teams across Sydney, Melbourne and Hyderabad (India) oversaw implementing the new "loans system" that replaced the older legacy systems. This phase of the project was already deemed a huge success in terms of the iron triangle (cost, scope, time and quality). We were interested to explore what network structures existed during the project in such high-performing teams. Thus, a SNA was conducted to evaluate the project team network structure in terms of who sought whom for advice, who relied on whom to get work done, and who socialised with whom during nonworking times (e.g. during lunch breaks or after work hours).

The data collection included asking for demographic data such as gender, level of education, role in the organisation and in the project and years in the organisation and in the project. The second part of the data collection presented a roster of names. As the sample size was 56 members in total, the respondent was thus presented with 55 names in the roster. For each name, the respondent was then asked to indicate (1) whether the respondent sought advice from the contact; (2) if so, the frequency of advice-sharing that occurred; and (3) the extent to which he or she felt "comfortable" in seeking advice from that person. The same questions were asked for the "Reliance" network (i.e. to elicit who relied on whom) and the "socialising" network (i.e. who hung out with whom). Once all the 56 members answered these questions, a "whole network" was then constructed using Netdraw (Borgatti, Everett, and Freeman 2002). An example of the figure (named de-identified for confidentiality reasons) is shown in Figure 18.3.

There are limitations and challenges for conducting network data collection using a whole network approach. Let's say you have an organisation where there are thousands of employees. Let's also assume that you are interested in a network of who knows whom. In this case, having all employees go through a roster of thousands of names and to recall whether one knew someone else would be highly impractical. Previous studies suggest that scrutinising through long lists of names and identifying the multiple types of ties with each person on the roster cause fatigue and recall problems (Bernard, Killworth, and Sailer 1982). Given these difficulties, the egocentric network approach is an alternative strategy for data collection.

EGOCENTRIC NETWORK APPROACH

In the SN parlance, the person we are interested in is referred to as the "ego" and the people referred to by the "ego" as his affiliate, advisor, friend, or relative are known as "alters".

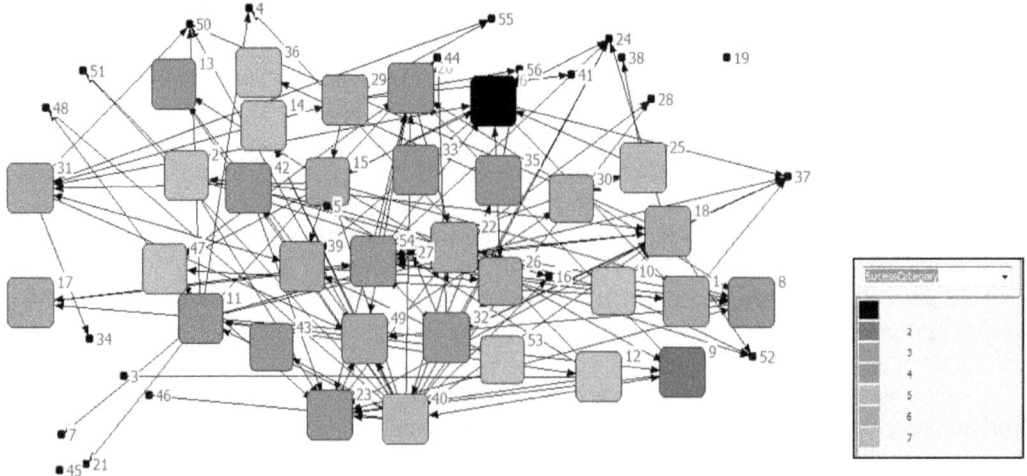

FIGURE 18.3 Sample advice-seeking network using sociocentric network data collection (size of node = closeness centrality; high colour code = high perceived project success by team member).

In a study to develop a theoretical model for understanding the effects of SN properties on individual performance and technology use, Chung (2011) and Chung and Hossain (2009) sampled 109 general practitioners (GPs) in rural New South Wales (NSW), Australia and used the following name-generator question:

> By "professional network", we mean professional people whom you associate, interact or work with for the provision of care to patients (e.g. nurses, admin staff, specialists, pathologists, doctors etc.) Looking back over the last six months, please identify people (up to 15 maximum) who are important in providing you with information or advice for providing care to patients.

Once the respondent provided the list of people, the name interpret questions were asked. These questions elicited information about each of the people named. For instance, the occupation and the proximity within which the respondent worked with the person were asked. Other name-interpreter items solicited were strength of each tie, measured by "time known the person", "frequency of interaction", "type of relationship" and "degree of closeness".

Finally, to complete the (ego) network of the respondent GP, each respondent was asked to determine how the members of their professional network relate to each other based on a five-point degree of closeness scale ranging from "especially close" to "do not know each other". In other words, for each alter nominated, the GP would determine a closeness scale for every other alter.

A sample sociogram showing the egocentric professional network of the GP is shown in Figure 18.4.

In the following section, I demonstrate some case studies on how SNs can be used for stakeholder engagement analysis and as a visually useful tool for project evaluation.

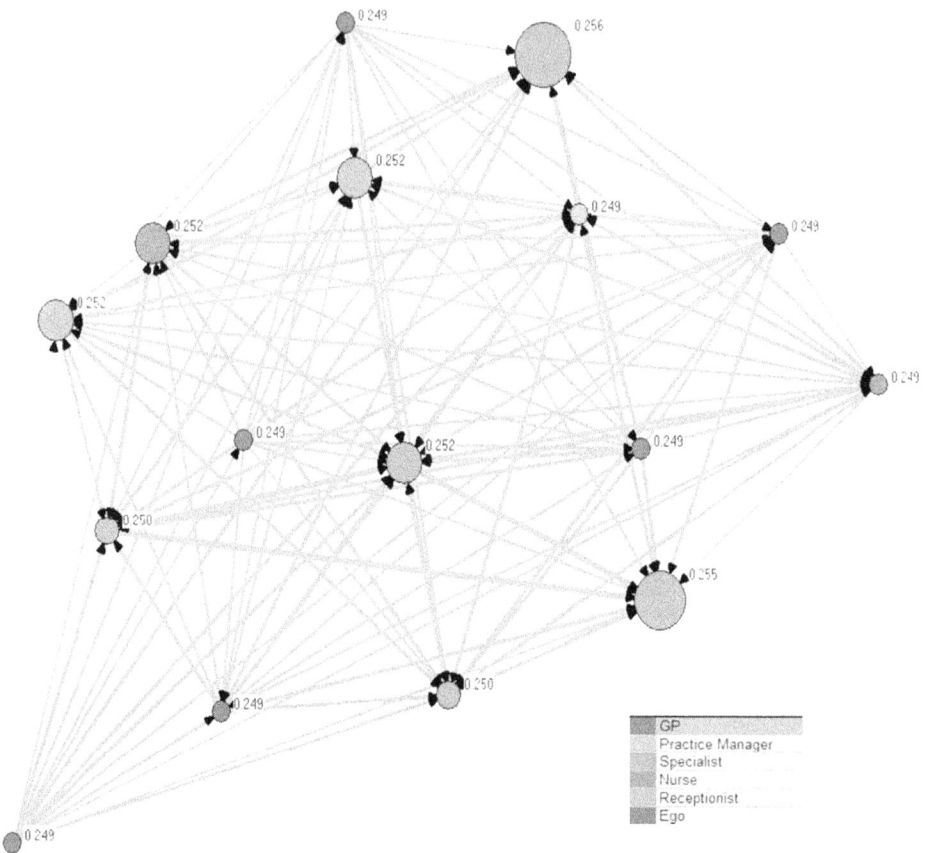

FIGURE 18.4 Professional network of respondent. Larger node size = larger constraint within network.

SAMPLE CASE STUDY 1: THEORY AND APPLICATION OF SNS IN PROJECT TEAM ENGAGEMENT ANALYSIS

In this case study (Chung, Du, and Crawford 2017), I demonstrate how theories and analytics in SN offer a rather useful and rich insight for understanding stakeholder engagement from a bottom-up, organic perspective. I make several assertions about traditional models for stakeholder analysis (Mitchell, Agle, and Wood 1997, Eskerod and Jepsen 2013):

i. They are often not capable of identifying influential stakeholders from the viewpoint of all other stakeholders.
ii. Its method of assessment as to which stakeholders are helpful or harmful are often static in nature and not aligned with the dynamic nature of stakeholder relationships which influences each other's attitudes.
iii. Traditional models of stakeholder prioritisation fail to focus attention on stakeholders who can truly make the project go or stop or who can influence the direction of the project significantly.

I then propose an improvised network model for stakeholder analysis (Chung and Crawford 2016) originally inspired by Rowley (1997).

Rowley suggests the need for moving beyond the dyadic ties analysis that was recurrent in most of the contemporary stakeholder management approaches. He mapped multiple and interdependent interactions that simultaneously exist in stakeholder environments, thus holistically capturing the complex nature of stakeholder interactions for both the focal organisation and its stakeholders, and its stakeholders' stakeholders. He theorised that how stakeholders affect the focal organisation and how the focal organisation responds to these influences depends on the network of stakeholders surrounding the relationship. To do this, he used the notion of density and (betweenness) centrality as key factors for stakeholder analysis.

While stakeholder network density indicates the nature of coalitions or shared behaviour, thus increasing the power of stakeholders to pressure or govern expectations of the focal organisation, centrality of the focal organisation (CFO) confers power, in its ability to resist stakeholder pressures. In effect, Rowley (1997) proposes a four-way structural classification of stakeholder influences accounting for organisational responses to stakeholder pressures, shown in Table 18.2.

- **Compromiser:** When density of the stakeholder network (DSM) and the CFO is high, it means that the high DSM facilitates stakeholder communication and coordination to form an influential collective force. However, because the CFO is also high, it can influence the formation of expectations. Therefore, the strategy here would be to pacify and balance expectations with a view to create win-win situations.
- **Commander:** When DSM is low, it means that stakeholders are rather sparse or isolated leaving them in a position where they do not communicate or collaborate with each other to form a coalition. Coupled against the high CFO, it means the focal organisation is now in a commanding position to stipulate expectations and exercise high levels of discretion.

TABLE 18.2
Structural Classification of Stakeholder Influence (Rowley 1997)

DSM		CFO	
		High	*Low*
	High	Compromiser	Subordinate
	Low	Commander	Solitarian

- **Subordinate:** The reverse of the "commander" scenario applies here as the CFO is low and the DSM is high. This means stakeholders enjoy a power advantage and have higher access to information flows, leaving the focal organisation no choice but to accede stakeholder expectations and pressure.
- **Solitarian:** In this scenario, there is low CFO and low DSM. Neither the focal organisation nor its stakeholders are well connected, and therefore, the power difference remains trivial. Information flow is impeded in such a scenario.

The context of this study is a small ICT company (telco) based in Tasmania, Australia, which was established in 2008 when the founders saw a gap in the market for quality Internet services provision. There was a need for more pragmatic, service-focused providers who were willing to partner and grow with businesses and take on a role of a trusted business advisor. Since then, the local telco grew significantly in operations and now works with over 100 small and medium-sized enterprises (SMEs) in Australia across all industries. Employing 32 employees at the time of the study, the telco's primary objective is to maintain the upmost levels of service for their customers and strive to place the local telco company at the forefront of Internet and Cloud Services within the ICT industry. This telco was selected as they were in the process of announcing company structure changes that would impact all the teams. The deployment of the structural changes provided an opportunity for the research study to review the pre- and post-deployment impacts to change management in projects. We will address the telco firm as "ACME telco".

DEMOGRAPHIC AND SN DATA COLLECTION

Ethics application was successfully obtained for this study. All 32 employees including team members, team leaders and business unit managers in ACME telco were invited to participate in an online survey and closed a month after. With support from the top management, a total of 27 employees responded achieving a response rate of 84%.

Demographic items in the survey included gender, birth year, highest level of education, role in ACME telco, years worked and the department they belonged to. Respondents were also asked to consider the most recent project they were involved in and to rate themselves on a scale of 1 (low) to 5 (high) their degree of (1) influence and (2) interest in the project as a stakeholder.

The second section of the survey pertained to SN data. As the entire list of the employee names was available, a sociocentric approach (Chung, Hossain, and Davis 2005) was utilised where each respondent was asked if they had communicated with the other employees in the list. Using the following name generator, a communication network of each respondent was elicited:

> Looking over the past three months, please tell us who you have communicated with for work-related matters.

Respondents could then choose the name of the person they had communicated with, followed by another set of questions, which elicited the strength and nature of the relationship. This includes frequency of the communication (quarterly to daily) and emotional closeness (ranging from "not close at all" to "very close"). Respondents were then asked to rate the degree to which each person they nominated had influence and interest in the project as a stakeholder. Unlike the self-reported rating of influence and interest described above, this provides a measure of the person's stakeholder influence and interest from the perspective of others.

As each respondent completed their survey, we could obtain a whole communication network of the organisation. Even though there were five respondents who did not participate in the survey, others have nominated them during the name-generator component of the communication network question. Therefore, all the 32 employees appeared in the network for analysis, and for confidentiality purposes, their names were converted to unique IDs that are non-identifiable.

SN Measures: Betweenness Centrality and Ego Density

To operationalise the model proposed above, we used betweenness centrality and ego density, mathematical details of which are available in Chung and Crawford (2016).

It is sufficient here to recall that betweenness centrality measures the extent to which a node (person) lies in between the shortest path of all other nodes (persons) in the network. Mathematically, it is expressed as the ratio of the number of shortest paths between two nodes passing through a particular node over the total number of shortest paths from one node to the other. High betweenness centrality means more information will flow through that node. Hence, it will have more control over the network and more likely to be the information broker or bottleneck of the network.

Density or ego density in this case is calculated as the ratio of actual number of ties over maximum possible number of ties in the network. The higher the density, the more members in the network connecting with each other. From egocentric perspective, ego density represents how dense other nodes that one specific ego communicates with are connected. An ego density of 1 means all members that the ego communicates with are connected with each other, which forms a clique.

Using median scores from both variables, which were plotted on a two-dimensional graph (x and y axes), a grid can be obtained so that each team member can be plotted on a profile.

STAKEHOLDER ASSESSMENT MEASURE: INTEREST AND INFLUENCE

To operationalize *interest* and *influence*, we considered the average of all ratings from others rather than the self-reported ratings to keep the measures of interest and influence objective. The ranges of x and y axes were then determined by the minimums and maximums of the variables – interest and power. Since both samples are normally distributed, means of interest and power rated by others were chosen to be the cut points for x and y axes, respectively, in order to produce the grids in the graph that depicts the PM stakeholder analysis model.

RESULTS: PM MODEL VERSUS THE NETWORKS MODEL FOR STAKEHOLDER IDENTIFICATION AND ENGAGEMENT ANALYSIS

For the power (influence) and interest grid (see Figure 18.5), the mean scores for interest and power are 3.73 and 3.36, respectively (scale out of 5). Results are summarised as follows:

- Thirteen out of 32 members are suggested to be "managed closely" (IDs 3, 4, 5, 6, 8, 9, 13, 15, 24, 28, 29, 30, 31), indicating both high power and interest as stakeholders of the project.
- Five are rated to be "kept satisfied" (IDs 1, 7, 14, 19, 20), which means they are rated to have high power but less interest as stakeholders in the project.
- Three are classified to be kept informed (IDs 2, 27, 23).
- Eleven of all members should be just monitored (IDs 10, 11, 12, 16, 18, 21, 22, 25, 26, 32, 33) according to the grid.

In addition, interest and power are very significantly and strongly positively correlated according to the data ($r = 0.888$, $p < 0.01$), which means in the opinions of majority employees in the company, one person rated with high interest will most likely also be rated to have high power/influence in the project.

On the other hand, according to the betweenness centrality and ego density grid (see Figure 18.6), the medians for centrality and ego density are 0.192 and 0.649, respectively (out of 1). Among all 32 employees,

- No one is classified as a *compromiser* or *solitarian*, who has both high or low centrality and ego density, respectively.

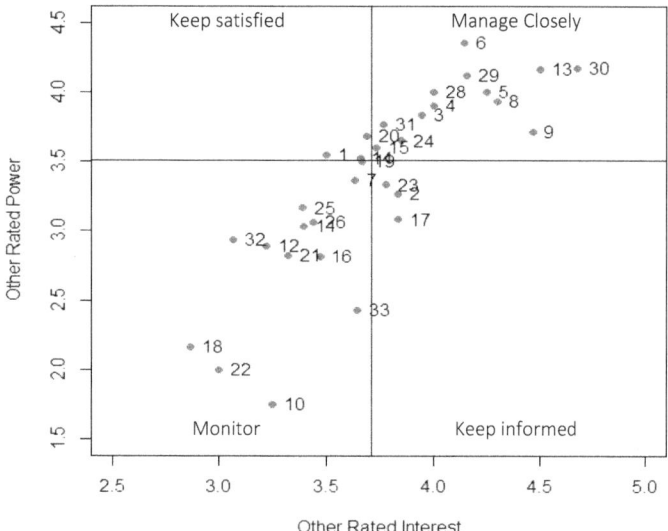

FIGURE 18.5   Power (influence) and interest grid (PM model).

- Sixteen are *commanders* in the communication network structure (IDs 2, 3, 4, 5, 7, 8, 9, 12, 14, 15, 16, 18, 21, 23, 24, 26) indicating their ability to resist pressure from other stakeholders and their ability to execute or broker information (Chung and Crawford 2015)
- The remaining sixteen are classified as *subordinates* in the communication network (IDs 1, 6, 10, 11, 13, 17, 19, 20, 22, 25, 28, 29, 30, 31, 32, 33), indicating their relatively high probability to form a coalition but the inability to influence the wider members of the network because of their low individual centrality scores.

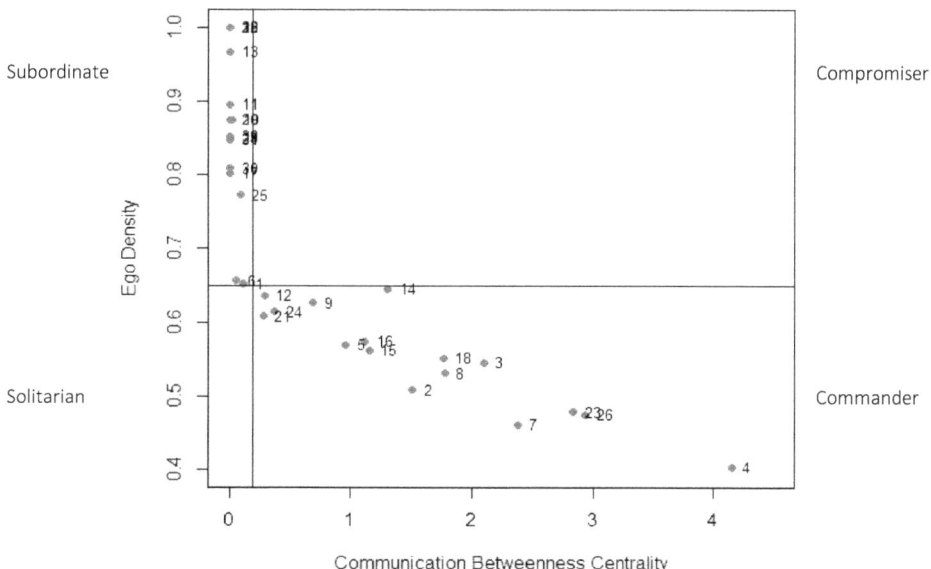

FIGURE 18.6   Betweenness centrality and ego density grid (SN model).

There is a very significant and strong negative correlation between betweenness centrality and ego density ($r = -0.943$, $p < 0.01$), and this validates the nature of the SN property where it is expected.

Among the four groups in the power and interest grid (PM model), members that need to be "managed closely" are more likely to have significant impact on the project, in contrast to members that are being 'monitored', who have both low interest and power. Similarly, in the centrality and ego density grid (SN model), information is more likely to flow through members with high betweenness centrality; hence, those who have higher centrality in the network are probably more influential.

After comparison, it is obvious that there are differences between two grids. For instance, employees 6, 13, 29, 30 and 31 who have been rated to have both high power and interest and are suggested to be 'managed closely' according to the power and interest grid have been grouped into subordinates based on the betweenness centrality and ego density grid. Another significant difference is that employees 12, 16, 18, 21 and 26 who have been classified to be in a commanding position only need to 'monitored' in the power and interest grid.

In order to visualise the changes of positions between two grids, some of the members with significant shifts who have been mentioned above are depicted in the chart (see Figure 18.7) showing key movements between the two different classifications of the PM model and the SN model. In fact, only a few number of all 32 members have shown their consistent importance or less in impact on the project in both grids. Hence, it is reasonable to deduce that managing stakeholders simply using classification models such as the PM models or using SN models only cannot summarise the complex diversity of stakeholders in the context of a real organisation; rather, a more accurate and comprehensive model needs to be proposed to manage stakeholders appropriately.

To improve the previous two models, Chung and Crawford's model (2016) combines power and interest grid with the SN model, which enables the visualisation of stakeholder management strategies suggested by power and interest grid, attribute data like roles as well as the relationships between stakeholders. The model utilises different colours of nodes indicating management strategies, sizes of nodes indicating betweenness centrality and shapes indicating organizations that stakeholders belong to. In addition, tie strength between stakeholders is represented by the thickness of the lines. Combining with SN, relationships between stakeholders can be understood better and easier to propose proper management approaches.

We do not contend that one model is superior over the other. As stated above, it is useful to form an amalgamation of both models as there are clear advantages in doing so.

Firstly, the PM model is useful at the outset of the project life cycle – particularly in planning phases. The SN model would not be useful – particularly in the instance of communication networks, where much communication of the project has not occurred as yet. Where the SN model would be useful would be in the project execution phase, where much communication has occurred, and team members have settled in and have worked on significant phases of the project.

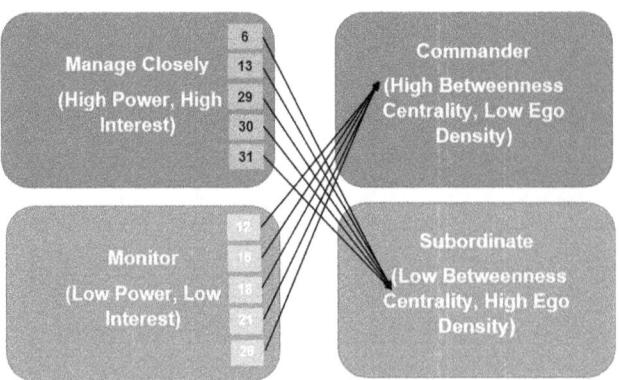

FIGURE 18.7   Key movements highlighted between classifications in the PM model and the SN model.

Secondly, the PM model is useful to navigate the organisational hierarchy. It does help to identify those who hold high organisational authority, salience, currency and influence in the project. These could be stakeholders who literally hold the highest stakes – ones who could stop or continue to fund the project. The PM model, however, is less useful when one needs to identify how influence occurs, or how work really gets done, or in identifying who the real commanders and leading groups are within the organisation in an informal sense. This is where the SN model can address each of these areas.

Combining both models together with the SN metrics that allow for statistical testing and association with project team and project outcomes appear to be extremely promising, because by doing so, one gets a complete picture of stakeholder identification, engagement and analysis in both the formal and informal networks.

SAMPLE CASE STUDY 2: USING SNA AS A VISUAL EVALUATION TOOL

During 2014–2017, integration of care has been considered a major priority in Australia. The NSW Integrated Care Strategy is a state-wide ministry of health initiative that has been locally interpreted. This case study examines the integration of care within a particular local health district (LHD) in NSW (Fares et al. 2018). The project, from the government's perspective, aimed to improve the integration of care between all health services within a defined geographical area, in order to improve patient experience and outcomes, as well as to reduce duplication of services and improve efficiencies.

The evaluation of whether integration within this LHD area, however, was being explored at the time. Since the success of this integrated care project was determined by how different stakeholder groups, such as health care services and providers, perceive integration, therefore, integrated care required a multi-stakeholder and PM approach with the application of SNA.

Therefore, the research sought to use SNA to capture a snapshot of how stakeholders (health care services) interact and provide professional advice to each other after the implementation of an integrated health care project. By exploring how health care services are connected, we aimed to identify (1) areas of strength to be capitalised on and areas of weakness to be improved in the integrated network and (2) key and marginal stakeholders that need to be engaged in order to increase communication and facilitate the integration of services to provide efficient ongoing care for patients. Therefore, it was necessary to explore the actual structure of the stakeholder network in terms of team care collaborations, advice seeking and referrals, the key stakeholders that play an influential role and whether there was an inherent nature in the relationship between social-professional networks and health care integration.

The research team identified 68 health care services within the LHD to be included in the study. An email that carried information on the integrated care project was sent to all the services identified. Out of the 68 services priory identified, 53 services consented for their name to be on the list in the SN survey. This list included the most relevant health care services that provided mental and physical services.

STUDY DESIGN

Through interviews, a sociocentric questionnaire was provided, listing all health care providers (HCPs) within the LHD (who consented), and the respondents were asked which HCP they made referrals to, had team care arrangements with and sought advice from. The provider provided information orally on the advice network relationship, and the researchers completed the survey instrument with this information. Each health service was represented by one or two providers (clinicians or professionals) working within the service. Demographic data collected included questions such as the name of health care services, whether the service provides physical or mental health and the number of years/months in current position.

As an example for the advice network, the name-generator question was:

Advice to: Please identify those services who have given you advice related to your work in the last 6 month, then identify the frequency of interaction.

The study also measured perceived integrated care from the respondent's perspective using the Rainbow Model (Valentjin et al. 2013) for measuring integration. The integrated care instrument consisted of 29 questions and a seven-point Likert scale ranging from strongly agree to not applicable.

Out of the 53 services that consented for their name to be on the list of services, 49 services participated in the study. The number of providers interviewed was 56.

Using Netdraw and UCINet, which are one of the most widely used tools in SNA (others include Pajek, Gephi, NetMiner, and ORA), the following sociogram was produced:

In Figure 18.8, the density of the entire network is 0.13, which means it is relatively sparse and centralisation being 0.63 indicates that the density of the network revolves around 63% of particular HCPs. A visual investigation of the sociogram shows that community-based services, outreach services and hospital-based services seem to be at the heart of the advice network within the LHD. The size of the nodes represents out-degree centrality. The larger the size, the more frequent the stakeholder provides advice. On the other hand, betweenness centrality (calculated by UCINet but not shown in the sociogram) measures the extent to which an actor lies on the shortest path and has a brokerage position between other nodes in the network. These two centrality measures are used to identify the top five stakeholders (Table 18.3) that are responsible for advice sharing and are considered to be the most influential within the LHD.

As noted above, SNA provides a very useful visual as to the mapping of the structure of HCPs within the LHDs. Without such powerful visualisations, the richness of interactions such as advice seeking and provision cannot be captured. With sociograms produced from respondents who are at the coal face of actual health care provision, the network is thus organic, realistic and truly formed bottom-up. Although networks are dynamic and changing, such a visual provides a very insightful snapshot and a clear X-ray into the community organisation at large. At the time of writing, updated data showing the advice network of the HCPs within the LHD was not available; otherwise, a useful comparison of before and after would be highly ideal.

For illustrative purposes, consider the work of Cross, Borgatti and Parker (2002). They researched into the advice provision network of a boutique consultancy firm that pride itself in the provision of thought leadership through technology-driven and business-focused solutions. There was clear

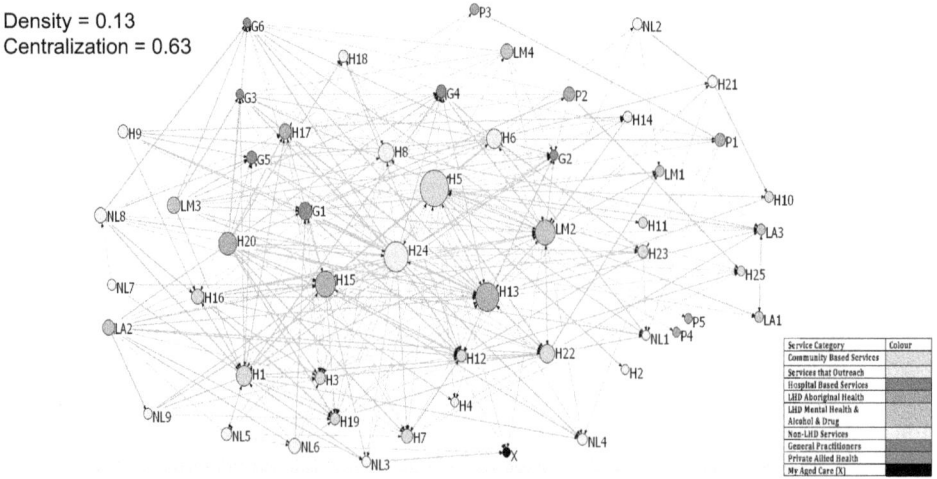

FIGURE 18.8 Sociogram showing advice network of HCPs in the integration care project.

TABLE 18.3
Centrality Measures for Key HCPs

Stakeholder ID	Stakeholder Group	Out-Degree Centrality	Betweenness Centrality
H5	Community-based services	39	540
H24	Services that outreach	29	63
H13	Hospital-based service	28	501
H15	Hospital-based service	23	196
LM2	LHD mental health and drug and alcohol	21	270

synergy of adept business analysts and expert technical specialists. What the researchers noted was that the very reason for which the consultancy was formed was in fact, also the reason why the firm was segregated into two distinct coalitions (Figure 18.9).

As per Cross et al. (2002),

> The group on the left side of the network was skilled in the "softer" issues of strategy or organizational design, often focusing on cultural interventions or other aspects of organizations to help improve knowledge creation and sharing. The group on the right was composed of people skilled in "harder" technical aspects of knowledge management, such as information architecture, modeling, and data warehousing.
>
> Over time, members of these two sub-groups had gravitated to each other based on common interests. These people often worked on projects together and just as importantly shared common work-related interests in terms of what they read, conference attendance, and working groups within the organization. The problem was that each sub-group had grown to a point of not knowing what people in the other sub-group could do in a consulting engagement or how to think about involving them in their projects. Thus, even when there were opportunities in client engagements to incorporate each other's skill sets, this was often not done because neither group knew what the other knew or how to apply their skill sets to new opportunities. This was despite the fact that the group's strategic charter was to integrate these unique skill sets and that all aspects of formal organizational design supported this mission (e.g., reporting structure, common performance metrics and incentives).

As a result, an intervention took place consisting of the numerous strategies. Firstly, new projects would require the representation of both sides and they had to be jointly staffed. Secondly, remuneration via sales goals was modified such that managers from each group would be accountable to sell projects that included both technical and organisational aspects. Thirdly, communication

Pre-Intervention

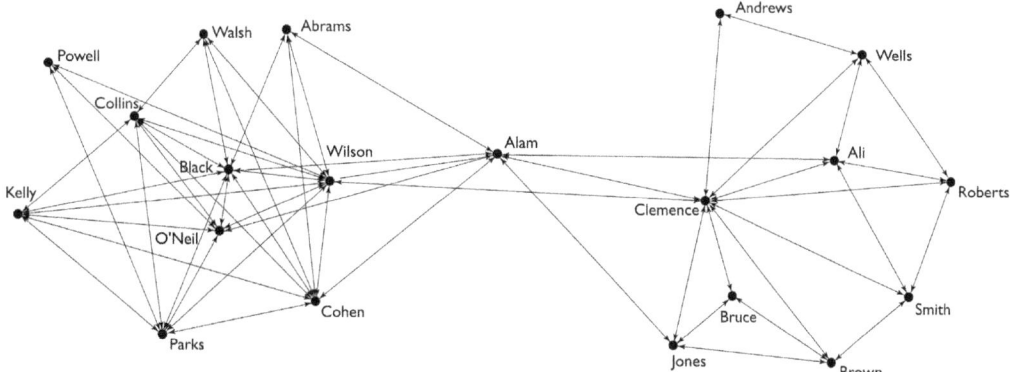

FIGURE 18.9 Information sharing network (pre-intervention) (Cross, Borgatti and Parker 2002).

modes were changed. These included weekly status calls, weekly brief email updates, and a project-tracking database that allowed members of both groups to be updated as to what each group was working on. After 9 months, the information sharing became a lot more cohesive and the overall project-based organisation achieved high-performance efficiency (Figure 18.10).

TOOLS AND TECHNOLOGIES FOR SNA IN PROJECT WORK

For beginners, perhaps the easiest and most accessible tool for SNA is UCINet (Borgatti, Everett, and Freeman 2002). Developed first at the University of California Irvine, by professors Stephen Borgatti, Martin Everett and the late Linton Freeman, UCINet is fairly robust in terms of analysing SN data and has a comprehensive library of SN analytics for evaluating network properties at the network, node and tie levels.

UCINet is also packaged with another tool called "Netdraw" (a freeware) that allows one to visualise sociograms. Netdraw allows the customisation of the nodes, ties and the location of each node in through the use of colours, sizes, labels and symbols in order to better represent the sociogram for meaningful inferences. It also contains a suite of visualisation algorithms that allow the representation of the network based on a number of criteria (e.g. clustering of nodes based on where the bulk of connections lie). At the analytic level, Netdraw offers calculation of network metrics such as density, centralisation, centrality, structural holes, factions, cliques, keyplayer, cut points and blocks.

The drawback of UCINet and Netdraw is that they were programmed in the late 1990s for the Windows environment and is available in 32-bit and 64-bit versions. As a result, it cannot handle large datasets (e.g. with over millions of nodes) and runs only on Windows operating systems (OS). Non-Windows operators such as Mac users are able to use UCINet and Netdraw by installing third-party applications such as Wine (better known as a "compatibility layer") or by installing Windows OS on the Mac through third-party emulation software such as Parallels and VMWare.

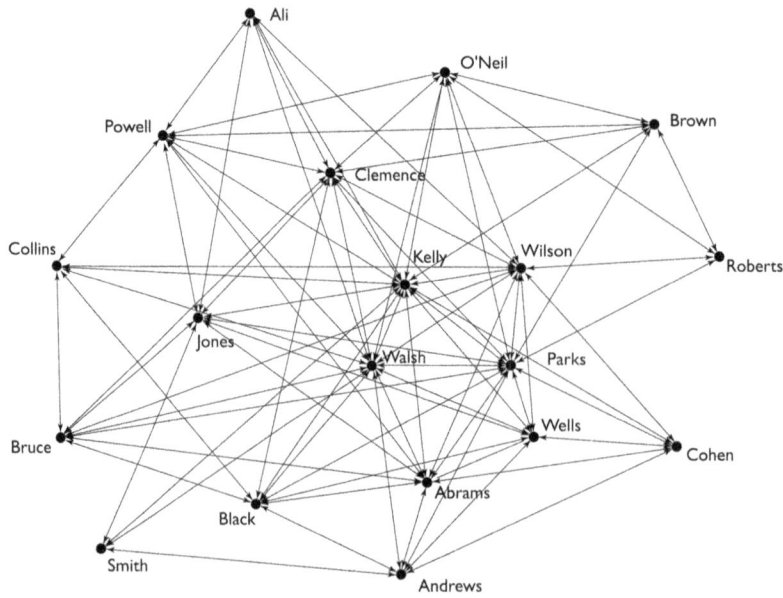

Post-Intervention (Nine Months Later)

FIGURE 18.10   Information sharing network (post-intervention) (Cross, Borgatti and Parker 2002).

Gephi is another freeware for SNA that is often hailed as the "Photoshop of social network maps". It also has a suite of SN analytics but is often used to display professionally and aesthetically pleasing sociograms.

For SNA of huge datasets (e.g. networks with millions of nodes and ties), Pajek is a good recommendation. While there are a number of other noteworthy tools for SN data visualisation and analysis such as NetMiner, ORA, NodeXL and SN Visualizer, there are also a number of technology companies that specialise in the planning, designing, collection and analysis of social or organisational network data.

To cite a few examples, consider Onasurveys.com.au and Polinode.com. Both platforms focus on organisational network analysis (ONA) and is particularly useful where a defined set of actors are possible – i.e. a whole network approach to data collection. For instance, when planning for relocation of physical office space, a network analysis of who collaborates with whom is useful in order to plan and design physical co-location and collaborative spaces. Here, the entire occupants of the building are surveyed, and so, the respondents are a priori known. As another example, a project team could be surveyed to understand who sought advice from whom in order to produce a network map of informal leaders that is otherwise not gleanable from the project organisational chart.

Similar to Onasurveys.com.au, SWOOP Analytics (swoopanalytics.com) is also an ONA platform that works seamlessly with Yammer and Workplace (organisational social media platform developed by Microsoft and Facebook, respectively) networks. Given that most large organisations use Yammer or Workplace for collaboration and work, Swoop Analytics provides numerous profiles based on network analytics. The categories in the profile include Engagers, Responders, Observers, Broadcasters and Catalysts. While the intention of the profile categories is not to name and shame, the idea is that project leaders and managers can use this as a framework within which to operate, while showing indicators and guidelines for what success looks like in a collaborative project work. Project workers can then decide what it is that they want to achieve within the project context and how they want to collaborate.

CLASSROOM DISCUSSION QUESTIONS AND ACTIVITIES

1. Consider the meaning of social media as opposed to SNs. How are they different? In particular, what does one mean by the study of SNs?
2. Imagine you are trying to explain the value of network literacy to your project sponsor. You are convinced that you can use it to evaluate the collaborative nature of the project team. What are some of the key concepts you would discuss? Can you explain the most commonly used concepts at the network level, actor level and relational level?
3. **Making sense of stakeholder networks:** In a recent capital project, the question was asked to project team members: *Select the key stakeholders (max 5–10) in relation to Communication and Change Management that you regularly collaborate with in relation to your Capital Project.*

   Collaboration includes helping one think through difficult problems, sharing of insights and experience, providing approvals for activities to go ahead and coordination of project resources.

   The six sociograms shown in Figure 18.11 were then drawn up based on the team members' response.

QUESTIONS

1. Explain your observations of what is going on in each of the six sociograms.
2. Which project network is collaborating most well?
3. What would you do to intervene in each of the six networks so that project team collaboration may be improved?

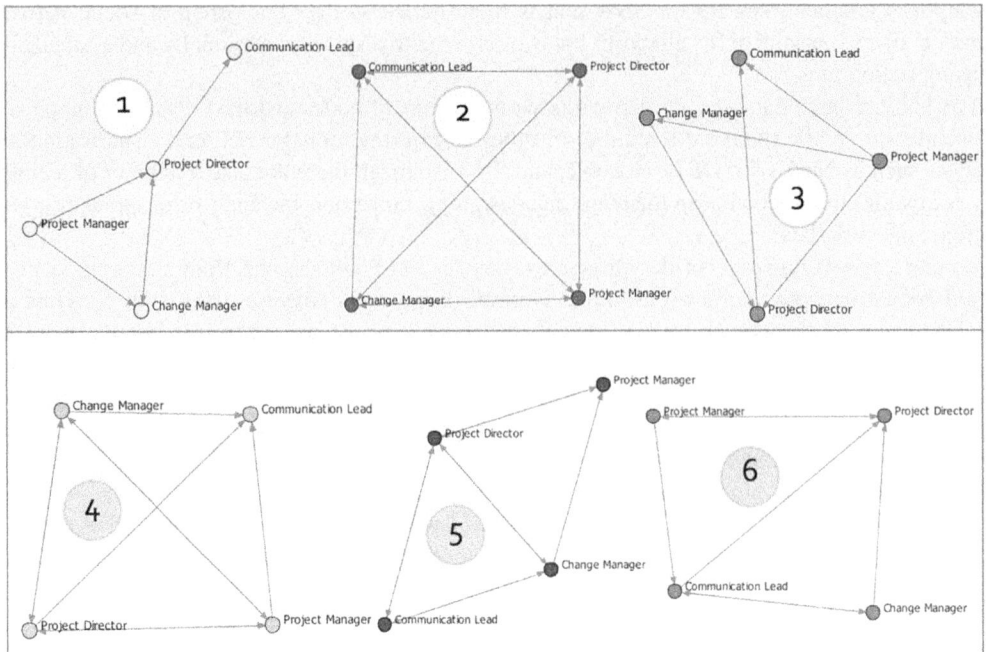

FIGURE 18.11   Six structures of stakeholder networks (courtesy of Cai Kjaer from Swoop Analytics.)

CONCLUSION

In this chapter, I have introduced the notion of SN thinking as a toolset for complex PM. SN theory and methodology both offer a unique but very powerful lens for understanding project-based behaviour at the network, individual and relational level.

SN studies are increasingly rapid in the academic field; however, the level of literacy among project professionals and practitioners is still relatively low. It is envisaged that this book will help lift the level of network literacy and the application of networking thinking in project environments.

While only introductory notions and concepts of networks are presented in this chapter, the examples are directly relevant to project scenarios – such as stakeholder engagement and analysis and project evaluation. As long as relational data exists, there will always be structural analysis possible and this is where networks prove useful.

As noted by Turner and Müller (2003), project organisations are indeed network organisations – and the growth and application of network thinking in project networks will only continue to increase (Manning 2017). In a world of complex megaprojects and the management of such projects, consider the importance of understanding the interrelatedness and the interdependence of many parts within the project in order to fully understand the whole. Imagining being able to visualise the shifting sentiments of the community as a key stakeholder and being able to understand how key stakeholders shift from a help to harm paradigm, would not the network literate project leader and manager understand then the best way to intervene? That is the power of understanding and harnessing invisible networks – the dynamic and shifting relationships which are so complex, often discounted and ignored.

REFERENCES

Anichenko, E, KSK Chung, and L Crawford. 2016. Formal organisational networks and informal project networks: Implications for project performance. *AIPM Conference 2016*, Hilton Sydney, Sydney, 16–19 October 2016.

Bavelas, A. 1950. Communication patterns in task-oriented groups. *Journal of Acoustical Society of America* 22 (6):725–730.

Bernard, HR, PD Killworth, and L Sailer. 1982. Informant accuracy in social-network data V. An experimental attempt to predict actual communication from recall data. *Social Science Research* 11:30–66.

Bhagat, S, M Burke, C Diuk, IO Filiz, and S Edunov. 2016. Three and a half degrees of separation. Accessed 12th February 2018.

Borgatti, S, and X Li. 2009. On social network analysis in a supply chain context. *Journal of Supply Chain Management* 45 (2):5–22.

Borgatti SP, MG Everett, and LC Freeman. 2002. *Ucinet for Windows: Software for Social Network Analysis.* Analytic Technologies, Harvard, MA.

Breiger, RL. 2004. The analysis of social networks. In *Handbook of Data Analysis*, edited by M Hardy and A Bryman, 505–526. London: Sage Publications.

Burt, RS. 1992. *Structural Holes: The Social Structure of Competition.* Cambridge, MA: Harvard University Press.

Burt, RS. 1997. A note of social capital and network content. *Social Networks* 19:355–373.

Burt, RS. 2000. Structural holes versus network closure. In *Social Capital: Theory and Research*, edited by N Lin, KS Cook and RS Burt, 31–56. New York: Aldine de Gruyter.

Burt, RS. 2005. *Brokerage and Closure: An Introduction to Social Capital.* New York: Oxford University Press.

Chinowsky, PS, J Diekmann, and J O' Brien. 2010. Project organizations as social networks. *Journal of Construction Engineering and Management* 136 (4):452–458.

Chung, KSK. 2011. *Understanding Attitudes to Performance in Knowledge-Intensive Work: The Influence of Social Networks and Information and Communication Technologies Use.* Saarbrücken: Lambert Academic Publishing.

Chung, KSK, and L Crawford. 2015. The role of social networks theory and methodology for project stakeholder management. *29th World Congress International Project Management Association (IPMA) 2015*, IPMA WC, Westin Playa Bonita, Panama City, 28–30 September.

Chung, KSK, and L Crawford. 2016. The role of social networks theory and methodology for project stakeholder management. *Procedia - Social and Behavioral Sciences* 226 (2016):372–380.

Chung, KSK, X Du, and L Crawford. 2017. Improving stakeholder engagement: Looking at the unseen. IRNOP, Boston, MA, 11–14 June 2017.

Chung, KSK, and L Hossain. 2009. Measuring performance of knowledge-intensive workgroups through social networks. *Project Management Journal* 40 (2):34–58.

Chung, KSK, L Hossain, and J Davis. 2005. Exploring sociocentric and egocentric approaches for social network analysis. *International Conference on Knowledge Management Asia Pacific*, Victoria University Wellington, Wellington, 27–29 November.

Cross, R, SP. Borgatti, and A Parker. 2002. Making invisible work visible: Using social network analysis to support strategic collaboration. *California Management Review* 44 (2):25–46.

de Montjoye, Y-A, A Stopczynski, E Shmueli, A Pentland, and S Lehmann. 2014. The strength of the strongest ties in collaborative problem solving. *Scientific Reports (Nature Publishing Group)* 4 (5227):6. doi:10.1038/srep05277.

Eskerod, P, and AL Jepsen. 2013. *Project Stakeholder Management.* 1 ed. New York: Routledge Taylor and Francis Group.

Fares, J, KSK Chung, M Passey, J Longman, and P Valentijn. 2018. Analysing stakeholder advice networks: An Australian integrated healthcare project. Project management research and practice. *Project Management Institute Australia Conference 2017*, UTS ePress, Sydney: NSW, pp. 1–14.

Frank, O, and D Strauss. 1986. Markov graphs. *Journal of the American Statistical Association* 81 (395):832–842.

Freeman, LC. 1978. Centrality in social networks: Conceptual clarification. *Social Networks* 1 (3):215–239.

Freeman, LC. 2004. *The Development of Social Network Analysis: A Study in the Sociology of Science.* Vancouver BC: Empirical Press.

Garton, L, CA Haythornthwaite, and B Wellman. 1997. Studying online social networks. *Journal of Computer Mediated Communication* 3 (1).

Granovetter, MS. 1973. The strength of weak ties. *The American Journal of Sociology* 78 (6):1360–1380.

Kilduff, M, and W Tsai. 2003. *Social Networks and Organizations.* London: Sage Publications.

Killen, CP, and C Kjaer. 2012. Understanding project interdependencies: The role of visual representation, culture and process. *International Journal of Project Management* 30 (5):554–566.

Li, Y, Y Lu, YH Kwak, Y Le, and Q He. 2011. Social network analysis and organizational control in complex projects: Construction of EXPO 2010 in China. *Engineering Project Organization Journal* 1 (4):223–237. doi:10.1080/21573727.2011.601453.

Lin, N. 2000. Building a network theory of social capital. In *Social Capital: Theory and Research*, edited by N Lin, KS Cook and RS Burt, 3–29. New York: Aldine de Gruyter.

Manning, S. 2017. The rise of project network organizations: Building core teams and flexible partner pools for interorganizational projects. *Research Policy* 46 (8):1399–1415. doi:10.1016/j.respol.2017.06.005.

McPherson, JM, and L Smith-Lovin. 1987. Homophily in voluntary organizations: Status distance and the composition of face-to-face groups. *The American Journal of Sociology* 52:370–379.

Milgram, S. 1967. The small-world problem. *Psychology Today* 1:62–67.

Mitchell, RK, BR Agle, and DJ Wood. 1997. Towards a theory of stakeholder identification and salience: Definining the principle of who and what really counts. *Academy of Management Review* 22:853–886.

Monaghan, A, and KSK Chung. 2016. Advice, reliance, work and social networks for project success. ANZAM (Australian & New Zealand Academy of Management, QUT Business School, Brisbane (accepted 5 September 2016), 6–9 December 2016.

Pentland, AS. 2012. The new science of building great teams. *Harvard Business Review* 90 (4):60–70.

Pryke, SP. 2017. *Managing Networks in Project-Based Organisations*. Oxford, UK: Wiley-Blackwell.

Putnam, R. 1993. *Making Democracy Work: Civic Traditions in Modern Italy*. Princeton, NJ: Princeton Univesity Press.

Putnam, R. 1995. Bowling along: America's declining social capital. *Journal of Democracy* 6 (1):65–78.

Reagans, R, and B McEvily. 2003. Network structure and knowledge transfer: The effects of cohesion and range. *Administrative Science Quarterly* 48 (2):240–267.

Robins, G, L Bates, and P Pattison. 2011. Network governance and environmental management: Conflict and cooperation. *Public Administration* 89 (4):1293–1313. doi:10.1111/j.1467-9299.2010.01884.x.

Robins, G, P Pattison, Y Kalish, and D Lusher. 2007. An introduction to exponential random graph (p*) models for social networks. *Social Networks* 29:173–191.

Rowley, TJ. 1997. Moving beyond dyadic ties: A network theory of stakeholder influences. *The Academy of Management Review* 22 (4):887–910.

Scott, J. 2000. *Social Network Analysis: A Handbook*. London: Sage Publications.

Toomey, L. 2012. Social networks and project management performance: How do social networks contribute to project management performance? *PMI® Research and Education Conference*, Limerick, Munster.

Tortoriello, M, and D Krackhardt. 2010. Activating cross-boundary knowledge: The role of simmelian ties in the generation of innovations. *Academy of Management Journal* 53 (1):167–181.

Turner, JR, and R Müller. 2003. On the nature of the project as a temporary organization. *International Journal of Project Management* 21 (1):1–8.

Valentjin, PP, SM Shcepman, W OPheij, and MA Bruinjnzeels. 2013. Understanding integrated care: A comprehensive conceptual framework based on the integrative functions of primary care. *International Journal of Integrated Care* 13: 655–679.

Wasserman, S, and K Faust. 1994. Social network analysis: Methods and applications. In *Structural Analysis in the Social Sciences*, edited by M Granovetter, 1–819. New York: Cambridge University Press.

Wasserman, S, and P Pattison. 1996. Logit models and logistic regressions for social networks: I. An introduction to Markov graphs and p*. *Psychometrika* 60:401–426.

Wellman, B. 1926. The school child's choice of companions. *Journal of Educational Research* 14:126–132.

19 Modeling and Simulation Toolset

Sergey Suslov
The AnyLogic Company

Dmitry Katalevsky
Russian Presidential Academy of National Economy
and Public Administration (RANEPA)

CONTENTS

INTRODUCTION

This chapter is devoted to contemporary modeling and simulation techniques applied to complex project management. It is divided into two main parts:

- In the *first part*, we provide a theoretical overview of project management from the perspective of a systemic approach applying casual-loops diagrams, a methodology frequently used to produce a conceptual view of complex systems. We reflect typical causes of frequent project failures. We use a case study of a software development company to illustrate a common behavior when complex project spirals out of control. We provide a brief literature overview of project management simulations studies and give some practical advice for project management practitioners.
- In the *second part*, we provide an overview of three most widespread modeling techniques (system dynamics, agent-based, and discrete event modeling) using some

simple models of project management as examples. The second part of the chapter might be of interest to those who are interested to make their first steps in application of simulations to project modeling. AnyLogic software is used to illustrate case studies.

MODELING AND SIMULATION

Modeling plays a very significant role in our lives, yet people often underestimate how frequently they apply models. By simply calculating the time and cost of getting to the city center from the hotel while traveling, you are making use of a mental model. This model might even include an equation, for example, when you multiply the ticket price by the size of your family. Models can be observed everywhere in our daily life. They can come in a variety of forms, such as mental models, formulas, organizational charts, and physical models of buildings.

Modeling in a digital environment with the help of computer software tools has become a standard in business and engineering. Many vertical software solutions are used for complex project management in various industries. A good example of modeling in the construction industry is building information modeling (BIM). Currently, before any construction work can be started, a digital model of the building is created, using tools, for instance, Autodesk Revit. Such models include all the architectural and engineering aspects of the potential building (Figure 19.1 is an example of such a model).

It simplifies the construction project, because the technical side becomes transparent to all involved professionals (architects, structural engineers, construction professionals) and people who will use the building in future, such as industrial engineers that automate the factory when it is ready, or healthcare researchers who will organize processes in the future hospital.

Why do people create models? First of all, they might create a model because it is easier to experiment with a model than with real-life objects. Real-life experiments are often too expensive or even impossible to do, for example, the planning horizon for the construction of a new seaport will take tens of years. It is physically impossible to test which of the several processes in this port is best if we have several options; thus, a model makes a great alternative.

We provide overview of three the most useful simulation approaches – system dynamics (SD), discrete event simulation (DES) and agent-based modeling (ABM). While system dynamics is based on holistic process description, the ABM system uses a bottom-up approach that is described as interacting objects with their own behaviors. System behavior emerges as a summary of individual agents' actions. The main idea of DES is to consider the system as a process, that is, a sequence of operations being performed across entities. Depending on project' and organizational specifics, one can choose and apply the most useful simulation method case-by-case.

FIGURE 19.1   A building information model of oil refinery in Autodesk Revit software tool.

Unlike real-world projects, the virtual modeling world is completely risk-free. Managers, researchers, and engineers can create and test models of various system designs, and answer hundreds of what-if questions by experimenting virtually in a risk-free environment (Figure 19.2) (Borshchev 2013). Modeling is a very efficient instrument to improve people communication during the initial steps of a complex project, because it allows for objective discussion. There is no need to think that one person's opinion is better than the other's opinion because of differences in experience; managers and engineers can get modeling results to support their opinions. If we model the suggested scenarios, modeling will show which scenario has the lowest technical risks and the net present value (NPV) that fits our budget limitations.

> In this chapter, we provide universal theoretical foundation of complex projects dynamics. Unlike real-world projects, the virtual modeling world is completely risk-free. Managers, researchers and engineers can create and test models of various system designs, and answer hundreds of what-if questions by experimenting virtually in a risk-free environment.

Simulation is unique among other modeling tools and technologies because it allows the user to add dynamics into models. The simulation model evolves over time, discretely or continuously changing its state. To create a simulation model is to define a set of rules that will drive the system over time.

Every simulation is a model, but not every model is a simulation. For example, the most widely used modeling tool is Microsoft Excel spreadsheets. Is a project's budget estimation in Excel a simulation? Definitely not, as it has no time component, it just includes slices of data on particular moments in time.

Simulation modeling is a very powerful instrument that is widely used for solving various business challenges. This is mostly because it is very practical and easy to understand, and there is no need to guess when you can just run a simulation model and see what happens. Animation is a significant advantage that simulation models can give, compared to static models such as those which are Excel-based. When people watch an animation like a video showing where a system, for example, a port is working with given equipment configuration and parameters, they find it easier to visually validate the model and trust its results.

Stochastics is another big advantage of simulation; it is possible to set up any characteristics of the model as a probability distribution with given parameters, let's say a project step duration. Each time the model is executed, a new random value is used as the parameter's value. If the model is executed many times, we can see the model result (e.g., total project completion time)

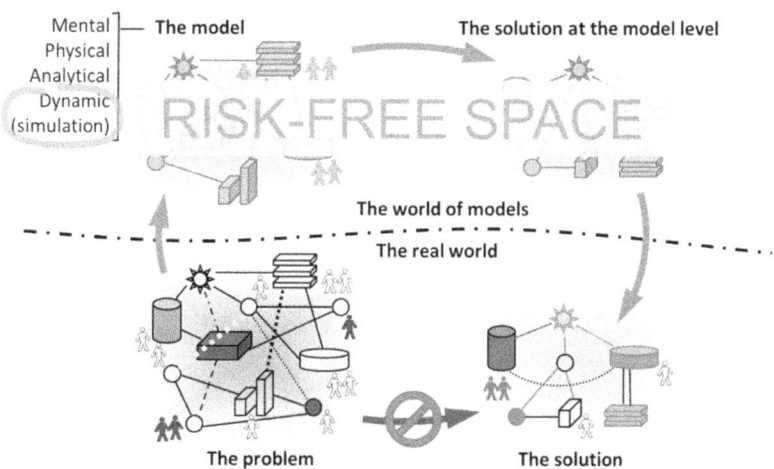

FIGURE 19.2 Risk-free virtual "world" of models.

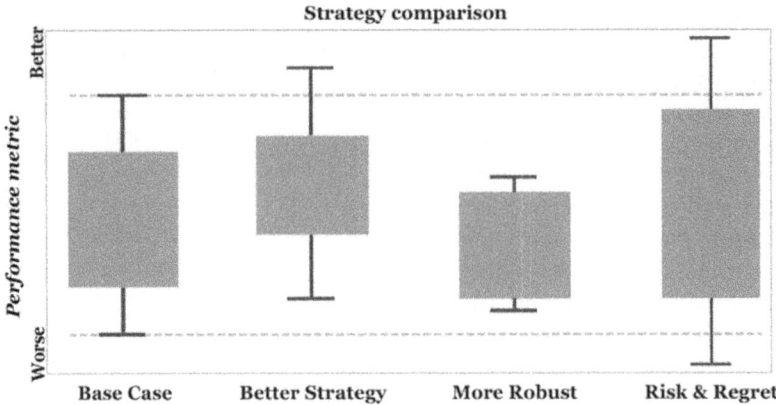

FIGURE 19.3   Box and whisker plot with simulation results of four alternative strategies.

as a stochastic value—create a histogram or box and whisker plot, and calculate mean and deviation. This allows you to take into account deviation, which represents the risks of being over project time or budget.

The box and whisker plot in Figure 19.3 shows the simulation results of four scenarios representing four different strategies. This plot allows you to compare them, taking into account not only average performance metrics (NPV, project duration, etc.) but also a deviation. "Box" represents a range with 50% results; "whisker" represents a whole range of simulation results (The AnyLogic Company 2016).

Time and stochasticity are the two key features which make the simulation model especially effective for use in complex project management. That is why we will continue to write about simulation only.

SIMULATIONS IN PROJECT MANAGEMENT

Simulation science has a *proven track record of studying complex projects*. For instance, the first research in simulation dates back to the 1960s, but only in the 1980s were the first project-specific modeling applications established. Since the 1990s, many publications have appeared, documenting usage of simulation approach in project management. Many studies of project management and simulations tried to grasp major reasons behind project failures and significant cost overruns being typical for this field of practice.

The first systematic studies of Roberts applied simulations (system dynamics, in particular) to project management by introducing flows of project work (job units) and concepts of perception gaps (differences between perceived progress and real progress). Together with underestimation of required scope and effort to complete a project, Roberts pointed out that these project errors were responsible for managements' distorted project perception, which in turn inevitably leads to resource misallocation (Roberts 1964, 1974). The first models produced by Roberts, Kelly referred to research and development projects and studied *perceived* versus *real progress* in projects as well as

Simulation science has a proven track record of studying complex projects. Practitioners of simulations believe that complex feedback structures around managerial perception of projects and managers' decision-making are the elements key to project disruption. Since projects (especially complex) are usually tightly coupled systems, it becomes increasingly important to assess potential ripple effects. Due to multiple interactions of non-linear feedback loops with unintended and counterintuitive consequences and their compound nature, complex projects often spiral out of control.

development of R&D dynamics over time and effects of multi-project management. These were the early attempts to investigate the impact of **managerial decisions** on project execution, based on the assumption that *perception of the project's state may be different from reality.*

Later on, Richardson and Pugh added the concept of Rework, undiscovered Rework, perceived progress, and real productivity (currently the classics of project management models in system dynamics). The model introduced by Richardson and Pugh (1981) concentrated on studying several key domains typical for any complex projects (monitoring and control cycle, Rework generation, and staff hiring).

Based on the achieved progress in project simulations, Pugh-Roberts Associates created the "Program Management Modeling System" (a set of complex system dynamics models used as a project management tool to support managerial decision-making). The tool was successfully applied in a number of management consulting projects for large construction and even dispute resolutions (delay and disruption (D&D) cases). One of the early successes was the Ingalls Shipbuilding case against the U.S. Navy in the 1970s. Ingalls Shipbuilding won a contract to build a number of warships for the Navy in 1969–1970. The contract price was fixed. The project resulted in about USD 500m cost overrun; however, the Navy agreed to reimburse only USD 150m direct costs, blaming the rest of the overruns for Ingalls' mismanagement. Ingalls sued the Navy, claiming that constant design changes caused the D&D. Pugh-Roberts Associates created a complex model, replicating in detail a shipbuilding project to quantify the cost of disruption resulting from delays and design changes from the customer. The case was settled out of court for USD 447m. In an investigation of the case, Cooper identified that about USD 200m–300m of the settlement could be attributed to the model-produced analysis (Cooper 1980).

Many researchers have made other important contributions to the field of simulations in project management over the years, to mention a few:

- Abdel-Hamid (1993) (integrative model of a software management project): explicit incorporation of managerial functions of planning and staffing linked to the process of software development.
- Cooper and PA Consulting (Cooper 1980, 1993a,b,c; Cooper et al. 2002; Cooper and Reichelt 2004): project management as a complex system, non-linear feedback, and quantification of ripple effects on cost overruns.
- Williams et al. (1995): the effects of design changes and delays on project costs (compounding effects).
- Ford and Sterman (1998): multiple phase project model (aging chain structure) explicitly portraying iteration over four distinct development activities to describe average development processes.
- Williams (1999): investigation into what constitutes complexity in project management.
- Graham (2000): insights from system dynamics modeling of complex programs.
- Lyneis et al. (2001): strategic management of complex projects.
- Eden et al. (2005): claim analysis of complex project failures (comparison of system dynamics modeling and "measured mile" methods).
- Taylor and Ford (2006): strategies for managing projects near tipping point.
- For example, Lee et al. (2007) investigated the interaction of resource allocation delays and different amounts of control imposed by managers and made some counter-intuitive conclusions (i.e., because delays are inevitable, optimal delays with minimum timing are preferable).

And many more.

As a result of cumulative studies of project management failures, the simulation practitioners have developed a unified theory of a *typical project management disruption mechanism.* We summarize it below using a case study of a software developing company to illustrate a step-by-step typical project management disruption dynamics.

A CASE STUDY: A MECHANISM OF A PROJECT DISRUPTION

There is an important difference between traditional approaches to project management and simulation modeling (i.e., system dynamics, agent-based modeling (ABM), etc.). In traditional project management tools, critical path method (CPM)-based tools describe a project as a networked sequence of discrete technical tasks and events, and portray a project as the sum of discrete work segments. Such tools and systems can be badly misleading by failing to portray that projects *really do not work* in a straight line of tasks started and ended but in an iterative process of accomplishment. Further, they encourage the view of projects as projectiles, hurtling toward an outcome on which human intervention has little effect (Cooper 1994).

Practitioners of simulations and, in particular, of system dynamics widely believe that complex feedback structures around managerial perception of projects and managers' decision-making are the elements key to project disruption. It's usually our decisions and actions that work through the multiple non-linear cause-and-effect relationships constantly involving delays that affect project execution. In this section, we will review the typical mechanism of project disruption and propose several remedies and recommendations to mitigate negative effects and prevent project failures.

Let's imagine that ABC Company is trying to create a complex software project. We assume that in order to launch the software successfully, ABC Company staff needs to develop about 100 thousand lines of source code.[1] Using typical system dynamics language in VenSim software tool,[2] we can propose a simple structure (Figure 19.4).

Stock **Work to be Done** represents an initial projected work (i.e., 100 thousand code lines) that the ABC management thinks they need to implement to finish the software. The stock **Work Really Done** represents a finished piece of work that does not need any further work. Both **Work to be Done** and **Work Really Done** are measured in the number of program code lines. As the software development progresses, the **Work to be Done** is depleted and **Work Really Done** increases. **Work Being Done** is a flow of work at any given time over the software development period. It can be measured in code lines added per time period, that is, the number of lines of code added per day.

Typically, **People** and **Productivity** directly influence the speed of **Work Being Done** (Figure 19.5). One would assume that usually the more technical staff (software developers) are hired to work on the software and that the more productive they are at adding code per time period, the quicker the stock **Work Really Done** is growing. However, it is reasonable to assume that not all **Work Being Done** is useful and meets our quality criteria. The variable **Quality** should be included into our system, measuring a fraction of **Work Being Done** (between 0 and 1) that is actually entering the pool of **Work Really Done** and does not require any further work. For simplicity reasons, we may assume that **Quality** is linked to the number of generated errors in program code. In reality, there are many factors both internal and external that directly and indirectly influence Quality and Productivity.

As noted previously, a huge advancement in understanding project management dynamics was achieved by adding the concept of a **Rework Cycle** (see Figure 19.7). Errors are detected via the testing process. As errors are detected (depending on the error discovery rate), the Rework is identified, increasing the amount of work remaining. Typically, system dynamics practitioners use the stock **Known Rework** to mark the amount of work that needs to be redone. There is also a notion of **Undiscovered Rework**—the work that contains existing but undiscovered errors. For the purpose of simplification in our chapter, we will operate with the only **Rework** stock in this case. **Rework** is added by **Rework Discovery** flow (in our case, a level of code lines classified as needing a rewrite per time) and depleted by **Rework Done** flow (a level of code lines fixed per time period). **Rework Discovery** can be tricky. Sometimes Rework is quickly discovered; however, this might not

FIGURE 19.4   Main stocks and a flow in the model.

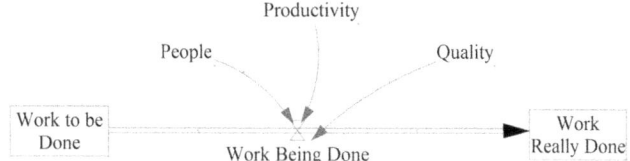

FIGURE 19.5 Influencing variables.

be always the case. The NASA study of flight software complexity shows (NASA Office of Chief Engineer 2009: 47) that many defects are inserted during the design and coding stages of the software development process (see Figure 19.6 for illustration). As Holzmann claims,

> The amount of control software needed to, say, fly a space mission is rapidly approaching a million lines of code. If we go by industry statistics, a really good—albeit expensive—development process can reduce the number of flaws in such code to somewhere in the order of 0.1 residual defects per 1,000 lines. (A residual defect is one that shows up after the code has been fully tested and delivered. The larger total number of defects hiding in the code is often referred to as the latent defects.) Thus, a system with one million lines of code should be expected to experience at least 100 defects while in operation.
>
> *Holzmann (2007)*

However, even **Rework Done** flow might contain errors that will result in another cycle of the Rework.

As Rodrigues and Bowers observe, the Rework cycle identifies four key factors that are partially under management control and should be treated with care: resource level, productivity, quality, and error discovery time (Rodrigues and Bowers 1996). Usually, project managers focus on resources and productivity—in ABC Company's case, people and their programming skills—treating them as fundamental prerequisites for project success or failure. However, the system dynamics experience suggests that it is equally important to monitor both quality and *error discovery rate*.

Lyneis and Ford consider the Rework cycle to be the most important feature of project disruption:

> The rework cycle is, in our opinion, the single most important feature of system dynamics project models. The rework cycle's recurrent nature in which rework generates more rework that generates more rework, etc., creates problematic behaviors that often stretch out over most of a project's duration and are the source of many project management challenges.
>
> *Lyneis and Ford (2007)*

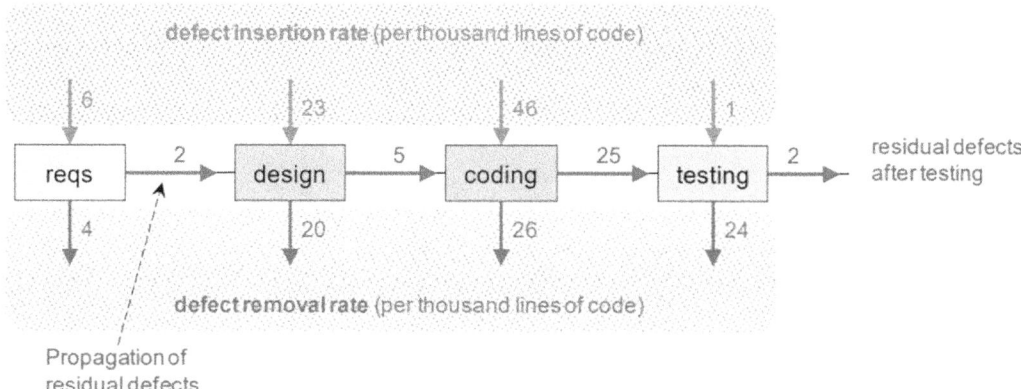

FIGURE 19.6 Defects propagation through various stages of software development (after testing, two defects remain per 1,000 lines of code) (Eick et al. 1992).

Rework cycle can easily be accountable for over 50% of the total time for project development (Cooper 1994) (Figure 19.7).

Further development of the ABC software project development case will add several cause–effect chains and feedback loops describing project lifecycle dynamics.

The management of ABC tries to control the project by evaluating its progress periodically. Primarily, they assess the necessity of getting additional personnel to finish the remaining work (including discovered Rework) on time. **Perceived Progress** translates into **Expected Completion Time**. If **Expected Completion Time** significantly deviates from the **Scheduled Completion Time**, the management needs to take action.

The most common approach reflecting conventional wisdom would be *to add more personnel* to increase the rate of **Work Being Done** (the cycle *Undiscovered Rework → Perceived Progress → Expected Completion Time → Scheduled Completion Time → Time Remaining → Staff needed → Staff Requested → Hiring → Staff on Project* in Figure 19.8).

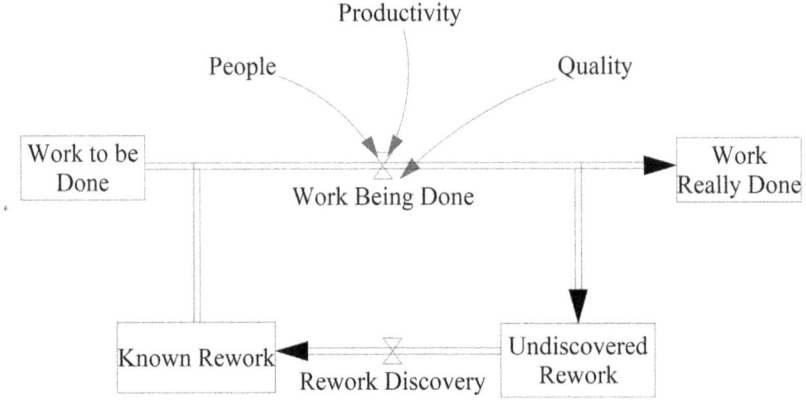

FIGURE 19.7   Typical Rework cycle.

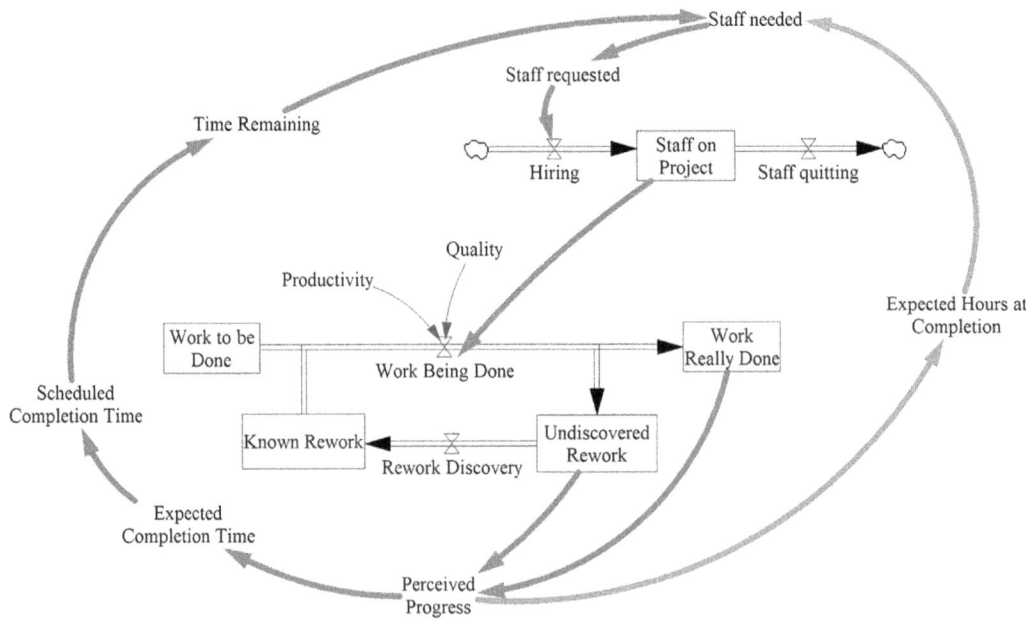

FIGURE 19.8   "We-need-more-staff!" cycle.

If this is not an option—for instance, due to resource constraints or unavailability of candidates with relevant qualifications and skills—the management will respond with a requirement of **extended work hours** for their software development team (overtime). This is the most common option to avoid the additional cost and commitment of bringing in additional people.

Overtime usage quickly becomes the new "normal" for the Company employees. The perceived gap between **Work Really Done** and **Work to be Done** still remains. Constant overtime with some delay effect reduces productivity and increases error rate (fatigue effect). More errors push the Rework cycle further, in turn increasing the amount of work that needs to be done and delaying work completion. The full cycle repeats itself. The **Overtime** feedback loop quickly becomes *self-reinforcing* (Figure 19.9a). Usually, the Overtime cycle and its

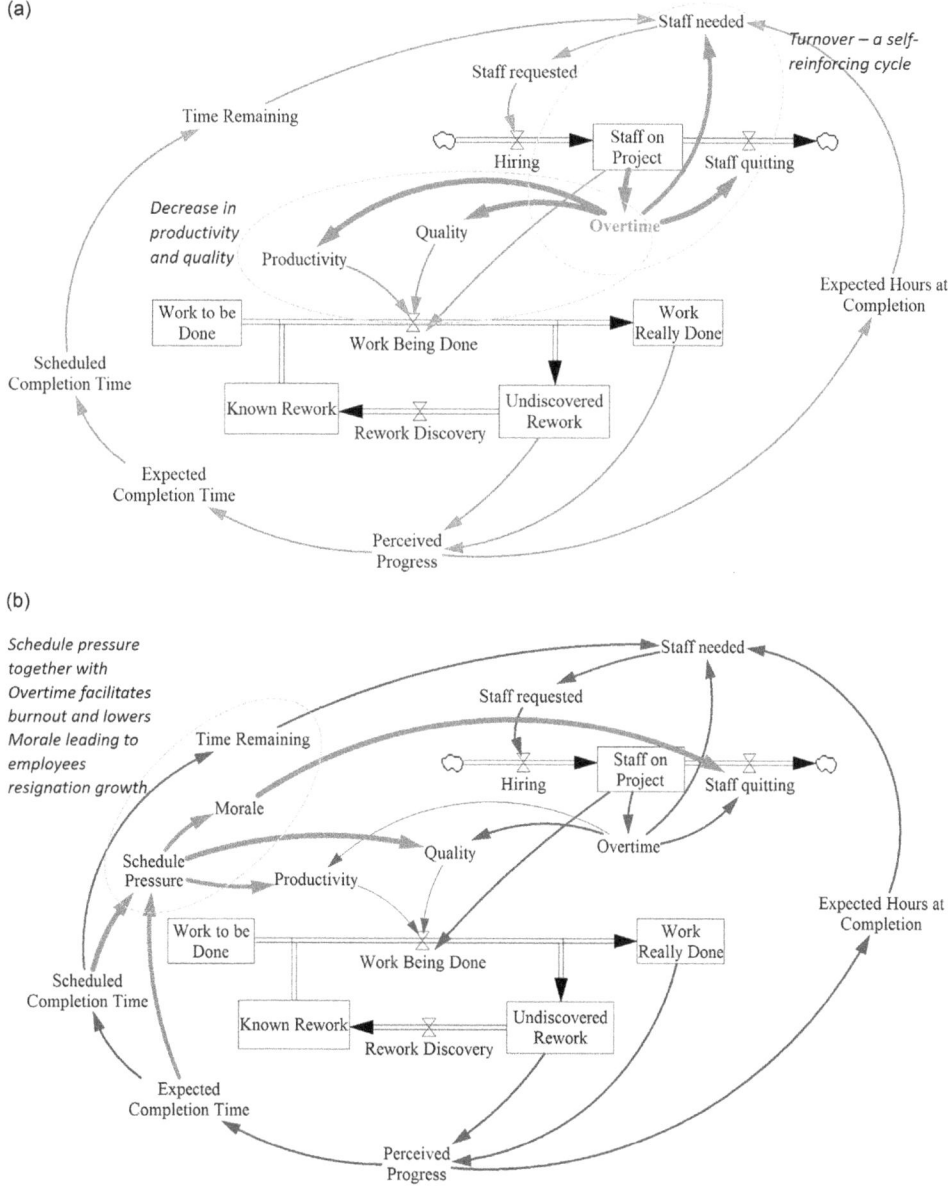

FIGURE 19.9   (a) Overtime cycle. (b) Schedule pressure and deteriorating morale added to the turnover cycle.

impact on fatigue and eroded personnel morale are significantly underestimated by management (Chan 2011; Neves et al. 2016).

After a few weeks, several software developers (the key staff, in fact) decide to quit ABC Company due to constant **Schedule Pressure** and deteriorating **Morale** (Figure 19.9b). Often, the highly qualified personnel leave first, since it's easier for them to find better jobs due to their qualifications and credentials. In addition to the fact that the project is experiencing a brain drain, the abandoned **Work to be Done** is redistributed on the personnel left further, exacerbating their performance and further pushing the self-reinforcing feedback loop of **Overtime**.

Now, management is committed to hiring or transferring additional personnel to the software project. However, it takes time to find and hire the right people (it is a process with a delay). Therefore, the overloaded remaining personnel further experience deteriorating morale and decreasing **Quality**.

However, bringing in new personnel has another counter-intuitive effect[3] which is usually underestimated or ignored. As new hires or specialists are transferred to supplement the current software team progress, new people with less experience or skill than those already on board enter the project. This is especially true if skills required are rare or in high demand on the labor market. The more constrained the labor market is, the lower the level of newcomers joining the project (Figure 19.10a). Therefore, it takes time for the newcomers to get themselves familiar with the project and go along the learning curve that the remaining staff has already been through.

There is a complex interaction of the feedback loops arising from the newcomers. Newly hired individuals might, intentionally or not, contribute to the project team turmoil by being in some cases less loyal, demonstrating higher attrition rates, being less skillful than expected (oversold), having unrealistic project expectations, etc. This is another vicious (self-reinforcing) cycle of *new hiring–higher departure rates–more hiring*.

New staffs require more supervision, and precise task-setting and control. Therefore, the experienced personnel's attention is now more and more diverted from the software development tasks to interacting with the newcomers, training, mentoring, and supervising them (Figure 19.10b). The management of ABC Company realizes that the software development process is now even more prone to the coding errors which increase the Rework cycle. The counterintuitive result of a substantial team expansion policy is *skill dilution*. Skill dilution has direct consequences such as lower productivity and further deteriorating quality (Figure 19.10a,b). The **Work Being Done**, Quality and Productivity are decreasing, while error rate, **Rework Discovery**, and **Rework are** increasing; the gap between **Work Really Done** and **Perceived Progress** is widening, which in fact suggests *to add even more project personnel* (!).

The management team cannot keep up with rising project costs. The growing organization becomes dysfunctional. It generates many meetings and discussions, while the responsibility is blurred. Another paradox is that despite the growing number of the project team, the availability of each person may even reduce, since personnel becomes sub-divided into smaller groups and much more time is now taken for internal discussions and reviews of design changes. As a result, the project team feels increasingly mentally and physically exhausted. This contributes to burnout, lower morale, and all kind of negative consequences associated with it.

It can still get worse. Problems that occur in the early stages of the project quickly propagate to downstream work—poor requirements (unclear, incomplete, too general), software tests/evaluations lacking rigor and breadth, etc. can trigger significant Rework further (NASA Office of Chief Engineer 2009: 57). Classical examples include large construction, manufacturing, new product design, and other projects where changes in the project design and engineering stages affect construction and manufacturing stages (Figure 19.11).

It is possible that errors are discovered months and years later—a typical case in large and complex projects (construction, infrastructure, software, etc.). Some errors discovered may have

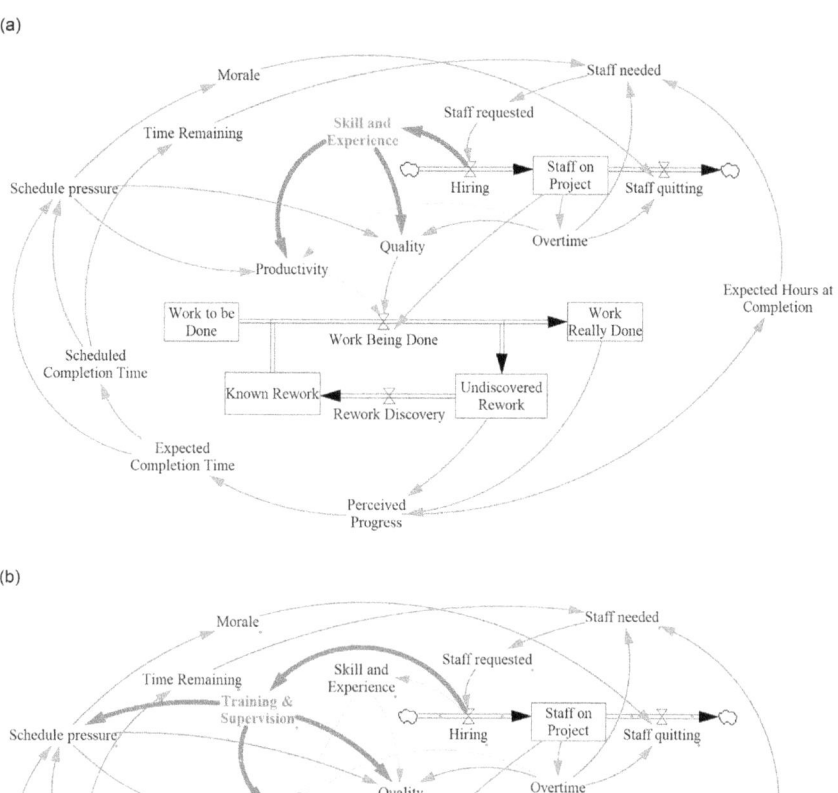

FIGURE 19.10   (a) Skill dilution from the new hires. (b) Training and supervision of newcomers diverts time from the project adding to schedule pressure.

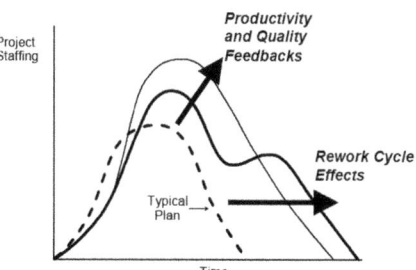

FIGURE 19.11   Typical rework cycle and productivity/quality effects on project staffing dynamics. The rework cycle tends to delay project staffing, pushing the actual staff profile to the right, as errors are created, discovered, and reworked. This creates more errors which are discovered and reworked. Vicious circle feedbacks triggered by management responses to project problems tend to increase staffing, pushing project staffing (and effort applied through overtime) up, and to the right.[4]

devastating impacts on the project performance, necessitating changes that eliminate a substantial part of the project progress.

SOURCES OF RISKS FOR A TYPICAL PROJECT EXECUTION

Analysis of complex projects suggests that key sources of risks in terms of impact on delivery deadline and costs are as following:

- Late information or changes in initial design
- Constraints in resource availability (i.e., slow project start, insufficient skill mix, forced cost cuts to meet financial constraints, etc.)
- New processes, materials, and team members
- Management and organizational changes
- Initially aggressive project assumptions (compressed timing, inadequate budget, misperception of overall project complexity).

The degree of cost escalation found in software development processes was addressed by a number of researchers, including Boehm (1981), Larson and Wertz (1993) and Stecklein et al. (2004). Larson and Wertz (1993) conducted a study on satellite costs, estimating error costs over each lifecycle phase of the satellite production and developed the guidelines for NASA on estimation of the impact of errors. The NASA study (2004) distinguishes four main phases (requirement preparation, design, code, and testing phases) in the software development project and forecasts that the overall error cost factor would increase dramatically from phase to phase. Let's assume the cost of an error in the requirement preparation phase is 1 unit of cost. Then, the cost of fixing that error in the design phase increases to 5–7 units; in the coding phase, 10– 25 times; and in testing phase, more than 50 times (the NASA study compared different methods to calculate escalation of costs and revealed that cost growth rises exponentially over the project lifecycle). Software projects as well as many large-scale complex projects are prone to so-called "killer errors".

Since projects (especially complex) are usually tightly coupled systems, it becomes increasingly important to assess potential ripple effects. Due to multiple interactions of non-linear feedback loops with unintended and counterintuitive consequences and their compound nature, complex projects often spiral out of control.

Even insignificant by itself, residual mistakes in software development or small independent failures in a complex project combined can cause a major disruption. NASA's expert in complex software development G. Holzmann concludes that adding backup and fault protection translates into increase in a system's size and complexity with "unplanned couplings between otherwise independent system components..... Given the magnitude of the number of possible failure combinations, there simply isn't enough time to address them all in a systematic software testing process. For example, just hundred residual defects might occur in close to ten thousand different combinations" (Holzmann 2007). A combination of minor failures was directly responsible for the failure of NASA's Mars Global Surveyor mission (Holzmann 2009).

Conclusion. Complex projects are prone to errors and disruptions. Humans are typically not very good at working with tightly coupled systems while the cost of error sometimes is too high (i.e., human life). Ability to predict various scenarios of the future development of the project and to quantify potential outcomes and impact of managerial decisions on project outcomes becomes a critical skill. Therefore, simulations start to play an increasingly important role.

In the next part of the chapter, we will provide some insight into simulation science reviewing three key approaches to project simulations coupled with simple models and will discuss advantages of each approach.

OVERVIEW OF KEY MODELING TECHNIQUES

The simulation technics applicable to complex project management is dominated by three paradigms of thinking: system dynamics, discrete event, and agent-based modeling (ABM). We should say that we see applications of the same three modeling approaches across business system simulations. Let's go into details and see what these approaches are.

SYSTEM DYNAMICS

System dynamics is the oldest simulation approach, with roots in work done by MIT Professor J. Forrester in 1950. When joining MIT Sloan School of Management in 1956, J. Forrester was already famous for his work for the U.S. military on the first U.S. air defense system (Semi-Automatic Ground Environment, SAGE), random access memory (RAM) for computer industry, computer numerical control (CNC) machines for manufacturing. Forrester applied this significant experience in technical system design to business and social simulations.

Initially, J. Forrester called the new approach industrial dynamics. He later reviewed its name to system dynamics to highlight that economics consists of complex systems with non-linear non-obvious behaviors which are interconnected in dynamics.

From the very beginning, system dynamics stayed focused on the management and organizational layer, abstracting from the engineering and manufacturing sides of the business. This was established because of the idea that most problems are caused by the management and organizational side of the business.

Let's try to build a model of a software development project, which is a good example of a complex project involving a lot of people. System dynamics is supported by three to four tools: VenSim, PowerSim, iThink, and AnyLogic that use one and the same notation for model building. We will learn its basic element—stock and flow diagram in this exercise.

A stock is a digital representation of something that is modeled, for example, a number of people, a set of requirements, and a number of products. Any process in system dynamics is modeled as a flow between stocks. In our example, software development is a flow between *requirements* and *developed software* (Figure 19.12). You can imagine that any production process is just a raw material flow to finished goods.

This is already the simplest model that will work if we know the *software development rate*, but it is not a constant; it depends on the number of *staff*, their *productivity*, and even their *communication overhead*. Adding such variables to our model makes it a bit more complicated (Figure 19.13).

The model is graphical and easy to understand, but it cannot be executed until we add numbers and formulas to connected variable flows.

Productivity is just a constant; *staff* is a variable that we can set up by a value at this stage. You may see that stocks, flows, variables, and constants are graphically represented differently. It is a standard system dynamics notation supported by a variety of tools. The screenshots you see in our chapter were captured from AnyLogic software.[5] We will continue illustrating our model

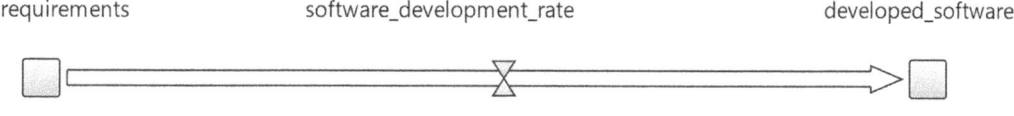

requirements   software_development_rate   developed_software

FIGURE 19.12   Simplest system dynamics diagram of the software development project.

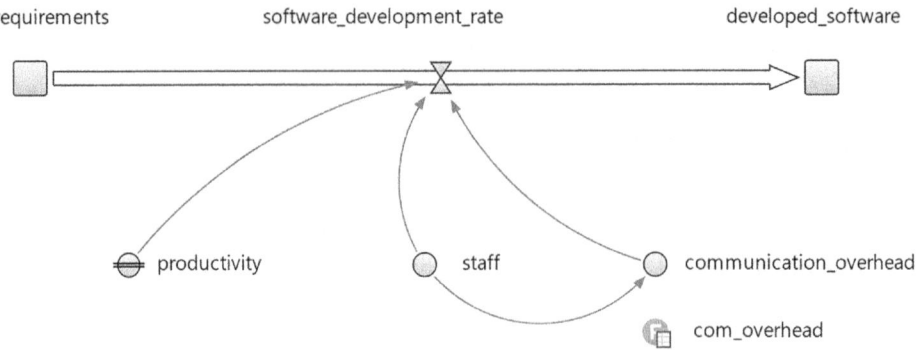

FIGURE 19.13   Modified diagram of the same model.

development process with screenshots from the AnyLogic software, since it supports all major simulation approaches. Some elements may be new to you, but never mind, we will explain all significant elements of the model.

We will define *communication overhead* with a table function that is also typical of system dynamics models. Such table functions can be a result of real system observations. Table function includes pairs of arguments and values. It allows you to set up a function when you do not know a formula, the function will be defined and data approximated.

The horizontal axis on Figure 19.14 represents team size, and the vertical axis shows overhead percentage. When we have less than five people, we have no overhead on their communication; but if we start with five and later on the number grows, the more people are working on the project, the bigger overhead they generate.

In AnyLogic, we call functions by function names with parameters in parentheses. We will define *communication overhead* as it is shown as follows:

$$communication\_overhead = com\_overhead(staff)$$

The software development rate is equal to productivity multiplied by the number of staff and the communication overhead.

$$software\_development\_rate = productivity * \left(1 - \frac{communication\_overhead}{100}\right) * staff$$

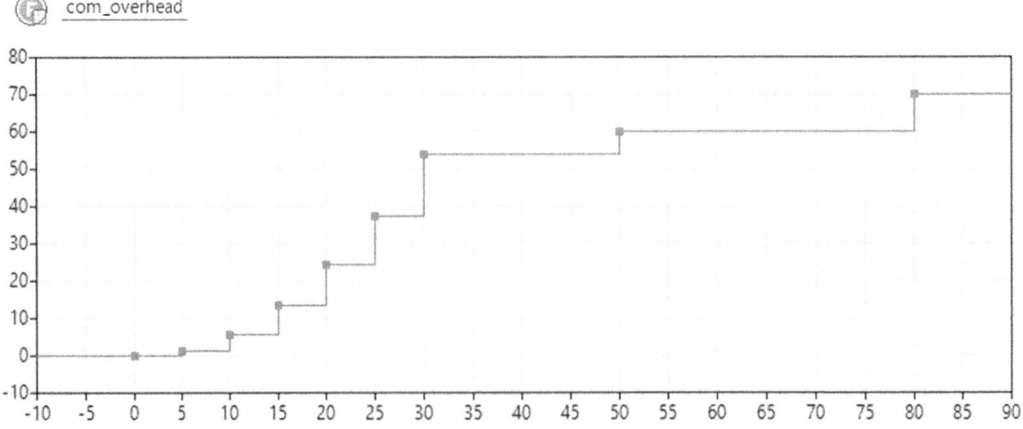

FIGURE 19.14   A table function that defines communication overhead.

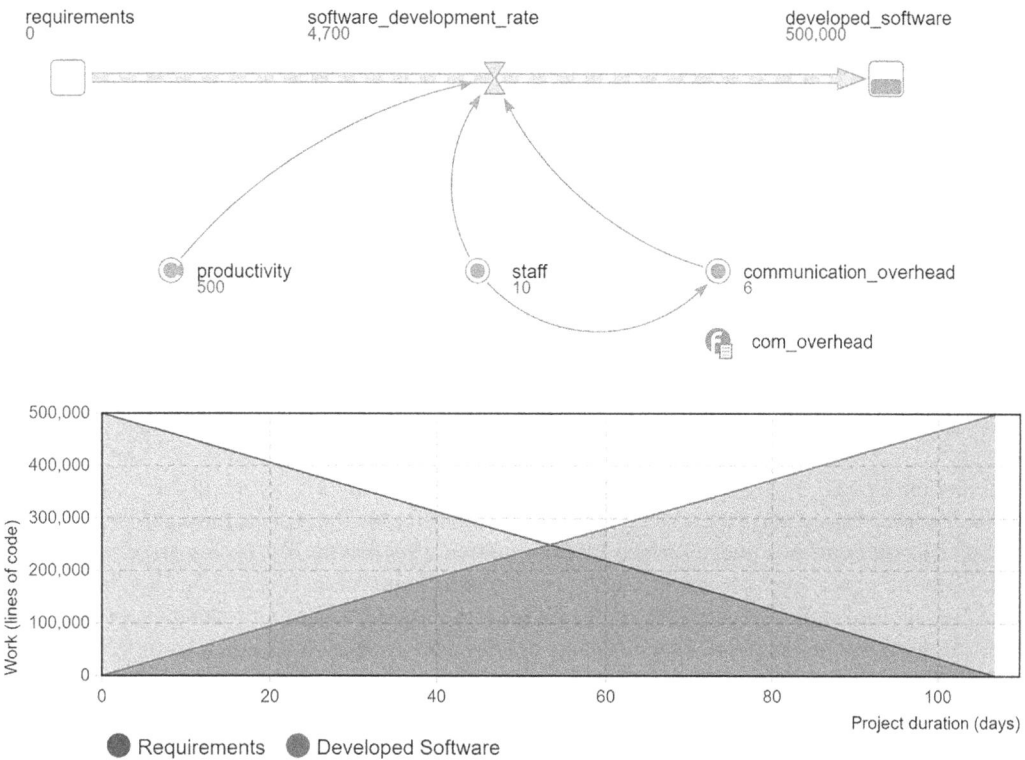

FIGURE 19.15 Modeling results, ten people, 500,000 lines of code.

Let's set up model parameters and run it. Initial *requirements* include half a million lines of code, *staff* number is 10 people, and each one's productivity includes 500 code lines per day. In this case, the project will be completed in 106 days (Figure 19.15). We can easily understand this result; since each developer produces 500 lines of code per day and all of them produce 5,000 per day, they would finish in 100 days, but we have 6% overhead on their communication.

We can get the same results by static modeling in Excel. The more details we add, the more difficult it is to get the same result by spreadsheet-based modeling. To expand the model further by adding a staff-hiring process that strongly affects total software development project time, let's continue thinking in terms of stocks and flows.

A major enhancement in system dynamics that was added in the first decade of the 21st century is hierarchical modeling. It allows you to create reusable components and use them on the main system dynamics diagram. This makes the system dynamics model easier to read, understand, and manage. Let's add a new object representing our development team into the model and use it as a subcomponent of the model.

The structure of our sub-model that represents a development team is shown in Figure 19.16. We hope you got used to system dynamics notation and AnyLogic.

Looking at the model in detail, there are two stocks "*new project staff*" and "*experienced staff*", and two flows representing new developers added into the project and their assimilation. By assimilation, we mean the process of training people to understand project details, scope, and boundaries.

Staff *assimilation rate* depends on the number of new people, which we should specify in the corresponding variable value. After 1 month of work, *new project staff* become experienced.

$$assimilation\_rate = \frac{new\_project\_staff}{month}$$

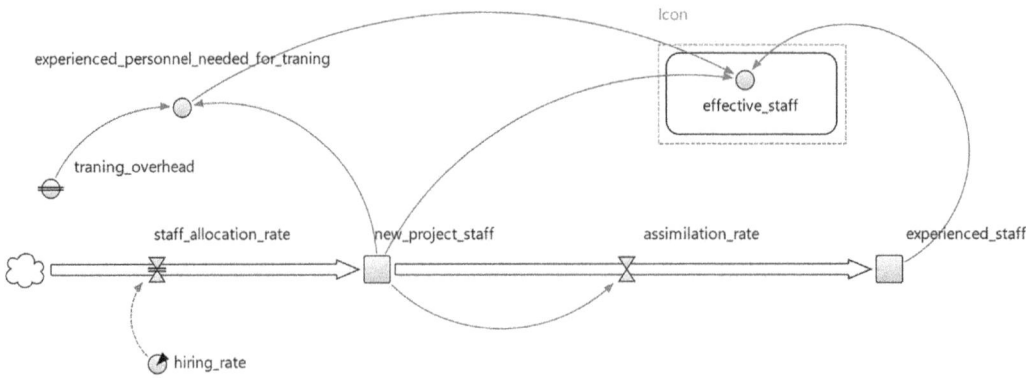

FIGURE 19.16   New staff allocation and assimilation—a subcomponent stock and flow diagram.

Experienced staff is used to train new employees, so there is *training overhead*. Let's assume that in our model, we need one experienced person working full time to train four newcomers.

Our sub-diagram has one interface variable called *effective staff*. The interface variable is visible on the upper level when you use the component. In our case, it represents a value of full-time experienced staff that can work on the project. We use the assumption that new people have 80% productivity and that only experienced people who are not involved in training activities can develop.

$$effective\_staff = 0.8 * new\_project\_staff$$

$$+ \left(experienced\_staff - experienced\_personnel\_needed\_for\_training\right)$$

Let's combine two models by adding the object "development team" onto the main diagram and connecting *effective staff* interface variable to *staff* variable. After the connection, the *staff* variable is driven by the *effective staff* variable (Figure 19.17).

Let's assume that in the same team of ten people, three new developers can be hired per month; to support hiring, we define *staff allocation rate* value.

$$staff\_allocation\_rate = 3.0/month$$

If we run the model, we see that the project is now finished in 83 days (Figure 19.18).

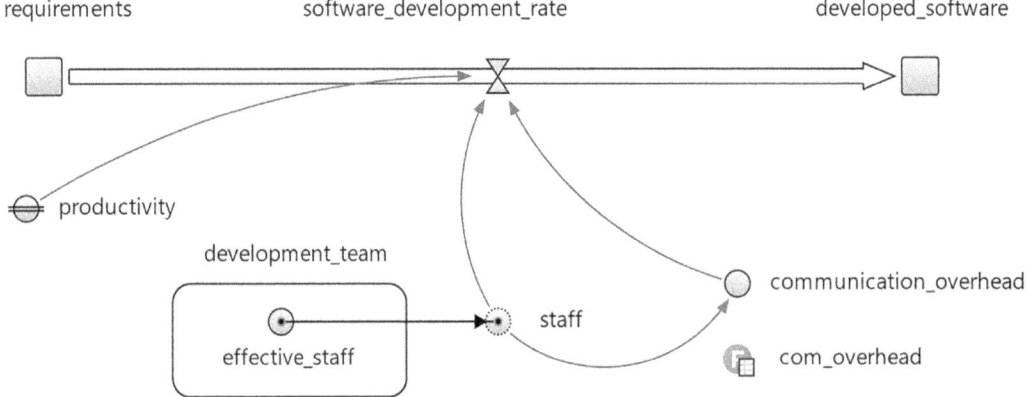

FIGURE 19.17   *Development team* sub-model embedded on to the main stock and flow diagram.

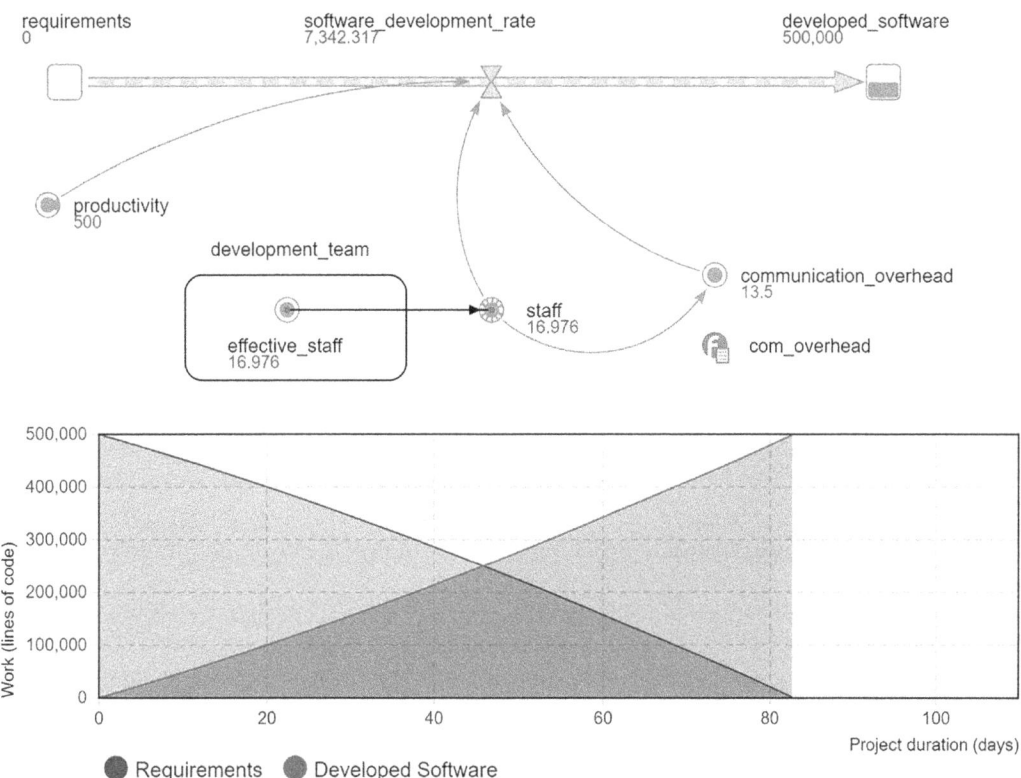

FIGURE 19.18   Modeling results, ten people, three new developers per month, 500,000 lines of code.

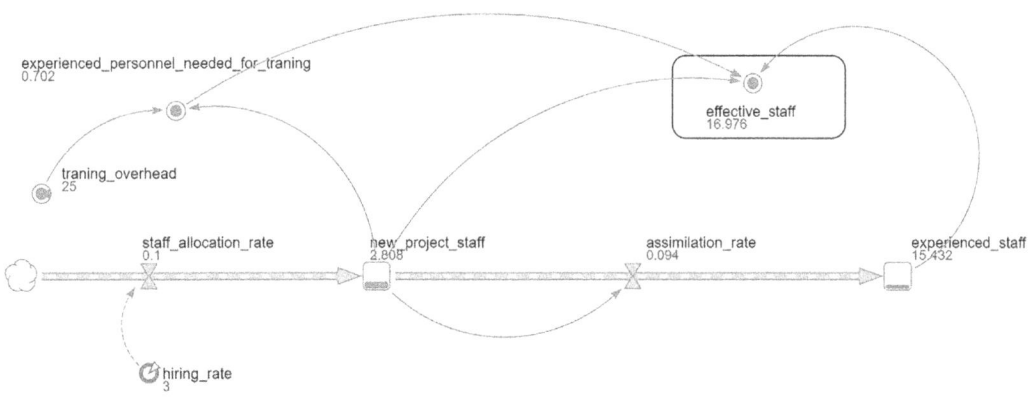

FIGURE 19.19   Modeling results, *development team* sub-model.

At the end of the project, we have a team of 15.432 experienced developers and 2.808 new employees, and their effective number is 16.976 (Figure 19.19). This is the typical situation for system dynamics, when we have fractional variable numbers representing people or other indivisible items. So we need to decide what having 2.808 new people on the team means. The reasonable assumption for this model would be to round this number up to 3, but for some other models, we just have to drop the fractional part of the figure.

Now the project manager has to understand everything about how a decision to add more people into the project influences project terms and costs. For simplicity, let's run another experiment to vary *hiring rate* from zero to twenty and create a plot of completion time versus hiring rate. As you can

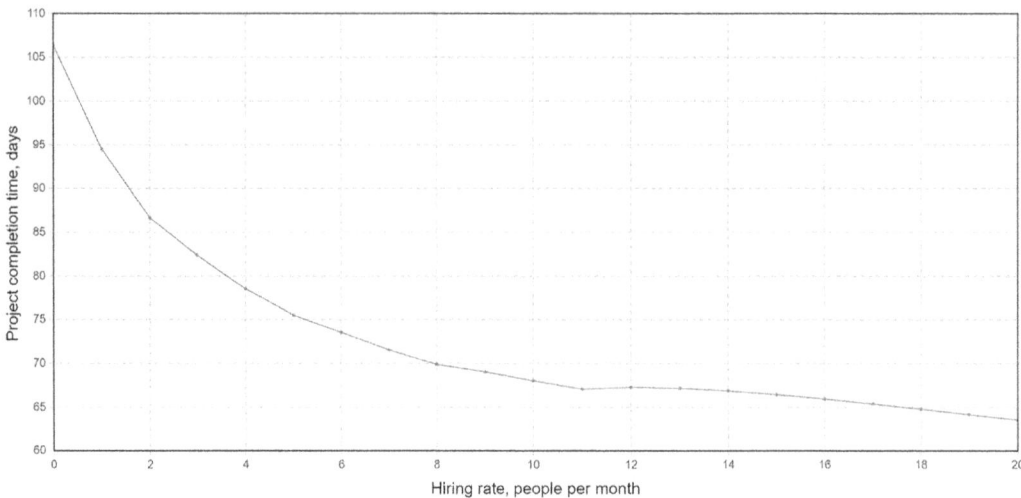

FIGURE 19.20   Hiring rate variation.

see in the plot in Figure 19.20, project completion time shows non-linear dynamics when we add new people to the project.

At this stage, we have not touched stochastics, which is the key benefit of simulation in project risk analysis. Let's do an experiment with our model setting *staff allocation rate* as a probability triangular distribution (Figure 19.21).

If we run the model 1,000 times and draw project completion time on a histogram, we will be able to analyze the risk of deadline failure. Assuming our deadline is 90 days, we know from simulation that if we allocate three new developers per month to our team, we finish the project in 83 days, a week ahead of schedule. When we add stochastic allocation rate, we see on the histogram that there is about a 28% probability that we do not meet the 90-day deadline; 27% is the sum of two last bars in Figure 19.22. The project manager should probably consider allocating more resources to the project, and we can continue experimentation with the model to support his decision-making.

To create this model, we used a general-purpose simulation software called AnyLogic. The model is publicly available in AnyLogic Cloud, along with the experiments we described and the model source code (it might be found by the name "Software development process dynamics" or by the author Sergey Suslov).[6]

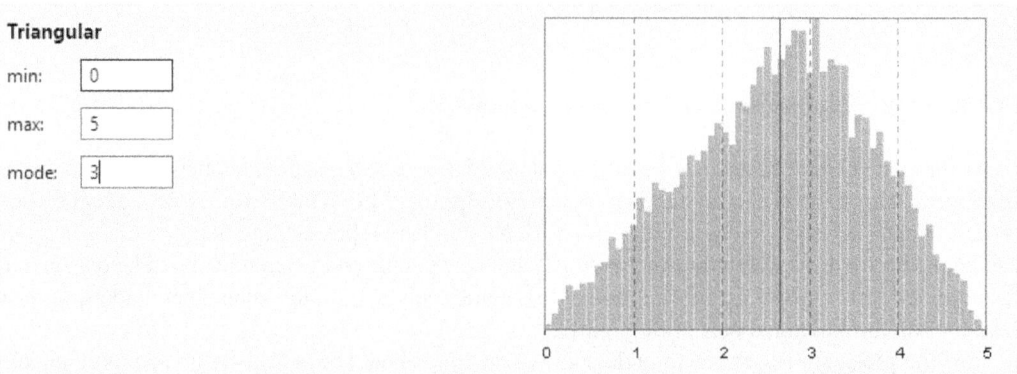

FIGURE 19.21   Triangular probability distribution.

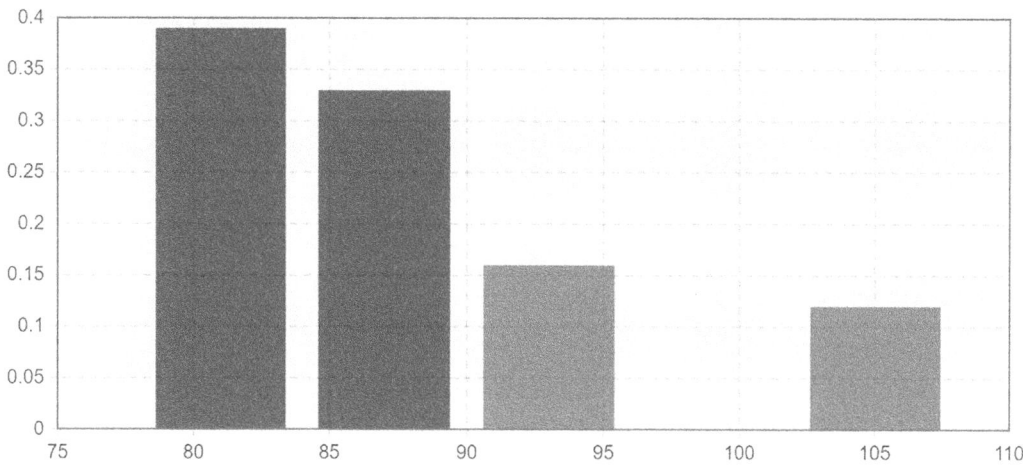

FIGURE 19.22 Histogram of project completion times in the case of stochastic *hiring rate.*

AGENT-BASED MODELING

Agent-based modeling (ABM) is the most recent major simulation method. Nevertheless, some models can be found in old papers and books like Thomas Shelling's 1978 book *Micromotives and Macrobehavior* (Schelling 1978). Such papers may be treated as the beginning of the ABM concept; for decades, ABM was limited in practical application by the state of computer hardware. In 1990–2000, ABM stayed a purely academic topic, but the 21st century with its booming computer hardware made ABM commercially applicable for solving high-scale business tasks; moreover, it now shows the fastest application growth compared with other simulation methods.

ABM and system dynamics are on different poles. While system dynamics is based on holistic process description, the ABM system uses a bottom-up approach that is described as interacting objects with their own behaviors. System behavior emerges as a summary of individual agents' actions. Scientific literature concentrates on the problems, such as what we can call an agent, whether or not it should have individual behaviors, memory, goals, etc. For the purpose of this chapter, we will concentrate only on the applied aspects of ABM.

In the past two decades, the ABM community has developed several practical ABM toolkits. In authors' opinion, the most popular tools are AnyLogic, Swarm, RePast, and NetLogo. Each toolkit has a variety of characteristics, its strong and weak points.[7]

To understand ABM in detail, let's try to reconceptualize our system dynamics software development model using ABM. We will continue using AnyLogic simulation software to develop new models.

The main object of an agent-based model looks very simple, as it includes only a couple of agent populations. A population is a set of agents, sometimes also called an array or collection. In our case, they are *developers* and *trainees*. The *Allocation* event periodically adds new *Trainees* to the system (Figure 19.23).

Since there is no place in an agent-based model where a global behavior is defined, let's start from the bottom with a *Developer* behavior. A statechart is a powerful tool to describe agent states and behaviors. Our *developer* and *trainee* agents' behavior can be graphically represented by the following statecharts (Figure 19.24).

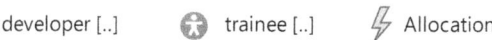

FIGURE 19.23 Agents collections and the event.

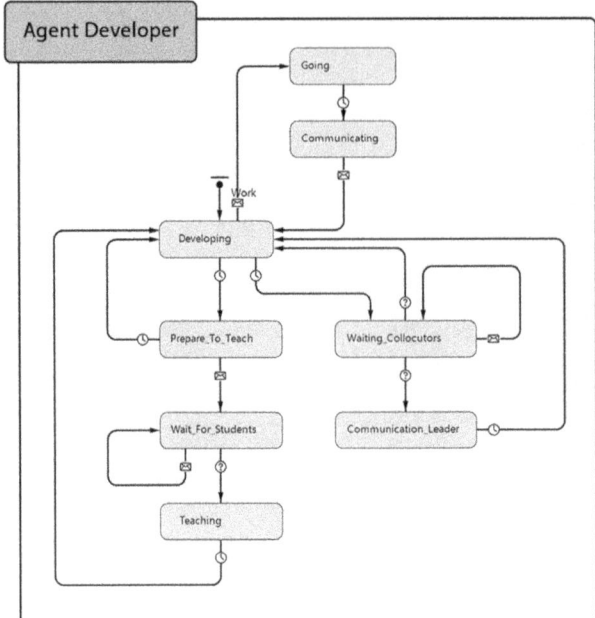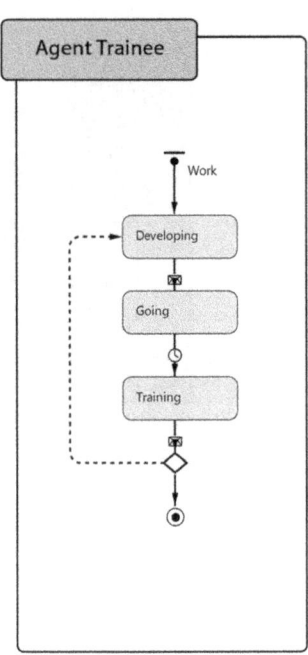

FIGURE 19.24 Statecharts that define agents' behavior.

Developing is the main state that the agent is in when starting its life in the model. Being in *Developing* state, the agent is producing the software product. *Teaching* is a state where a *Developer* agent does not develop but spends his time teaching new employees (represented by *Trainee* agent). There are two intermediate states between *Development* and *Teaching*, and they represent preparation for teaching and waiting for students. Agents communicate by sending messages, for example, in *Wait for students* state, a developer is periodically sending signals to all the trainees that it is in the room and ready to start the training.

Trainee logic is much simpler, there are only three states: *Developing* and *Training* which are clear, and *Going* which represents a break between development and training (it may be different in time depending on where the training room is located). One and the same model shows different results, depending on the spatial environment (whether all the developers and trainees are in one room, floor, building, city, country, etc.). To support a particular case, we should set up a timeout and determine how long the agent is spending in *Going* state. After each training, we check the experience of the trainee and decide whether to convert him to a developer. In this case, *Trainee* agent is disposed of and a new *Developer* agent is created.

Communication is also supported in the *Developer* agent's statechart. On a timeout, a developer decides to initiate a discussion and goes to *Waiting Collocutors* state, where it sends messages to its colleagues. When everybody is here, the agent becomes a *Communication leader*, leaving this stage after a given random timeout. Any *Developer* agent in *Developing* state switches to *Going* state on a corresponding signal from another agent who initiates the communication. When the agent becomes communication leader, it switches to *Communicating* state.

Agent-based models are usually full of code scripts, and in our case, Java language is provided by AnyLogic simulation software. A lot of small actions such as sending messages and agents' movements are done by adding a few lines of code into statecharts' transitions or states. For example, in our model we add new trainees by periodically executing three code lines.

If we execute this model, we get approximately the same results we would have gotten from the system dynamics model. You might wonder why we need ABM if we already have system dynamics. It is a very reasonable question, to which we have a set of answers:

- **Building agent-based models is more of a descriptive process.** We just describe how system components behave and communicate to get results. In a lot of cases, it is easier to describe the system components than to identify stocks, flows, and especially feedback loops.
- **ABM has a wider application range.** System dynamics is usually used for high-level management and organizational models, while ABM provides much better support for mid-level challenges such as supply chain simulation—where attention is paid to vehicles, distances, particular product stock levels on warehouses, etc.
- **ABM is a good tool to utilize company data.** Nowadays, companies have tons of data, and an agent-based model can use this data to get very precise model outputs. For instance, mobile network providers have their clients' full information (age, income, gender, etc.). They can model each client as an agent and get a precise digital twin of their client base or even a market segment in a given region.

AN AGENT-BASED CASE STUDY OF A SOFTWARE DEVELOPMENT COMPANY

There are many good case studies of ABM application for complex project management. For example, one big software development company used an agent-based model build with the approach described above to test its internal human resource management policies. They analyzed three alternative strategies to manage a growing global team:

- Internal education and coaching
- Replacing low-rated employees with new ones
- Combination of the strategies above.

This company does outsource projects, so the human resource part of the project is significantly important. They should always plan whether and how to grow the team, and develop its skills to have project managers, team leaders, and senior developer in place, along with regular staff that can be hired much easily. The agent-based model supports them in this planning.

Originally, most ABM applications assumed that the agent is a person: employee, customer, patient, etc. With growths in technology, the situation is rapidly changing; we see much more agent-based models where agents are objects, vehicles, warehouses, and even non-physical assets like projects. An agent-based case study of a logistics provider (see below) can help us to expand this idea, since the model includes such agents as vehicles, sites, etc.

AN AGENT-BASED CASE STUDY OF A LOGISTICS PROVIDER

Increasing competition in the logistics services market drives a demand for tools and solutions that make logistics operations more effective and efficient. A big 3PL logistics operator created an agent-based simulation model of over-sized cargo delivery over thousands of kilometers, to support a construction project in oil and gas industry.

The construction project was already planned, and a proposal was organized between logistics operators. The logistics operator sees calculating the precise cost of delivery and planning the logistics project as a big task. This is because there is a lot of freedom and small decisions, like what vehicles to use, what routes to take, and how to combine rail, road, and river transportation.

An agent-based model created in AnyLogic included such agent types as vehicle, base (intermediate point where the cargo can be stored), and construction plants (destination of the cargo). The model made allowance for what-if scenarios and also predicted how many vehicles were needed, where, and at what points in time.

The simulation model created with AnyLogic software supported logistics experts in

- Choosing the optimal form of multimodal transportation and considering lead times, transportation costs, and risks.
- Mitigating the financial risks and understanding the probability of satisfying expected lead time criteria.
- Forecasting the future state of the vehicles in short- and midterm perspectives.
- Budgeting and providing end customer with better contract conditions than other 3PL-operators
- Financial calculation in a short period for operational decision-making.

We hope that the case studies above helped show the diversity of agent-based applications for the management of complex projects. ABM can be applied at a high level of abstraction where most top management challenges occur, as well as at the middle abstraction level, where project details play a bigger role in emergent behavior.

Discrete Event Simulation

Discrete event simulation (DES) is almost as old as system dynamics. In October 1961, IBM engineer Geoffrey Gordon introduced the first version of GPSS (General Purpose Simulation System, originally Gordon's Programmable Simulation System), which is considered to be the first method of software implementation of discrete event modeling (Borshchev 2013). These days, discrete event modeling is supported by many software tools. From a practitioner's point of view, we should comment that when people apply simulation, the model is a discrete event or includes a discrete event subpart in over 50% of cases.

The main idea of DES is to consider the system as a process, that is, a sequence of operations being performed across entities. The operations include delays, service by various resources, choosing the process branch, splitting, combining, and so on. As long as entities compete for resources and can be delayed, queues are present in virtually any discrete event model. The model is specified graphically as a process flowchart, where blocks represent operations. The flowchart usually begins with "source" blocks that generate entities and inject them into the process and ends with "sink" blocks that remove entities from the model. This type of diagram is familiar to the business world as a process flowchart and is ubiquitous in describing their process steps.

All major DES tools support the same set of basic flowchart blocks such as source, queue, delay/process/service, and sink. There are more than 50 different DES tools[8], some of them are general purpose such as Arena, AnyLogic, Simul8, and ExtendSim; some are industry-specific, for example, Siemens Plant Simulation, FlexSim, and Automod. Any general-purpose software tool can be easily applied for solving project management challenges using DES.

Let's think about how we can create a model of our software project management using DES. First of all, we need a source for our requirements and sink for developed software. As soon as we have requirements, we need to seize a developer from *experienced staff*. Then, we process a piece of the requirements and release a developer (Figure 19.25).

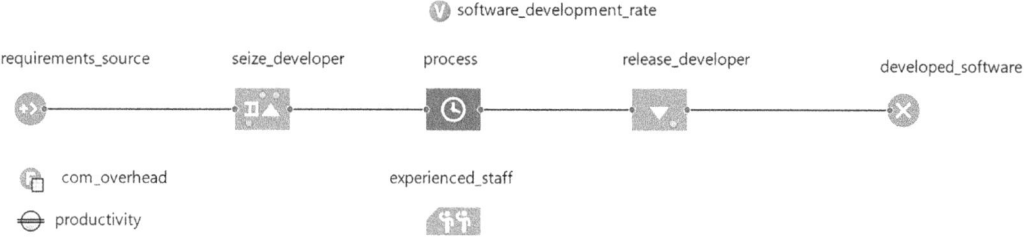

FIGURE 19.25 Software development model implemented as a DE flowchart.

Let's feed the model with data. Requirements source should produce 500,000 lines of code, each of them an entity. Let's introduce a new variable called *software development rate* and assign it the following value:

$$software\_development\_rate = \left(\frac{1 - com\_overhead\left(experienced\_staff.size()\right)}{100} \right) * productivity$$

By doing this, we decrease the software development rate on communication overhead, exactly as we did in system dynamics. We can reuse the table function *com_overhead*.

We should set up the *process* time as 1/*software_development_rate*, because this is a time in days that one developer needs for one line of code. We have a team of ten developers so that our *process* can process up to ten lines of code simultaneously.

Running the model, we get the same 106 days for project completion time as we got by applying system dynamics (Figure 19.26).

Let's think about how to implement new developer allocation. We can start by adding a new, initially empty resource pool called *new project staff*. We should specify that we can use this pool when we *seize developer* (Figure 19.27).

A scripting language is a necessary evil of the professional model-building process that some simulation practitioners like the most and some others struggle with the most. Scripting makes modeling tools very powerful and flexible, but not so easy to create and manage. In our model, we use Java code expression to set up unique *software_development_rate* for each line of code processed in process block. This expression is executed each time we seize a developer.

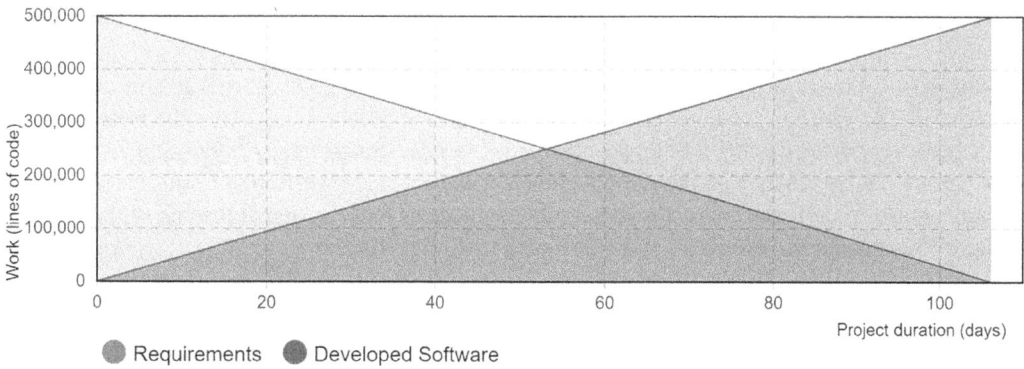

FIGURE 19.26 DE model results, 500,000 lines of code, ten people.

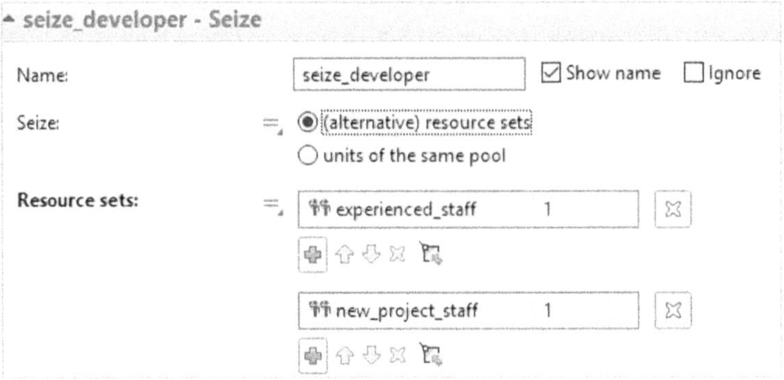

seize_developer - Seize

Name:	seize_developer	☑ Show name	☐ Ignore
Seize:	⦿ (alternative) resource sets ○ units of the same pool		
Resource sets:	👥 experienced_staff 1	☒	
	⊕ ⇧ ⇩ ☒ 🔖		
	👥 new_project_staff 1	☒	
	⊕ ⇧ ⇩ ☒ 🔖		

FIGURE 19.27   Seize block settings in AnyLogic software tool.

Then, let's modify the Java code that we execute on seizing the developer:

```
software_development_rate = productivity*
(1-com_overhead(experienced_staff.size()+new_project_staff.size())/100)*
(((Developer)unit).experienced? 1 : 0.8)*
(1-training_overhead/100*new_project_staff.size()/experienced_staff.size());
```

Software development rate is assigned a new value that equals productivity decreased by three multipliers:

- The first one to support communication overhead
- The second one to change productivity in case the seized developer is a new employee
- The last multiplier to support that we have to allocate experienced people time for training.

By this expression, we added *new project staff* into communication overhead, supported training overhead calculations, and changed productivity for new developers. Processing speed will be unique or each particular entity representing a line of code, depending on the developer and current team size and structure.

The last and most important part is the allocation process for new developers (Figure 19.28). As the most important part of the processes, it starts with *Source* block and ends with *Sink* block. New developers leave *staff allocation* source block with *hiring rate* per month. As soon as a new developer appears, we increase the capacity of *new project staff* resource pool. After that, new developers enter *training* service block, where they immediately start monthly training. After the *training* ends, we increase *experienced_staff*, adding new resources and decreasing *new_project_staff*.

When we run the model with the same parameters, we get slightly different results caused by the discrete nature of *staff allocation*. Project completion time is 88 days, and at the end of the project, we have 15 experienced developers and 3 new people (Figure 19.29).

FIGURE 19.28   DE model results, 500,000 lines of code, ten people.

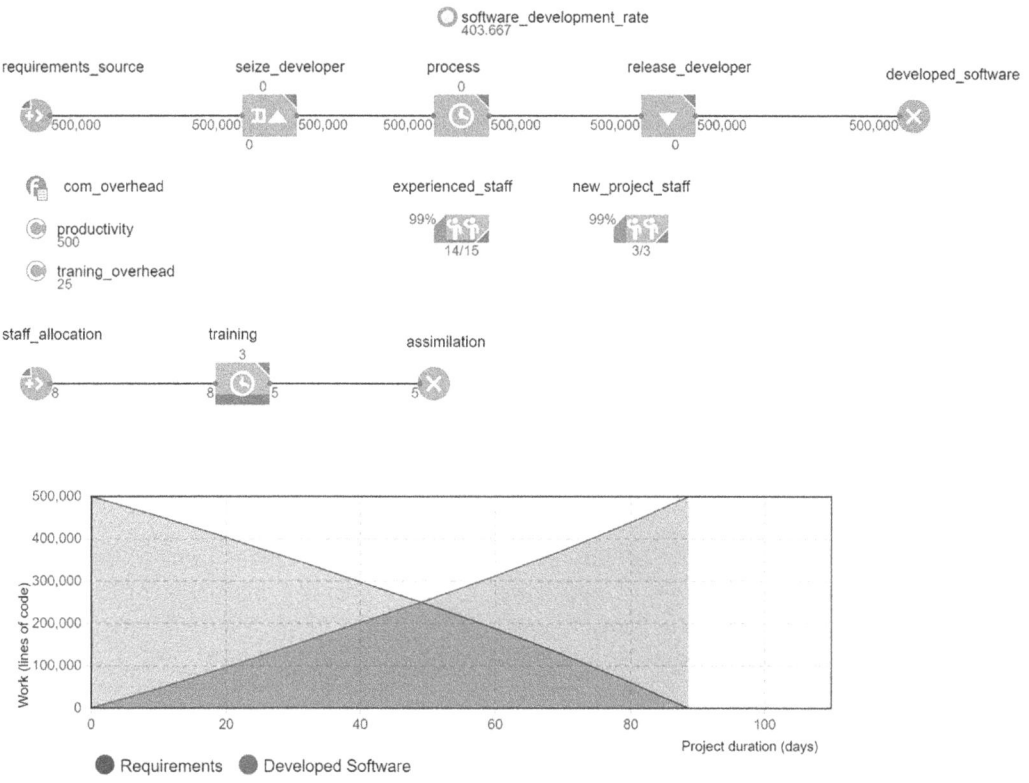

FIGURE 19.29   DE model results, 500,000 lines of code, ten people, three new developers per month.

The model along with its source code is published in AnyLogic Cloud,[9] and you can find it by searching by the name "Software Development Process Dynamics DE" or by the author name Sergey Suslov.

You can use free AnyLogic PLE[10] to see the full model internals or even continue the development.

DESs are usually used for the engineering part of project management, where managers and engineers try to predict the completion time of every project stage and mitigate the risk of being out of schedule due to technical reasons. In comparison with system dynamics or ABM, DES is used more on low to middle abstraction levels, where detailed business or technology processes play key roles. The case study below should help to visualize how operational constraints influence the feasibility of the project goals.

A CONSOLIDATED CONTRACTORS COMPANY CASE STUDY

Let's look at the case study of Consolidated Contractors Company (CCC), which is the largest construction company in the Middle East and ranks #18 internationally. CCC has offices and projects in over 40 countries and a workforce of over 130,000 employees. Its portfolio includes oil and gas plants, refineries and petrochemical facilities, pipelines, power and desalination plants, light industries, water and sewage treatment plants, airports and seaports, heavy civil works, dams, reservoirs and distribution systems, road networks, and skyscrapers. Just after signing a contract for a site preparation project (earthmoving scope of approximately four million cubic meters), the client and local authorities placed new, more restrictive constraints on the operation. These constraints included

- Trucks were allowed to move at a maximum of 10 kmph within the construction project site instead of the original 20 kmph.
- Trucks from/to the dump location were instructed to follow a specific route full of traffic lights, intersections, roundabouts, and security gates. Accordingly, the original assumption of the average 40 kmph truck speed on route from and to dump site could not be maintained.
- Truck sizes/loads were brought down from the maximum allowable of 32–15 m^3.
- The number of truck trips restricted to a maximum of 100/h.
- Only one of the original four site access points was granted after signing the contract, thus restricting all traffic on site to one gate.
- Only one work shift (10 h/day) was allowed for dumping at the dump site in contrast to the original two-shift (20 h/day) schedule.

The newly placed constraints essentially meant that the project schedule would be severely impacted, as would the total cost of completing it. The number and complexity of the constraints made it quite difficult to manually evaluate the impact of the constraints on time to complete equipment requirements. As a result of the newly added complexity, the simulation team was asked to help quantify the impact of the new constraints and substantiate a claim by CCC for a time extension. CCC already had a well-established construction simulation model made in AnyLogic software. This model was used for basic construction optimization, to forecast equipment and time requirements for earthworks operations. Using the earthworks simulator requires multiple parameter inputs, including the assumed average speed the trucks will travel on their haul routes and back. With a new truck route imposed on the project, it was very difficult to manually estimate the average speed the trucks would run at.

As such, a more complex construction simulation model was built in AnyLogic software to mimic the trucks traversing the route segments, while both loaded and empty. Using optimization techniques in AnyLogic was helpful for risk-free experimentation and for effective construction management on this project. In the construction simulation model, each route segment was modeled with a stochastic distribution for the total time to traverse the segment. Then, the trucks were made to run 10,000 times each way in the simulator to arrive at an average speed for each route and loaded/empty combination.

Why use simulation in construction? DES with AnyLogic was specifically selected for construction modeling because this software allowed CCC to

- Very quickly build a construction optimization model representing the route and its segments.
- Add a map of the routes and superimpose an animation of the trucks traversing the routes to make it easier to visually explain the work to stakeholders.

The average speeds deduced from the truck route simulator were then fed into the earthworks construction simulation model along with the remainder of the new constraints. These included

- Truckload sizes and the number of work shifts (working hours) per day, to produce multiple scenarios showing the original time and equipment requirement forecast to complete the operations.

- The current time and equipment requirement forecast to complete the operations (impact quantification in time and resource requirements).
- Proposed mitigation scenario.

The results of the two-step process of using the truck route simulator to summarize the route and then feeding it into the earthworks simulator enabled CCC to quantify very quickly the impact of the new more restrictive constraints and to build mitigation scenarios to aid the claim for extension of time. Both of the construction simulation models were created in AnyLogic software, giving them the flexibility to consider all the new constraints. DES has proven to be good assistants for construction management. Based on the construction simulation model results, the client agreed to extend the total duration of the earthworks operation by 50% on top of the original schedule duration and to allow two work shifts per day. This essentially saved the project an estimated additional cost equivalent to 18% of the original total contract value.

COMBINED APPROACH: KEY CHALLENGES AND ADVANTAGES

Even though we were able to create one and the same model of software project management using three different simulation methods, it is rare in practice. Often you can choose a modeling method following practical recommendations that can be found in the Internet:

- If there are many independent objects, use an agent-based approach.
- If there is only information about global dependencies, use system dynamics.
- If a system is easily described as a process, use a discrete event approach (Figure 19.30).[11]

Most real-world cases are complex, and it is convenient to describe different parts of a system with different methods. In using only one method, your ability to capture business systems with their real complexity and interactions may be seriously limited. Some system elements will have

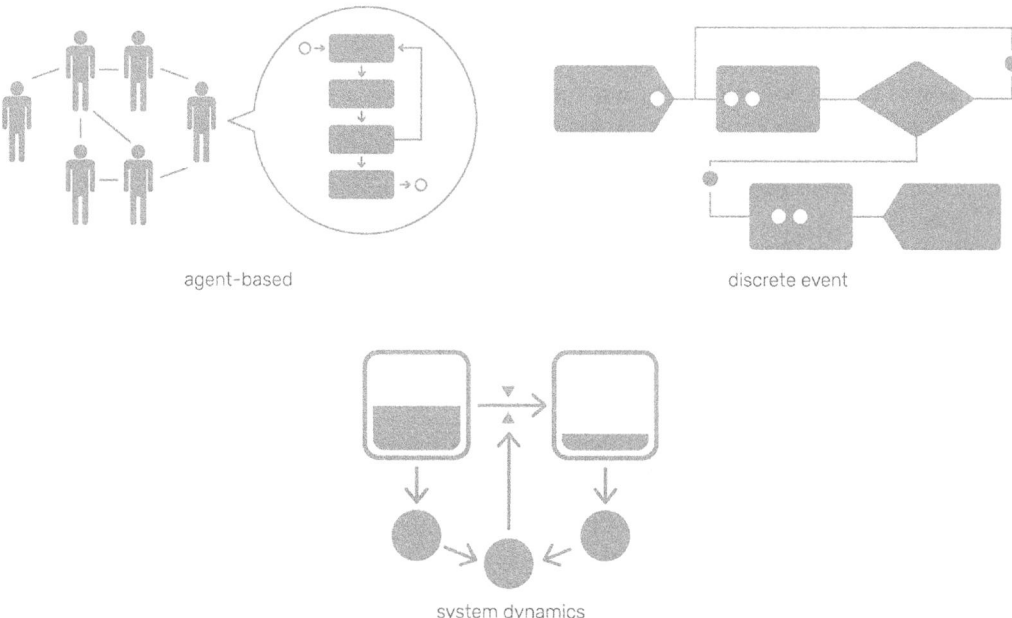

agent-based   discrete event

system dynamics

FIGURE 19.30   Simulation methods.

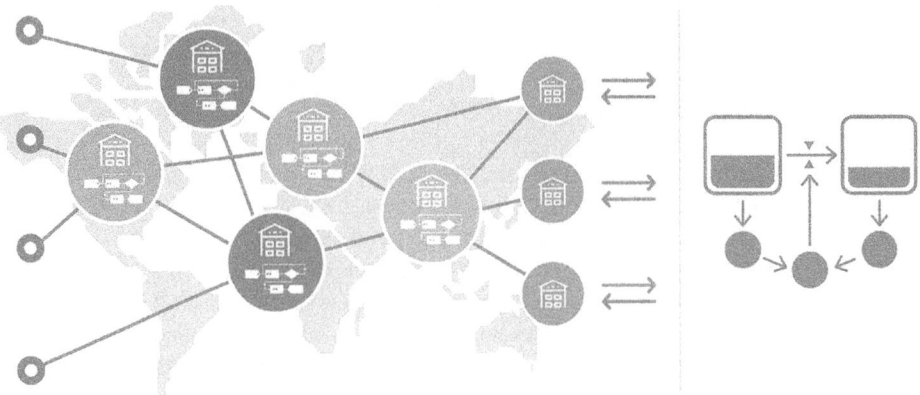

FIGURE 19.31   Supply chain multimethod model.

Here, in Figure 19.31 you can see how production, distribution, and the market can be combined in one model, using different techniques. A discrete event model describes the processes within each warehouse. The warehouses then appear as agents on the distribution network. Finally, the market, which drives the system, is modeled with system dynamics. Everything is captured without compromise.

to be excluded or a workaround developed. In such cases, it makes sense to consider mixing different simulation approaches in one model, which is called multimethod or hybrid simulation.

Most of the actively developing simulation vendors have started implementation of the support for at least some multimethod modeling architectures. It is likely that moving forward only multimethod simulation will survive as an industry standard for general-purpose simulation software tools. In *The Big Book of Simulation Modeling* (Borshchev 2013), Andrei Borshchev suggested a set of typical multimethod architectures including

- **Agents interacting with a process model.** For instance, equipment defined as an agent influences the process of a conveyor system.
- **A process model linked to a system dynamics model.** A business process flowchart can change the variables of a system dynamics diagram for a company's profit and loss statement.
- **System dynamics inside agents.** Production processes inside factories can be defined with system dynamics (SD) and used as agents in a supply chain model.
- **Processes inside agents.** A good example are the warehouses in Figure 19.31.
- **Agents temporarily act as entities in a process.** For instance, a regional healthcare model, where people defined as agents go to hospital and follow a discrete event flowchart.

We recommend you follow technical recommendations given in this book during project management challenges conceptualization phase.

QUESTIONS

1. Why is it beneficial to use simulation for complex projects? What advantages does simulation provide?
2. Why do projects fail from the simulation experience perspective?

3. What is counter-intuitive in the system behavior of complex projects?
4. What are ineffective remedies that managers typically use in practice? Why do these measures fail to fix the project?
5. How would you suggest fixing project management problems applying a systemic perspective?
6. What are the key sources of risk in the management of complex projects?
7. What are the main modeling approaches for the simulation of complex projects? What are their differences, advantages, and drawbacks?
8. What practical recommendations would you provide to a project manager?
9. What in your opinion prevents project managers from widespread use of simulations? Why? Please, explain your point of view.

CONCLUSION. SIMULATIONS: WHY BOTHER?

Simulation modeling offers a holistic view of the project. The overall complexity of even not-so-large projects quickly exceeds human capacity to rationally assess dynamics of the system. Herbert Simon, an economics Nobel Prize laureate, introduced a concept of "bounded rationality" to emphasize this phenomenon (Simon 1972, 1979; Dörner 1996). Simulations help to produce counter-intuitive conclusions. For instance, Lee et al. (2007) investigated and found out that contrary to the widespread assumption that everything should be done as quickly as possible to minimize delays, projects may have optimal delay sizes that are inevitable but still do not significantly hurt the overall project performance. Experienced project managers intuitively account for the possibility of small project delays.

Simulations are the right tool to make such an assessment holistically and simulate various scenarios. Unlike traditional approaches, simulations can incorporate many subjective factors such as staff morale, productivity, motivation, fatigue, deterioration of quality, and many others that are usually not explicitly analyzed. Over the years, system dynamics has developed a language and methodology for expressing various tangible and non-tangible variables (factors) that can be quantified (numerical estimates can be provided) and added to the model.

A number of studies document a successful use of simulations to support various projects. For example, naval shipbuilding (Cooper 1980), highway construction (Ford et al. 2004), semiconductor cheap development process (Ford and Sterman 1998), flour production (Godlewski et al. 2012), military projects (Lyneis et al. 2001), software project management (Abdel-Hamid and Madnick 1982), and construction projects (Pena-Mora and Park 2001; Love et al. 2002).

The advances in simulation software now help building models in the presence of management easier and faster. MIT professor John Sterman explains:

> Traditionally, formal modeling tools were complex and inaccessible to all but the trained analyst. As a result, models were developed by experts without direct involvement in the modeling process .... The inability of the managers [who were expected to use the results of the model] to participate in the process created a dilemma. If the modelers built a simple model, they were criticized for ignoring important relationships. If they built a complex model, they were criticized for creating a black box no one could understand. .... The solution to the dilemma is the intensive involvement of the management team in the modeling process. New software tools now make it possible for managers to participate as full partners in model development. Sophisticated interfaces and intuitive designs allow managers and employees throughout a firm to use, test, and revise complex models.

Sterman (1992: 11–12)

A group model building (GMB) methodology was developed by Richardson and Andersen (1995, 2010), Vennix (1996), Lane (2000), Andersen et al. (2007), and others. Originally, such workshops were conducted as 1- to 2-day seminars where experienced simulation practitioners together with top-management representing the client company used to discuss the conceptual structure of the system and build a model afterwards. Lately, the practice evolved into what is currently known as "participatory system modeling" (Van den Belt 2004, Stave 2010, and others) "… which encompasses GMB and mediated modeling, but places more emphasis on seeing participation along a continuum of model building and formulation, from no involvement to high involvement" (Hovmand 2014: 17).

Group model-building workshops can be very different in terms of the type of model to be developed—from a simple map providing a first glance into the project to a comprehensive computer simulation model. Even simple maps typically used at the early stages of a modeling process prove to be effective in conceptualizing a problem and providing useful insights into the key feedback structure of the modeled project. GMB workshops proved to be very useful in terms of higher quality of the model, since a model is built with the help of facilitation teams—who are usually experienced simulation practitioners—and deep expertise of industry practitioners—who are not necessarily experts in simulation modeling. In addition, GMB with early involvement of clients or key stakeholders helps to get a buy-in from stakeholders and propagate practical model application by end-users at a later stage.

When prepared, a simulation can be used at various stages of the project in a number of ways. For instance, Lyneis et al. (2001) suggest using simulations for project management practitioners:

- **At a pre-project stage** for bid or plan analysis, risk assessment, and competitor analysis (to estimate using information from public sources, if possible, what the program can cost competitors)
- **During the project stage** for risk management, change management (analysis of potential impact in terms of costs and scheduling specification and work scope changes from customer), and evaluation of process changes (total impact of changes to evaluate trade-off between short-term costs and long-term benefits realized)
- **After the project is finished** for benchmarking, training, and developing personnel (as a management flight simulator to communicate project experience and lessons learnt to the management).

RECOMMENDATIONS FOR PROJECT MANAGEMENT PRACTITIONERS

- It might be rather useful to view project management, not as a sequence of discrete tasks but *flows of work*, much of which consists of a Rework cycle. It's important to have a *rigorous* and *timely error-discovery process* to reveal errors soon enough that their correction is done quickly and with minimal effort. Early resolution significantly cuts disruption.
- Managers can influence project performance; however, they should take into account the *complex* and *counter-intuitive nature of major feedback loops* involved. Avoid conventional wisdom (i.e., "Let's hire more") and apply system thinking.
- Beware of *self-reinforcing feedback loops* (overtime, rapid hiring, conflicting deadline pressures, etc.) which quickly spiral a project out of control. It is important to understand the sources of the vicious circles, to maintain control over project costs.
- Usually, it is much cheaper to *move schedule pressure* (if possible) than to try to meet deadlines.
- More and later changes do not just create more impact; they create them disproportionately.
- Indirect project disruption costs are far more dangerous, typically resulting in costs up to ten times higher than direct cost overruns.
- *Try to budget for 10%–50% overspending* to be able to mitigate early problems quickly—major project disruption can easily lead to extra costs *times* the original project budget.
- *Try to model and simulate:* the theory and practice are already here.

Simulation is a powerful tool for project managers. With the advancements in simulation tools and dissemination of knowledge in theory and practice, it becomes easier and cheaper to simulate complex projects.

As project practitioners get more experience in the modeling and simulation of complex projects, we will experience less project failures and cost overruns, making the project management industry less prone to errors.

Simulation is a powerful tool for project managers. With the advancements in simulation tools and dissemination of knowledge in theory and practice, it becomes easier and cheaper to simulate complex projects thus considerably saving money and efforts of project practitioners.

ENDNOTES

1 Productivity of a software development project can be measured in various ways, that is, lines of source code produced per certain amount of time, number of bugs per 1,000 of code lines, active days (the time spent by a programmer to develop a code, not including time for planning and other minor activities), tasks scope (a volume of code a programmer can deliver yearly), and others. Some also use cycle time (order is taken for production and complete) and lead time (order is received from customer and delivered to him) as productivity metrics. For convenience purpose, in this publication, we will measure progress in *lines of source code produced per certain amount of time* (although industry practitioners generally do not believe this is an effective metric to measure software development productivity).

2 http://vensim.com/ (accessed March 12, 2019).

3 The term suggested by Jay Forrester, the founder of the system dynamics method.

4 "The rework cycle tends to delay project staffing, pushing the actual staff profile to the right, as errors are created, discovered, and reworked. This creates more errors which are discovered and reworked. Vicious circle feedbacks triggered by management responses to project problems tend to increase staffing, pushing project staffing (and effort applied through overtime) up, and to the right. This is due to productivity losses and increasing errors as a consequence of delays in recognizing and responding to project problems and the rework cycle" (Ford et al. 2007: 9).

5 www.anylogic.com (accessed March 12, 2019).

6 The direct link to the model https://cloud.anylogic.com/model/14482b77-2be0-4a80-bfb9-2f8c20ea627f?mode=SETTINGS (accessed March 12, 2019).

7 Comparison of agent-based modeling software: https://en.wikipedia.org/wiki/Comparison_of_agent-based_modeling_software (accessed March 12, 2019).

8 Simulation Software Survey, ORMS Today: www.informs.org/ORMS-Today/OR-MS-Today-Software-Surveys/Simulation-Software-Survey (accessed March 12, 2019).

9 The direct link to the model: https://cloud.anylogic.com/model/3fbdf014-76f2-4393-b194-cad66dd12381?mode=SETTINGS (accessed March 12, 2019).

10 www.anylogic.com/downloads/personal-learning-edition-download/ (accessed March 12, 2019).

11 www.anylogic.com/use-of-simulation/multimethod-modeling/ (accessed March 12, 2019).

FURTHER READING

Cooper, K. G. 1980. Naval ship production: A claim settled and a framework built. *Interfaces* 10, no. 6 (December): 20–36. doi:10.1287/inte.10.6.20.

Cooper, K. G. 1994. The $2,000 hour: How managers influence project performance through the rework cycle. *Project Management Journal* 25, no. 1: 11–24.

Godlewski, E., G. Lee and K. Cooper. 2012. System dynamics transforms fluor project and change management. *INFORMS Journal on Applied Analytics* 42, no. 1: 17–32. doi:10.1287/inte.1110.0595.

Howick, S. 2003. Using system dynamics to analyse disruption and delay in complex projects for litigation: Can the modelling purposes be met? *The Journal of the Operational Research Society* 54, no. 3: 222–229. doi:10.1057/palgrave.jors.2601502.

Lyneis, J., O. de Weck and S. Eppinger. 2003. Course notes for MIT course ESD.36J: System and project management (Fall 2003). https://dspace.mit.edu/handle/1721.1/80702 (accessed March 12, 2019).

Lyneis, J. M., K. G. Cooper and S. A. Els. 2001. Strategic management of complex projects: A case study using system dynamics. *System Dynamics Review* 17, no. 3 (Autumn): 237–260. doi:10.1002/sdr.213.

Lyneis, J. M. and D. N. Ford. 2007. System dynamics applied to project management: A survey, assessment, and directions for future research. *System Dynamics Review* 23, no. 2–3 (Summer–Autumn): 157–189. doi:10.1002/sdr.377.

Reichelt, K. and J. Lyneis. 1999. The dynamics of project performance: Benchmarking the drivers of cost and schedule overrun. *European Management Journal* 17, no. 2 (April): 135–150. doi:10.1016/S0263-2373(98)00073-5.

Repenning, N. and J. Sterman. 2001. Nobody ever gets credit for fixing problems that never happened: Creating and sustaining process improvement. *California Management Review* 43, no. 4 (Summer): 64–88. doi:10.2307/41166101.

Rodrigues, A. and J. Bowers. 1996. System dynamics in project management: A comparative analysis with traditional methods. *System Dynamics Review* 12, no. 2 (Summer): 121–139. doi:10.1002/(SICI)1099-1727(199622)12:2<121::AID-SDR99>3.0.CO;2-X.

REFERENCES

Abdel-Hamid, T. K. 1993. A multiproject perspective of single-project dynamics. *Journal of Systems Software* 22, no. 3: 151–165. doi:10.1016/0164-1212(93)90107-9.

Abdel-Hamid, T. and E. Madnick. 1982. *Modeling the Dynamics of Software Project Management.* Working paper, Alfred P. Sloan School of Management (February 1988, #WP 1980–88). https://dspace.mit.edu/bitstream/handle/1721.1/48430/modelingdynamics00abde.pdf?sequence=1 (accessed March 12, 2019).

Andersen, D. F., J. A. M. Vennix, G. P. Richardson and E. A. J. A. Rouwette. 2007. Group model building: Problem structuring, policy simulation and decision support. *The Journal of the Operational Research Society* 58, no. 5 (May): 691–694.

Boehm, B. W. 1981. *Software Engineering Economics.* Upper Sadle River, NJ: Prentice-Hall PTR.

Borshchev, A. 2013. *The Big Book of Simulation Modeling: Multimethod Modeling with AnyLogic 6.* La Vergne, TN: Lightning Source Inc.

Chan, M. 2011. Fatigue: The most critical accident risk in oil and gas construction. *Construction Management and Economics* 29, no. 4: 341–353. doi:10.1080/01446193.2010.545993.

Cooper, K. G. 1980. Naval ship production: A claim settled and a framework built. *Interfaces* 10, no. 6 (December): 20–36. doi:10.1287/inte.10.6.20.

Cooper, K. G. 1993a. The rework cycle: Why projects are mismanaged. *PM Network* 7, no. 2: 5–7.

Cooper, K. G. 1993b. The rework cycle: How it really works and reworks. *PM Network* 7, no. 2: 25–28.

Cooper, K. G. 1993c. The rework cycle: Benchmarks for the project manager. *Project Management Journal* 24, no. 1: 17–21.

Cooper, K. G. 1994. The $2,000 hour: How managers influence project performance through the rework cycle. *Project Management Journal* 25, no. 1: 11–24.

Cooper, K. G., J. M. Lyneis and B. J. Bryant. 2002. Learning to learn, from past to future. *International Journal of Project Management* 20, no. 3 (April): 213–219.

Cooper, K. G. and K. S. Reichelt. 2004. Project changes: Sources, impacts, mitigation, pricing, litigation, and excellence. In *The Wiley Guide to Managing Projects*, ed. P. W. G. Morris and J. K. Pinto, 743–772. Hoboken, NJ: Wiley. doi:10.1002/9780470172391.ch31.

Dörner, D. 1996. *The Logic of Failure: Recognizing and Avoiding Error in Complex Situations*, trans. R. and R. Kimber. New York: Basic Books.

Eden, C., T. Williams and F. Ackermann. 2005. Analysing project cost overruns: Comparing the "measured mile" analysis and system dynamics modelling. *International Journal of Project Management* 23, no. 2 (February): 135–139. doi:10.1016/j.ijproman.2004.07.006.

Eick S. G., C. R. Loader, M. D. Long, L. G. Votta and S. V. Wiel. 1992. Estimating software fault content before coding. In *ICSE '92 Proceedings of the 14th international conference on Software engineering (Melbourne, Australia—May 11–15, 1992)*, 59–65. New York: ACM. doi:10.1145/143062.143090.

Ford, D. N., S. Anderson, A. Damron, R. de Las Casas, N. Gokmen, and S. Kuennen. 2004. Managing constructability reviews to reduce highway project durations. *Journal of Construction Engineering and Management* 130, no. 1 (February): 33–42. doi:10.1061/(ASCE)0733-9364(2004)130:1(33).

Ford, D. N., J. M. Lyneis and T. R. B. Taylor. 2007. Project controls to minimize cost and schedule overruns: A model, research agenda, and initial results. In *Proceedings of the 25th International Conference of the System Dynamics Society and 50th Anniversary Celebration*, ed. J. Sterman, R. Oliva, R. S. Langer, J. I. Rowe and J. M. Yanni, 1–27. Albany, NY: System Dynamics Society. http://citeseerx.ist.psu.edu/viewdoc/download?doi=10.1.1.488.4883&rep=rep1&type=pdf (accessed March 12, 2019).

Ford, D. N. and J. Sterman. 1998. Dynamic modeling of product development processes. *System Dynamics Review* 14, no. 1 (Spring): 31–68. doi:10.1002/(SICI)1099-1727(199821)14:1<31::AID-SDR141>3.0.CO;2-5.

Godlewski, E., G. Lee and K. Cooper. 2012. System dynamics transforms fluor project and change management. *INFORMS Journal on Applied Analytics* 42, no. 1: 17–32. doi:10.1287/inte.1110.0595.

Graham, A. K. 2000. Beyond PM 101: Lessons for managing large development programs. *Project Management Journal* 31, no. 4: 7–18. doi:10.1177/875697280003100403.

Holzmann, G. 2009. Appendix D – Software complexity. In *NASA Office of Chief Engineer. NASA Study of Flight Software Complexity: Final Report*, www.nasa.gov/pdf/418878main_FSWC_Final_Report.pdf (accessed March 12, 2019).

Holzmann, G. J. 2007. Conquering complexity. *Computer* 40, no. 12 (December): 111–113. doi:10.1109/MC.2007.419.

Hovmand, P. S. 2014. *Community Based System Dynamics*. New York: Springer Science + Business Media. doi:10.1007/978-1-4614-8763-0_2.

Lane, D. C. 2000. Diagramming conventions in system dynamics. *The Journal of the Operational Research Society* 51, no. 2 (February): 241–245.

Larson, W. J. and J. R. Wertz. 1993. *Space Mission Analysis and Design*. 2nd ed. Torrance, CA: Microcosm.

Lee Z. W., D. N. Ford and N. Joglekar. 2007. Resource allocation policy design for reduced project duration: A systems modeling approach. *Systems Research and Behavioral Science* 24, no. 6 (November/December): 551–566. doi:10.1002/sres.809.

Love, P. E. D., G. D. Holt, L. Y. Shen, H. Li, and Z. Irani. 2002. Using systems dynamics to better understand change and rework in construction project management systems. *International Journal of Project Management* 20, no. 6 (August): 425–436. doi:10.1016/S0263-7863(01)00039-4.

Lyneis, J. M., K. G. Cooper and S. A. Els. 2001. Strategic management of complex projects: A case study using system dynamics. *System Dynamics Review* 17, no. 3 (Autumn): 237–260. doi:10.1002/sdr.213.

Lyneis, J. M. and D. N. Ford. 2007. System dynamics applied to project management: A survey, assessment, and directions for future research. *System Dynamics Review* 23, no. 2–3 (Summer–Autumn): 157–189. doi:10.1002/sdr.377.

NASA Office of Chief Engineer. 2009. NASA study of flight software complexity: Final report. www.nasa.gov/pdf/418878main_FSWC_Final_Report.pdf (accessed March 12, 2019).

Neves, F. G., H. Borgman and H. Heier. 2016. Success lies in the eye of the beholder: The mismatch between perceived and real IT project management performance. In *49th Hawaii International Conference on System Sciences*, 5878–5887. Los Alamitos, CA: IEEE Computer Society.

Pena-Mora, F. and M. Park. 2001. Dynamic planning for fast-tracking building construction projects. *Journal of Construction Engineering and Management* 127, no. 6 (December): 445–456. doi:10.1061/(ASCE)0733-9364(2001)127:6(445).

Richardson, G. P. and D. F. Andersen. 1995. Teamwork in group model building. *System Dynamics Review* 11, no. 2 (Summer): 113–137. doi:10.1002/sdr.4260110203.

Richardson, G. P. and D. F. Andersen. 2010. Systems thinking, mapping, and modeling in group decision and negotiation. In *Handbook of Group Decision and Negotiation*, ed. D. M. Kilgour and C. Eden, 313–324. Dordrecht: Springer Verlag.

Richardson, G. P. and A. Pugh III. 1981. *Introduction to System Dynamics Modeling with Dynamo*. Cambridge, MA: MIT Press.

Roberts, E. B. 1964. *The Dynamics of Research and Development*. New York: Harper & Row.

Roberts, E. B. 1974. A simple model of R&D project dynamics. *R&D Management* 5, no. 1: 1–15. doi:10.1111/j.1467-9310.1974.tb01217.x.

Rodrigues, A. and J. Bowers. 1996. The role of system dynamics in project management. *International Journal of Project Management* 14, no. 4 (August): 213–220. doi:10.1016/0263-7863(95)00075-5.

Schelling, T. C. 1978. *Micromotives and Macrobehavior*. New York: W. W. Norton & Co.

Simon, H. A. 1972. Theories of bounded rationality. In *Decision and Organization: A Volume in Honor of Jacob Marschak*, ed. C. B. McGuire and R. Radner, 161–176. Amsterdam: North-Holland.

Simon, H. A. 1979. Rational decision making in business organizations. *The American Economic Review* 69, no. 4 (September): 493–513.

Stave, K. 2010. Participatory system dynamics modeling for sustainable environmental management observations from four cases. *Sustainability* 2, no. 9: 2762–2784. doi:10.3390/su2092762.

Stecklein, J. M., J. Dabney, B. Dick, B. Haskins, R. Lovell and G. Moroney. 2004. Error cost escalation through the project life cycle. Conference paper. NASA Johnson Space Center. https://ntrs.nasa.gov/search.jsp?R=20100036670 (accessed March 12, 2019).

Sterman, J. 1992. System dynamics modeling for project management. http://scripts.mit.edu/~jsterman/docs/Sterman-1992-SystemDynamicsModeling.pdf (accessed March 12, 2019).

Taylor, T. and D. N. Ford. 2006. Tipping point failure and robustness in single development projects. *System Dynamics Review* 22, no. 1 (Spring): 51–71. doi:10.1002/sdr.330.

The AnyLogic Company. 2016. Developing disruptive business strategies with simulation. www.anylogic.com/resources/white-papers/developing-disruptive-business-strategies-with-simulation/ (accessed March 12, 2019).

Van den Belt, M. 2004. *Mediated Modeling: A System Dynamics Approach to Environmental Consensus Building.* Washington, DC: Island Press.

Vennix, J. A. M. 1996. *Group Model Building: Facilitating Team Learning Using System Dynamics.* Chichester: Wiley.

Williams, T., C. Eden, F. Ackermann and A. Tait. 1995. The effects of design changes and delays on project costs. *Journal of the Operational Research Society* 46, no. 7 (July): 809–818.

Williams, T. M. 1999. The need for new paradigms for complex projects. *International Journal of Project Management* 17, no. 5 (October): 269–273. doi:10.1016/S0263-7863(98)00047-7.

20 Adaptability, Agility, and Resilience Toolset

Nil H. Kilicay-Ergin
Pennsylvania State University

CONTENTS

INTRODUCTION

As systems continue to grow in complexity, they pose increasingly greater design and management challenges. Designing to achieve primary functional requirements is important for initial fielding of a system. Afterwards, system properties also known as ilities gain importance as a determinant of competitiveness and means to maximize the lifecycle value of systems. For example, as discussed in de Weck et al. (2011), first automobiles were motorized versions of horse-drawn carriages. Many incremental improvements were made over the long lifetime of automobiles. Initial four-wheel car brakes were replaced by hydraulic four-wheel brakes which were then replaced by dual hydraulic breaks. These changes in cars were necessary to address safety, an important system property for many engineering systems, in more demanding operating environments. Other ilities emerged over the evolution of systems. Safety, quality, reliability, usability, and maintainability are some of the core system properties that are important in any traditional engineering system including cars, airplanes, communication systems, and many more.

The evolution of ilities continues as independent systems are interconnected to form system of systems (SoS) to address uncertainty and changes in operational environment, social values, and technological advances. A special type of SoS, cyber–physical system, is increasingly engineered to integrate cyber and physical systems using computation, communication, and control technologies. Many granting agencies emphasize that cyber–physical systems will be more transformative than the IT revolution due to several advances in adaptive computing, pervasive sensing and actuating capabilities, advanced real-time control, cloud computing, and cognitive capabilities (Muller, 2018). Other systems that are closely related to cyber–physical systems such as Internet of Things, adaptive systems, and smart systems are also becoming key enablers for many applications. For example, growing human populations and growing demand for limited natural resources necessitate sustainable cities that require efficient and effective energy generation and transportation infrastructures. Smart grid, connected and autonomous vehicles within a networked infrastructure, and other smart city applications are evolving to address these societal needs. Healthcare systems are changing as well. Smart medical devices, wearable sensors, and connected home monitoring systems are all part of the modern healthcare systems. Proliferation of all these cyber–physical systems has given rise to a new generation of system properties including flexibility, adaptability, resilience, agility, and sustainability. This expanded set of ilities are systemic and emerged from the interactions between systems. Resilience in large-scale infrastructures is a system property that has emerged as a response to observed cascading failures in networked systems. Agility emerged as a response to fiercely competitive product development environments. Adaptability has emerged as a response to instability and uncertainties in design and operating environments.

Project managers of today's complex systems need to understand the implications of these new generation of system properties on complex project management. Refactoring these properties later is not feasible as any changes to address these properties after implementation phase have a domino effect on constituent systems and cost may be prohibitive. Therefore, project managers need to spend time up front and account for these ilities during project planning and work allocation. This necessitates clear definition of these properties so that project managers can track and evaluate them throughout development and implementation. In addition, project managers should be familiar with strategies and methods used to implement ilities along with functional requirements during system architecture generation. This chapter aims to provide a basis for project managers and engineers of today's complex systems in terms of understanding three important system properties:

> This chapter aims to provide a basis for project managers and engineers of today's complex systems in terms of understanding three important system properties: adaptability, resilience, and agility.

> While there is agreement that these properties provide value to stakeholders in dynamic and uncertain operating environments, there is a plethora of definitions for these -ilities which makes it difficult to measure and evaluate these properties in design and operation.

adaptability, resilience, and agility. While there is agreement that these properties provide value to stakeholders in dynamic and uncertain operating environments, there is a plethora of definitions for these -ilities which makes it difficult to measure and evaluate these properties in design and operation. This chapter provides an overview of definitions for these properties in the "Adaptability", "Resilience", and "Agility" sections. Since these ilities are interrelated, relationship among these properties is discussed in the "Relationships among Adaptability, Resilience, and Agility" section considering the current research thrusts in these areas. The "Methods and Tools for Design and Management of Adaptability, Resilience, and Agility" section provides a review of the design principles, techniques, and tools for project managers/engineers to design, analyze, and manage these system properties. The "Summary" section concludes this chapter with a summary of key concepts.

ADAPTABILITY

One of the major challenges in design of complex systems is the management of change in the system itself as well as in its operational environment. Adaptability is a system property that deals with these changes and uncertainty in operational environments. Adaptability expectations can vary depending on the system context. For example, smart home applications require adaptability to temperature changes by self-adjusting its heating and air-conditioning. In a different system such as platooning autonomous vehicles, adaptability means ability to self-organize and respond to changing traffic conditions. Therefore, project managers should have a clear definition of adaptability for their system context to be able to manage it effectively. To help project managers, this section provides a review of various definitions of adaptability for complex engineering systems.

Adaptability is defined in different ways. Fricke and Schulz (2005) define it as the ability of a system to change itself under uncertain or changing operating environments. Other definitions emphasize ability of a system to self-modify itself to fit to changes in its environment (Crawley et al., 2004). Adaptable systems use sensors, control algorithms, and human operators to accommodate changes in its operating environment. Autonomy is a key enabler for detecting changes and undesirable system states.

Other definitions of adaptability include ability to exhibit self-organizing and emergent behavior, ability to change to satisfy changing requirements, and ability to respond to predictable changes in operational environment (Chalupnik et al., 2013). Others view adaptability as a different form of reliability (Chalupnik et al., 2013). While there is no consensus on the type of changes an adaptive system responds, to the concept of change within the system is an accepted term in the definition of adaptability.

Ross et al. (2008) provide a context-free definition of adaptability, flexibility, modifiability, and robustness by describing "changeability", a term they coined to describe change events. In their definition, a change event is described by three elements: the agent of change, the mechanism of change, and the effect of change. Change agent which is the reason for initiating change is used to make a distinction between adaptability and flexibility. Based on this element, if the change agent is external to the system, then flexibility is considered. If the change agent is internal to the system, then adaptability is considered. The distinction between flexibility and adaptability can be made once the system boundary is explicitly defined. An adaptable system undergoes internal change autonomously without human intervention, whereas a flexible system is modified by an external system such as humans. A thermostat controlling the heating of a room undergoes self-modification, whereas a flexible manufacturing system is reconfigured by production engineers to meet changing demands. Other elements of changeability are useful to define other related system properties including robustness, scalability, and modifiability. All these ilities deal with change effect which is the difference in states before and after a change has occurred. Robustness is the ability of a system to remain constant in parameters under changes. Scalability is the ability of a system to change the level of a parameter. Modifiability is the ability of a system to change the membership of the parameter set. Change mechanism is another element of changeability which is the path the system must take to transition from its prior state to its post-state. A change path details the necessary resources and constraints for the change.

When considering adaptability, several concepts should also be considered to deal with unpredictable events. Principles of self-organization is a key concept for engineering of adaptiveness. Systems that have the capability to reorganize its components through transitioning from any state to an equilibrium state require proper selection of states. This means systems should be designed with plasticity and self-awareness (Sanchez-Escribano and Sanz, 2014). Autonomy is another enabler for the design of adaptability and dealing with unpredictable internal and external events. Therefore, autonomous and self-organization capabilities are key enablers of adaptability. The adaptability of a system strongly influences the resilience and agility properties of a system. The relationship among these will be clarified after a review of resilience and agility definitions.

RESILIENCE

Like adaptability, resilience is a critical property in today's complex engineering systems. Project managers need to have a clear definition of this property to be able to track and manage it through various measures. Resilience is a system property used in various disciplines. In the engineering domain, resilience is an important feature of any interconnected system. Like other ilities, it is defined in different ways (Hosseini et al., 2016):

- The ability of a system to return to its operation after disturbances
- The ability of a system to function or survive when disrupted

- The ability of a system to respond to off-nominal conditions
- The ability of a system to recover from disturbances rapidly with minimal impact on its stability.

While there is no unique definition of resilience, there are some similarities across these definitions. Many of these definitions focus on the ability of the system to absorb and adapt to disruptions. These definitions emphasize recovery from unexpected disturbances in the operational environment. Some systems are expected to recover after a shock and return to its steady-state performance level (dynamic resilience). Other systems such as infrastructure systems are not expected to return to its steady state performance level but rather survive or continue to function when disrupted (static resilience). In these definitions, survivability and recoverability are key elements. Other definitions define resilience in terms of a system's proactiveness, how well a system is prepared for disturbances and discard any post-disturbance recovery roles.

Haimes (2009) points to the multidimensional nature of resilience and indicates that certain states of a system are more resilient than others because resilience is context dependent. The architecture of the system, its operational environment, and types of disruptions play a role on system resilience. A system can be resilient in some disruptions but not in others. For example, infrastructure systems are SoS distributed geographically. As these power, transportation, and communication networks grow in complexity, maintaining and improving resilience over the SoS constituent systems becomes critical for providing uninterrupted service. However, infrastructure networks may be resilient to specific disruptions, but over time, the network becomes less resilient due to the emergence of new types of threats. The heterogeneous characteristic of SoS provides opportunities to improve resilience, but interdependencies increase the risks of failures cascading throughout the network. Traditional reliability and risk approaches are not sufficient to manage resilience. Studies focused on designing resiliency into systems indicate that resiliency cannot be achieved by a single principle or method due to its systemic characteristic.

Other systems that are vulnerable to disruptions include defense, banking, insurance, and healthcare systems. These systems operate by complex processes within a larger SoS, and consequences of disruptions can be a source of ripple effect. Process resilience becomes vital in these systems which are highly regulated. For example, a disruption in banking services directly impacts its customers. However, there might be more severe consequences if a certain bank is a major provider of inter-bank transactions. In that case, the impact would ripple to nationwide customers (Cohen et al., 2018). Another challenge in these highly networked systems is challenge in testing new applications. Since most applications are closely linked to telecom service providers, testing is not an internal activity anymore. An organization can implement resilience into its processes by first identifying its critical processes and then by determining the risk preference for the organization. Then, depending on the risk preference and operational environment, processes can be redesigned or reconfigured.

It is also necessary to discuss system properties that are closely related to resilience. These include robustness, survivability, reliability, flexibility, agility, and safety. Chalupnik et al. (2013) define these attributes based on the design changes required or not for a product or process to respond to off-nominal conditions. A more systematic analysis of the relationship between these ilities is discussed in the "Relationships among Adaptability, Resilience, and Agility" section.

AGILITY

The most important role of a project manager is to manage the project from inception to deployment by ensuring that the project stays within budget and is delivered on time and with high quality. This becomes challenging in highly competitive and uncertain environments. For example, software development projects carry many technical uncertainties due to rapid evolution of technologies and changing stakeholder expectations. Agility is now an essential part of both the system development process and a property of the system itself. Agility consists of two properties:

1. Flexibility
2. Speed.

This means that agile processes and systems can be rapidly changed to respond to changes in the operational environment. An agile process can be characterized as adaptive and responsive to unexpected information that unfolds during system development. As opposed to traditional development processes, requirements and design solutions are not frozen but continuously evolve to respond to changes.

While agility is embraced widely in software development, other industries also apply agile principles. The manufacturing industry defines agility in terms of flexibility and leanness. A manufacturing organization should respond flexibly and rapidly while eliminating waste to the extent possible without hindering the ability to respond to change. In the manufacturing setting, agility incorporates all elements of leanness including economic dimension, high quality, and simplistic components into the rapid change process (Conboy and Fitzgerald, 2004). Other related industries include agile supply chains, agile decision support systems, military forces, and command and control systems (C2). For some of these applications, agility means applying the concepts of flexibility throughout different parts of the organization. For example, military forces look at agility in individuals (commanders), in supporting information systems (C2 systems), as well as in the overall military forces. In that perspective, agility is a multidimensional concept which has elements of system attributes including robustness, resilience, responsiveness, flexibility, innovation, and adaptation (Albert and Hayes, 2003). This holistic perspective is supported by others (Nagel and Dove, 1991) with the suggestion that agility should be viewed throughout the enterprise in a business-wide context.

Apart from the processes, agility is also an important property of today's complex systems. Agile systems can be characterized as (Haberfellner and de Weck, 2005)

- Flexible, reconfigurable, and extensible
- Scalable in terms of capacity
- Flexible in terms of performance levels and functions.

Thus, agile systems can be reconfigured or modified rapidly by adding, removing components, or changing its performance levels or capacity. Adaptability is also used in the definition of agility. Dove (2005) defines agile systems in terms of response ability which consists of four metrics:

- Response time
- Response quality
- Response cost
- Response scope.

Response quality refers to the suitability and effectiveness of the selected changes. Response scope refers to the range of response alternatives and can be categorized as proactive or reactive response. Some examples of proactive response include the development of a new system to respond to a new need, upgrading an existing system to increase its performance, switching to a more efficient process, and expanding the system capabilities by integrating new subsystems with new capabilities. Reactive response examples include fixing a failed subsystem or resolving an unexpected event, increasing or decreasing capacity, and reconfiguring components of subsystems.

There are trade-offs for incorporating agility into systems including increased system complexity through the addition of interfaces and other components. The increased complexity eventually impacts the cost, mass, and other performance properties of systems. Therefore, it is important to understand when agile systems are needed and the value of agility for different stakeholders such as users, system operators, production, maintenance personnel, and system owners. Agile systems are necessary when

- Systems are large scale and have an operational life greater than 10 years where user requirements and operational environment are likely to change significantly.
- Cost of building a new system is not economically feasible each time requirements change.

Agile systems in such cases have a set of embedded sensors to monitor and alert decision-makers when change is necessary as well as flexible systems/components that allow it to be changed easily and rapidly. A decision mechanism to evaluate the cost-benefit of change is also necessary to ensure most effective strategy is used to respond to changes and agility is used when needed. Process agility is necessary when

- There is high uncertainty in the operational environment in terms of demand.
- Time to market is critical, and products are outdated quickly.

For example, in the toy industry, there is high uncertainty in terms of predicting the demand for toys, and here, agility is part of the product design process. In other settings where software applications are outdated quickly with new releases, agile processes are essential part of software development.

RELATIONSHIPS AMONG ADAPTABILITY, RESILIENCE, AND AGILITY

Adaptability, resilience, and agility are not mutually exclusive. There are interrelations between these properties. Resilience definitions vary depending on the context. For some systems, resilience includes returning to the original steady state; for other systems, resilience means rapidly maintaining an acceptable level of opera-

> Adaptability, resilience, and agility are not mutually exclusive. There are interrelations between these properties.

tion after a disturbance; and for other systems, resilience means self-organization and adaptation. We can see adaptability and agility properties discussed within the resilience definitions. A computer network must be resilient to cyber-attacks in an agile fashion and being adaptable is one way of responding to disruptions rapidly. For project managers and system designers, a variety of different definitions and interrelations among them creates a challenge in terms of managing interdisciplinary teams. Therefore, it is important to be aware of the relationships among system properties.

Generic definitions provided by standardization institutes are not helpful in terms of evaluating these system properties, and most importantly, these definitions do not consider various stakeholder value expectations. Also, SoS and other large-scale systems continuously evolve to respond to changing requirements and emergent behavior. As seen in the definition of resilience, the definition of system properties is context dependent and varies by environmental conditions, operational scenarios, and multi-stakeholder expectations.

Various studies on system properties investigated the interrelationships between different ilities from different perspectives to address the drawbacks of single view definitions. de Weck et al. (2012) studied subset–superset interrelations between ilities using means-end models. Semantic sets are formed based on the identification of closely related ilities. Within these semantic sets, a hierarchy of ilities are identified where lower levels serve as enablers for higher level ilities. Most often modularity and interoperability appear at lower levels and serve as means to achieve higher level ilities. Some of the higher level ilities include resilience, robustness, and flexibility which provide value to system stakeholders.

System Quality Definition Structure (SQDS) is another study that investigates the relationship among system ilities comprehensively by mapping subclasses of ilities to satisfy a parent class ility.

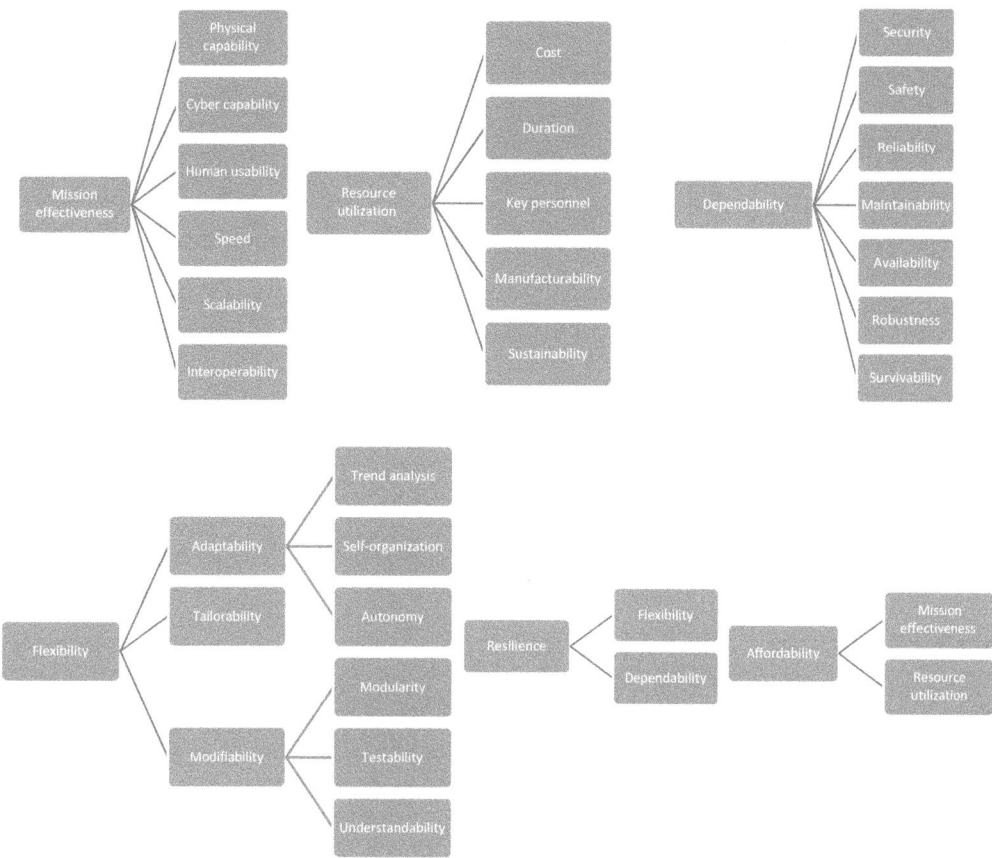

FIGURE 20.1 Means-ends hierarchy of system qualities. (Adapted from Boehm and Kukreja (2015).)

This ontology uses a modified IDEF5 ontology framework, with ility classes represented in a stakeholder value-based, means-end hierarchy (Boehm and Kukreja, 2015). Figure 20.1 illustrates these relationships in a means-end hierarchy. In this semantic ontology model, there is no overlap between subclasses within a parent class, and all the subclasses combine to satisfy the parent class system ility. For example, to achieve flexibility, modifiability, tailorability, and adaptability should be considered. System properties such as understandability, testability, and modularity are means to achieve modifiability. Trend or anomaly analysis, self-organization, and autonomous modification of system structure or state are means to achieve adaptability.

Design principles we choose to implement can very often have a negative impact on the system qualities they do not target specifically. SQDS also helps to identify conflicts among system qualities. For example, flexibility can have a negative impact on affordability, or sub-qualities such as modifiability and performance can have negative impact on each other. It is important to consider these conflicts during alternative solution generation.

METHODS AND TOOLS FOR DESIGN AND MANAGEMENT OF ADAPTABILITY, RESILIENCE, AND AGILITY

Project managers need to understand the design choices that affect adaptability, resilience, and agility of a system as these properties are critical to the success of today's systems. Several areas of research propose methods for design of these system properties at the architectural level:

a. Design principles
b. Trade space studies
c. Network analysis models
d. Computational intelligence models and other models.

> Project managers need to understand the design choices that affect adaptability, resilience, and agility of a system as these properties are critical to the success of today's systems.

Project managers also need techniques and tools to manage these system properties during implementation and system evaluation. Several methods and tools are useful in that perspective:

e. Quality function deployment (QFD) (house of quality)
f. Technical debt.

> Project managers also need techniques and tools to manage these system properties during implementation and system evaluation.

These methods, techniques, and tools are reviewed in the following subsections. While this is not a comprehensive list, it serves as a guide for project managers, system architects, and design engineers of complex engineering systems. Detailed information about these methods can be found in the cited references.

Design Principles

A design principle (also referred to as heuristic) is an abstraction of experience that has proven to be useful and can be used to guide the design process. Several studies present design principles to guide the design for adaptability, resilience, and agility. Tables 20.1–20.3 summarize some of the design principles for these system properties. A cyber–physical system, disaster assistance system (PEAK), for US military is used to explain some of the examples provided in the tables. A brief description of PEAK system is provided below. Detailed information can be found in Neill et al. (2017).

The Pre-positioned Expeditionary Assistance Kit (PEAK) is a system for providing essential services to a region in crisis as a part of a disaster relief operation. In March 2010, the Office of the

---

TABLE 20.1

Design Principles for Adaptability (Fricke and Schulz, 2005)

Design Principle	Description and Enablers	Example
Autonomy is key to achieving adaptability	Control theory principles and algorithms (feedback/control loops) supports autonomy. Object orientation with inheritance of attributes or properties and independence among objects enables autonomy.	Unmanned air vehicle (UAV) system going into safe mode or performing predefined actions at specific events. Autonomy is important in applications where user or operator access is restricted.
Decentralization	Decisions are made at local nodes that have the best information, while information is accessible to the entire system for system-level decisions. System properties can be allocated to lower levels within the system.	PEAK system situational awareness is decentralized. The first responders make decisions based on situational awareness information on the field in a timely fashion. Information is made available to US-based headquarters through satellite communication link.
Self-organization	Reorganize system components through transitioning from any state to an equilibrium state. Requires plasticity and self-awareness	PEAK water purification system can self-adjust chemicals for the purification of various water resources including blackish or salt water.

TABLE 20.2
Design Principles for Resilience

Design Principle	Description and Enablers	Example
Physical redundancy	Employ redundant hardware as a backup when systems fail	PEAK power system utilizes redundant solar panels as a backup.
Functional redundancy	Leverage heterogeneity in the SoS to provide redundant capabilities without adding new systems	PEAK power system utilizes renewable energy sources to generate power. A gasoline generator is included as a functional backup.
System-level properties	Improve system-level properties such as flexibility, adaptability of the constituent systems within SoS	The situational awareness and communication subsystem consists of wireless mesh network that sends data to a remote coordination station at the US Command headquarters through a base station that uses a very small aperture terminal (VSAT) station to create a reach back communication link via a satellite. By breaking the system into these distinct components, the individual components can address their specific system-level properties.
Repairability	Decrease total time to recovery, ensure availability of resources and personnel to reduce disruption impact	Commercial off-the-shelf (COTS) products are utilized whenever possible to reduce total time to recovery.
Internode interaction	Constituent systems in the SoS should communicate and collaborating with each other	Water, situational awareness, power, and communication nodes share system information with each other.
Localized capacity	If a node in the SoS failed, the remaining nodes should continue to function	In the water system, a pump can be replaced with one that can be operated manually to draw water in case no power is available.
Human-in-the loop	For rapid response and creative option generation, humans should be in the loop	Situational awareness UAV system is controlled by human operators.
Drift correction	Initiate resilience measures before the disruption so that mitigation steps may be initiated before actual disruption	Communication system protects data from unauthorized access over exposed communication links.
Improved communication at organizational level	Facilitate real-time information sharing between stakeholders	First responders at the disaster can communicate with each other on the field as well as reach back to US headquarters.

Source: Adapted from Uday and Marais (2015) and Jackson and Ferris (2013).

TABLE 20.3

Design Principles for Agility

Design Principle	Description and Enablers	Examples
Reusable	Encapsulated modularity—modules are separable and distinct Plug compatibility—modules can be easily added or removed due to shared interaction and interface standards Facilitated reuse—modules are reusable	The modular nature of the PEAK system allows checking and swapping components that are not compliant with export regulations.
Reconfigurable	Flat interaction—modules communicate directly on a peer-to-peer relationship Deferred commitment—module relationships are transient when possible Distributed control Self-organization—module interaction is self-adjusting	The first responders use this mesh network to connect their mobile devices and transmit situational awareness data in the form of voice, video, images, and text to the base station.
Scalable	Evolving standards—communication between modules is standardized Elastic capacity—modules can be increased or decreased Redundancy—capacity sizing can be achieved through duplicate modules	In the water system, multiple treatment chambers can be used depending on the water need or type of water—one for desalination, another for pH correction, etc.

Source: Adapted from Dove (2005).

Secretary of Defense (OSD), US Southern Command (USSOUTHCOM), and the National Defense University (NDU) partnered to form the PEAK JCTD (Joint Capability Technology Demonstration) to address the following problem statement: USSOUTHCOM's capability for promoting security and enhancing stability within its geographic area of responsibility is constrained by a limited capacity for enabling scalable critical services during time-sensitive events. The goal of the PEAK system is to provide effective, low-cost and sustainable crisis services that support and build capacity in partner nations to promote security and stability in the theater. To meet these objectives, the system must have the following broad functional capabilities:

- Provide water purification and electrical power services
- Allow communication and sharing of time-sensitive information among its users
- Local situational awareness and information sharing on threats, local populace, services, environment, infrastructure, and other support personnel.

TRADE SPACE STUDIES

Several investigators have developed techniques to help architects to interactively explore the trade space that system attributes form. Crawley's combinatorial design generation and efficient frontier computation (Selva and Crawley, 2010) derive alternative solutions and evaluate these solutions based on a combination of system property variables. Chung et al. (2000) review different approaches to addressing system attributes in software engineering: product- and process-oriented approaches. Product-oriented approaches aim to find measurable metrics for quality attributes to measure how architectural solutions meet target values. Process-oriented approaches aim to determine the interrelationships between quality attributes to support trade-off analysis in architectural design. Wang and de Neufville (2005) formulated a theoretical approach to valuing flexibility using real options.

Engel and Browning (2008) developed a quantitative approach to valuing system flexibility using architecture options. Keeney's value functions (1996; 2007) are derived from stakeholder expectations and use decision theory and trade-off analysis to address quality attributes.

NETWORK ANALYSIS STUDIES

System networks including process networks can be analyzed using network analysis tools (Braha and Bar-Yam 2006). All process charts including project management tools such as PERT charts can be analyzed using network analysis. Some of the network analyses include the slope of power laws associated with in-degree and out-degree, the effects of disruptions on the network of broken links, and the time it takes for the network to return to a steady state or to a converged state once all activities are completed (Sheard and Mostashari, 2009).

Nagurney and Qiang (2009) review network equilibrium models for large-scale systems and apply these for the analysis of transportation networks, supply chain networks with disruption risks, as well as dynamic networks such as Internet and electric power. They extend their work on assessing the network performance measure and identification of important network components.

Dependency graph networks are another useful tool in the analysis of system properties. Garvey and Pinto (2009) model system as a directed graph where nodes represent the direction, strength, and criticality of supplier–provider dependency relationships. Their Functional Dependency Network Analysis (FDNA) method determines the criticality of nodes and dependencies to understand ripple effects of service failures. Guariniello and DeLaurentis (2013) adapted the FDNA method to SoS to identify critical capabilities in operational SoS networks.

Identifying important nodes of a SoS is essential for resilience analysis. Lu et al. (2016) provide a systematic review of the metrics for the identification of important nodes of a network. Srinivasan et al. (2018) propose a discrete-time Markov chain-based ranking algorithm to rank important activities in a variety of process networks.

COMPUTATIONAL INTELLIGENCE AND OTHER METHODS

Researchers have studied the use of computational intelligence methods for searching system architectures that reflect stakeholder preferences for certain -ilities (Dagli et al., 2009). A fuzzy-genetic optimization model is constructed in Pape et al. (2013) to search for alternative SoS architectures, and fuzzy associate memory is used to evaluate the fitness of architectures based on preferred stakeholder quality attributes such as flexibility, affordability, robustness, and performance.

Other studies apply quality attribute workshops and attribute-driven design that has been used successfully in software-centric systems in the systems domain (Neill et al., 2017). Unlike most system design approaches, the quality-based approach is driven from systemic quality attributes that are determined from both the business and technical goals of the system rather than just its functional requirements. This ensures that the architecture of the system clearly, and traceably, reflects the most important goals for the system. Quality attribute-driven design is integrated with other system architecting methods such as Object Process Methodology to capture functions and desired quality attributes early in conceptual design (Kilicay-Ergin et al., 2016). Quality-based design utilizes analytically derived tactics to inform the architects design choices about system architecture.

QUALITY FUNCTION DEPLOYMENT

QFD is a tool to address quality attributes by providing traceability from customer expectations to design solutions. The tool is useful especially during implementation and production phases. Relationship matrices provide structure to translate customer expectations into technical requirements. Customer needs, technical requirements, priorities, and evaluation of competing products are captured in relationship matrices which are also referred to as the "house of quality". QFD supports the management of system qualities and avoids loss of information. However, there are some drawbacks. This tool does not provide design principles to address quality attributes or how to solve bottlenecks that may arise from the interaction of different quality attributes. Further information about this tool can be found in Akao (1990).

TECHNICAL DEBT

Technical debt is a concept used in agile software development to understand the trade-off between short-term benefit of releasing software rapidly versus the additional rework caused by selecting easier solutions instead of better solutions that would take longer. Kruchten et al. (2012) discuss ways to tackle technical debt and the decision process of balancing cost and value. Seaman and Guo (2011) provide processes and tools to measure and manage technical debt in software systems.

SUMMARY

Adaptability, resilience, and agility are inherent properties of today's complex engineering systems. While the most important role of a project manager is to ensure that the project stays within budget and is delivered on time and with high quality, it is crucial to manage these inherent properties throughout system development and implementation. For example, SoS engineering aims to maintain dynamic stability through self-adjustment and evolution. There should be a balance between maintaining order and sufficient chaos for growth. Finding this balance is not trivial because SoS capabilities can conflict with constituent system objectives. Besides, individual systems can belong to more than one SoS leading to conflicts. Managing adaptability within SoS and other large-scale systems requires an understanding of the elements of adaptability. This chapter provided an overview of definitions for adaptability where autonomy, self-organization, decentralization, and learning are some of the key enablers for this system property. Adaptability is also a key enabler for other system-level properties including resilience and agility.

Resilience is another important system property that provides value to system stakeholders. The architecture of the system, its operational environment, and types of disruptions play a role on system resilience. A system can be resilient in some disruptions but not in others.

Agility is now an essential part of both the system development process and a property of the system itself. Agility consists of two properties: flexibility and speed. Adaptability, resilience, and agility are not mutually exclusive. There are interrelations between these properties. Various methods and tools are available for analysis and design of system properties including the application of design principles, trade space studies, network analysis, and model-based approaches including computational intelligence methods. QFD and technical debt are some of the tools that can support project managers in terms of managing these system properties effectively. Ricci et al. (2014) provide a systemic process that can guide project managers and architects in the quest for designing adaptability, resilience, and agility in SoS. While the process is focused on SoS engineering, it can be adapted to individual systems as well. The method steps are summarized:

1. Identify SoS value propositions and design constraints.
2. Identify possible sources of uncertainty that could jeopardize value delivery and parameterize these into perturbations.
3. Identify desired ilities using analytical tools such as semantic ontology models and means-end models. Certain ilities can be more important than others depending on the perturbations.

4. Generate alternative architectures based on the concept of operations and associated design variables. This step generates high-level architecture alternatives using traditional system architecting methods.

5. Generate ility driving architecture alternatives. This time the focus is on generating alternative solutions that enable desired ilities identified in Step 3. Here, design principles for specific ilities (discussed in the "Design Principles" section) or other methods such as quality attribute-driven design or evolutionary search-based architecture models and network models can be utilized (discussed in the "Network Analysis Studies" and "Computational Intelligence and Other Methods" sections).

6. Evaluate ility driving architecture alternatives using various metrics such as cost, performance, and other ility metrics.

7. Analyze architecture alternatives using various trade space analysis approaches (discussed in the "Trade Space Studies" section).

8. Select the best architecture with ilities is the final step of the process and involves selecting the best architecture considering the analysis done in Step 5. Once this selection is documented, iterations can continue to get more insights or the conceptual design outcome; architecture with desired ilities can proceed to the next phase.

This process shares common themes with quality attribute-driven design methodology discussed in the "Computational Intelligence and Other Methods" section. Both emphasize business and stakeholder value view throughout architecture generation. Both emphasize using design principles and tactics to generate ility driving architecture alternatives. Both emphasize importance of trade-off analysis. While these process models provide a systematic guide to address and design ilities including adaptability, resilience, and agility, there are still many open research areas that need to be explored. For example, there are cost/benefit imbalances for the constituent systems within a SoS. It is important to find ways to resolve these imbalances; otherwise, independent systems will not cooperate in design of ilities within their system. As discussed in the previous sections, there are interdependencies between ilities. As stakeholder value propositions evolve, it is important to understand the co-evolution of multiple ilities. This also necessitates a need for the visualization of SoS ilities to improve decision-making. Finally, there is now a strong need for validated formal models of system properties and link to design principles. This will facilitate the use of Ricci's process and quality attribute-driven design within a wide variety of system applications including cyber–physical systems, Internet of Things, and smart city applications.

DISCUSSION QUESTIONS

1. Look at the disaster assistance system (PEAK) presented in this chapter. Discuss the definitions of adaptability, resilience, and agility for the disaster assistance system (PEAK). Alternatively identify a complex engineering system of interest to you and discuss what adaptability, resilience, and agility mean in that context.

2. Most often modularity and interoperability serve as means to achieve higher level ilities. Discuss how modularity and interoperability can serve as means to achieve adaptability, resilience, and agility for the given PEAK system or a complex engineering system of interest to you.

3. It is impossible to optimize a system for all system qualities that are important for its stakeholders. Therefore, it is necessary to identify the interrelationships between system qualities and determine whether one quality would positively or negatively affect any other system qualities. Construct a trade-off matrix for the PEAK system (or an alternative system of interest to you), and identify positive and negative relations among system qualities relevant for the PEAK system. For example, flexibility positively affects maintainability but negatively affects reliability.

REFERENCES

Akao, Y. 1990. *Quality Function Deployment: Integrating Customer Requirements into Product Design.* Productivity Press, Cambridge, MA.

Albert, D. S. and R. E. Hayes. 2003. Power to the edge. DoD command and control research program. www.dtic.mil/dtic/tr/fulltext/u2/a457861.pdf (accessed October 15, 2018).

Bar-Yam, Y. 2006. Engineering complex systems: Multiscale analysis and evolutionary engineering. In *Complex Engineered Systems: Science Meets Technology*, D. Braha, A. A. Minai, and Y. Bar-Yam (Editors), pp. 22–39, Springer, Cambridge, MA.

Beesemyer, J. C., A. M. Ross, and D. H. Rhodes. 2012. An empirical investigation of system changes to frame links between design decisions and ilities. *Procedia Computer Science* 8:31–38. Conference on Systems Engineering Research.

Boehm, B. and N. Kukreja. 2015. An initial ontology for system qualities. *INCOSE International Symposium* 25:1:341–356, Seattle.

Chalupnik, M. J., D. C. Wynn, and J. Clakson. 2013. Comparison of ilities for protection against uncertainty in system design. *Journal of Engineering Design* 24:12:814–829.

Chung, L., B. A. Nixon, E. Yu, and J. Mylopoulos. 2000. *Non-Functional Requirements in Software Engineering.* Kluwer, Boston, MA.

Cohen, M., P. Pathak, and A. Samuel. 2018. Process resilience is becoming a business imperative. WIPRO, 2014, http://knowledge.wharton.upenn.edu/article/process-resilience-becoming-business-imperative/ (accessed November 1, 2018).

Conboy, K. and B. Fitzgerald. 2004. Toward a conceptual framework of agile methods: A study of agility in different disciplines. *Proceedings of the 2004 ACM Workshop on Interdisciplinary Software Engineering Research*, Newport Beach, CA, pp. 37–44.

Crawley, E., O. D. Weck, E. Eppinger, C. Magee, J. Moses, W. Seering, J. Schindall, D. Wallace, and D. Whitney. 2004. Engineering systems monograph, Massachusetts Institute of Technology, Technical Report. http://strategic.mit.edu/docs/architecture-b.pdf (accessed June 26, 2019).

Dagli, C. H., A. Singh, J. P. Dauby, and R. Wang. 2009. Smart systems architecting: Computational intelligence applied to trade space exploration and system design. *Systems Research Forum* 3:2:101–119.

de Neufville, R., O. de Weck, D. Frey, D. Hastings, R. Larson, D. Simchi-Levi, K. Oye, A. Weigel, and R. Welsch. 2004. Uncertainty management for engineering systems planning and design. MIT Engineering Systems Symposium, Cambridge, MA.

de Weck, O. L., D. Roos, and C. L. Magee. 2011. *Engineering Systems: Meeting Human Needs in a Complex Technological World.* The MIT Press, Cambridge, MA.

de Weck, O. L., A. M. Ross, and D. H. Rhodes. 2012. Investigating relationships and semantic sets amongst system lifecycle properties (ilities). *Third International Engineering Systems Symposium*, Delft University of Technology, Delft, The Netherlands, 18–20 June 2012.

Dove, R. 2005. Fundamental principles for agile systems engineering. *Conference on Systems Engineering Research*, Hoboken, NJ.

Engel, A. and T. R. Browning. 2008. Designing systems for adaptability by means of architecture options. *Systems Engineering Journal* 11:2:125–146.

Fricke, E. and A. Schulz. 2005. Design for changeability: Principles to enable changes in systems throughout their entire lifecycle. *Systems Engineering Journal* 8:4:342–359.

Garvey, P., and C. Pinto. 2009. Introduction to functional dependency network analysis. *Second International Symposium on Engineering Systems*, MIT, Cambridge, MA.

Guariniello, C. and D. DeLaurentis. 2013. Dependency analysis of system-of-systems operational and development networks. *Conference on Systems Engineering Research*, Atlanta, GA.

Haberfellner, R. and O. de Weck. 2005. Agile systems engineering versus agile systems engineering. *15th Annual International Symposium of the INCOSE*, 10–15 July.

Haghnevis, M. and R. G. Askin. 2012. A modeling framework for engineered complex adaptive systems. *IEEE Systems Journal* 6:3:520–530.

Haimes. Y. 2009. On the definition of resilience in systems. *Risk Analysis* 29:4:498–501.

Hosseini, S., K. Barker, and J. E. Ramirez-Marquez. 2016. A review of definitions and measures of system resilience. *Reliability Engineering and System Safety* 145:47–61.

Jackson, S. and T. L. J. Ferris. 2013. Resilience principles for engineered systems. *Systems Engineering* 16:2:152–164.

Keeney, R. L. 1996. Value-focused thinking: Identifying decision opportunities and creating alternatives. *European Journal of Operational Research* 92:537–549.

Keeney, R. L. and D. Winterfeldt. 2007. Practical value models. In *Advances in Decision Analysis*, W. Edwards, R. F. Miles Jr., and D. von Winterfeldt (Editors), pp. 232–252, Cambridge University Press, New York.

Kilicay-Ergin, N. H., C. J. Neill, and R. S. Sangwan. 2016. Integrating object-process methodology with attribute driven design. *Annual IEEE Systems Conference*, Orlando, FL, pp. 1–8.

Kruchten, P., R. L. Nord, and I. Ozkaya. 2012. Technical debt: From metaphor to theory and practice. *IEEE Computer Society* 29:6:18–21.

Madni, A. M. and S. Jackson. 2009. Towards a conceptual framework for resilience engineering. *IEEE Systems Journal* 3:2:181–191.

Muller, H. 2018. The rise of intelligent cyber-physical systems. *Computing Edge*, IEEE Computer Society, pp. 31–33.

Nagel, R. and R. Dove. 1991. *21st Century Manufacturing: Enterprise Strategy.* Iacocca Institute, Lehigh University, Bethlehem, PA.

Nagurney, A. and Q. Qiang. 2009. *Fragile Networks: Identifying Vulnerabilities and Synergies in an Uncertain World*. John Wiley & Sons, Hoboken, NJ.

Neill, C., R. Sagwan, and N. Kilicay-Ergin. 2017. A prescriptive approach to quality focused system architecture. *IEEE Systems Journal* 11:4:1994–2005.

Lu, L., D. Chen, X. Ren, Q. Zhang, Y. Zhang, and T. Zhou. 2016. Vital nodes identification in complex networks. *Physics Reports* 650:1–63.

Pape, L., K. Giammarco, J. Colombi, N. Kilicay-Ergin, and G. Rebovich. 2013. A fuzzy evaluation method for system-of-systems meta-architectures. *Procedia Computer Science* 16:245–254. Conference on Systems Engineering Research.

Ricci, N., M. E. Fitzgerald, A. M. Ross, and D. H. Rhodes. 2014. Architecting systems of systems with ilities: An overview of the SAI method. *Procedia Computer Science* 28:322–331. Conference on Systems Engineering Research.

Ross, A. M., D. H. Rhodes, and D. E. Hastings. 2008. Defining changeability: Reconciling, flexibility, adaptability, scalability, modifiability, and robustness for maintaining system lifecycle value. *Systems Engineering Journal* 11:3:246–262.

Sanchez-Escribano, M. G. and R. Sanz. 2014. Emotions and the engineering of adaptiveness in complex systems. *Proceedia Computer Science* 28:473–480. Conference on Systems Engineering Research.

Seaman, C. and Y. Guo. 2011. Measuring and monitoring technical debt. *Advances in Computers* 82:25–45.

Selva, D. and E. F. Crawley. 2010. Integrated assessment of packaging architectures in earth observing programs. *IEEE Aerospace Conference*, Big Sky, MT, pp. 1–17.

Sheard, S. A. and A. Mostashari. 2009. Principles of complex systems for systems engineering. *Systems Engineering Journal* 12:4:295–311.

Srinivasan, S. M., N. Kilicay-Ergin, R. S. Sangwan, and C. J. Neil. 2018. Ranking critical activities in process architectures. *Procedia Computer Science* 140:46–55. Complex Adaptive Systems Conference.

Takeuchi, H. and I. Nonaka. 1986. The new new product development game. *Harvard Business Review*, pp. 1–11.

Uday, P. and K. Marais. 2015. Designing resilient system-of-systems: A survey of metrics, methods, and challenges. *Systems Engineering Journal* 18:5:491–510.

Valverde, M. S. and R.V. Sole. 2007. Hierarchical small worlds in software architecture. Special Issue on Software Engineering and Complex Network Dynamics of Continuous, Discrete, and Impulsive Systems Series B.

Wang, T. and R. de Neufville. 2005. Real options in projects. *Proceedings of the 9th Real Options Annual International Conference*, Paris, France.

21 Cyber-Systemic Toolset

Fredmund Malik
Malik International AG

Nam Nguyen
Malik Australia and SE Asia
Adelaide Business School

CONTENTS

THE CHALLENGE OF THE GREAT TRANSFORMATION21 FROM THE OLD WORLD TO A NEW WORLD

Globally, economies and societies are going through the most fundamental change in history. We are experiencing the displacement of the "Old World", as we have come to know it, by the "New World", which is still largely unknown. It is the origin of a new order and a new societal functioning—a new kind of societal revolution.[1]

This process has been called the "Great Transformation21" since 1997.[2] This transformation will change almost everything: *What* we do, *how* we do it and *why* we do it—and, maybe also, *who we are*. In just a few years, almost everything will be new and different: How we manufacture, transport, finance and consume; how we educate, learn, do research and innovate; how we share information, communicate and cooperate; and how we work and live. Most probably that will also change *who* we are.[3]

A new dynamic order is forming, and—more importantly—so is a new mode of functioning of society and its organizations. The all prominent digitalization is but one of the most powerful and most visible of several major driving forces.[3] Digitalization as such is not really new, however, on its present and still growing capacity and speed it leads to ever more interconnectivity: Potentially everything is going to be interconnected to everything else like in natural ecosystems.

Other challenges are the worldwide demographic changes, and the global ecology and economies. Together and through their mutual interconnectivities, they produce a new property of reality—which is exponentially exploding systemic complexity. For most people, this is an entirely new kind of reality of which most of them have never heard of and which they have never experienced. Complexity, on the one hand, can be seen as the "Great Destroyer" and, on the other hand, also as the "Great Creator"—depending on how we cope with it.

Its full potential will be exploited by an equally powerful force which is a new kind of management, governance and leadership.[4] It is the entirely new and innovative kind of managing complexity itself, because complexity is the very source of intelligence, of creativity and of novel solutions. It is the kind of management which makes use of the complexity sciences like system theory and above all of the science of cybernetics which is the basis for communication, control and navigation. It is the cyber-systemic kind of managing complex systems. As we will later show, both digitalization and cybernetic management have much in common since they had their birth at the very same time and in the same place.[4]

> Complexity, on the one hand, can be seen as the Great Destroyer and, on the other hand, also has the Great Creator.

The Great Transformation is the reason why ever more organizations—in the business sector as well as in the non-business sector—operate in a zone of excessive and accelerating changes. The origin of bureaucratic paralysis and ossification lies in the obsolete methods of conventional mechanistic management. Today's organizations prevent solutions and even contribute to the intensification of crises with their growing inability to master complexity. Digitization, on the one hand, is aggravating the difficulties, and, on the other hand, it will be one of the major solutions if it is properly applied.[5]

All the social mechanisms that make organizations function will change fundamentally and irreversibly—worldwide. Millions of organizations of every kind and size will have to adapt and be rebuilt, as they no longer meet the new standards. Across the generations, people will be required to rethink and relearn.[1]

THE MAP AND MODEL OF GROWTH, OF CREATIVE DESTRUCTION AND OF EVOLUTIONARY INNOVATION

Change in itself—even big change—is not unusual. There is always improvement, adaptation, innovation, and also disruption. Here, we are talking not about any kind of change but about a very specific kind of change, the kind that will replace the existing with something new according to a pattern. We are talking about *substitution*. The famous Austrian economist Joseph Schumpeter called this kind of change "creative destruction". With that, he put in words the basic law of change that also governs evolution in nature.[6]

The favorite paradigm of the Great Transformation is two overlapping s-curves (Figure 21.1) because most people grasp it immediately and rarely forget it once they have seen it. The research into this pattern goes back to the late 1970s.[5] They are s-shaped because they represent growth processes, and there is no such thing as linear growth processes. The graph represents the "Old World", the "New World" and the foundations of tomorrow's existence.[1]

Between the curves, we see an area of increasing turbulences, as the old is replaced by the new. This is the critical decision zone; this is also where disruptions can take place; this is where the "Old World" starts dissolving and the "New World" begins to take shape. It is the kind of process which in biology is called metamorphosis, an example of which is the transformation of a caterpillar into a butterfly. In the societal organizations, it is predominantly large and complex projects by which change will be brought about.[1]

This is the zone where the really difficult questions and risks of navigation and management occur. Previous key resources become increasingly meaningless; they have to be shifted or newly created.

In search of answers, many reflexively still cling to the old methods, although they are becoming increasingly weak and useless clashing against the walls of complexity for which they have never been designed.[5] To a large part, it is these old methods themselves which are causing the troubles in the first place when one continues to apply them where they have to fail by their very nature.

Among other things, one key question is whether people in a "red" business will also be able to contribute to a "green" business, and all of a sudden, it is doubtful as to whether you will have use for even your best people in the future. This zone of shifts is a black box—as it is named in the science of cybernetics. It is a system which, due to its immense complexity, is incomprehensible, unpredictable and incomputable. However, with the new kind of management, we are able to cope with it—but not with conventional methods.[7]

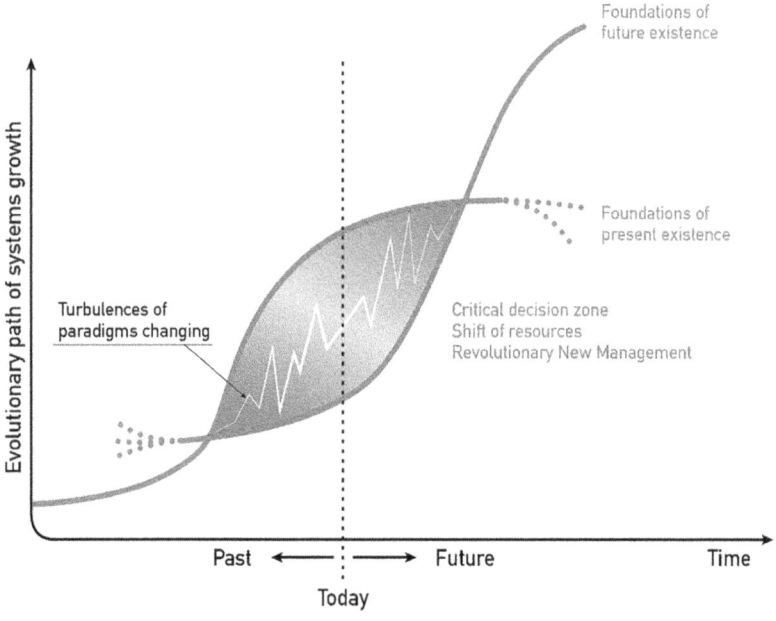

FIGURE 21.1 The paradigm of the Great Transformation[1]: growth, creative destruction, and evolutionary innovation.

NAVIGATING INTO THE UNKNOWN BY SYSTEMATICALLY MISLEADING SIGNALS

One basic rule of change is: *Whatever exists will be replaced.* Looking at it with hindsight, we know what the pattern looks like. We also know then what would have been the right decision at any given point. Standing in the here and now (Figure 21.2), the existing information systems in our present organizations deliver systematically misleading signals as their output.[1]

In today's world, the signals tell us to continue on the red curve. To the extent that—even if we notice the green curve at all and take it seriously—our old compass warns us not to pursue that route. Only when it is very late or too late, do our old systems sound the alarm.[1] However, by knowing the entire pattern up to a certain degree, the risk of making wrong decisions can be circumvented or at least reduced.

Needed: Three different strategies and systemic leadership[1]

There is yet another important challenge. We need not only one but three different strategies: The first strategy is to take advantage of the red curve as long as possible. We need a second strategy to build the green curve in time to have it when we need it. And we need a third strategy to make the metamorphic transition from red to green.[1]

At this point, it also becomes evident where we need real leadership and what it must entail. True leadership is needed for the start into an unknown future, when all visible signs seem to indicate that we should stay in the past. There are such organizations that have mastered change several times in the course of their history, mainly by making it happen themselves. Examples include Siemens, Bosch and for a long time also General Electric, but not Kodak, to give a contrarian example. For instance, nothing could be more useless than having the world's best chemists in the photo industry when the substituting technology is digital. Virtually overnight Kodak's most valuable asset—the knowledge of its top people—had become worthless. What is worse, apart from having become "useless", these same people offered the strongest resistance to digitization for the very reason that they were the best chemists in their field.[1]

Being ahead of change by substituting yourself

Just as there is a substitution pattern, there is also a strategic principle that successful organizations adhere to: *Be ahead of change!* They actively make change happen instead of waiting for it to

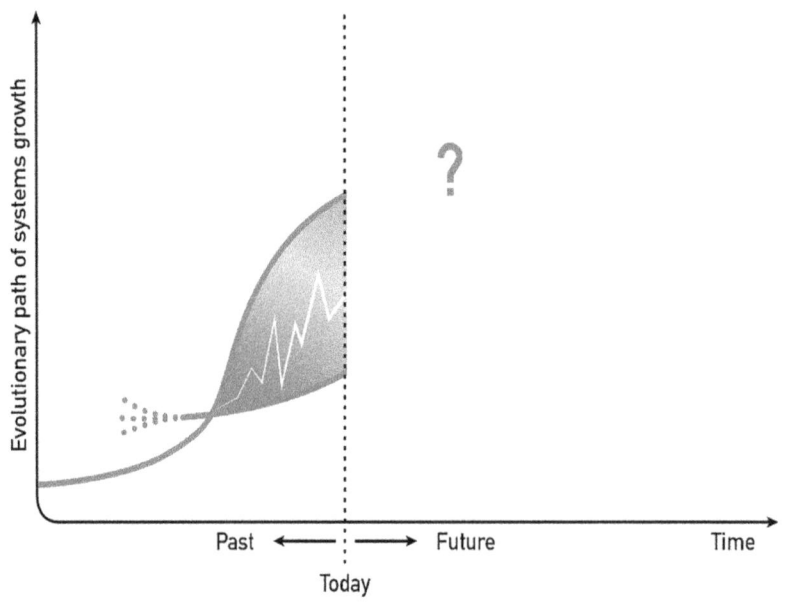

FIGURE 21.2   Here and now: navigating into the unknown with systematically misleading signals.[1]

hit them. They take advantage of the forces of this relentless law of business—and not only business—to start into a new dimension of performance rather than fight it. They keep the initiative and invent the rules. Hence, change is not a must but a want to them. The organization itself determines what happens instead of drifting along. By outgrowing itself and its previous limits, it in effect substitutes itself. *If we don't do this, others will. It will happen one way or another*—that is their maxim.[1]

RADICAL NEW GOVERNANCE THINKING: FROM THE MECHANISTIC MANAGEMENT TO CYBER-SYSTEMIC MANAGEMENT OF COMPLEXITY

Information technology is one of the major drivers of the Great Transformation. However, digitization alone would soon freeze within the labyrinths of obsolete organizations and mechanistic management processes. Most probably, it would just reinforce already existing bureaucracy. Successful digitization needs cyber-systemic complexity management. Only together will they create the ability and willingness of people and organizations for fast change.[5]

Properly managed digitization primarily means the possibility of rapidly growing interconnectedness of hitherto separated systems. And it also means doing things simultaneously which so far could only be done sequentially. As a consequence, this means an exponential increase in intelligence, adaptively, speed and productivity. This is what challenges millions of organizations in our modern complexity society. Without functioning organizations, a collapse of societal functioning looms.[4]

The danger is real because the origin of most of today's organizations' morphology and principles of functioning reach far back into the last century. Therefore, they are ever less equipped to deal with the challenges of today's complexity and speed. They are too slow for the transformation, not efficient enough and not adaptable enough.[1]

Decision-making processes seem paralyzed and block themselves. Collective intelligence, creativity, innovation and ability for change are lacking, as well as self-coordination, self-regulation and self-organization. If we were to stick to conventional ways of thinking and methods, a social disaster would be inevitable. On the other hand, a historically unique period of prosperity on a global scale could be reached if we rethink now. A new societal order—facilitating a *humane as well as functioning way of living together*—could thus be created, beyond the more than 200-year-old and gridlocked political ideologies.[5]

DIGITIZATION AND CYBER-SYSTEMIC MANAGEMENT ARE TWINS

Digitization and the cyber-systemic management of organizations have the same hour and place of birth. The origin of cyber-systemic—or simply cyber management—lies exactly in the time and place that also saw the development of today's computer technology. The foundations for modern computer technology are the same as for the proper management of societal organizations. Some of the pioneers realized early on that the same principles apply to both areas.[5]

To name but a few: the mathematician Norbert Wiener, the founder of modern cybernetics, with his book *Cybernetics: Control and Communication in the Animal and the Machine*,[8] (also wrote *The Human Use of Human Beings*).[9] The British neurophysiologist Ross W. Ashby with his revolutionary book *Design for a Brain*.[10] Added to these were John von Neuman, the inventor of the modern computer, Claude Shannon's *Information Theory*, Heinz von Foerster's *Causal Circularity*, and also Warren Mc Culloch's *Embodiments of Mind*.[11]

With their early theories on information and communication, algorithms and heuristics and on the design and navigation of dynamically complex systems, they have created the prerequisites for today's real "cyberspace" as well as for the cybernetics of the management of complex systems. It was still too early for this new cyber-systemic management though. For decades, the industrialized society's mechanistic notion of management would continue to dominate, and it would be taught at thousands of universities and business schools—until today.[5]

ORGANIZATIONS AS LIVING ORGANISM

The current challenge of the Great Transformation21 needs and forces the renunciation of mechanistic management. The prevalent notion of the last decades that the company is a machine which can be steered with the linear principles of cause and effect blocked the necessary progress.[12]

The more helpful notion is the organization as a living organism in its evolutionary environment. The new kind of management based on the findings of cybernetics then is enabling organizations to self-organize and self-regulate wherever numerous and ever more people work together to reach common goals. Contrary to widespread fears, this is exactly what allows people the freedom for the first time to unfold their intelligence and creativity in the digital world in a new and better way.[6]

The new goal is the adaptive viability of a flexible organization that far surpasses the notion of sustainability. *What* needs to be done has been identified—and cyber-systemic management provides the *How* and *Whereby*. Hence, digital interconnectivity and cyber-systemic management of complexity will become the very societal functions that enable people to effectively exploit the new possibilities and opportunities of the New World as it evolves.[5]

THE TWO LAWS OF PROPERLY USING COMPLEXITY

The proper use of complexity is governed by two laws of nature. They are all present in the natural ecosystems: the Law of Interconnectivity and the Law of Simultaneity.[5] Once recognized, they are easy to understand, because examples abound. However, the invention of the proper methods and tools to apply them in the context of societal organizations took time and was not that easy.

The Law of Interconnectivity
What with old methods can be done only separately can be done interconnectedly with cyber tools.
The Law of Simultaneity
What could be done only sequentially with the old methods with cyber tools can now be done simultaneously.

These are the laws which produce intelligence, speed, harmony, consensus and joint self-organizing action.

The most obvious system which operates on those two laws are living brains and the most effective one so far is the human brain.

We are now going to show two breakthrough cyber technologies which *amplify speed and effectiveness of complex project management* by factors of 60–100. The first cyber-system technology is called *Syntegration* and the second *Cyber Sensitivity Modelling*. Together, they demonstrate that higher competencies only result from higher complexity.

BIG CHANGE WITH SYNTEGRATION TECHNOLOGY— COMPLEX PROJECT AND STAKEHOLDER MANAGEMENT

WHAT IS SYNTEGRATION?

The term "Syntegration" is a combination of "synergy" and "integration". It stands for our innovative system-technological process to master highly complex challenges and transform organizations.

The basis of our work is our results on the functioning of complex systems gained in over 30 years of research and development of organization and management systems suitable for dealing with complexity.[5] The scientific bases of our system technologies and social techniques are the complexity sciences. These comprise systems theory, cybernetics and bionics, information and communication theory, algorithmics and heuristics as well as the geometry of interconnection as given by complex hyper polyhedrons.[13] Our understanding of governance and management far exceeds the notions of business administration here. It is verifiably wrong to simply reduce complexity. This destroys all efficient and intelligent functions. This is why complexity is new resource for a better functioning of our society's organizations.[1]

Syntegration is an innovative system-technological process to master highly complex challenges and transferal organizations as traditional change methods are too weak and too slow for the profound change required in this day and age. The developed solutions have a maximum chance of implementation due to the common consensus on the cultural level.

Syntegration is by far more effective and efficient than other change methods. It is an altogether new way to communicate, orchestrate, synchronize, innovate and collaborate. The effect is an unparalleled increase of management effectiveness and an amplification of institutional leadership. It applies in particular when there are large numbers of stakeholders.

> We can't master the challenges we are facing with conventional methods anymore…

This is one of the results of a survey that we recently conducted. The participants were 250 top executives of companies in the German-speaking area, CEOs, and Heads of HR. We have been conducting such surveys already for a long time. But we experienced this kind of unambiguousness for the first time. Until now, the opinions were usually mixed: "Somehow we can make it work … it may be difficult, but we can manage it".

The executives do not say that there is no way to manage the change. They do say, however, that it can't be managed with conventional methods. So what are the conventional methods for change management? Count among them, the "team" and team work, small group workshops and conventional management meetings. However, they still apply where complexity is low.

The 250 executives are right in their conclusion and are very aware of the new situation. In comparison with the new process, traditional change methods are too weak and too slow for the profound change required in this day and age.

What the Syntegration Methodology Accomplishes

The Syntegration methods are innovative, high-performance and high-speed procedures for mastering the most complex challenges and big change. During a Syntegration, the 40 most knowledgeable and important people together develop the solutions to a key topic—which has been determined by the corporate management—in a 4-day intensive closed conference (large group workshop) on a level playing field. The Syntegration process is superior to any other method in terms of effectiveness and speed of operation by factors between 60 and 100.

At the end of the Syntegration, the following results will be present:

1. Solutions to the key topic and to its necessary and interconnected sub-subjects.
2. The objectives and measures required for implementation.
3. A comprehensive analysis and diagnosis of the existing management systems as well as a necessary concept with objectives and measures.
4. The implementation concept for the compiled solutions for points 1 and 3.
5. Allotted responsibilities to individuals for implementation.
6. An innovative reinforcement of knowledge and intelligence through a simultaneous, interconnected and cross-organizational involvement of the selected people.
7. A measureable improvement of corporate culture.

SAMPLE APPLICATIONS

The Syntegration method can be applied to those challenges that have their origins in the ongoing "Great Transformation". They are problems whose solution calls for profound and frequently also fast change:

1. Complex innovation projects, in connection with a strategic new alignment.
2. Cost reduction and improvements in efficiency, often in connection with a restructuring in the organization and a realignment of the lines of business.
3. Growth strategies, also in connection with cross-organizational partnerships.
4. The development of new business models, strategic positioning connected to personnel and organizational problems.
5. Post-merger integration interconnecting different cultures, developing a common picture of the future and integration of the two companies in the merger.
6. Transformation of organizational structures, in connection with questions of corporate culture and a reform of the management and personnel systems together with the necessary rules of management and leadership.
7. Large and complex digitalization projects in connection with interventions into the organizational structures and business processes.

The Syntegration framework has been applied more than 1,000 times since the early 2000s in all kinds of organizations: business companies from medium-sized to large corporations, organizations from public administration, scientific research, politics, sports and more. The confidentially written rating of satisfaction by participants (always the top 30–40 persons of the respective organization) has been more than 90% sum of very good and good. The oral comments are from very good to enthusiastic since people have never experienced something similar. Written testimonials by top executives of respective organizations are excellent without exception. Enclosed with the submission of this chapter are several example cases and feedback from these applications.

INDEPENDENT FROM THE CONTENT OF A CHALLENGE

The Syntegration applications have the big advantage of being independent from the specific contents on the task level of a challenge. Meaning that methodically, it isn't decisive for the application of the Syntegration whether it is, for example, about cost problems, questions of strategy, changes in the organization, tapping into new markets, IT projects, product innovations or corporate culture.

It is, however, much more decisive for the application of the Syntegration procedures whether the challenges are complex or simple, whether they are heavily interconnected or can be worked on individually, whether they change rapidly or whether they remain relatively stable because the content/expertise requirements in a narrower sense are being guaranteed by choosing the right selection of participants.

A SUPERIOR SOLUTION METHODOLOGY FOR COMPLEX CHALLENGES

A huge number of specialists and managers, who simultaneously bring and interconnect their knowledge, experience and intelligence, to ensure that optimal solutions can be achieved, are almost always needed for mastering complex challenges today. This works best when the participants can agree on a level

The best solutions are created when such challenges are dealt with in a holistic, collective, simultaneous and interconnected way.

playing field and with equal participation. The best solutions are created when such challenges are dealt with in a holistic, collective, simultaneous and interconnected way.

Simultaneous Effect on Several Levels

The Syntegration procedures are simultaneously effective on the task level, the cultural level as well as on the management level. On the task level, optimal and innovative solutions, objectives and measures are being created. On the cultural level, the Syntegration procedures release a strong social energy. Both a strong feeling of togetherness and individual commitment are created by this kind of cooperation. During work on the solutions, a culture of common strength and common trust as well as high consensus is being created. The developed solutions have a maximum chance of implementation due to the common consensus on the cultural level. On the management level, the catalyzers for implementing the best solutions are being released.

Solutions on the Management, Governance and Leadership Level

The Syntegration events are optimal opportunities for a comprehensive diagnosis of a company's management and regulatory systems. The strengths and weaknesses of the management systems become apparent due to the special communication processes of the Syntegration. This also includes the corporate strategy, structure and culture, the corporate governance, corporate mission and corporate policy leadership as well as management effectivity and efficiency. At the end of a 4-day Syntegration, reliable diagnoses for the management level are clear, and also—if need be—recommendations for necessary measures for improvement.

How Does the Syntegration Work?

The Syntegration methodology consists of a combination of some 30 different kinds of communication processes that are dynamically interconnected with each other. Following clear rules, these processes function reliably and consistently, because they are self-regulating and work in real time. Therefore, no "idling circle" exists, which is typical for many of the traditional methods (Figure 21.3).

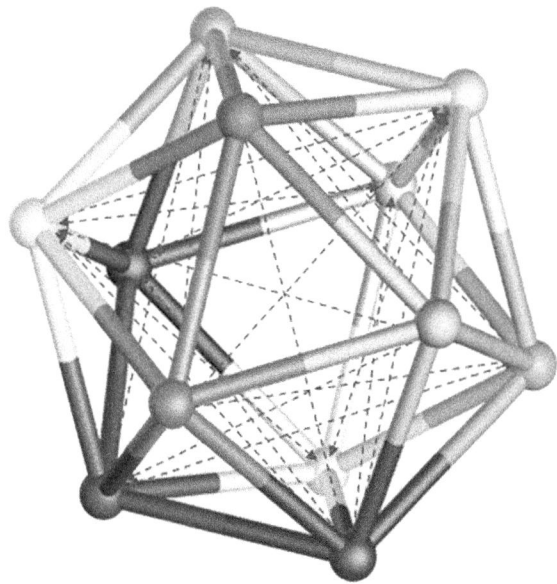

FIGURE 21.3 The "Icosahedron", the most complex Platonic solid.[14] Its complex geometric structure serves as the logic control for the Syntegration processes.

All participants are always active during Syntegrations in different roles and contribute to the common success. The process architecture ensures that nobody can remain passive. The participants often work on their topics voluntarily and with major commitment until the late evening hours. An intensity prevails that can hardly be ever observed during usual business routines.

All processes are being supported and accompanied by experienced facilitators.

Procedural Sequence

The application of Syntegration procedures follows clearly defined steps for which the client gets thorough support.

1. Determination of the key topic which will be solved with the Syntegration. It serves as the opening and guiding mission of the syntegrative event.
2. Formulating the key topic in the form of a leading action message along the lines of the question "What do we have to do in order to...". This question is the navigational goal for starting the process and the measure for the solutions.
3. Selection of participants that are necessary for the optimal solution of the opening question. There are two basic criteria: first, concerning the participants' knowledge and, second, their importance in the given corporate culture.
4. Determination of date and place.
5. The further stages of the process will follow a clearly structured protocol which regulates the preparation phase, the execution phase as well as the implementation phase.

IMPLEMENTATION OF SOLUTIONS

The subsequent implementation process is determined on the organizational and personnel level in advance of the Syntegration event, so that the implementation can be started without any delay. This succeeds very quickly because the key executives themselves will have participated in drafting the measures and solutions in consensus.

THE SENSITIVITY MODELING TECHNOLOGY FOR UNDERSTANDING AND MANAGING COMPLEX SYSTEMS

During the Syntegration process, the huge interconnectivity of 12 topics and 40 persons in three different roles produces an enormous amount of information in the realm of terabytes. It is not just data but information proper because it is interconnected and therefore it comes in patterns.

These can be explored and skimmed by specially designed processes. We gave that methodology the name of *Total Immersion Exploration*.

One of our most effective cybernetic tools of the kind of exploration is the Cybernetic *Sensitivity Model (SensiMod)*.[14] It was pioneered by Prof. Dr. Frederic Vester. The Sensitivity Model is a tool for the depiction of apparently unconnected parts of reality as an interconnected system. One then moves on to find out cybernetic properties of the system and the most effective ways of understanding and handling it.

> The Sensitivity Model is a tool for the depiction of apparently unconnected parts of reality as an interconnected system.

Table 21.1 provides a brief comparison of several systems tools.

Our main comparison/consideration is between the first three systems tools/models. In addition to the features above, SensiMod has other additional advantages, for example,

- With SensiMod, simulation is only one of nine largely independent procedural tools. It is a complementing tool to the other tools of a SensiMod, and going through each of the nine steps already provides a lot of insights and understanding of the system under consideration.

TABLE 21.1

A Brief Comparison—Systems Tools/Models

Features		1 Systems tools (SensiMod)	2 Systems dynamics	3 BBN	4 Consideo modeler	5 Meta models	6 Coupled complex models	7 Agent-based models	8 Expert systems
Model purpose	Prediction			X	X	X	X		X
	Forecasting	X	X		X	X	X		X
	Decision-making	X	X	X	X	X	X		X
	System understanding	X	X		X		X	X	X
	Social learning	X	X		X		X	X	X
Input data type	Qual. and quan.	X		X	X				X
	Quan. only		X			X	X	X	
Focal range	Focused and in-depth						X		
	General and broad		X						
	Compromise					X			X
	Both	X		X	X			X	
Express uncertainty	Yes	X		X					X
	No		X		X	X	X	X	
Model output	Individual							X	
	Aggregated	X	X	X	X	X	X		X
User-friendly	Easy to handle	X		X	X				
	Nice look at the surface	X	X		X	X		X	X

Source: Added in and adapted from Refs. [14,15].

- Fuzzy logic is the foundation of SensiMod.
- Interactive operation.
- Feedback loops are fully taken into account (i.e., in the "Effect System"), while Bayesian belief networks (BBNs) do not allow for feedback loops.
- Inclusion of both quantitative and qualitative data (which is very important in dealing with complex problems), while system dynamics is not good at working with qualitative data.
- SensiMod is easier to use/apply—can be used by "non-modelers", social scientists, etc.; system dynamics would require some level of modeling skills; compared with the "conditional probability tables" of BBNs, the "impact matrix" of SensiMod is much easier to understand and filled in by workshop participants.
- Results can be visually presented and explained easily to decision-makers, managers and leaders.

MODELING A PROJECT AS A CYBERNETIC SYSTEM. FINDING THE VARIABLES

In Figure 21.4, we see the factors/variables that were identified as being important from the intense discussions in the course of a Syntegration process as described above. In this case, the number of variables is 18. We know that somehow they have to belong together; otherwise, they

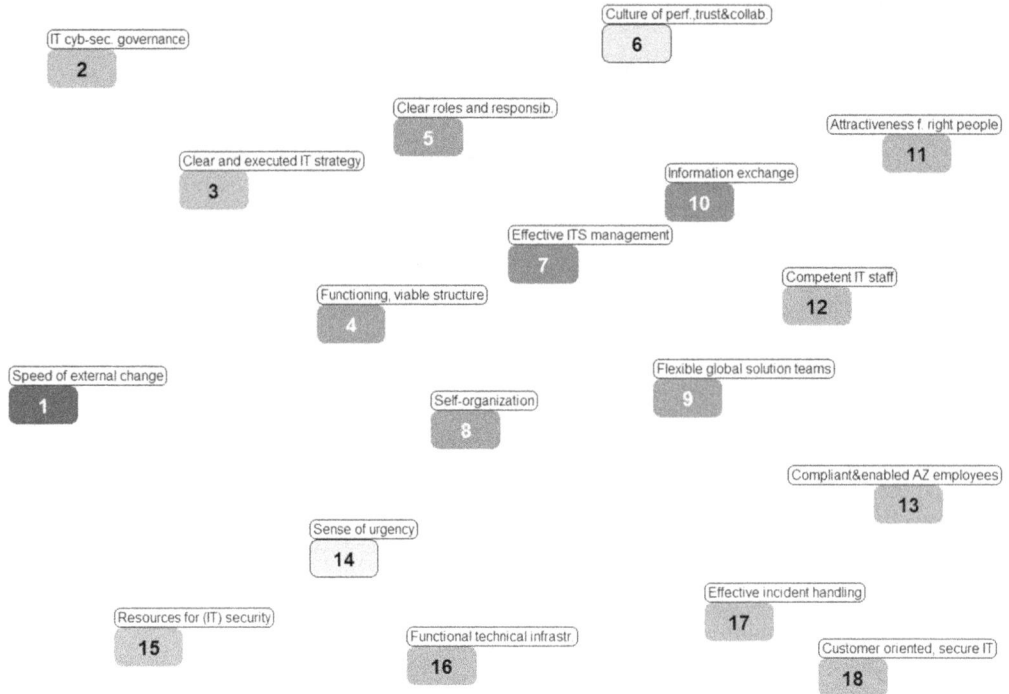

FIGURE 21.4   The 18 key factors for developing the interconnected system model.

TABLE 21.2
Variable Names

The Variables Are

1. Speed of external change IT cyber-security governance	2. IT cyber-security governance
3. Clear and executed IT strategy	4. Functioning, viable structure
5. Clear roles and responsibilities	6. Culture of performance, trust and collaboration
7. Effective ITS management	8. Self-organization
9. Flexible global solutions teams	10. Information exchange
11. Attractiveness for the right people	12. Competent IT staff
13. Compliant and enabled employees	14. Sense of urgency
15. Resources for (IT) security	16. Functional technical infrastructure
17. Effective incident handling	18. Customer-oriented, secure IT

would not have come up during the Syntegration process. But how? Eighteen variables are quite a large number for treating them with conventional methods. Figure 21.4 shows the 18 key factors spread out over the surface. The names of the variables are shown in Table 21.2. Since all of the participants were high-positioned, experienced executives and experts in cyber-security issues, it was highly probable that all the important variables have been identified.

DISCOVERING THE INVISIBLE CYBERNETICS OF A STRATEGY PROJECT TO TRANSFORM A COMPANY

Just as we cannot see natural forces, such as gravitational forces, we are also unable to see cybernetic forces such as control circuits. Both kinds of forces can only be observed from their effects on objects and variables and their consequent behaviors.

What we see in the following pictures is the project model of a large corporation that was challenged continuously by hacking and other cyber-attacks.

People there were instantly intrigued by our approach, as they could see, with our tools and ways of thinking, how we were approaching the core of their own expertise. Management realized that we were applying their existing expertise on the company itself and its functionality, thus raising their own management system to a new and higher level of functioning, just as they had done before for their customers when they worked on control systems for machines, vehicles and aircraft.

In Figure 21.5, we see the interconnections between the variables, thus forming a cybernetic network, the control circuits in which these variables are embedded. There is actually no hierarchy but a heterarchy. Just as the brain has no boss, networks in general have no bosses in the conventional meaning. Rather both may have many bosses, since every circuit can take command depending on the situation with which the system has to cope. Now we have no longer just variables but a closely knit dynamic *system*.

It is the invisible interconnections that turn a set of variables into a dynamic system. And since these are not just arbitrary kinds of interconnections but regulating and influencing ones, we have modeled the "cybernetics of the system". Just as one might speak of the "physics of a body" or the "statics of a building", we speak of the "cybernetics of a system", referring to its self-regulating interrelations, which control its functionality. It is one of the purposes of the Sensitivity Model to make the invisible visible and tangible.

Let us consider the complexity of the system which can be found out by studying the system: The systemic facts are quite interesting and also challenging: 18 key factors, 87 direct effects, 2,684 feedback loops, 2,240 reinforcing loops, and 444 stabilizing and inhibiting loops.[14]

Strong interconnection and the dominance of reinforcing feedback loops show that this is the type of *system which can be activated*. For gaining active control of the strongly interconnected system, single measures are not sufficient. *Concerted, simultaneous intervention* at different parts of the system is needed.

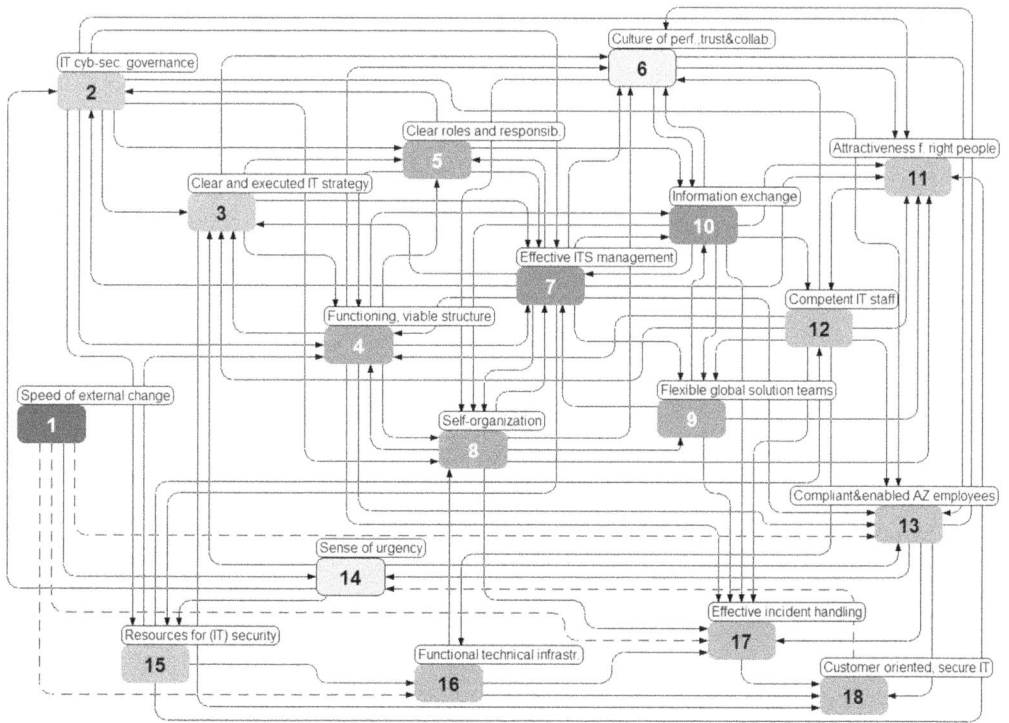

FIGURE 21.5 How the 18 variables are interconnected by invisible control circuits.

Master Controls and Control Circuits

Figure 21.6 shows a screenshot of the system with the first analysis results with respect to its control circuits and control dynamics. The systemic diagnosis shows two really decisive levers which act as master controls: The two variables "Effective IT-Security Management" and "Functioning, Viable Structures" are embedded in 1808 (81%) circuits and in 1538 (69%) of all of the 2,240 reinforcing feedback loops. Therefore, if one of these two *systemically essential factors is deactivated*, the number of reinforcing feedback loops is reduced to 19% which means 432 instead of 2,240 and 31% meaning 702 instead of 2,240, respectively.

This shows the system can only be activated, if these two key drivers are activated. Effective ITS management and a functioning, viable structure are essential for achieving cyber security in the organization for which the Sensitivity Model was designed.

Sensitivity Impact and Risk Map

Humans lack several natural prerequisites for understanding and steering complex systems. For instance, as George A. Miller has discovered long ago, the human brain can only grasp a small number of variables at any given time to steer its actions. He called it "The Magical Number 7 ± 2". Only this small number of variables can be observed independently of each other and even fewer interconnections.[16]

This insight is of course used in the design of aircraft cockpits and similar devices. With extended training, specialists may be able to increase the number by one or two additional variables but rarely more. After that "magical number", there is fading out of control and may be breakdown of our sensomotoric organization. This goes to prove that life, based on biological systems, has so far only been able to master a limited extent of complexity and that, if complexity is to develop further, this

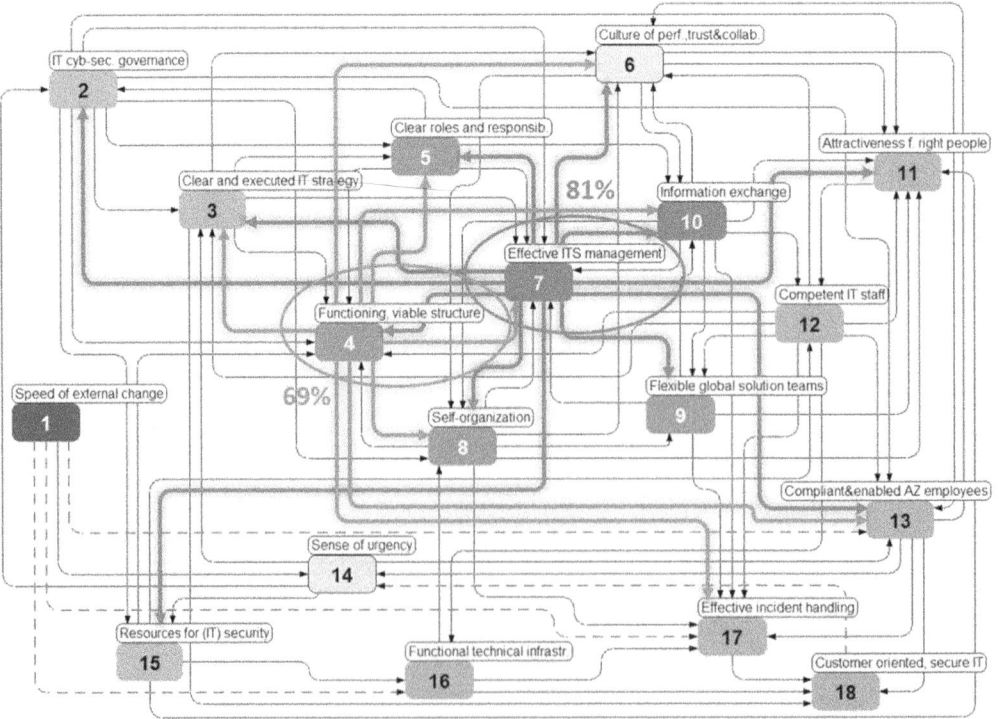

FIGURE 21.6 Master controls and control circuits.

can only happen in socio-cultural and socio-technical ways, by enhancing adaptation capabilities and intelligence—for example, by digitization.

The dynamics of the system are then made visible in the *Sensitivity Impact and Risk Map*[14] as shown in Figure 21.7. This helps to find out which variable will have what effect, that is, whether a variable is active or passive, and whether it acts as a buffer or has critical impact.

Systems Diagnosis:

- Most effective lever: IT cyber-security governance (2) is the most effective lever to initiate change in the system. It has a strong effect on the system (high active sum, vertical axis).
- Active-critical drivers: Effective ITS management (7) is even more active (highest active sum) but is already quite strongly influenced by other factors (medium passive sum, horizontal axis). Effective ITS management, clear and executed IT strategy (3), functioning, viable structure (4), self-organization (8) and flexible global solution teams (9) are active-critical drivers that have an essential influence on the development of the system. However, simultaneous interventions at different parts of the system are needed to activate these factors.

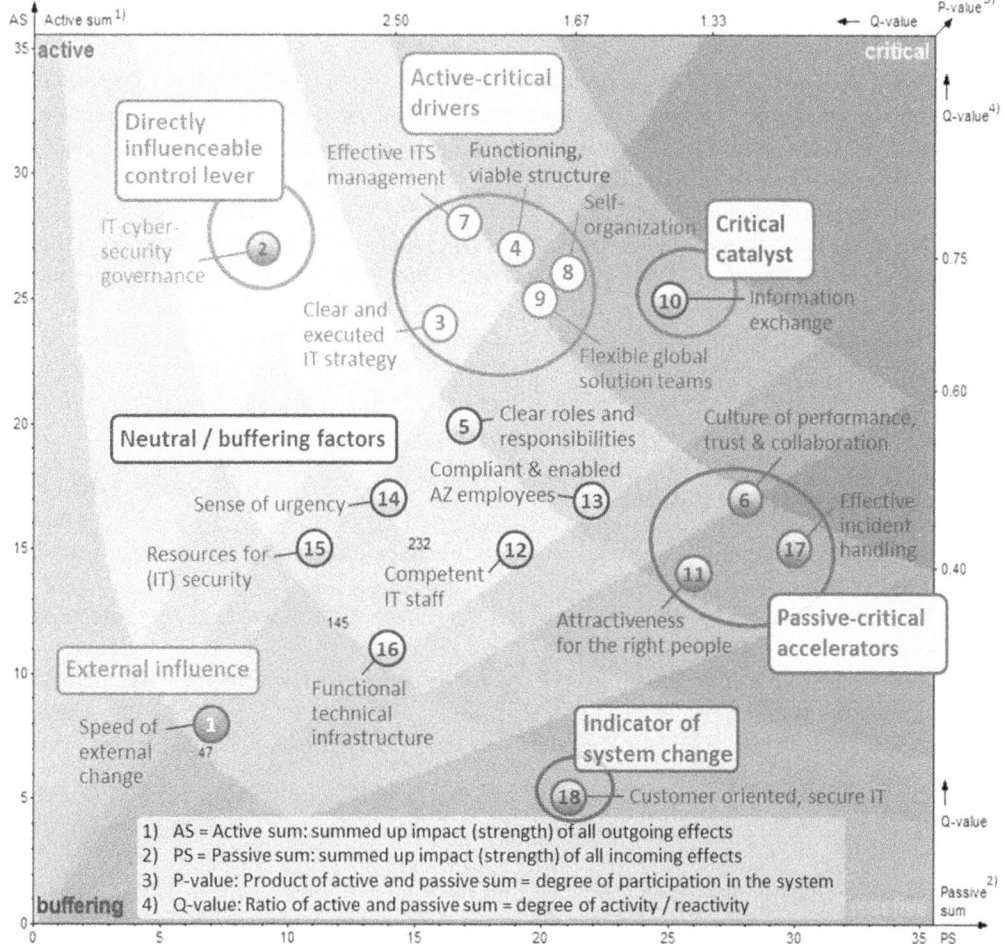

FIGURE 21.7 How to move, change and control the system: the Impact/Risk Map as a tool for action. Simultaneous intervention from several angles is needed.

- Critical catalyst: Information exchange (10) is most critical. Improvement of this factor will have a strong influence on the system, but it is also strongly influenced by many other factors. It requires simultaneous action from several angles to achieve effective information exchange.
- Passive-critical accelerators: Culture of performance, trust and collaboration (6), attractiveness for the right people (11) and effective incident handling (17) are even more dependent on change in other parts of the system but give essential accelerating impulses, if activated.
- Indicator of system change: Customer-oriented, secure IT (18) is an indicator of the successful development of the cyber security.
- Neutral/buffering factors: The other factors are not especially effective levers, though an unintended status of these factors can block off system transformation.

Cyber-Systemic Governance and Management

- "Effective ITS management" (7), followed by "IT cyber-security governance" (2) and Functioning, viable structure (4), has the *strongest overall effect* on the system (highest active sum, vertical axis).
- Effective management, a clear and executed strategy (3) and the structural factors, including self-organization and flexible global solution teams (4, 8, 9) forms a self-reinforcing *nucleus of effectiveness.*
- Embedded in an effective IT cyber-security governance, a *mutually reinforcing dynamic* can boost the system to a *new level of functioning*: The right cyber-systemic governance forms the stable and adaptable basis on which the viable system structure can be implemented and managed effectively.
- As managerial professionalism, a functioning structure and an effective governance are mutually reinforcing, they should be improved simultaneously and immediately. Together, they can act as the *nucleus for system transformation*, as they are influencing all other system elements directly.
- Governance, management, strategy and structure are rather *weakly effected by other factors* (relatively low passive sum, horizontal axis). This means improving these elements of effectiveness simultaneously and in a harmonized, interconnected way is the *optimal starting point for sustainable change* in the system (Figure 21.8).

QUESTIONS FOR DISCUSSION

1. Please discuss and map the current situations, complex challenges of your organization, joint-project, etc. on the double S-curves.
2. What would you think are the benefits of the Syntegration process in engaging stakeholders in complex projects?
3. How do you see the application of the Syntegration framework within your organization, your complex project?
4. How do you see the application of the Sensitivity Model for your organization, your complex project?

CONCLUSION

So if we need to observe many more variables and their interconnections to really understand the system and be able to steer it, we need special tools—such as the Sensitivity Model. It would not suffice to say "computer-based" models as is often the case. The important thing is the particular kind of model and philosophy of modeling systems. Executives of all levels can now obtain just the kind of information—as distinguished from data—needed to understand the system, and its inherent

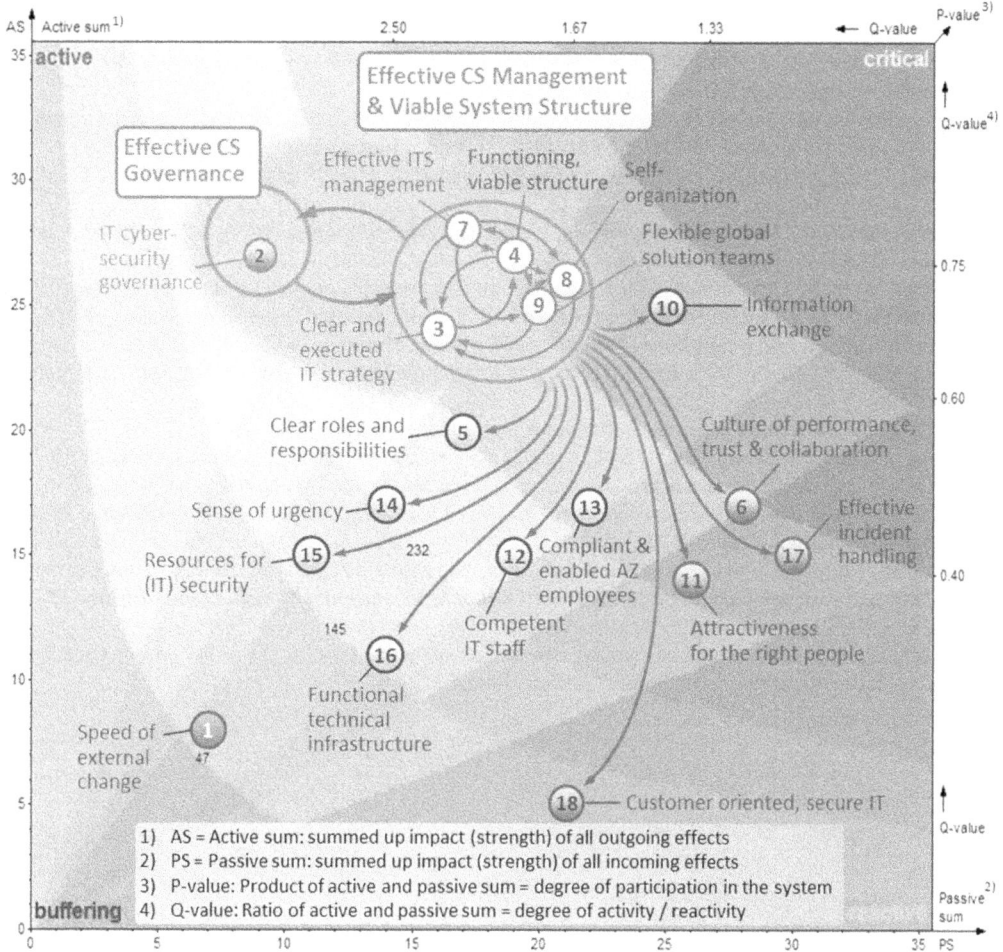

FIGURE 21.8 Cyber-systemic governance, management and a viable system structure can form the self-reinforcing nucleus of effectiveness for fast and sustainable system transformation.

functional laws, and the systems controls both for the organization as a whole and for individual operational units. With Sensitivity Models for each level of the organization, intelligence and control can be implanted in and out, up and down to the capillaries of each business unit, thus reinforcing itself throughout the system to an extent that would not be possible without organizational cybernetics.

Think of not just one single Cybernetic Sensitivity Model but of layers of networks of many such models each one designed to the peculiarities of their variables, all of them interconnected, we would in fact end up in an organization with the equivalent of a nervous system as the regulatory mechanism of the organization.

Together with the master controls of the "double S-curve" and the high-performance Syntegration methodology for big change combined with networks of Sensitivity Models, it is quite a powerful system for effective stakeholder engagement and complex project management.

REFERENCES

1. Malik, Fredmund. *Navigating into the Unknown: A New Way for Management, Governance and Leadership.* Frankfurt, NY: Campus Verlag, 2016.
2. Malik, Fredmund. *Governance.* Frankfurt, NY: Campus Verlag, 1997.

3. Malik, Fredmund. *Strategy: Navigating the Complexity of the New World.* 2nd ed. Frankfurt, NY: Campus Verlag, 2016.

4. Malik, Fredmund. *The Right Corporate Governance: Effective Top Management for Mastering Complexity.* Frankfurt, NY: Campus Verlag, 2012.

5. Malik, Fredmund. *Strategy for Managing Complex Systems: A Contribution to Managerial Cybernetics for Evolutionary Systems.* 11th ed. Frankfurt, NY: Campus Verlag, 2016.

6. Malik, Fredmund. *Corporate Policy and Governance: How Organizations Self-Organize.* Frankfurt, NY: Campus Verlag, 2011.

7. Malik, Fredmund. *Managing Performing Living: Effective Management for a New World.* 2nd ed. Frankfurt, NY: Campus Verlag, 2016.

8. Wiener, Norbert. *Cybernetics: Control and Communication in the Animal and the Machine.* Paris: Hermann & Cie & Cambridge, MA: MIT Press, 1948, 2nd revised ed. 1961.

9. Wiener, Norbert. *The Human Use of Human Beings.* Paris: Hermann & Cie, 1950.

10. Ross. Ashby W. *Design for a Brain. The Origin of Adaptive Behavior.* 2nd ed. London: Chapman and Hall, 1960.

11. Mc Culloch, Warren. *Embodiments of Mind.* Cambridge, MA: MIT Press, 1988.

12. Malik, Fredmund. *Management: The Essence of the Craft.* Frankfurt, NY: Campus Verlag, 2010.

13. Beer, Stafford. *Beyond Dispute: The Invention of Team Syntegrity.* Chichester: John Wiley, 1994.

14. Vester, Frederic. *The Art of Interconnected Thinking: Ideas and Tools for a New Approach to Tackling Complexity.* Stuttgart: MCB-Verlag, 2012.

15. Jakeman, Anthony J, and Rebecca A Letcher. Integrated assessment and modelling: Features, principles and examples for catchment management. *Journal of Environmental Modelling & Software*, 2003, 18 (no. 4): 491–501.

16. Miller, George A. The magical number seven, plus or minus two: Some limits on our capacity for processing information. *Psychological Review*, 1956, 63 (2): 81–97.

22 Systemic Risk Toolset
Another Dimension

Fran Ackermann
Curtin University

CONTENTS

WHY DOES SYSTEMIC RISK MATTER?

This chapter as its name suggests is based on a focus on risk and in particular *systemic* risk. However, taking a systems perspective (Reynolds and Holwell 2010) is relevant for all elements of project management, not just risk. For example, when considering any project, there is almost without exception a recognition that there will be the need to attend to technical systems with electrical designs taking cognizance of mechanical, ventilation, environmental, etc. considerations. These technical systems will also impact and be impacted by financial systems (depending on the type of project, e.g., PPP contracts and DBOM contracts) and potentially human factor systems (staffing, supervision, suppliers, etc.)

As early as the 1970s, this requirement to take a systems thinking approach when considering engineering-oriented projects emerged particularly through the work of Peter Checkland (1980). Checkland recognized the need when working on projects to attend to a wider range of interlocking systems including socio-political systems and realized that a systems engineering perspective would benefit from augmentation. As a consequence, he developed Soft Systems Methodology (Checkland 1980, 2001; Checkland and Poulter 2010). One example Peter Checkland used to discuss this need was the development of Concorde, noting that whilst there were clear engineering system challenges to overcome, these were also impacted by financial systems (it was a costly venture), political systems (Concorde was a joint project between France and the UK), and environmental systems (needing to be over the Channel before passing through the sound barrier). Each of these systems had implications on the success of the project (both in their own right and in their impact on one another). As such, understanding the different perspectives (legal, governmental, financial, and engineering) was critical.

Another more recent example is a growing attention on the decommissioning of oil platforms as it is increasingly becoming recognized that paying attention to stakeholders (Fowler et al. 2013; Laraia 2017) and the environment (McHaina 2001) is as important as paying attention to the engineering considerations. There is also considerable research illustrating that safety systems and rework systems have significant implications for one another (Love et al. 2018), further reinforcing the importance of taking a systemic perspective.

That said, for the most part projects tend to be 'distilled' into their components and whilst this has worked well for many projects and traditionally is the accepted way of managing projects, the very fact that projects are continuing to experience considerable cost and time overruns suggests that whilst these techniques are beneficial, they are not sufficient. For example, we have seen projects such as Denver's US$5 billion airport that was 200% overspent (Nutt 2002), the UK's Scottish Parliament anticipated to be ten times the original budget (Scottish Parliament 2003), and more recently the Perth Children's Hospital which experienced significant delays, that is 4 years overrun – the hospital was due to be completed in 2014 and began opening in 2018 a 40% time overrun (Special Inquiry into Government Programs and Projects (2018); Public Accounts Committee (PCH)). These examples straddle a period from the 1970s to today – reflecting and reinforcing the need to consider additional techniques. The cost and time overruns suggest we have not taken all considerations into account and attending to the systemic impacts of risks on one another provides a potentially important avenue for research and practice.

Browning and Ramasesh (2015) in their MIT article 'Reducing unwelcome surprises in project management' explicitly note that 'projects operate as systems' (2015 p. 54) and argue that there are at least five key subsystems experienced in projects. These include the results, process, organizational, goals, and tools subsystems. Moreover, whilst each of these subsystems is made of many elements, each subsystem and its elements impact the others increasing complexity and generating many possible outcomes. Projects thus comprise systems each interacting with the others.

- *Thus, an important consideration is thinking systemically when managing projects.*

Projects have always been risky, and therefore, understanding both the entirety and nature of risks is important. Risks create uncertainties in projects, and uncertainties create risks. Likewise, as project complexity increases so too do risks and this will increase the complexity of the project. Finally, complexity and uncertainty are interrelated with an increase in complexity giving rise to additional uncertainties and uncertainties compounding the complexity. Consequently, it is important to recognize that a comprehensive understanding not only of risk but also of complexity and uncertainty is necessary. For example, complexity can be structural (comprising the technical aspects of the project including but not limited to construction, wiring, heating, ventilation, and air conditioning (HVAC), it can be processual (encompassing considerations such as contracts, human resources, procurement), and it can be relational (acknowledging the impact of joint venture partners, suppliers, etc.), each form generating project-specific risks. For a good discussion of forms of complexity in projects, see Geraldi et al. (2011) who define the complexity of projects according to five dimensions, namely, structural, dynamics, pace, socio-political, and uncertainty (reinforcing the link between uncertainty and complexity). Uncertainty can relate to the environment within which the project is taking place ranging from consideration of political uncertainties (and thus possibly funding uncertainties) to environmental and legislative. Likewise, uncertainty can be technical particularly where state-of-the-art technologies are being adopted. As such, the link between uncertainty and complexity is clearly revealed.

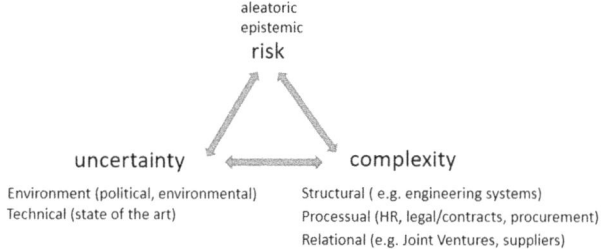

FIGURE 22.1   An illustration of the interconnectivity of risk, uncertainty, and complexity.

Furthermore, when we come to consider risk, it is worth noting that there are different sorts of risk – e.g., aleatoric and epistemic. Probabilistic analysis is the predominant method used to handle the uncertainties involved in risk analysis, both aleatory (representing variation) and epistemic (due to lack of knowledge). For

> It is important to recognize that a comprehensive understanding not only of risk but also of complexity and uncertainty is necessary.

aleatory uncertainty, there is broad agreement about using probabilities with a limiting relative frequency interpretation. However, for representing and expressing epistemic uncertainty, the answer is not so straightforward. As such, risk and uncertainty are interconnected. Not only do risks exist within projects demanding a need for systemic consideration, but they also exist between projects, that is where there are portfolios of projects. This particular consideration will be explored further later on in the chapter (Figure 22.1).

Overtime uncertainties can be reduced as more information is captured. As such, good risk management aims to reduce uncertainty over time as new information comes to light, and further analyses are undertaken. As such, the concept of 'Left Shift' (Winch 2010) provides a useful conceptual lens to consider projects. This conceptual lens recognizes that over the course of the project – time – information will increase starting with relatively little at the commencement of the project and ideally completing the project with a complete understanding of the facts. The faster the information can be ascertained and absorbed/analyzed, the faster uncertainty is diminished – giving rise to the name 'left shift'. Finding processes to aid with this left shift aspiration early on in the project is therefore important, and some of the systemic methods noted in this chapter aim to do this (Figure 22.2).

As a key aspect of eliciting information to help with managing the uncertainty (typically, this is focused on the technical aspects of a project), attention is needed to ensure that a continuous review of

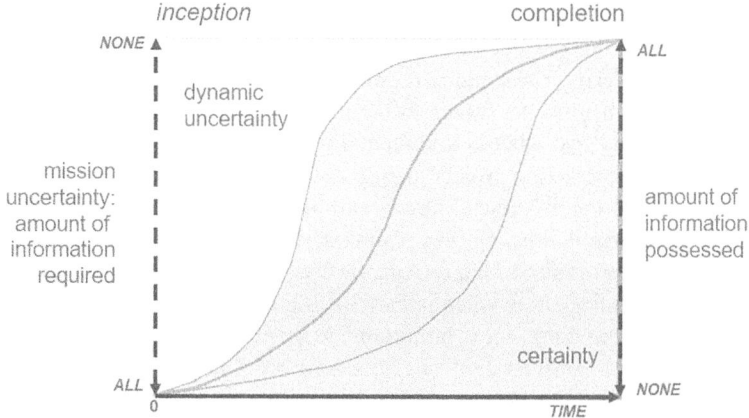

FIGURE 22.2   Left shift. [Figure 1.3 in: Winch, G. M. (2010) *Managing Construction Projects: An Information Processing Approach* (2nd ed.) Oxford, Wiley-Blackwell.]

the risks is undertaken as risk salience and occurrence will change over time. Some risks will cease to be relevant and therefore 'drop off the radar', and other risks will emerge or gain importance (due to changing conditions). For example, as the project progresses, some of the technical uncertainties will be resolved and thus no longer pose a risk to the project; however, changes in the environment might occur raising the significance of other risks, e.g., changes in government and thus funding patterns.

However, in addition to these changes are the changes in terms of the impact the risk might have to other risks around it. Thus, at the commencement of the project, a risk might be relatively peripheral and therefore not requiring a great deal of attention and energy. However, over the duration of the project, the risk may gain salience and impact as changing circumstances result in it having the potential to impact a wide range of other risks. Consequently, its position in a network of risks becomes more central and significant. It is this reason that understanding the impact of risks as a system becomes so important.

Systemic risk increases with the increase in project complexity. This frequently arises due to projects involving a wider range of stakeholders (see, e.g., new nuclear build projects where there are extremely high numbers of sub-contractors with different cultures, languages, and expertise). In addition, complexity increases with the changing nature of the external environment through changing legislation, political inclinations, etc. Another contributor to complexity is through projects spanning longer life cycles due to a proliferation of PPP, PFI, etc. type projects which bring in operation and maintenance risks. Each of these increases not only is a challenge in itself but also impacts one another, thus generating intra- and interdynamics through the project's life.

- *Thus, an important consideration is seeing risk, complexity, and uncertainty as interrelated and changing over time.*

Moreover, there is a tendency to take a top-down approach starting with a set of well-understood risk categories rather than taking a bottom-up approach and considering what are the possible risks for this specific project. This is understandable – as the list provides a relatively easy starting point. However, it exposes the risk assessment process to a potentially myopic examination. By only considering those experienced on other projects (and synthesized into a generic list), those particular to the project are missed. This consideration resonates with Browning and Ramasesh's paper (2015) noted above regarding unwelcome surprises, where they note that it is determining the known knowns, the unknown knowns, the known unknowns, and more critically the unknown unknowns that is so significant. Starting with the generic list risks focusing on the known knowns and the known unknowns potentially misses the other two forms of information uncertainty.

Exacerbating this situation is the fact that in many projects, once the bid has been won, there is a rush to get started due to time pressures (particularly the case when the project has significant liquidated damages for project overrun being imposed). There is a sense that "we're already behind. We know what we need to do. Let's get started" (Browning and Ramasesh 2015 p. 54). Therefore, instead of carefully considering risks and determining what is known about the project and what is knowable, project work begins. As shown in Figure 22.3, careful attention to and consideration of the risks in their entirety and what is known and what is assumed known are important early on. Linking with Left Shift assisting project managers to assess not only those known knowns and known unknowns, but also the unknown knowns and unknown unknowns will increase the probability of project success. An interesting observation derived from the literature is that often when considering the information required for a project, staff when not knowing something assume other staff are in receipt of this information when in fact this is not the case. As such, understanding what is known, assumed known, and present is important (Ackermann et al. 2007).

The attention to and the consideration of risk are nicely illustrated by Thamhain's statement that good risk assessment "requires broad involvement and collaboration across all segments of the project team and its environment, and sophisticated methods for assessing feasibilities and usability early and frequently during the project life cycle" (2013 p. 20). Note that Thamhain also stresses the need for early identification of risks.

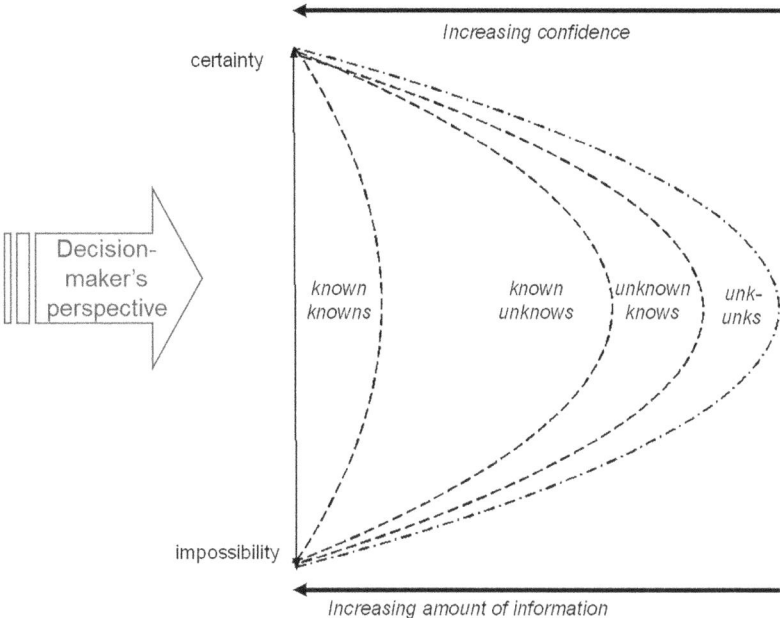

FIGURE 22.3 Adapted from the information space: a cognitive model for managing risk and uncertainty on projects (Winch 2010 Figure 13.2, p. 359).

Helping project managers take a more comprehensive view of unknowns and risks can be achieved by actively increasing the viewpoints taken. For example, "a complete risk management analysis must include not only the technical factors but also a realistic assessment of environmental and social risks" (De Lemos et al. 2004 p. 63). As such, it is important to move beyond technical and financial (which are important) but also consider stakeholders (internal and external). "Currently risk management has been commonly applied across the construction sector, however it very rarely includes the effects of human factors" (Thevendran and Mawdesley 2004 p. 131). This can be very significant where inter-personal considerations exist as "Conflicts between project management team and project owner are often neglected" (Krane et al. 2012 p. 54) and projects have come unstuck due to this not being considered. Other aspects that are worth considering are the environmental aspects.

As can be seen, this attention to taking a wide and comprehensive view of risk resonates with the earlier discussion regarding the need to see risk alongside complexity and uncertainty and further illustrates the point made at the beginning of the chapter regarding Concorde and the need to consider technical, financial, and environmental issues.

Unless a thoughtful consideration of risks is undertaken, effective allocation of resources to manage the risks is likely to be challenging. This is aligned with Simon's four stages of decision-making (Simon 1977), namely, intelligence, design, choice and review. Getting a good understanding of the wealth of risks and uncertainties is necessary before doing any evaluation. After all, "Project managers do not have unlimited resources for interacting with stakeholders. You must decide carefully how to spend the time and resources which you have available for this task" (Eskerod and Jepsen 2013 p. 7). This will include thinking about project outcomes, goals, decommissioning at the same time as construction, etc. For example, in a recent paper on the costs of decommissioning, it was noted that "new development project teams need to spend as much effort preparing for decommissioning in the design stage as they do preparing for the maintenance of the platform" (Carpenter 2015 p. 113).

Finally, the well-known book *Black Swan* (Taleb 2007) reinforces this need to consider risks widely with the call to consider black swans, that is very unlikely events (akin to unknown unknowns) (Aven 2015). This resonates with experience from litigation projects where small probability events, e.g., the most severe winter for 100 years, can have massive impacts on a project. Aspects of weak signal management theory (Rossel 2013; Kaivo-oja 2011) are clearly relevant in this domain.

Thus, an important consideration is taking a comprehensive view of risks

Returning to the topic of systemic risk, a project can be viewed as organic – growing like Amoeba (Eden et al. 2005). "Project risks are not always independent, yet current risk management practices do not clearly manage dependencies between risks" (Kwan and Leung 2011 p. 635).

For example, when viewing the life of a project, it is possible to see how one risk impacts on another and causes spiraling overruns as noted in the introduction. Understanding risks as systems allows for better management. This can be achieved through identifying those risks that have significant impacts on the project, that is, those with lots of potential to affect other risks. They are central to the risk network (which returns to the earlier point regarding shifting saliency). However, in addition, risks and their management can also cause counterintuitive consequences and feedback behavior which can escalate out of control.

For example, a construction company had bid and won a contract to build a state-of-the-art drill ship. The contract had within it significant liquidated damages imposed for any late delivery. However, not long into the project the client changed their mind about the design and did so a number of times. As a consequence, engineering drawings needed to be redone to reflect these changes, and the scope of the project increased. This had a knock on effect on the number of hours engineering required. Given that this increase in hours had a deleterious impact on the schedule, the first action of management was to increase overtime (providing additional hours). This can be seen in Figure 22.4 showing the trigger in bold, and the management action to control the situation illustrated through thick links (generating a negative-feedback loop).

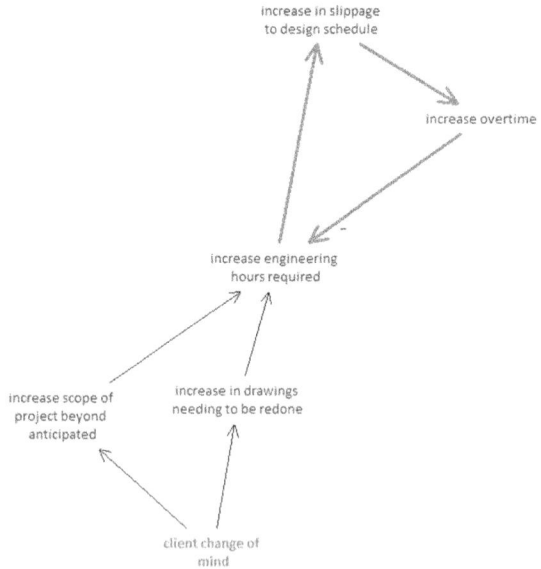

FIGURE 22.4 The impact caused by a client change of mind. Note: The statement in bold represents the trigger, and the other statements are consequences emanating from the trigger. The arrows are causal and thus can be read "client change of mind" may lead to "increase in drawings needing to be redone". The thick arrows represent statements comprising a feedback loop; that is, an increase in engineering hours required may lead to an increase in slippage to design schedule" which, in turn, may result in "increase in overtime". This then has a controlling impact on the "increase in engineering hours required" reflecting a negative or control feedback loop. A small negative symbol is attached to the head of the link between increase overtime and increase engineering hours required recognizing that an increase in overtime will lead to a decrease in hours required.

However, this was only a temporary solution. In addition to changing their mind, the client was late in providing information and in approving drawings. These actions resulted in a number of further changes to the design activity. For example, engineering, due to the pressures imposed by the liquidated damages penalty, was compelled to start working out of order (another management action). Unsurprisingly, working out of order alongside the fatigue caused by overtime began to generate mistakes in the engineering effort increasing the rework required and subsequently the engineering hours required. As this demand for engineering hours spiraled, engineers were 'borrowed' from other projects increasing the numbers of engineers working on the project (and thus the engineering hours available) but increasing communication challenges further leading to the potential for mistakes. The project was now experiencing considerable stress as illustrated in Figure 22.5.

Not surprisingly the challenges being experienced in engineering spilled over into construction with drawings being released out of order, due to the time pressures being experienced. This, in turn, meant that construction could not do the pre-outfitting originally anticipated increasing inefficiency and thus increasing construction hours. As with engineering, construction also opted for overtime as an initial solution, but again, this resulted in errors through fatigue and thus rework, further increasing the hours. Moreover, the client furbished equipment which was promised to be delivered on schedule in the contract arrived late causing construction to have to work out of order and again experience rework. Finally, the managerial action of bringing in more staff to manage the hours resulted in overcrowding as ships are a finite space and thus resulted in inefficiencies.

As can be seen from this simplified situation (the real project was far more complex), a number of vicious feedback cycles were experienced on the project resulting in massive overrun. The small example comprises 21 feedback loops. Not only did the project become severely impacted but so too did other projects as staff were 'borrowed' from them in the attempt to avoid running over time.

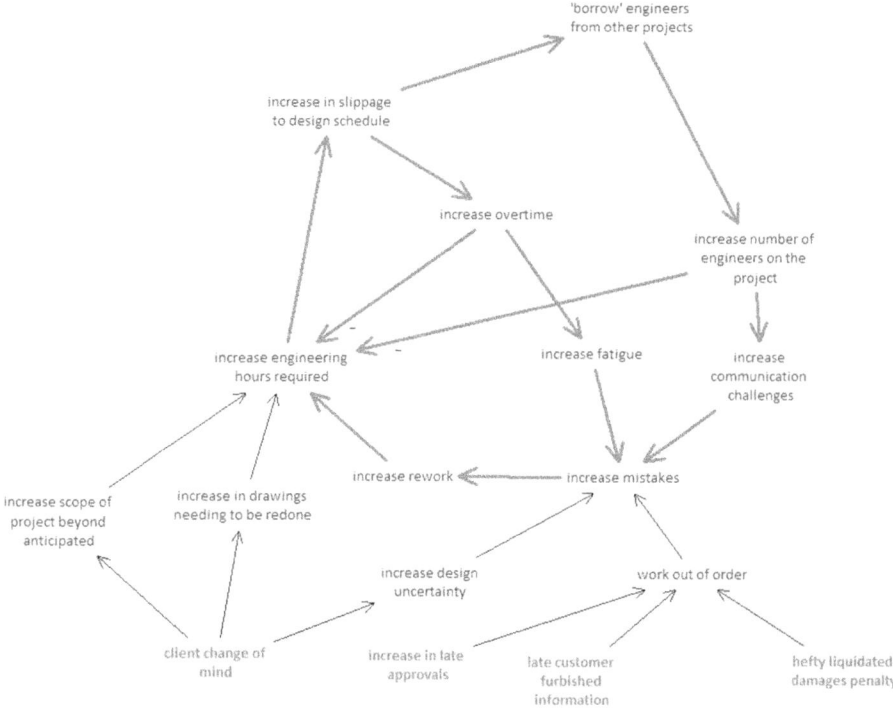

FIGURE 22.5 Escalating project dynamics. Note: The four triggers (bold statements) cause multiple consequences which produce positive (vicious) feedback loops as well as negative (control) feedback loops. The impact is escalating.

This results in the amoebic life of a project unfolding. As noted in the Eden et al.'s (2005) paper, the growth of a project can be triggered by a number of factors. For example, the initial project (the circle) grows due to a range of factors. For example, as illustrated in Figure 22.7, factors include A. giveaways (the contractor does work for free because the change is innovative and fun), the mix of B. under and C. overestimating contingency required, D. different interpretations of the contract, E. interrupting the client through changes of mind, through late information, or through approval delays. Finally, there is of course F. inefficiency of the contractor. Each of these factors pulls the project away from the original design estimate (the central circle) and schedule and, in doing so, has impact on other areas and so the total overrun in costs grows amoebic like.

As noted in the drill ship example, many of the managerial actions appear sensible when taken in isolation, but when the project is viewed systemically, their counterintuitive ramifications become apparent. This is not unusual. Another example, slightly adapted, illustrating why it is important to take a systemic view comes from experience when working on modelling disruption and delay to support litigation. The project

> Many of the managerial actions appear sensible when taken in isolation, but when the project is viewed systemically, their counterintuitive ramifications become apparent.

was the design and build of the Channel Tunnel train that transports cars, buses, trucks, etc. from the UK to France and vice versa. The train comprising a number of carriages connected together through couplers was designed so that different information requirements could flow up and down the train length. As such, the original design allowed for X number of pins in the coupler connecting X number of cables. Over the course of the project, the client requested more and more information be made available increasing the coupler pins by over 100%. The contractor when accounting for these extra requirements was able to calculate the cost of the extra cabling and the hours required to fit the cabling. However, what the contractor did not account for was the fact that this extra cabling now resulted in the carriage exceeding the weight limit and thus led to a considerable effort to redesign the carriage. Materials were changed (to lighten the carriage) and repositioned causing changes to the ventilation system and thus to the wiring system. Thus, changes in one area of the train led to massive changes in another. Indirect costs are hard to identify and add to changes in scope – as they may be experienced as death by a 1,000 cuts, each increase in information not obviously contributing to a problem but when combined together causes the weight problem.

As such, as can be seen from Figures 22.4–22.7, the impact of events can be escalatory rather than additive – experienced through a number of illustrative feedback loops. For more examples showing the impact of systemicity on a project straddling engineering, construction, and beyond, see Williams et al. (1995a,b). As such, this dynamic behaviour illustrates William's (1997 p219)

> The impact of events can be escalatory rather than additive – experienced through a number of illustrative feedback loops.

point namely 'developing project risk management frameworks that inform teams about likely cross impacts' is a critical task.

Another consideration is as noted above the changing nature of contracts. For many researching or working on a project, it comprises the design and build stages with the construction company handing over responsibility to the client at the commission stage. However, with the advent of contracts such as DBOM, BOOT, and PPP, this is changing. First, the project now can be envisaged as comprising not only design and build but also operate and maintain (Alexander et al. 2019). It might also include decommissioning as is being argued for in recent oil and gas projects (see earlier in this chapter). Consequently, when thinking about risks a longer time horizon is now required. This is due to risks emerging through two avenues.

The first focuses on the particular project itself and aspects such as the selection of materials or the design of infrastructure. In managing the project costs, project managers might decide

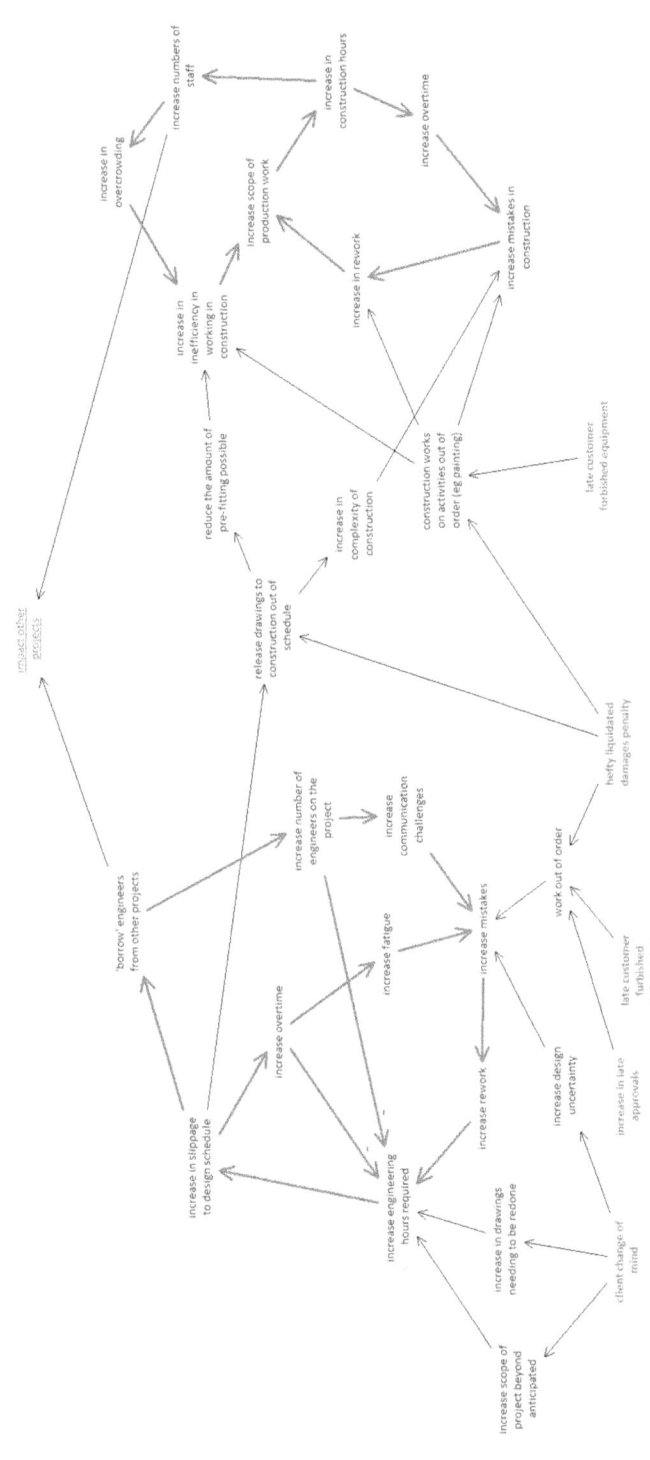

FIGURE 22.6 Disruptions in engineering impacting construction. Note: Additional trigger (late customer furbished equipment) plus the ramifications of triggers experienced in engineering impact construction creating further vicious cycles and further impacting overruns. Also noted are the consequences of these dynamics in engineering and construction impacting on other projects being undertaken as shown by the statement in underlined text.

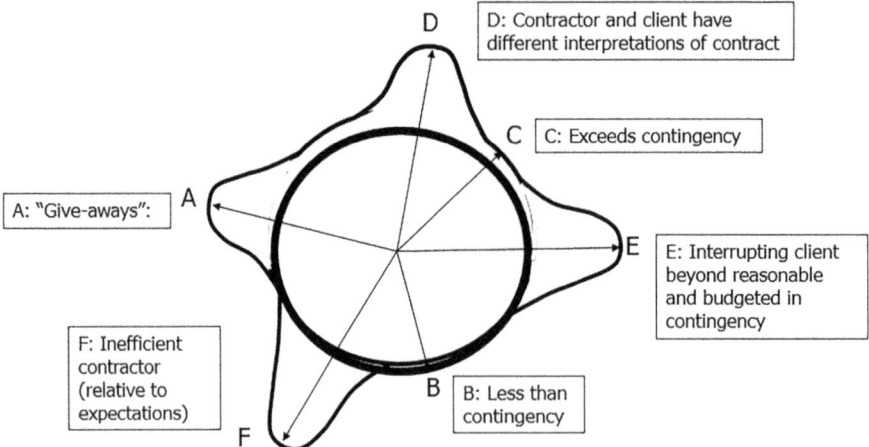

FIGURE 22.7   Actual cost build-up (post-hoc view). Adapted from Eden et al. (2005 p. 22).

to use a particular component which initially does the job; however, its durability is question-able, and therefore, there is a risk that within a short period of time in operation maintenance is required costing the organization. Alternatively, in designing a product, e.g., a locomotive, then there is a risk that maintenance issues are not considered and relatively trivial considerations, e.g., the removal and replacement of an air filter, can cause a risk of incurring significant costs when the design requires a virtual strip down of the locomotive rather than a relatively straight-forward operation.

The second avenue of risk caused by these longer duration projects is that they may also be part of a portfolio, e.g., a series of schools. If something happens to one of the schools, as was experi-enced in Scotland (Carrell 2016, 2017; Cole 2017), then there is a risk that all of the schools within that portfolio require adaptation. Thus, risk systems straddle across the entire cradle to grave spec-trum of a project's life.

- *Thus, an important consideration is taking a systemic view of risks.*

This chapter next provides a brief review of the development of some risk tools, particularly those that acknowledge that risk interact, explores one approach in depth providing an illustration of where it has been used (forensically and proactively) along with some brief guidelines and some further reading. The chapter concludes with some interesting questions for students and a summary.

MODELLING SYSTEMIC RISK

COMPARING SYSTEMIC RISK APPROACHES WITH THE TRADITIONAL PARADIGM

Taking a systems perspective to risk modelling does demand a different paradigm from that of the more traditional forms and therefore before exploring systemic risk analysis tools, some compari-son of these two paradigms will be presented so as to provide an understanding of the differences. That said, it is worth noting that these two paradigms should not necessarily be seen as being in competition with one another. It is entirely feasible to use a systemic approach to ensure a wide cov-erage of both risks themselves and their impact on one another and then use this analysis to populate a risk register. Furthermore, it is worth noting that for some projects, traditional methods provide the necessary analysis and management support, whereas for others, a wider (systemic) view is necessary. For example, Williams (2005 p. 497) notes that for projects that are "complex, uncertain and time limited, conventional methods might be inappropriate".

One of the biggest differences between the paradigms is the assumptions underpinning them. Traditional methods are predominantly viewed as "rationalist and normative. Project management in this paradigm presents itself as self-evidently correct … and provides a normative set of techniques" (Williams 2005 p. 498). In addition, the traditional paradigm is essentially positivist predicating itself on an objective, fact-based ontology (rather than one more socially construed as is the systemic risk paradigm). Traditional methods for risk assessment see projects broken down "into their constituent parts – in scope (work breakdown structure), time (critical path networks) cost (budgets), risk (risk registers), and so on" (Williams 2017 p. 56). However, this 'breakdown' has limitations. For example, Merrow (2011 p. 327) notes "in projects, bad things tend to happen in groups, not individually.… Events that affect projects in major ways … tend to go together. Even when one of those things occurs individually, it tends to trigger a cascade of problematic effects". Thamhain echoes this view when he notes "undesirable events (contingencies) are often caused by a multitude of problems … these problems often cascade, compound and become intricately linked" (2013 p. 29) – all arguments for the systems paradigm.

In addition, and related to the assumptions, traditional methods also tend to take a 'parsimonious' view to what is considered a risk. For example, understanding other parties' reactions to actions/events is not seen as a legitimate part of risk; however, these behaviors can result in a massive impact on the project's weal. A risk may take the form of lobbying behavior or a demand for benchmarking. So too can changes in client's strategy. As such, when analyzing which risks are likely to have the most impact, the conventional risk analysis focus on a subset of risks is potentially missing those whose impact is going to be the greatest.

In many instances, it could be argued that traditional risk assessment focuses on those risks that appear to be more controllable and to challenge this is to challenge the dominant paradigm – risks are – as noted by Gephart, Van Maanen, and Oberlechner (2009 p. 143) – "identifiable through scientific measurement and calculation, and (can) be controlled using such knowledge". As such, there are a number of stakeholders who see retaining conventional processes to be in their best interests. Furthermore, the original development of project management approaches are built on work done in the late 1980s, and as such, these traditional methods are well established and understood – they are part of the project management vernacular and are the accepted approach (e.g., Project Risk Registers may be mandated by the contract).

Systemic risk thus confronts the established way of working causing potential discomfort which is reinforced by project managers needing to be comfortable with 'opening Pandora's box'. Taking a systemic view to risk management demands working with complexity – looking at risk structures that might exceed 100 events each with different impacts on one another. This is in many ways contrary to their training and experience. Moreover, whilst many project managers are at least aware of the systemic nature of risks, as Williams (2017) points out, the vast majority of project management literature touches only briefly on methods to deal with the impacts of risks, and not only key texts such as PMBOK but also many other standards are silent on either the need or methods for managing systemicity in risk. This leaves project managers with little to work with.

> Taking a systemic view to risk management demands working with complexity – looking at risk structures that might exceed 100 events each with different impacts on one another.

EXISTING RISK TOOLS FOR SYSTEMIC RISK ANALYSIS

Despite the paucity of detail in texts such as PMBOK, there exist a range of tools and techniques that have been developed to take a systemic approach to risk. Each has its own benefits and shortcomings. Some of them stimulate conversation (and changing mental models and/or effectively engaging with stakeholders), and others are more for quantitative modelling. Moreover, selection of the tools and techniques to be used is dependent on the project's requirements. For example, systems dynamics (SD) can be applied in a purely qualitative form when quantification is fraught with challenges due to high levels of uncertainty. It is also worth being mindful of the fact that understanding the results and their significance is important – the more mathematical the form of modelling, the more challenging understanding (as noted by Taleb in Black Swans, blind application is fraught with danger). This is illustrated well by Coyle's comment regarding intelligent consumption "in system dynamics practice the uncertainties are usually justified by the argument that one is concerned with general patterns of behavior rather than precise numbers" (Coyle 2000 p. 227). Finally, it is worth noting that "the way we understand and describe risk strongly influences the way risk is analyzed and hence it may have serious implications for risk management and decision making" (Aven 2016 p. 4). As such, each of the methods presented has an impact on how it assists project managers as each will have advantages and disadvantages depending on the type of project, knowledge of the project manager, number of parties involved, etc.

Project Risk Register

The register (often required as part of regulatory requirements) allows for the calculation of the impact of a risk considering its probability (on an integer scale) and comprises the following steps: (1) identify, (2) analyze, (3) plan, and (4) monitor and review. The register can either use qualitative probabilities, e.g., low/medium/high, or more quantitative ones, e.g., 65% probability. Given its focus on understanding risks and consequences, the approach illustrates some network properties (e.g., through the inclusion of potential responses), although very limited and typically in a table form. As a consequence of frequently being mandated as part of the governance process of the project, Project Risk Registers are probably the most common tool used for risk assessment. However, it is not without its weaknesses. The most common limitation is that the process of constructing the register is at best given superficial consideration (ticking the box) and doesn't impact on decision-making. A second limitation is the illusion of control (Lyytinen 2011).

Bow Ties

Bow ties are a pictorial representation which allows for a systemic view of risks to be portrayed in an easy to understand manner (Jones and Israni 2012; Saud et al. 2014). The bow-tie structure illustrates how potential causes, and options for prevention and/or mitigation are linked following/ considering a significant event. Due to their accessibility, experience from a range of staff is able to be captured and structured. In the center of the tie is positioned the event – the hazard – with consequences (different scenarios) linking out and triggers/threats linking in. The diagram allows for multiple triggers to be displayed along with proactive mechanisms to control or prevent these events from occurring and multiple consequences with reactive controls displayed. They have typically been for hazard identification although their elements of (1) identify the hazard, (2) prevent or mitigate and (3) assess lend themselves to risk assessment.

Fishbone Diagrams

Fishbone diagrams, also referred to as Ishikawa diagrams (Ishikawa 1976), provide a method that enables the structured identification and categorization of causes. Initially used for managing quality problems in product design, the diagrams/method has been used to great effect in risk identification. The defect is noted with the causes being displayed as 'fishbones' with each bone representing a major cause and explanatory causes, that is root causes, branching off these major cause bones – as such provides a systemic representation. Correlations can then be determined regarding an event and possible causes. In addition, the causes and sub-causes can provide a structure for considering risk management options – ensuring a sustainable outcome. Where appropriate further analysis can be undertaken. Fishbone diagrams have also been referred to as cause-and-effect diagrams (Watson 2004).

Bayesian Networks

Bayesian Networks are a form of network modelling, using a chain rule (sequence) to represent causal relationships among a set of random variables considering local dependencies (Jensen and Neilsen 2007). As such, they form a directed acyclic graph. They have been used for risk analysis and safety assessment. The modelling approach enables probability updating using Bayes' theorem; thus, as new information becomes available, the network is updated aiding decision-making and providing the opportunity for learning. They are considered to be superior over fault trees (a similar process) through a relaxing of limitations providing a more flexible structure and their ability to handle uncertainty and thus being able to address a wider range of applications (Khakzad et al. 2011). Bayesian networks can, when reflecting and supporting decision-making, be referred to as influence diagrams. However, they are not able to handle feedback (Marle and Vidal 2011).

System Dynamics

SD is a continuous simulation modelling technique that is based on the work of Jay Forrester (1961). The models comprise stocks and flows along with delays and auxiliary variables which impact the rate of flow. As such, they are designed to model the entire project rather than focus on one particular event or hazard. A key characteristic of SD models is their ability to capture and subsequently analyze feedback behavior. SD modelling typically commences with the drawing of a causal loop diagram (or influence diagram) and can subsequently be modeled using sophisticated software such as Stella™, Powersim™, or Vensim™. As such using SD, it is possible to model a project as anticipated and compare this with what was experienced. SD has been used to understand project risks both prospectively (Sternam 1992; Gray and Shahidi 2011) and forensically (Cooper 1981; Howick et al. 2009).

Hindcasting

Hindcasting (Klein 2007) is a technique that occurs at the beginning of a project (assisting with the importance noted earlier of paying sufficient attention early in the project's life) and aims to elicit cause and effect from a futuristic perspective (as such, it can be conceived as being the opposite of a post-mortem). The technique commences with an assertion (possibly by the project manager) that the project has failed and team members individually note down possible causes. These causes are then shared and reviewed in order to adapt the plan to avoid their occurrence. This technique is particularly good for managing some of the conformity pressures that can exist and enabling those taking part to share their expertise and others to learn. It also enables team members to be more alert to a range of possibilities and be watching out for them. As such, it can be seen as a form of scenario planning where different futures are imagined and managed.

Cynefin

Another framework that can provide a useful structure to support project management is that of Snowden and Boone (2007), namely, the Cynefin framework. This framework "helps leaders determine the prevailing operative context so that they can make the appropriate choices" (Snowden and Boone 2007 p. 72). Snowden and Boone divide the decision-making landscape into four domains, each of which demands different actions. These domains include the simple (known knowns), the complicated (known unknowns), the complex (unknown unknowns), and the chaotic (unknowables). In each of these domains, they provide suggestions. The framework explicitly recognizes the need to take a holistic (systemic) approach when considering projects along with appreciation that projects are dynamic. Snowden and Boone also see the framework as helping to alleviate/reduce biases that are common in decision-making (e.g., the optimism bias which preferences the benefit of the doubt, the anchoring bias which restricts adjusting from an initial guess, the framing bias which notes the significance of individual perspectives, and the representation bias which deals with base rate ignorance). The developers also note the challenges encouraged with counterintuitive dynamics (Table 22.1).

Other techniques that recognize the value of a systemic approach to considering risk include The Cascade model (Ellinas 2019), Activity-on-Node networks (Project Management Institute 2013), Cybernetics Risk Influence Diagramming (Vinnakota 2011), Analytical Hierarchy Processes – AHP (Beauchamp-Akatova and Curran 2013), root cause analysis (Gangidi 2018), and the tailored business card (Yiannaki 2012).

In summary and as noted in the earlier quote regarding perspective taken, each technique has different contributions. It is worth noting that for either bow-tie or fishbone/Ishikawa diagrams, they focus on events separately rather than looking at the interaction between events. This reflects a key consideration namely the need to balance the complexity of modelling with accessibility – that is ensuring those using the techniques understand the modelling and are not overwhelmed by the learning demands. It is also worth noting that some of the methods are very mathematical and as such opaque to many. For example, the *Handbook on Systemic Risk* (Fouque and Langsam 2013) presents a range of mathematical/statistical forms of analyses which for many present significant challenges. It is also worth noting that some of the methods are explicitly noted as not being effective on complex projects suggesting care is taken to ensure fit for purpose usage.

TABLE 22.1

A Selection of the Benefits and Disadvantages of Systemic Tools

Tool	Benefits	Disadvantages
Project Risk Register	• Well known and often mandatory	• Does not allow for risks to be linked (Williams 2017)
	• Practical – enables time, cost, and technical risk analyses to be carried out	• Tends to focus on technical (and sometimes financial) risk and therefore misses risks (Williams 2017)
	• Can act as a central repository	• Does not consider social effects (De Bakker 2011)
	• Forms basis for the required outputs for risk work	• Provides an illusion of control (Lyytinen 2011)
	• Provides an audit trail (Williams 1994)	
Bow Tie	• Graphical and easy to use	• Does not allow for risks to be connected together
	• Helps with identifying hazards along with their causes and options for prevention/mitigation in an all-inclusive manner	• Does not quantify risks the likelihood of any barrier failing/risk occurring (McLeod and Bowie 2018)
	• Good for incident analysis (post-project)	• Requirement to acquire bow-tie software
	• Can include human	
	• Encourages group participation and group knowledge (Saud et al. 2014)	• Need to have a robust risk-assessment matrix to identify representative set of bow-tie diagrams (Saud et al. 2014)
	• Simple diagrammatic way of describing and analyzing risk – gives clear road map/overview in a single picture	• Generally limited to qualitative measures
	• Versatile	• Not effective in addressing potential synergetic effects of hazards/risks
	• Clear graphic representation of problem – readily understood (Popov et al. 2016)	• Can sometimes oversimplify complex situations
		• Additional risk assessment methods required if purpose is to model complex relationships (Popov et al. 2016)

(Continued)

TABLE 22.1 (*Continued*)

A Selection of the Benefits and Disadvantages of Systemic Tools

Tool	Benefits	Disadvantages
Fishbone	• Simple and effective tool – aids investigating and identifying numerous different causes of a problem	• Cannot link two risks together
	• Provides a structured means for determining causes of problems	• Time consuming – to analyze problems individual fishbone diagrams must be drawn for each problem
	• Helps focus on the causes of problems (Bilsel and Lin 2012)	• Possible to miss interrelationships between different problems and causes (Bilsel and Lin 2012)
	• Encourages group participation and group knowledge (Ilie and Ciocoiu 2010)	• Cannot provide quantified evaluation of risk (Luo et al. 2018)
Bayesian Belief Nets	• Able to handle incomplete data sets • Enables learning about causal relationships (Heckerman 2008) • Flexible graphical structure	• Significant challenges regarding assessing probabilities within feedback loops. Probabilities are subjective measures
	• Possible to analyze domino effects through probabilistic framework, thus considering synergistic effects	• Computational difficult of exploring a previously unknown network: maybe too costly to perform or impossible given the number and combination of variables
	• Benefits from both qualitative and quantitative modelling techniques (Kahkzad et al. 2013)	• Depending on prior knowledge being reliable (excessively optimistic or pessimistic assessments will distort network and impact results (Niedermayer 2008)
System Dynamics	• Recognizes counterintuitive dynamics namely feedback • Valid tool for large, complex projects (systems)	• Fully working model takes time and money and is complicated to build, and ensure reliability • Coyle (2000)
	• Able to incorporate concepts of quantitative and qualitative analysis into the model (Coyle 2000)	• Incorporating uncertainties on both inputs and outputs may result in misleading interpretations from the results
	• Enables hard quantitative effects to be captured along with the 'human' effects (Howick 2003)	• Assumes agreed/shared objectives
Hindcasting	• Helps teams to identify potential problems early on	• Doesn't provide any quantification
	• Reduces the 'damn the torpedo, full speed ahead' attitude of people overinvested in the project –this refers to attitude of continuing to go ahead with course of action despite known risks/reckless attitude (Klein 2007)	• Doesn't have any formal mechanism for structuring the outcomes • Doesn't model risks directly

(Continued)

TABLE 22.1 (*Continued*)

A Selection of the Benefits and Disadvantages of Systemic Tools

Tool	Benefits	Disadvantages
	• Prevents groupthink as team members are asked to come up with different reasons for problems • Encourages creative thinking and imaginative solutions (Kahneman 2011)	
Cynefin	• Dialectic approach • Help executives effectively consider their context (Snowden and Boone 2007) • Gives decision-makers new constructs allowing them to break out of old ways of thinking (Kurtz and Snowden 2003) • Provides an orderly way to evaluate the interaction of organizational systems, the external environments, etc. (Dettmer 2011)	• Doesn't provide any quantification • Doesn't have any formal mechanism for structuring the outcomes • Doesn't model risks directly

Possible videos for further information

- Cynefin Framework: www.youtube.com/watch?v=N7oz366X0-8
- Bow tie: www.youtube.com/watch?v=dpGKHncw-d8
- Fishbone diagram: www.youtube.com/watch?v=ZS8Re23Z_4k.

AN ILLUSTRATION OF A PARTICULAR TECHNIQUE'S CONTRIBUTION TO SYSTEMIC RISK MODELLING

Engineering Australia in association with the Risk Engineering Society, University of Sydney, and Australian Cost Engineering Society produced a white paper (2015) focusing on master complex projects. The paper highlighted a number of key considerations, namely,

- The importance of identifying and treating critical risks through *workshopping risk* – "an effective approach is to provide a forum that encourages participants to combine creative and logical thinking" (p. 18)
- The importance of taking a comprehensive view through ensuring "*differing perspectives* of project stakeholders should be taken into account, and a 'group think' mentality avoided" (p. 3)
- The importance of understanding the critical risks "for complex projects an exclusive reliance on the standard [analyzing the likelihood and consequences] means that the *early identification and treatment of low likelihood 'show stoppers'* may not occur" (p. 18).

Thus, taking a collaborative approach that ensures a comprehensive consideration and also attends to systemic risk emerges as critical. One particularly method – causal mapping – (Ackermann and Alexander 2016) has been used extensively in the arena of risk and project management. Initially to help with modelling disruption and delay (Ackermann et al. 1997; Williams et al. 2003), the technique has been used to support risk workshops (Ackermann et al. 2014), to underpin the management of projects schools of thought (Edkins et al. 2007), identifying project risk (Maytorena et al. 2004) and risk filters (Ackermann et al. 2007).

The technique allows for the capture and modelling of a range of perspectives (Bryson et al. 2004). This is important as no one person can have an appreciation of all of the potential project risks and is akin to blind men exploring the elephant and only having a partial understanding. Mapping not only allows for the elicitation of risks but also the modelling of the impacts of one risk on another and thus develops a systemic approach to risk management. This helps in a number of ways. The first is that in capturing the chains of argument (risk chains), it is possible to help each participant understand better the risks. The second is that in sharing the risks, a common understanding can be developed with each contributor being able to 'piggy back' off one another's contributions increasing coverage. The third benefit is that the resultant network – directed graph – can then be analyzed. This allows for a thoughtful assessment of where to focus energy rather than a more intuitive and potentially flawed appreciation. The analysis can point out feedback loops, along with giving insights into those that are central and have a significant impact compared with those that are relatively peripheral.

Causal mapping uses a set of well-established guidelines (Bryson et al. 2004; Ackermann et al. 2005) in order to ensure rigor in modelling (see Appendix A). Maps can be produced through one-to-one interviews, through manual workshops, and through computer-assisted/-supported workshops (Ackermann and Eden 2004). Once the maps have been created, they are amendable to a range of analyses (Eden et al. 1992) allowing for the identification of potential priority areas. Software is available for both the modelling element (Decision Explorer) along with providing facilities to support group interaction (Group Explorer Ackermann and Eden 2010).

EXAMPLE: UNDERSTANDING RISKS FOR ENERGY

Scottish Hydro Electric Power Distribution (SHEPD) was looking at designs for the powering of the Shetland Islands, a remote set of islands at the northern end of the United Kingdom. The power station supplying the islands required replacement due to age and changes in emission regulations. A key consideration in the design was meeting the UK and Scottish governmental aims to increase the use of renewal power and the fact that the Shetland Islands, given their geographic location, experience regularly strong to very strong winds. There was also the recognition that any new power system needed to be able to run independently of any other power generation infrastructure (the Shetland Islands lack grid connection to the rest of the United Kingdom). As such, the organization was considering options. Given the range of power generation options available (wind, oil burning, and waste burning), storage through batteries and the need to attend to a range of domestic and organizational demands (e.g., factories, hospitals, schools), SHEPD through the Northern Isles New Energy Solutions (NINES) project was designed to carry out a comprehensive risk assessment exercise for the different options available. As such, the organization engaged with researchers to carry out a range of risk analyses.

One particular component of the risk assessment exercise was a series of risk workshops. These workshops allowed a range of different perspectives to be elicited ranging from those working in the energy company, to Shetland Islanders and also academics (power engineers knowledgeable about smart grids). It also ensured that stakeholders felt included in the project and their views appreciated. Each of the workshops involved around eight to sixteen participants. It was believed that incorporating this range of perspectives would ensure the greatest diversity and coverage. They were held at locations convenient to the participants and lasted for half a day.

The workshops sought to both capture a comprehensive number of risks (eliciting divergence) and structure the risks so that (1) further insights and shared understanding could be elicited and (2) the resultant structure could be analyzed. As such, the process adopted the cause mapping technique supported by a group support system (see Figure 22.8). The group support system allowed for simultaneous capture and also anonymity – a valuable feature given that participants can be uncomfortable raising potential risks.

The workshops commenced with a brief introduction to the work (see the guidelines noted in Appendix B) and then asked participants to type into the system those risks that they saw potentially occurring. The group was also asked for opportunities so as to capture both the positive and negative

FIGURE 22.8   Example of a group using the group support system to surface and review risks. Adapted from Ackermann et al. (2014 p. 292).

aspects facing the project allowing for maximum effectiveness. This stage allowed risks to be identified from the bottom up ensuring breadth and recognizing the importance of context.

As participants generated the risks and opportunities, the facilitator sought to cluster the contributions into content-oriented clusters. This was for two reasons: the first to help with the mass of information being generated (and thus reduce cognitive overload) and the second so that a review of the clusters could help reveal any missing areas for consideration. The surfacing period lasted around 15–20 min before a review of the clusters was undertaken and a second brief contribution period provided.

The group then began to explore, with the facilitator's help, the clusters particularly paying attention to the interaction between the risks and opportunities. As noted above in this chapter, risks can impact on other risks, and this interactivity was captured through entering causal links between the risks/opportunities. From this work, the original content-oriented clusters morphed into structural clusters each having a number of interactions with other clusters forming a network.

The facilitator was able to undertake various analysis, e.g., identification of those risks/opportunities that were central or extremely busy (lots of links in and out) and also feedback (which as noted above can have serious ramifications for project success). This analysis provided a first pass at determining priority areas. A small section of the map with a feedback loop is shown in Figure 22.9. The loop is a potentially vicious feedback loop focusing on staffing and funding. Starting at the bottom left, the loop recognizes the lack of engineers resulting in insufficient knowledge and thus negatively impacting on a successful project. This, in turn, has ramifications on the regulator's perception of the organization and thus negatively impacts future project which subsequently affect recruitment – illustrating how impacts on one project can ricochet onto others.

Following this exploration, the group was given the opportunity to prioritize the risks according to the potential impact and probability using a 0–100 scale. Risks that were seen to have the least impact were positioned at 0 and risks seen as having the most impact at 100 with the remainder positioned relative to these two anchor points. A second prioritization activity had participants determining which of the risks, if occurring, would be in the short term (6 months) and which were more long term. Where there were discrepancies in views further discussion allowed for a deeper understanding to emerge.

The workshops also allowed for the examination of goals/objectives and revealed significant differences with some of the organizational staff noting that the goal was *to contribute to renewal targets*, another noting *keep the lights on* (i.e. no black outs), and a third noting *keep costs at a minimum*. This difference is not untypical – the Engineers Australia paper (2015) noted how important it was to "achieve genuine alignment of the stakeholders" (pp. 4 and 9).

The process was seen as very beneficial to the organization as it revealed a number of risks previously not considered. For example, running a smart grid power system demands considerable engineering skills, and staff with these skills are fairly scarce. Given the remoteness of Shetland,

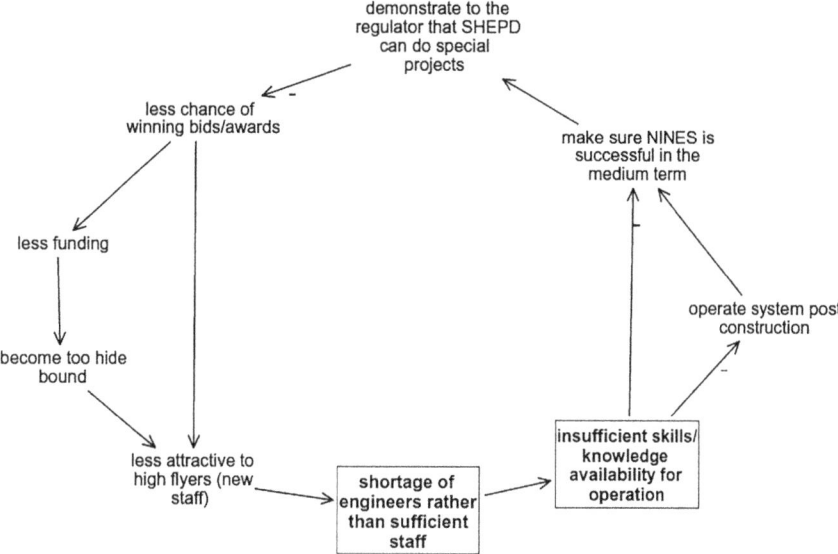

FIGURE 22.9 Extract from the model showing one of the feedback loops. Note: Each of the arrows should be read as 'may lead to' except for those arrows with a negative sign at the arrowhead which should be read as, 'may not lead to'. Adapted from Ackermann et al. (2014 p. 295).

along with fairly extreme weather in winter, and a small close-knit community, attracting and retaining these skilled staff under such conditions was likely to be a real challenge. The process was also used again on further projects.

QUESTIONS FOR DISCUSSION

Q. What are the benefits and disadvantages of taking a systemic approach to risk assessment?

A. Benefits include engaging stakeholders through the process of ensuring comprehensive capture of risks, being able to prioritize risks in a systematic fashion (rather than 'gut feel') and ensure management options are sustainable and gain greatest traction (through tailoring each option to address potentially more than one risk). Disadvantages include difficulty in learning the methods/approaches particularly those that demand mathematical/modelling skills and accessibility of the results (both in opening up the risk identification space – which can feel overwhelming – and in managing the resultant material).

Q. When might undertaking a risk workshop make sense (and where might it not make sense)?

A. Where projects involve a number of different stakeholders, e.g., joint ventures, or where there are lots of suppliers – tapping into the wealth of risks from different perspectives. This relates to the case study example where the Shetland Islanders were able to provide the energy company with an insight into particular localized risks. In addition, where projects involve challenging conditions, e.g., the environment (from all dimensions, ecological, political, economic).

Q. When should traditional risk assessment techniques be used and when should systemic ones be used?

A. Probably, a combination of both is best with one augmenting the other. Traditional methods such as Project Risk Registers are often necessary due to compliance requirements, but these can be enhanced by taking a systemic review. Projects that have considerable complexity will definitely benefit from using systemic methods but again probably alongside traditional techniques to ensure a degree of comfort and accessibility.

SUMMARY

To conclude, as this chapter has noted, there are a number of good reasons to adopt a systemic approach to risk assessment and management, namely, the four below considerations:

- *Think systemically when managing projects*
- *See risk, complexity, and uncertainty as inter-related and changing over time*
- *Take a comprehensive view of risks*
- *Take a systemic view of risks.*

Attending to each of these considerations will reduce the chance of unwelcome surprises (unknown unknowns) through increasing the chance of surfacing 'black swans' as both the comprehensiveness and systemic approach teases out further insights.

Allocating sufficient time to consider risks at the commencement of the project and at regular intervals (thinking systematically) will also help increase the chance of project success particularly if a bottom-up and bespoke approach is taken allowing for a comprehensive view of risks.

Keeping the risk management plan up to date can transform it from a door stop into a vital project management tool. "Remember what you don't know can kill your project", Bruce Pitman (Senior Vice President and Senior Operating Officer of the US National Space Society).

Involving stakeholders not only ensures a wider consideration of risks but also builds relationships and ensures shared understanding of the project. This is not only valuable in terms of ensuring that the 'project team' are all on the 'same page' but also helps manage context if others such as regulators, local authorities, and communities are involved as once again a more comprehensive set of risks can be identified (as in the NINES example case) but also potentially a greater chance of project success through 'buy-in'.

> Involving stakeholders not only ensures a wider consideration of risks but also builds relationships and ensures shared understanding of the project.

Finally, systemic risk means thinking wider than the stages of design and build and considering projects in terms of their operations and maintenance. "Most of the time, your risk management works. With a systemic event such as the recent shocks following the collapse of Lehman Brothers, obviously the risk-management system of any one bank appears, after the fact, to be incomplete. We ended up where banks couldn't liquidate their risk, and the system tended to freeze up" (Myron Scholes, Nobel Laureate in Economic Sciences).

APPENDIX A: GUIDELINES FOR MAPPING USING A MANUAL PROCESS

PRE-WORKSHOP

- Identify a room with a large blank (preferably cream or white) wall and enough space to position six to ten chairs in a semi-circle around the wall space.
- Position 12 sheets of flipchart paper to the wall – portrait layout – 6 × 2.
- Have plenty of large post-it pads, flip chart pens (multiple colors), and single color, thick-nib pens for participants
- Agree on the focus question – e.g., 'what are the opportunities, risks and concerns facing xx'.

WORKSHOP

- Provide a clear introduction to the workshop including the benefits of being inclusive and not self-censoring, writing each risk as a single item with six to eight words to describe it, building on one another's ideas, etc.
- Provide pends and post-its to participants.
- Encourage contribution of risks and opportunities.
- Cluster the risks and opportunities into content oriented clusters – and be prepared to change these as more contributions surface and suggest alternatives. Number the post-its as you go.
- Review the clusters with the group to stimulate new ideas and check contents.
- Review each cluster in terms of relationships – best to position the post-its first and then draw in the links. Where the links straddle clusters consider just drawing in a small link and noting the risk/opportunity number at the end rather than drawing long arrows.
- Capture new risks/opportunities as the group discusses each of the clusters and weave these into the structure.
- Provide participants with sticky dots for prioritization (reflecting on the structure first).
- Review the outcome and discuss next steps.

POST-WORKSHOP

Provide participants with photos of the clusters/wall map and list the prioritized risks/opportunities.

APPENDIX B: MAPPING GUIDELINES ADAPTED FROM ACKERMANN ET AL. (2005)

WORDING

- Action oriented
- Six to eight words
- Think about 'who'/'what'/'when'
- Avoid 'should'/'ought'/'need'/'must', etc. (as these are prescriptive and do not add value)
- Avoid 'in order to'/'due to'/'through', etc. (as these are implicit links)
- Separate sentences into phrases (potentially with different consequences).

DIRECTION OF THE ARROW

- Options to outcomes; means to ends
- Not a chronology/flow chart
- Avoid double-headed arrows – tease out the meaning between both statements, e.g., how does A link to B, and B link to A. This will assist in determining the direction of the link or elaborate a feedback loop.

ADDITIONAL CONSIDERATIONS

- Link specific activities into generic activities, e.g., privatize food into privatize hospitals.
- Position assertions/facts at the bottom of the hierarchy and goals/objectives at the top.
- Do not paraphrase; keep the contributor's own language.
- Encourage participants to build the map together – a shared artefact.

REFERENCES

Ackermann, Fran, and James Alexander. 2016. Researching complex projects: Using causal mapping to take a systems perspective. *International Journal of Project Management* 34, no. 6: 891–901. doi:10.1016/j. ijproman.2016.04.001.

Ackermann, Fran, and Colin Eden. 2004. Using causal mapping: Individual and group; traditional and new. In *Systems Modelling: Theory & Practice*, edited by M. Pidd, pp. 127–145. Chichester: Wiley.

Ackermann, Fran, and Colin Eden. 2010. The role of group decision support systems: negotiating safe energy. In *Handbook of Group Decision and Negotiation*, edited by D. Kilgour, and C. Eden, pp. 285–299. Dordrecht: Springer,

Ackermann, Fran, Colin Eden, and Ian Brown. 2005. *The Practice of Making Strategy*. London: Sage.

Ackermann, Fran, Colin Eden, and Terrence Williams. 1997. A persuasive approach to delay and disruption using 'mixed methods'. *Interfaces* 27, no. 2: 48–65.

Ackermann, Fran, Colin Eden, and Terrence Williams et al. 2007. Systemic risk assessment: A case study. *Journal of the Operational Research Society* 58, no. 1: 39–51. doi:10.1057/palgrave.jors.2602105.

Ackermann, Fran, Susan Howick, John Quigley et al. 2014. Systemic risk elicitation. Using causal maps to engage stakeholders and build a comprehensive view of risks. *European Journal of Operational Research* 238, no. 1: 290–299. doi:10.1016/j.ejor.2014.03.035.

Alexander, James, Fran Ackermann, and Peter Edward D. Love. 2019. Taking a holistic exploration of the project life cycle on public private partnerships: Considering issues relating to quality, risk and HRM. *Project Management Journal*: 1–13. doi:10.1177/8756972819848226.

Aven, Terje. 2015. Implications of black swans to the foundations and practice of risk assessment and management. *Reliability Engineering & System Safety* 134: 83–91. doi:10.1016/j.ress.2014.10.004.

Aven, Terje. 2016. Risk assessment and risk management: Review of recent advances on their foundation. *European Journal of Operational Research* 253, no. 1: 1–13. doi:10.1016/j.ejor.2015.12.023.

Beauchamp-Akatova, Elena, and Richard Curran. 2013. From initial risk assessments to system risk management. *Journal of Modelling in Management* 8, no. 3: 262–289. doi:10.1108/JM2-01-2011-0008.

Bilsel, R. Ufuk, and Dennis Lin. 2012. Ishikawa cause and effect diagrams using capture recapture techniques. *Quality Technology & Quantitative Management* 9, no. 2: 137–152. doi:10.1080/16843703.201 2.11673282.

Browning, Tyson R., and Ranga Ramasesh. 2015. Reducing unwelcome surprises in project management. *MIT Sloan Management Review* 56, no. 3: 53. https://sloanreview.mit.edu/article/reducing-unwelcome-surprises-in-project-management/.

Bryson, John Michael, Fran Ackermann, Colin Eden et al. 2004. *Visible Thinking: Unlocking Causal Mapping for Practical Business Results*. Chichester: John Wiley & Sons.

Carpenter, Chris 2015. Decommissioning costs can be reduced. *Journal of Petroleum Technology* 67, no. 1: 113–116. doi:10.2118/0115-0113-JPT.

Carrell, Severin 2016. Building flaws found in all 17 closed Edinburgh schools. *The Guardian*: April 14, 2016. www. theguardian.com/uk-news/2016/apr/14/building-flaws-found-in-all-17-closed-edinburgh-schools.

Carrell, Severin 2017. Damning report attacks firms which built fault-ridden Scottish schools. *The Guardian*: February 11, 2017. www.theguardian.com/uk-news/2017/feb/09/damning-report-slams-firms-who-built-fault-ridden-scottish-schools.

Checkland, Peter 1980. *Systems Thinking, Systems Practice*. Chichester: Wiley.

Checkland, Peter 2001. Soft systems methodology. In *Rational Analysis for a Problematic World Revisited*, edited by J. Rosenhead, and J. Mingers, pp. 61–90. Chichester: Wiley.

Checkland, Peter, and John Poulter. 2010. Soft systems methodology. In *Systems Approaches to Managing Change: A Practical Guide*, edited by M. Reynolds, and S. Holwell, pp.191–242. London: Springer.

Cole, John. 2017. Report of the independent inquiry into the construction of Edinburgh schools. *The City of Edinburgh Council*: February 9, 2017. www.edinburgh.gov.uk/news/article/2245/independent_report_into_school_closures_published.

Cooper, Robert. G. 1981. The components of risk in new product development: Project new prod. R&D *Management* 11, no. 2: 47–54. doi:10.1111/j.1467-9310.1981.tb00449.x.

Coyle, Geoff. 2000. Qualitative and quantitative modelling in system dynamics: Some research questions. *System Dynamics Review: The Journal of the System Dynamics Society* 16, no. 3: 225–244. doi:10.1002/1099-1727(200023)16:3<225::AID-SDR195>3.0.CO;2–D.

De Bakker, Karel Franciscus Christiaen. 2011. *Dialogue on Risk: Effects of Project Risk Management on Project Success*. Netherlands: University of Groningen.

De Lemos, Teresa, David Eaton, Martin Betts et al. 2004. Risk management in the Lusoponte concession—A case study of the two bridges in Lisbon, Portugal. *International Journal of Project Management* 22, no. 1: 63–73. doi:10.1016/S0263-7863(03)00013-9.

Dettmer, William. 2011. *Systems Thinking and the Cynefin Framework: A Strategic Approach to Managing Complex Systems*. Port Angeles: Goal Systems International.

Eden, Colin, Fran Ackermann, and Steve Cropper. 1992. The analysis of cause maps. Journal of *Management Studies* 29, no. 3: 309–324. doi:10.1111/j.1467-6486.1992.tb00667.x.

Eden, Colin, Fran Ackermann, and Terrence Williams. 2005. The amoebic growth of project costs. *Project Management Journal* 36, no. 2: 15–27. doi:10.1177/875697280503600203.

Edkins, Andrew, Esra Kurul, Eunice Maytorena-Sanchez et al. 2007. The application of cognitive mapping methodologies in project management research. *International Journal of Project Management* 25, no. 8: 762–772. doi:10.1016/j.ijproman.2007.04.003.

Ellinas, Christos. 2019. The domino effect: An empirical exposition of systemic risk across project networks. *Production and Operations Management* 28, no. 1: 63–81. doi:10.1111/poms.12890.

Eskerod, Pernille, and Anna Lund Jepsen. 2013. *Project Stakeholder Management*. London: Gower Publishing.

Forrester, Jay. 1961. *Industrial Dynamics*. Waltham, MA: Pegasus Communications.

Fouque, Jean-Pierre, and Joseph A. Langsam. 2013. *Handbook on Systemic Risk*. Cambridge: Cambridge University Press.

Fowler, Ashely, Peter Macreadie, Daniel Jones et al. 2013. A multi-criteria approach to decommissioning of offshore oil and gas infrastructure. *Ocean & Coastal Management* 87: 20–29. doi:10.1016/j.ocecoaman.2013.10.019.

Gangidi, Prashant. 2018. A systematic approach to root cause analysis using 3 × 5 why's technique. *International Journal of Lean Six Sigma*. doi:10.1108/IJLSS-10-2017-0114.

Gephart Jr, Robert P., John Van Maanen, and Thomas Oberlechner. 2009. Organizations and risk in late modernity. Organization Studies 30, no. 2–3: 141–155. doi:10.1177/0170840608101474.

Geraldi, Joana, Harvey Maylor, and Terry Williams. 2011. Now, let's make it really complex (complicated): A systematic review of the complexities of projects. *International Journal of Operations & Production Management* 31, no. 9: 966–990. doi:10.1108/01443571111165848.

Government of Western Australia. 2018. *Special Inquiry into Government Programs and Projects*. Western Australia: The Public Sector Commission. https://publicsector.wa.gov.au/sites/default/files/documents/special_inquiry_into_government_programs_and_projects_volume_2.pdf.

Gray, Mark, and Azin Shahidi. 2011. Applying the principles of system dynamics to project risk management or "the domino effect". *Conference paper presentation at the Project Management Institute Global Congress*, Dallas, TX.

Heckerman, David. 2008. A tutorial on learning with Bayesian networks. In *Innovations in Bayesian Networks: Theory and Applications*, edited by D.E. Holmes, and L.C. Jain, pp. 33–82. Berlin: Springer.

Howick, Susan. 2003. Using system dynamics to analyse disruption and delay in complex projects for litigation: Can the modelling purposes be met? *Journal of the Operational Research Society* 54, no. 3, 222–229. doi:10.1057/palgrave.jors.2601502.

Howick, Susan, Fran Ackermann, Colin Eden et al. 2009. System dynamics and disruption and delay in complex projects. In *Encyclopaedia of Complexity & System Science*, edited by R. Meyers, pp. 1845–1864. New York: Springer.

Ilie, Gheorghe, and Carmen Ciocoiu. 2010. Application of fishbone diagram to determine the risk of an event with multiple causes. *Management Research and Practice* 2, no. 1: 1–21.

Ishikawa, Kaoru. 1976. *Guide to Quality Control*. Tokyo: Asian Productivity Organization.

Jensen, Thomas, and Finn Jensen. 2007. *Bayesian Networks and Decision Graphs*. New York: Springer.

Jones, Fredrick, and Kumar Israni. 2012. Environmental risk assessment utilizing bow-tie methodology. *Conference paper presentation at the International Conference on Health, Safety and Environment in Oil and Gas Exploration and Production*, Perth.

Kahneman, Daniel. 2011. *Thinking, Fast and Slow*. New York: Farrar, Straus and Girouz.

Kaivo-oja, Jari. 2011. Weak signals analysis, knowledge management theory and systemic socio-cultural transitions. *Futures* 44, no. 3: 206–217. doi:10.1016/j.futures.2011.10.003.

Khakzad, Nima, Faisal Khan, and Paul Amyotte. 2011. Safety analysis in process facilities: Comparison of fault tree and Bayesian network approaches. *Reliability Engineering & System Safety* 96, no. 8: 925–932. doi:10.1016/j.ress.2011.03.012.

Khakzad, Nima, Faisal Khan, Paul Amyotte et al. 2013. Domino effect analysis using Bayesian networks. *Risk Analysis: An International Journal* 33, no. 2: 292–306. doi:10.1111/j.1539-6924.2012.01854.x.

Klein, Gary. 2007. Performing a project premortem. Harvard *Business Review* 85, no. 9: 18–19. doi:10.1109/EMR.2008.4534313.

Krane, Hans Petter, Nils Olsson, and Asbjørn Rolstadås. 2012. How project manager-project owner interaction can work within and influence project risk management. *Project Management Journal* 43, no. 2: 54–67. doi:10.1002/pmj.20284.

Kurtz, Cynthia, and David Snowden. 2003. The new dynamics of strategy: Sense-making in a complex and complicated world. *IBM Systems Journal* 42, no. 3: 462–483. doi:10.1147/sj.423.0462.

Kwan, Tak Wah, and Hareton Leung. 2011. A risk management methodology for project risk interdependencies. *IEEE Transactions on Software Engineering* 37, no. 5: 635–648. doi:10.1109/TSE.2010.108.

Laraia, Michele. 2017. New and unexpected stakeholders in decommissioning projects. In *Advances and Innovations in Nuclear Decommissioning*, edited by M. Laraia, pp. 131–151. Cambridge: Elsevier Science & Technology.

Love, Peter Edward, Pauline Teo, Fran Ackermann et al. 2018. Reduce rework, improve safety: An empirical inquiry into the precursors of error in construction. *Production Planning & Control* 29, no. 5: 353–366. doi:10.1080/09537287.2018.1424961.

Luo, Tongyuan, Chao Wu, and Lixiang Duan. 2018. Fishbone diagram and risk matrix analysis method and its application in safety assessment of natural gas spherical tank. *Journal of Cleaner Production* 174: 296–304. doi:10.1016/j.jclepro.2017.10.334.

Lyytinen, Kalle. 2011. MIS: The urge to control and the control of illusions – Towards a dialectic. *Journal of Information Technology* 26, no. 4: 268–270. doi:10.1057/jit.2011.12.

Marle, Franck and Ludovic-Alexandre Vidal. 2011. Project risk management processes: Improving coordination using a clustering approach. *Research Engineering Design* 22: 189–206.

Maytorena, Eunice, Graham Winch, Sharon Clarke et al. 2004. Identifying project risks: A cognitive approach. In *Innovations: Project Management Research*, edited by D.I. Cleland, J.K. Pinto, and D.P. Slevin, pp. 465–479. Newtown Square, PA: Project Management Institute.

McHaina, David. 2001. Environmental planning considerations for the decommissioning, closure and reclamation of a mine site. *International Journal of Surface Mining, Reclamation and Environment* 15, no. 3: 163–176. doi:10.1076/ijsm.15.3.163.3412.

McLeod, Ronald, and Paul Bowie. 2018. Bowtie analysis as a prospective risk assessment technique in primary healthcare. *Policy and Practice in Health and Safety* 16, no. 2: 177–193. doi:10.1080/14773996.2018.1466460.

Merrow, Edward. 2011. Industrial *Megaprojects*: Concepts, *Strategies*, and *Practices* for *Success*. Hoboken, NJ: John Wiley.

Niedermayer, Daryle. 2008. An introduction to Bayesian networks and their contemporary applications. In *Innovations in Bayesian Networks: Theory and Applications*, edited by D.E. Holmes, and L.C. Jain, pp. 117–130. Berlin: Springer.

Nutt, Paul. 2002. *Why Decisions Fail: Avoiding the Blunders and Traps that Lead to Debacles*. San Francisco, CA: Berrett-Koehler Publishers.

Popov, Georgi, and Bruce Lyon. 2016. Bow-tie risk assessment methodology. In *Risk Assessment: A Practical Guide to Assessing Operational Risks*, edited by G. Popov, B.K. Lyon, and B. Hollcroft, pp. 181–208. Hoboken, NJ: Wiley.

Project Management Institute. 2013. A *Guide to the Project Management Body of Knowledge (PMBOK)*. Newtown Square, PA: Project Management Institute.

Reynolds, Martin, and Sue Holwell. 2010. *Systems Approaches to Managing Change: A Practical Guide*. London. Springer.

Rossel, Pierre. 2013. Weak signals as a flexible framing space for enhanced management and decision making. In *Foresight for Dynamic Organisations in Unstable Environments*, edited by S. Mendonça, and S. Bartolomeo, pp. 31–44. Abingdon: Routledge.

Saud, Yaneira, Kumar Israni, and Jeremy Goddard. 2014. Bow-tie diagrams in downstream hazard identification and risk assessment. *Process Safety Progress* 33, no. 1: 26–35. doi:10.1002/prs.11576.

Scottish Parliament. 2003. Corporate body issues August update on Holyrood. Media Release, Parliament House, 24 November. www.parliament.scot/VisitorInformation/PressReleases2003.pdf.

Simon, Herbert. 1977. *The New Science of Management Decision*. Upper Saddle River, NJ: Prentice-Hall.

Snowden, David, and Mary Boone. 2007. A leader's framework for decision making. Harvard Business Review 85, no. 11: 68.

Sternam, John. 1992. System dynamics modelling for project management. Unpublished manuscript, Cambridge: Massachusetts Institute of Technology, Sloan School of Management.

Taleb, Nassim Nicholas. 2007. *The Black Swan: The Impact of the Highly Improbable*. London: Penguin.

Thamhain, Hans. 2013. Managing risks in complex projects. *Project Management Journal* 44, no. 2: 20–35. doi:10.1002/pmj.21325.

Thevendran, Vicknayson, and Michael Mawdesley. 2004. Perception of human risk factors in construction projects: An exploratory study. *International Journal of Project Management* 22, no. 2: 131–137. doi:10.1016/S0263-7863(03)00063-2.

Vinnakota, Tirumala. 2011. Systemic assessment of risks for projects: A systems and cybernetics approach. *2011 IEEE International Conference on Quality and Reliability*, Bangkok, Thailand.

Watson, Greg. 2004. The legacy of Ishikawa. *Quality Progress* 37, no. 4: 54.

Williams, Terrence. 1994. Using a risk register to integrate risk management in project definition. *International Journal of Project Management* 12, no. 1: 17–22. doi:10.1016/0263-7863(94)90005-1.

Williams, Terrence. 1997. Empowerment vs risk management? *International Journal of Project Management* 15, no. 4: 219–222. doi:10.1016/S0263-7863(96)00074-9.

Williams, Terrence. 2005. Assessing and moving on from the dominant project, management discourse in the light of project overruns. *IEEE Transactions on Engineering Management* 52, no. 4: 497–508. doi:10.1109/TEM.2005.856572.

Williams, Terrence. 2017. The nature of risk in complex projects. *Project Management Journal* 48, no. 4: 55–66. doi:10.1177/875697281704800405.

Williams, Terrence, Fran Ackermann, and Colin Eden. 2003. Structuring a delay and disruption claim: An application of cause-mapping and system dynamics. *European Journal of Operational Research* 148, no. 1: 192–204. doi:10.1016/S0377-2217(02)00372-7.

Williams, Terrence, Colin Eden, Fran Ackermann et al. 1995a. The effects of design changes and delays on project costs. *Journal of the Operational Research Society* 46, no. 7: 809–818. doi:10.1057/jors.1995.114.

Williams, Terrence, Colin Eden, Fran Ackermann et al. 1995b. Vicious circles of parallelism. *International Journal of Project Management* 13, no. 3: 151–155. doi:10.1016/0263-7863(95)00034-N.

Winch, Graham. 2010. *Managing Construction Projects: An Information Processing Approach.* Oxford: Wiley-Blackwell.

Yiannaki, Simona. 2012. A systemic risk management model for SMEs under financial crisis. *International Journal of Organizational Analysis* 20, no. 4: 406–422. doi:10.1108/19348831211268607.

23 Systemic Innovation Toolset

Gerrit Muller, Kristin Falk, and Marianne Kjørstad
University of South-Eastern Norway

CONTENTS

BACKGROUND: THE HOLY GRAIL OF SYSTEMIC INNOVATION—CHALLENGES AND OPPORTUNITIES

In this chapter, we will discuss methods, tools, and the mindset that may help to get closer to the Holy Grail of systemic innovation. A major challenge of innovation is that the organization inherently enters new territory, which implies that it does not know the lay of the land. Innovation brings uncertainty and unknowns. In today's world, we expand this notion further in VUCA, an acronym standing for volatility, uncertainty, complexity, and ambiguity. The volatility relates to the fast pace in society and technology. Complexity stems from the integration of systems and organizations, resulting in many emerging dependencies and interactions. The human and organizational context with a wide variation of properties as emotional, cultural, social, and political brings ambiguity.

> In this chapter, we will present methods and tools that may help to get closer to the Holy Grail of systemic innovation. Tools covered include visual ConOps, A3 Architecture Overviews (A3AO), and co-creation sessions using ideation, and methods like pacing and agile fast iterations.

Project management works often closely together with systems engineers or architects. The systems engineers are among others responsible for a sound solution design:

- The decomposition of the solution in parts
- The definition of interfaces

- The design of the dynamic behavior of the parts (the interaction between parts resulting in functionality)
- The allocation of quality attributes to functions and parts.

A challenge of systems design, also in less innovative projects, is to orchestrate that the emerging dynamic behavior delivers the desired functionality and performance. A threat is that undesired behavior emerges, causing quality problems related to attributes such as reliability, safety, and security. The worst situation is that *unforeseen* emergence results in undesired behavior. Innovative solutions have an increased risk of unforeseen emergence, due to a lack of experience and knowledge of the solution. Weaknesses of the systems design show up during systems integration or later in the life cycle.

Traditional project management has the focus on organizing and managing projects such that the projects conclude successfully; that is, deliver the pre-defined results within the given time and with the available resources. Most activities are control oriented, such as planning, resource allocation, progress monitoring, and risk mitigation. For mature projects, where the objectives are clear, the solution technology is ready, the design is a variant of previous designs, and the deployment and business context are existing, a control mindset in project management may work well. A major challenge for project management of innovation is that a pure control mindset may kill the innovation before it has started. Figure 23.1 shows the essential tasks of project management for systemic innovation.

> A major challenge for project management of innovation is that a pure control mindset may kill the innovation before it has started.

Innovation requires the recognition of an opportunity, creation of a fitting solution, successful deployment of this solution with a viable business, and an innovation culture. A major challenge is that initially, the innovating organization lacks knowledge of the opportunity (the customer context), potential solutions (technology, the supply chain, and potential technology and business partners), the deployment (the operational context from manufacturing to commissioning to support), and the business context (competition, regulations, finance). This knowledge is dynamic and may change fast. Other stakeholders (customers, partners, competitors, regulators, suppliers, investors, etc.) may lack knowledge and perspective, resulting in noisy and ambiguous inputs, such as needs and concerns, to the project.

A challenge for innovators is to distinguish the problem space and solution space (see Figure 23.2). The problem space is the context where specific problems occur. These problems are opportunities for entrepreneurs, if they can find (technical) feasible and (business) viable solutions. Part of the challenge is that most engineers "live" and think in the solution space, which obfuscates their understanding of the actual problem. Vice versa, many stakeholders may live and think in the problem space.

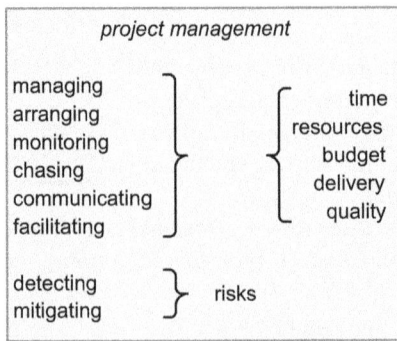

FIGURE 23.1   Essential tasks of project management for systemic innovation.

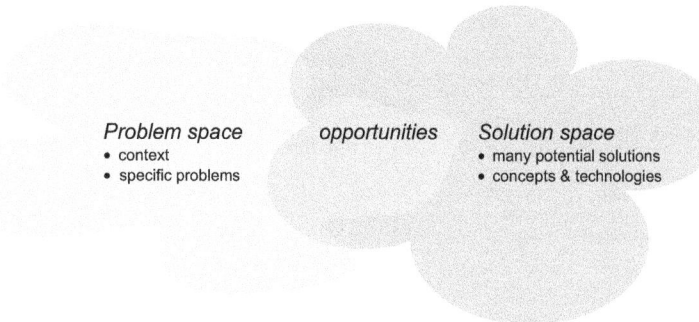

FIGURE 23.2 Innovation projects need to explore and understand both the problem and the solution space.

Project management in this context needs tools to explore the problem and solution space. These tools help by facilitating understanding, reasoning, communication, and decision-making. Examples are

- Visual ConOps to capture the envisioned operations
- A3AO providing multiple views at various abstraction levels to capture the architecture overview
- Co-creation sessions using ideation.

> These tools help by facilitating understanding, reasoning, communication, and decision-making.

Another class of tools serve the planning, organizing, and controlling roles in the VUCA innovation setting. These tools must facilitate fast learning and fast detection of knowledge gaps, to ensure that the solution results only in desired functionality and performance. Examples of such tools are

- Pacing and agile fast iterations
- Planning for integration

The "Towards the Holy Grail of Systemic Innovation – Methods and Tools" section describes the tools further.

FRAMEWORK FOR PROJECT MANAGEMENT FOR SYSTEMIC INNOVATION

FRAMEWORK FOR PROJECT MANAGEMENT FOR SYSTEMIC INNOVATION

Figure 23.3 shows the project stakeholders and main project artifacts in typical projects. Project management should be planning, organizing, staffing, leading, and controlling in this project landscape. The figure shows that the project management job extends significantly beyond the project boundaries. Many stakeholders including investors, legislators, customers, and life cycle organization operate at a distance from the project. These stakeholders interact with the project via a wide variety of artifacts.

Project management for systemic innovation needs a prime focus on making progress by learning fast. Fast learning helps to reduce knowledge gaps and, in that way, helps projects coping with VUCA. A key aspect of fast learning is failing early (Kelley and Kelley 2013), which helps in identifying obstacles, limitations, and needs. In this landscape, we elaborate some of the tasks in Figure 23.1 (project management tasks):

Composing the project team with a proper balance of capabilities such as leadership skills, level and type of proficiency, creativity, ability to think outside the box, and doers. Attention points are team dynamics and the environment they work in.

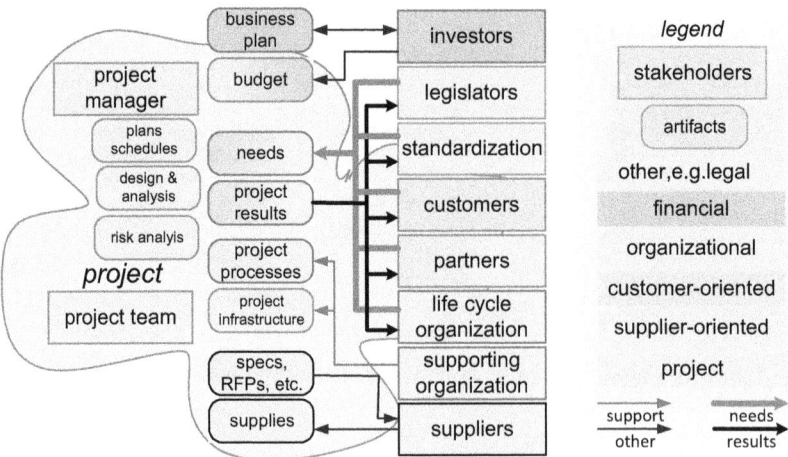

FIGURE 23.3　The project landscape with stakeholders and artifacts.

Organizing and facilitating project members such that they cooperate early and interact with external stakeholders early. Many project members in innovation teams may tend to work locally, unable to see the impact of their work at system level, e.g., emerging behavior. In practice, problems often arise when multiple parts interact causing undesired emerging behavior. Similarly, interaction with the context typically causes undesired side effects, besides the desired emerging functionality and performance.

Orchestrating solution design and analysis to form a basis for planning, identifying required competences, risk analysis, business plan, and communication with external stakeholders.

Organizing the project infrastructure and processes such that they suit the dynamics of the innovation project. Project infrastructure and processes consist among other of the IT tools supporting engineering, such as configuration and version management, documentation repositories, and workflow support. It also facilitates practicalities such as housing and test equipment. In general, processes need to be "just enough", so rather lightweight to avoid bureaucratization. General opinion is that innovation takes place at the edge of chaos; however, complete chaos is a risk for progress of the project. This is a delicate balancing act for project management.

Managing budget and business and project plans with the focus on making progress (including failing early and learning) and a secondary focus on economic constraints. Project management will plan and make schedules to make progress. However, in this VUCA context, project plans and schedules need high adaptability, and the project as a whole needs responsiveness. In innovation, project plans are even more a means rather than goals as in mature solution development.

Ensuring and monitoring progress, using plans and schedules, and taking action when falling behind. In general, taking action means following-up project members or external parties. In some cases, the action may be redirecting the project and adapting to a new direction. This is a challenging task, especially when tasks may change or are ambiguous.

Managing external contacts (potential customers, partners, suppliers, legislators, standardization bodies, investors) and internal contacts (such as the life cycle and supporting organizations) ensuring that all stakeholders are involved, well informed, and know what the project expects from them. A significant attention point is making sure that internal stakeholders interact timely and properly with external stakeholders.

Detecting and mitigating risks in an agile manner. The VUCA in the context translates into a continuous state of high alert. A challenge for project management is to catch major risks early and mitigate these.

Project management has to cooperate closely with leaders in the organization, both technical (technical leads, systems engineers, and systems architects) and commercial (marketing and sales). Both technical and commercial leaders contribute to the above project management tasks.

Positioning Systemic Innovation in the Body of Knowledge

Many incumbent companies are searching for the Holy Grail: a recipe for systemic innovation. These companies have many reasons to be caught in existing solutions, in existing markets, and in supply chains, causing them to fall victim to disruptive technologies (Christensen 1997). The recipe for the Holy Grail is not easily gained.

Taking the uncertain steps into innovation can be perceived as both chaotic and complex. Apparently, this uncertain phase is such a general and necessary step towards innovation that has been given a separate name. The very early phase of innovation has been given the name "Fuzzy Front End". Koen et al. (2001), however, propose to change this name into "Front End of Innovation" (FEI), because the word "fuzzy" suggests that this innovative phase is inherently unmanageable due to its inherent uncertainty and unpredictability. The New Concept Development Model that Koen et al. advocate offers five core elements, in a circular iteration: idea genesis, idea selection, concept and technology development, opportunity identification, and opportunity analysis.

These five core elements (the so-called engine) allow for systemic approaches by using methods and tools. According to Koen et al., the leadership and culture of the organization "fuel" the engine. This illustrates that the mindset of the leadership plays a crucial role in realizing the innovation. The final part of the model is the influencing factors (the environment), consisting among others of business strategies, competitive factors, organizational capabilities, and technology readiness.

O'Reilly III and Tushman (2004) provide the term "ambidextrous" that explains the tension between the terms "systemic" and "innovation". They assert an ambidextrous organization as part of the solution. In such an organization, they state that the traditional businesses and the innovative starters are "organizationally distinct units that are tightly integrated at senior executive level". This indicates that fostering innovation requires a different skillset than traditional management.

Taylor and Levitt (2004) showed that as outsourcing of specialized skills increased, innovations with potential to improve overall productivity significantly such as supply chain management, enterprise resource planning, or component prefabrication often required multiple interdependent firms to change their processes. Data showed that these systemic innovations diffused slowly in project-based industries such as the construction industry. This was a fact even if the innovations held the promise of significant increases in productivity and profitability.

Dettmer (2011) relates to the Observe-Orient-Decide-Act (OODA) method developed by John Boyd. OODA promotes agility over raw power when dealing with human's opponents. Dettmer makes the following observation that is relevant for systemic innovation:

> Regardless of the nominal OODA cycle time, however, in a competitive environment the party with the faster OODA loop cycle—often referred to as the decision cycle—gains a decisive advantage over its competitor.

Dettmer (2011) building on Fayol stated that the following functions, in general, are the core management functions:

- Planning
- Organizing

- Staffing
- Directing (or leading)
- Controlling.

Dettmer (2011) uses the Cynefin framework explaining the challenges of managing complex projects. The Cynefin framework (pronounced ku-*nev*-in), developed by David J. Snowden, is a decision-making framework that differs between ordered and unordered systems: the ordered systems being in the simple and complicated domain, and the unordered being in the chaotic and complex domains (due to unpredictable phenomena such as chance and choice). "Unknown unknowns" characterize the complex domain; that is, a typical challenge to project management of innovations. According to Snowden and Boone, each domain requires different types of decision-making, and effective leadership is dependent on the ability to recognize the operational context and act accordingly.

A multiple case study of six "innovation-friendly" infrastructure projects within the Swedish Transport Administration showed how challenges often related to the fact that even small innovations become systemic, affecting multiple stakeholders (Larsson et al. 2017). Client and contractors often had different views on these challenges and to what extent the clients' procurement strategies and project management practices have given opportunities for innovation. This is true also for other domains.

Eling and Herstatt (2017) support the use of the term Front End of Innovation as applied by Koen, when they discuss the earliest phase of innovation. They state "that the FEI consists of activities such as:

- opportunity/problem identification,
- analysis, and matching;
- market and technology analysis;
- idea generation,
- evaluation, and screening;
- concept development, evaluation, and testing;
- requirement definition;
- and project planning and risk analysis (Khurana and Rosenthal 1998; Reid and de Brentani 2004).

Midgley and Lindhult (2017) described several interpretations of the term "systemic innovation". Some literature has interpreted it as a type of innovation where value can only be derived when the innovation is synergistically integrated with other complementary innovations, going beyond the boundaries of a single organization. In this chapter, we follow Midgley's fourth interpretation where the term 'systemic innovation' concerns how people are acting to bring about an innovation engaging in a process supporting systemic thinking.

COMPLEX PROJECT MANAGEMENT OF SYSTEMIC INNOVATIONS

Innovation adds specific challenges that make project management more complex, and unhinges all aspects that make innovation projects controllable. In innovation projects

- The direction (what opportunity do we pursue) is vague and moving.
- Technology may be immature with a low technology readiness level (TRL).
- Previous designs may be non-existing or not applicable.
- The deployment and business context may be new and hence unknown.
- The architecture, e.g., how the solution will fulfill the value proposition, the business proposition, and how to make concept and technology choices needs significant development.

Nevertheless, successful innovations require strong management, especially since many contributors such as inventors, engineers, and designers typically lack a management focus. The project manager role needs to focus on achieving results within time and resource constraints, with a complex interaction between many players. In other words, systemic innovation requires organizational competences that ensure resources and time work properly together to achieve results.

A specific challenge related to many innovations is that the success of the innovation depends on successful development of all formal and informal stakeholder relations. Professor Kanter stated that "Lack of consensus among players in a complex system is one of the biggest barriers to innovation. One subgroup's innovation is another subgroup's loss of control" (Taylor and Levitt 2004).

Currently, there is a proliferation of information-related innovations, characterized by keywords such as *big data*, *deep learning*, *cloud services*, and *Artificial Intelligence (AI)*. These services promise a wide variety of capabilities such as *condition-based maintenance*, *computer-assisted diagnosis and treatment*, and *optimization of logistics and production*. Specifically, any information-related innovations depend on among others

- The integration of communication, storage, and processing platforms
- A multitude of services, such as identification and payments
- Standards for interoperability
- Compliance with regulations that are still crystalizing out.

In other words, the project scope is significantly more than a technical solution and a user application. Innovation projects tend to be complex projects.

In this chapter, we collect methods and tools for *orchestrating solution design and analysis*. A tool such as pacing helps to avoid that progress stalls due to the VUCA context. Tools like T-shaped presentations and A3AO help to align the wide variety of stakeholders. Both VUCA and the wide variety of stakeholders contribute to the complexity of project management. Orchestrating solution design and analysis, the basis for most other activities, help to cope with VUCA and the variety of stakeholders.

The specific systemic innovation challenge is the lack of a priori design knowledge in a context where there are also knowledge gaps. Therefore, we also include tools for *understanding and exploration of the problem and solution space*. Tools such as divergence–convergence, co-creation, and Pugh Matrix are tools to facilitate teams with a variety of participants to gather understanding and to explore problem and solution space. Tools such as Illustrative ConOps and A3AOs help to capture and communicate complex operational and architectural knowledge in an accessible way. The core to managing complex projects is the ability to understand, reason, discuss, and communicate complex issues at an abstraction level that fits human capabilities.

Finally, we focus on tools for *identifying and mitigating risk*. The driving force in achieving this is the mindset to fail early. We advocate that the solution integration primarily drives the project planning, since risks typically materialize during this integration. Core to complex project management is the ability to plan the integration and to prioritize integration sufficiently in the planning force field of suppliers, partners, resource owners, and other necessary facilities. Figure 23.4 shows how the combination of these tools helps to make innovation systemic, as far as the VUCA context allows this.

TOWARDS THE HOLY GRAIL OF SYSTEMIC INNOVATION—METHODS AND TOOLS

In this chapter, we focus on tools for coping with a VUCA context in the interaction between project management and technical and commercial leaders. This interaction is a cornerstone in enabling systemic innovation.

Figure 23.5 shows planning tools suitable for innovation projects. The figure divides the tools into three classes represented by the three boxes in the figure. The three classes are planning tools that focus on planning, on decision timing, and on long-term outlook.

In the figure, methods and tools that focus on the actual planning include PERT (Program Evaluation Review Technique), Pacing, Last Planner, agile and wall planning, and planning

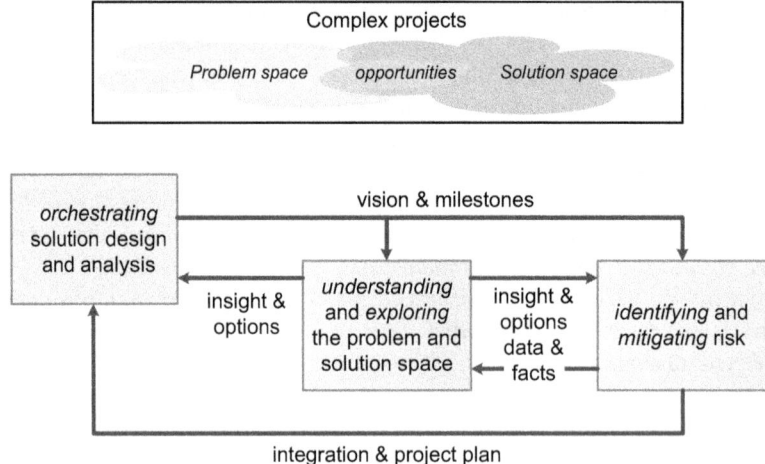

FIGURE 23.4   The value created by the tools for project management of complex projects.

FIGURE 23.5   Planning Methods for innovation projects.

for integration. PERT planning is useful since it emphasizes relations between activities with less emphasis on the timing. In VUCA circumstances, timing tends to change, while planning relations are more stable. Pacing is a method suited to get pressure in the team and progress transparency towards investors. See the next subsection for an elaboration. Lean and agile approaches offer methods and tools such as *Last Planner* and *wall planning. Planning for integration* is a planning approach; the prime focus of this approach is to integrate the parts into a working solution within its context.

The second box in Figure 23.5 shows methods focusing on timely decisions. The Lean methodology advocates *late decision-making* and using *set-based design*. These methods relate to the *real option theory* from the financial world. The main idea is to delay decisions to a moment that the team has more knowledge. By delaying a decision, the team keeps more options open.

The third box in Figure 23.5 relates to getting a long-term perspective. In a VUCA world, teams may use a long-term perspective for a better understanding of the context. These methods help to understand what *may* happen. Roadmapping and Foresight are methods to explore what the future may bring. Gigamapping is a broad method to explore the context by visualizing it on the wall. Scenario Planning helps in exploring various options for the future.

Figure 23.6 shows methods and tools used to communicate with stakeholders, and to explore and understand the problem and solution space. The problem space includes customer context as well as business context. This context includes, among others, ways of working, stakeholders' concerns, needs and interests, economic and financial relations, legislation and standardization, and competition and partners. The solutions space is the superset of potential solutions. The purpose with the solution-space tools is to help understanding potential solutions and their characteristics.

Stakeholder communication is easier if we have something to show. The tools listed in Figure 23.6 include paper illustrations, PowerPoint presentations, and mockups. The A3AOs tool supports communicating the resulting architecture, and mutual learning about the chosen problem throughout the creation process. The goal of the T-shaped presentations is to provide a wide variety of stakeholders with a clear and to-the-point overview. By offering depth in the core of the innovation, it ensures that stakeholders not only get glossy superficial insights. Demonstrators, such as physical objects, illustrations, or virtual reality, are beneficial, since these types of tangible items enable the creators and other stakeholders to challenge ideas and assumptions.

A common trap is to work on the solution space alone. In innovation, we need to understand our customers by working in the problem space, but also look into the solution space in order to validate our thoughts and understanding. We promote the use of several of these tools in conjunction, since they provide complementary insights.

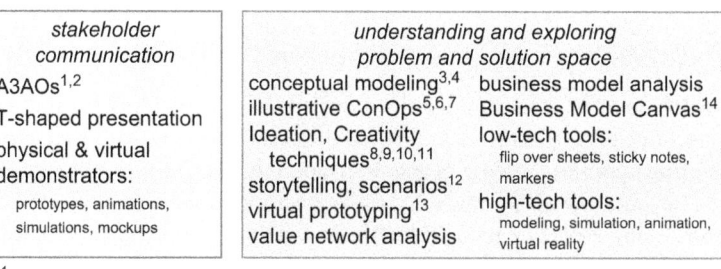

1 Borches D, 2010 A3 architecture overviews: a tool for effective communication in product evolution.

2 https://gaudisite.nl/BorchesCookbookA3architectureOverview.pdf

3 Muller, G. Challenges in Teaching Conceptual Modeling for Systems Architecting, ER 2015

4 Muller, G. Teaching conceptual modeling at multiple system levels using multiple views, CIRP 2014

5 https://gaudisite.nl/INCOSE2016_Solli_Muller_VisualConOps.pdf

6 https://gaudisite.nl/INCOSE2015_MullerEtAl_SubseaOverviewA3.pdf

7 ISO/IEC 2011. Systems and software engineering - Life cycle processes - Requirements engineering.

8 https://www.ideou.com/pages/ideation-method-mash-up

9 Skjelten, E.B.: Complexity & other beasts a guide to mapping workshops. The Oslo School of Architecture and Design, Oslo (2014).

10 Young, J.W. 2016: A Technique for Producing Ideas, Stellar Editions.

11 Bhattacharya, Hemerling & Waltermann, 2010, Competing for Advantage; How to Succeed in the new Global Reality. Boston Consulting Group https://www.bcg.com/documents/file37656.pdf

12 Muller, G., 2011, Systems Architecting; a Business Perspective, CRC Press

13 http://www.esi.nl/innovation-support/documents/symposium-2016/2-PT_Virtual-Prototyping-Interventional-X-Ray-Systems.pdf

14 Osterwalder, A. et al., 2004, The business model ontology: A proposition in a design science approach.

FIGURE 23.6   Stakeholder methods applicable for innovation projects.

THE METHODS AND TOOLS IN THE TOOLSET OF SYSTEMIC INNOVATION

This section describes eight methods and tools that project managers and their team can apply to improve innovation projects.

1. **Pacing**: It forces a project into a rhythm of showing progress regularly and thereby facilitating failing early.
2. **T-shaped presentation**: It provides a blueprint to present the innovation top-down. The blueprint will help in finding the content for the T-shaped presentation.
3. **Divergence/Convergence**: Early in projects, there is a need for diverging perspectives exploration, while later in projects, there is a need for converging goal orientation.
4. **Co-creation**: It is an approach for creating a collaborative environment and a mutual understanding of the problem- and solution space.
5. **Illustrative ConOps**: It captures the sequence of operations as visualizations.
6. **A3AO**: It is a tool designed for effective communication of architectural knowledge by providing overview in a compact way.
7. **Pugh matrix**: It is a multi-criteria decision-making method that allows for the comparison of a number of design concepts.
8. **Planning for integration**: It strives to fail as early as possible such that major uncertainties, unknowns, and ambiguity get visible as early as possible.

PACING

Pacing is an approach where a set of high-level milestones, related to specific achievements, sets the pace of the innovation team. The milestones serve the investors' need to monitor progress; often achievement of such milestone is a prerequisite for paying the next installment. Each milestone is stretching the team's ability to keep the team sharp. The project team gets a financial and a technical incentive for making progress. Figure 23.7 shows an example of such pacing milestones.

T-SHAPED PRESENTATIONS

T-shaped presentations are presentations for a variety of stakeholders, combining breadth and depth. The goal of these presentations is to provide these stakeholders with a clear and to-the-point overview. By offering depth in the core of the innovation, it ensures that stakeholders not only get glossy superficial insights, but also know what the innovation is about, and where challenges in the project may occur. Using a presentation addressing a variety of stakeholders facilitates alignment of stakeholders.

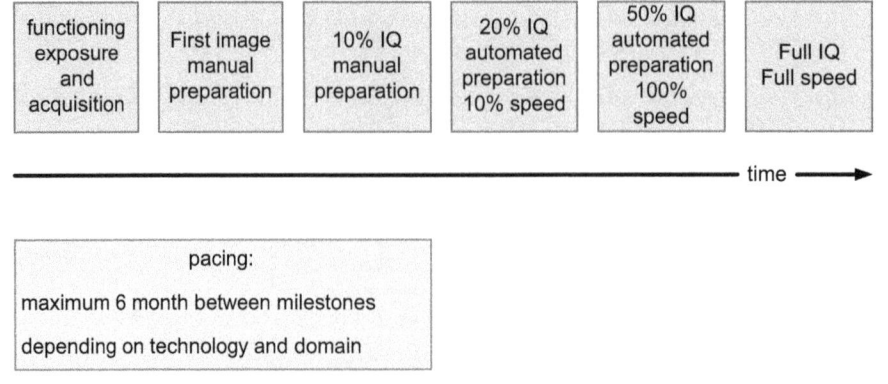

FIGURE 23.7   Example of pacing milestones.

Making such T-shaped presentation forces the project team members to create a clear storyline. The presentation is connecting the external world and market to the envisioned company, product, and business model, and connecting that to the core challenges of the solution. In practice, project teams tend to "live" in the solution space and this type of presentation forces them to connect to the broader customer and business context.

Figure 23.8 shows the content of a T-shaped presentation. Horizontally, it is showing the width of the presentation and vertically the technical depth in the core solution. Such presentation starts with a broad perspective on current societal trends, opportunities, problems, and needs. It narrows down to the envisioned business and market position, and the competition. Then, it zooms in on foreseen customers and stakeholders with their key drivers, concerns, and application. From this point onward, the presentation dives into depth, starting with the solution, e.g., the product, system, or service, and its functionality and key performance parameters. The solution is then further elaborated via concepts and design to specific aspects of critical design, and all the way down to critical or new technologies.

Once the presentation reaches its full depth, the presentation guides the audience back by summarizing and concluding how the proposed design choices fit the solution specification, and how that, in turn, answers the stakeholders' needs and concerns. The presentation finalizes by providing the project and business plans to conclude with recommendations.

DIVERGENCE–CONVERGENCE

A spiral can describe an innovation process where the team gradually obtains more knowledge within design, business, and engineering disciplines (see Figure 23.9). The team should include aspects of technology (systems engineering), value creation (business development), and human aspects (human systems integration), instead of working only at one aspect at a time. In different companies, and at different stages in a development project, one or two of these three aspects are often lost. Our suggestion is to address the contradictions and differences between the theories, and apply what is suited for each phase in the development process.

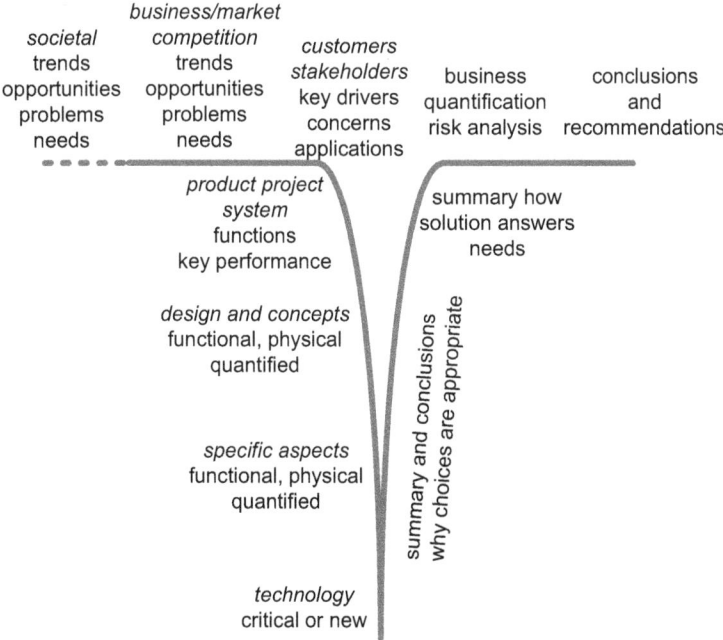

FIGURE 23.8   A T-shaped presentation, combining breadth and depth.

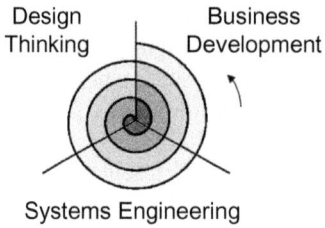

FIGURE 23.9   The development spiral.

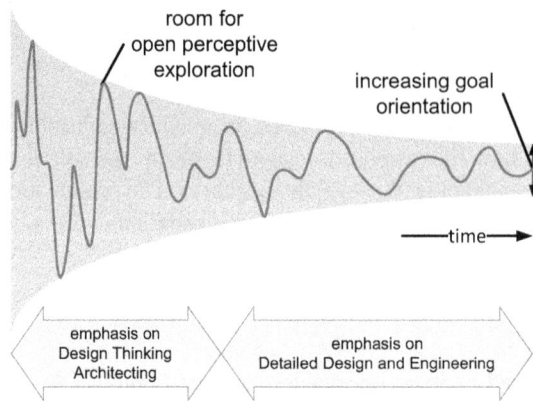

FIGURE 23.10   The mindset of the project team shifts over time.

Although the state of the art of existing theories has proven to deliver competitive advantages in a VUCA world, the theories have their distinct weaknesses. Traditional systems engineering works very well to narrow down and deliver solutions, especially combined with the Lean Thinking of knowledge-based development. However, systems engineering tend to be too stringent to foster disruptive human-centric innovations. Design Thinking, on the other hand, spans out solutions adapted to humans but is only applied to the early phase of a development and not in the later engineering stages.

Figure 23.10 illustrates this shift of mindset during the project. Early in projects, there is a need for open perceptive exploration, where approaches like Design Thinking shine. Later in projects, there is a need for increased goal orientation, where systems engineering can offer effective tools.

CO-CREATION SESSION FOR IDEATION

Running co-creation sessions with customers and other relevant stakeholders is a good approach for creating a collaborative environment and a mutual understanding of the problem- and solution space early on. Co-creation sessions with relevant participants, carefully selected methods and tools, and experienced facilitator(s) are often a good starting point to understand opportunities, potential solutions, and the business case. The duration of a co-creation session could be from 1 to 3 days.

The purpose of co-creation sessions may vary. In this section, we focus on co-creation sessions with the purpose of developing a set of early-phase innovative concepts with ownership from the customers. The nature of this early phase will reflect in ambiguous concepts. Sub-sequent concept development will combine and expand on these, rather than choosing a "winning" concept.

Selection of participants for the co-creation session will be based on roles, proficiencies, as well as personalities. It might also be beneficial to bring in other participants than customers, such as third parties to add different perspectives. It can be beneficial to include potential users of the system under design, with the aim of gaining insight in users' perspectives and unmet needs. The purpose of a co-creation session is to enable thinking creatively. This requires participants with a good group dynamic, a trusting environment, and a location that facilitates various mapping tools. Facilitating the session outside the workplace might be beneficial to remove the participants from their everyday controlled working environment.

The co-creation session needs one or two facilitators with experience and ability to guide, inspire, and lead the participants through the session. The session will need to be planned upfront, choosing the relevant tools for the particular group of people and opportunity at hand. However, it is not necessarily important to give the participants too much information before the session. Too much directions and micromanagement might limit the creativity.

A playful warm-up exercise as a start will help to remove self-induced limitations and create a trusting environment (Kelley and Kelley 2013). Such an exercise could be that the participants are acting as an expert to describe a random object or pretending to be a superhero and describe one's powers. A variety of such methods and tools can be found in literature and on the Internet.

A prerequisite to the co-creation session is to gain insight in the problem space, user need, and enabling technology up-front. The session itself will build upon this insight and enable a mutual understanding of the problem space seen from the perspective of each of the participants. There are various methods and tools to understand the problem space better, such as using mapping tools to map the current view of a situation or to perform a service journey (Skjelten 2014). As participants tend to jump into solution space, effort should be made on holding the participants in the problem space in the first part of the session. This is to avoid being trapped into solving the wrong problem.

Building the problem landscape will provide insight needed to explore innovative ideas and find new relations. Ideation tools such as brainstorming, mapping better view of a situation, identifying pains and gains, or mash-up tool are useful in this phase. The ideation process is quite open in the beginning and then converging to a set of tangible concepts in the last half of the co-creation session. Voting with stickers is an effective way of narrowing down the number of concepts, categorizing the ideas based on effort and value may help to determine "low-hanging fruits" and "diamonds". Methods like rapid prototyping are used to communicate and test out ideas in the workshop, applying low-cost equipment such as tape, carton and paint, and Lego. A sum-up in plenum of the final set of concepts is an effective way of closure of the co-creation session. The session can be documented using pictures of the mapping and prototypes made. This documentation will also act as input to the next phase of further concept development.

ILLUSTRATIVE CONOPS

Traditional Concept of Operations (ConOps) documents tend to be entirely textual. Illustrative ConOps captures the sequence of operations as visualizations.

Figure 23.11 shows an example of an illustrative ConOps in the oil and gas domain. This figure shows the "workover" operation; this is an operation to perform maintenance on a subsea oil and gas well. For this operation, a vessel has to bring the workover equipment to the well, assemble the equipment, and connect to the subsea equipment to access the well. After the maintenance, the reverse sequence is followed: disconnecting the equipment, disassembling it, and moving to the next well, leaving only the subsea well equipment that was there in the first place.

The benefit of an illustrative ConOps is that its representation is close to the mental world of a varied set of stakeholders. The visual format engages the stakeholders more than (detailed) text. Consequently, it works better to get feedback and to validate early concepts.

Solli and Muller (2016) add to the individual steps more information about this step, such as resources, environmental conditions, stakeholders, and constraints and opportunities. This enrichment facilitates further discussion among the different stakeholders.

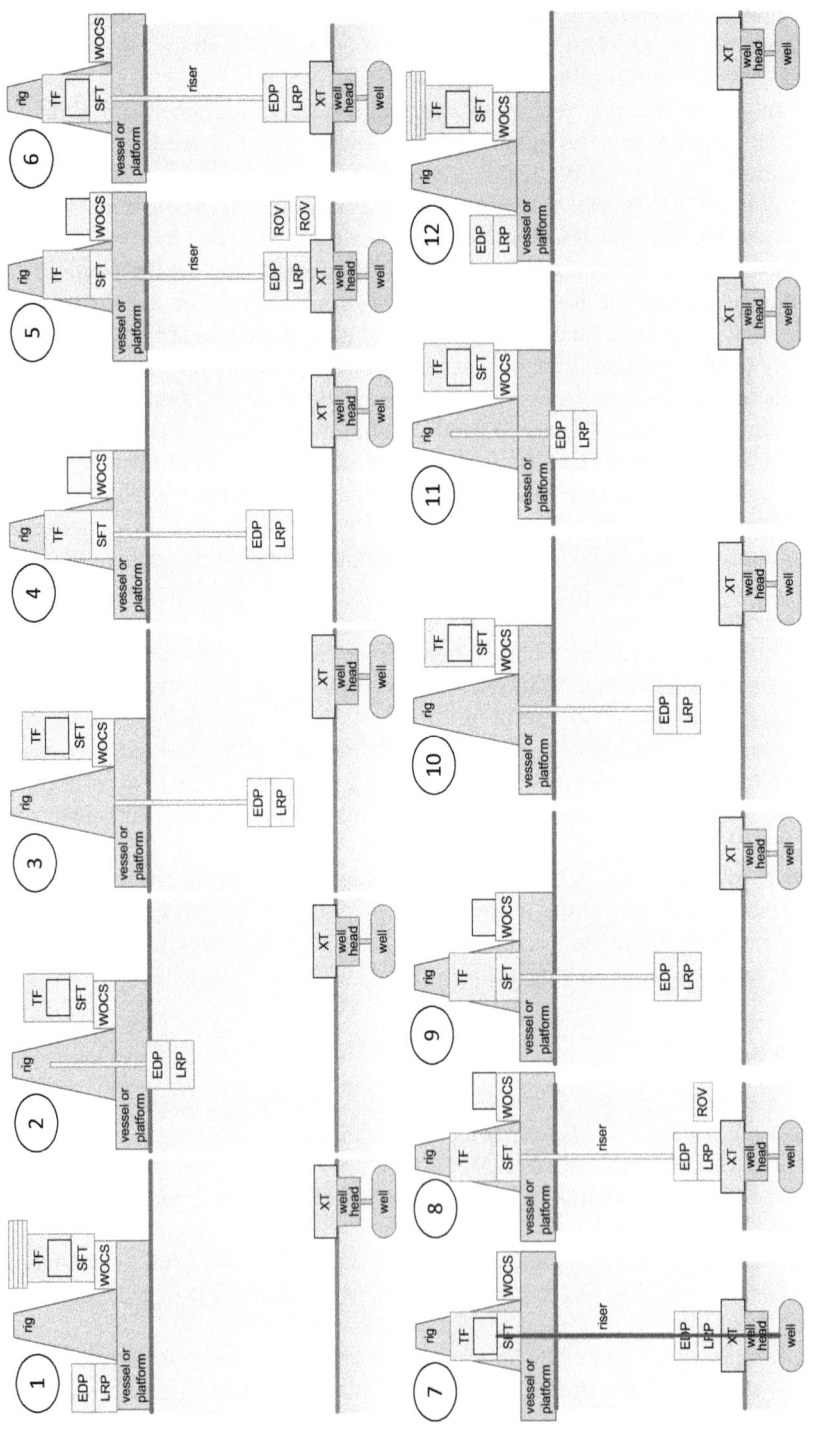

FIGURE 23.11 Example of an illustrative ConOps.

A3 ARCHITECTURE OVERVIEW

The A3AO is a tool designed for effective communication of architectural knowledge (Borches Juzgado 2010). The original A3 tool, developed by Toyota, is simply based on a piece of paper with the standard European size 297×420 mm. It is sufficient in size for communication purposes, big enough to include both visual and textual information.

A number of elements are standard content on A3AOs: physical models, functional models (dynamics), and quantification. An A3AO is normally based on a user story. That is, it has the operational viewpoint and can thus be compared to a conceptual ConOps. As such, it can partly replace written documents, at least in an early stage of a project.

A selection of benefits by applying A3AOs is given below based on (Solli and Muller 2016, Løndal and Falk 2018):

- Improve decision-making
  - Contribute positively in creating a shared understanding by improving the perceived learning and visualization capabilities.
  - Support the selection of systems in a System of Systems.
- Improve communication
  - Help the discussions to stay focused, dynamic, and inclusive.
  - Improve and increase cross-boundary communication in the organization.
  - Handle the communication challenges in lean product development teams.
  - Improve the communication and system overview across the organization.
  - Cross-boundary communication and early validation.
- Improve knowledge and knowledge management
  - Capture design and system knowledge.
  - Share knowledge in an engineering context.
  - Share understanding across the organization.
  - Learn about a complex system by using it for reversed architecting.
  - Replace heavy text communication of requirements in combination with model-based systems engineering.
  - Increase the R&D product system knowledge (by interactive dynamic A3AO).
  - Give a systems overview, easiness to locate information, readability, and well presentation of knowledge.
  - Save up to 95% time on finding product and project specific information.

The tool is not just about communicating after creation, but also learning about the chosen problem throughout the creation process of A3AOs. The effort creating the architectural overview as well as the novelty of the tool are factors that may prevent people from using and adapting A3AOs in their work.

Figure 23.12 shows an example A3AO (Muller et al. 2015). The figure provides an explanation of the cost of workover operations and all underlying models. Benefit of having all these models concurrently is that stakeholders can point to complementary models when reasoning about improvement options.

Stakeholders in the oil and gas industry gave an enthusiastic response to a number of characteristics of this specific A3.

- The "cartoon" relates immediately to problems they experienced in the past. They recognized that they could have prevented some past problems if they would have followed such approach.
- The A3 connects the technical system to the business interests in terms of time and costs.

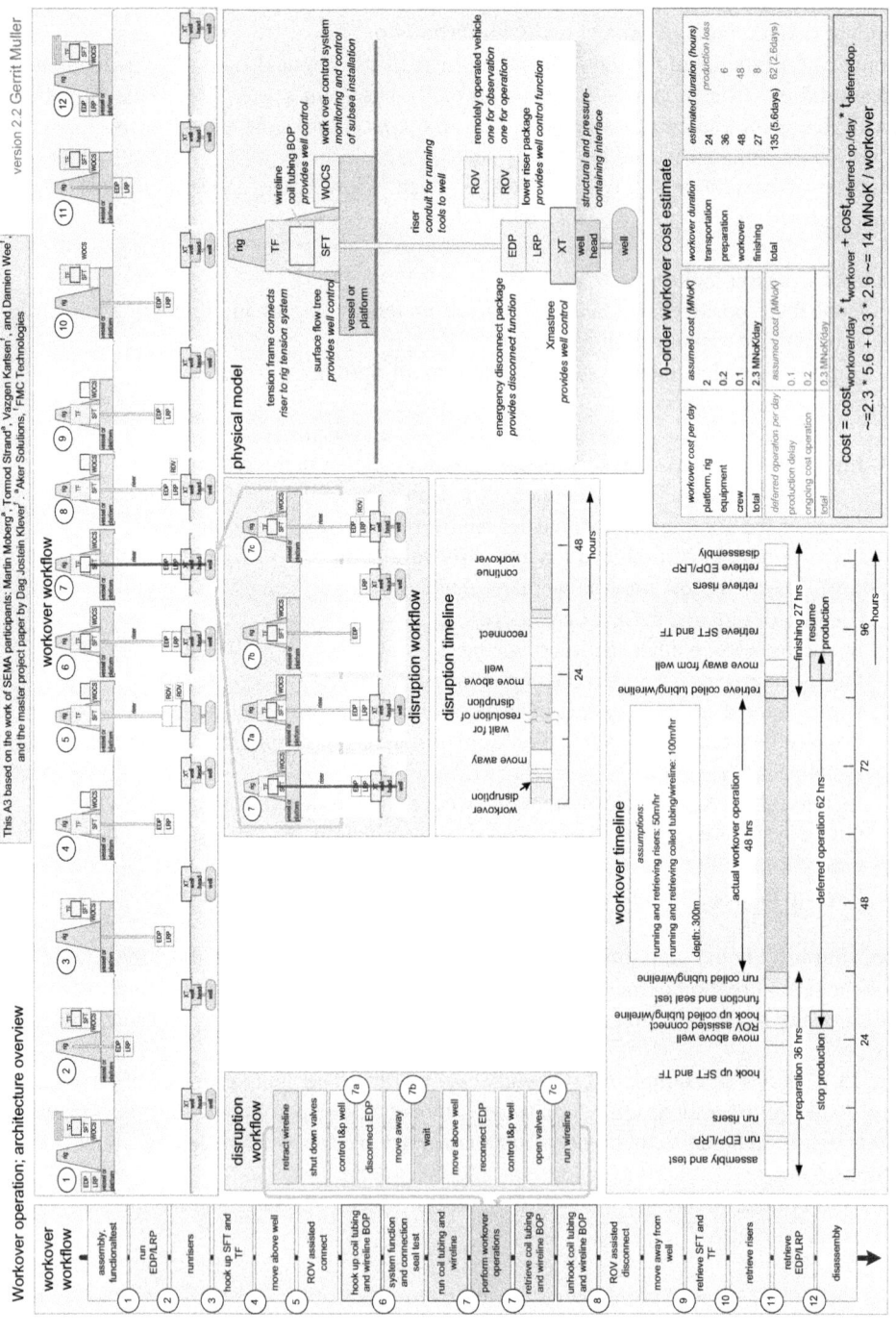

FIGURE 23.12 An example A3AO from offshore energy.

- The A3 approach is pragmatic. It fits and complements the current way of working that has emerged under a combination of high cost and time pressure and high demands for safety, reliability, and lifetime at the same time. Approaches that are more formal seem yet to be beyond current domain culture; they are perceived as time-consuming and not applicable (Muller et al. 2015). For example, attempts at using IDEF0 and SysML typically meet skepticism and resistance.

CONCEPT EVALUATION USING PUGH MATRIX

Pugh Matrix is a multi-criteria decision-making method that allows for the comparison of a number of design concepts leading ultimately to better understanding of the concepts. It is a systems engineering tool to extract the knowledge and experience from the team, and to display the complexity of the interwoven factors in a comprehensible way.

The Pugh Matrix helps in communicating qualities of conceptual solutions between project members and stakeholders. Project engineers can utilize this to determine the concept most suitable for the scope. Drawbacks of the method may be increased workload on the responsible engineers. Another aspect worth mentioning is the risk of the method being used as pure documentation of a predetermined choice. Mitigating actions such as clear communication of the benefits, and upper management commitment, may ease implementation.

The Pugh Matrix consists of columns listing the proposed concepts, and rows with evaluation criteria. The evaluation criteria are typically linked to stakeholder requirements. This setup forms a matrix with cells ranking the concept per criteria. The output from the Pugh Matrix is an overall score of the concept performance based on the evaluation criteria. The Pugh Matrix is expandable with weighing and prioritizing of criteria, and it is possible to show the output with percentages and advanced graphs. However, the quality of output from the Pugh Matrix is highly dependent on the input. Incorrect, incomplete, or inadequate evaluation criteria or ranking will corrupt the value of the performance score. On the other hand, the Pugh Matrix gives the involved users a greater view of the strengths and weakness of the concepts, and where to focus improvements.

Figure 23.13 shows a Pugh Matrix from Solli and Muller (2016). In this matrix, Solli used four categories of criteria (cost, design, installability and retrievability, and operability). Each individual criterion is ranked on a Likert scale from 1 (unfavorable performance) to 5 (excellent performance). The use of traffic light colors from red (1) to bright green (5) in the original paper helps to see weak and strong points quickly. Solli adds a visual summary of the concepts based on the clusters of criteria.

Pugh (1981) explains that the process of concept selection works best when used in a successive set of divergence and convergence steps. In the discussion about concepts, criteria, and scores, typically new concepts pop up. Successive iterations using the decision matrix evolve the design, while increasing the insight of the designers.

PLANNING FOR INTEGRATION

During the integration of the solution, many of the uncertainties, unknowns, and ambiguity will show up and disrupt the progress. Planning for integration is an approach, where the integration plan strives to fail as early as possible, such that major uncertainties, unknowns, and ambiguity get visible as early as possible. This is a risk detection and mitigation strategy. Core idea behind the approach is that early failure allows for recovery from the cause of the failure.

For example, when a new sensor behaves unexpectedly, e.g., it has too much noise or it has unexpected discontinuous behavior, the component that uses the sensor can search for solutions. It may search for the root cause of the problem, such as poor shielding, look for another sensor, or compensate for the behavior in the control logic.

from:Solli, H., and Muller G. 2016, Evaluation of illustrative ConOps and Decision Matrixas Tools in concept selection, INCOSE 2016 in Edinburgh, GB

User guide:
The concepts listed are ranked on a scale from 1-5 based on their attributes for each criteria. 3 is the mean value and describes a good enough performance to the criteria. A higher number shows a better performance, while a lower number shows aworse performance on the criteria listed.
The priority setting enables you to prioritize individual criteria to a higher or lower importance. If the priority is set to low for a criteria, that criteria will count less compared to a standard or higher prioritized one.

Rating	Description
1	Unfavorable performance
2	Less than satisfactory performance
3	Satisfactory performance
4	More than satisfactory performance
5	Excellent performance

User input

	Criteria	Priority setting	Concepts			
			A Simplified	B.1 PGB with Toast rack	B.2 PGB with GP's	C Satellite XT on WH
Cost	Hardware Cost	High	2	3	4	5
Cost	Installation Cost	Standard	2	2	3	4
Cost	Operational Cost	Standard	3	3	3	3
Design	Engineering hours (Amount of new engineering, re-use, analysis)	Standard	5	3	3	2
Design	Design familiarity (Is the design known in AkSo? Previously delivered?)	Standard	4	2	3	3
Design	Requirement compliance	Standard	5	4	3	2
Design	Deliverytime from call-off (Long lead items, fabrication time)	High	3	3	3	4
Design	Amount of new qualifications (TQP's)	High	5	2	2	2
Design	On-shore Testability (Availability of necessary equipment and procedures)	Standard	4	3	3	4
Installability & Retrievability	Number of installation runs required	Standard	1	2	2	5
Installability & Retrievability	Installation time	Standard	1	2	3	4
Installability & Retrievability	Weather vulnerability (Metocean constraints, Hs)	Low	2	4	4	4
Installability & Retrievability	Need for special tools	Low	4	3	3	3
Installability & Retrievability	Guide system robustness	High	4	4	3	2
Installability & Retrievability	Size of vessel required (Rig, heavy lift vessel, installation vessel)	Standard	1	2	3	5
Installability & Retrievability	Weight & Size	Standard	1	3	4	5
Installability & Retrievability	Retrieval flexibility of equipment	Standard	3	4	4	2
Operability	ROV access	Standard	3	4	4	4
Operability	Flow assurance (Hydrate/Scale, pipeline friction, pressure bleed-off)	Standard	3	3	3	3
Operability	Dewatering & start-up (Service access, injection points, etc.)	Standard	3	4	4	4
Operability	Reliability	Standard	3	4	4	4
Operability	Interchangeability	Standard	5	2	2	1
	Indicating summary:		78	74,5	78,5	84,5

Executive Summary

Legend: ■ Cost ▨ Design ▨ Installability & Retrievability ▨ Operability

X-axis: Simplified A-, PGB with Toast B.1-rack, PGB with GP's B.2-, Satellite on WH C-

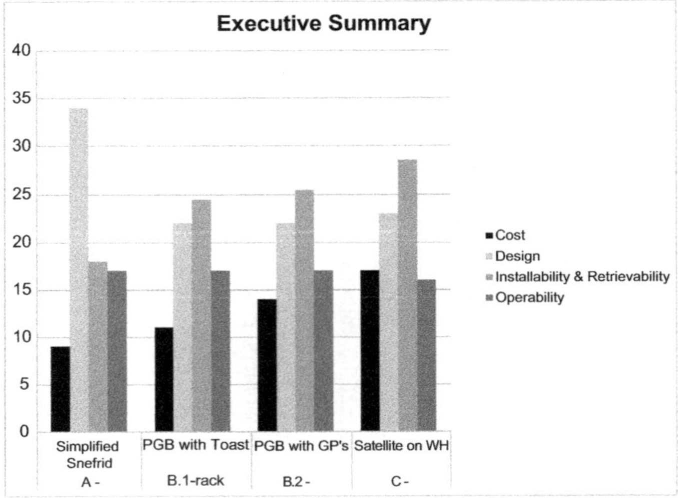

FIGURE 23.13   Example Pugh Matrix from Solli and Muller (2016).

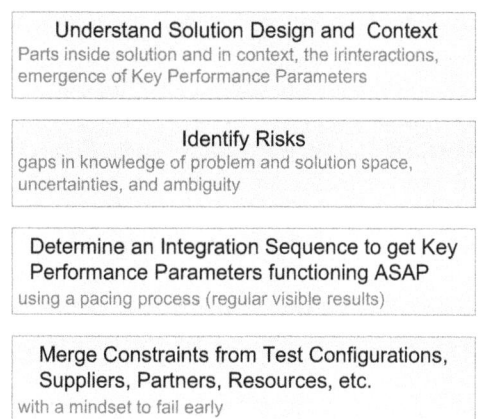

FIGURE 23.14 Stepwise approach to planning for integration.

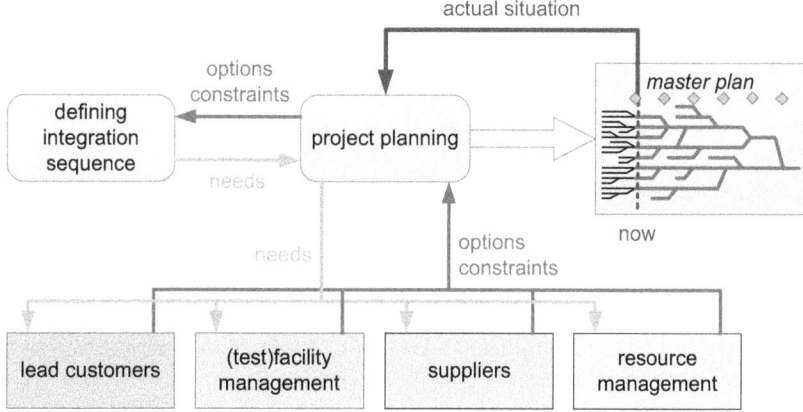

FIGURE 23.15 The transformation of an integration sequence into a project master plan with pacing milestones.

Figure 23.14 shows a stepwise approach to planning for integration. Project management needs understanding of the solution design and its context (what are the parts inside and outside the solution, how do they interact, how do the key performance parameters emerge from the interaction). The project team identifies risks, such as gaps in knowledge of problem and solution space, uncertainties, and ambiguity.

Typically, there are unknowns where there are gaps in the knowledge. Interaction of parts and emergence of behavior and performance tend to be areas of unknown unknowns. Hence, an integration plan needs an *integration sequence* that may uncover unknown unknowns as early as possible. This integration sequence is core to the planning activity; it strives to have an operational solution as early as possible, such that key performance parameters can be measured and analyzed early.

In practice, many other project constraints get in the way of this preferred *integration sequence*. The planning challenge for project management and project team is to create an integration plan that strives for early failure, considering the constraints. In this over-constrained puzzle, the team needs creativity to stay close to the fail early mindset. For example, prototype components or functions may enable early functioning. Modeling, e.g., hardware in the loop and software in the loop, may also enable early functioning of critical functions.

Figure 23.15 shows how the *integration sequence* is transformed into a project *master plan* with pacing milestones. The project planning links directly to *suppliers, customers, (test) facility, and resources*, imposing the needs from the project and receiving constraints back.

As indicated by the red arrow in the figure, the continuously updated master plan feeds the actual project status (indicated by "now" in the figure) back to the project planning. The project planning activity is a continuous activity. This is specific to systemic innovation projects since all of the input changes during the project, and since the integration itself brings new insights that demand adaptation of specification and design of the solution, and hence of the plan.

Many projects use a static master plan, rather than continuously updating it. Project members may oppose continuous update, since it feels uncomfortable; innovation projects inherently are shooting on a moving target, which may feel inherently uncomfortable. There are several methods for continuous updating plans, such as wall planning. These methods make the plan visual for everyone, and the planning process is inclusive to improve the involvement of all project members.

QUESTIONS RELATED TO THE TOOLSET FOR DISCUSSION TO BE USED IN THE CLASSROOM

Question 1. Select an innovative project as a topic for this assignment, and apply the following questions on that project:

- Make an inventory of stakeholders and their concerns.
- Use this inventory to analyze the current knowledge level and expectations of these stakeholders related to your innovation. Look at a broad set of aspects, such as the business, technical, legal, organizational, legal, social, and environmental aspects.
- Use the same list and aspects to determine the desired knowledge and expectation level for these stakeholders.
- What tools can you use to capture and communicate missing knowledge to these stakeholders?
- Explore the solution space and determine an initial solution concept.
- Identify gaps, uncertainties, and ambiguity in the knowledge of the problem and the solution space. Use this for a risk assessment, in terms of probability and severity.
- Use the understanding of the problem and solution space and the risk assessment to create an initial integration sequence.
- Identify the constraints affecting the transformation of the integration sequence into a project master plan.
- Make an initial project master plan, striving for early failure.

Question 2: Select the A3AO tool and apply it on the same project as Question 1.

- Explain the main conclusions as part of the A3AO.
- What will be the main discussion from a general manager's perspective?
- Who would be interested in viewing the physical view?
- How could you as a project manager apply this sheet to improve the services?

Question 3: Select the co-creation tool and apply it on the same project as Question 1.

- Determine the purpose of the co-creation session related to your innovation.
- Select the participants that you need to include in this session to achieve the purpose.
- Plan the agenda of the co-creation session, and identify techniques needed in the session to achieve the purpose.

SUMMARY AND FUTURE CHALLENGES

Systemic innovation requires project management tools to cope with VUCA. In such environment, learning fast is crucial, which translates into strategies and a mindset of failing fast and early. We presented tools that

- Help in directing and monitoring progress with a tool such as pacing that helps to avoid that the progress stalls due to the VUCA context.
- Facilitate the exploration of the problem and solution space, with tools such as divergence–convergence, co-creation, and Pugh Matrix. Both VUCA and the wide variety of stakeholders contribute to the complexity of project management. These tools facilitate teams with a variety of participants to gather understanding and to explore problem and solution space. Facilitating the exploration of the problem and solution space help to cope with VUCA and the variety of stakeholders.
- Facilitate communication with a wide variety of stakeholders and create overview in chaotic circumstances with tools such as T-shaped presentations and A3AOs. Tools such as Illustrative ConOps and A3AOs help to capture and communicate complex operational and architectural knowledge in an accessible way.
- Facilitate planning in these circumstances with planning systems integration as tool. These tools help in *identifying and mitigating risk*. The driving force in achieving this is the mindset to fail early. Core to complex project management is the ability to plan the integration and to prioritize integration sufficiently in the planning force field of suppliers, partners, resource owners, and other necessary facilities.

> Systemic innovation requires project management tools to cope with VUCA. In such environment, learning fast is crucial, which translates into strategies and a mindset of failing fast and early.

The core to managing complex projects is understanding, reasoning, discussing, and communicating complex issues at an abstraction level that fits human capabilities. In innovation, the added challenge is coping with VUCA and a wide variety of stakeholders. The interplay of the tools presented is the key to cope with the complexity and to make the innovation systemic.

Scalability of project management tools and the broader project execution is a major challenge in today's world. This challenge will increase further in the near future, when future innovations will rely on integrating even more existing capabilities. At the same time, we see that many stakeholders expect faster delivery with the same or higher reliability. The project teams face a multitude of complicating factors (higher interoperability, more partners and stakeholders, less time, more competition and globalization, higher quality expectations, more difficulties to distinguish from the "crowd", and other business aspects), which increases the project management complexity.

REFERENCES

Borches Juzgado, P.D. 2010. A3 architecture overviews. A tool for effective communication in product evolution, University of Twente, Enschede, The Netherlands. doi:10.3990/1.9789036531054

Christensen, C. 1997. *The Innovator's Dilemma – When New Technologies Cause Great Firms to Fail.* Boston, MA: Harvard Business Review Press.

Dettmer, H.W. 2011. *Systems Thinking and the Cynefin Framework: A Strategic Approach to Managing Complex Systems.* Goal Systems International, retrieved from https://static1.squarespace.com/static/578d0f8459cc6877481865ef/t/57ec5284579fb363a5b9a18c/1475105417186/Systems-Thinking-and-the-Cynefin-Framework-Final.4.pdf on August 11, 2018.

Eling, K., and C. Herstatt. 2017. Managing the front end of innovation—Less fuzzy, yet still not fully understood. *The Journal of Product Innovation Management*, 34, 6.

Kelley, T., and D. Kelley. 2013. *Creative Confidence: Unleashing the Creative Potential Within Us All.* New York: Crown Business.

Khurana, A., and S.R. Rosenthal. 1998. Towards holistic "front ends" in new product development. *Journal of Product Innovation Management*, 15 (1), 57–74.

Koen, P., G. Ajamian, R. Burkart, A. Clamen, J. Davidson, R. D'Amore, C. Elkins, K. Herald, M. Incorvia, A. Johnson, R. Karol, R. Seibert, A. Slavejkov, and K. Wagner. 2001. Providing clarity and a common language to the "Fuzzy Front End". *Research-Technology Management*, 44 (2), 46–55. doi:10.1080/08956308.2001.11671418.

Larsson, J., P. Eriksson, and A. Udén. 2017. Challenges in implementing systemic innovation in transport infrastructure projects. *Proceedings of 13th International Conference on Organization, Technology and Management in Construction*, Zagreb, Croatia.

Løndal, S., and K. Falk. 2018. Implementation of A3 architectural overviews in Lean Product Development Teams. A case study in the Subsea Industry, *INCOSE 2018* in Washington, DC.

Midgley, G., and E. Lindhult. 2017. What is system innovation, Research Memorandum 99, Hull University Business School, retrieved from www.researchgate.net/publication/315692364_What_is_Systemic_Innovation on October 18, 2018.

Muller, G., D. Wee, and M. Moberg. 2015. Creating an A3 architecture overview a case study in SubSea systems, *INCOSE 2015* in Seattle, WA.

O'Reilly III, C.A., and M.L. Tushman. 2004. The ambidextrous organization. *Harvard Business Review*, April 2004.

Pugh, S. 1981. Concept selection: A method that works. In: Hubka, V. (ed.), *Review of Design Methodology. Proceedings International Conference on Engineering Design*, March 1981, Rome. Zürich: Heurista, blz. 497–506.

Reid, S.E., and U. de Brentani. 2004. The fuzzy front end of new product development for discontinuous innovations: A theoretical model. *Journal of Product Innovation Management*, 21 (3), 170–184.

Singer, J.D., N. Doerry, and M.E. Bukley. 2010. What is set-based design? *Naval Engineers Journal*, 121, 31–43. doi:10.1111/j.1559–3584.2009.00226.x.

Skjelten, E.B. 2014. *Complexity and Other Beasts. A Guide to Mapping Workshops*. Oslo, Norway: The Oslo School of Architecture and Design.

Solli, H., and G. Muller. 2016. Evaluation of illustrative ConOps and Decision Matrix as Tools in concept selection, *INCOSE 2016* in Edinburgh, GB.

Taylor, J., and R. Levitt. 2004. Understanding and managing systemic innovation in project-based industries. *Innovations: Project Management Research*, 83–99.

Index

Lightning Source UK Ltd.
Milton Keynes UK
UKHW030804101219
355076UK00010B/403/P